*Introduction to Probability and Statistics*

# Introduction to Probability and Statistics

by Malcolm Goldman, 1970

*New York University*

Harcourt, Brace & World, Inc.

*New York | Chicago | San Francisco | Atlanta*

*Introduction to Probability and Statistics*

ISBN: 0-15-543595-7

LIBRARY OF CONGRESS CATALOG CARD NUMBER: 77–105694

PRINTED IN THE UNITED STATES OF AMERICA

*To Ruby*

# Preface

This book is the product of my experience with courses taught at New York University. It is primarily for undergraduates with enough interest in mathematics to require a course in probability and statistics above the "cookbook" level—one emphasizing concepts as well as problems—but one which does not require extensive mathematical preparation or great technical ingenuity. It is not oriented toward any particular field of application and would be appropriate for mathematics majors as well as science and some social science majors.

In order to use this book successfully the reader should be acquainted with the material of a traditional calculus course through multiple integrals, power series, and elementary differential equations. This text is suitable for a one-year course for juniors and seniors. However, a one-semester or two-quarter course may be fashioned from Chapters 1–7 and 12–14 together with sections 10.1 and 10.2.

The main topics are: combinatorial and geometric probability problems (Chapters 1, 3, and 5), random variables and their expectations (Chapters 6, 7, and 8), limit theorems via characteristic functions (Chapters 9 and 10), introduction to stochastic processes (Chapter 11), estimation (Chapter 13), hypothesis testing (Chapter 14), and mathematical models (Chapter 4).

Solution of counting problems by means of "code words" is emphasized. One advantage of this procedure is that it applies to more types of problems than does a tree diagram. Furthermore, it allows for an easier explanation of the errors which beginners frequently make.

The intuitive device of a "payoff book" is used in the context of random variables. The concept of a "standard form" for a random variable (or vector) is used to bridge the main definitions of random variable.

I have elected to use characteristic functions because of my success with them in the classroom. Even students of modest ability have been able to use the uniqueness theorem after considering a number of examples based on this technique. The body of material on characteristic functions includes some mathematical topics which many instructors of my acquaintance would regret to see slip from the mathematics curriculum.

Chapter 11 treats elementary aspects of stochastic processes with emphasis on the joint distribution rather than on the underlying measure. Admittedly, I have sacrificed zero-one laws, etc. to maintain the level of the book. However, first passage of random walks, an intuitive arcsine law, a potential-theoretic approach to recurrence of multidimensional simple random walks, and transformations of Gaussian processes are included.

The material on statistics is quite classical. Loss and risk functions have not been mentioned. I have been careful to point out which parts of a problem require a nonstatistical judgment; for example, what significance level to use. Each chapter also contains a section of motivative and exploratory material to support the basic intent of clarifying the significance of statistical inference rather than its range of application. The latter is left to other courses in statistics.

Basic courses in probability should emphasize problem-solving and concepts. Thus, even though I have stated many important results formally as theorems, listed the Kolmogorov axioms, and made some careful deductions for countable sample spaces, I have also included an abundance of illustrative examples and exercises designed to encourage the student's own exploration of this subject. These exercises range from routine drills for practice of computational methods to more penetrating problems containing information essential to the development of the theory. The latter are marked with a diamond (♦) and should be treated as "must" reading. Although it is not necessary that such exercises be worked out in every case, it is important that their assertions be understood. Optional sections and their related exercises are marked with a star (★) and can be omitted on first reading. These portions of the text have been so marked because of their relative difficulty.

A few guiding remarks are due the reader. Chapters 1–5 may lull an unwary student into a false sense of security because the material contained in them does not require much attention to terminology. However, terminology is very important in the remaining chapters and there it must be studied carefully. There also are a few places in which some unreasonable probabilities or techniques are offered "reasonably" to illustrate the pitfalls of the subject and to encourage the student to analyze a situation rather than just to leap ahead mechanically. Since the correct assertions usually follow within a page, it is always worthwhile to read a bit further after hitting a snag.

A book of this type is invariably influenced by one's students, whom I thank as a group. My colleagues Morris Meisner, John Mineka, and Richard Pollack read parts of the manuscript and were helpfully critical. The elegant proof of Theorem 14.2.1 was supplied by Professor William Studden. I am indebted to the Literary Executor of the late Sir Ronald A. Fisher, F.R.S., and to Oliver & Boyd Ltd., Edinburgh, for their permission to reprint Tables 2 and 3 from their book *Statistical Methods for Research Workers*. Miss Marilyn Davis' expeditious editorial work was a great help. For her constant encouragement, patience, and representation of the potential reader, it is a great pleasure to thank my wife.

Malcolm Goldman

# Contents

# 3

## *Probability Systems*

# 4

## *Mathematical Models*

# 5

## *More on Probability Systems*

# 6

## *Random Variables*

# 7

## *Mathematical Expectation*

# 8

## *Random Vectors*

# 9

## *Characteristic Functions*

# 10

## *Limit Theorems*

# 11

## *Random Processes*

# 12

## *Preliminaries of Statistics*

# 13

## *Estimation*

# 14

## *Hypothesis Testing*

# 15

## *Some Topics in Sampling*

# Introduction to Probability and Statistics

CHAPTER **1**

# *Counting*

## 1.0  Prelude

Probability theory relates to questions such as the following:

(a)  What is the probability that it will rain tomorrow?

(b)  What is the probability of winning in a lottery?

(c)  What is the probability of filling an inside straight?

(d)  What is the probability of winning in a dice game?

(e)  What is the probability that 500 or more of 1000 people insured by company X will die before age 65?

(f)  What is the probability that 600 or more heads will show in 1000 tosses of a coin?

(g)  What is the probability that a particle undergoing Brownian motion will be outside of a certain region at some time?

We will not treat each of these questions directly. Instead, we will set up a much more general framework in which each of these questions is a special case. This is customarily done in mathematics because experience has shown that we get more insight into the questions at hand as well as more technique to attack new questions. We may even be led to develop new approaches in mathematics and its applications.

The material on probability theory begins in Chapter 3. Chapters 1 and 2 treat certain background material with which many readers are acquainted in varying degrees. Chapter 1 deals with elementary counting techniques. These techniques enable us to get numerical answers to numerous problems. Chapter 2 deals with set theory. Readers who are familiar with these topics may skim these chapters in order to adjust their terminology and their notation to that of this book* and to see if their background is sufficient. Readers who are unfamiliar with these topics should read Chapters 1 and 2 carefully. They are very important to a thorough understanding of probability.

# 1.1 Lists, collections, and sets

Many basic problems in probability involve counting certain collections. In this chapter we take up some techniques of counting collections which are formed according to some given pattern.

First of all, we must distinguish between a list and a collection. For example, a grocery list might be "chicken, tomatoes, milk" or "chicken, milk, tomatoes" or possibly even "milk, chicken, tomatoes, milk." These are different *lists* but the collection of items is the same (providing that one does not buy the milk twice merely because it appears twice on the list). That is, the order of entries *is* at issue in the formation of a list in addition to the collection of items on the list. Thus two lists are the same only if the items on the list are the same and appear in the same order on both lists. We call lists of letters "words" since the order of letters is important in a word. For example, TWO and TOW can both be formed with three blocks labeled with the obvious letters but the *words* are different because the order of the letters is different.

Since it is difficult to write (and have printed) standard English letters without suggesting order, an artificial device is used to denote the *collection* of certain items rather than the list of those items. Braces { } are used for this purpose. Thus {tomatoes, chicken, milk} denotes the collection of the enclosed items. The notation is read as "the collection (consisting) of tomatoes, chicken, and milk." That collection is the same as {milk, chicken, tomatoes} and is the same as {chicken, milk, tomatoes, milk}. If there are too many items to conveniently list, other devices are used. For example, {1, 2, 3, . . . , 26} is the collection of numerals from 1 to 26 inclusive. In some cases it would be the collection of numbers from 1 to 26 (recall that numerals are just symbols). Which it is, is usually clear from the context. Sometimes a descriptive phrase is used, such as {citizens of the U.S.} or {real numbers > 3} or {$n^2$; $n$ an integer} or {$n^2$ | $n$ an integer}.* The last two types of notation are sometimes called set builders. Each may be read as "the collection of all $n^2$ such that $n$ is an integer." Similarly,

---

* Different authors use different separating marks in the braces notation. One might see {. . . | . . .} or {. . . ; . . .} or {. . . : . . .}.

{citizens of the U.S.} is read as "the collection of citizens of the U.S." It should be observed that

$$\{n^2; n \text{ an integer}\} = \{m^2; m \text{ an integer}\} = \{n^2; n \text{ a non-negative integer}\}$$

since we are collecting squares of integers, not letters, and because repetitions in the description do not change the collection. The set builder notation is utilized in the next chapter where we discuss matters more fully. For the present, various examples should serve to explain the terminology.

The concept of a collection of things is fairly natural and for certain purposes it makes little difference whether we have a collection of numbers, a collection of apples, a collection of fruit, a collection of letters of the alphabet, or a collection of blocks labeled with letters of the alphabet. Certainly, as far as counting is concerned one has the same problem with {A, B, C} as with {a banana, an apple, an orange} and the same problems with {1, 2, 3, 4, 5, ...} as with {2, 4, 6, 8, 10, ...}.

For the present we refer to the "objects" in a collection as *elements* or *members* of the collection. The word "element" is somewhat less suggestive of a concrete object and most objects of mathematics are not very concrete. Thus A, B, and C are the elements of {A, B, C}; any even whole number is an element of {2, 4, 6, 8, 10, ...}; and all real numbers greater than $-1$ and less than $+1$ are elements of

$$\{x; x \text{ is a real number and } x^2 < 1\}$$

Instead of the words collection and *subcollection* we use the words *set* and *subset*, respectively. The latter words are shorter and are more commonly used in this context.

We say that two sets are *disjoint* if they have no common elements. The *union* of a number of sets is the set consisting of those elements which are in one or more of the given sets.

At this point, we can consider some examples. In these examples, we follow the mathematical practice of making no restrictions unless restrictions are specifically called for.

### Example 1.1.1

How many different 2-letter "words" can be made with the letters of the English alphabet? The term "word" merely signifies a list of letters. Our words need not be words in any language.

We can have

$$AA, AB, AC, AD, \ldots, AZ$$
$$BA, BB, BC, BD, \ldots, BZ$$
$$CA, CB, CC, CD, \ldots, CZ$$
$$\vdots$$
$$ZA, ZB, ZC, ZD, \ldots, ZZ$$

It is convenient to think of 26 disjoint sets of words. In the first set are all words which have A as the first letter. In the second are all words which have B as the first letter. In the third set are all words which have C as the first letter, and so on. Since there are 26 letters available as the second letter each set contains 26 words. Thus the total number of words in the union of the sets is $26 \cdot 26 = (26)^2$.

### Example 1.1.2
How many 2-letter words may be made with the English alphabet if no letter is used twice in the same word?

The obvious answer is $(26)^2 - 26$ since exactly 26 words from Example 1.1.1 have repeated letters, but let us look at the problem in another way. This way, though complicated, leads to a pattern which enables us to handle more general situations.

We consider

$$AB, AC, AD, \ldots, AZ$$

$$BA, BC, BD, \ldots, BZ$$

$$CA, CB, CD, \ldots, CZ$$

$$\vdots$$

$$ZA, ZB, ZC, \ldots, ZY$$

Again we envision 26 disjoint sets of words. All words in the first set have A as the first letter. All words in the second set have B as the first letter, etc. However, there are only 25 words in each set since the words in the first set can have any letter but A as the second letter. Similarly, the words in the second set can have any letter but B as the second letter. Since we have 26 sets with 25 words in each set we have a total of $26 \cdot 25$ words of the type called for.

### Example 1.1.3
How many 2-letter words may be formed from the English alphabet if the first letter is a vowel and the second letter is a consonant which occurs later in the alphabet than the vowel in the first position? We define the vowels to be A, E, I, O, U for this example.

We could imagine five sets—the A set, E set, I set, O set, and U set. There are 21 consonants occurring beyond A in the alphabet, 18 such consonants beyond E, 15 beyond I, 10 beyond O, and 5 beyond U. Thus, the A set has 21 words. The E set has 18 words, etc. The total number of words in the union of our sets is $21 + 18 + 15 + 10 + 5 = 69$. It might be useful for the student to write all of these words and place them in appropriate sets.

The foregoing examples suggest the following fundamental principle:

If there are $k$ disjoint sets and if the first set contains $m_1$ objects, the second set contains $m_2$ objects, the third set contains $m_3$ objects, ..., the $k$th set contains $m_k$ objects then the total number of objects in the union of the sets is $m_1 + m_2 + m_3 + \cdots + m_k$.

**Corollary**     If $m_1 = m_2 = \cdots = m_k = n$, the total number of objects in the union of the sets is $kn$.

These rather innocent remarks will take us quite a way in counting various collections.

### Example 1.1.4
How many 3-letter words may be formed from the English alphabet?

We could form a set of words all of which have AA as the first *two* letters. The second set could be composed of all words which have AB as the first two letters. The words of the third set would have AC as the first two letters. Continuing this way, e.g., in the 53rd set all words would have CA as the first two letters. In view of Example 1.1.1 there would be $(26)^2$ sets since there are as many sets as there are 2-letter words. Since there are 26 letters available for the third letter each set would have 26 words. By the corollary to the fundamental principle with $k = (26)^2$ and $n = 26$ there are $(26)^3$ 3-letter words.

The student should convince himself that there are $(26)^3 \cdot 26 = (26)^4$ possible 4-letter words which may be made with the English alphabet since we could form $(26)^3$ sets of words where each set consists precisely of all words with the same first *three* letters. Thus there are as many sets as there are 3-letter words. Each set can have 26 words so the above corollary applies with $k = (26)^3$ and $n = 26$.

It should be quite clear that $(26)^r$ $r$-letter words may be formed with the English alphabet. More generally, if an "alphabet" has $n$ different symbols, the above arguments would show that there are $n \cdot n = n^2$ 2-symbol words, $n^2 \cdot n = n^3$ 3-symbol words, and, in general, $n^r$ $r$-symbol words which may be formed with that alphabet if all symbols may be repeated arbitrarily.

### Example 1.1.5
How many 3-letter words may be formed with the English alphabet if no letter is repeated in a given word?

Following Example 1.1.2 we could arrange for the first set to have all words with AB as the first two letters. All words in the second set could have AC as the first two letters. Then, continuing this way, the 26th set would consist of all words with BA as the first two letters. The number of sets is the same as the number of 2-letter words where repetitions are not allowed, namely $26 \cdot 25$.

Each of the above sets contains 24 words since any letter except the first two can be used in the third position. By the corollary there are $26 \cdot 25 \cdot 24$ words satisfying the conditions of the example.

In a similar fashion we could show that there are $26 \cdot 25 \cdot 24 \cdot 23$ 4-letter words which may be made with our alphabet if no letter is repeated in a word. In general if an alphabet contains $n$ distinct symbols and if $r \leq n$ we may make $n(n-1)(n-2) \cdots (n-r+1)$ $r$-symbol words from such an alphabet if symbols may not be repeated in a given word. The condition $r \leq n$ is imposed because such a word cannot have more "letters" than there are symbols in the alphabet.

### Example 1.1.6

How many words of four letters or fewer may be made with our alphabet?

Suppose we put all 1-letter words in one set, all 2-letter words in the second set, all 3-letter words in the third set, etc. By previous work we have $m_1 = 26$, $m_2 = (26)^2$, $m_3 = (26)^3$, and $m_4 = (26)^4$. Therefore, according to the fundamental principle, the answer is $26 + (26)^2 + (26)^3 + (26)^4$.

Although the fundamental principle lies at the heart of our counting procedures it is hoped that the reader will develop techniques of obtaining answers to counting problems without referring to elaborate systems of sets. Some of the ensuing exercises and text material should help to bring this about.

**E X E R C I S E S  1 . 1**  ────────────────────────────────

Throughout this set of exercises, "word" will mean list of symbols from some "alphabet." It is not implied that the words be words in any language. Exercises 9–13 contain ideas which are essential to the subsequent material and should be carefully studied.

**1**  Consider the alphabet consisting of W, X, Y, Z.

(a)  How many 2-letter words can be made?

(b)  How many 3-letter words can be made?

(c)  How many 4-letter words can be made?

(d)  How many 2-letter words can be made if no letter can be repeated in one word?

(e)  How many 4-letter words can be made if no letter can be repeated in one word?

(f)  List the 2-letter words in sets where two words are in the same set if their second letters are the same. How many sets are there? How many words are in each set?

(g)  List the 3-letter words in sets where two words are in the same set if their first two letters form the same 2-letter word.

**2**  (a)  How many 2-digit numerals can be made with the digits 2, 3, 5, 7?

 (b)  How many 3-digit numerals can be made with these digits?

 (c)  Could problems analogous to those of exercise 1(a)–(g) be posed with essentially the same results?

**3**  How many 2-digit numerals have 2, 3, 5, or 7 as the first digit and any of the digits 2, 3, 4, 5, 6, 7, 8, or 9 which are no greater than the first digit in the second position? List the numerals in sets according to the first digit.

**4**  How many different 6-letter words can be made with the English alphabet? How many such words have no letter repeated in the same word?

**5**  In how many different ways can 14 people be seated in a row of 14 seats? How does this problem differ from the problem of making 14-letter words with a 14-letter alphabet if no letter is repeated in a single word?

**6**  Consider the problem of counting 3-letter words with the regular alphabet. Suppose we argued as follows. Put all words having the same second and third letters in the same set. There are 26 words in a set because there are 26 possible first letters. There are $(26)^2$ sets because the second and third letters form a 2-letter word and each determines a set. What is wrong with such an argument?

**7**  How many different 2-letter words can be formed if the first letter is a consonant and the second letter is a vowel which occurs earlier in the alphabet than the first letter of the word?

**8**  How many zeros appear at the right-hand side of the decimal representation of the following numbers?

 (a)   $3! = 3 \cdot 2 \cdot 1$

 (b)   $5! = 5 \cdot 4 \cdot 3 \cdot 2 \cdot 1$

 (c)   $8! = 8 \cdot 7 \cdot 6 \cdots 2 \cdot 1$

 (d)   $10! = 10 \cdot 9 \cdot 8 \cdots 2 \cdot 1$

 (e)   $26! = 26 \cdot 25 \cdot 24 \cdot 23 \cdots 2 \cdot 1$

 (f)   $137! = 137 \cdot 136 \cdot 135 \cdots 2 \cdot 1$

[*Hint:* Collect the factors into sets containing numbers divisible by 2, by 5, by 25, by 125, and by neither 2 nor 5.]

♦ **9**  How many different 2-letter words may be formed if the first letter must be a consonant and the second must be a vowel?

♦ **10**  How many different 3-letter words may be formed if the first letter is a consonant, the second letter is a vowel, and third letter is one of the letters A, H, I, M, O, T, U, V, W, X, Y?

♦ **11**  How many different 2-symbol words can be formed if the first symbol must be chosen from among $m_1$ symbols and the second symbol must be chosen from among $m_2$ symbols?

♦ **12**   How many different 3-symbol words may be formed if the first symbol must be chosen from among $m_1$ different symbols, the second symbol must be chosen from among $m_2$ different symbols, and the third symbol must be chosen from among $m_3$ symbols?

♦ **13**   (Generalization of 10–12.) How many different $r$-symbol words can be formed if the first symbol must be chosen from among $m_1$ different symbols, the second symbol must be chosen from among $m_2$ different symbols, ..., the $r$th symbol must be chosen from among $m_r$ different symbols? (Naturally, $r$ is a positive integer.)

**14**   How many different 3-symbol words may be formed if the first symbol is one of the symbols {AE}, {AI}, {AO}, {AU}, {EI}, {EO}, {EU}, {IO}, {IU}, {OU}, the second symbol is an English alphabet consonant, and the third symbol is a standard numeral from 1 to 999 inclusive?

**15**   How many license plates may be made which have the appearance ☆□△–⊙, where ☆ is a vowel, □ is a vowel beyond ☆ in the alphabet, △ is a consonant, and ⊙ is a standard numeral between 1 and 999 inclusive? Two such license plates would be EOW–912 , AUK–71 .

**16**   Suppose $f(n)$ is a non-negative integer-valued function of the integer-valued variable $n$ where $1 \le n \le 17$. How many different integer-valued functions $g(n)$ with $1 \le n \le 17$ are there satisfying $0 \le g(n) \le f(n)$? How many such $g(n)$ satisfy $|g(n)| \le f(n)$?

**17**   How many different non-negative integer divisors does $9000 = 2^3 \cdot 3^2 \cdot 5^3$ have? [*Hint:* $2^0 \cdot 3^2 \cdot 5^1$ is one divisor.]

**18**   How many 4-letter words made with the English alphabet have repeated letters?

**19**   How many 5-letter words can be made with the English alphabet in such a way that vowels and consonants alternate?

**20**   In how many ways may balls numbered $1, 2, 3, \ldots, n$ be placed into $k$ distinguishable boxes?

# 1.2   Permutations and combinations

Certain collections occur frequently and it is desirable to have notation for the number of objects in those collections.

In the discussion following Example 1.1.5, we observed that the number of $r$-letter words made from an alphabet of $n$ symbols and having no symbol repeated in the same word is

$$n(n - 1)(n - 2) \cdots (n - r + 1), \quad \text{where } 0 < r \le n$$

Such words which are just lists of symbols without repetitions are often called *permutations* of the symbols; i.e., arrangements of some symbols. They occur frequently and for convenience we define

$$_nP_r = n(n - 1)(n - 2) \cdots (n - r + 1)$$

The symbol $_nP_r$ is read as "the number of permutations of $n$ symbols taken $r$ at a time." Furthermore we set $_nP_n = n(n - 1)(n - 2) \cdots 2 \cdot 1 = n!$ (read "$n$ factorial"). Other notation for $_nP_r$ is $n_r$ and $P_r^n$.

Since $(n + 1)! = (n + 1)(n)(n - 1)(n - 2) \cdots 2 \cdot 1 = (n + 1)n!$, we may define $0! = 1$ and have a value consistent with $(n + 1)! = (n + 1)n!$ in the case $n = 0$. For $0 < r < n$, $n! = n(n - 1) \cdots (n - r + 1)(n - r)(n - r - 1) \cdots 2 \cdot 1$ so we have $n! = {_nP_r}(n - r)!$. This can be verified separately for $n = r$. Thus for $0 < r \leq n$ we have

**(1.2.1)**    $$_nP_r = \frac{n!}{(n - r)!}$$

### Example 1.2.1

Five cards from a standard deck are dealt one at a time, face up. How many different arrangements are possible?

Presumably the order is relevant since if the order of the cards were different the appearance would be different. Assuming this we have a list of five symbols (cards) with no repetitions. Hence there are $_{52}P_5 = 52 \cdot 51 \cdot 50 \cdot 49 \cdot 48$ such arrangements.

### Example 1.2.2

Five pieces of paper are inscribed with the letters A, B, C, D, E—one letter on each piece of paper. How many 5-letter words may be formed with these slips? How many of these words have B occurring immediately after A? How many words begin with A and end with E?

Assuming that we may only arrange the slips in a row, the answer to the first part is 5! since a piece of paper cannot occupy two positions. To answer the second part we imagine *four* slips of paper—$\boxed{AB}$ $\boxed{C}$ $\boxed{D}$ $\boxed{E}$. A word of the sort we seek is merely an arrangement of these four slips in a row. There are 4! words of this type. For part three, we need only count the number of arrangements of the three slips labeled B, C, D in the second, third, and fourth positions since A is fixed in the first position and E is fixed in the fifth position. There are 3! arrangements (permutations) of B, C, D so there are 3! words beginning with A and ending with E.

Suppose we have a basic collection of distinct objects $S = \{s_1, s_2, \ldots, s_n\}$. In the context of counting, a subcollection containing $r$ of the objects is frequently called a *combination* of the $n$ objects taken $r$ at a time.

To be sure, a combination of elements is a set. However, the terms "combination" and "set" come from different mathematical traditions and both remain in use. The fields of probability and statistics have acquired a peculiar and nonmathematical terminology as a result of their history.

**Example 1.2.3**

Suppose $n = 4$. List all of the combinations of the four objects of $S$ taken two at a time.

One answer is $\{s_1, s_2\}, \{s_1, s_3\}, \{s_1, s_4\}, \{s_2, s_3\}, \{s_2, s_4\}$, and $\{s_3, s_4\}$. There are obviously other *lists*.

**Example 1.2.4**

Suppose $n = 5$. List all of the combinations of the five objects of $S$ taken three at a time.

The required combinations are

$$\{s_1, s_2, s_3\} \quad \{s_1, s_4, s_5\}$$
$$\{s_1, s_2, s_4\} \quad \{s_2, s_3, s_4\}$$
$$\{s_1, s_2, s_5\} \quad \{s_2, s_3, s_5\}$$
$$\{s_1, s_3, s_4\} \quad \{s_2, s_4, s_5\}$$
$$\{s_1, s_3, s_5\} \quad \{s_3, s_4, s_5\}$$

We now determine the total number of combinations of $n$ objects taken $r$ at a time if $0 < r \leq n$. Let $k$ be the number of (different) combinations of the $n$ symbols taken $r$ at a time. Suppose we take one such combination of $s_1, s_2, \ldots, s_n$; e.g., if $n = 7$ and $r = 3$ one such combination is $\{s_1, s_3, s_6\}$. We can form $3!$ permutations of these symbols—namely, $s_1, s_3, s_6$; $s_1, s_6, s_3$; $s_3, s_1, s_6$; $s_3, s_6, s_1$; $s_6, s_1, s_3$; $s_6, s_3, s_1$. In general there are $r!$ permutations of the symbols in any specific combination of $r$ symbols. Thus we have a division of the $_nP_r$ permutations of $n$ symbols $r$ at a time into $k$ sets of $r!$ permutations. Any two permutations in the same set are different arrangements of the same combination of $r$ symbols. By the corollary to the fundamental principle we have $_nP_r = kr!$. Therefore the number of combinations

**(1.2.2)** $$k = \frac{n!}{(n-r)!\,r!}$$

Note that $k = \binom{n}{r}$, the coefficient of $x^r$ is the binomial expansion of $(1 + x)^n$. The appearance of the binomial coefficient $\binom{n}{r}$ is no coincidence (such things rarely are) but we will save further elaboration until section 1.5. It should be pointed out that other books use $C_r^n$ or $_nC_r$ as notation for $\binom{n}{r}$.

The expression $\dfrac{n!}{n!0!} = 1$ (since $0! = 1$) is a natural value for the number of combinations of $n$ symbols taken zero at a time. Thus we can think of $\dbinom{n}{r}$ for $r = 0$ as well as for $0 < r \le n$ as giving the number of subsets (combinations) of $n$ distinct objects taken $r$ at a time.

The binomial formula states that

(1.2.3) $\quad (1 + x)^n = x^n + \dbinom{n}{1}x^{n-1}\cdot 1 + \dbinom{n}{2}x^{n-2}\cdot 1^2 + \dbinom{n}{3}x^{n-3}\cdot 1^3 + \cdots$

$$+ \dbinom{n}{n-1}x\cdot 1^{n-1} + 1^n$$

Since $\dbinom{n}{0} = \dbinom{n}{n} = 1$ and since $1^r = 1$ we can write (1.2.3) as

$$(1 + x)^n = \dbinom{n}{0}x^n + \dbinom{n}{1}x^{n-1} + \dbinom{n}{2}x^{n-2} + \cdots + \dbinom{n}{n-1}x + \dbinom{n}{n}$$

If we set $x = 1$ we obtain

(1.2.4) $\quad 2^n = \dbinom{n}{0} + \dbinom{n}{1} + \dbinom{n}{2} + \cdots + \dbinom{n}{n-1} + \dbinom{n}{n}$

Thus there are $2^n$ different subsets of a set of $n$ objects if one includes the subset having no objects as a legitimate subset.

In the following examples we use the words "hand of cards" to mean a collection of cards, i.e., order is irrelevant. A "kind" is a level such as king, four, ace, eight, etc. A standard deck has four cards of each kind and 13 cards in each suit. The suits are clubs, diamonds, hearts, and spades.

### Example 1.2.5

How many hands of five cards have cards of one suit only?

Suppose we form four sets; in the first set we put all such hands which contain only clubs, in the second set we put all such hands containing only diamonds, etc. Since there are 13 club cards, the number of different hands containing five club cards is $\dbinom{13}{5}$. The answer to our problems is therefore $4\dbinom{13}{5}$.

### Example 1.2.6

How many hands of five cards contain three cards of one kind and two cards of a different kind?

It should be fairly clear that there are $13\cdot 12 \cdot \dbinom{4}{3}\cdot\dbinom{4}{2}$ such hands. But if further clarification is needed we argue as follows: Any such hand can be described by a word of the type first kind, another kind, combination of three suits, combination of two suits. For example, ace, eight, $\{\heartsuit, \diamondsuit, \spadesuit\}, \{\heartsuit, \clubsuit\}$ means that the hand is composed of the aces of hearts, diamonds, and spades and the eights of hearts and clubs. Since there are $\binom{4}{3}$ combinations of four things three at a time and $\binom{4}{2}$ combinations of four things two at a time and since $_{13}P_2 = 13 \cdot 12$ is the number of ways of describing a first kind and another kind, the answer follows.

The discussion of Example 1.2.6 uses the technique of "code words." We use this technique in some of the subsequent examples.

### Example 1.2.7

How many hands of five cards contain two cards of one kind, two cards of a second kind, and one card of a third kind?

The number $13 \cdot 12 \cdot 11 \cdot \binom{4}{2} \cdot \binom{4}{2} \cdot \binom{4}{1}$ is *not* the correct answer. It would be correct if among others $J\heartsuit$, $J\diamondsuit$, $K\diamondsuit$, $K\clubsuit$, $4\heartsuit$ is different from $K\diamondsuit$, $K\clubsuit$, $J\heartsuit$, $J\diamondsuit$, $4\heartsuit$. However, since two jacks and two kings in a *hand* are no different from the same two kings and the same two jacks, the presence of $13 \cdot 12$ is erroneous. The correct answer is

$$\binom{13}{2} \cdot 11 \cdot \binom{4}{2} \cdot \binom{4}{2} \cdot \binom{4}{1}$$

since one needs only a combination of 13 (kinds) taken two at a time in order to select two kinds from which the two cards are to come. In other words, we form a code word as follows: combination of two kinds, third kind, combination of two suits, combination of two suits, combination of one suit. As in Example 1.2.6, such a code word represents a hand containing two cards of the kinds given by the first symbol of the word and of the suits specified by the third and fourth symbols of the word and containing one card of the kind given by the second symbol and the suit given by the fifth symbol.

The idea of forming what one might call code words to describe the cards in a hand as in Examples 1.2.6 and 1.2.7 might be puzzling at first sight. The fact remains that it is more in keeping with the style of the book to count words of diverse alphabets. Furthermore, counting hands of cards which fit various descriptions is not so easy. Experience shows that people often get 2, 6, or 24 times the correct answer because they count the same thing repeatedly. We feel that careful study and a few examples will show how to set up code words that are in one-to-one correspondence with the hands at issue. It is hoped that this method reduces the chances of error.

Another example of this type is as follows.

### Example 1.2.8

Twenty universities sent a physicist, a chemist, a biologist, an astronomer, and a mathematician (from each university) to a scientific meeting. How many committees of ten which can be formed from the 100 scientists have each field represented by two scientists? How many such committees have two scientists from each field and ten different universities represented?

In essence we must choose two universities to send (two) physicists, two universities to send (two) chemists, two universities to send (two) biologists, etc. Another way to look at this problem is to imagine an alphabet each of whose symbol is a *combination* of two universities. There are $\binom{20}{2} = \dfrac{20 \cdot 19}{2}$ such combinations. We want combinations because Harvard's physicist and Chicago's physicist would give the same representation as Chicago's physicist and Harvard's physicist. We now want to form a 5-letter code word using that alphabet. Repetitions are allowed because both Chicago and Harvard could have a physicist and a biologist on the committee. The answer is $\binom{20}{2}^5 = (190)^5$.

The second part is somewhat more complicated. In essence, we wish to form 5-letter words that look like

$$\{1, 2\}, \{7, 9\}, \{3, 20\}, \{6, 4\}, \{8, 13\}$$

if the universities are numbered from 1 to 20. The first symbols can be formed in $\binom{20}{2}$ ways and then the second symbol can be formed in $\binom{18}{2}$ ways after which the third symbol can be formed in $\binom{16}{2}$ ways, etc. The answer is

$$\frac{20 \cdot 19}{2} \cdot \frac{18 \cdot 17}{2} \cdot \frac{14 \cdot 13}{2} \cdot \frac{12 \cdot 11}{2}$$

This answer can be deduced from other approaches with more ease.

### Example 1.2.9

How many of the committees of ten formed as in Example 1.2.8 have three scientists from each of three universities and one scientist from a fourth university?

We use a 6-letter code word of the form

$$\star \quad n \quad \triangle \quad \square \quad \bigcirc \quad f$$

The $\star$ represents a combination of three universities from the universities numbered $1, 2, 3, \ldots, 20$, $n$ is a number not in $\star$ but $1 \leq n \leq 20$, $\triangle$ is a combination of three of the five fields telling which scientists come from the lowest number in ☆, □ is a combination of three fields for the next number in ☆, ○ is a combination of three fields for the largest number in ☆, and $f$ is the field of university $n$'s scientist.

It should be clear that any code word of the form indicated determines and is determined by one and only one committee. We illustrate how the committee $\{p_1, b_1, c_1, a_9, b_{12}, a_{12}, c_9, m_9, b_{18}, m_{12}\}$ leads to a code word. Here $p, c, b, a, s$ are the first letters of the fields and the subscripts are for the universities numbered 1 to 20. The code word for that committee is

$$\{1, 12, 9\}, 18, \{p, b, c\}, \{a, c, m\}, \{a, b, m\}, b$$

We see that there are $\binom{20}{3} \cdot 17 \cdot \binom{5}{3} \cdot \binom{5}{3} \cdot \binom{5}{3} \cdot 5$ such words and therefore that many such committees.

**EXERCISES 1.2** _____

The answers to most of the combinatorial problems which follow are given in the form $_6P_3$ or 4! rather than 120 or 24 since the latter numbers are relatively meaningless as far as the patterns of the problem are concerned. Naturally, there are situations where two expressions which look different give the same answer.

**1**  Evaluate $_{10}P_7$, $_6P_4$, and 7!.

**2**  Evaluate $\binom{6}{4}$, $\binom{10}{3}$, $\binom{6}{2}$, and $\binom{10}{7}$. Do you notice anything? If not, evaluate $\binom{7}{3}$ and $\binom{7}{4}$ as well as $\binom{8}{3}$ and $\binom{8}{5}$.

◆ **3**  Show that $\binom{n}{r} = \binom{n}{n-r}$ if $0 \leq r \leq n$.

**4**  Show that if $r \leq \frac{1}{2}(n-1)$, then $\binom{n}{r} \leq \binom{n}{r+1}$.

**5**  How many 4-flag signals may be made with six different flags? A "4-flag signal" consists of running up four flags on a single pole.

**6**  Five boys and five girls go on a hike. How many single file arrangements are possible if a certain boy, Courtney, is to lead? How many arrangements if Courtney leads and boys and girls alternate?

**7**  A barrel of 10 000 screws contains 20 defective screws. How many different batches of six screws containing no defectives are possible? How many batches of six screws containing at least one defective screw are possible?

**8**  A local political organization has 60 men and 40 women members. How many different 4-person delegations to the national convention are possible? How many of the delegations are not constituted entirely of members of the same sex? [*Hint:* Find out how many delegations are composed of men only, of women only, and subtract intelligently.]

**9**   In accordance with Example 1.2.9, form code words for the following committees.

(a)   $\{p_3, b_3, c_3, p_8, m_8, c_8, m_{11}, b_{11}, a_{11}, c_{16}\}$

(b)   $\{c_7, a_4, b_3, m_4, m_7, p_7, b_4, a_{12}, c_{12}, m_{12}\}$

(c)   $\{a_4, a_8, b_8, c_5, c_8, c_{20}, m_5, m_{20}, p_5, p_{20}\}$

Form the committee corresponding to the code word

(d)   $\{3, 6, 15\}, 9, \{a, c, m\}, \{a, m, p\}, \{a, b, c\}, p$

**10**   How many different hands of five cards from a standard deck contain three cards of one kind, one card of a second kind, and one card of a third kind?

**11**   How many different hands of seven cards from a standard deck contain five spade cards and two diamond cards? How many hands of seven cards contain five cards from one suit and two cards from another suit?

**12**   How many different hands of seven cards from a standard deck contain two cards from each of three kinds and one card from a fourth kind? [*Hint:* See Example 1.2.7.]

**13**   How many different hands of ten cards contain two cards from each of two kinds and three cards from each of two other kinds?

**14**   How many 7-letter words may be made from blocks labeled A, B, C, D, E, F, G, if A, B, C must occur in that order and be adjacent, and D, E must occur in that order and be adjacent?

**15**   When one talks about a combination lock, is the word "combination" used in the same way as we use the word "combination" in this section? Explain.

**16**   A campus organization contains six professors and 98 students. How many different delegations of two professors and four students can be formed?

**17**   In order to shoot off a hypothetical missile one must have three keys which must be put simultaneously into three holes. The outward appearance of the keys are the same. If a person finds 20 of such keys, among them the right keys, how many arrangements of three keys might he have to try before he successfully launches the missile?

**18**   How many different hands of five cards from a standard deck have all cards coming from kinds ten, jack, queen, king, ace?

**19**   In how many ways can balls numbered $1, 2, 3, \ldots, n$ be put in $k$ distinguishable boxes? If $k \leq n$, in how many such arrangements are there balls in each box?

**20**   If $n \geq 1$ show that

$$\binom{n}{0} + \binom{n}{1}\frac{1}{2} + \binom{n}{2}\left(\frac{1}{2}\right)^2 + \binom{n}{3}\left(\frac{1}{2}\right)^3 + \cdots + \binom{n}{n-1}\left(\frac{1}{2}\right)^{n-1} + \binom{n}{n}\left(\frac{1}{2}\right)^n = \left(\frac{3}{2}\right)^n$$

**21**   If $n \geq 1$ show that

$$\binom{n}{0} - \binom{n}{1} + \binom{n}{2} - \binom{n}{3} + \cdots + (-1)^n\binom{n}{n} = 0$$

# 1.3 More counting problems

Let us look at another counting problem.

### Example 1.3.1
How many different 6-letter words may be made with slips of paper labeled A, A, A, B, B, C? The slips marked A are considered indistinguishable from one another as are those marked B.

We consider some associated symbols, say, $A_1$, $A_2$, $A_3$, $B_1$, $B_2$, C which are taken as distinguishable. There are 6! permutations of these distinguishable symbols. We form $k$ sets where two permutations are in the same set if they form the same word when the subscripts are ignored, e.g., $A_1A_2B_1A_3B_2C$, $A_2A_3B_2A_1B_1C$, and $A_1A_3B_1A_2B_2C$ are in the same set. There are as many sets as there are words of the type called for in the example. Now each set contains $3! \cdot 2!$ permutations of the distinguishable symbols because there are 3! permutations of the subscripts 1, 2, 3 for the A's and 2! permutations of the subscripts 1, 2 for the B's. We thus have $6! = {}_6P_6 = k(3!2!)$ by the corollary to the fundamental principle. Thus $k = 6!/3!2!$ is the answer we seek.

### Example 1.3.2
How many different 11-letter words may be formed with slips labeled M, I, S, S, I, S, S, I, P, P, I?

The reasoning of Example 1.3.1 is applicable. There are 11! permutations of 11 distinguishable symbols. Moreover a word such as SMISSIPIPIS leads to 4!4!2! permutations of the distinguishable symbols M, $I_1$, $I_2$, $I_3$, $I_4$, $S_1$, $S_2$, $S_3$ $S_4$, $P_1$, $P_2$. We therefore have $11! = k4!4!2!$, where $k$ is the number of sets as in Example 1.2.8. However, it is preferable to write

$$k = \frac{11!}{4!4!2!1!}$$

in order to display it as a specialization of the general formula.

The preceding two examples suggest the following theorem.

**Theorem 1.3.1**   The number of $n$-symbol words which may be formed with $k_1$ indistinguishable copies of a first symbol, $k_2$ indistinguishable copies of another symbol, ..., $k_r$ indistinguishable copies of an $r$th symbol is

$$\frac{n!}{k_1!k_2!\cdots k_r!}$$

if $k_1 + k_2 + \cdots + k_r = n$ and the symbols of the different kinds are distinguishable.

We will not give a formal proof of this theorem.

## Example 1.3.3

(The classical occupancy problem) In how many different ways may $n$ indistinguishable balls be put into $k$ distinguishable boxes?

We imagine the boxes arranged as a row of pigeon-holes. We denote by W the wall between two adjacent boxes (pigeon-holes) and by B a ball. Suppose we arrange $n$ indistinguishable letters B and $k - 1$ indistinguishable letters W to form a word, e.g., if $n = 5$ and $k = 4$, a word could be BWWBBBWB. We interpret that word as follows: All B's to the left of the first W denote balls in the first box. All B's between the first and second W's denote balls in the second box. All B's between the second and third W's denote balls in the third box, and so on. Finally, all B's to the right of the $(k - 1)$st W denote balls in the $k$th box. Thus BWWBBBWB signifies one ball in the first box, no balls in the second box, three balls in the third box, and one ball in the fourth box. It is clear that to each arrangement of balls in the boxes there is one and only one word of the kind mentioned and to each such word there is one and only one such arrangement of $n$ balls in $k$ boxes. There are

$$\frac{(n + k - 1)!}{n!(k - 1)!} = \binom{n + k - 1}{n}$$

such words by Theorem 1.3.1 with $k_1 = n$ and $k_2 = k - 1$.

## Example 1.3.4

In how many different ways may $n$ or fewer indistinguishable balls be put into $k$ distinguishable boxes?

Suppose we consider putting $n$ indistinguishable balls into $k + 1$ distinguishable boxes where the first $k$ boxes are thought of as being the $k$ boxes indicated in the problem and the remaining box as being a kind of refuse box for those balls not arranged in the first $k$ boxes. Since $0, 1, 2, 3, \ldots, n$ balls may be put into the refuse box we will be arranging $n$ or fewer indistinguishable balls in $k$ boxes for each arrangement of $n$ balls in $k + 1$ boxes. By Example 1.3.3, there are $\binom{n + k}{n}$ of the latter arrangements.

## EXERCISES 1.3

1   How many different 10-letter words can be made with slips labeled AAABBBCCDD? How many of them have the three A's next to one another?

2   Paul has blocks labeled AAABBCCD. How many different 8-letter words can he make? How many 7-letter words can he make?

3   Explain why the expansion of $(x + y + z)^n$ has $\dfrac{n!}{n_1! n_2! n_3!}$ as the coefficient of $x^{n_1} y^{n_2} z^{n_3}$ (with $n_1 + n_2 + n_3 = n$).

4   How many different 8-flag signals can be made with four identical red flags, two identical blue flags, and two identical yellow flags?

5    Twelve levers are in a row. Each may be in three positions which we will call 1, 2, and 3. How many arrangements involve five levers in position 1, four in position 2, and three in position 3?

6    Show that $\dfrac{13!}{4!5!3!}$ is a whole number.

7    A medical experimenter proposes to give six doses of drug A, three of drug B, and four of drug C in some order. How many different orders are possible?

8    Ten balls are numbered 1–10 and there are ten other unnumbered balls which are otherwise indistinguishable. In how many different ways can all of these balls be placed in 12 distinguishable boxes?

9    In how many different ways can some, all, or none of the balls of exercise 8 be placed in 12 distinguishable boxes?

10    How many different expressions of the form

$$\frac{\partial^n}{\partial x_1{}^{n_1} \, \partial x_2{}^{n_2} \, \partial x_3{}^{n_3} \cdots \partial x_k{}^{n_k}}$$

where $n_1 + n_2 + n_3 + \cdots + n_k = n$, are there? [*Hint:* This is like putting indistinguishable balls into distinguishable boxes.]

11    How many arrangements of 15 indistinguishable balls in eight distinguishable boxes have no empty boxes?

12    How many arrangements of 20 indistinguishable balls in eight distinguishable boxes have at least two balls in each box?

13    In how many different ways can one arrange $m$ red but otherwise indistinguishable and $n$ blue but otherwise indistinguishable balls in $k$ boxes?

14    How many ways may 15 indistinguishable A's and 12 indistinguishable B's be arranged in a row so that there are five groups of B's, four groups of A's, and the groups alternate? Two such arrangements are BBAAAABBAAAA-BBAAAABBBAAABBB and BAAAAAAAAABBAABAABAABBBBBBB.

[*Hint:* In essence one is arranging 15 A's in four boxes and 12 B's in five other boxes with no empty boxes in each case.]

15    In how many ways can $m$ red balls be arranged in $r$ distinguishable boxes and $n$ green balls in $s$ other distinguishable boxes?

# 1.4   Stirling's formula

The examples and exercises of the previous pages made extensive use of $n!$ for various integers $n$. It is not convenient to calculate $n!$ for large $n$. Values of $n!$ for $n > 15$ are extremely tedious to calculate by hand and although electronic computers can do amazing things, each machine has its limits.

To estimate $n!$ for large $n$ we introduce Stirling's formula without proof.*

**(1.4.1)** $\quad n! \sim \sqrt{2\pi n}\, n^n e^{-n}$

The symbol "$\sim$" is read as "is asymptotic to" and (1.4.1) means that

$$\lim_{n \to \infty} \frac{n!}{\sqrt{2\pi n}\, n^n e^{-n}} = 1$$

Generally speaking we write $f(x) \sim g(x)$ as $x \to a$ if $\lim_{x \to a} [f(x)/g(x)] = 1$. Now if $g(x) \to \infty$ as $x \to a$ it is quite possible that $|f(x) - g(x)| \to \infty$ as $x \to a$. (See exercises 1.4.5 and 1.4.6.) It is true that $|n! - \sqrt{2\pi n}\, n^n e^{-n}| \to \infty$ as $n \to \infty$. However, if one approximates (by Stirling) $n!/m!$ for large $n$ and $m$, the errors compensate one another to some extent.

We use the symbol "$a \approx b$" to mean that $a$ is approximately equal to $b$.

**Example 1.4.1**

Estimate $_{1000}P_{200}$ by Stirling's formula. $_{1000}P_{200} = 1000!/800!$.

We estimate by computing

$$\frac{\sqrt{2\pi \cdot 1000}}{\sqrt{2\pi \cdot 800}} \frac{1000^{1000} e^{-1000}}{(800)^{800} e^{-800}}$$

The common logarithm of the last expression is

$$\tfrac{1}{2} \log 1000 - \tfrac{1}{2} \log 800 + 1000 \log 1000 - 800 \log 800 + 800 \log e - 1000 \log e$$
$$= \tfrac{1}{2}(1 - \log 8) + 200 \log 1000 + 800(1 - \log 8) - 200 \log e$$

where we have used $\log 1000 - \log 800 = \log(1000/800) = \log 10 - \log 8$. Since $\log 1000 = 3$ and $\log e = 0.43429$, we find that

$$\log_{10}{}_{1000}P_{200} \approx 590.675$$

so

$$_{1000}P_{200} \approx 4.73 \times 10^{590}$$

**Example 1.4.2**

Estimate $\binom{2n}{n}$ for large $n$.

Again we apply (1.4.1) to $\binom{2n}{n} = \dfrac{(2n)!}{n!n!} = \dfrac{(2n)!}{(n!)^2}$. This yields

$$\frac{\sqrt{2\pi 2n}(2n)^{2n} e^{-2n}}{(2\pi n)(n^n)^2 (e^{-n})^2} = \frac{\sqrt{2}}{\sqrt{2\pi n}} \frac{2^{2n} n^{2n} e^{-2n}}{n^{2n} e^{-2n}}$$

$$= \frac{1}{\sqrt{\pi n}} 2^{2n}$$

---

* For proof of Stirling's formula see J. V. Uspensky *Introduction to Mathematical Probability* (McGraw-Hill, New York, 1937); R. C. Buck, *Advanced Calculus* (McGraw-Hill, New York, 1965).

**EXERCISES 1.4**

1   Estimate $_{100}P_{60}$ by Stirling's formula.

2   What is the percentage error made in estimating 6! and 8! by (1.4.1)?

3   Estimate $\binom{4n}{n}$ for large $n$.

4   Estimate the coefficient of $x^{25}\, y^{25}\, z^{25}\, w^{25}$ in $(x + y + z + w)^{100}$.

5   Show that if $\lim\limits_{x \to a} g(x) = b$ and $b \neq 0$ and "$b \neq \infty$" then $f(x) \sim g(x)$ implies that $\lim\limits_{x \to a} |f(x) - g(x)| = 0$.

6   Give an example of functions $f(x)$ and $g(x)$ such that $f(x) \sim g(x)$ as $x \to \infty$ but $f(x) - g(x) \to \infty$ as $x \to \infty$.

7   Use the ratio test on the power series $\sum\limits_{n=1}^{\infty} \dfrac{n^n}{n!}\, x^n$. For what values of $x$ does the series converge? Does the result seem surprising in view of (1.4.1)? Why?

# ★ 1.5   Generating functions

In section 1.2 it was noted that the coefficient of $x^r$ in the usual binomial expansion of $(1 + x)^n$ is $\binom{n}{r}$, the number of combinations of $n$ things taken $r$ at a time. We now suggest a way of looking at this which throws light on similar questions.

An elementary computation gives

$$(x_1 + y_1)(x_2 + y_2)(x_3 + y_3) = x_1 x_2 x_3 + x_1 x_2 y_3 + x_1 y_2 x_3 + y_1 x_2 x_3$$
$$+ x_1 y_2 y_3 + y_1 x_2 y_3 + y_1 y_2 x_3 + y_1 y_2 y_3$$

Similar computations, such as that of $(x_1 + y_1)(x_2 + y_2)(x_3 + y_3)(x_4 + y_4)$, reveal the following pattern: The answer is a sum of products which are composed of one term from each bracketed quantity. All possible terms of this form occur and no term involving two terms from the same bracketed quantity occurs, e.g., we never have $x_1 y_1 \cdots$. Similarly

$$(x_1 + y_1 + z_1)(x_2 + y_2 + z_2) = x_1 x_2 + x_1 y_2 + x_1 z_2 + y_1 x_2 + y_1 y_2 + y_1 z_2$$
$$+ z_1 x_2 + z_1 y_2 + z_1 z_2$$

If we consider $(x + y)^n = \underbrace{(x + y)(x + y) \cdots (x + y)}_{n \text{ factors}}$ and expand as above,

there are $2^n$ terms before simplification.

Since there are $n$ factors $(x + y)$, in general we select an $x$ from $r$ factors and a $y$ from the remaining $n - r$ factors. The $r$ factors from which we select an $x$ are a combination of $n$ things taken $r$ at a time. Thus there are as many terms involving $r$ factors $x$ (and $n - r$ factors $y$) as there are combinations of $n$ things (factors) taken $r$ at a time. Thus $(x + y)^n = \sum_{r=0}^{n} \binom{n}{r} x^r y^{n-r}$. If we set $y = 1$ we get the desired result.

If we consider $(x + y + z)^n = \underbrace{(x + y + z)(x + y + z) \cdots (x + y + z)}_{n \text{ factors}}$

and let $k_1 + k_2 + k_3 = n$ where $k_1$, $k_2$, $k_3$ are non-negative integers, the coefficient of $x^{k_1} y^{k_2} z^{k_3}$ is $\dfrac{n!}{k_1! k_2! k_3!}$. To see this one could imagine placing an $X$ over any factor from which an $x$ is chosen, a $Y$ over any factor from which a $y$ is chosen, and a $Z$ over any factor from which a $z$ is chosen. Thus there are as many terms in $(x + y + z)^n$ with $k_1$ $x$'s, $k_2$ $y$'s, and $k_3$ $z$'s as factors as there are $n$-letter words with $k_1$ identical $X$'s, $k_2$ identical $Y$'s, and $k_3$ identical $Z$'s.

Similarly, if $k_1$, $k_2$, $k_3$, $k_4$ are non-negative integers with $k_1 + k_2 + k_3 + k_4 = n$, the coefficient of $x^{k_1} y^{k_2} z^{k_3} w^{k_4}$ in the expansion of $(x + y + z + w)^n$ is

$$\frac{n!}{k_1! k_2! k_3! k_4!}.$$

### Example 1.5.1

How many positive integers are divisors of $3528 = 2^3 \cdot 3^2 \cdot 7^2$?

This type of problem was handled in exercise 1.1.17 by a simple method. Here, we use a more complicated procedure which may illustrate other possibilities. We form $(1 + 2 + 2^2 + 2^3)(1 + 3 + 3^2)(1 + 7 + 7^2)$. The number of terms in the expansion is $4 \cdot 3 \cdot 3 = 36$. Each term is clearly a divisor of $3528$ and there are no others.

Consider the polynomials

$$P(x) = a_0 + a_1 x + a_2 x^2 + \cdots + a_m x^m$$

and

$$Q(x) = a_0 + b_1 x + b_2 x^2 + \cdots + b_n x^n$$

If we define $a_k = 0$ for $k > m$ and $b_l = 0$ for $l > n$ and let $P(x)Q(x) = R(x) = \sum_{r=0}^{m+n} c_r x^r$ it is easy to verify that

(1.5.1)    $c_r = a_0 b_r + a_1 b_{r-1} + a_2 b_{r-2} + \cdots + a_r b_0$

$$= \sum_{k=0}^{r} a_k b_{r-k}$$

since $a_k x^k b_l x^l$ contributes to $c_r x^r$ if and only if $k + l = r$.

### Example 1.5.2

Define $\binom{n}{k} = 0$ if $n$ and $r$ are positive integers with $k > n$. Show that

$$\binom{m + n}{r} = \sum_{k=0}^{r} \binom{m}{k}\binom{n}{r - k} \quad \text{if } r \geq m \text{ and } r \geq n$$

Since $(1 + x)^m(1 + x)^n = (1 + x)^{m+n}$ and

$$(1 + x)^m = \sum_{j=0}^{m} \binom{m}{j}x^j, \qquad (1 + x)^n = \sum_{k=0}^{n} \binom{n}{k}x^k$$

we may use the previous remarks with $P(x) = (1 + x)^m$, $Q(x) = (1 + x)^n$, $R(x) = (1 + x)^{m+n}$. Let

$$a_k = \binom{m}{k}, \qquad b_l = \binom{n}{l}, \quad \text{and} \quad c_r = \binom{m + n}{r}$$

Equation (1.5.1) gives $c_r = a_0 b_r + a_1 b_{r-1} + \cdots + a_r b_0 = \sum_{k=0}^{n} a_k b_{r-k}$. But

$a_k = 0$ for $k > m$ and $b_l = 0$ for $l > n$ essentially restates $\binom{m}{k} = 0$ if $k > m$

and $\binom{n}{l} = 0$ if $l > n$. Thus

$$\binom{m + n}{r} = c_r = \sum_{k=0}^{r} \binom{m}{k}\binom{n}{r - k}$$

### Example 1.5.3

Find the sum of the reciprocals of all positive integers whose prime factors are 2, 3, 7, 19.

This problem calls for $A = \sum \frac{1}{k}$ where $k = 2^{j_1}3^{j_2}7^{j_3}19^{j_4}$ and all possible combinations of non-negative integers $j_1$, $j_2$, $j_3$, $j_4$ are used except that we exclude $j_1 = j_2 = j_3 = j_4 = 0$. Obviously $A$ is an infinite series and it may not converge. First of all we recall from the theory of infinite series that

$$1 + x + x^2 + x^3 + \cdots = \frac{1}{1 - x}, \quad \text{if } |x| < 1$$

Thus

$$1 + \frac{1}{2} + \frac{1}{2^2} + \frac{1}{2^3} + \cdots = \frac{1}{1 - \frac{1}{2}}$$

$$1 + \frac{1}{3} + \frac{1}{3^2} + \frac{1}{3^3} + \cdots = \frac{1}{1 - \frac{1}{3}}$$

$$1 + \frac{1}{7} + \frac{1}{7^2} + \frac{1}{7^3} + \cdots = \frac{1}{1 - \frac{1}{7}}$$

$$1 + \frac{1}{19} + \frac{1}{(19)^2} + \frac{1}{(19)^3} + \cdots = \frac{1}{1 - \frac{1}{19}}$$

We expect that the product of the series

$$\left(1 + \frac{1}{2} + \frac{1}{2^2} + \frac{1}{2^3} + \cdots\right)\left(1 + \frac{1}{3} + \frac{1}{3^2} + \cdots\right)\left(1 + \frac{1}{7} + \frac{1}{7^2} + \cdots\right)$$

$$\times \left(1 + \frac{1}{19} + \frac{1}{(19)^2} + \cdots\right)$$

should behave like the product of finite sums and should be a sum of terms $a_i b_j c_k d_l$ where $a_i$ is in $\left\{1, \frac{1}{2}, \frac{1}{2^2}, \frac{1}{2^3}, \cdots\right\}$, $b_j$ is in $\left\{1, \frac{1}{3}, \frac{1}{3^2}, \cdots\right\}$, $c_k$ is in $\left\{1, \frac{1}{7}, \frac{1}{7^2}, \cdots\right\}$, $d_l$ is in $\left\{1, \frac{1}{19}, \frac{1}{(19)^2}, \cdots\right\}$, and each possible arrangement is used once and only once. Since $1 = \frac{1}{2^0} = \frac{1}{3^0} = \frac{1}{7^0} = \frac{1}{(19)^0}$ the only term which we do not want in $A$ is $1 = \frac{1}{2^0 3^0 7^0 (19)^0}$. Therefore

$$A = \frac{1}{1 - \frac{1}{2}}\frac{1}{1 - \frac{1}{3}}\frac{1}{1 - \frac{1}{7}}\frac{1}{1 - \frac{1}{19}} - 1$$

$$= 2 \cdot \frac{3}{2} \cdot \frac{7}{6} \cdot \frac{19}{18} - 1 = \frac{133}{36} - 1$$

### Example 1.5.4

Explain the relationship between the quantity $\binom{n + k - 1}{n}$ which occurs in Example 1.3.3 and the coefficient of $x^n$ in the MacLaurin series for $(1 - x)^{-k}$.

We write

$$(1 - x)^{-k} = \underbrace{\frac{1}{1 - x}\frac{1}{1 - x}\cdots\frac{1}{1 - x}}_{k \text{ factors}}$$

As noted in Example 1.5.4 this is

**(1.5.2)**  $\underbrace{(1 + x + x^2 + \cdots)(1 + x + x^2 + \cdots)\cdots(1 - x + x^2 + \cdots)}_{k \text{ factors}}$

We get $x^n$ in multiplying out these series if we choose $x^{r_1}$ from the first bracket in (1.5.2), $x^{r_2}$ from the second bracket, $\ldots$, $x^{r_k}$ from the $k$th bracket, and $r_1 + r_2 + \cdots + r_k = n$. The number of different ways of doing this must be the coefficient of $x^n$ in a series of the form $\sum_{n=0}^{\infty} a_n x^n$ equal to (1.5.2) for $|x| < 1$. The theory of power series guarantees that this is the MacLaurin series for (1.5.2) which is $(1 - x)^{-k}$. Now we may interpret $r_j$ as the number of balls in the $j$th box where $j = 1, 2, \ldots, k$ and $1 = x^0$. Since we have $r_1 + r_2 + \cdots + r_k = n$ when we study the coefficient of $x^n$ we have a total of $n$ balls in our $k$ boxes. Conversely any arrangement of $n$ indistinguishable balls in $k$ distinguishable boxes determines a choice of $x^{r_j}$ from the $j$th bracket. Therefore, the number of arrangements of this kind must equal the coefficient of $x^n$ in the series.

The ideas of this section may seem frightening since there are no procedures given for dealing with a definite class of questions. One may very well feel that he could not make the correct choice of polynomial or series to solve these problems neatly. This is true of other mathematical problems at all levels. The remarks of this section were designed to suggest certain possibilities. It is hoped that these remarks together with the associated exercises will furnish some insight into neat solutions of a certain type. If not, one may use other techniques to obtain not-so-neat solutions.

### Example 1.5.5

Consecutive rows of dots have the following property. The first row has one dot, the second row has two dots, and if $n > 3$ the $n$th row has two times the number of dots in the $(n - 1)$st row plus the number of dots in the $(n - 2)$nd row. Find an expression for the number of dots in the $n$th row. Specifically, find the number of dots in the 10th row.

Let $a_n$ be the number of dots in the $n$th row. Then $a_1 = 1$, $a_2 = 2$, and $a_n = 2a_{n-1} + a_{n-2}$.

Let $R(x) = \sum_{n=1}^{\infty} a_n x^n$. Now if $n \geq 2$, $a_n x^n = 2a_{n-1}x^n + a_{n-2}x^n$ or $a_n x^n = x2a_{n-1}x^{n-1} + x^2 a_{n-2}x^{n-2}$. If we sum over $n = 3, 4, 5, \ldots$ we obtain

$$\sum_{n=3}^{\infty} a_n x^n = 2x \sum_{n=3}^{\infty} a_{n-1}x^{n-1} + x^2 \sum_{n=3}^{\infty} a_{n-2}x^{n-2}$$

By writing out the terms, it is easy to see that

$$\sum_{n=3}^{\infty} a_{n-2}x^{n-2} = \sum_{n=1}^{\infty} a_n x^n$$

Also

$$\sum_{n=3}^{\infty} a_n x^n = R(x) - x - 2x^2$$

and

$$\sum_{n=3}^{\infty} a_{n-1}x^{n-1} = R(x) - x$$

Therefore,

$$R(x) - x - 2x^2 = 2x[R(x) - x] + x^2 R(x)$$

and

$$-x - 2x^2 = -R(x) + 2xR(x) - 2x^2 + x^2 R(x)$$

or

$$-x = (x^2 + 2x - 1)R(x)$$

Thus

$$-\frac{x}{x^2 + 2x - 1} = R(x)$$

Now $x^2 + 2x - 1 = (x + 1 - \sqrt{2})(x + 1 + \sqrt{2})$ and

$$-\frac{x}{(x + 1 - \sqrt{2})(x + 1 + \sqrt{2})} = \frac{x}{(\sqrt{2} - 1 - x)(1 + \sqrt{2} + x)}$$

$$= \frac{A}{\sqrt{2} - 1 - x} + \frac{B}{1 + \sqrt{2} + x}$$

where $A(1 + \sqrt{2}) + B(\sqrt{2} - 1) + (A - B)x = x$ by partial fraction methods. Thus $A - B = 1$ and $A(1 + \sqrt{2}) + B(\sqrt{2} - 1) = 0$. Routine calculations give

$$R(x) = -\frac{1 - \sqrt{2}}{2\sqrt{2}} \frac{1}{\sqrt{2} - 1 - x} - \frac{(1 + \sqrt{2})}{2\sqrt{2}} \frac{1}{1 + \sqrt{2} + x}$$

$$= \frac{1}{2\sqrt{2}} \left[ \frac{1}{1 - x/(\sqrt{2} - 1)} \right] - \frac{1}{2\sqrt{2}} \left[ \frac{1}{1 + x/(1 + \sqrt{2})} \right]$$

Since if $|z| < 1$, $\sum_{n=0}^{\infty} z^n = \frac{1}{1 - z}$ we have

$$R(x) = \frac{1}{2\sqrt{2}} \left[ \sum_{n=0}^{\infty} \left( \frac{x}{\sqrt{2} - 1} \right)^n - \sum_{n=0}^{\infty} \left( -\frac{x}{\sqrt{2} + 1} \right)^n \right]$$

$$= \frac{1}{2\sqrt{2}} \sum_{n=0}^{\infty} \left[ \left( \frac{1}{\sqrt{2} - 1} \right) - (-1)^n \left( \frac{1}{\sqrt{2} + 1} \right)^n \right] x^n$$

Therefore,

$$a_n = \frac{1}{2\sqrt{2}} \left( \frac{1}{\sqrt{2} - 1} \right)^n - (-1)^n \left( \frac{1}{\sqrt{2} + 1} \right)^n$$

If we rationalize denominators we get

$$a_n = \frac{1}{2\sqrt{2}} \left[ \left( \frac{\sqrt{2} + 1}{1} \right)^n - (-1)^n \left( \frac{\sqrt{2} - 1}{1} \right)^n \right]$$

Thus for example if $n = 10$, we use the binomial expansion of $(\sqrt{2} \pm 1)^n$ to obtain

$$a_{10} = \frac{1}{2\sqrt{2}} \left[ \sum_{k=0}^{10} \binom{10}{k} 2^{k/2} - \sum_{k=0}^{10} \binom{10}{k} 2^{k/2} (-1)^{10-k} \right]$$

$$= \frac{1}{2\sqrt{2}} \left[ 2\binom{10}{1} 2^{1/2} + 2\binom{10}{3} 2^{3/2} + 2\binom{10}{5} 2^{5/2} + 2\binom{10}{7} 2^{7/2} + 2\binom{10}{9} 2^{9/2} \right]$$

$$= 10 + \frac{10 \cdot 9 \cdot 8}{3 \cdot 2 \cdot 1} 2 + \frac{12 \cdot 9 \cdot 8 \cdot 7 \cdot 6}{5 \cdot 4 \cdot 3 \cdot 2 \cdot 1} 4 + \frac{10 \cdot 9 \cdot 8}{3 \cdot 2 \cdot 1} 8 + 10 \cdot 16$$

The preceding example displays the method of generating functions. Specifically if $a_0, a_1, a_2, \ldots$ is a sequence of numbers then the function $\sum_{n=0}^{\infty} a_n x^n$ is called the *generating function* of the numbers if the series converges for some $x = x_0 \neq 0$ (and hence for at least all $x$ with $|x| < |x_0|$). A function to which the series converges on an interval is also called the generating function.

In many problems one can determine the generating function from relationships between the numbers $a_0, a_1, a_2, \ldots$ and manipulate the generating function to get certain results about the numbers themselves. Generating functions will be used in section 11.2.

### Example 1.5.6

How many $n$-letter words which can be made with the alphabet {A, B} do not have two (or more) consecutive A's?

If $n = 1$, the words are A, B. If $n = 2$, the words are AB, BA, BB. If $n = 3$, the words are B AB, B BA, B BB, AB A, and AB B. We write the 3-letter words in a way to suggest a general procedure. Consider a word of the type we seek which has $n + 1$ letters. If it starts with B it must have as its last $n$ letters an $n$-letter word with no two consecutive A's. If it starts with A, the next letter must be B and then it must have as its last $n - 1$ letters an $(n - 1)$-letter word with no two consecutive A's.

Let $a_n$ be the number of words having $n$ letters with no two consecutive A's. We have $a_1 = 2$ and $a_2 = 3$. However, the above argument says that

**(1.5.3)** $\qquad a_{n+1} = a_n + a_{n-1}, \quad$ for $n = 2, 3, 4, 5, \ldots$

where $a_n$ is the number of words of the form

$$\text{B} \;\boxed{n\text{-letter word no two consecutive A's}}$$

and $a_{n-1}$ is the number of words of the form

$$\text{AB} \;\boxed{(n-1)\text{-letter word no two consecutive A's}}$$

If we let $R(x) = \sum\limits_{n=1}^{\infty} a_n x^n$ and if we multiply both sides of (1.5.3) by $x^{n+1}$ we have

$$a_{n+1}x^{n+1} = xa_n x^n + x^2 a_{n-1}x^{n-1}$$

We now sum over $n = 2, 3, 4, 5, \ldots$ and obtain

$$\sum_{n=2}^{\infty} a_{n+1}x^{n+1} = \sum_{n=2}^{\infty} xa_n x^n + \sum_{n=2}^{\infty} x^2 a_{n-1}x^{n-1}$$

$$= x \sum_{n=2}^{\infty} a_n x^n + x^2 \sum_{n=2}^{\infty} a_{n-1}x^{n-1}$$

Now

$$\sum_{n=2}^{\infty} a_{n+1}x^{n+1} = a_3 x^3 + a_4 x^4 + a_5 x^5 + \cdots$$

$$= R(x) - a_1 x - a_2 x^2 = R(x) - 2x - 3x^2$$

Similarly

$$\sum_{n=2}^{\infty} a_n x^n = R(x) - 2x \qquad \sum_{n=2}^{\infty} a_{n-1}x^{n-1} = R(x)$$

We now can write

$$R(x) - 2x - 3x^2 = x[R(x) - 2x] + x^2 R(x)$$

and

$$-2x - x^2 = (x^2 + x - 1)R(x)$$

As in Example 1.5.5 it can be shown that

$$R(x) = \frac{x^2 + 2x}{\sqrt{5}}\left[\frac{2}{\sqrt{5}-1}\sum_{n=0}^{\infty}\frac{(2x)^n}{(\sqrt{5}-1)^n} + \frac{2}{\sqrt{5}+1}\sum_{n=0}^{\infty}\frac{(-2x)^n}{(\sqrt{5}+1)^n}\right]$$

and so $a_n$, the coefficient of $x^n$ in $R(x)$, is given by

$$\frac{1}{\sqrt{5}}\left\{\frac{2}{\sqrt{5}-1}\left[\left(\frac{2}{\sqrt{5}-1}\right)^{n-2} + 2\left(\frac{2}{\sqrt{5}-1}\right)^{n-1}\right]\right.$$

$$\left. + \frac{2}{\sqrt{5}+1}\left[\left(\frac{-2}{\sqrt{5}+1}\right)^{n-2} + \left(\frac{-2}{\sqrt{5}+1}\right)^{n-1}\right]\right\}$$

for $n = 2, 3, 4, 5, \ldots$.

Another way of writing $a_n$ is

$$a_n = \frac{1}{\sqrt{5}}\left[\left(\frac{\sqrt{5}+1}{2}\right)^{n-1}\left(2\frac{\sqrt{5}+1}{2}+1\right) - \left(-\frac{\sqrt{5}-1}{2}\right)^{n-1}\left(-2\frac{\sqrt{5}-1}{2}+1\right)\right]$$

$$= \frac{1}{\sqrt{5}}\left[\left(\frac{\sqrt{5}+1}{2}\right)^{n-1}(\sqrt{5}+2) - \left(-\frac{\sqrt{5}-1}{2}\right)^{n-1}(-\sqrt{5}+2)\right]$$

**EXERCISES 1.5**

1  Recall that $e^u = u^0/0! + u^1/1! + u^2/2! + u^3/3! + \cdots$ Compute

$$\sum_{k=0}^{n} \frac{1}{k!(n-k)!} z^k w^{n-k}$$

Show that $e^{zx}e^{wx} = e^{(z+w)x}$ by comparing the coefficients of $x^n$ in the expressions

$$\frac{(z+w)^0 x^0}{0!} + \frac{(z+w)^1 x^1}{1!} + \frac{(z+w)^2 x^2}{2!} + \frac{(z+w)^3 x^3}{3!} + \cdots$$

and

$$\left(\frac{(zx)^0}{0!} + \frac{(zx)^1}{1!} + \frac{(zx)^2}{2!} + \cdots\right)\left(\frac{(wx)^0}{0!} + \frac{(wx)^1}{1!} + \frac{(wx)^2}{2!} + \cdots\right)$$

2  Let $a_n$, $n = 0, 1, 2, 3, \ldots$, be the number of ways of putting $n$ or fewer indistinguishable balls into $k$ distinguishable boxes. Find the generating function for $a_0, a_1, a_2, a_3, \ldots$.

3  Let $s > 1$. Show, intuitively, that

$$\sum_{n=1}^{\infty} \frac{1}{n^s} = \left(1 - \frac{1}{2^s}\right)^{-1}\left(1 - \frac{1}{3^s}\right)^{-1}\left(1 - \frac{1}{5^s}\right)^{-1}\left(1 - \frac{1}{7^s}\right)^{-1}\cdots$$

where the factors on the right-hand side involve $p^s$ for all prime numbers $p$. Why do we assume $s > 1$?

4  Find the MacLaurin series for $(1 - x)^{-r}$ where $r > 0$. Denote the coefficient of $x^n$ in that series by $(-1)^n \binom{-r}{n}$. Consider the assertion

$$\sum_{k=0}^{n} \binom{-r}{k}\binom{-s}{n-k} = \binom{-r-s}{n}.$$

Check it for $n = 1, 2$, and 3. Prove it in general.

5  Let $a_0, a_1, a_2, a_3, \ldots$ satisfy $a_0 = 0$, $a_1 = 1$, and $a_{n+2} = a_{n+1} + a_n$ for all integers $n \geq 0$. Evaluate $a_n$ for $2 \leq n \leq 7$. Let $\gamma(x) = \sum_{n=0}^{\infty} a_n x^n$. Show that

$$\gamma(x) = \frac{1}{1 - x - x^2}, \quad \text{if } |x| \leq \tfrac{1}{2}$$

The numbers $a_1, a_2, a_3, \ldots$ are the Fibonacci numbers.

6  Find the generating function for the sequence $a_0, a_1, a_2, a_3, \ldots$ where $a_0$ and $a_1$ are assumed to be known and for some constants $A$, $B$, and $C$ with $A \neq 0$, $Aa_{n+2} = Ba_{n+1} + Ca_n$ for all integers $n \geq 0$. An equation of this type is called a linear difference equation.

7  In reference to Example 1.5.6 estimate as roughly as seems reasonable $(1/\sqrt{5})|-\tfrac{1}{2}(\sqrt{5} - 1)|^{n-1}|-\sqrt{5} + 2|$. Show that $a_n$, the number of words of the type indicated, is the nearest whole number to $(1/\sqrt{5})[\tfrac{1}{2}(\sqrt{5} + 1)]^{n-1} \cdot (\sqrt{5} + 2)$.

**8**   Find the generating function for the number of $n$-letter words using the alphabet A, B, C which do not have two or more consecutive A's.

**9**   Find the generating function for the number of $n$-letter words using the alphabet A, B, C which do not have three or more consecutive A's.

Exercises 10–15 are keyed to a certain property of generating functions that make them very useful in probability problems. These ideas are used in section 11.2.

**♦ 10**   Let   $A(x) = \sum\limits_{n=0}^{\infty} a_n x^n$   and   $B(x) = \sum\limits_{n=0}^{\infty} b_n x^n$.   Let   $c_n = a_0 b_n + a_1 b_{n-1} + a_2 b_{n-2} + \cdots + a_{n-1} b_1 + a_n b_0 = \sum\limits_{k=0}^{n} a_k b_{n-k}$.   Find, intuitively, $\sum\limits_{n=0}^{\infty} c_n x^n$ in terms of $A(x)$ and $B(x)$.

**♦ 11**   Show, intuitively, that

$$\sum_{n=1}^{\infty} \sum_{k=1}^{n} r_k s_{n-k} = (r_1 + r_2 + r_3 + \cdots)(s_0 + s_1 + s_2 + \cdots)$$

by verifying that the same term appears once and only once on each side of the equation.

**♦ 12**   Let $a_0 = 1$. Suppose that $a_n = \sum\limits_{k=1}^{n} 2^{-k} a_{n-k}$ for $n = 1, 2, 3, \ldots$. Find $a_1, a_2, a_3, a_4$ by successive substitution for $n$.

**♦ 13**   Suppose that $a_j = \binom{2j}{j} 2^{-2j}$ for $j = 0, 1, 2, 3, \ldots$ and $a_n = \sum\limits_{k=1}^{n} b_k a_{n-k}$ for $n = 1, 2, 3, \ldots$. Find $b_1, b_2, b_3, b_4$ by successively substituting and solving for the $b_n$'s with $n = 1, 2, 3, 4$.

**♦ 14**   Show that if $a_0, a_1, a_2, a_3, \ldots$ and $b_1, b_2, b_3, \ldots$ are sequences satisfying $a_n = \sum\limits_{k=1}^{n} b_k a_{n-k}$ for $n = 1, 2, 3, \ldots$ and if $A(x) = \sum\limits_{n=0}^{\infty} a_n x^n$, $B(x) = \sum\limits_{n=1}^{\infty} b_n x^n$ then $A(x) - a_0 = A(x)B(x)$. [*Hint:* Use exercise 11 with $r_k = b_k x^k$ and $s_j = a_j x^j$.]

**♦ 15**   Use exercise 14 to find the generating functions for the sequences in exercises 12 and 13.

# CHAPTER 2

# *Sets*

## 2.1 Notation

In Chapter 1 we had occasion to count certain collections. Historically speaking, many important results in probability theory were obtained with no organized study of collections. But a systematic treatment is convenient for our purposes and is essential in studying the deeper results of modern probability.

If $s$ is one of the elements (members) of the set $M$ we write $s \in M$ and read "$s$ is an element of the set $M$." We also (more commonly) read $s \in M$ as "$s$ belongs to $M$" or "$s$ is in $M$." We write $s \notin M$ if $s$ is not one of the elements of $M$, and read this as "$s$ does not belong to $M$" or "$s$ is not in $M$."

**Example 2.1.1**
Let $S = \{1, 2, 3, 4, 5, 6\}$ and $M = \{1, 3, 4, 6\}$. Discuss the membership of 3, 4, 5, 6 in $M$.

$$3 \in M, \quad 4 \in M, \quad 5 \notin M, \quad 6 \in M$$

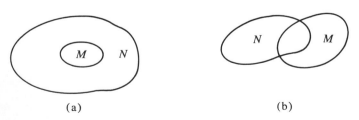

(a)                                                (b)

**Figure 1**

*Example 2.1.2*

Let $S = \{\text{positive integers}\}$, $N = \{\text{even integers}\}$, $P = \{\text{prime numbers}\}$. Discuss membership of 1, 2, 3, 4, 7, 9, $\sqrt{2}$ in $S$, $N$, and $P$.

$$1 \in S, \quad 1 \notin N, \quad 1 \notin P \text{ (by convention)}$$
$$2 \in S, \quad 2 \in N, \quad 2 \in P$$
$$3 \in S, \quad 3 \notin N, \quad 3 \in P$$
$$4 \in S, \quad 4 \in N, \quad 4 \notin P$$
$$7 \in S, \quad 7 \notin N, \quad 7 \in P$$
$$9 \in S, \quad 9 \notin N, \quad 9 \notin P$$
$$\sqrt{2} \notin S, \quad \text{so } \sqrt{2} \notin N, \quad \sqrt{2} \notin P$$

Other notation can be defined using the idea of set membership. We write $M \subset N$ if whenever $s \in M$, $s \in N$. [See Figure 1(a).] The notation $M \subset N$ is read as "$M$ is a subset of $N$" or "$M$ is contained in $N$." We write $M = N$ if $M$ and $N$ have the same elements, i.e., $s \in M$ if and only if $s \in N$. Note that $M = N$ if and only if $M \subset N$ and $N \subset M$. Furthermore, we write $N \supset M$ if and only if $M \subset N$. We read $N \supset M$ as "$N$ contains $M$." The denial of $M \subset N$ is $M \not\subset N$ which is read as "$M$ is not contained in $N$." In other words, $M \not\subset N$ means that there is an element $s$ such that $s \in M$ and $s \notin N$. The case of $M \not\subset N$ is illustrated in Figure 1(b). It should be noted that if $M = N$, then $M \subset N$. The symbol $\varnothing$ denotes the empty set, that is, the set which contains no elements. Hence for any element $s$, $s \notin \varnothing$. Furthermore $\varnothing \subset M$ for all $M$. The concept of the empty set is notationally convenient even if a bit disconcerting.

Under most circumstances, we usually start out with a given set $S$ and investigate properties of its subsets. We have the following notation for the subsets which can be constructed from subsets of $S$. If $M \subset S$ and $N \subset S$ we define

$$M \cup N = \{s \in S; s \in M \text{ or } s \in N\}$$
$$M \cap N = \{s \in S; s \in M \text{ and } s \in N\}$$
$$M - N = \{s \in S; s \in M \text{ and } s \notin N\}$$

The subset $M \cup N$ is called the *union* of $M$ and $N$; $M \cap N$ is called the *intersection* of $M$ and $N$; and $M - N$ is called the (set-theoretic) *difference* of $M$ and $N$. The set $S - M$ is called the *complement* of $M$ and is frequently denoted by $\overline{M}$ or $M'$ (in this book we use the latter). Diagrams (called Venn diagrams) for these sets are given in Figure 2.

We will assume that $(M')' = M$, i.e., if $s \in S$ and $s \notin M'$, then $s \in M$. We further assume that $M \cap M' = \varnothing$.

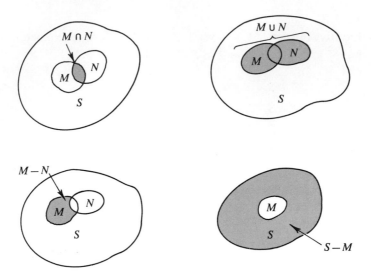

**Figure 2**

### *Example 2.1.3*

Prove that $M \cap (N \cup P) = (M \cap N) \cup (M \cap P)$. We show that $s \in M \cap (N \cup P)$ if and only if $s \in (M \cap N) \cup (M \cap P)$. If $s \in M \cap (N \cup P)$, then $s \in M$ and $s \in (N \cup P)$. Since $s \in (N \cup P)$, $s \in N$ or $s \in P$. But then $s \in (M \cap N)$ or $s \in (M \cap P)$ by definition. Therefore $s \in (M \cap N) \cup (M \cap P)$ by definition. If $s \in (M \cap N) \cup (M \cap P)$ then $s \in (M \cap N)$ or $s \in (M \cap P)$. If $s \in (M \cap N)$, $s \in M$ and $s \in N$ and hence $s \in (N \cup P)$. Similarly if $s \in (M \cap P)$, $s \in M \cap (N \cup P)$. In either case $s \in M \cap (N \cup P)$.

Before proceeding, the reader should convince himself that $M \cup N = N \cup M$, $M \cap N = N \cap M$, $M \subset M \cup N$, $M \supset M \cap N$, $M \cup \varnothing = M$, $M \cap \varnothing = \varnothing$ as well as that $M = M \cap N$ if and only if $M \subset N$ and $M = M \cup N$ if and only if $M \supset N$.

An interesting set equation is

**(2.1.1)**    $(Q - M) \cap (Q - N) = Q - (M \cup N)$

which is one of De Morgan's laws. To prove (2.1.1) we will show that $s \in (Q - M) \cap (Q - N)$ if and only if $s \in Q - (M \cup N)$.

Suppose $s \in (Q - M) \cap (Q - N)$. Then $s \in (Q - M)$ and $s \in (Q - N)$ so $s \in Q$, $s \notin M$, and $s \notin N$. Thus $s \notin M \cup N$ for if $s \in (M \cup N)$ we would have $s \in M$ or $s \in N$, both of which are denied to us. Therefore $s \in Q - (M \cup N)$.

If $s \in Q - (M \cup N)$ we have $s \in Q$ and $s \notin (M \cup N)$. Therefore $s \notin M$ *and* $s \notin N$. We now can assert that $s \in Q$ and $s \notin M$ so $s \in (Q - M)$ and that $s \in Q$ and $s \notin N$ so $s \in (Q - N)$. Therefore $s \in (Q - M) \cap (Q - N)$ and (2.1.1) is proved.

Another De Morgan law is

**(2.1.2)**    $(Q - M) \cup (Q - N) = Q - (M \cap N)$

We leave the proof of (2.1.2) as an exercise for the reader.

It is desirable to allow for unions and intersections of more than two sets. If $M_1, M_2, \ldots, M_k$ are subsets of $S$ we let

$$M_1 \cup M_2 \cup \cdots \cup M_k = \{s \in S; s \in M_i \text{ for at least one integer } i \text{ with } 1 \le i \le k\}$$

Similarly

$$M_1 \cap M_2 \cap \cdots \cap M_k = \{s \in S; s \in M_i \text{ for each integer } i \text{ with } 1 \le i \le k\}$$

If $k = 3$, Figure 3 illustrates $M_1 \cup M_2 \cup M_3$ and $M_1 \cap M_2 \cap M_3$ in that the darkly shaded region is $M_1 \cap M_2 \cap M_3$ and the lightly shaded region is $M_1 \cup M_2 \cup M_3$. If we have infinitely many subsets where the subscripts take all positive-integer values, say $M_1, M_2, M_3, M_4, \ldots$, the notation $M_1 \cup M_2 \cup M_3 \cup \cdots$ and $M_1 \cap M_2 \cap M_3 \cap \cdots$ are the union and intersection respectively.

However, a general notation for all of these cases is available. Suppose $I$ is an (index) set, i.e., to each $\alpha \in I$, there is a subset $M_\alpha$. Let

$$\bigcup_{\alpha \in I} M_\alpha = \{s \in S; s \in M_\alpha \text{ for at least one index } \alpha \in I\}$$

Let

$$\bigcap_{\alpha \in I} M_\alpha = \{s \in S; s \in M_\alpha \text{ for all } \alpha \in I\}$$

These are the union and intersection, respectively, of the sets\ $M_\alpha$ for $\alpha \in I$. In the case that $I$ is a suitable set of integers we might write $\bigcup_{j=1}^{\infty} M_j$ or $\bigcap_{j=-\infty}^{\infty} M_j$. Thus $\bigcup_{j=1}^{k} M_j = M_1 \cup M_2 \cup \cdots \cup M_k$ and $\bigcap_{j=3}^{\infty} M_j = M_3 \cap M_4 \cap M_5 \cap \cdots$.

**Example 2.1.4**
Let $S = \{1, 2, 3, 4, 5, 6, 7, 8\}$. Let $M = \{1, 2, 3\}$, $N = \{2, 4, 5, 6, 7, 8\}$. Find $M \cup N$, $M \cap N$, and $M'$.

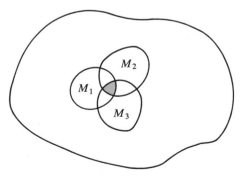

**Figure 3**

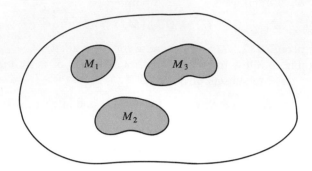

**Figure 4**

By the definitions $M \cup N = \{1, 2, 3, 4, 5, 6, 7, 8\} = S$,   $M \cap N = \{2\}$, and $M' = \{4, 5, 6, 7, 8\}$.

### *Example 2.1.5*

Let $S = \{x; x \text{ is a real number}\}$. Let $I = \{-1, 0, 1, 2, 3, 4, \ldots\}$ considered as a subset of $S$. Let $M_\alpha = \{x; x^2 \leq \alpha\}$. Describe $M_{-1}$, $M_2 \cap M_3$, $\bigcup\limits_{\alpha=1}^{\infty} M_\alpha$.

The set $M_{-1} = \{x; x^2 \leq -1\} = \varnothing$ since $x^2 \geq 0$ for any real number $x$; $M_2 \cap M_3 = \{x; x^2 \leq 2\} \cap \{x; x^2 \leq 3\} = \{x; x^2 \leq 2\} = M_2$ by definition of the intersection of two sets; $\bigcup\limits_{\alpha=1}^{\infty} M_\alpha = \bigcup\limits_{\alpha=1}^{\infty} \{x; x^2 \leq \alpha\} = \{x; x^2 \leq \alpha \text{ for at least one integer } \alpha = 1, 2, 3, \ldots\}$. Now given any real number $x$, there is at least one positive integer $\alpha$ such that $x^2 \leq \alpha$. Therefore $\bigcup\limits_{\alpha=1}^{\infty} M_\alpha = S$, the set of all real numbers.

If $\{M_\alpha\}$ is a collection of sets where $\alpha \in I$, an index set, we say that the sets in $\{M_\alpha\}$ are *disjoint* or *mutually exclusive* if for each $\alpha \in I$, $\beta \in I$, $\alpha \neq \beta$ we have $M_\alpha \cap M_\beta = \varnothing$. This is illustrated in Figure 4 for $M_1, M_2, M_3$.

### EXERCISES 2.1

**1**  Let $S = \{1, 2, 3, \ldots, 10\}$,   $M = \{1, 2, 4, 5, 7, 9\}$, and $N = \{1, 3, 4, 7, 8\}$. Describe $M \cap N$,   $M \cup N$,   $M - N$,   $N - M$, and $M' = S - N$ in braces notation.

**2**  Let $M_j = \{x; x \text{ is a real number and } j < x \leq j + 1\}$ where $j \in \{1, 2, 3, 4, 5, \ldots\}$. Show that the sets in $\{M_j\}$ are disjoint. Would the sets be disjoint if $M_j = \{x; x \text{ is a real number and } j \leq x \leq j + 1\}$?

3 With $j \in \{1, 2, 3, \ldots\}$, describe $\bigcap_{j=1}^{\infty} \{x; x \geq j\}$.

4 Draw a Venn diagram illustrating

   (a) $M \cap (N_1 \cup N_2) = (M \cap N_1) \cup (M \cap N_2)$

   (b) $M \cup (N_1 \cap N_2) = (M \cup N_1) \cap (M \cup N_2)$

♦ 5 Prove the equalities of exercise 4.

6 What can be said about $M - (M - N)$ if $N \subset M$?

♦ 7 (a) Draw a Venn diagram illustrating equation (2.1.2).

   (b) Prove equation (2.1.2) in the same manner that (2.1.1) was proved.

   (c) Prove the generalized form of (2.1.1); namely,

     **(2.1.3)** $\qquad \bigcap_{\alpha \in I} (Q - M_\alpha) = Q - \bigcup_{\alpha \in I} M_\alpha$

   (d) Draw a Venn diagram for equation (2.1.3) in the case $I = \{1, 2, 3, 4\}$.

♦ 8 State a generalized form of (2.1.2) which is analogous to (2.1.3).

9 Let $M = \{(x, y); x$ and $y$ are real numbers such that $x^2 + 2y^2 = 5\}$ and let $N = \{(x, y); x$ and $y$ are real numbers such that $6x^2 - y^2 = 4\}$. Find $M \cap N$.

10 Let $S = \{1, 2, 3, \ldots, 20\}$, $\quad Q = \{2, 3, 5, 7, 11, 13, 17, 19\}$, $\quad M = \{1, 2, 3, \ldots, 12\}$, and $N = \{7, 8, 9, 10, \ldots, 20\}$.

   (a) Work through equation (2.1.1) with these sets by direct examination of the elements in the sets.

   (b) Work through $M \cap (N \cup Q) = (M \cap N) \cup (M \cap Q)$ in the same way.

♦ 11 Show that (2.1.3) with $Q = S$ states that $s \notin \bigcup M_\alpha$ if and only if $s \in M_\alpha'$ for each $\alpha$.

12 Let $S = \{$polynomials $p(x) = ax^3 + bx^2 + cx + d\}$, $\quad K = \{p \in S; p(0) = 3\}$, $L = \{p \in S; p'(0) = -2\}$, $\quad M = \{p \in S; p''(0) = 1\}$, and $N = \{p \in S;$ $\int_0^1 p(x)\,dx = 0\}$. Describe $K \cap L$, $\quad K \cap L \cap M$, and $K \cap L \cap M \cap N$ in the notation $\{p \in S; \cdots\}$.

# 2.2 Countable sets

A set $S$ is said to be *finite* if $S = \varnothing$ or if there is a positive integer $n$ such that $S = \{s_1, s_2, \ldots, s_n\}$, that is, the elements of the set $S$ can be indexed by the integers $i$ such that $1 \leq i \leq n$. A set $S$ is said to be *denumerable* if we have $S = \{s_1, s_2, s_3, \ldots\}$, which means that the elements of $S$ can be indexed by the set of all positive integers or, in other words, there is a one-to-one correspondence between the elements of $S$ and the positive integers. A set $S$ is said to be *countable* if it is finite or denumerable. One also says that such a set has *countably many* elements. (It can be shown that a set cannot be both finite and denumerable but we shall not prove it here and we use the result only indirectly.) An *enumeration* of a denumerable set is a procedure giving a one-to-one correspondence between the elements of the set and the positive integers; i.e., a list $s_1, s_2, s_3, \ldots$.

Some denumerable sets are

$$E = \{m; m = 2n \text{ for } n = 1, 2, 3, 4, \ldots\}$$

$$Z = \{m; m \text{ is an integer}\}$$

$$R^+ = \{\text{positive rational numbers}\}$$

Of these sets, the description of $E$ gives an indexing required by the definition of denumerable. As for $Z$, let

$$0 = s_1, \quad 1 = s_2, \quad -1 = s_3, \quad 2 = s_4, \quad -2 = s_5, \quad 3 = s_6, \quad -3 = s_7, \quad \ldots$$

In general $s_{2n} = n$ for $n > 0$, $s_{2n+1} = -n$ for $n > 1$, and $s_1 = 0$. Since any positive integer is 1 or of the form $2n$ for $n > 0$ or $2n + 1$ for $n > 1$ we have our indexing. The denumerability of $R^+$ is best seen from a diagram (Figure 5). We let $s_1 = 1/1$, $s_2 = 2/1$, $s_3 = 1/2$, $s_4 = 3/1$, $s_5 = 1/3$ (we omit 2/2 since it is the same rational number as 1/1), $s_6 = 4/1$, $s_7 = 3/2$, $s_8 = 2/3$, $s_9 = 1/4$, $s_{10} = 5/1$, $s_{11} = 1/5$ (since 4/2, 3/3, 2/4 are already accounted for) and we continue to trace out the diagonal lines and omit what has already been accounted for. It is clear that any positive rational number will be $s_k$ for some positive integer $k$ and it is clear that all positive integers $k$ must eventually be used as subscripts.

This same procedure can be used to show that if $M_1, M_2, M_3, \ldots$ are denumerable sets, then $\bigcup_{n=1}^{\infty} M_n$ is a denumerable set.

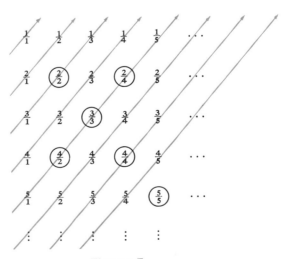

**Figure 5**

Once it has been shown that the set of positive rational numbers is denumerable it is easy to apply the method used to show that $Z$ is denumerable to show that the set of all rational numbers is denumerable.

Some properties of countable sets are as follows:

(a)  Any subset of a countable set is countable.

(b)  The union of countably many countable sets is countable.

(c)  If a countable set has a denumerable subset, the set is denumerable.

We close this section on countable sets by showing that the set $U$ of real numbers $x$ such that $0 < x \leq 1$ is not denumerable. Any such $x$ is the sum of an infinite series (its decimal representation) and we have

$$x = \frac{n_1}{10} + \frac{n_2}{10^2} + \frac{n_3}{10^3} + \frac{n_4}{10^4} + \cdots$$

where the numbers $n_i \in \{0, 1, 2, \ldots, 9\}$. Certain numbers have two such representations, for example,

$$\frac{1}{2} = \frac{1}{10} + \frac{9}{10^2} + \frac{9}{10^3} + \frac{9}{10^4} + \frac{9}{10^5} + \cdots$$

Similarly

$$\frac{1}{25} = \frac{0}{10} + \frac{4}{10^2} + \frac{0}{10^3} + \frac{0}{10^4} + \frac{0}{10^5} + \frac{0}{10^6} + \cdots$$

$$= \frac{0}{10} + \frac{3}{10^2} + \frac{9}{10^3} + \frac{9}{10^4} + \frac{9}{10^5} + \cdots$$

To relieve us of this type of ambiguity we will elect to use the series where $n_i = 9$ from some point on as a convention. Naturally, numbers like

$$\frac{1}{3} = \frac{3}{10} + \frac{3}{10^2} + \frac{3}{10^3} + \frac{3}{10^4} + \cdots$$

have only one representation and we leave them alone.

Suppose the set $U$ is denumerable. Suppose all of the elements of $U$ have been written as

$$s_1 = \frac{n_1^{(1)}}{10} + \frac{n_2^{(1)}}{10^2} + \frac{n_3^{(1)}}{10^3} + \frac{n_4^{(1)}}{10^4} + \cdots$$

$$s_2 = \frac{n_1^{(2)}}{10} + \frac{n_2^{(2)}}{10^2} + \frac{n_3^{(2)}}{10^3} + \frac{n_4^{(2)}}{10^4} + \cdots$$

$$s_3 = \frac{n_1^{(3)}}{10} + \frac{n_2^{(3)}}{10^2} + \frac{n_3^{(3)}}{10^3} + \frac{n_4^{(3)}}{10^4} + \cdots$$

$$\vdots$$

Let

$$m_k = \begin{cases} 7, & n_k^{(k)} = 0, 1, 2, 3, 4 \\ 2, & n_k^{(k)} = 5, 6, 7, 8, 9 \end{cases}$$

Note that $m_k \neq n_k^{(k)}$.

Let

$$s = \frac{m_1}{10} + \frac{m_2}{10^2} + \frac{m_3}{10^3} + \frac{m_4}{10^4} + \cdots$$

The number $s \in U$ since $0 \le m_k \le 9$ and $s$ cannot have all $m_k = 0$ beyond some place because none of the $m_k = 0$. We assert that $s \ne s_k$ for $k = 1, 2, 3, 4,$ $\ldots$ . This follows easily because

$$s_k = \frac{n_1^{(k)}}{10} + \frac{n_2^{(k)}}{10^2} + \cdots + \frac{n_k^{(k)}}{10^k} + \cdots$$

and

$$s = \frac{m_1}{10} + \frac{m_2}{10^2} + \cdots + \frac{m_k}{10^k} + \cdots$$

By definition of $m_k$, $m_k \ne n_k^{(k)}$ and since these representations are unique, $s \ne s_k$ for $k = 1, 2, 3, \ldots$ . Thus the assumption that $U$ is denumerable leads to a contradiction since we supposed that $U = \{s_1, s_2, s_3, \ldots\}$ yet we found an $s \in U$ and $s \notin \{s_1, s_2, s_3, \ldots\}$.

## EXERCISES 2.2

1  What are $s_{18}$, $s_{19}$, and $s_{20}$ in the enumeration of $R^+$ given by Figure 5? Find $n$ for which $s_n = \frac{3}{8}$.

2  Using the enumeration of $R^+$ given by Figure 5 show that if $s_k = n/m$ then $k \le \sum_{j=1}^{n+m} j = \frac{1}{2}(n + m)(n + m + 1)$.

3  Write the following numbers as $n_1/10 + n_2/10^2 + n_3/10^3 + \cdots$ so that none end in zeros: (a) 1/2, (b) 1/6, (c) 1/8, (d) 1/40.

4  In the discussion which shows that $U$ is not countable suppose that $s_1 = 0.13721\ldots$, $s_2 = 0.38815\ldots$, $s_3 = 0.98765\ldots$, and $s_4 = 0.71368\ldots$. Find the first four decimal places of the number which is claimed not to be in the enumeration.

5  Show that the set $\{k/2^n;\ k$ an integer, $n$ a positive integer$\}$ is denumerable without using the denumerability of the rationals. Numbers of the form $k/2^n$ are called *dyadic rationals*.

6  Show that the set of points $(j, k, l)$ in 3-dimensional space, where $j$, $k$, and $l$ are integers, is countable.

7  Give an example of a denumerable subset of $\{0, 1, 2, 3, \ldots\}$ whose complement is denumerable.

♦ 8  Show that the union of countably many countable sets is countable.

9  Show that if $M$ is an uncountable set of positive real numbers, there is a positive integer $n$ such that infinitely many elements of $M$ are greater than $1/n$.

# 2.3  **Cartesian products**

Suppose $S_1, S_2, \ldots, S_n$ are nonempty sets. We define the set $S_1 \times S_2 \times \cdots \times S_n$ to be the set of all lists $l = (s_1, s_2, \ldots, s_n)$ which can be formed from $s_1 \in S_1$, $s_2 \in S_2, \ldots, s_n \in S_n$. We write parentheses (  ) around the entries of the list to set them off from other symbols which may appear near them.

**Definition**       The set $S_1 \times S_2 \times \cdots \times S_n$ defined above is called the *Cartesian product* of the sets $S_1, S_2, \ldots, S_n$.

The name Cartesian is in honor of the French mathematician Descartes and the motivation for that name should be clear from Example 2.3.2 below.

### Example 2.3.1
Let $S_1 = \{H, T\}$ and $S_2 = \{1, 2, 3, 4, 5, 6\}$. Describe $S_1 \times S_2$.

$S_1 \times S_2 = \{(H, 1), (H, 2), (H, 3), (H, 4), (H, 5), (H, 6),$
$$(T, 1), (T, 2), (T, 3), (T, 4), (T, 5), (T, 6)\}$$

### Example 2.3.2
Let $S_1 = S_2 = \{x; x \text{ is a real number}\}$. Describe $S_1 \times S_2$.

The set $S_1 \times S_2 = \{(x, y); x \text{ and } y \text{ are real numbers}\}$. We use $(x, y)$ rather than $(x_1, x_2)$ to suggest that $S_1 \times S_2$ is essentially the coordinatized Cartesian plane.

### Example 2.3.3
Let $S_1 = S_2 = S_3 = \{x; x \text{ is a real number}\}$. Describe $S_1 \times S_2 \times S_3$.

The set $S_1 \times S_2 \times S_3 = \{(x, y, z); x, y, z \text{ are real numbers}\}$. It is convenient to think of $S_1 \times S_2 \times S_3$ as coordinatized 3-dimensional Euclidean space.

### Example 2.3.4
Let $S_1 = \{x; x \text{ is a real number}\}$ and $S_2 = \{p; p \text{ is a point on a circle of radius 1}\}$. Describe $S_1 \times S_2$.

In our formal mathematical notation $S_1 \times S_2 = \{(x, p); x \in S_1, p \in S_2\}$ but a more intuitive description is useful. Imagine an infinite right circular cylinder of radius 1 (see Figure 6) in a coordinatized 3-dimensional space.
Any point $q$ on the (surface of the) cylinder can be described by specifying one point $p$ on a fixed cross-section circle and a real number $x$ giving the (directed) distance from the plane of the cross-section circle to the point $q$.

### Example 2.4.5
Let $S_1 = \{p; p \text{ is on a circle of radius 1}\}$ and $S_2 = \{q; q \text{ is on a circle of radius 4}\}$. Describe $S_1 \times S_2$.

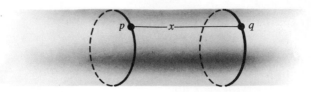

**Figure 6**

Again we have the formal description $S_1 \times S_2 = \{(p, q); p \in S, \quad q \in S_2\}$ but again there is an interesting interpretation. Suppose we imagine a copy of the circle $S_1$ being swung around the circle $S_2$ in such a way as to make the surface of a torus (doughnut or anchor ring—see Figure 7).

It is convenient to imagine that the center of the copy of $S_1$ lies on $S_2$ and that the plane of $S_1$ is perpendicular to the plane of $S_2$. Any point $r$ on the surface of such a torus can be described by a pair $(p, q)$ where $q$ specifies the point on $S_2$ at which to place the plane of the copy of $S_1$ and $p$ gives the point on $S_1$ needed to produce $r$ on the copy of $S_1$.

The geometric objects indicated in the above examples are not the only geometric representations of the Cartesian products indicated but they do furnish us with an intuitive view of the Cartesian product set which can be very useful.

Some questions connected with the successive formation of Cartesian products seem troublesome at first glance. For example, let $S_1 = S_5 \times S_6$ and $S_2 = S_7$. Let $S_3 = S_1 \times S_2$ and $S_4 = S_5 \times S_6 \times S_7$. According to the definitions,

$$S_3 = \{((s_5, s_6), s_7); s_5 \in S_5, \quad s_6 \in S_6, \quad s_7 \in S_7\}$$

and

$$S_4 = \{(s_5, s_6, s_7); s_5 \in S_5, \quad s_6 \in S_6, \quad s_7 \in S_7\}$$

What is the difference between $S_3$ and $S_4$? There is no essential difference between $S_3$ and $S_4$ since there is a simple one-to-one correspondence between

**Figure 7**

lists of the form $((s_5, s_6), s_7)$ and $(s_5, s_6, s_7)$. We will therefore treat $S_3$ and $S_4$ as basically the same. Similarly, for our purposes, $S_1 \times S_2$ and $S_2 \times S_1$ can be thought of as the same.

**EXERCISES 2.3** ———————————————————————

**1** Suppose $S_1$ has $n_1$ elements, $S_2$ has $n_2$ elements, and $S_3$ has $n_3$ elements. How many elements has $S_1 \times S_2$? How many elements has $S_1 \times S_2 \times S_3$? Generalize. (Note why we call it the Cartesian *product*.)

♦ **2** Suppose $M_1 \subset S_1$, $M_2 \subset S_1$, and $N \subset S_2$. Show that $(M_1 \cup M_2) \times N = (M_1 \times N) \cup (M_2 \times N)$. Also, draw a diagram.

**3** Suppose $M_1 = \{x; a \leq x \leq b\}$ and $M_2 = \{x; c \leq x \leq d\}$. Describe $M_1 \times M_2$ in reference to $S_1 \times S_2$ of Example 2.3.4.

**4** Suppose $S_1 = \{p; p$ is a point on the circumference or in the interior of a circle of radius 1$\}$ and $S_2 = \{x; 0 \leq x \leq 1\}$. Give a geometric interpretation of $S_1 \times S_2$.

**5** Suppose $S_1 = S_2 = \{x; x$ is a real number$\}$ and $S_3 = \{x; 0 \leq x \leq 1\}$. Interpret $S_1 \times S_2 \times S_3$ geometrically.

**6** Suppose $S_1 = S_2 = \{x; x$ is a real number$\}$ and $S_3 = \{p; p$ is a point on a circle of radius 1$\}$. Give a geometric representation for $S_1 \times S_2 \times S_3$.

**7** Suppose $S_1 = \{1, 3, 5, 7\}$. Is it possible for there to be a set $S_2$ such that $S_1 \times S_2$ has 15 elements?

**8** Give an example of a set with infinitely many elements in the $xy$ plane which is not the Cartesian product of a set $M_1$ on the $x$ axis with a set $M_2$ on the $y$ axis.

**9** Draw a rectangle in the $xy$ plane with sides parallel to the axes. Draw a smaller rectangle of the same kind inside the first rectangle. Look at the portion of the larger rectangle which is outside of the smaller rectangle. Show that if $M_1 \subset S_1$, $M_2 \subset S_2$ then $S_1 \times S_2 - M_1 \times M_2$ is the union of two sets each of which is a Cartesian product. The latter sets may not be disjoint.

# 2.4 Functions

In calculus a function is described as a particular kind of correspondence between two sets (usually sets of real numbers). A useful intuitive analog of a function $f$ from a (nonempty) set $S$ (called the domain of $f$) to a set $T$ (called the range space) of $f$ is a (hypothetical) computing machine (see Figure 8) which "turns out" a definite element of $T$ corresponding to any element $s \in S$ which is put into the machine. It is assumed that this machine turns out some

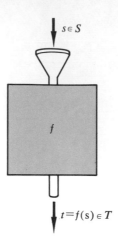

**Figure 8**

element $t \in T$ for each $s \in S$ and that the element $t$ which is turned out when $s$ is put in is uniquely determined by $s$. Furthermore if $f$ is the symbol for the function at issue, we write $t = f(s)$ in the present circumstances. Occasionally, for emphasis, such functions are called single-valued functions.

An alternative rigorous way of viewing a function is as a certain type of subset of a Cartesian product. Hence, a set $G \subset S \times T$ specifies a function $f$ with domain $S$ and range space $T$ if for each $s \in S$ there is one and only one $t \in T$ such that $(s, t) \in G$. We denote that $t$ corresponding to $s$ by $t = f(s)$.

If one interprets the computing machine sufficiently loosely it is fairly clear that these two views are equivalent in the sense that a computing machine $f$ can be used to determine any point $(s, t) \in G$ if $s$ is given. Also, if $G$ is given we could construct our machine to find the $t \in T$ which is such that $(s, t) \in G$ for a given $s \in S$. That $t$ is then turned out of the machine.

To designate a function $f$ with domain $S$ and range space $T$ we will write

$$f: S \rightarrow T$$

which is read as "$f$ goes from $S$ to $T$" or "$f$ maps $S$ into $T$." It should be noticed that we can have $f(s_1) = f(s_2)$ for $s_1 \neq s_2$, $s_1 \in S$, $s_2 \in S$. What we cannot have is $f(s_1) = t_1$ and $f(s_1) = t_2$ for $t_1 \neq t_2$. Note also that we can draw a graph (see Figure 9) which, in this case, illustrates a set $G$ when

$$S = \{x; 0 \leq x \leq 2, x \text{ real}\}, \quad T = \{x; x \text{ real}\}$$

We frequently use the phrase "the function $f(x)$." We will understand this to mean "the function $f$ whose value at $x$ is $f(x)$."

If $f: S \rightarrow T$ is such that to each $t \in T$ there is *at most* one $s \in S$ such that $f(s) = t$, $f$ is said to be a *one-to-one function*. If $f$ is such that

$$\{t \in T; t = f(s) \text{ for some } s \in S\} = T$$

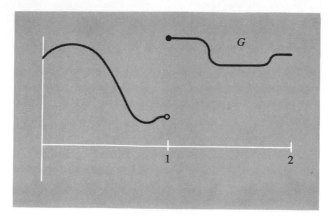

**Figure 9**

then $f$ is said to be *onto* $T$. If $f$ is one-to-one and onto $T$, we can define $g: T \twoheadrightarrow S$ by the relationship

$$g(t) = s \text{ if (and only if) } f(s) = t$$

In that case $g$ is called the *inverse* function of $f$ and is frequently written $g = f^{-1}$.

We will not belabor the topic of functions. To a great extent, this material was inserted here because it is a natural topic to study after the topic of Cartesian products. However, in later chapters there will be a number of functions in use simultaneously. To facilitate the reader's understanding of those functions we have mentioned three ways to view functions.

**EXERCISES 2.4** _____

1   Let $S_1 = \{x; x \text{ is a real number}\}$, $S_2 = \{x; x \text{ is a real number and } x \geq 0\}$ and $G = \{(x, y) \in S_1 \times S_2; y = x^2\}$. Does $G$ specify a function $f: S_1 \to S_2$? Does $G$ specify a function $g: S_2 \to S_1$?

2   Let $S_1 = S_2 = \{x; x \text{ is a real number and } -1 \leq x \leq 1\}$. Let $G = \{(x, y) \in S_1 \times S_2; x^2 + y^2 = 1\}$. Does $G$ specify a function from $S_1$ to $S_2$? Suppose $S_3 = \{x; x \text{ is a real number and } 0 \leq x \leq 1\}$. Let $H = \{(x, y) \in S_1 \times S_3; x^2 + y^2 = 1\}$. Does $H$ specify a function from $S_1$ to $S_3$? From $S_3$ to $S_1$?

3   Suppose $f: S \to T$ and $g: T \to S$ are functions. Show that $f^{-1} = g$ if and and only if for each $s \in S$, $g(f(s)) = s$ and for each $t \in T$, $f(g(t)) = t$.

4   A *function of two variables* from a set $S$ to a set $T$ is a function $f: S \times S \to T$. Suppose $S = \{x; x \text{ real and } -1 \leq x \leq 1\}$ and $T = \{x; x \text{ real}, x \geq 0\}$. Does $G = \{(x, y, z) \in S \times S \times T; x^2 + y^2 + z^2 = 1\}$ specify a function of two variables (from $S$ to $T$)? Does $H = \{(x, y, z) \in S \times S \times T; x^2 + y^2 + 2z^2 = 4\}$ specify a function of two variables (from $S$ to $T$)?

5 Suppose $S$ has $n$ elements and $T$ has $m$ elements. How many different functions $f: S \to T$ are there? Two functions $f_1: S \to T$ and $f_2: S \to T$ are the same if $f_1(s) = f_2(s)$ for each $s \in S$.

6 Let $S_1 = \{n; n$ is an integer greater than 3$\}$. Let $S_2 = \{m; m$ is a positive integer$\}$. Let $S = S_1 \times S_2$. Does the set satisfying $3m = 3n + 3$ determine a function from $S_1$ to $S_2$? Does the set satisfying $m - n = -2$ determine a function from $S_1$ to $S_2$?

7 Let $S_1 = \{1, 2, 3, 4, 5, 6\}$. Let $S = S_1 \times S_1 \times S_1$. For $\alpha = (s, t, u) \in S$, let $f(\alpha) =$ the position of the largest number among $s$, $t$, and $u$. Does this determine a function from $S$ to the set $\{$first position, second position, third position$\}$?

8 Let $S_1 = \{x; x$ is a positive real number$\}$ and let $S_2 = \{x; x$ is a real number$\}$. Does the graph of $x^2 + y^2 = 1$ for $(x, y) \in S_1 \times S_1$ determine a function from $S_1$ to $S_1$? Does the graph of $x^2 + y^2 = 1$ for $(x, y) \in S_2 \times S_2$ determine a function from $S_2$ to $S_2$?

9 Let $S_2 = \{x; x$ is a real number$\}$. Describe the usual domains for the function into $S_2$ whose values are,

(a) $\cos x$                          (g) Arccos $x$

(b) $\sin x$                          (h) $\tan x$

(c) $e^x$                             (i) $\dfrac{x + 2}{x^2 - 5x + 4}$

(d) $\log x$                          (j) $\displaystyle\sum_{n=1}^{\infty} \frac{1}{n^x}$

(e) $\sqrt{1 - x^2}$                  (k) $\displaystyle\int_0^{\infty} t^x e^{-t}\, dt$

(f) Arcsin $x$

10 Describe a function of three variables from a set $S$ to a set $T$ in a fashion similar to exercise 4.

11 Give a more precise statement for the assertion, "The function $f(x) = x^2$ is continuous."

# 2.5  Systems of subsets

Instead of using the phrases "set of subsets" or "collection of subsets" we shall use the phrases "system of subsets" or "system of subsets of $S$." Presumably these phrases can be understood without a general definition. However, we give a few examples. (Henceforth, we denote the set of real numbers by $R$.)

### Example 2.5.1

Let $S = \{1, 2, 3, 4\}$. List the sets in the system of 3-element subsets of $S$.

The sets $\{1, 2, 3\}, \{1, 2, 4\}, \{1, 3, 4\}, \{2, 3, 4\}$ comprise the system of 3-element subsets of $S$.

### Example 2.5.2

Let $S = \{1, 2, 3, 4\}$. Find some systems $\mathscr{S}$ of subsets of $S$ having the property that if $M \in \mathscr{S}$ and $N \in \mathscr{S}$ then $M \cap N \in \mathscr{S}$.

One such system is $\mathscr{S} = \{\varnothing, \{1\}, \{2\}, \{3\}\}$. Another such system is $\mathscr{S} = \{$all subsets of $\mathscr{S}\}$. Still another is $\mathscr{S} = \{\varnothing, \{1\}, \{2\}, \{3\}, \{4\}, \{1, 2\}, \{1, 3\}, \{1, 4\}, \{2, 3\}, \{2, 4\}, \{3, 4\}\}$.

### Example 2.5.3

Let $S = \{1, 2, 3, 4\}$. Find the smallest system $\mathscr{S}$ of subsets of $S$ having the property that $\{1, 2\}, \{2, 3\}$, and $\{3, 4\}$ are in $\mathscr{S}$ and that whenever $M \in \mathscr{S}$, $N \in \mathscr{S}$ then $M \cap N \in \mathscr{S}$ and $M \cup N \in \mathscr{S}$.

Suppose $\mathscr{S}$ consists of $\{\varnothing, \{2\}, \{3\}, \{1, 2\}, \{3, 4\}, \{1, 2, 3\}, \{2, 3, 4\}, S\}$. Certainly $\{1, 2\}, \{2, 3\}$, and $\{3, 4\}$ are in $\mathscr{S}$. Intersections of those sets give $\varnothing$, $\{2\}, \{3\}$ and unions of those sets give $\{1, 2, 3\}, \{2, 3, 4\}$, and $S$. Since $S$ has four elements there are $2^4 = 16$ subsets of $S$ by (1.2.4). We have accounted for nine of them in $\mathscr{S}$. An examination of each of the remaining sets shows that none of them is obtainable from unions and intersections of the above sets. Therefore they are not in the *smallest* such system.

In Example 2.5.3, as well as in other cases, we may interpret the phrase "smallest system of subsets" having certain properties as meaning the system $\mathscr{S}$ consisting of those subsets $M$ which are in *all* systems $\mathscr{S}_\alpha$ having the given properties, i.e., $M \in \mathscr{S}$ if and only if $M \in \bigcap_{\alpha \in I} \mathscr{S}_\alpha$ where $I$ is a suitable index set.

It must be verified that $\mathscr{S}$ also has the required properties. Thus in reference to Example 2.5.2, certain of the systems which satisfy the conditions but which have more subsets than does $\mathscr{S}$ (systems which contain $\mathscr{S}$), are
$$\mathscr{S}_1 = \{\varnothing, \{1\}, \{2\}, \{3\}, \{1, 2\}, \{2, 3\}, \{3, 4\}, \{1, 3\}, \{1, 2, 3\}, \{1, 3, 4\}, \{2, 3, 4\}, S\}$$
and
$$\mathscr{S}_2 = \{\text{all subsets of } S\}$$
A system $\mathscr{S}_3$ which contains $\mathscr{S}$ but which does not contain $M \cup N$, $M \cap N$ whenever $M, N \in \mathscr{S}_3$ is
$$\mathscr{S}_3 = \{\varnothing, \{2\}, \{3\}, \{4\}, \{1, 2\}, \{2, 3\}, \{2, 4\}, \{1, 2, 3\}, \{1, 2, 4\}, \{1, 3, 4\}, \{2, 3, 4\}, S\}$$
A very important system of subsets of $R$ is $\mathscr{B}$, the smallest system such that

**(2.5.1a)**   if $M_1, M_2, M_3, \ldots$ are in $\mathscr{B}$, then $(\bigcup_{k=1}^{\infty} M_k) \in \mathscr{B}$

**(2.5.1b)**   if $M \in \mathscr{B}$,   $N \in \mathscr{B}$,   $M - N \in \mathscr{B}$

**(2.5.1c)**   if $a \in R$,   $b \in R$, and $a < b$ then $\{x \in R; a < x \le b\} \in \mathscr{B}$

This system is called the system of *Borel sets of R*. Any set in $\mathscr{B}$ is called a *Borel set*.

### Example 2.5.4

Show that $R \in \mathscr{B}$.

Let $n$ be a positive integer. Let $M_n = \{x \in R; -n < x \le n\}$. By condition (2.5.1c) $M_n \in \mathscr{B}$. For any real number $x$, there is a positive integer $n$ such that $|x| < n$. Therefore $-n < x \le n$, so each real number is in $M_n$ for some $n$. Therefore $R = \bigcup_{n=1}^{\infty} M_n$. But by condition (2.5.1a), $R = \bigcup_{n=1}^{\infty} M_n \in \mathscr{B}$.

From Example 2.6.4 and condition (2.5.1b) with $M = R$ we see immediately that $N' = R - N \in \mathscr{B}$ whenever $N \in \mathscr{B}$. Furthermore by the generalized De Morgan law (exercise 2.1.7) with $Q = S$ we have $(\bigcap_{n=1}^{\infty} M_n)' = \bigcup_{n=1}^{\infty} M_n'$.

Thus if $M_1, M_2, M_3, \ldots$ are in $\mathscr{B}$, then $M_1', M_2', M_3', \ldots$ are in $\mathscr{B}$ so $\bigcup_{n=1}^{\infty} M_n' \in \mathscr{B}$ and hence $(\bigcap_{n=1}^{\infty} M_n)' \in \mathscr{B}$. But $(N')' = N$ so we have $\bigcap_{n=1}^{\infty} M_n \in \mathscr{B}$. This shows that if $M_1, M_2, M_3, \ldots$ are in $\mathscr{B}$ then

$$(2.5.2) \qquad \bigcap_{n=1}^{\infty} M_n \in \mathscr{B}$$

Since $S \in \mathscr{B}$, $\varnothing = S' \in \mathscr{B}$ by condition (2.5.1b). Therefore if $N_1, N_2, \ldots, N_k$ is a finite list of sets in $\mathscr{B}$ we can set

$$M_1 = N_1, \qquad M_2 = N_2, \qquad \ldots, \qquad M_k = N_k$$
$$M_{k+1} = \varnothing, \qquad M_{k+2} = \varnothing, \qquad M_{k+3} = \varnothing, \qquad \cdots$$

Thus $N_1 \cup N_2 \cup \cdots \cup N_k = \bigcup_{n=1}^{\infty} M_n \in \mathscr{B}$. In a similar way we can show that the intersection of finitely many sets of $\mathscr{B}$ is in $\mathscr{B}$. It is sometimes said that $\mathscr{B}$ is *closed under* countable union and difference (as well as countable intersection and complementation).

### Example 2.5.5

Let $H = \{x \in R; x > a\}$. Show that $H \in \mathscr{B}$.

If we can show that $K = \{x \in R; x \le a\} = (R - H) \in \mathscr{B}$ we can apply Example 2.5.4 and condition (2.5.1b) to give $H \in \mathscr{B}$. Let $M_1 = \{x \in R; a - 1 < x \le a\}$. Let $M_2 = \{x \in R; a - 2 < x \le a - 1\}$. In general, let

$$M_n = \{x \in R; a - n < x \le a - n + 1\}$$

These sets link up as in Figure 10.

Now if $x \le a$ then $a - x \ge 0$ and there is an integer $n \ge 1$ such that $n - 1 \le a - x < n$. The last double inequality is equivalent to the pair of

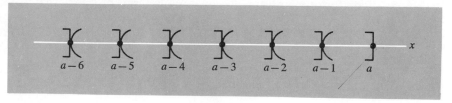

**Figure 10**

inequalities $a - n < x$ and $x \le a - n + 1$. Therefore any $x \in K$ is in some $M_n$. Thus $K \subset \bigcup\limits_{n=1}^{\infty} M_n$. But each $M_n$ is clearly contained in $K$. Since each $M_n \in \mathscr{B}$ by (2.5.1c), $K = \bigcup\limits_{n=1}^{\infty} M_n \in \mathscr{B}$ by (2.5.1a). By the above remark, $H = R - K \in \mathscr{B}$.

### Example 2.5.6

Show that for any real number $a$, the set $\{a\}$, consisting only of $a$, is in $\mathscr{B}$.

For any positive integer $n$, let $M_n = \{x \in R; a - 1/n < x \le a\}$. By condition (2.5.1c) $M_n \in \mathscr{B}$. Consider $\bigcap\limits_{n=1}^{\infty} M_n = A$; by (2.5.2) $A \in \mathscr{B}$. Now for each $n$, $a \in M_n$; so $a \in A$. If $x > a$, then $x \notin M_1$. So $x \notin A$ by the definition of intersection. Also, if $x < a$, then $a - x > 0$. So for some positive integer $n$, $1/n < a - x$ or $x < a - 1/n$. Then for that $n$, $x \notin M_n$ and $x \notin A$. Thus $A = \{a\}$. By the above remark, $\{a\} \in \mathscr{B}$.

These examples, together with some of the exercises to follow, indicate that many, if not all, subsets of $R$ are in $\mathscr{B}$. It can be shown that not all subsets of $R$ are in $\mathscr{B}$. However, all "nice" subsets of $R$ are in $\mathscr{B}$. We proceed as though any set which would arise in an elementary level book, such as this, is in $\mathscr{B}$.

The reader may wonder why the basic set in $\mathscr{B}$ is a "half-open" interval $\{x \in R; a < x \le b\}$. We find, in later chapters, that it is very convenient to have disjoint sets readily available. Note that if $a < b < c$, then

$$\{x \in R; a < x \le b\} \cap \{x \in R; b < x \le c\} = \varnothing$$

We could have just as easily used basic intervals open on the right. That is a matter of convention.

**Definition**   A set $M$ in $n$-dimensional space is said to be an *open* set if for each point $x = (x_1, x_2, \ldots, x_n) \in M$ there is a number $\epsilon > 0$ (which may vary with $x$) such that the generalized square

$$\{(y_1, \ldots, y_n); x_i - \epsilon < y_i < x_i + \epsilon, \qquad i = 1, \ldots, n\} \subset M$$

A set $N$ in $n$-dimensional space is said to be *closed* if its complement $N'$ is an open set.

The condition for openness can also be given in terms of generalized disks, i.e., for any $x = (x_1, \ldots, x_n) \in M$ there is a number $\delta > 0$ (depending on $x$) such that $\{(y_1, \ldots, y_n); (y_1 - x_1)^2 + (y_2 - x_2)^2 + \cdots + (y_n - x_n)^2 < \delta^2\} \subset M$. This follows substantially because any generalized square contains a generalized disk and any generalized disk contains a generalized square.

In 2-dimensional space the following sets are open.

**(2.5.3a)**   the interior of a disk $\{(x_1, x_2); (x_1 - a_1)^2 + (x_2 - a_2)^2 < r^2\}$ with center $(a_1, a_2)$ and radius $r > 0$

**(2.5.3b)**   the exterior of the disk in (a)

**(2.5.3c)**   the interior of a rectangle $\{(x_1, x_2); a_1 < x_1 < b_1, \quad a_2 < x_2 < b_2\}$

**(2.5.3d)**   the exterior of the rectangle in (c)

**(2.5.3e)**   the interior of a parabola $\{(x_1, x_2); x_2 > cx_1^2\}$ for $c > 0$

In 1-dimensional space, generalized disks and squares reduce to intervals. Figure 11 gives some insight into the appearance of open sets. The boundary is not in $M$.

**Theorem 2.5.1**   An open set $M$ in 2-dimensional Euclidean space is a union of countably many rectangles of the form

$$\{(x_1, x_2); a_1 < x_1 \leq b_1; a_2 < x_2 \leq b_2\}$$

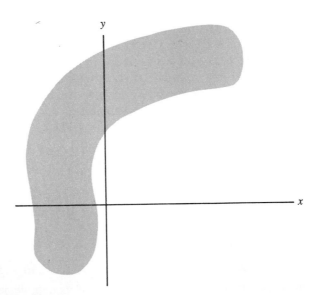

**Figure 11**

PROOF    Consider the squares

$$Q_{nm}^{(0)} = \{(x_1, x_2); n < x_1 \leq n + 1, \quad m < x_2 \leq m + 1\}$$

for $n$, $m$ chosen from among all integers, There are countably many such squares because there are only countably many pairs of integers $(n, m)$.

Now consider the squares

$$Q_{nm}^{(1)} = \left\{(x_1, x_2); \frac{n}{2} < x_1 \leq \frac{n + 1}{2}, \quad \frac{m}{2} < x_2 \leq \frac{m + 1}{2}\right\}$$

for $n$, $m$ chosen from among all integers. There are countably many of these. Let

$$Q_{nm}^{(2)} = \left\{(x_1, x_2); \frac{n}{4} < x_1 \leq \frac{n + 1}{4}, \quad \frac{m}{4} < x_2 \quad \frac{m + 1}{4}\right\}, \; n, m \text{ integers}$$

In general

$$Q_{nm}^{(k)} = \left\{(x_1, x_2); \frac{n}{2^k} < x_1 \leq \frac{n + 1}{2^k}, \quad \frac{m}{2^k} < x_2 \leq \frac{m + 1}{2^k}\right\}, \; n, m \text{ integers}$$

The length and width of the square $Q_{nm}^{(k)}$ are $\dfrac{n + 1}{2^k} - \dfrac{n}{2^k} = \dfrac{1}{2^k} = \dfrac{m + 1}{2^k} - \dfrac{m}{2^k}$

so that no two points in $Q_{nm}^{(k)}$ are farther apart than $\sqrt{2}\,\dfrac{1}{2^k}$.

There are only countably many squares of the form $Q_{nm}^{(k)}$ where $k = 0, 1, 2, 3,$ $\ldots$ and $n$, $m$ range separately through the integers. For any $k$ and any fixed point $(x_1, x_2)$ there is a square (and only one) of the form $Q_{nm}^{(k)}$ such that $(x_1, x_2) \in Q_{nm}^{(k)}$ because if $x_1$ and $x_2$ are real numbers, $2^k x_1$ and $2^k x_2$ are real numbers and, obviously, there must be integers, say $n_0$ and $m_0$, such that $n_0 < 2^k x_1 \leq n_0 + 1$ and $m_0 < 2^k x_2 \leq m_0 + 1$. Thus $(x_1, x_2) \in Q_{n_0 m_0}^{(k)}$.

Now let $N$ be the union of all of the $Q_{nm}^{(k)}$ which are entirely contained in $M$, i.e., $\bigcup_{Q_{nm}^{(k)} \subset M} Q_{nm}^{(k)}$. There are only countably many $Q_{nm}^{(k)} \subset M$. Obviously $N \subset M$.

Suppose $x = (x_1, x_2) \in M$. By definition of an open set there is a real number $\epsilon > 0$ such that

$$L = \{(y_1, y_2); x_1 - \epsilon < y_1 < x_1 + \epsilon, \quad x_2 - \epsilon < y_2 < x_2 + \epsilon\} \subset M$$

Choose $k$ such that $1/2^k < \epsilon/2$. By the foregoing argument there is a rectangle $Q_{nm}^{(k)}$ such that $x \in Q_{nm}^{(k)}$. Now all points whose distance from $x$ is less than $\epsilon$ are in $L$ since $L$ is a square with center $x$ and side $2\epsilon$. As we previously observed, no point in $Q_{nm}^{(k)}$ is farther from $x$ than $\sqrt{2}/2^k < \sqrt{2}(\epsilon/2) < \epsilon$. Therefore, all points of $Q_{nm}^{(k)}$ are in $L$ and hence $Q_{nm}^{(k)} \subset M$. But, therefore, $x \in Q_{nm}^{(k)} \subset N$ so we have $M \subset N \subset M$ and $N = M$ and $M$ is a union of countably many sets of the form indicated.

## EXERCISES 2.5

◆ 1 Let $S$ be any set. Show that the system $\mathscr{S}$ of finite subsets of $S$ is closed under $\cup$ and $\cap$.

◆ 2 Let $S$ be denumerable. Let $\mathscr{S}$ be the system of finite subsets of $S$ and their complements. Show that $\mathscr{S}$ is closed under $\cup$, $\cap$, and complementation.

3 Let $S$ be a set with $n$ elements. Let $0 < k < n$. Let $\mathscr{S}$ be the system of subsets of $S$ which have $k$ or more elements. Under which, if any, of $\cup$, $\cap$, and complementation is $\mathscr{S}$ closed?

4 Show that $\mathscr{S}$ of exercise 2 is not closed under countable union.

5 Show that $\mathscr{S}$ of exercise 1 is closed under countable intersection.

◆ 6 Let $\mathscr{S}_\alpha$ be systems of subsets of $S$. Suppose that each $\mathscr{S}_\alpha$ is closed under countable union. Show that $\mathscr{S} = \cap \, \mathscr{S}_\alpha$ is closed under countable union by showing that if $M_n \in \mathscr{S}_\alpha$ for all $\alpha$ and $n = 1, 2, 3, \ldots$ then $\bigcap_{n=1}^{\infty} M_n \in \mathscr{S}_\alpha$ for all $\alpha$.

7 Let $\mathscr{S}_n$ be the system of sets $M \subset R$, the real numbers, such that $M$ has infinitely many elements and if $x \in M$, $y \in M$ then $|x - y| < 1/n$. Show

$$\mathscr{S}_1 \cap \mathscr{S}_2 \cap \cdots \cap \mathscr{S}_n \neq \varnothing \text{ but } \bigcap_{n=1}^{\infty} \mathscr{S}_n = \varnothing.$$

8 Sketch figures to indicate why the sets of (2.5.3) are open.

9 Let $f: R \to R$ be a continuous function. Let $M$ be an open set in $R$. Show that $f^{-1}(M)$ is an open set.

10 Show, by example, that the intersection of countably many open sets may be open or it may not. What about intersections of finitely many open sets?

11 Let $\mathscr{B}$ be the system of Borel sets of $R$. Show that $\mathscr{B}$ contains all sets of the form
(a) closed sets (complements of open sets)
(b) union of countably many closed sets (called $F_\sigma$ set)
(c) intersection of countably many open sets ($G_\delta$ set).

12 Show that any open rectangle in the $xy$ plane with sides parallel to the axes is the union of countably many open squares whose centers have rational coordinates and whose side length is a rational number. [*Hint:* Given any real number $x < y$, there is a rational number $r$ such that $x < r < y$.]

## SUPPLEMENTARY EXERCISES FOR CHAPTER 2

◆ 1 (a) How many 1st-degree polynomials have integer coefficients?
(b) How many 2nd-degree polynomials have integer coefficients?
(c) How many polynomials of arbitrary degree have integer coefficients?

A real number is called an *algebraic* number if it is a zero of a polynomial with integer coefficients.

(d)   What is the maximum number of real zeros which a polynomial of degree $n$ can have?

(e)   How many algebraic numbers are there?

◆  **2**   Let $S$ be a set with subsets $M_1, M_2, M_3, \ldots$. Show that

(a)   $F' = \bigcup\limits_{n=1}^{\infty} (\bigcap\limits_{k=n}^{\infty} M_k)^*$ consists of all elements of $S$ each of which belongs to all but a finite number of the $M_j$.

(b)   $K = \bigcap\limits_{n=1}^{\infty} (\bigcup\limits_{k=n}^{\infty} M_k)$ consists of all elements of $S$ each of which belongs to infinitely many of $M_j$.

**3**   Recall that a function $f: R \to R$ is continuous at $x_0$ if for any $\epsilon > 0$ there is a $\delta > 0$ such that if $|x - x_0| < \delta$ then $|f(x) - f(x_0)| < \epsilon$. Let $\Omega$ be the set of functions from $R$ to $R$. Let

$$M_{jk} = \{f \in \Omega : |f(x) - f(x_0)| < 1/k \quad \text{if} \quad |x - x_0| < 1/j\}$$

Show that $\bigcap\limits_{k=1}^{\infty} (\bigcup\limits_{j=1}^{\infty} M_{jk})$ is the set of functions in $\Omega$ which are continuous at $x_0$.

---

\* The notation $\bigcup\limits_{n=1}^{\infty} (\bigcap\limits_{k=n}^{\infty} M_k)$ signifies $\bigcup\limits_{n=1}^{\infty} N_n$ where $N_n = M_n \cap M_{n+1} \cap M_{n+2} \cap \cdots$.

CHAPTER **3**

# *Probability Systems*

## 3.1 **Background**

Many standard English words have meanings different from their usual meanings when they are used in a mathematical context, e.g., ring, integrate, and differentiate. The standard usage gives little or no intuitive help in understanding the mathematical usage of these words. However, the mathematical and standard usage of the word "probability" are quite similar, although the word probability will have a much more precise meaning in the mathematical context. This meaning is developed in this and later chapters.

Much earlier work in probability was stimulated by games of chance, that is to say, gambling. At present, gambling in the traditional sense is a minor stimulus of work in probability. Nevertheless, it is convenient to use certain ideas of gambling and certain games of chance to provide intuition. Let us consider a variety of examples.

*Example 3.1.1*
A coin is to be tossed in such a way that it spins many times in the air and then lands.

One frequently says, "There is a fifty-fifty chance that heads will show" or "It is equally probable that a head or a tail will show" or "The probability that a head will show is $\frac{1}{2}$."

### Example 3.1.2
A card is to be drawn from a well-mixed standard deck.

One frequently says, "The odds are 3 to 1 that it is not a spade" or "The probability that it is not a spade is $\frac{3}{4}$."

### Example 3.1.3
The names of all of the residents in a town are written on cards which are put in a barrel, mixed well, and one card is chosen.

One may say, "The probability that a woman's name will appear on the card is $\frac{1}{2}$."

### Example 3.1.4
A red die and a green die are to be rolled.

One might say, "The odds are 10 to 1 against a total of seven dots showing on the top faces when the dice come to rest" or "The probability is $\frac{10}{11}$ against a seven showing."

### Example 3.1.5
A stick is to be broken at a random place.

Under certain conditions one might say, "The probability that the longer piece is more than twice as long as the shorter piece is $\frac{1}{2}$."

The above examples have certain common characteristics. In the first sentence, an act or experiment is described. We may not be able to predict in detail the exact outcome of the act or experiment in advance of its completion. In the second sentence of the examples, certain numbers are associated with a situation which might arise under the conditions of the experiment or act described in the first sentence.

Before proceeding further, we admit that the numbers in Examples 3.1.3, 3.1.4, and 3.1.5 might be questioned. At present, the exact numbers are not important. We are merely pointing to the fact that the assignment of some number seems appropriate and satisfactory although the number given in the text might be somewhat unsettling. If the reader prefers some other numbers in the above examples, he should feel free to insert them.

In view of these and other similar examples, both mathematicians and nonmathematicians have idealized and abstracted the essential aspects and have developed general mathematical systems to cover the above problems and many more.

As the basic ingredient we have certain sets pertaining to the experiment. For Example 3.1.1, we may consider {H, T}, the set consisting of the two indicated letters of the alphabet (H is for heads, T is for tails). For Example 3.1.2, we could consider

$$\{1_C, 1_D, 1_H, 1_S, 2_C, 2_D, 2_H, 2_S, \ldots, 13_C, 13_D, 13_H, 13_S\}$$

a set consisting of the numerals $1, 2, 3, \ldots, 13$ with subscripts C, D, H, S (standing for clubs, diamonds, hearts, and spades) and $1, 2, 3, \ldots, 13$ standing for ace, two, three, $\ldots$, king. For Example 3.1.3, consider {cards each having the name of a resident in the town}. For Example 3.1.4, consider

$$\{(m, n); m \in \{1, 2, \ldots, 6\} \text{ and } n \in \{1, 2, 3, \ldots, 6\}\}$$

the set of ordered pairs $(m, n)$ with the indicated possible values for $m$ and $n$. Here we may take $m$ and $n$ to represent the number of dots on the top face of the red die and of the green die, respectively, when the dice stop rolling. In Example 3.1.5, consider

$$\{x; x \text{ is a real number and } \tfrac{1}{2} \leq x \leq 1\}$$

The number $x$ may be thought of as the length of the longer piece of the stick if we think of the whole stick as having length of one unit.

The above sets, which have been suggested in connection with the examples, index the possible outcomes which one might expect to occur in the course of the experiment. The fact that there is more than one outcome reflects our uncertainty of the outcome at the beginning of the experiment. The fact that the above sets are not completely unrestricted reflects the fact that we have some idea of the range of the possible outcomes.

Let us concentrate for the moment on

$$\{(m, n); 1 \leq m \leq 6, \quad 1 \leq n \leq 6; m, n \text{ are integers}\}$$

which we will call $D_2$ (for two dice). It is clear that $D_2$ consists of 36 ordered pairs $(m, n)$. In Example 3.1.4 we are interested in the subset $\{(1, 6), (2, 5), (3, 4), (4, 3), (5, 2), (6, 1)\}$ whose elements correspond to the cases in which the total number of dots showing on the top faces is seven. We could have asked for the probability that the total showing was five. This would lead us to the subset $\{(1, 4), (2, 3), (3, 2), (4, 1)\} \subset D_2$. The subset we select for consideration depends on the problem and other probability questions arising in connection with the rolling of a red die and a green die could lead to the subsets

**(3.1.1)**    $\{(5, 6), (6, 5), (1, 6), (2, 5), (3, 4), (4, 3), (5, 2), (6, 1)\}$

or

**(3.1.2)**    $\{(1, 1), (2, 2), (3, 3), (4, 4), (5, 5), (6, 6)\}$

or

**(3.1.3)**    $\{(1, 1), (1, 2), (1, 3), (1, 4)\}$

Now from expression (1.2.4), we know that the number of subsets of $D_2$ is $2^{36}$ (including $\varnothing$). Mathematicians have learned to feel at home with all of the $2^{36}$ subsets of $D_2$ even if not equally at home with each. By this, we mean to suggest that one should take a broad view and be prepared to assign probabilities lavishly rather than restricting one's attention to "natural" subsets. Of course, in the present example we have very few difficulties. In other problems it may be necessary to narrow one's viewpoint in order to get good results, as we shall see later.

# 3.2   Sample spaces and events

At this point, it is convenient to make some definitions.

**Definition**     A set $S$ whose elements can be used to index the outcomes of an experiment, or equivalently, whose elements are in one-to-one correspondence with the outcomes of the experiment is called a *sample space* (for the experiment). The elements $s \in S$ are called *sample points* (for the experiment).

The word "sample" reflects the role which data gathering and statistics played in the development of probability theory and the words "point" and "space" are standard mathematical terms which need have no special connection with geometrical or physical points and space.

In the above examples we were led to the sample spaces $S = \{H, T\}$ for Example 3.1.1,

$$S = \{1_C, 1_D, 1_H, 1_S, 2_C, 2_D, 2_H, 2_S, \ldots, 13_C, 13_D, 13_H, 13_S\}$$

for Example 3.1.2, $S = \{$cards each of which has the name of a resident of the town$\}$ for Example 3.1.3, $S = D_2$ for Example 3.1.4 and $S = \{x; x$ is a real number and $\frac{1}{2} \le x \le 1\}$ for Example 3.1.5. In each of these we may consider the elements as indexes for the outcomes of the experiment. For Example 3.1.2 we could just as easily have used $\{1, 2, 3, \ldots, 52\}$ if we have a procedure which assigns to each such number a card in the deck. One may ask, "Why not use $S_{11} = \{2, 3, 4, \ldots, 12\}$ for Example 3.1.4?" There is no reason not to use this set as the sample space for Example 3.1.4. There are always many legitimate sample spaces for an experiment. In many cases one set is more useful than another because of ease of computation in that framework, because it allows for more problems to be solved at one stroke, or because it is intuitively more appealing.

**Definition**     Subsets of a sample space $S$ are called *events* when they are used in a probabilistic context.

Events will usually be denoted by "$M$" and "$N$."

In Example 3.1.1, we were especially interested in the event $M = \{H\}$. In Example 3.1.2, $M = \{1_C, 1_D, 1_H, 2_C, 2_D, 2_H, \ldots, 13_C, 13_D, 13_H\}$ was of special interest. In Example 3.1.3, the relevant event was {cards having names of female residents of the town}. In Example 3.1.5, $M = \{x; \frac{2}{3} < x \leq 1\}$ was of interest because the longer piece must be longer than $\frac{2}{3}$ of a unit if it is to be more than twice as long as the shorter piece. In Example 3.1.4, the desired event was $\{(1, 6), (2, 5), (3, 4), (4, 3), (5, 2), (6, 1)\} \subset D_2$ but in other similar problems the event could be $\{(1, 4), (2, 3), (3, 2), (4, 1)\}$ or the expression in (3.1.1). Note that if $S_{11} = \{2, 3, 4, \ldots, 12\}$ is desired as the sample space for dice tossing (and certain problems), the events in $S_{11}$ corresponding to the above events in $D_2$ are $\{7\}$, $\{5\}$, and $\{7, 11\}$. It should also be noted that the events in expressions (3.1.2) and (3.1.3) have no corresponding events in $S_{11}$, hence making this sample space less desirable for use.

The above definitions of sample space and event are very general since the concept of an experiment is very general. Mathematicians are not frightened by generality, since it often leads to their most significant advances.

It was suggested that one should deal with many subsets (events) of a sample space unless it is very inconvenient. In Examples 3.1.1–3.1.5 there was only one event at issue in the statement of each example. Usually, however, if we can determine the probability of a particular event in $S$ we can determine the probability of other events in $S$.

### Example 3.2.1
A red and a green die are to be rolled. Assuming that the dice are symmetrical, which event would pose special difficulty in determining its probability?

No event seems especially difficult after it is described.

### Example 3.2.2
A stick is to be broken at a "random" place. What is the probability that the ratio of the length of the longer piece to the length of the shorter piece is between $r$ and $s$ where $r$, $s$ are real numbers greater than 1?

Here we are not asking for the probabilities of all subsets but certainly for the probabilities of many of them.

Examples 3.2.1 and 3.2.2 suggest that we can deal with a system of events as well as with merely one event. Since the words "or," "and," and "not" represent basic logical operations which may link or operate upon the descriptions of events and lead to other events (seven *or* eleven, in dice), it is reasonable to require the following properties of the system $\mathscr{A}$ of events.

**E1** The sample space $S$ is an event in $\mathscr{A}$.

**E2**   If $M$ and $N$ are events in $\mathscr{A}$, then $M \cup N$ and $M' = S - M$ are events in $\mathscr{A}$.

**E3**   If $M_1, M_2, M_3, \ldots$ is a list of countably many events in $\mathscr{A}$, then $\bigcup_{n=1}^{\infty} M_n$ is an event in $\mathscr{A}$.

We require E3 for certain technical reasons that will be apparent later.

A system $\mathscr{A}$ satisfying E1, E2, E3 is called an *admissible system* of events. Such a system will be in the background of all our probability problems even if it is not mentioned explicitly. Some of the problems to follow will request the student to use E2 to show that $M \cap N$ and $M - N$ are in $\mathscr{A}$ if $M \in \mathscr{A}$ and $N \in \mathscr{A}$. In other books admissible systems are called sigma fields.

### Example 3.2.3
Show that the system $\mathscr{A}$ consisting of $S$ and $\varnothing$ satisfies E1, E2, and E3.

Condition E1 is satisfied by the definition of $\mathscr{A}$. Since $S \cup S = S$, $\varnothing \cup \varnothing = \varnothing$, $S \cup \varnothing = \varnothing \cup S = S$, $S' = S - S = \varnothing$, $\varnothing' = S - \varnothing = S$, it follows that E2 is satisfied. As for E3, suppose $M_1, M_2, M_3, \ldots$ are in $\mathscr{A}$. Each is either $S$ or $\varnothing$. If at least one of the $M_n$ is $S$, then $\bigcup_{n=1}^{\infty} M_n = S$. If all of the $M_n$ are $\varnothing$ then $\bigcup_{n=1}^{\infty} M_n = \varnothing$. In either case $\bigcup_{n=1}^{\infty} M_n \in \mathscr{A}$.

It is not always so easy to verify E2 and E3 and it is rarely so simple to describe all the sets of an admissible system. In many cases we use the system of *all* subsets of a sample space $S$. This is obviously an admissible system of events. In other cases it is not feasible to do this. In those cases we use the smallest admissible system containing certain subsets of special interest. The fact that there is such a smallest admissible system follows from the definition of smallest system and from E1, E2, E3 as the defining conditions of the system. A basic admissible system of events in $R$ is the system of Borel sets $\mathscr{B}$ (see section 2.5). A detailed study of that system would take us far afield. To a fair extent one can ignore the admissible system in a first course in probability. We will not pursue the topic very much further but we will close by recommending that the reader give some attention to the role of the admissible system in the ensuing material. Although it would be dishonest to claim that the admissible system is the primary concept in probability theory, a study of admissible systems is essential in advanced topics.

**EXERCISES 3.2** _____

**1**   A regular dodecahedron is a regular geometric solid which has 12 congruent faces each of which is a regular pentagon. Suppose we mark the faces so that they are distinguishable. Describe a sample space for the experiment of rolling the dodecahedron on a flat table and seeing on which face it rests when it stops rolling.

2   The chairman of an international body consisting of one representative from each of 12 countries is to be chosen by lot from among the representatives. Describe a sample space for this experiment.

3   Describe a sample space for the experiment of choosing one card from a standard deck and recording the suit of the chosen card.

4   A coin is tossed six times and the number of heads which show is tabulated. Describe a sample space for this experiment.

5   A flashbulb is chosen at random from a batch of ten flashbulbs which have serial numbers from 21786 to 21795 inclusive. Describe a sample space for this experiment. How many subsets does this sample space have?

6   Two flashbulbs are chosen at random from the batch described in exercise 5. Describe a sample space for this experiment. How many subsets does this sample space have?

7   From the batch of flashbulbs described in exercise 5 a flashbulb is chosen at random. Its serial number is noted and it is returned to the batch which is then mixed thoroughly. A flashbulb is again chosen at random and its serial number is noted. Describe a sample space for this experiment. How many sample points does this sample space have?

8   Flashbulbs numbered 21786 and 21787 of those described in exercise 5 are glued together. Describe a sample space for the experiment in which exactly two flashbulbs are chosen from the batch. How many sample points are in the sample space?

9   (a)   Let $S$ be any nonempty set. Let $A$ be a fixed subset of $S$ with $\varnothing \neq A$ and $S \neq A$. Show that $\mathscr{A} = \{\varnothing, A, S - A, S\}$ is an admissible system of events; i.e., satisfies E1, E2, and E3.

(b)   Is $\mathscr{A}_0 = \{\varnothing, A, S\}$ an admissible system?

10   Let $S = \{1, 2, 3, 4, \ldots\}$. Let $\mathscr{A}_0$ consist of $S$ and all finite subset of $S$. Which of E1, E2, E3 does $\mathscr{A}_0$ satisfy?

11   Let $S = \{1, 2, 3, 4, \ldots\}$. Let $\mathscr{A}_0$ consist of $\varnothing$ and all denumerable subsets of $S$. Which of E1, E2, E3 does $\mathscr{A}_0$ satisfy?

12   In $D_2$ describe the events which correspond to the following descriptions using the bracket notation.

(a)   The total number of dots showing is six.

(b)   The total number of dots showing is more than nine.

(c)   The total number of dots showing is an even number.

13   Describe a sample space which would be useful for studying the sex of the children in families having fewer than 100 children.

**14** Describe a sample space which would be useful in studying the experiment in which five cards are chosen at random from a standard deck. How many sample points are there in this sample space? How many sample points are there in the event in which all cards of the five chosen are of different kinds? How many sample points are there in the event in which all cards of the five chosen are of the same suit?

♦ **15** Show that if $M$ and $N$ are events in $\mathscr{A}$ which satisfies E1, E2, E3, then $M \cap N$ and $M - N$ are in $\mathscr{A}$.

♦ **16** Show where "or," "and," and "not" relate to E1, E2, and E3.

**17** Give some examples of the use of "or," "and," and "not" in forming events.

**18** Imagine that a red die and a green die have coordinate axes imbedded in them. What would be the disadvantage of using a sample space whose sample points state the relationship of these coordinate axes to certain fixed external coordinate axes when we consider such problems as rolling a seven or eleven, or having less than four spots on the red die?

# 3.3 Equal likelihood

Suppose an experiment has a finite number $n$ of possible outcomes. Furthermore, assume there is a great deal of symmetry (rolling dice) or that due to some data gathered (sex of offspring in certain species) or due to simplicity (day of birth of people in a room) it appears that there is no probabilistic preference for one outcome as opposed to another outcome. This problem was the traditional object of intense study and we will call it the case of *equally likely outcomes*. It should be recognized that if simplicity is the only reason for choosing this case, we may pay the price in highly inaccurate predictions.

In this case the probability of an event $M$, which we denote by $P(M)$, is defined as the ratio

$$P(M) = \frac{\text{number of sample points in } M}{n}$$

If $M$ and $N$ are events and if $M \cap N = \varnothing$

**(3.3.1)**   the number of sample points in $M \cup N = $ (the number of sample points in $M$) + (the number of sample points in $N$)

If we divide both sides of equation (3.3.1) by $n$ we get $P(M \cup N) = P(M) + P(N)$ in the case $M \cap N = \varnothing$. We could further deduce that $P(M \cup N) = P(M) + P(N) - P(M \cap N)$ regardless of the events $M$ and $N$ but this is not germane at present.

In the case of equally likely outcomes virtually all problems are counting problems.

### Example 3.3.1
Two standard fair dice of different colors are rolled. What is the probability of the event that the total number of dots showing on the top faces is seven?

If we interpret the problem to mean that $D_2$ is a correct sample space and if the symmetry of the dice imply that the outcomes are equally likely, then

$$P(\text{dice show seven}) = \tfrac{6}{36}$$

since the event is $\{(1, 6), (2, 5), (3, 4), (4, 3), (5, 2), (6, 1)\}$ and $D_2$ has 36 sample points. Note that this answer differs from that offered in Example 3.1.4. The number $\tfrac{6}{36} = \tfrac{1}{6}$ is considered the correct answer.

### Example 3.3.2
Two people are to be selected at random from a group consisting of eight men and two women. What is the probability that at least one woman is selected? Suppose three people are selected.

If we take the words "at random" to mean that the sample space can be {all pairs of people from ten people} with equal likelihood we have $n = \binom{10}{2}$. The event of having at least one woman in the pair that is selected is the complement of the event in which only men are chosen. The number of ways of choosing a pair, both of whom are men, is $\binom{8}{2}$. Therefore, there are $\binom{10}{2} - \binom{8}{2}$ pairs having at least one woman. Therefore,

$$P(\{\text{at least one woman in the pair}\}) = \frac{\binom{10}{2} - \binom{8}{2}}{\binom{10}{2}} = 1 - \frac{\binom{8}{2}}{\binom{10}{2}}$$

$$= 1 - \frac{8 \cdot 7}{10 \cdot 9} = \frac{34}{90} \approx 0.378$$

If three people are selected, the probability of at least one woman being chosen is

$$\frac{\binom{10}{3} - \binom{8}{3}}{\binom{10}{3}} \approx 0.533$$

if one uses the same reasoning.

### Example 3.3.3
Five cards are dealt from a well-shuffled deck. What is the probability of getting a full house? (A full house is three of one kind and two of another kind.)

For convenience, we will first interpret the problem as though the order of the cards is irrelevant. We may therefore choose as sample space the set of combinations of 52 cards taken five at a time. We interpret the words "well-shuffled" to mean that any hand of five cards (sample point) is treated the same as any other hand, i.e., equally likely outcomes. The event of a full house was essentially studied in section 1.2 where we found that there are $13 \cdot 12 \cdot \binom{4}{3} \cdot \binom{4}{2}$ hands containing three cards of one kind and two cards of another kind. Therefore

$$P(\text{hand is a full house}) = 13 \cdot 12 \cdot \binom{4}{3} \cdot \binom{4}{2} \Big/ \binom{52}{5}$$

If it is felt that the word "dealt" means that the order of the cards is relevant, but "well-shuffled" still means equally likely outcomes, the number of lists of five cards from a regular deck is $_{52}P_5$. The number of lists of cards which form a full house is $13 \cdot 12 \cdot \binom{4}{3} \cdot \binom{4}{2} \cdot 5!$ since from any collection of five cards which form a full house one may form $5!$ lists. In this context the answer is $13 \cdot 12 \cdot \binom{4}{3} \cdot \binom{4}{2} \cdot 5! / _{52}P_5$ but it is easy to show that this result is the same as the previous answer.

### Example 3.3.4

A fair coin is to be tossed five times in such a way that it spins many times before it lands. What is the probability that three or more heads show?

If one assumes that the appropriate sample space consists of the $2^5 = 32$ 5-letter words using the alphabet $\{H, T\}$ and that "fair coin" and "spinning many times" suggest that all sample points are equally likely, we have $\binom{5}{3}$ words with three H's and two T's, $\binom{5}{4}$ words with four H's and one T, $\binom{5}{5}$ words with five H's and 0 T's. Therefore,

$$P(\{\text{three or more heads occur}\}) = \left[ \binom{5}{3} + \binom{5}{4} + \binom{5}{5} \right] \div 2^5 = \tfrac{1}{2}$$

Henceforth, we omit such phrases as "spins many times," "well-shuffled," etc. We assume that such simple experiments are executed well and our theory is readily applicable.

**EXERCISES 3.3**

1   A batch of ten radio tubes contains two defective tubes which are not distinguishable by sight from the good tubes. Two tubes are to be chosen at random from the batch. What is the probability that at least one will be defective?

**2**   A fair coin is to be tossed 30 times. What is the probability that heads will show on exactly 14 tosses?

**3**   A fair die is to be rolled 30 times. What is the probability that one dot or two dots will show on exactly 14 of the rolls?

**4**   A community contains 5400 children who live in the town and 3600 children who live in the outlying districts. 1000 children are chosen at random to be put into a new school. What is the probability that there will be fewer than 200 rural children chosen?

**5**   What is the probability of getting two cards in each of two different kinds and one card of a third kind in a hand of five cards selected from a standard deck?

♦ **6**   Using only the fact that $P(M \cup N) = P(M) + P(N)$ if $M \cap N = \varnothing$, show that if $M_1, M_2, \ldots, M_k$ are events such that $M_i \cap M_j = \varnothing$ for $i \neq j$, then $P(M_1 \cup M_2 \cup \cdots \cup M_k) = P(M_1) + P(M_2) + \cdots + P(M_k)$. [*Hint:* Use mathematical induction on $k$.]

**7**   Verify that the second viewpoint in Example 3.3.3 gives the same answer as the first viewpoint.

**8**   What is the probability that a hand of five cards from a standard deck has all cards of the same suit? What assumptions are being made?

**9**   A room contains eight adults and two children. Two people are to be selected at random from the room. What is the probability that at least one is a child? What is the difference between this problem and exercise 1?

**10**   A 3-letter word (as in Chapter 1) made from the symbols of the English alphabet is to be chosen at random. What is the probability that it is made up only of consonants? What is the probability that it has exactly one vowel? What is the probability that all the letters are different? Discuss the case where repetitions are allowed and the case where no repetitions are allowed.

**11**   Find the probability that a hand of five cards selected from a standard deck has a ten, a jack, a queen, a king, and an ace.

**12**   Find the probability that a hand of five cards selected from a standard deck has a two, three, four, five, and six. Compare with exercise 11.

**13**   All cards except the 13 spade cards are removed from a deck. Five cards are chosen at random from the spade cards. What is the probability that the hand contains the two, three, four, five, and six of spades? Compare with exercise 12.

**14**   A political meeting contains six people from each of 52 countries. Five people are selected at random. What is the probability that the five people are from different countries? Compare with exercises 11, 12, and 13.

**15**   Ten unlabeled cans of peas are on a shelf. Two cans contain very good peas. The other eight cans contain ordinary peas. Two cans are chosen at random. Find the probability that at least one contains very good peas.

16 Ten new senators have been elected. Six are liberals and four are conservatives. Find the probability that a committee of five of them selected at random (unlikely though that may be) has a majority of conservatives.

17 What is the probability that a word selected at random from 30-letter words using the alphabet A, E, I, O, U, Y has exactly 14 of its positions filled by A or E?

18 (17 continued.) What is the probability that a 30-letter word of our current type has five of each of A, E, I, O, U, Y?

19 A box contains 30 red balls, 40 white balls, and 50 green balls. Ten balls are selected at random. What is the probability that exactly two are red, three are white, and five are green?

20 Three fair dice are rolled. What is the probability that a total of ten dots show?

21 An integer between 10 and 49 inclusive is selected at random. What is the probability that it is divisible by seven or is odd? What is the probability that it is either odd or it is even and divisible by 7?

# 3.4 Additivity

In section 3.3 we defined the probability $P(M)$ of the event $M \subset S$ where $S$ is a finite sample space with $n$ sample points. There we wanted to treat the outcomes of an experiment having $S$ as a sample space as equally likely. For this purpose we let $P(M)$ be the number $1/n$ (number of sample points in $M$). With this definition the following numerical properties are obvious.

**M1** $P(M)$ is a real number and $0 \leq P(M) \leq 1$.

**M2** $P(S) = 1$.

**M3** $P(\varnothing) = 0$.

**M4** If $M$ and $N$ are events such that $M \cap N = \varnothing$, then $P(M \cup N) = P(M) + P(N)$.

Also true and almost as obvious is the following additional property.

**M5** If $M_1, M_2, M_3, \ldots$ is a countable system of events in $S$ such that $M_i \cap M_j = \varnothing$ for $i \neq j$, then

$$P(\bigcup_{j=1}^{\infty} M_j) = \sum_{j=1}^{\infty} P(M_j)$$

where we interpret the "infinite" sum on the right as the infinite series whose $j$th term is $P(M_j)$.

The main reason for the validity of M5 is its essentially finite character in this case. One cannot have an infinite system of disjoint nonempty subsets of a finite sample space, that is, there must be a whole number $l$ such that $M_j = \varnothing$ for $j \geq l$. Otherwise there would be infinitely many nonempty sets $M_j \subset S$ which, since they are disjoint, would imply that $S$ is infinite. Therefore,

$$\bigcup_{j=1}^{\infty} M_j = \bigcup_{j=1}^{l} M_j$$

and

$$\sum_{j=1}^{\infty} P(M_j) = \sum_{j=1}^{l} P(M_j)$$

since

$$P(M_j) = P(\varnothing) = 0, \quad \text{for } j \geq l$$

By exercise 3.3.6, $P(\bigcup_{j=1}^{l} M_j) = \sum_{j=1}^{l} P(M_j)$ so (M5) holds in our current case.

Although M5 implies M4 it is convenient to list both conditions in the present discussion.

We now make an inductive leap and define, on a general sample space, mathematical concepts which can be specialized to the above concepts in the case of equal likelihood on finite sample spaces. For this purpose we let $S$ be a sample space, $\mathscr{A}$ an admissible system of events, and $R$ the real numbers.

**Definitions**    A *probability measure* on $S$ (actually on $\mathscr{A}$) is a function $P: \mathscr{A} \to R$ which satisfies conditions M1 through M5 above.

Condition M5 is called *countable additivity*. A real-valued function $P: \mathscr{A}_0 \to R$, where $\mathscr{A}_0$ is a system of sets satisfying only E1 and E2, is called a *finitely additive probability measure* if $P$ satisfies M1 through M4.

If $P$ is a probability measure, then the *probability* of an event $M$ (with respect to $P$) is the number $P(M)$.

This definition of "probability" is to be interpreted as the official mathematical usage of this word for the purposes of this book. Hence, we cannot rigorously talk about the probability of an event unless we have the entire framework of $S$, $\mathscr{A}$, and $P$. We will occasionally call $P$ a measure.

One may justifiably ask why finitely additive probability measures are not the exclusive object of study. Every probability measure satisfying E1–E3 and M1–M5 satisfies E1 and E2 and M1–M4. However, M5 enables us to determine the probability of "infinite events" from the probabilities of finite events in a certain sense. Without M5 there may be a certain kind of indeterminacy. This, together with certain other technical reasons, is why we restrict ourselves to probability measures satisfying M1–M5.

In an orderly study of our material we may wish to establish that certain functions from $\mathscr{A}$ to the reals are probability measures. The earlier material of this section established that equal likelihood on a finite sample space gives a probability measure. In the case of sample spaces with infinitely many sample points, we can give a careful, but complicated proof that probability measures determined in a certain way satisfy M1–M5. However, to show that probability measures determined in other ways satisfy M1–M5 is much beyond this book. We will state that certain functions can be shown to satisfy M1–M5 and we will use the consequences of M1–M5 for these functions.

After equal likelihood we may consider unequal likelihood. This can be done in countable as well as finite sample spaces.

### Example 3.4.1

An experiment (or game) consists of tossing a fair coin until heads appears for the first time. Describe a sample space, a system of events, and a probability measure for the problem. Calculate the probability that heads appears for the first time on an even-numbered toss. Calculate the probability that heads appears for the first time on a toss evenly divisible by 3.

We may take $S = \{1, 2, 3, 4, \ldots\}$ with the convention that the sample point $n$ represents the case in which $H$ appears for the first time at the $n$th toss. We can interpret "fair coin" as meaning that the various outcomes in tossing the coin $n$ times are equally likely; there are $2^n$ such outcomes possible, one of which consists of one H following $(n - 1)$ T's. It seems plausible to award to the event $\{n\}$ the probability $2^{-n}$; i.e., $P(\{n\}) = 2^{-n}$. This fits in very nicely because

$$2^{-1} + 2^{-2} + 2^{-3} + 2^{-4} + \cdots = 1$$

in the sense of sums of infinite series. What can we use as our $\mathscr{A}$? A natural choice is the system of all subsets of $S$. How do we define $P(M)$ for a typical event $M$? Why not take $2^{-j_1} + 2^{-j_2} + 2^{-j_3} + \cdots$ if $M = \{j_1, j_2, j_3, \ldots\}$? Thus the probability that H occurs for the first time on an even-numbered toss is

$$
\begin{aligned}
P(\{2, 4, 6, 8, \ldots\}) &= 2^{-2} + 2^{-4} + 2^{-6} + 2^{-8} + \cdots \\
&= 1/2^2 + 1/2^4 + 1/2^6 + 1/2^8 + \cdots \\
&= \frac{1/2^2}{1 - 1/2^2} = \frac{1/4}{1 - 1/4} = \frac{1}{3}
\end{aligned}
$$

by a well-known formula for the sum of a geometric series. Similarly, the probability of the coin showing H for the first time on a toss divisible by 3 is

$$
\begin{aligned}
P(\{3, 6, 9, 12, \ldots\}) &= 2^{-3} + 2^{-6} + 2^{-9} + 2^{-12} + \cdots \\
&= \frac{1/2^3}{1 - 1/2^3} = \frac{1/8}{1 - 1/8} = \frac{1}{7}
\end{aligned}
$$

It seems clear that the function $P$ defined in the example is a probability measure but we have not proved this. We will state and give an intuitive proof of a theorem which includes this. Since a careful proof is difficult we save such a proof until the end of the section. This will leave us clear to follow up some of the consequences of the assumption of countable additivity.

Let $S = \{1, 2, 3, \ldots\}$, $\mathscr{A} = \{\text{all subsets of } S\}$, and $p_1, p_2, p_3, \ldots$ be non-negative real numbers with $\sum_{j=1}^{\infty} p_j = 1$. If $N \neq \varnothing$ is a finite set and $N \subset S$, we let $\sum_{j \in N} p_j$ be the sum of those $p_j$ for $j \in N$. We assume that we know all of the properties of such sums. If $N = \varnothing$, we define $\sum_{j \in N} p_j = 0$. For $M \subset S$, where $M$ can be finite or infinite, we define

$$\textbf{(3.4.1)} \qquad P(M) = \underset{\substack{N \subset M \\ N \text{ finite}}}{\text{l.u.b.}} \sum_{j \in N} p_j$$

In (3.4.1) l.u.b. means least upper bound and it is taken over all finite sets $N \subset M$. Note that for such sets $\sum_{j \in N} p_j$ is defined.

**Theorem 3.4.1**    The function $P: \mathscr{A} \to R$ defined by (3.4.1) is a probability measure.

Intuitive argument for Theorem 3.4.1:

**M1**    $P(M)$ is obviously real and non-negative,

$$\sum_{j \in N} p_j \leq \sum_{j=1}^{\infty} p_j = 1$$

so

$$\underset{\substack{\text{finite set} \\ N \subset M}}{\text{l.u.b.}} \sum_{j \in N} p_j = P(M) \leq 1$$

**M2**    Obvious.

**M3**    A definition.

**M4**    Subsumed under M5.

**M5**    Let $M = \bigcup_{k=1}^{\infty} M_k$. Since the $M_k$ are disjoint every $j \in M$ belongs to one and only one $M_k$. Thus every $p_j$ used in computing $P(M)$ appears once and only once in $\sum_{k=1}^{\infty} P(M_k)$ and vice versa.

Therefore, M5 is proved.

Since $\{1, 2, 3, \ldots\}$ can be used to index any denumerable set, it can be used as the most general denumerable sample space although there may be many cases when another denumerable set is more convenient. At any rate, the important concept in Theorem 3.4.1 is that the most general probability measure on the admissible system $\mathscr{A}$ consisting of all subsets of a denumerable sample space $S = \{s_1, s_2, s_3, \ldots\}$ is given by some collection $\{p_1, p_2, \ldots\}$ of non-negative real numbers with $\sum\limits_{n=1}^{\infty} p_n = 1$ and $P(M) = \sum\limits_{s_n \in M} p_n$ for $M \subset S$. This

can be seen by identifying $n \in \{1, 2, 3, \ldots\}$ with $s_n \in \{s_1, s_2, s_3, \ldots\}$. Consider

any convergent series $\sum\limits_{n=1}^{\infty} a_n$ with $a_n \geq 0$ such that $\sum\limits_{n=1}^{\infty} a_n = A > 0$. Let $p_n = \dfrac{a_n}{A}$

Thus

$$\sum_{n=1}^{\infty} p_n = \frac{A}{A} = 1$$

Thus there are many probability measures on a denumerable sample space. Naturally, certain sequences $p_1, p_2, p_3, \ldots$ are more interesting than others. Some of the more interesting probability measures on denumerable sample spaces are studied in later sections.

We now give an example of a genuine finitely additive measure.

### Example 3.4.2

Let $S = \{1, 2, 3, \ldots\}$. Let $\mathscr{A}_0$ consist of all finite sets (including $\varnothing$) and the complements of finite sets. (By exercise 2.6.1 we know that $\mathscr{A}_0$ satisfies E1 and E2.) For $M \in \mathscr{A}_0$ let $P^*(M) = 0$ if $M$ is finite and let $P^*(M) = 1$ if $M' = S - M$ is finite, i.e., if $M$ is the complement of a finite set. Show that $P^*$ satisfies M1–M4 but not M5 in the restricted sense that there are $M_1, M_2, \ldots$

in $\mathscr{A}_0$, $M_i \cap M_j = \varnothing$ for $i \neq j$ such that $\bigcup\limits_{n=1}^{\infty} M_n$ is in $\mathscr{A}_0$ but

$$P(\bigcup_{n=1}^{\infty} M_n) \neq \sum_{n=1}^{\infty} P(M_n)$$

That $P^*$ satisfies M1–M3 follows from the definition of $P^*$ and the fact that $S = \varnothing'$ is the complement of a finite set. As for M4, we examine various cases.

If $M$ and $N$ are finite sets, $M \cup N$ is a finite set and we have $P^*(M \cup N) = 0 = 0 + 0 = P^*(M) + P^*(N)$. If $M$ is the complement of a finite set $Q$, i.e., $M = S - Q$ and if $M \cap N = \varnothing$, it is easy to see that $Q \supset N$ so $N$ must be a finite set. Moreover $(M \cup N)' = M' \cap N' \subset Q$ so $M \cup N$ is the complement of a finite set. Therefore $P^*(M \cup N) = 1 = 1 + 0 = P^*(M) + P^*(N)$. If $N$ is the complement of a finite set, the roles of $M$ and $N$ should be interchanged.

Finally, suppose $M_n = \{n\}$, the set having as its sole element the integer $n$.

By definition $\bigcup\limits_{n=1}^{\infty} M_n = \{1, 2, 3, \ldots\} = S \in \mathscr{A}_0$. However, $\sum\limits_{n=1}^{\infty} P^*(M_n) = 0 + 0$

$+ 0 + \cdots = 0 \neq 1 = P^*(S) = P^*(\bigcup\limits_{n=1}^{\infty} M_n)$.

It is somewhat unfortunate that the full effect of M5 cannot be discussed at this level. Suffice it to say that we will be using the properties of probability measures satisfying M5 without giving adequate acknowledgement.

We now join the sample space, the admissible system of events, and the probability measure.

**Definition**     A triple consisting of a sample space $S$, an admissible system of events $\mathscr{A}$ (satisfying E1–E3), and a probability measure $P$ on $\mathscr{A}$ (satisfying M1–M5) is called a *probability system*.

Such a system is in the background of any result which we obtain, even though it may not be explicitly mentioned on all occasions. (It may be useful to look for the probability system when it is not explicit.) The letter $\mathscr{S}$ is used to denote a probability system.

**Theorem 3.4.2**     Let $\mathscr{S}$ be a probability system. If $M \in \mathscr{A}$ and $N \in \mathscr{A}$ and $M \subset N$, then $P(N - M) = P(N) - P(M)$.

PROOF     Since $N - M = N \cap M'$ it is clear that $N = M \cup (N - M)$ and $M \cap (N - M) = \varnothing$. Therefore, by M4, $P(N) = P(M) + P(N - M)$ so $P(N - M) = P(N) - P(M)$.

**Corollary**     $P(M') = 1 - P(M)$.

PROOF     Obvious.

**Corollary**     Under the hypothesis of Theorem 3.4.2 $P(M) \leq P(N)$.

PROOF     By M1, $P(N - M) \geq 0$ so $P(M) \leq P(N)$.

**Theorem 3.4.3**     Let $\mathscr{S}$ be a probability system. If $M$ and $N$ are in $\mathscr{A}$ then

$$P(M \cup N) = P(M) + P(N) - P(M \cap N)$$

PROOF     Using the various set theoretic definitions we obtain $M \cup N = M \cup (N - M)$ and $M \cap (N - M) = \varnothing$. Therefore, $P(M \cup N) = P(M) + P(N - M)$. But $N - M = N - (N \cap M)$ and $N \cap M \subset N$. Therefore, by Theorem 3.4.2, with $M$ replaced by $N \cap M$, we have $P(M \cap N) = P(N) - P(N - (N \cap M))$ or $P(N - M) = P(N - (N \cap M)) = P(N) - P(M \cap N)$. Therefore, $P(M \cup N) = P(M) + P(N) - P(M \cap N)$.

**Corollary**     $P(M_1 \cup M_2 \cup M_2) = P(M_1) + P(M_2) + P(M_3) - P(M_1 \cap M_2) - P(M_2 \cap M_3) - P(M_1 \cap M_3) + P(M_1 \cap M_2 \cap M_3)$ whenever $M_1$, $M_2$, $M_3$ are events in a probability system $\mathscr{S}$.

PROOF     Let $M = M_1$ and $N = M_2 \cup M_3$ in Theorem 3.4.3. Then

$$P(M_1 \cup M_2 \cup M_3) = P(M_1) + P(M_2 \cup M_3) - P(M_1 \cap (M_2 \cup M_3))$$
$$= P(M_1) + P(M_2) + P(M_3) - P(M_2 \cap M_3)$$
$$- P(M_1 \cap (M_2 \cup M_3))$$

But by Example 2.1.3 $M_1 \cap (M_2 \cup M_3) = (M_1 \cap M_2) \cup (M_1 \cap M_3)$ so

$$P(M_1 \cap (M_2 \cup M_3)) = P(M_1 \cap M_2) + P(M_1 \cap M_3) - P(M_1 \cap M_2 \cap M_1 \cap M_3).$$

But $M_1 \cap M_2 \cap M_1 \cap M_3 = M_1 \cap M_2 \cap M_3$, so we have

$$P(M_1 \cup M_2 \cup M_3) = P(M_1) + P(M_2) + P(M_3) - P(M_2 \cap M_3)$$
$$- [P(M_1 \cap M_2) + P(M_1 \cap M_3) - P(M_1 \cap M_2 \cap M_3)]$$

and further simplification gives the desired result.

These are very simple results and since they did not use M5, they are true for finitely additive probability measures. We now pass to theorems which are valid for probability measures.

**Theorem 3.4.4**   Suppose $\mathscr{S}$ is a probability system and $M_1 \subset M_2 \subset M_3 \subset \cdots$ are events in $\mathscr{A}$. Then

$$P(\bigcup_{n=1}^{\infty} M_n) = \lim_{n \to \infty} P(M_n)$$

PROOF    The key to the proof is to go from the sets $M_1, M_2, M_3, \ldots$ as in Figure 1 to disjoint sets $N_1, N_2, N_3, \ldots$ where $N_1 = M_1$, $N_2 = M_2 - M_1$, $N_3 = M_3 - M_2, \ldots$, $N_k = M_k - M_{k-1}$ for $k > 1$. To show that such sets are actually disjoint we assume $k > j$ and have $N_j \subset M_j \subset M_{k-1}$. Therefore $N_j \cap N_k = N_j \cap (M_k - M_{k-1}) = \varnothing$ since $N_j \subset M_{k-1}$. Figure 1 also suggests, and we prove it by mathematical induction, that $N_1 \cup N_2 \cup \cdots \cup N_n = M_n$. For $N_1 = M_1$ and if $N_1 \cup N_2 \cup \cdots \cup N_k = M_k$ then $N_1 \cup N_2 \cup \cdots \cup N_k \cup N_{k+1}$ $= M_k \cup N_{k+1} = M_k \cup (M_{k+1} - M_k) = M_{k+1}$ since $M_k \subset M_{k+1}$. Furthermore, $\bigcup_{k=1}^{\infty} M_k = \bigcup_{k=1}^{\infty} N_k$ so $P(\bigcup_{k=1}^{\infty} M_k) = P(\bigcup_{k=1}^{\infty} N_k) = \sum_{k=1}^{\infty} P(N_k)$ by M5 and the disjointness of the $N_k$. However, $\sum_{k=1}^{\infty} P(N_k) = \lim_{n \to \infty} \sum_{k=1}^{n} P(N_k) = \lim_{n \to \infty} P(\bigcup_{k=1}^{n} N_k)$ by

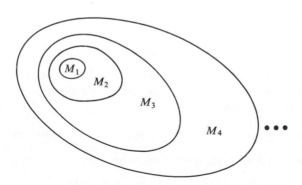

**Figure 1**

M4 and its implications for finite unions of disjoint sets. However, $\bigcup\limits_{k=1}^{n} N_k = M_n$, so $P(\bigcup\limits_{k=1}^{\infty} M_k) = \lim\limits_{n \to \infty} P(M_n)$ and the theorem follows since $\bigcup\limits_{k=1}^{\infty} M_k = \bigcup\limits_{n=1}^{\infty} M_n$.

**Corollary**     If $M_1 \supset M_2 \supset M_3 \supset \cdots$ are events in $\mathscr{A}$ then $P(\bigcap\limits_{n=1}^{\infty} M_n) = \lim\limits_{n \to \infty} P(M_n)$.

PROOF     By the first corollary to Theorem 3.4.2 $P(M') = 1 - P(M)$ for any event $M$. By the generalized De Morgan law (exercise 2.1.7) $(\bigcap\limits_{k=1}^{\infty} M_k)' = \bigcup\limits_{k=1}^{\infty} M_k'$ and it is immediate that $M_1' \subset M_2' \subset M_3' \subset \cdots$ since $M_1 \supset M_2 \supset M_3 \supset \cdots$. By Theorem 3.4.4 $P(\bigcap\limits_{k=1}^{\infty} M_k) = 1 - P((\bigcap\limits_{k=1}^{\infty} M_k)') = 1 - P(\bigcup\limits_{k=1}^{\infty} M_k')$ $= 1 - \lim\limits_{n \to \infty} P(M_n') = 1 - \lim\limits_{n \to \infty} [1 - P(M_n)]$. By the second corollary to Theorem 3.4.2 $P(M_1) \geq P(M_2) \geq P(M_3) \geq \cdots \geq 0$ so $\lim\limits_{n \to \infty} P(M_n)$ exists. Therefore, $\lim\limits_{n \to \infty} [1 - P(M_n)] = 1 - \lim\limits_{n \to \infty} P(M_n)$. If we complete the chain we see that $P(\bigcap\limits_{n=1}^{\infty} M_n) = P(\bigcap\limits_{k=1}^{\infty} M_k) = 1 - [1 - \lim\limits_{n \to \infty} P(M_n)] = \lim\limits_{n \to \infty} P(M_n)$.

Before we give a rigorous proof of Theorem 3.4.1, we prove some preliminary results.

**Lemma 3.4.1**     Let $I$ be a nonempty index set and $\{b_\alpha\}$ for $\alpha \in I$ a set of real numbers. Suppose that l.u.b. $b_\alpha = a$, a real number. Then for $\epsilon > 0$ there is an $\alpha \in I$ such that $b_\alpha > a - \epsilon$.

PROOF     Suppose that for some $\epsilon_0 > 0$, $b_\alpha \leq a - \epsilon_0$ for all $\alpha \in I$. Then $a - \epsilon_0$ is a smaller upper bound for the $b_\alpha$. This contradicts the definition of $a$, the smallest upper bound of the $b_\alpha$.

**Lemma 3.4.2**     Let $n_1, n_2, n_3, \ldots$ be an enumeration of $M$. (If $M$ is a finite set we assume that the enumeration terminates.) Then $P(M) = p_{n_1} + p_{n_2} + p_{n_3} + \cdots$. (This is a finite sum if $M$ is finite.)

PROOF     If $M$ is finite the lemma follows from the definition of l.u.b. and the properties of finite sums. Suppose that $M$ is infinite. Let $N$ be finite with $N \subset M$. For some integer $k$, $N \subset \{n_1, n_2, \ldots, n_k\} = N^*$. Therefore, $\sum\limits_{j \in N} p_j \leq \sum\limits_{j \in N^*} p_j = p_{n_1} + p_{n_2} + \cdots + p_{n_k} \leq p_{n_1} + p_{n_2} + p_{n_3} + \cdots$ by known properties of infinite series with non-negative terms. Therefore

$$P(M) = \underset{\substack{N \subset M \\ N \text{ finite}}}{\text{l.u.b.}} \sum\limits_{j \in N} p_j \leq p_{n_1} + p_{n_2} + p_{n_3} + \cdots$$

However, for all $k$, $p_{n_1} + p_{n_2} + \cdots + p_{n_k} \leq P(M)$ since $N^* \subset M$ and $N^*$ is finite. By known properties of infinite series with non-negative terms $p_{n_1} + p_{n_2} + p_{n_3} + \cdots \leq P(M)$. This, combined with the opposite inequality, proves the lemma.

PROOF OF THEOREM 3.4.1    M1, M2, M3 follow from the intuitive argument with Lemma 3.4.2 used to obtain M2. Furthermore, M4 is subsumed by M5.

Let $M = \bigcup\limits_{k=1}^{\infty} M_k$ where the $M_k$ are disjoint events. Let $N$ be a finite subset of $M$. There must be some $l$ such that $N \subset M_1 \cup M_2 \cup \cdots \cup M_l$. Thus $N = (N \cap M_1) \cup (N \cap M_2) \cup \cdots \cup (N \cap M_l)$. But $N \cap M_k$ is finite, so we have $\sum\limits_{j \in N \cap M_k} p_j \leq P(M_k)$. Furthermore, by the disjointness of the $M_k$ the sets $N \cap M_k$ are disjoint. Since we are familiar with the properties of finite sums

$$(3.4.2) \qquad \sum_{j \subset N} p_j = \sum_{k=1}^{l} \sum_{j \in N \cap M_k} p_j \leq \sum_{k-1}^{l} P(M_k) \leq \sum_{k=1}^{\infty} P(M_k)$$

The quantity on the far right of (3.4.2) is independent of $N$ so

$$P(M) = \text{l.u.b.}_{\substack{N \subset M \\ N \text{ finite}}} \sum_{j \in N} p_j \leq \sum_{k=1}^{\infty} P(M_k)$$

Let $n$ be a positive integer. By Lemma 3.4.1 and M2, for any $\epsilon > 0$ there are finite sets $N_k \subset M_k$, $k = 1, 2, \ldots, n$ such that $P(M_k) - \epsilon/n \leq \sum\limits_{j \in N_k} p_j$. Since the $M_k$ are disjoint so are the $N_k$. Let $N = N_1 \cup N_2 \cup \cdots \cup N_n$—a finite set. Then for $N \subset M$

$$(3.4.3) \qquad P(M) \geq \sum_{j \in N} p_j = \sum_{k=1}^{n} \sum_{j \in N_k} p_j \geq \sum_{k=1}^{n} [P(M_k) - \epsilon/n] = \sum_{k=1}^{n} P(M_k) - \epsilon$$

since $P(M) = \text{l.u.b.}_{M \supset N \text{ finite}} \sum\limits_{j \in N} p_j$. Thus $P(M) + \epsilon \geq \sum\limits_{k=1}^{n} P(M_k)$. Since $\epsilon > 0$ is arbitrary, $P(M) \geq \sum\limits_{k=1}^{n} P(M_k)$. However, $P(M)$ does not depend on $n$, so $P(M) \geq \text{l.u.b.}_{n \to \infty} \sum\limits_{k=1}^{n} P(M_k) = \sum\limits_{k=1}^{\infty} P(M_k)$. Hence we have $P(M) = \sum\limits_{k=1}^{\infty} P(M_k)$.

Lemma 3.4.2 is important in its own right because it shows that $P(M)$ can be computed from any enumeration of $M$.

Many of the techniques in the proof of Theorem 3.4.1 apply to non-negative series without the assumption that $\sum\limits_{n=1}^{\infty} p_n = 1$.

In the ensuing chapters we will use the symbol $P$ somewhat loosely, that is, if $s$ is a sample point (and if $\{s\}$ is an event) we will let $P(s) = P(\{s\})$. Also, if an event is described by a phrase of words we will let $P(\text{phrase}) = P(\{\text{phrase}\})$ instead of setting up a special letter. For example, we may write "find $P$ (full house in a hand of five cards)." Finally, we will write $P$ for most probability measures. This is somewhat like writing $f$ for most functions but it will actually cause less trouble than one might imagine. In certain cases where the distinction between probability measures is important we write $P_1$, $P_2$, etc.

**EXERCISES 3.4** _____

**1**   Let $S = \{1, 2, 3, 4, 5, 6\}$. Suppose that we wish to have $P(n)$ proportional to $n$ for each $n \in S$. Find $P(n)$ exactly.

**2**   Work exercise 1 in the cases that we wish $P(n)$ to be proportional to (a) $n^2$ (b) $1/n$.

**3**   Work exercise 1 if $S$ is changed to $\{1, 2, 3, \ldots, N\}$.

**4**   Let $S = \{0, 1, 2, 3, \ldots\}$ with the usual $\mathscr{A}$. Find, exactly, the probability measure $P$ such that

   (a)   $P(n)$ is proportional to $(\frac{1}{6})^n$ for all $n \in S$

   (b)   $P(n)$ is proportional to $3^n/n!$ for all $n \in S$

**5**   Give an example of events $M_1, M_2, M_3, \ldots$ such that $P(\bigcap\limits_{n=1}^{\infty} M_n) \neq \lim\limits_{n \to \infty} P(M_n)$.

**6**   Let $M_1, M_2, M_3, \ldots$ be events in $\mathscr{A}$. Show that $P(\bigcup\limits_{n=1}^{\infty} M_n) \leq \sum\limits_{n=1}^{\infty} P(M_n)$. Show first that $P(\bigcup\limits_{n=1}^{m} M_n) \leq \sum\limits_{n=1}^{m} P(M_n)$.

**7**   Let $M_1, M_2, M_3, \ldots$ be events in $\mathscr{A}$. Suppose that $\sum\limits_{n=1}^{\infty} P(M_n)$ converges to a finite limit. Show that $P(\bigcap\limits_{n=1}^{\infty} (\bigcup\limits_{k=n}^{\infty} M_n)) = 0$. [*Hint:* Note that $\lim\limits_{m \to \infty} \sum\limits_{n=m}^{\infty} a_n = 0$ for a convergent series; then use exercise 6 and the corollary to Theorem 3.4.4.]

**8**   Suppose that we try to build a finitely additive measure $P^*$ on $S = \{0,1,2,3,\ldots$ and $\mathscr{A}$, the set of all subsets of $S$, as follows: $P^*$(finite set) $= 0$,   $P^*$(infinite set) $= 1$. What goes wrong?

**9**   In reference to Theorem 3.4.1 show that $P(N \cup \varnothing) = P(N) + P(\varnothing)$ when $N$ is a finite set.

# 3.5   The continuous case

So far we have avoided a probability system for Example 3.1.5. The experiment of breaking a stick is very vague as to how the stick is to be broken. For the present, we suggest a specific type of chance breaking which may not be very credible from the point of view of how real world sticks break but which provides insight into problems of continuous chance phenomena.

We imagine the stick placed on a coordinate axis as in Figure 2. There is no loss of generality in assuming that the stick is one unit long. We then proceed to ignore the physical stick and consider $S = \{x; x \text{ a real number}, 0 \leq x \leq 1\}$. We now define the words "broken at a random place" for our convenience. We say that the probability that the break occurs between $a$ and $b$ is $b - a$ where $0 \leq a \leq b \leq 1$. In essence we are specifying probabilities of *certain* events, that is, $P(\{x \in S; a \leq x \leq b\}) = b - a$. Since $S = \{x; 0 \leq x \leq 1\}$, i.e., $a = 0$, $b = 1$; $P(S) = 1 - 0 = 1$. Notice that

$$P(\{x \in S; 0 \leq x \leq \tfrac{1}{4}\}) = P(\{x \in S; \tfrac{1}{4} \leq x \leq \tfrac{1}{2}\}) = P(\{x \in S; \tfrac{1}{2} \leq x \leq \tfrac{3}{4}\})$$
$$= P(\{x \in S; \tfrac{3}{4} \leq x \leq 1\}) = \tfrac{1}{4}$$

Thus there is a great deal of symmetry in our interpretation. Moreover, in order for the longer piece to be more than twice as long as the shorter piece, it is obvious that the (coordinate of the) break $x$ must satisfy $0 \leq x < \tfrac{1}{3}$ or $1 \geq x > \tfrac{2}{3}$ — that is, we cannot have $\tfrac{1}{3} \leq x \leq \tfrac{2}{3}$. Therefore,

$P$(longer piece is more than twice as long as the shorter piece)

$$= 1 - P(\{x \in S; \tfrac{1}{3} \leq x \leq \tfrac{2}{3}\}) = 1 - \tfrac{1}{3} = \tfrac{2}{3}$$

Note that this is at variance with the $\tfrac{1}{2}$ asserted in Example 3.1.5.

This seems to settle the problem, except, we have not specified a probability system which, supposedly, was to be the basis for every question in probability. We have a sample space $S$, and we have discussed certain events and their probabilities but we should have an admissible system $\mathscr{A}$ and $P$ should define a function from $\mathscr{A}$ to $R$ satisfying M1–M5. A good choice for our $\mathscr{A}$ is the sets of $S$ which are Borel sets in $R$.

The existence of $P: \mathscr{A} \to R$ such that $P$ satisfies M1–M5 and such that $P(\{x; a \leq x \leq b\}) = b - a$ is an important topic in the subject of measure theory. In this book we accept the result that such a measure exists and work with the implications of the existence of such a probability measure.

### Example 3.5.1
Within the present framework, what is the probability that the stick will be broken at a definite point $c$ such that $0 \leq c \leq 1$? What is the probability that the break will occur at an algebraic number?

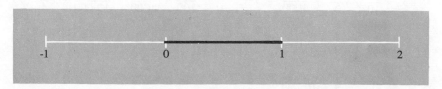

**Figure 2**

By Example 2.5.6 any $\{c\}$ is a Borel set in $R$. If $0 \le c \le 1$, $\{c\} \subset S$ so $\{c\} \in \mathcal{A}$. Furthermore $0 \le P(c) \le P(\{c - 1/n \le x \le c + 1/n\} \cap S) \le 2/n$ for all integers $n > 0$. This can be true if and only if $P(c) = 0$. Thus the probability of the break occurring at a fixed predetermined point is zero. This should not be surprising. There are infinitely many points in $S$ and by symmetry they should be equally probable. The only way for this to occur is for $P(c) = 0$.

In supplementary exercise 2.1(e), it was shown that the set $A$, of algebraic numbers, is a denumerable set. Therefore, $A = \bigcup_{n=1}^{\infty} \{a_n\}$, where $a_n$ is the algebraic number corresponding to $n$ in an enumeration. As noted in section 2.2, $A \cap S$ is countable since it is a subset of $A$. Thus $A \cap S = \bigcup_{n=1}^{\infty} \{a'_n\}$, where $a'_1, a'_2, a'_3, \ldots$ is an enumeration of the algebraic numbers between 0 and 1, inclusive. By the above argument for any $c$, $P(a'_n) = 0$. Therefore, $P(A \cap S) = \sum_{n=1}^{\infty} P(a'_n) = 0 + 0 + 0 + \cdots = 0$.

## Example 3.5.2

What is the probability that the break in our current stick will occur at a point $x$ where

$$x = \frac{1}{10} + \frac{n_2}{100} + \frac{n_3}{1000} + \frac{n_4}{10\,000} + \cdots$$

for $n_i \in \{0, 1, 2, \ldots, 9\}$ and the conventions of section 2.2 are applied to select the nonterminating expansion for real numbers $x$ with $0 < x \le 1$.

What is the probability that the break occurs at

$$x = \frac{1}{10} + \frac{n_2}{10} + \frac{7}{1000} + \frac{n_4}{10\,000} + \frac{n_5}{100\,000} + \cdots$$

in the above context?

Since the $n_i \ge 0$ and since we cannot have $n_i = 0$ for $i = 2, 3, 4, \ldots$ then

$$x = \frac{1}{10} + \frac{n_2}{100} + \frac{n_3}{1000} + \cdots > \frac{1}{10}$$

Furthermore since $n_2, n_3, n_4, \ldots \le 9$ we have

$$x = \frac{1}{10} + \frac{n_2}{100} + \frac{n_3}{1000} + \cdots \le \frac{1}{10} + \frac{9}{100} + \frac{9}{1000} + \frac{9}{10\,000} + \cdots$$

$$= \frac{1}{10} + \frac{9}{100}\left[1 + \frac{1}{10} + \frac{1}{100} + \frac{1}{1000} + \cdots\right] = \frac{1}{10} + \frac{9}{100}\left(\frac{1}{1 - \frac{1}{10}}\right)$$

$$= \frac{1}{10} + \frac{9}{100} \cdot \frac{10}{9} = \frac{1}{10} + \frac{1}{10} = \frac{2}{10}$$

Thus we have $\frac{1}{10} < x \le \frac{2}{10}$. Since any real number greater than zero and less than or equal to one has an expansion of the form

$$y = \frac{n_1}{10} + \frac{n_2}{100} + \frac{n_3}{1000} + \frac{n_4}{10\,000} + \cdots$$

and if $n_1 \ge 2$, $y > \frac{2}{10}$ and if $n_1 = 0$,

$$y \le \frac{9}{100} + \frac{9}{1000} + \frac{9}{10\,000} + \cdots = \frac{1}{10}$$

we find that for any $x$, $\frac{1}{10} < x \le \frac{2}{10}$ we have

$$x = \frac{1}{10} + \frac{n_2}{100} + \frac{n_3}{1000} + \cdots$$

Now $\left\{\frac{1}{10}\right\} \cup \left\{x; \frac{1}{10} < x \le \frac{2}{10}\right\} = \left\{x; \frac{1}{10} \le x \le \frac{2}{10}\right\}$ so we have $P\left(\left\{\frac{1}{10}\right\}\right) +$

$P(x$ of type we want$) = \frac{2}{10} - \frac{1}{10}$ or $0 + P(x$ is of the type we want$) = \frac{1}{10}$.

Therefore

$$P\left(x + \frac{1}{10} + \frac{n_2}{100} + \frac{n_3}{1000} + \frac{n_4}{10\,000} + \cdots\right) = \frac{1}{10}$$

As for those

$$x = \frac{1}{10} + \frac{n_2}{100} + \frac{7}{1000} + \frac{n_4}{10\,000} + \frac{n_5}{100\,000} + \cdots$$

we still have $\frac{1}{10} < x \le \frac{2}{10}$ but what other things are true of such $x$? For convenience, suppose $n_2 = 2$. Then $x > \frac{1}{10} + \frac{2}{100} + \frac{7}{1000}$ because we are not allowed $n_j = 0$ for $j \ge 4$. But also

$$x \le \frac{1}{10} + \frac{2}{100} + \frac{7}{1000} + \frac{9}{10\,000} + \frac{9}{100\,000} + \cdots$$

$$= \frac{1}{10} + \frac{2}{100} + \frac{7}{1000} + \frac{9}{10\,000}\left(1 + \frac{1}{10} + \frac{1}{100} + \cdots\right)$$

$$= \frac{1}{10} + \frac{2}{100} + \frac{7}{1000} + \frac{9}{10\,000} \cdot \frac{10}{9}$$

by a previous computation. Therefore

$$x \le \frac{1}{10} + \frac{2}{100} + \frac{7}{1000} + \frac{1}{1000}$$

Now we can argue, as before, that any real number $y$ with

$$\frac{1}{10} + \frac{2}{100} + \frac{7}{1000} < y \le \frac{1}{10} + \frac{2}{100} + \frac{8}{1000}$$

can be written as

$$y = \frac{1}{10} + \frac{2}{100} + \frac{7}{1000} + \frac{n_4}{10\,000} + \frac{n_5}{100\,000} + \frac{n_6}{1\,000\,000} + \cdots$$

Now

$$P\left(\left\{x \in S; \frac{1}{10} + \frac{2}{100} + \frac{7}{1000} < x \le \frac{1}{10} + \frac{2}{100} + \frac{8}{1000}\right\}\right) = \frac{8}{1000} - \frac{7}{1000} = \frac{1}{10^3}$$

since the absence of an equality on the left omits only $\frac{1}{10} + \frac{2}{100} + \frac{7}{1000}$ and

that point has probability measure zero. Thus when $n_2 = 2$ we have an interval of real numbers between 0.127 and 0.128 as in Figure 3. The same argument can be carried out for $n_2 = 0, 1, 3, 4, 5, 6, 7, 8, 9$ to yield, in succession, the real numbers between 0.107 and 0.108, between 0.117 and 0.118, between 0.137 and 0.138, between 0.147 and    148, etc. There are a total of ten such intervals of real numbers, each of them having probability $\frac{1}{1000} = \frac{1}{10^3}$. Since they are disjoint we find

$$P\left(x = \frac{1}{10} + \frac{n_2}{100} + \frac{7}{1000} + \frac{n_4}{10\,000} + \frac{n_5}{100\,000} + \cdots\right)$$

$$= \underbrace{\frac{1}{10^3} + \frac{1}{10^3} + \cdots + \frac{1}{10^3}}_{\text{ten terms}} = \frac{1}{10^2}$$

We will provide a number of exercises in this vein but at present we point out that we have strongly used the probability measure $P$ in which $P(a < x \le b) = b - a$ if $0 \le a < b \le 1$.

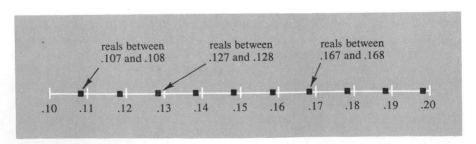

Figure 3

We now pose a slight variant of the stick-breaking problem which leads to difficulties.

### Example 3.5.3

A stick is broken at a random rational number $x$ between 0 and 1. Find the probability that the longer piece is more than twice as long as the shorter piece.

If we interpret rational number to mean the sample space $S = \{x; x$ is a rational number and $0 \leq x \leq 1\}$ and if we interpret "at a random rational number" to mean $P(\{x \in S; a \leq x \leq b\}) = b - a$ for $0 \leq a < b \leq 1$ and $a, b \in S$ we find ourselves in trouble. The procedures of Example 3.5.1 apply to this problem and show that $P(\{c\}) = 0$ for any $c \in S$. But if M5 is to hold, since $S$ is a union of countably many sets each of which contains exactly one rational number between 0 and 1, $P(S) = 0 + 0 + 0 + \cdots = 0$. Obviously, this is a violation of M2.

The conclusion of Example 3.5.3 is that we cannot have M5, M2, and $P(\{a < x \leq b\}) = b - a$ simultaneously in the context of the example. However, it was never said that we would be able to adapt a probability system to any group of sentences containing the word "probability." One may develop another theory for finitely additive measures (drop M5) or deal with infinite measures (drop M2). In this book, we will just pass up such problems entirely. The requirements of the problem are simply not compatible with our system.

### Example 3.5.4

A point is picked at random in a square. What is the probability that it lies inside the inscribed circle which is concentric with the square?

Here we may assume that the square has its sides of length one unit. Therefore we can take as the sample space $S = \{(x, y); 0 \leq x \leq 1, \; 0 \leq y \leq 1\}$. We shall interpret "at random" to mean that the probability that the chosen point lies in a rectangle $\{(x, y); a \leq x \leq b, \; c \leq y \leq d\} = (b - a)(d - c)$, the area of the rectangle (see Figure 4), if $0 \leq a < b \leq 1, \; 0 \leq c < d \leq 1$. It should be noted that this insures that $P(S) = \{(x, y); 0 < x \leq 1, \; 0 \leq y \leq 1\} = (1 - 0)(1 - 0) = 1$. If $P$ really is a legitimate probability measure, then any line segment of the form $\{(x, c); a \leq x \leq b\}$ or $\{(a, y); c \leq y \leq d\}$, (line segments parallel to the $x$ or $y$ axes) should have probability zero. This can be seen by observing that $\{(x, c); a \leq x \leq b\} = \bigcap_{n=1}^{\infty} \{(x, y); a \leq x \leq b, \; c - 1/n \leq y \leq c + 1/n\}$ and using an argument like that in Example 3.5.1 to show that $P(\{(x, c); a \leq x \leq b\}) \leq \lim_{n \to \infty} (b - a)(2/n) = 0$. Notice that this also shows that such a segment is an event if there is a genuine admissible system of events containing all rectangles of the type described. Since any point is the intersection of two such segments, all points are events in the same context. Therefore $P(\{(x, y); a < x < b, \; c < y < d\}) = P(\{(x, y); a \leq x \leq b, \; c \leq y \leq d\})$

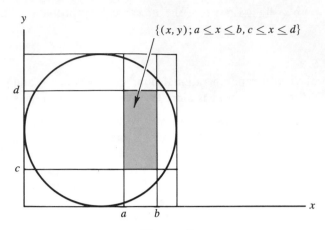

**Figure 4**

because the set on the right is the (disjoint) union of the set on the left and some segments (parallel to the axes) and some points. Similarly we have $P(\{(x, y);\ a < x \le b,\ c < y \le d\}) = (b - a)(d - c)$ for $0 \le a < b \le 1,\ 0 \le c < d \le 1$.

Suppose we consider a union of disjoint rectangles of this latter type (including right-hand and upper boundary segments) which is contained in the circle in question (see Figure 5). If we call these rectangles $M_1, M_2, \ldots, M_n$ and if we call the inside of the circle $C$, we see that if $C$ is an event, $P(C) \ge$

$$P(\bigcup_{k=1}^{n} M_n) = \sum_{k=1}^{n} P(M_n) = \text{sum of the areas of the rectangles} = \text{area of } (\bigcup_{k=1}^{n} M_n).$$

Now the least upper bound of the areas of such inscribed figures is $\pi(\frac{1}{2})^2$ since $C$ has radius $\frac{1}{2}$; therefore, $P(C) \ge \frac{1}{4}\pi$. In a similar fashion we may build a

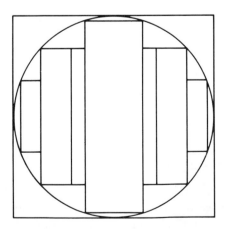

**Figure 5**

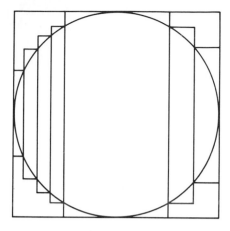

**Figure 6**

union of disjoint rectangles which contain $C$ as in Figure 6 and argue that $P(C) \le$ area of $(\bigcup_{k=1}^{r} N_k)$ and since the area of such figures have greatest lower bound $\frac{1}{4}\pi$ also, we get $\frac{1}{4}\pi \ge P(C)$ so $P(C) = \frac{1}{4}\pi$ if $C$ is an event. That $C$ is an event is not too difficult to prove but we omit the proof at this stage.

The lesson to be learned from Example 3.5.4 is that probability and area are defined in such a way that if the events are generated by rectangles and the probabilities of those rectangles are their areas, then the probabilities of other figures are the areas of those figures.

We now give an example where probability is not directly given by area, although the sample space is a subset of the plane.

***Example 3.5.5***

A point is selected in the interior of a circle of radius 1 in such a way that the radial polar coordinate is chosen at random between 0 and 1 and the angular polar coordinate is chosen at random between 0 and $2\pi$. Find the probability that the point lies in the inscribed square.

We interpret the conditions of the problem as follows: If $(r, \theta)$ are the polar coordinates of the chosen point,

$$P(\{(r, \theta); r_1 \le r \le r_2 \text{ and } \theta_1 \le \theta \le \theta_2\}) = (r_2 - r_1)\frac{1}{2\pi}(\theta_2 - \theta_1)$$

if $0 \le r_1 < r_2 \le 1$ and $0 \le \theta_1 < \theta_2 \le 2\pi$ (see Figure 7). This is not the area of the figure. This says that the radial coordinate is chosen in a symmetric way at random between 0 and 1 and that the angular coordinate is so chosen between 0 and $2\pi$. It is easy to show that $P(r = r_1) = 0$ and $P(\theta = \theta_1) = 0$ in the same

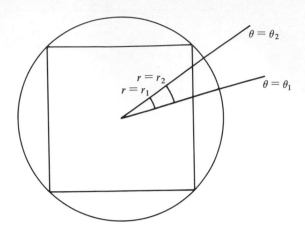

**Figure 7**

way as before. We calculate $P(0 \le r \le \frac{1}{2})$. Since there are no restrictions on $\theta$ we have

$$P(0 \le r < \tfrac{1}{2}) = P(\{(r, \theta);\ 0 \le r \le \tfrac{1}{2},\ \ 0 \le \theta < 2\pi\})$$

$$= (\tfrac{1}{2} - 0)\frac{1}{2\pi}(2\pi - 0) = \tfrac{1}{2}$$

Similarly

$$P(\tfrac{1}{2} \le r \le \tfrac{1}{2}) = (1 - \tfrac{1}{2})\frac{1}{2\pi}(2\pi - 0) = \tfrac{1}{2}$$

The areas of these figures are $\pi(\frac{1}{2})^2 = \frac{1}{4}\pi$ for the inner circle of radius $\frac{1}{2}$ and $\pi - \frac{1}{4}\pi = \frac{3}{4}\pi$ for the annular region $\{(r, \theta);\ \frac{1}{2} \le r \le 1\}$.

Thus we have two sets with equal probabilities but different areas, so probability is not directly proportional to area. We may view this probability as follows: Suppose that a very thin nonhomogeneous disk or plate which has a total mass of one unit rests on the region enclosed by the circle of Example 3.5.5. We think of the probability of a set $M$ as the mass of the portion of the plate which rests exactly on $M$. The exact value of that probability depends on $M$ and on the nature of the plate. More precisely, if $p$ is the function giving the mass density of the plate, it is known from calculus that the mass of the portion of the plate resting on the set $M$ is $\iint_M p\, dA$, where $dA$ is the "element of area."

In polar coordinates $dA = r\, dr\, d\theta$. According to our specifications we have

$$P(\{(r, \theta);\ \theta_1 \le \theta \le \theta_2 \text{ and } r_1 \le r \le r_2\}) = \frac{1}{2\pi}(r_2 - r_1)(\theta_2 - \theta_1)$$

$$= \int_{\theta_1}^{\theta_2} \int_{r_1}^{r_2} p(r, \theta) r\, dr\, d\theta$$

It can be seen that $p(r, \theta) = 1/2\pi r$ yields this result. It is the only "nice" function which satisfies this equation.

We now want to find the probability (mass) "over" the inscribed square. By symmetry we need only consider one-eighth of the square; let us use the one between $\theta = 0$ and $\theta = \frac{1}{4}\pi$. Therefore $P(\text{square}) = 8 \int_0^{\pi/4} \int_0^{\sqrt{2} \sec \theta/2} \frac{1}{2\pi r} r \, dr \, d\theta$,

since the right-hand boundary of the square is the line $x = \frac{1}{2}\sqrt{2}$ in rectangular coordinates or $r \cos \theta = \frac{1}{2}\sqrt{2}$ or $r = \frac{1}{2}\sqrt{2} \sec \theta$ in polar coordinates. Then

$$\int_0^{\pi/4} \int_0^{\sqrt{2} \sec \theta/2} \frac{1}{2\pi r} r \, dr \, d\theta = \int_0^{\pi/4} \frac{\sqrt{2}}{4\pi} \sec \theta \, d\theta$$

$$= \frac{2\sqrt{2}}{\pi} \log_e (\sqrt{2} + 1) \approx 0.792$$

since $\int \sec \theta \, d\theta = \log_e |\sec \theta + \tan \theta| + C$ by standard integral formulas.

We will not go into details of the system of events. Our sample space really is $S = \{(r, \theta); 0 < r \le 1, \quad 0 \le \theta < 2\pi\}$ (ignoring the center of our circle).

In Example 3.5.5 we obtained a number for the probability of a particular event by referring to mass and mass density. Is this justified? In the case of $S = \{1, 2, 3, \ldots\}$ with $P(\{n\}) = p_n$, $\sum_{j=1}^{\infty} p_j = 1$, we may regard the point $n$ as a point mass with mass $p_n$. Then the mass of an event $M$ is $P(M)$ as defined in (3.4.1). But Theorem 3.4.1 states that this gives a probability measure. A generalization of point masses is a continuous distribution of mass given by a mass density function $p$ which may have the form of mass per unit length, mass per unit area, mass per unit volume, etc. according to the dimensionality of the set $S$. Integrals of the form $\int_M p$ are analogs to $\sum_{n \in M} p_n$. Naturally, for probability, we would restrict ourselves to those mass distributions having the property that $\int_S p = 1$. Note that in the special case the mass density is constant and $S$ is a 2-dimensional region, so $p(x, y) = k$, $\int_M p = \int \int_M k \, dx \, dy = k$ (area of $M$). A similar result holds for a constant mass density when $S$ is a 3-dimensional set.

All we can really claim so far is that mass is a convenient intuitive device for visualizing probabilities. We discuss another such intuitive device in section 5.4 after we give some additional examples. However, a key issue which we cannot treat is whether probabilities assigned by $P(M) = \int_M p$ satisfy axioms M1–M5 for a suitable system of events if $p$ has the form of a mass density which gives mass of one unit to $S$. In advanced treatments of probability theory it is shown that such is the case. We, therefore, accept this result without proof and use its consequences.

### Example 3.5.6

Let $S = \{(x, y); x \geq 1 \text{ and } y \geq 1\}$ and let $p(x, y) = 1/x^2 y^2$. Verify that $P(S) = 1$, and find $P(y \leq kx)$ where $0 < k \leq 1$.

By definition

$$\iint_S p(x, y) \, dx \, dy = \lim_{\substack{a \to \infty \\ b \to \infty}} \int_1^a \int_1^b \frac{1}{x^2 y^2} \, dx \, dy = \lim_{\substack{a \to \infty \\ b \to \infty}} \int_1^a \frac{1}{y^2} \left[ -\frac{1}{x} \right]_1^b \, dy$$

$$= \lim_{\substack{a \to \infty \\ b \to \infty}} \int_1^a \frac{1}{y^2} \left[ 1 - \frac{1}{b} \right] \, dy = \lim_{\substack{a \to \infty \\ b \to \infty}} \left[ 1 - \frac{1}{b} \right] \int_1^a \frac{1}{y^2} \, dy$$

$$= \lim_{\substack{a \to \infty \\ b \to \infty}} \left[ 1 - \frac{1}{b} \right] \left[ -\frac{1}{y} \right]_1^a = \lim_{\substack{a \to \infty \\ b \to \infty}} \left[ 1 - \frac{1}{b} \right] \left[ 1 - \frac{1}{a} \right] = 1$$

For $k \leq 1$ the region in which $y \leq kx$ is shaded in Figure 8. Accordingly

$$P(y \leq kx) = \iint_{y \leq kx} \frac{1}{x^2 y^2} \, dx \, dy = \lim_{b \to \infty} \int_{1/k}^b dx \int_1^{kx} \frac{1}{x^2 y^2} \, dy$$

$$= \lim_{b \to \infty} \int_{1/k}^b \frac{1}{x^2} \left[ -\frac{1}{y} \right]_1^{kx} dx = \lim_{b \to \infty} \int_{1/k}^b \frac{1}{x^2} \left[ 1 - \frac{1}{kx} \right] dx$$

$$= \lim_{b \to \infty} \left( \left[ -\frac{1}{x} + \frac{1}{2kx^2} \right]_{1/k}^b \right) = \lim_{b \to \infty} \left[ k - \frac{1}{2k(1/k^2)} - \frac{1}{b} + \frac{1}{2kb^2} \right]$$

$$= k - \frac{k}{2} = \frac{k}{2}$$

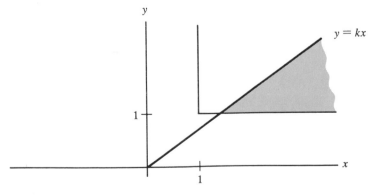

**Figure 8**

### Example 3.5.7

A point $(x, y)$ is chosen at random in the square $S = \{(x, y); 0 < x \leq 1,$
$0 < y \leq 1\}$ and the quotient $z = y/x$ is formed. Find the "mass" density on
the set $\{z; z > 0\}$ which gives the correct $P(a < z \leq b)$.

We first assume that $0 < a < b \leq 1$ and refer to Figure 9. We then interpret
"at random" to mean that the probability that the point $(x, y)$ lies in a rectangle
contained in $S$ is proportional to the area of this rectangle. Since the area of $S$
is 1, the probabilities of events must be equal to the areas of those events in
view of Example 3.5.4. Now

$$\left\{(x, y); a < z = \frac{y}{x} \leq b\right\}$$

consists of all points $(x, y)$ above the line $y = ax$ and below or on the line
$y = bx$ and inside the square. The area of this triangle is $\frac{1}{2}b \cdot 1 - \frac{1}{2}a \cdot 1 = \frac{1}{2}(b - a)$
since it is the set-theoretic difference of two right triangles. If $1 \leq c < d$,

$$\left\{(x, y); c < z = \frac{y}{x} \leq d\right\}$$

consists of all points $(x, y)$ in the square $S$ and between the lines $y = cx$ and
$y = dx$, including the line $y = dx$. This is a triangle and by similar argumenta-

tion, the area of this triangle is $\frac{1}{2} \cdot 1 \cdot \frac{1}{c} - \frac{1}{2} \cdot 1 \cdot \frac{1}{d} = \frac{1}{2}\left(\frac{d - c}{cd}\right)$ in view of Figure 9.

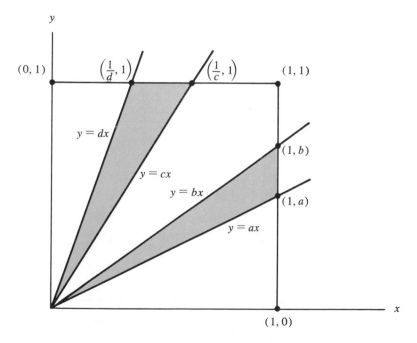

**Figure 9**

Therefore, if $0 < a < b \le 1$, $P(a < z \le b) = \frac{1}{2}(b - a)$ and if $1 < c \le d$,

$$P(c < z \le d) = \frac{1}{2c} - \frac{1}{2d} = \frac{1}{2}\left(\frac{d - c}{cd}\right).$$

Now what kind of density (of mass) on a line would make the mass of a line segment $\{z; a < z \le b\} = \frac{1}{2}(b - a) = \frac{1}{2}$ (length of the segment). The answer here is simple. The mass density must be constant and equal to $\frac{1}{2}$. If we call $p(z)$ the mass density at the point $z$ we get $p(z) = \frac{1}{2}$ for $0 < z \le 1$. This is effectively the only way that the mass of an interval is equal to half of its length. Now if $1 < c \le d$, we have a different story. The mass of the interval

$$\{z; c < z \le d\} = \frac{1}{2}\left(\frac{d - c}{cd}\right)$$

This is not the result of a constant mass density because, for example, the intervals with $c = 2$, $d = 3$ and $c = 3$, $d = 4$ have equal lengths but different masses in the present context. But what is the density of mass at a point $z_0(> 1)$? It is the limit of the mass of an interval (containing $z_0$) divided by the length of the interval as the length of that interval tends to zero. (Naturally, if this limit does not exist we have difficulties, but this will not concern us.) Consider the interval $\{z; z_0 - h < z \le z_0 + h\}$ where $h$ is such that $1 < z_0 - h$. Therefore $c = z_0 - h$, $d = z_0 + h$. The length of that interval is $2h$ and the mass is

$$\frac{(z_0 + h) - (z_0 - h)}{2(z_0 + h)(z_0 - h)} = \frac{h}{(z_0 + h)(z_0 - h)}$$

Therefore the density

$$P(z_0) = \lim_{h \to 0} \frac{\text{mass of interval}}{\text{length of the interval}} = \lim_{h \to 0} \frac{h/(z_0 + h)(z_0 - h)}{2h}$$

$$= \lim_{h \to 0} \frac{1}{2(z_0 + h)(z_0 - h)} = \frac{1}{2z_0^2}$$

Summarizing this as,

$$p(z) = \begin{cases} \frac{1}{2}, & 0 < z \le 1 \\ \frac{1}{2z^2}, & 1 < z \end{cases}$$

gives the density function. To check, one could integrate to see if $P(M) = \int_M p(z)\, dz$ since the integral of the density has to give the mass and we are thinking of the probability of a set as the mass of that set.

**EXERCISES 3.5**

1   A stick of unit length is to be broken at random. What is the probability that the longer piece is more than three times the length of the shorter piece? What assumptions are being made?

2   Under the assumptions of exercise 1, what is the probability that the larger piece is at least twice as long but no more than three times the length of the shorter piece?

3   A point is selected at random from inside a circle of radius $a$. What is the probability that it lies inside the concentric circle of radius $\frac{1}{2}a$? What assumptions are being made?

4   A point is selected at random from the interior of a sphere of unit radius. What is the probability that it lies inside the concentric sphere of radius $\frac{1}{2}$? What assumptions are being made?

5   Verify that $\int_c^d \frac{1}{2z^2} \, dz$ has the properties asserted in Example 3.5.7.

6   Under the assumptions of exercise 4, what is the probability that the point lies inside

(a)   the inscribed cube?

(b)   the inscribed regular tetrahedron?

7   Suppose that we want to describe a chance breaking of our stick in which it is more likely that the stick breaks close to the middle rather than at the ends. Which of the following gives highest probability to the middle half $\frac{1}{4} < x \leq \frac{3}{4}$, which gives next highest and which gives least? We assume $0 \leq a < b \leq 1$.

(a)   $P(\{x; a < x \leq b\}) = 3(b^2 - a^2) - 2(b^3 - a^3)$

(b)   $P(\{x; a < x \leq b\}) = \frac{1}{2}(\cos \pi a - \cos \pi b)$

(c)   $P(\{x; a < x \leq b\}) = 2[b - a - (b - \frac{1}{2})|b - \frac{1}{2}| + (a - \frac{1}{2})|a - \frac{1}{2}|]$

8   How would you choose one of the probability measures of exercise 7 to describe an experiment if you had a large number of sticks upon which you were exerting forces by means of a particular machine?

9   Find the probability mass density for the three probability measures in exercise 7. Graph these densities.

10   Suggest some probability mass densities for chance breaking of a stick under which it is more likely that the stick breaks near the ends rather than near the middle.

11   One unit of mass (to be thought of as probability) is distributed on the square $\{(x, y); 0 \leq x \leq 2, \ 0 \leq y \leq 2\}$ in such a way that $P(\{(x, y); a < x \leq b, c < y \leq d\}) = (b - a)(a^2 - c^2)/8$. Find the probabilities of the rectangles with

(a)   $a = c = 0, \quad b = d = 1$

(b)   $a = 0, \quad b = 1, \quad c = 1, \quad d = 2$

(c)   $a = 0, \quad b = 1, \quad c = 0, \quad d = 2$

(d)   $b = a + \epsilon, \quad d = c + \delta$

(e)   Find $\lim\limits_{\substack{\epsilon \to 0 \\ \delta \to 0}} \frac{1}{\epsilon\delta} P(\{(x, y); a < x \leq a + \epsilon, \ c < y \leq c + \delta\})$.

(f)   Find the mass density which gives the probabilities in question.

**12** Let $S = \{s; s$ is a point inside a circle of radius 1$\}$. Let $M_n = \{s; s$ is a point inside a circle of radius $1/n^2\} \subset S$ for $n = 2, 3, 4, \ldots$. Show that the probability that a point, selected at random from $S$ lies in infinitely many of the $M_n$, is zero. This is true regardless of the location of the $M_n$. [*Hint:* Exercise 3.4.7 and supplementary exercise 2.2.]

# 3.6 Summary

As a mathematical model for, or abstraction of, ideas about experiments or acts whose outcome is uncertain, we use a triple consisting of $S, \mathscr{A}, P$ where $S$ is a nonempty sample space, $\mathscr{A}$ is an admissible system of events (subsets) of $S$, and $P$ is a real-valued function on $\mathscr{A}$ called a probability measure. The conditions imposed on $S, \mathscr{A}, P$ are

**E1** $S \in \mathscr{A}$.

**E2** If $M$ and $N \in \mathscr{A}$, then $M \cup N \in \mathscr{A}$ and $M' = S - M \in \mathscr{A}$.

**E3** If $M_1, M_2, M_3, \ldots$ is a denumerable system of events in $\mathscr{A}$ then $(\bigcup\limits_{n=1}^{\infty} M_n) \in \mathscr{A}$.

**M1** $P(M)$ is a real number and $0 \leq P(M) \leq 1$.

**M2** $P(S) = 1$.

**M3** $P(\varnothing) = 0$.

**M4** If $M$ and $N \in \mathscr{A}$ and $M \cap N = \varnothing$ then $P(M \cup N) = P(M) + P(N)$.

**M5** If $M_1, M_2, M_3, \ldots$ is a denumerable system of events in $\mathscr{A}$ and if $M_i \cap M_j = \varnothing$ for $i \neq j$ then $P(\bigcup\limits_{n=1}^{\infty} M_n) = \sum\limits_{n=1}^{\infty} P(M_n)$.

The sample spaces which we use are

(a) finite sets

(b) denumerable sets

(c) certain subsets of $n$-dimensional Euclidean space (Borel sets).

More advanced topics in probability theory use more complicated sample spaces.

Borel sets in $n$-dimensional Euclidean space are the sets of the smallest system satisfying E1, E2, E3 which contain all "generalized rectangles" of the form $\{(x_1, x_2, \ldots, x_n); a_i < x_i \leq b_i$ for $i = 1, 2, 3, \ldots, n\}$. All open sets are Borel sets and Borel sets cover such a broad range of sets that it is not easy to directly describe a non-Borel set.

We always use the system of all subsets as the system of events when we use countable sample spaces. When $S$ is a Borel set we restrict our studies to the admissible system of events consisting of all Borel sets which are subsets of $S$.

If $S = \{s_1, s_2, s_3, \ldots\}$ (or finite), by our convention about $\mathscr{A}$, the probability measure $P$ (under our usage of all subsets as events) must be determined by a sequence $p_1, p_2, p_3, \ldots$ with $p_i \geq 0$ and $\sum p_i = 1$. In that case $P(M) = \sum_{i \in M} p_i$. When $S$ is a Borel set which is not countable, we restrict our attention to probability measures which arise from "mass" densities and which are related to integration in the way that summation is used in countable sample spaces. We use densities which are piecewise continuous and where the integration theory is simple. In more advanced works, much more complicated probability measures are studied.

# CHAPTER 4

# Mathematical Models

## 4.1 Models for deterministic phenomena

In Chapter 3 a mathematical model for chance phenomena was proposed. We shall use the term "model" instead of (mathematical) "theory." The concept of a mathematical model deserves further elaboration. However, in a book of this kind one must strike a balance between the discussion of models in general and the exposition of the theory of one particular model. Therefore, we will give a brief discussion of models in order to further a thorough understanding of our particular model for chance phenomena. It should be understood that a discussion of mathematical models is more philosophical than mathematical.

We first give some examples which point to relevant aspects of mathematical models. Consider a problem which is not normally thought of as a problem of chance. A large, well-sealed tank, from which the air has been evacuated has a device in the bottom which can shoot a steel ball straight upwards at various initial velocities (Figure 1).

We wish to describe the position of the ball at various times *after* it has been shot and *before* it hits a wall of the tank. A common mathematical model for this problem is the *equation*.

**(4.1.1)** $\qquad s = -\frac{1}{2}gt^2 + vt$

Here $t$ represents the elapsed time after the ball has been shot, $v$ represents the initial velocity which the projection device imparts, $s$ represents the position of the ball in a certain reference system, and $g$ is an appropriate constant which may depend on the location of the tank.

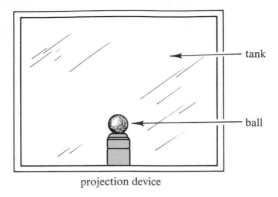

projection device

**Figure 1**

This mathematical model is usually considered to be a good model for this problem in the sense that it makes fairly accurate predictions of the position of the ball. Thus (with $g = 32$ ft/sec$^2$) the mathematical model (4.1.1) predicts that the highest point (apogee) of the ball is 4 ft above the release point if the initial velocity $v$ is 16 ft/sec (and the tank is more than 4 ft high). Furthermore, it predicts that 1 sec is required for the ball to return to the level of the release point under these conditions. We also find that the velocity of the ball is $-\frac{3}{4}g + v$ when the elapsed time $t = \frac{3}{4}$ providing that the ball has not struck a wall before $t = \frac{3}{4}$. These predictions are based on the assumption that the velocity is the derivative of the position function (calculus has been used freely). The predictions are in good agreement with experiments.

Another common mathematical model is that for the position of the weight on the end of a very light spring when that weight is making small oscillations from equilibrium (Figure 2). The model in question is the differential equation with initial conditions

**(4.1.2)**    $$\frac{d^2s}{dt^2} = -ks, \quad s(t_0) = s_0, \quad \frac{ds}{dt} t_0 = v_0$$

Here $t$ is again time, $s$ gives the displacement from equilibrium of the weight, $s_0$ and $v_0$ are the initial displacement and velocity at time $t = t_0$. For some purposes we can take $t_0 = 0$ but it is not necessary. One form of the general solution of the equation in (4.1.2) is $s = c_1 \cos \omega t + c_2 \sin \omega t$ where $\omega = k^{1/2}$. If we take "initial" conditions into account we obtain

**(4.1.3)**    $$s = s_0 \cos \omega(t - t_0) + v_0/\omega \sin \omega(t - t_0) = B \cos(\omega t + b)$$

for suitable $B$ and $b$. Note that (4.1.2) and (4.1.3) form a common model for a suitable pendulum undergoing small oscillations. This model is somewhat more interesting than that of the steel ball in the tank.

Let us make some deductions from this model.

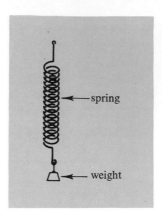

**Figure 2**

## Example 4.1.1

A weight which hangs from a spring oscillates after an initial displacement of 0.003 cm and an initial velocity of 0.5 cm/sec. The weight passes downward through the equilibrium position every 2 sec. Fit this information into equation (4.1.3) and determine the maximum velocity of the weight.

We can take $t_0 = 0$ in our time scale. The initial conditions lead us to the special form of (4.1.3)

$$s = 0.003 \cos \omega t + 0.5\omega^{-1} \sin \omega t = B \cos (\omega t + b)$$

Using a trigonometric identity we obtain

$$\cos (\omega t + b) = \cos \omega t \cos b - \sin \omega t \sin b$$

Therefore

$$B \cos b = 0.003, \quad B \sin b = -0.5\omega^{-1},$$
$$B^2 = B^2 \cos^2 b + B^2 \sin^2 b = (0.003)^2 + (-0.5\omega^{-1})^2$$

Since the equilibrium position is $s = 0$ and we have downward passage through that position every 2 sec we must have

**(4.1.4)**    $\omega(t + 2) + b = 2\pi + \omega t + b$

This follows from the periodicity of the cosine function. That is, $B \cos (\omega t + b)$ must repeat itself every 2 sec. Repetition means that $\omega t + b$ changes by $2\pi$ every 2 sec. From (4.1.4) we find that $2\omega = 2\pi$ so $\omega = \pi$. Thus $B^2 = (0.003)^2 + (-1/2\pi)^2$. Now the velocity $\dfrac{ds}{dt} = -\omega B \sin (\omega t + b)$ from (4.1.3). Since the sine function varies from $-1$ to $1$ the maximum of $\dfrac{ds}{dt}$ is

$$\omega B = \pi\sqrt{(0.003)^2 + 1/4\pi^2} = \sqrt{(0.003\pi)^2 + 0.25}$$

*Example 4.1.2*

A weight on a spring oscillates after an unspecified initial displacement and velocity. The maximum displacement from the equilibrium position is 0.04 cm and the weight returns to this position every 0.5 sec. What is the maximum velocity of the weight according to our present model?

Since $s = B \cos(\omega t + b)$, $\cos(\omega t + b)$ varies between $+1$ and $-1$ and 0.04 is the maximum displacement, $B = 0.04$. Since the weight returns to the maximum position every 0.5 sec, we must have

$$\omega(t + \tfrac{1}{2}) + b = \omega t + b + 2$$

Thus $\tfrac{1}{2}\omega = 2$ and $\omega = 4$. As we saw in Example 4.1.1,

$$\frac{ds}{dt} = -\omega B \sin(\omega t + b)$$

Therefore the maximum velocity is 4 (0.04) or 0.16 cm/sec.

Example 4.1.1 differs from Example 4.1.2 in that the latter contains no initial values. However, observed or experimental values, namely the maximum displacement and the period of oscillation, can be used instead of the initial values $v_0$ and $s_0$ to make predictions. This is very important because, frequently, initial data are difficult or impossible to obtain. The observed values can select the specific model for a specific real world problem from a number of similar models just as the initial values make this selection.

We may consider another model for the displacement of the weight on the spring. Suppose we let

**(4.1.5)**      $s = 0$   for all $t > t_0$

The reader may balk at such a model; it seems crude if not wrong. On the other hand, it may seem crude only because the reader knows a better model, namely (4.1.3). A more sophisticated reader may feel that (4.1.3) is unsuitable. Let us consider (4.1.5) to be a model which may have some value but which is more elementary than (4.1.3).

*Example 4.1.3*

A weight on a spring is given an initial displacement of 0.002 cm and an initial velocity of $-0.4$ cm/sec. What are the predictions of the displacement and velocity at time $t > t_0$ which are made by the model (4.1.5)?

The model predicts $s = 0$ and $\dfrac{ds}{dt} = 0$ for all $t > t_0$. Obviously these are not impressive predictions, but for some purposes they suffice. They could describe the spring as seen from a distant star.

We see then, that a mathematical model is a mathematical system which is associated with a real-world system in such a way that certain aspects of the real-world system correspond to certain aspects of the mathematical system. It is as though we have a dictionary which enables us to translate from one to the other. Mathematical models are used because a limited amount of real-world information gathered by experimentation together with a certain amount of mathematical work can lead to additional information about the real-world system, if the model is good. This additional information might otherwise be obtained by experiment or by shrewd guessing, but the model may be capable of providing unlimited amounts of additional information. Naturally, if the mathematical work is difficult or economically expensive one need not use the model. However, experiments also cost money and they are often dangerous.

It is important to note the relationship between mathematics and experiment or observation. Most good mathematical models involve a family of particular systems which can treat a family of possible real-world problems. The choice of the appropriate particular system for a particular real-world problem is made as a result of experimentation. Our model for chance phenomena also has this generality. It, too, requires observation or experimental data if it is to be applied to a specific problem.

**EXERCISES 4.1** _____

1   Using model (4.1.1) and $g = 32$ ft/sec$^2$, $s = 24$ ft when $t = 1$ sec, find $v$ and $s$ when $t = 2$, assuming that the ball has not hit a wall before $t = 2$.

2   Given the information of exercise 1 find the value of $t$ at which the apogee is reached. Find the apogee.

3   What is unusual about the consequences of model (4.1.3) in terms of your observation of weights on springs?

4   A somewhat improved model for the displacement of a weight on the end of a spring is given by the differential equation

$$\frac{d^2s}{dt^2} + 2a\frac{ds}{dt} + ks = 0$$

where $a > 0$ and $k > 0$. Show that $s = e^{-at} \cos(\omega t + b)$ is a solution of the differential equation if $\omega = (k - a^2)^{1/2}$ and $a^2 < k$. Find $\lim_{t \to \infty} e^{-at} \cos(\omega t + b)$ and compare with exercise 3. What can be said about the behavior of $e^{-at} \cos(\omega t + b)$ when $a$ is very small and $t$ is not much larger than $(100a)^{-1}$.

# 4.2   How do mathematical models come about?

We can answer this question only superficially since it is as much philosophy as mathematics.* In some cases it is a matter of a scientist with the proper insight realizing that a phenomenon which he is studying has parallels in a body of known mathematics and that he can form a "dictionary" between the details of his phenomenon and certain of the details of the mathematics. In such a case mathematical deductions from the latter can be translated into predictions about the phenomenon. Such predictions are frequently valuable. For example, suppose one wants some idea of the temperature of an environment into which he is sending a spacecraft.

It should not be imagined that the mathematical model can always be found in an area of mathematics which is already developed. In some cases the phenomenon suggests certain mathematical concepts and some mathematical deductions actually occur *after* the corresponding behavior of the phenomenon is observed. We know that the elliptical motion of the planets about the sun was observed before it was predicted by Newton's Laws. In fact, the elliptical motion is part of the evidence from which one obtains the inverse square law of gravitational attraction. It would be unfortunate if one could not then rederive the evidence. However, many additional predictions follow from the inverse square law.

It is fair to say that the relationship between "real-world" phenomena and mathematical models is complicated. Scientists with strong personalities may influence the choice of mathematical models for the same phenomenon. That is, there are personal and historical factors. However, normally a model which predicts more accurately prevails over one which predicts less accurately unless the latter is much easier to manipulate than the former. Intuitive appeal of a model may override predictive (or manipulative) disadvantages in a model, although in some sense, intuitive appeal is some sort of manipulative advantage. Aspects of a model which tend to unify large areas of phenomena are favorable to the acceptance of the model although we would include this in the category of intuitive appeal. Some of this intuitive appeal is shown in the next example.

### Example 4.2.1
The "law" $F = ma$ does not depend on the location of the coordinate axes of a fixed rectangular coordinate system.

We will demonstrate this only in two dimensions. If the origin is shifted to the point $(-h, -k)$ in the original coordinate system, but axes are kept parallel to the old axes, we need only observe that

$$\frac{d^2x}{dt^2} = \frac{d^2(x + h)}{dt^2}, \qquad \frac{d^2y}{dt^2} = \frac{d^2(y + k)}{dt^2}$$

---

* For a more extensive discussion of this question, see James B. Conant, *On Understanding Science: An Historical Approach*, Terry Lecture (Yale University, New Haven, 1947).

If the axes are rotated through an angle $\theta$, we need only consider

$$\frac{d^2}{dt^2}(x\cos\theta + y\sin\theta, -x\sin\theta + y\cos\theta)$$

$$= \left(\frac{d^2x}{dt^2}\cos\theta + \frac{d^2y}{dt^2}\sin\theta, -\frac{d^2x}{dt^2}\sin\theta + \frac{d^2y}{dt^2}\cos\theta\right)$$

which is just the vector acceleration $\left(\frac{d^2x}{dt^2}, \frac{d^2y}{dt^2}\right)$ in the new coordinate system.

The equation $F = ma$ is used both in Newtonian and relativistic (Einsteinian) physics. In the Newtonian model $m$, the mass, is assumed to be constant whereas in the special theory of relativity we have $m = m_0(1 - v^2/c^2)^{-1/2}$, where $m_0$ is the so-called rest mass, $v$ is the velocity of the body, and $c$ is the speed of light. Normally, differential equations of the form

$$F(t) = m_0\left[1 - \frac{1}{c^2}\left(\frac{dx}{dt}\right)^2\right]^{-1/2}\frac{d^2x}{dt^2}$$

are more difficult to solve than those of the form

$$F(t) = m_0\frac{d^2x}{dt^2}$$

However, if $\left(\frac{dx}{dt}\right)$ is small in comparison to $c$, the velocity of light,

$$1 - \frac{1}{c^2}\left(\frac{dx}{dt}\right)^2 \approx 1$$

and the solutions of the equations are approximately equal. In some sense, Newtonian physics is a limiting case of relativistic physics. Or, alternatively, for small velocities the two theories (models) make essentially the same predictions. Therefore, for small velocities the Newtonian model is used because it is easier to work with and its predictions are essentially the same as those of the relativistic model.

This case shows how a "better" (relativistic) model replaced a "good" (Newtonian) model in its acceptance as the "true" description of nature. This is not to say that the Newtonian model is not used; it is just used with certain reservations. However, if one wishes to replace a "good" model, he will have to supply a much better model and normally will have to show that the "good" model is some sort of special or limiting case of the "better" model.

It should be evident that much more can be understood about mathematical models by studying the philosophy and history of science. Much of the learning is indirect because the term "mathematical model" is a recent one. Often one must read between the lines because it is common to write as though the real-world phenomena *is* the mathematical model. Phrases such as "physical space is Euclidean" or "is non-Euclidean" are common. It is our opinion that this style of writing is reasonable as long as one is aware of the limitations. Such locutions may suggest that there is an ultimate mathematical model for a phenomenon and may tend to cause a retention of the current model when the time may be ripe for a new one.

# 4.3   The model for chance phenomena

It is not the object of this book to present a number of possible models for chance phenomena, but rather to study one such model. This model is somewhat complicated so it seems appropriate to discuss its background and its advantages.

Let us consider the differential equation and initial conditions

$$(4.3.1) \qquad F = m\frac{d^2s}{dt^2}, \qquad s(0) = s_0, \qquad \frac{ds(0)}{dt} = v_0$$

as analogous to our probability system as follows.

The force $F$ is analogous to the sample space $S$ with its system of events $\mathscr{A}$. We may have various values $s_0$ and $v_0$ with the same force function just as we may have various probability measures $P$ on a given $\mathscr{A}$. Thus, in definite problems we will have definite values of $s_0$ and $v_0$ just as we have definite probability measures on $S$ in definite probability problems.

## Example 4.3.1
Compare the experiment "a fair die is rolled" to a deterministic experiment.

The deterministic experiment could be "the steel ball of Figure 1 is shot off in the tank with initial velocity 3 ft/sec." Here, $-g = \frac{d^2s}{dt^2}$ (which is essentially equivalent to $s = -\frac{1}{2}gt^2 + v_0t + s_0$) is analogous to the sample space $\{1, 2, \ldots, 6\}$ with its system of events. The initial velocity $v = 3$ is analogous to the equal-likelihood measure if we take $s_0 = 0$ and $g = 32$ as conventions.

## Example 4.3.2
Compare the experiment "a possibly biased die is rolled" with a deterministic experiment.

We might use "the above-mentioned steel ball is shot with some initial velocity." Here the words "possibly biased" do not specify the probability measure just as we have no specification of the initial velocity.

These two examples are only analogies and they should not be taken too literally. They are designed to put the probability system into a reasonable framework as a mathematical model and to aid the student in understanding probability rather than in conducting a number of manipulations in analogy with manipulations in the text.

### Example 4.3.3
Assume our steel ball is released with a certain initial velocity. At time $t = 0.2$ sec, the position $s = 3$ ft. Find $s$ when $t = 0.5$ sec. Construct a problem in chance phenomena analogous to this problem.

The nature of the deterministic problem is as follows. Certain initial conditions are not specified but subsequent data enable one to derive the initial conditions. From the derived initial conditions and the complete specification of the mathematical model, make a prediction.

Thus, an analogous chance problem could be: $A$ (possibly biased coin) is tossed 100 times and 25 heads and 75 tails appear. Find the probabilities for various numbers of heads appearing in the next 40 tosses.

There may be some discomfort in the "pat" way in which the problem in chance phenomena was offered. After all, do 100 tosses suffice to determine the bias? Could not there be an unusual situation in the 100 tosses that mislead us about the extent of the bias? Yes, this is indeed possible. There are more risks involved in prediction of chance phenomena as against deterministic phenomena; however, there is very little that we can do about it.

How then, do we test the validity of the probabilistic model if there are these inherent difficulties? Before we answer this let us be more precise about the correspondence between the model and real-world phenomena. The role of the sample space is clear. The system of events is a mathematical device only. However, what real-world situation corresponds to $P(M) = \frac{1}{3}$ or $P(M) = b$ for any $b$ between 0 and 1? Most of us have an immediate response to this. However, in many ways this is a complicated philosophical problem which is still under discussion. Nevertheless the predictions of both viewpoints are substantially the same.

Suppose we repeat a specific experiment $n$ times in such a way that outcomes do not affect other outcomes. For a specific event $M$ we let $f_r(M) = n^{-1}$ (number of times the event occurs in the $n$ trials).

The number $f_r(M)$ is called the *relative frequency* of $M$ in these trials. It should be thought of (at present) as purely empirical. It is a quite common feeling that if $n$ is large, $f_r(M)$ should be close to $P(M)$ in most situations. As a matter of fact, one can describe the deviations of $f_r(M)$ from $P(M)$ within the mathematical model.

Now, we can say that we test the validity of a probabilistic model by comparing the predictions through the use of the approximation of $P(M)$ by experimental values of $f_r(M)$.

Workers who apply probability theory claim that the general probability model is well confirmed. However, this is the same type of assurance that those of us who do not perform basic experiments have for most deterministic models.

**EXERCISE 4.3** ───────────────────────────────────

1 How would you verify the probabilistic model by testing $P$(fair coin shows 6 H's in a row) $= 2^{-6}$?

# CHAPTER 5

# *More on Probability Systems*

## 5.1 Compound Experiments

In the preceding chapters we have used the word "experiment" loosely and we will continue to do so. If the term was used more precisely, the range of application of the results would be unduly limited. Thus we may have a complicated experiment which is composed of a number of simpler experiments. Hence a *compound experiment* is one which is composed of component experiments.

The simplest case of a compound experiment is one which has two component experiments. Thus if $S_1$ and $S_2$ are the sample spaces of the component experiments we may use

$$S = S_1 \times S_2 = \{(s, t); s \in S_1, \quad t \in S_2\}$$

for the sample space of the compound experiment. The sample point $(s, t)$ denotes the occurrence of the outcome corresponding to $s$ in the first experiment *and* the outcome corresponding to $t$ in the second experiment. Since $S_1 \times S_2$ contains any possible pair $(s, t)$, we can account for any outcome in the first experiment paired with any outcome in the second experiment. If there are three component experiments we will normally use $S_1 \times S_2 \times S_3$ for the sample space of the compound experiment, if the component experiments have sample spaces $S_1, S_2, S_3$.

**EXERCISES 5.1** _____

1 An experiment consists of two components. First, a person is chosen from a group of 55 people with colds. Then one of the following is chosen; remedy

1, remedy 2, or a placebo. Describe a sample space for this experiment. Is this sample space useful even if the choices influence one another?

2    A card is chosen from the spade suit of a standard deck. A die is rolled. A coin is tossed. Describe a useful sample space for the compound experiment consisting of these three component experiments.

3    A real number is selected from $\{x; 0 \leq x \leq 1\}$. A point is chosen from the disk $\{(y, z); y^2 + z^2 \leq 1\}$. Describe a sample space for the compound experiment consisting of these two components. As in section 2.4, give a geometric description of the resulting sample space.

# 5.2  Independence

In reference to compound experiments, it is possible that the outcome in one component has no effect on the outcome in the other component. We now address ourselves to this case. It is the beginning of the most successfully developed area of probability theory. We treat two special examples.

### *Example 5.2.1*

A standard die is rolled and a card is selected from a well-shuffled standard deck. What is the probability that the die shows a number (of dots) which is divisible by 3 and that the card is a spade? The card choice is assumed to be unaffected by the die.

We choose $S_1 = \{1, 2, \ldots, 6\}$ as the sample space for the die and $S_2 = \{1, 2, 3, \ldots, 52\}$ for the card choice. For convenience, we assume that the spade cards are numbered $1, 2, 3, \ldots, 13$. The standard probability measures on $S_1$ and $S_2$, as a result of our assumptions, are the equal-likelihood measures.

As above, we take $S_1 \times S_2 = S$ to be the sample space for the compound experiment of picking a card and rolling a die. Since we have no preferred die face and no preferred card (equal-likelihood assumption) and since the card choice is unaffected by the die, no sample point $(s, t)$ should be preferred. We thus elect to use equal likelihood in $S$. If $M_1 = \{3, 6\} \subset S_1$ and $N_1 = \{1, 2, 3, \ldots, 13\} \subset S_2$ the event whose probability we seek is

$$\{(s, t); s \in M_1 \text{ and } t \in N_1\} = M_1 \times N_1 \subset S_1 \times S_2 = S$$

since any spade card can be paired with either die face 3 or die face 6. As observed in Chapter 2, $M_1 \times N_1$ contains $2 \cdot 13$ sample points so

$$P(M_1 \times N_1) = \frac{2 \cdot 13}{6 \cdot 52} = \frac{1}{3} \cdot \frac{1}{4} = \frac{1}{12}$$

We should make some further observations from the experiment described in Example 5.2.1. There is nothing special about $M_1 = \{3, 6\}$ and $N_1 = \{1, 2, \ldots, 13\}$.

If $M_2$ and $N_2$ are arbitrary subsets of $S_1$ and $S_2$, respectively, and if $M_2$ has $m_2$ sample points and $N_2$ has $n_2$ sample points then $M_2 \times N_2$ has $m_2 n_2$ sample points so $P(M_2 \times N_2) = \dfrac{m_2 n_2}{6 \cdot 52}$. Furthermore, if $P_1$ and $P_2$ are the probability measures on $S_1$ and $S_2$, respectively, then $P(M_2 \times N_2) = \dfrac{m_2 n_2}{6 \cdot 52} = P_1(M_2) P_2(N_2)$.

We can even do better. Let $M = M_2 \times S_2$ and $N = S_1 \times N_2$. Then $P(M) = P(M_2 \times S_2) = P_1(M_2) P(S_2) = P_1(M_2) \cdot 1 = P_1(M_2)$. Similarly $P(N) = P_2(N_2)$. Since (see exercise 2.4.2) $M_2 \times N_2 = (M_2 \times S_2) \cap (S_1 \times N_2)$,

$$P(M \cap N) = P(M_2 \times N_2) = P_1(M_2) P_2(N_2) = P(M)P(N)$$

in our present context of die rolling and coin tossing.

It is clear that a similar argument can be carried out in the case of any compound experiment when equal-likelihood probability measures are used throughout.

The equation $P(M \cap N) = P(M)P(N)$ does not hold for all pairs of events $M$ and $N$ since one could take $N = M'$ with $0 < P(M) < 1$ and hence, since $P(N) = 1 - P(M)$, $0 < P(N) < 1$, so $P(M)P(N) > 0$ whereas $P(M \cap N) = P(M \cap M') = P(\varnothing) = 0$. Nevertheless, it is important to gain insight into those situations where $P(M \cap N) = P(M)P(N)$. In fact, the greatest success of modern probability theory has been its extremely thorough understanding of cases of this sort.

### Example 5.2.2

A standard red die and a standard green die are rolled. Let $M$ be the event that the red die shows 3 and $N$ the event that the total number of dots showing is an even number. Show that $P(M \cap N) = P(M)P(N)$.

That $P(M) = \frac{1}{6}$ and $P(N) = \frac{1}{2}$ can be seen by referring to sample space $D_2$ of section 3.1. But $M \cap N = \{(3, 1), (3, 3), (3, 5)\}$ so $P(M \cap N) = \frac{3}{36} = \frac{1}{12} = \frac{1}{6} \cdot \frac{1}{2} = P(M)P(N)$.

We should note that $D_2 = \{1, 2, \ldots, 6\} \times \{1, 2, \ldots, 6\}$ and $M = \{3\} \times \{1, 2, \ldots, 6\}$ but $N$ is not of the form $\{1, 2, \ldots, 6\} \times N_2$. Therefore we may find $P(M \cap N) = P(M)P(N)$ in cases which are not strictly Cartesian product situations.

So far we have assumed that the component experiments have no influence upon one another, i.e., they are independent of one another in some philosophical sense. We have seen that a certain multiplicative relationship holds in that $P(M \cap N) = P(M)P(N)$ if $M$ deals with one component and $N$ with the other.

**Definition**    The events $M$ and $N$ of a probability system are called (*statistically*) *independent* if

**(5.2.1)**    $P(M \cap N) = P(M)P(N)$

The modifier "statistically" is falling into disuse (at least among mathematicians) and we will not use it. Apparently, it was inserted to distinguish those cases where we only know (5.2.1) in contrast to those cases where we expect independence from philosophical considerations. However, it is important to stress that, mathematically, independence of events means only what is contained in equation (5.2.1); philosophical matters are excluded except for those which are indicated in connection with Theorem 5.2.2.

Note that disjoint events are independent if and only if their probabilities are 0 or 1.

**Theorem 5.2.1**     Let $M$ and $N$ be independent events in the sense of the above definition. Then $M$ and $N'$ are independent events.

PROOF     $$M = M \cap S = M \cap (N \cup N') = (M \cap N) \cup (M \cap N')$$

$$(M \cap N) \cap (M \cap N') = M \cap M \cap N \cap N' = \varnothing$$

so

$$P(M) = P(M \cap N) + P(M \cap N') \quad \text{and} \quad P(M) - P(M \cap N) = P(M \cap N')$$

But by independence of $M$ and $N$,

$$P(M \cap N) = P(M)P(N)$$

so

$$P(M \cap N') = P(M) - P(M)P(N) = P(M)[1 - P(N)] = P(M)P(N')$$

and the theorem is proved.

It can be similarly shown that if $M$ and $N$ are independent, $M' = S - M$ and $N' = S - N$ are independent.

The following theorem proves (in the case of countable sample spaces) that one can construct a mathematical model for the compound experiment composed of two component experiments which are to be thought of as philosophically independent. The probabilities in the mathematical model for the compound experiment are consistent with those of the components and events relating to the separate components are (statistically) independent as in Example 5.2.1 and the succeeding comments. In the case where the outcomes in the components are equally likely, we get equal likelihood in the model for the compound experiment. The theorem is stated for two components only, but the case of more than two components can be handled in the same way.

**Theorem 5.2.2**     Let $S_1$ and $S_2$ be countable sample spaces with probability measures $P_1$ and $P_2$, respectively (on the admissible system of all subsets). Suppose $P_1(s_m) = p_m$ and $P_2(t_n) = q_n$ for $s_m \in S_1$ and $t_n \in S_2$. Let $S = \{(s_m, t_n); s_m \in S_1, \ t_n \in S_2\}$ (so $S$ is countable). Let $P((s_m, t_n)) = p_m q_n$. Then $P$ is a probability measure on (the admissible system of all subsets of) $S$ and the events $M = M_1 \times S_2$ and $N = S_1 \times N_1$ are independent for any $M_1 \subset S_1$ and $N_1 \subset S_2$ and $P(M \cap N) = P(M)P(N) = P_1(M_1)P_2(N_1)$.

PROOF    If $A$ and $B$ are any subsets of $S_1$ and $S_2$, respectively, we observe that

$$\sum_{(s_m,\, t_n) \in A \times B} p_m q_n = (\sum_{s_m \in A} p_m)(\sum_{t_n \in B} q_n)$$

Secondly, by exercise 2.3.2, $(M_1 \times S_2) \cap (S_1 \times N_1) = M_1 \times N_1$ so

$$P(M \cap N) = \sum_{(s_m,\, t_n) \in M \cap N} P((s_n, t_n)) = \sum_{(s_m,\, t_n) \in M_1 \times N_1} p_m q_n$$

$$= (\sum_{s_m \in M_1} p_m)(\sum_{t_n \in N_1} q_n) = P_1(M_1)P_2(N_1)$$

Now if $M_1 = S_1$ and $N_1 = S_2$ we see that $P(S) = P_1(S_1)P_2(S_2) = 1$. That $P$ satisfies the other conditions of a probability measure follows from Theorem 3.4.1 and the observation that the positive integers can be used in place of any denumerable sample space.

If we now take $N_1 = S_2$ we get $P(M_1 \times S_2) = P_1(M_1)P_2(S_2) = P_1(M_1) \cdot 1 = P_1(M_1)$. Similarly $P(S_1 \times N_1) = 1 \cdot P_2(N_1) = P_2(N_1)$. Therefore $P(M \cap N) = P_1(M_1)P_2(N_1) = P(M)P(N)$ and $M$ and $N$ are independent events.

### Example 5.2.3

Let $S = \{1, 2, \ldots, 12\}$ with equal likelihood. Let $M_1 = \{1, 2, 4, 12\}$, $M_2 = \{1, 2, 5, 6, 7, 8\}$, $M_3 = \{1, 5, 6, 7, 9, 12\}$. Show that $P(M_1 \cap M_2 \cap M_3) = P(M_1)P(M_2)P(M_3)$, $P(M_1 \cap M_2) = P(M_1)P(M_2)$, $P(M_1 \cap M_3) = P(M_1)P(M_3)$ but $P(M_2 \cap M_3) \neq P(M_2)P(M_3)$.

We will show only that $P(M_1 \cap M_2 \cap M_3) = P(M_1)P(M_2)P(M_3)$. $M_1 \cap M_2 \cap M_3 = \{1\}$, $P(\{1\}) = \frac{1}{12} = \frac{1}{3} \cdot \frac{1}{2} \cdot \frac{1}{2} = P(M_1)P(M_2)P(M_3)$. The remaining calculations are left to the reader.

### Example 5.2.4

Let $S = \{1, 2, \ldots, 30\}$ with equal likelihood. Let $M_1 = \{2, 4, 6, 8, \ldots, 30\}$, $M_2 = \{3, 6, 9, 12, \ldots, 30\}$, and $M_3 = \{5, 10, 15, \ldots, 30\}$. Show that $P(M_1 \cap M_2 \cap M_3) = P(M_1)P(M_2)P(M_3)$, $P(M_1 \cap M_2) = P(M_1)P(M_2)$, $P(M_1 \cap M_3) = P(M_1)P(M_3)$, and $P(M_2 \cap M_3) = P(M_2)P(M_3)$.

Here we leave all calculations to the reader.

### Example 5.2.5

Let $S$ and $P$ be the same as in Example 5.2.3. Let $M_1 = \{1, 2, 4, 12\}$, $M_2 = \{4, 12, 5, 6, 7, 11\}$, $M_3 = \{1, 2, 5, 6, 7, 8\}$. Show that any pair of events from $M_1, M_2, M_3$ are independent but $P(M_1 \cap M_2 \cap M_3) = 0 \neq P(M_1)P(M_2)P(M_3)$.

Here, too, we leave all calculations to the reader.

Examples 5.2.3, 5.2.4, and 5.2.5 show what can occur with more than two events. We extend the definition of independence so that the three events of Example 5.2.4 are independent but those of Examples 5.2.3 and 5.2.5 are not.

**Definition**    Suppose $\{M_\alpha\}$ is a family of events in the family of admissible events of a probability system. The events in $\{M_\alpha\}$ are said to be *independent* if for each finite subfamily of $\{M_\alpha\}$ having at least two events, the probability of the intersection of the events in the subfamily is the product of the probabilities of those events.

The analogs of Theorem 5.2.1 and 5.2.2 can be proved for many events in reference to this definition.

We defer the problem of independence in Euclidean sample space for a few sections.

**E X E R C I S E S  5.2** ———————————————

1   Prove Theorem 5.2.2 directly, in the case of finite sample spaces $S_1$ and $S_2$ with equal likelihood.

2   In Example 5.2.2, are the events {dice show < 5} and {dice show an even number} independent?

3   As in Example 5.2.2, are the events {red die shows 1} and {total number of dots is 4} independent?

4   A fair coin is tossed ten times. Let $L = \{$head occurs for the first time on the fifth toss$\}$, $M = \{$even number of heads occurs in ten tosses$\}$, and $N = \{$tenth toss is heads$\}$. Are $L$ and $M$ independent? Are $L$ and $N$ independent? Are $M$ and $N$ independent?

♦ 5   Prove that if $M$ and $N$ are independent, then $M'$ and $N'$ are independent.

6   A card is chosen from a standard deck. Determine whether {the card is a club} and {the card is a three} are independent events.

7   Find the probability that in six independent tosses of a fair die, each face will show exactly once.

8   Toss a pair of real coins 20 times and record the outcomes. Is there any indication that the coins are not independent? Assuming that two coins are fair and independent and that the tosses are independent, what are the probabilities that (T, T) occurs fewer than four times in 20 tosses and fewer than ten times in 60 tosses?

9   Finish Example 5.2.3.

10   Finish Example 5.2.4.

11   Finish Example 5.2.5.

**12**   Find three events $M_1$, $M_2$, $M_3$ in the context of Example 5.2.3 such that $P(M_1 \cap M_2 \cap M_3) \neq P(M_1)P(M_2)P(M_3)$ but any pair of which are independent.

**13**   Suppose $M_1$, $M_2$, and $N$ are events with $M_1 \cap M_2 = \varnothing$, $M_1$ and $N$ independent, and $M_2$ and $N$ independent. Are $M_1 \cup M_2$ and $N$ independent? Justify your conclusion.

# 5.3  Conditional probability

We begin this section with two examples.

### *Example 5.3.1*
A dime and a quarter (both fair) are tossed. The experiment at issue consists of recording the outcome if there is at least one head and ignoring the whole toss if both coins show tails. Describe a sample space $S_1$ for the experiment and a probability measure.

A suitable sample space is $S_1 = \{(H, H), (H, T), (T, H)\}$ where the outcome of the dime is recorded in the left-hand position of the ordered pair. The question of a probability measure is more complicated. If the act of ignoring the case of (T, T) has no effect on the relative weights of (H, H), (H, T), (T, H), one would feel that since (H, H), (H, T), (T, H), (T, T) are treated equally in the unrestricted tossing of two coins, then the sample points of the above $S_1$ should be treated as equally likely, namely $P((H, H)) = P((H, T)) = P((T, H)) = \frac{1}{3}$.

### *Example 5.3.2*
A hand of three cards from a well-shuffled standard deck is dealt. You accidentally see the ace of spades. What is the probability that the hand contains three aces?

Normally there are $\binom{52}{3}$ hands of three cards, but since the ace of spades was seen there are really $\binom{51}{2}$ possibilities. Of these, $\binom{3}{2}$ are composed of three aces. If the ace of spades had not been seen, no particular hand would be preferred probabilistically to any other hand, i.e., we have equal likelihood. It seems reasonable to retain equal likelihood in the new sample space. Thus the answer is $\binom{3}{2} \Big/ \binom{51}{2}$

In both of the above examples we could have thought of starting with a sample space $S$ (associated with the unrestricted tossing of two coins or the choice of three cards without knowing that one is the spade ace) and then passing

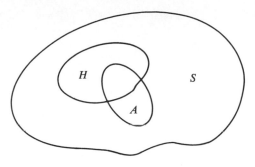

**Figure 1**

to a subset (event) of $S$ when the additional requirements "ignore (T, T)" or "one of the three chosen cards is the spade ace" are imposed.

Referring to Figure 1, we start with a sample space $S$. We are given some additional requirement $H \subset S$ and then ask what is the probability of the event $A$ in reference to the reduced sample space $H$. The letter $H$ is used because the additional requirement [such as "ignore (T, T)" or "the spade ace is seen"] is called the hypothesis. In the above two problems we obtained answers $\frac{1}{3}$ and $\binom{3}{2} / \binom{51}{2}$. We observe that

**(5.3.1)**    $$\frac{1}{3} = \frac{1}{4} \Big/ \frac{3}{4} \quad \text{and} \quad \frac{\binom{3}{2}}{\binom{51}{2}} = \frac{\binom{3}{2} \div \binom{52}{3}}{\binom{51}{2} \div \binom{52}{3}}$$

Thus if $A = \{\text{both coins show heads}\}$ or $A = \{\text{hand contains three aces}\}$ in the coin toss and three card hand choice, respectively, we can interpret the compound fractions on the right-hand sides of the equations in (5.3.1) as $\dfrac{P(A \cap H)}{P(H)}$, where $P$ is the probability measure of the system $S, \mathscr{A}, P$.

**Definition**    *The conditional probability of an event $A \subset S$ given the hypothesis* (event) $H \subset S$, written $P(A \mid H)$ and read as "the conditional probability of $A$ given $H$" is the number $\dfrac{P(A \cap H)}{P(H)}$ if $P(H) > 0$. If $P(H) = 0$ we do not define the conditional probability.

If $A_1 \subset H$,   $A_1 \cap H = A_1$ so

$$P(A_1 \mid H) = \frac{P(A_1)}{P(H)}$$

Thus the conditional probabilities of subevents of $H$ given $H$ are proportional to the probabilities of those events in the original sample space. The factor of proportionality is $1/P(H)$. It is clear that $P(H \mid H) = 1$. Finally, if $P(H) > 0$ then

**(5.3.2)**    $P(A \cap H) = P(A \mid H)P(H)$

One may ask, "Why do we not make the restriction $A \subset H$ in the definition of conditional probability?" This is answered indirectly by the exercises as well as by Example 5.3.3. Briefly, the reason is that a verbal description of $A$ may be more easily given in terms which do not include the conditions of $H$ and since this situation can be handled easily we do not require $A \subset H$. It should be observed that $P(A \mid H)$ makes perfectly good sense in any case where $P(H) > 0$. There is no need to assume equal likelihood or a countable sample space $S$.

It should also be noted that if $A$ and $H$ are independent and $P(\text{H}) > 0$ then

$$P(A \mid H) = \frac{P(A \cap H)}{P(H)} = \frac{P(A)P(H)}{P(H)} = P(A)$$

This relationship is sometimes described by saying "the probability of $A$ is not affected by knowledge that $H$ occurred."

Formula (5.3.2) is frequently used as follows: $P(H)$ and $P(A \mid H)$ are known in advance from a variety of considerations; $P(A \cap H)$ is then computed by (5.3.2). We see this technique in some of the examples which follow.

### Example 5.3.3

A die is tossed and if $n$ shows on the top face, a fair coin is tossed $n$ times. Find a sample space and probability measure for $(n, m)$ where $n$ is the number on the top face of the die and $m$ is the number of heads in the subsequent coin tosses.

A reasonable choice for the sample space is

$$S = \{(n, m); n = 1, 2, 3, 4, 5, 6 \text{ and } 0 \leq m \leq n\}$$

since there can be no fewer than zero heads and no more than $n$ heads in $n$ tosses. There are 27 sample points and here equal likelihood seems totally wrong. If we look at conditional probabilities it seems clear what measure should be given:

$$P(m \text{ heads} \mid n \text{ shows on die}) = \binom{n}{m}\left(\frac{1}{2}\right)^m\left(\frac{1}{2}\right)^{n-m} = \frac{1}{2^n}\binom{n}{m}$$

Furthermore we would want $P(n \text{ shows on die}) = \frac{1}{6}$ if the die is symmetrical. Thus

$$P(m \text{ heads} \mid n \text{ shows on die}) = \frac{P((m \text{ heads}) \text{ and } (n \text{ shows on die}))}{P(n \text{ shows on die})}$$

$$= \frac{P((n, m))}{P(n \text{ shows on die})}$$

and we obtain

$$\frac{1}{2^n}\binom{n}{m} = \frac{P((n, m))}{\frac{1}{6}} \quad \text{and} \quad P((n, m)) = \frac{1}{6 \cdot 2^n}\binom{n}{m}$$

**Example 5.3.4**

(Random walk.) A marker is moved one unit to the left or one unit to the right (on an axis) according to whether a fair coin shows T or H. Assuming the marker starts at zero, what can be said about the probability $q(n)$ that it *returns to zero for the first time* after the $n$th toss.

According to the conditions of the problem, the marker can be at zero after the $n$th toss if and only if there were equal numbers of heads and tails in the $n$ tosses since the marker starts at zero. That is, $n = 2m$. We change notation slightly. Let $A = \{$marker is at zero after the $2n$th toss$\}$. Let $B_{2m} = \{$marker returns to zero for the first time after the $2m$th toss$\}$. If $2l \neq 2m$, $B_{2l} \cap B_{2m} = \varnothing$ because the marker cannot return to zero for the *first* time both after the $2l$th toss and after the $2m$th toss if $2l \neq 2m$. Now $A \cap B_{2m} = \{$marker returns to zero for the first time after the $2m$th toss and is again at zero after the $2n$th toss$\}$. Furthermore $P(A) = P(n$ heads and $n$ tails in $2n$ tosses$)$ by our first observation that the number of heads and tails in the $2n$ tosses must be equal. Therefore

$$P(A) = \binom{2n}{n} \frac{1}{2^n} \frac{1}{2^n}$$

Let $P(B_{2m}) = q(2m)$. By (5.3.2),

$$P(A \cap B_{2m}) = P(B_{2m})P(A \mid B_{2m})$$

But

$$P(A \mid B_{2m}) = P(\text{marker is at zero after } 2n\text{th toss} \mid \text{marker returns to zero for the}$$
$$\text{first time after the } 2m\text{th toss})$$

For the marker to be at zero after the $2n$th toss given that it returned to zero for the first time after the $2m$th toss it is necessary and sufficient that there be equal numbers of heads and tails in the $2m + 1$, $2m + 2$, $2m + 3, \ldots, 2n$th tosses. The probability of this last eventuality is, clearly, $\binom{2n - 2m}{n - m} \frac{1}{2^{n-m}} \frac{1}{2^{n-m}}$ because the outcome of the $2m + 1$, $2m + 2, \ldots, 2n$th tosses are not affected by the first $2m$ tosses. It is also clear that $A = \bigcup_{m=1}^{n} (A \cap B_{2m})$ since if the marker is at zero after the $2n$th toss it must have returned to zero for the first time after either the 2nd, 4th, 6th, $\ldots$, or $2n$th toss. Since the events $B_{2m}$ are disjoint we obtain

$$P(A) = \sum_{m=1}^{n} P(A \cap B_{2m}) = \sum_{m=1}^{n} P(B_{2m})P(A \mid B_{2m})$$

or

$$\binom{2n}{n} \frac{1}{2^{2n}} = \sum_{m=1}^{n} q(2m) \binom{2(n - m)}{n - m} \frac{1}{2^{2n - 2m}}$$

This can be rewritten as

**(5.3.3)**   $$\binom{2n}{n}2^{-2n} = \sum_{m=0}^{n-1} q(2n - 2m)\binom{2m}{m} 2^{-2m}$$

if we replace $m$ by $n - m$. We shall do more with this formula in another section but the reader could try $n = 1, 2, 3$ in succession and find $q(2)$, $q(4)$, $q(6)$.

Sections 5.1–5.3 constitute a brief introduction to independent events, conditional probability, and some of their implications. A high percentage of the following material of this book is closely related to independent events. One can almost say that the consequences of independent events form the backbone of modern probability theory.

We will not do much with conditional probability. Although it is a most vital topic, the next level of implications of conditional probability is largely beyond the scope of this book.

**EXERCISES 5.3** _____

**1**   Two fair dice are rolled. Find $P$(dice show 7 | dice show at least 4).

**2**   A fair dime and a fair quarter are tossed independently. Find $P$(both coins show heads | quarter shows heads) and $P$(both coins show heads | at least one of the coins shows heads). Explain why the results are different.

**3**   In the situation described in Example 5.3.3 find $P$(die shows 5 | exactly three heads and unspecified number of tails occur).

**4**   Suppose instead of rolling a single die and then tossing a coin as many times as shown on the die, two dice are rolled and the coin is tossed as many times as is shown on the dice. Find $P$(dice show 7 | four heads and unspecified number of tails occur).

**5**   Medieval cities sometimes chose their government from their leading families by lot. Suppose that a city has five leading families each composed of two senior members and eight junior members. A governing council is chosen as follows: First, two senior men are chosen by lot. Then seven others are chosen by lot from the 48 who were not chosen on the first round. Find $P$(exactly one senior member of family A is chosen on the first round and exactly four members of family A are chosen on the second round).

**6**   A club consists of 50 couples. Six people are selected at random. Find $P$(three couples are selected | at least one couple is selected).

**7**   Show that if $P(H) > 0$ then the function $P(A \mid H)\colon \mathscr{A} \to$ reals is a probability measure.

**8**   Suppose that a coin and a die are magnetized and thrown together so that $P(1 \mid H) = \frac{1}{3}$,  $P(2 \mid H) = \frac{1}{6}$,  $P(3 \mid H) = \frac{1}{8}$,  $P(4 \mid H) = \frac{1}{8}$,  $P(5 \mid H) = \frac{1}{8}$,

$P(6 \mid H) = \frac{1}{8}$, where 1, 2, 3, 4, 5, 6 refer to the die; H, T refer to the coin. Are these results compatible with $P(H) = \frac{1}{2}$, $P(T) = \frac{1}{2}$, i.e., could a probability measure be put on $\{H, T\} \times \{1, 2, \ldots, 6\}$ which would lead to the above values?

**9**   Compute $q(2)$, $q(4)$, and $q(6)$ of Example 5.3.4.

♦ **10**   Let $L$, $M$, and $N$ be events with positive probability and with $P(L \cap M) > 0$. Show that

$$P(L \cap M \cap N) = P(N \mid L \cap M)P(M \mid L)P(L)$$

**11**   Generalize the result of exercise 10 to the case of four events.

**12**   A communicable disease has the following characteristics for a family of three in which the child attends school.

$P$(child will catch the disease) $= 0.60$

$P$(mother will catch the disease $\mid$ child catches it) $= 0.50$

$P$(father will catch it $\mid$ mother and child catch it) $= 0.40$

Find $P$(father, mother, and child all catch the disease). and $P$(mother and child catch the disease but father does not).

# 5.4   The continuous case

A thorough treatment of the analog of Theorem 5.3.2 for Euclidean sample spaces is beyond the scope of this book. We merely state that it is possible to form sample spaces for compound experiments whose component experiments have Euclidean sample spaces by using Cartesian products, and, a suitable probability measure $P$ can be put on the Cartesian product so that if $S_1$ and $S_2$ are the Euclidean sample spaces of the components and if $M = M_1 \times S_2$, $N = S_1 \times N_2$ then $P(M \cap N) = P(M)P(N) = P_1(M_1)P_2(N_2)$.

### Example 5.4.1

Two positions $x$ and $y$ on a stick of unit length are chosen independently and chosen so that the probability $P(a < x \le b) = b - a$ for $0 \le a \le b \le 1$ and $P(a < y \le b) = b - a$. What is the probability that a triangle can be formed from the pieces if the original stick is broken at the positions $x$ and $y$? (See Figure 2.)

It is convenient to think of the sample space for this problem as the square $\{(x, y); 0 \le x \le 1, \ 0 \le y \le 1\}$. Since the choices of $x$ and $y$ are independent we expect

$$P(a < x \le b \text{ and } c < y \le d) = P(a < x \le b)P(c < y \le d)$$

(See Figure 3.)

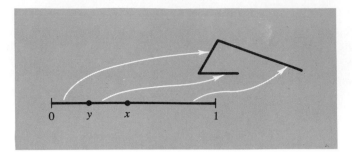

**Figure 2**

Suppose, temporarily, that $x < y$. The lengths of the pieces are then $x$, $y - x$, $1 - y$. A triangle can be formed from three line segments if and only if the sum of the lengths of any two segments is greater than the length of the third. We thus obtain the conditions

$$\begin{cases} x < y - x + 1 - y \\ y - x < x + 1 - y \\ 1 - y < x + y - x \end{cases} \text{ or } \begin{cases} 2x < 1 \\ 2y - 2x < 1 \\ 1 < 2y \end{cases}$$

If we shade the region satisfying $x < y$, $2x < 1$, $2y - 2x < 1$, and $1 < 2y$ we get the interior of the heavily shaded triangle in Figure 4. By symmetry we would obtain the lightly shaded triangle if we took $y < x$ and followed up the necessary and sufficient conditions for a triangle. So much for geometry; now we return to probability. According to our assumptions

$$P(a < x \le b \text{ and } c < y \le d) = P(a < x \le b)P(c < y \le d) = (b - a)(d - c)$$

Thus the probability that $(x, y)$ lies in a rectangle (with some sides deleted) is the area of the rectangle (see Figure 3). But then the probability of the union

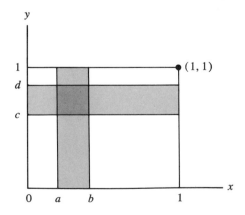

**Figure 3**

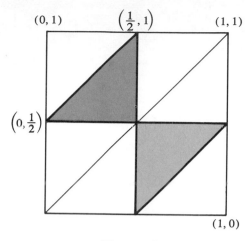

**Figure 4**

of finitely many disjoint rectangles is also the area of the figure. Since the indicated triangles of Figure 4 can be "approximated from within" by figures as indicated in Figure 5, it is quite clear that $P$(triangle) $\geq$ lim (area of inside figures). Similarly $P$(triangle) $\leq$ lim (area of outside figures). Thus $P$(triangle) $=$ area of triangle and thus the answer to our problem is $2(\frac{1}{2}\cdot\frac{1}{2}\cdot\frac{1}{2}) = \frac{1}{4}$ if there is a genuine probability measure for this problem.

### Example 5.4.2

Two positions $x$ and $y$ are selected independently on a stick of unit length but in this case $P(a < x \leq b) = b^n - a^n$ and $P(a < y \leq b) = b^m - a^m$ for $0 \leq a \leq b \leq 1$, where both $m, n > 0$. What is the probability that the pieces obtained by breaking the stick at the positions $x$ and $y$ can form a triangle?

In this problem the same sample space as in Example 5.4.1 can be used and the geometric aspects remain the same. The question is, "What is the probability of the union of the two triangles in Figure 4, when the $P(a < x \leq b$ and $c < y \leq d) = (b^n - a^n)(d^m - c^m)$?" First we observe that

**(5.4.1)**     $\displaystyle\int_a^b dx \int_c^d nx^{n-1}my^{m-1}dy = (b^n - a^n)(d^m - c^m)$

(Notice that we use the assumed independence.) We are really asking, "What density gives probability (mass) of $(b^n - a^n)(d^m - c^m)$ to the rectangle with opposite corners $(a, c)$ and $(b, d)$?" If $p(x, y)$ is the density at the point $(x, y)$ we should have

$$p(a, c) = \lim_{\substack{b \to a \\ d \to c}} \frac{P(a < x \leq b \text{ and } c < y \leq d)}{(b - a)(d - c)}$$

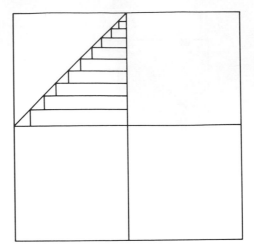

**Figure 5**

i.e., the density is the limit of the (probability) mass of a rectangle divided by the area of the rectangle as that area tends to zero.
But

$$\lim_{\substack{b \to a \\ d \to c}} \frac{P(a < x \le b \text{ and } c < y \le d)}{(b - a)(d - c)} = \lim_{\substack{b \to a \\ d \to c}} \frac{(b^n - a^n)(d^m - c^m)}{(b - a)(d - c)}$$

$$= \lim_{\substack{b \to a \\ d \to c}} (b^{n-1} + b^{n-2}a + b^{n-3}a^2 + \cdots + a^{n-1}) \cdot$$

$$(d^{m-1} + d^{m-2}c + d^{m-3}c^2 + \cdots + c^{m-1})$$

$$= na^{n-1}mc^{m-1}$$

That is, $p(a, c) = nma^{n-1}c^{m-1}$ or $p(x, y) = nmx^{n-1}y^{m-1}$. This brings us full circle to (5.4.1). By referring to Figure 4 we see that our answer is

$$\int_0^{1/2} dx \int_{1/2}^{(1/2)+x} nx^{n-1}my^{m-1} \, dy + \int_{1/2}^1 \int_{(-1/2)+x}^{1/2} nx^{n-1}my^{m-1} \, dy$$

Both terms are needed in the case $n \ne m$. We do not give the answer in a more simplified form.

Up to this stage we have interpreted probabilities obtained by integration such as $\int_a^b \int_c^d p(x, y) \, dx \, dy$ as the mass which a geometric region (a rectangle in this case) would have if there was a mass density $p(x, y)$ at the typical point $(x, y)$ or we have just decided to use integrals because integrals are the continuous analogs to discrete sums and $\sum_{s_n \in M} P(s_n) = P(M)$ is the probability of the event $M$. Naturally if the sample space $S$ is a subset of $R$, we will imagine a linear mass density, i.e., a function $p: S \to R$ such that $p(x) \geq 0$ for each $x \in S$, $\int_S p(x)\, dx = 1$ and $p$ is smooth enough to integrate on $S$. Thinking of probability as mass where the total mass is one unit can be very valuable, but other intuitive devices can be useful too, e.g., $\int_a^b p(x)\, dx$ is usually thought of as the area underneath the curve $y = p(x)$ [assuming $p(x) \geq 0$] and between the two vertical lines $x = a$ and $x = b$. Similarly, $\int_M p(x)\, dx$ may be thought of as the area of the region underneath the curve $y = p(x)$ and above the set $\{(x, 0); x \in M\}$. Thus if $M = \{x; x^2 \geq 2\}$ and $p(x) = 1/\pi(1 + x^2)$ we have that $P(M)$ is the area of the shaded region in Figure 6. The event $M$ may have many unconnected pieces when viewed as a subset of $R$ but we still have $\int_M p(x)\, dx$ as the sum of the integrals over the various pieces of $p$. It should be noted that in this case $P(M)$ is not the area of $M$ because $M$ is a 1-dimensional set. We actually attach another axis against which we plot $p(x)$. Similarly if $S \subset E^2 = R \times R$ and if $p(x, y)$ is the density of probability mass at $(x, y) \in S$ then $\iint_M p(x, y)\, dx\, dy = P(M)$ suggests finding the volume of the 3-dimensional set $\{(x, y, z); (x, y) \in M$ and $0 \leq z \leq p(x, y)\}$, i.e., the set under the surface $z = p(x, y)$ and above the points in the plane of the form $(x, y, 0)$ for $(x, y) \in M$. Here, too, we set up an auxiliary axis against which we plot $p(x, y)$. (See Figure 7.) The curve $y = p(x)$ and the surface $z = p(x, y)$ are frequently called the *probability curve* and the *probability surface*, respectively, when they arise.

There are many interesting questions which follow naturally from the above ideas. We defer consideration of these questions until Chapters 6 and 8. There we formulate new concepts which allow for a more effective treatment.

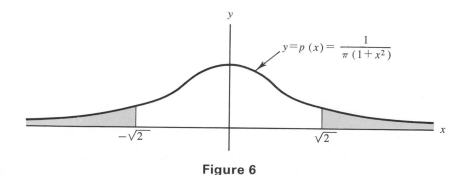

$$y = p(x) = \frac{1}{\pi(1 + x^2)}$$

**Figure 6**

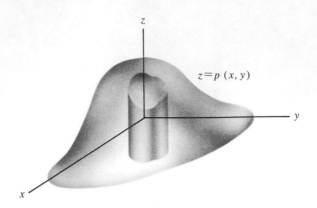

$z = p\,(x, y)$

**Figure 7**

**EXERCISES 5.4** ————————————————————————

**1**  A stick is marked as in section 3.5. For

$$x = \frac{n_1}{10} + \frac{n_2}{10^2} + \frac{n_3}{10^3} + \cdots$$

let $M = \{x; n_1 = 2\}$,  $N = \{x; n_2 = 5\}$, and $Q = \{x; n_3 = 5\}$. Show that $M$, $N$, $Q$ are independent. Make a more general conjecture.

**2**  A stick is broken as in exercise 3.5.7(a). For $M$ and $N$ as defined in exercise 1 find $P(M \mid N)$ and show that $M$ and $N$ are not independent.

**3**  A point $(x, y)$ is chosen in the square $S = \{(x, y); 0 \leq x,\ y \leq 1\}$ in such a way that $P(a < x \leq b$ and $c < x \leq d) = (b - a)(d - c)$. Let $\theta$ be the angle from the positive $x$ axis to the line segment joining $(0, 0)$ and $(x, y)$. Are the events $M = \{(x, y); \frac{1}{2} < x < 1\}$ and $\{(x, y); 0 < \theta \leq \frac{1}{6}\pi\}$ independent? Are $M$ and $N = \{(x, y); 0 < \theta \leq \frac{1}{4}\pi\}$ independent? Are $N$ and $\{(x, y); x^2 + y^2 \geq 1\}$ independent?

**4**  Would the point $(\frac{1}{3}, \frac{2}{3})$ in Example 5.4.1 signify that a triangle could be formed? What about the point $(\frac{1}{4}, \frac{2}{3})$? Show that points $(\frac{1}{4}, \frac{51}{100})$ and $(\frac{27}{100}, \frac{54}{100})$ signify triangles whose largest angle is greater than 90°. What is the probability that the triangle has its largest angle greater than 90° given that a triangle can be formed?

**5**  Let $S_1$ and $S_2$ be bounded or unbounded intervals in $R$. Let $S = S_1 \times S_2$. We take $\mathscr{A}$ to be the Borel sets in $R \times R$ which are contained in $S$. Suppose $p_1$ and $p_2$ are mass densities on $S_1$ and $S_2$, respectively, such that $\int_{S_1} p_1(x)\,dx = 1$ and $\int_{S_2} p_2(x)\,dx = 1$. Let $p(x, y) = p_1(x)p_2(y)$ be a mass density for $S$. Let $A_1 \subset S_1$ and $A_2 \subset S_2$ be intervals. Show that $A_1 \times S_2$ and $S_1 \times A_2$ are independent events.

6   Let $S = \{x; 1 \le x \le e\}$. Let $p(x) = 1/x$. Show that this gives a legitimate probability measure on $S$. Let $M = \{x \in S; (x - 2)^2 \le 1/36\}$. Find $P(M)$.

7   Let $S = \{(x, y); 0 \le x^2 + y^2 \le 1\}$. Let $p(x, y) = K(x^2 + y^2)$. Find $K$ such that $p$ is a probability mass density on $S$. Find $P$(the square inscribed in $S$). Are the events $M = \{(x, y); x > 0\}$ and $N = \{(x, y); y > 0\}$ independent? If $Q = \{(x, y); y < \frac{1}{2}\}$, are $M$ and $Q$ independent? Compare with Example 3.5.5. [*Hint:* Work in polar coordinates to find $K$.]

8   A point $(x, y)$ is chosen from the square $\{(x, y); 0 \le x, \ y \le 1\}$ at random. Find the conditional probability that $(x, y)$ has $y \le \sqrt{x}$ given that $y \ge 2x$.

9   Two points are selected independently and at random on the interval $\{x; 0 \le x \le 1\}$. (See Example 5.4.1.) Find the conditional probability that the points are less than $\frac{1}{20}$ apart given that they are less than $\frac{1}{10}$ apart.

10   Three points $x$, $y$, and $z$ are selected independently and at random on the interval $\{x; 0 \le x \le 1\}$. Describe how to find the conditional probability that $x^2 + y^2 \le 2z$ given that $x + y \le 1$.

# 5.5  Bayes' formula

In exercise 5.3.1, we sought $P$(die shows 5 | exactly three heads and unspecified number of tails occur in coin tossing). Since we were committed to the sample space of Example 5.3.3 the answer is easy to determine. Let us see an example in which the sample space is not given explicitly.

### *Example 5.5.1*
A man backstage prepares a box containing four balls. These balls are chosen from a collection of eight red balls and eight white balls which are indistinguishable except for color. A blindfolded player onstage draws two balls from the box. Find $P$(all the balls in the box are red | both balls drawn are red).

Since the man backstage may put in 0, 1, 2, 3, or 4 red balls we can form a sample space as in Example 5.3.3. Let us not go through the details but make some inferences. Let $H_j = \{j$ red balls and $4 - j$ white balls are put in the box$\}$ where $j = 0, 1, 2, 3, 4$. Let $M = \{$two red balls are selected$\}$. We seek

$$P(H_4 \mid M) = \frac{P(H_4 \cap M)}{P(M)}$$

It is clear that the sample space $S = H_0 \cup H_1 \cup H_2 \cup H_3 \cup H_4$ can be arranged where the events $H_j$ are disjoint. Thus $M = M \cap S = M \cap (H_0 \cup \cdots \cup H_4) = \bigcup_{j=0}^{4} (M \cap H_j)$. By a trivial observation the events $M \cap H_j$ are disjoint. The conditional probabilities $P(M \mid H_j)$ are clear. We use (5.3.2) for $P(M \cap H_j)$. Therefore, if the box is prepared so that the events $H_0, H_1, \ldots, H_4$ have the same probability

$$P(H_4 \mid M) = \frac{P(H_4 \cap M)}{\frac{1}{5} \cdot \frac{1}{6} + \frac{1}{5} \cdot \frac{1}{2} + \frac{1}{5} \cdot 1} = \frac{P(M \mid H_4)P(H_4)}{\frac{1}{5} \cdot \frac{1}{6} + \frac{1}{5} \cdot \frac{1}{2} + \frac{1}{5} \cdot 1}$$

$$= \frac{\frac{1}{5} \cdot 1}{\frac{1}{5}(\frac{1}{6} + \frac{1}{2} + 1)} = \frac{1}{\frac{5}{3}} = \frac{3}{5}$$

Suppose the box is prepared by a blindfolded choice of four balls from the 16 balls (eight red, eight white). In that case $P(H_j) = \binom{8}{j}\binom{8}{4-j} \div \binom{16}{4}$. The same formula applies and we would have

$$P(H_4 \mid M) = \frac{\binom{8}{4} \cdot 1}{\binom{8}{2}\binom{8}{2} \cdot \frac{1}{6} + \binom{8}{3}\binom{8}{1} \cdot \frac{1}{2} + \binom{8}{4} \cdot 1}$$

after the denominators of $\binom{16}{4}$ have been cancelled. In this case, $P(H_4 \mid M) = \frac{30}{182} \sim \frac{1}{6}$.

We have obtained two answers because the example was not precisely formulated. We could have obtained many other answers, each depending on the manner in which the boxes are prepared. This is natural in problems which are incompletely formulated.

Example 5.5.1 suggests a theorem which we shortly prove. The theorem follows so simply from the example that the reader should glance at the statement and try to prove it before reading the proof. (Perhaps that should be done with all theorems.) The example also suggests a type of problem where it is easy to make a mistake. We should look at this and take heed.

We go to the theorem first. Equation (5.5.1) has the form

$$P(H_4 \mid M) = \frac{P(M \mid H_4)P(H_4)}{P(M \mid H_2)P(H_2) + P(M \mid H_3)P(H_3) + P(M \mid H_4)P(H_4)}$$

However, certain terms are missing because certain conditional probabilities are zero. We really want, in more general form,

$$(5.5.1) \qquad P(H_1 \mid M) = \frac{P(M \mid H_1)P(H_1)}{\sum\limits_{j=1}^{n} P(M \mid H_j)P(H_j)}$$

**Theorem 5.5.1**   Let $\mathscr{S}$ be a probability system. Let $H_1, H_2, \ldots, H_n$ be disjoint events in $\mathscr{A}$ whose union is $S$, with $P(H_j) > 0$ for each $j$. Let $M \in \mathscr{A}$ with $P(M) > 0$. Then equation (5.5.1) holds.

PROOF   Since the $H_j$ are disjoint so are the events $M \cap H_j$. Since $H_1 \cup \cdots$ $\cup H_n = S$, then $(H_1 \cap M) \cup (H_2 \cap M) \cup \cdots \cup (H_n \cap M) = M$. Thus

$$P(M) = P(H_1 \cap M) + P(H_2 \cap M) + \cdots + P(H_n \cap M)$$

Now $P(H_j \cap M) = P(M \mid H_j)P(H_j)$ where the conditional probability exists because $P(H_j) > 0$. Therefore, since $P(M) > 0$

$$P(H_1 \mid M) = \frac{P(H_1 \cap M)}{P(M)} = \frac{P(M \mid H_1)P(H_1)}{\sum\limits_{j=1}^{n} P(M \mid H_j)P(H_j)}$$

Since the order of the events $H_1, \ldots, H_n$ is irrelevant, any of the events could be denoted by $H_1$.

Formula (5.5.1) is called *Bayes' formula* and the theorem proving its validity is called Bayes' Theorem. Bayes' formula is frequently applied in circumstances in which the events $H_j$ refer to actions which occur before (in time) the outcomes associated with $M$ but, for various reasons, one is uncertain which of the $H_j$ occurred. Furthermore, the events $H_j$ are of the kind which are often used as hypotheses in computing conditional probabilities. Therefore, Bayes' formula was frequently called the formula for computing the probabilities of hypotheses. For many years it was regarded with great suspicion because it appeared to imply ridiculous conclusions. However, upon more careful examination, it is not Bayes' formula which implies the ridiculous conclusions, but it is the improper use of that formula. Usually, the improper use of the formula occurs when the numbers $P(H_j)$ are not known and when there are $n$ events $H_j$. It was then common to replace one's ignorance of the values of $P(H_j)$ by taking $P(H_j) = 1/n$. Such a value makes for easy computation and has certain intuitive appeal—after all, if you do not know $P(H_j)$, why should $H_1$ be more or less favored than $H_2$. It was common to justify the choice $P(H_j) = 1/n$ by invoking a "Principle of Insufficient Reason." However, to do this is essentially the same as assuming that whenever one does not know the probability measure on a finite sample space, then the probability measure *must* be the equal-likelihood measure. It is comparable to replacing one's ignorance by a convenient assumption about the situation rather than an inconvenient assumption, thus giving rise to an absurd prediction.

### Example 5.5.2

A man leaves work promptly at 5:00 P.M. and keeps extensive data about his time of arrival home according to whether he takes a bridge route or a tunnel route. He duly informs his wife that the conditional probabilities are as indicated.

| | RANGE OF MINUTES | | | | |
| --- | --- | --- | --- | --- | --- |
| | 35–39 | 40–44 | 45–49 | 50–54 | Over 55 |
| BRIDGE | 0.10 | 0.25 | 0.45 | 0.15 | 0.05 |
| TUNNEL | 0.30 | 0.35 | 0.20 | 0.10 | 0.05 |

One day he flips a fair coin to decide between the bridge and tunnel. He arrives home at 5:47. Find the probability that he took the tunnel based on this information.

By (5.5.1) with $H_1 = \{$took tunnel$\}$, $H_2 = \{$took bridge$\}$,

$$P(\text{took tunnel} \mid \text{arrives between } 5:45\text{--}5:49) = \frac{\frac{1}{2}(0.20)}{\frac{1}{2}(0.20) + \frac{1}{2}(0.45)} = \frac{20}{65}$$

since we know that the man decided by flipping a fair coin.

### Example 5.5.3

A hardware store owner buys nails of a particular size from three different factories. They come in identical kegs with only a label to indicate the factory. These factories manufacture the nails independently and the probabilities of manufacturing a defective nail are 0.02, 0.03, and 0.04 for factories A, B, and C, respectively. The store owner happens to be examining a keg from which the label has disappeared. He looks at 20 randomly chosen nails and finds six defectives. What is the conditional probability that the keg came from factory $C$?

In a section on Bayes' formula it may be tempting to give

$$\frac{\binom{20}{6}(0.04)^6(0.96)^{14} \cdot \frac{1}{3}}{\binom{20}{6}(0.02)^6(0.98)^{14} \cdot \frac{1}{3} + \binom{20}{6}(0.03)^6(0.97)^{14} \cdot \frac{1}{3} + \binom{20}{6}(0.04)^6(0.96)^{14} \cdot \frac{1}{3}}$$

as the answer. However, how do we know that the store owner buys equal numbers of kegs of nails from each factory? If he only rarely buys from factory C it would be unlikely that the keg at issue comes from C. That would have to be balanced against factory C's higher defective rate. We would do well to say that we cannot give a definite answer until more information is given.

The reader may feel cheated by the discussion of this example. However, because of the history of fallacious use of Bayes' formula we have attempted to give a dramatic example of the pitfalls associated with its use. Needless to say, there are many problems in which all of the required information is present. Then Bayes' formula is perfectly applicable.

### EXERCISES 5.5

1  Work Example 5.5.1 if the man backstage puts in 0, 1, 2, 3, or 4 red balls with probabilities 0.1, 0.2, 0.1, 0.1, 0.5, respectively.

2  Assume that due to the vagaries of nature and commerce, cases of canned goods arrive at the supermarket with the following probabilities of defective cans per case of 48 cans:

| NO. OF DEFECTIVES | 0 | 2 | 5 | 8 | 12 | 20 | 24 | 30 | 36 | 45 |
|---|---|---|---|---|---|---|---|---|---|---|
| PROBABILITY OF THAT NO. OF DEFECTIVES | 0.30 | 0.20 | 0.20 | 0.05 | 0.05 | 0.02 | 0.02 | 0.08 | 0.04 | 0.04 |

Four cans are removed (without replacement) at random from the case. Exactly two are defective. Find the resulting conditional probability that there are exactly five damaged cans in the case.

3   With the assumptions of exercise 2 find the conditional probability that the next can drawn from the case (at random) is defective.

4   Derive a general solution for the following problem: Any one of $n$ events (hypotheses) $H_j$ may occur. $M$ is an event. $P(M \mid H_j)$ are known, $P(H_j)$ are known. $N$ is an event, $P(N \mid M \cap H_j)$ are known. Find $P(N \mid M)$.

5   Can Bayes' formula be extended to the case of infinitely many disjoint events $H_1, H_2, \ldots$? What makes the theorem work?

Exercises 6–10 are based on the following information.

A communications system transmits and receives sequences of zeros and ones. Its characteristics are as follows:

(5.5.2)
    (1)  Probability of receiving 0 given that 0 is sent = 0.99
    (2)     ,,      ,,      ,,  1  ,,    ,, 0 ,,  ,,  = 0.01
    (3)     ,,      ,,      ,,  0  ,,    ,, 1 ,,  ,,  = 0.02
    (4)     ,,      ,,      ,,  1  ,,    ,, 1 ,,  ,,  = 0.98

6   Suppose that the correctness or incorrectness of receptions of different signals in a sequence are independent. Compute (a) $P(000$ was transmitted $\mid 000$ was received), (b) $P(000$ was transmitted $\mid 001$ was received), (c) $P(000$ was transmitted $\mid 101$ was received).

7   Assuming that the answers to exercise 6 were any three definite numbers, where is there a kind of fallacy? Note that if a telegraph operator receives the message "XDDRESS: 314 FRONT ST." he is virtually sure that there is an error because there is no reason to send the sequence "XDDRESS"; therefore, why claim that 000 is a meaningful sequence? If 000 is not a meaningful sequence what would the answers to parts (a) and (b) of exercise 6 be?

8   Work parts (a) and (b) of exercise 6 under the assumption that all 3-digit sequences of zeros and ones are equally likely to be transmitted.

9   Work parts (a) and (b) of exercise 6 under the assumption that the sequences are selected according to the probabilities

| SEQUENCE | 000 | 100 | 010 | 001 | 011 | 101 | 110 | 111 |
|---|---|---|---|---|---|---|---|---|
| PROBABILITY | 0.05 | 0.15 | 0.15 | 0.15 | 0.15 | 0.15 | 0.15 | 0.05 |

**10** Suppose that the probabilities of the situations in (5.5.2) are changed as follows: (1) 0.90, (2) 0.10, (3) 0.20, (4) 0.80. The probability of a mistake is extremely high. To offset this, any message—a particular sequence of three digits—is repeated three times. Thus if message $xyz$ is at issue, the sequence $xyz\ xyz\ xyz$ is transmitted. Assume that the messages are chosen according to the probabilities in exercise 9. Find (a) $P(000\ 000\ 000$ is transmitted $|\ 010\ 000\ 000$ is received), (b) $P(010\ 010\ 010$ is transmitted $|\ 010\ 000\ 000$ is received).

**11** Justify each step in the proof of Theorem 5.5.1.

## SUPPLEMENTARY EXERCISE FOR CHAPTER 5

**1** Suppose that our stick is broken at random at a point $x$. The longer piece is then broken at random at a point $y$. Show that

$$p(x, y) = \begin{cases} \dfrac{1}{1 - x}, & 0 \le x \le \tfrac{1}{2}, \quad \tfrac{1}{2} \le y \le 1 \\[2mm] \dfrac{1}{x}, & \tfrac{1}{2} \le x \le 1, \quad 0 \le y \le x \\[2mm] 0, & \text{otherwise} \end{cases}$$

gives a mass density on $\{(x, y);\ 0 \le x,\ \ y \le 1\}$. Discuss why it gives plausible probabilities.

CHAPTER **6**

# *Random Variables*

## 6.1 Payoffs

In this chapter we define and begin the study of what is probably the most fundamental entity in probability and statistics—the random variable. Roughly speaking, a random variable is a variable which varies over certain values with certain probabilities. For many purposes, such a definition is fairly adequate, but we feel that a more precise treatment is desirable so that the reader may get a broader introduction to probability and statistics.

In order to develop some insight it is now convenient to think of a probability system as a game of chance; somewhat like a large, possibly very complicated slot machine which one may play by pulling the handle (once). All kinds of things may go on inside the slot machine but after the handle is pulled, certain (generally unpredictable) outcomes occur.

Thus far we have not suggested anything new. The slot-machine idea can serve as intuition for some of the material in the previous chapters. However, in addition to a slot machine with its (random) outcomes we may consider a "book" of payoff instructions which states how much should be paid to the player when a certain outcome occurs on the slot machine. Naturally, if this "book" is to have any significance it must designate a payoff for each of the possible outcomes on the slot machine. The "payoff book" is the intuitive version of the *random variable.** A precise definition for a random variable is made later.

* Many authors use the abbreviation r.v. for random variable. Abbreviations are extremely common in books on probability and statistics.

Since we are thinking of a probability system as a slot machine whose outcomes are (indexed by) the sample points $s \in S$, a sample space, and since to each outcome the payoff book specifies a payoff which we call $X(s)$, we may call the payoff book $X$. In fact, $X$ is a function from $S$ to the real numbers (we will use real-number payoffs). Such an intuitive scheme is illustrated in Figure 1.

### Example 6.1.1
A slot machine rolls a fair die. The payoff book states that if $n$ dots show the payoff is $n$ units of money.

Here we may take $S = \{1, 2, 3, 4, 5, 6\}$ with the usual system of events and equal likelihood (fair die) as the mathematical version of the slot machine. We may take $X(n) = n$ for $n \in S$ as the mathematical version of the payoff book.

### Example 6.1.2
The slot machine chooses a card (at random) from a well-shuffled standard deck. The payoff book states that the payoff is 0 if the card is neither a heart nor the queen of spades. If a heart is chosen the payoff is 1. If the queen of spades is chosen the payoff is $-13$.

Here we take $S = \{1, 2, 3, \ldots, 52\}$ with equal likelihood and the usual system of events as the mathematical version of the slot machine. We number the 13 heart cards $1, 2, 3, \ldots, 13$ and use 14 to index the spade queen. For the payoff book we consider $X(n) = 0$ if $n > 14$, $X(n) = 1$ if $1 \leq n \leq 13$, and $X(14) = -13$.

In the remaining examples we generally discard reference to slot machines and payoff books. We will subsequently make a number of definitions and move away from the intuitive realm.

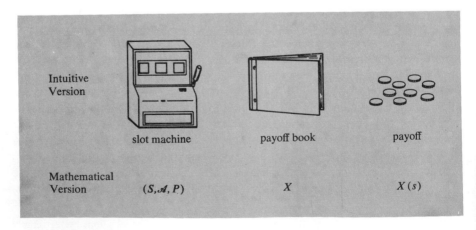

**Figure 1**

## Example 6.1.3

A standard die is rolled. If $n$ spots show the payoff is 0 if $n$ is even and 1 if $n$ is odd.

We could let $S = \{1, 2, \ldots, 6\}$ as in Example 6.1.1 and let

$$Y(n) = \tfrac{1}{2}[1 - (-1)^n] = \begin{cases} 1, & n \text{ odd} \\ 0, & n \text{ even} \end{cases}$$

(We use $Y$ to distinguish this payoff procedure from that in Example 6.1.1.)

Except for the payoffs in Example 6.1.2 being negative as well as positive, the above examples look fairly innocuous. Examples 6.1.1 and 6.1.3 suggest the fact that one may have various payoff procedures associated with the same underlying game. We will not feel obliged to restrict the payoffs to integers because payoff is only intuition. Random variables have much broader use than in gambling.

## Example 6.1.4

A fair coin is tossed and the payoff is 1 if the coin shows H and 0 if the coin shows T.

We can let $S = \{H, T\}$ and let $Y(H) = 1$ and $Y(T) = 0$.

## Example 6.1.5

A real number $x$ with $0 < x \leq 1$ is selected in such a way that $P(a < x \leq b) = b - a$ for $0 \leq a \leq b \leq 1$. The payoff is 1 if $\tfrac{1}{4} < x \leq \tfrac{3}{4}$ and 0 otherwise.

Let $S = \{x; 0 < x \leq 1\}$ and let

$$Z(x) = \begin{cases} 1, & \tfrac{1}{4} < x \leq \tfrac{3}{4} \\ 0, & 0 < x \leq \tfrac{1}{4} \\ 0, & \tfrac{3}{4} < x \leq 1 \end{cases}$$

It may be observed that in Examples 6.1.3–6.1.5 there is a certain similarity. Not only are the payoffs chosen from $\{0, 1\}$ in each case, but the probability of a payoff of 0 is $\tfrac{1}{2}$ and the probability of a payoff of 1 is also $\tfrac{1}{2}$ in each case. If one could really play such games for such payoffs there appears to be no monetary criteria to distinguish amongst them. It is true that the underlying games in Examples 6.1.3–6.1.5 are different and that tossing a coin may be less interesting than picking a real number, but the structure of the payoff books are such as to leave no apparent way to obtain a higher payoff by choosing one of Examples 6.1.3, 6.1.4, or 6.1.5 as compared to any other. These ideas are formalized in section 6.3.

The above examples show rather simple situations. Let us consider some more complicated examples.

## Example 6.1.6

From 6000 ears of corn, 2000 of which are top quality, 3500 of which are medium quality, and 500 of which are poor quality, a sample of 60 ears is selected at random (by the slot machine). If $u$, $v$, $w$ are the number of top, medium, and poor quality ears in the sample, the payoff is $7u + 5v + w$.

We would naturally let $S = \{$combinations of 60 objects chosen from 6000 objects of which 2000 are marked t, 3500 are marked m, and 500 are marked p$\}$. Probability is given by equal likelihood and all subsets are events. If $s$ is such a combination where $u(s)$ are marked t, $v(s)$ are marked m, and $w(s)$ are marked p, then $X(s) = 7u(s) + 5v(s) + w(s)$.

Example 6.1.6 is much less contrived than some of the previous examples. It is analogous to many industrial and commercial problems. For example, one has a carload of corn produced on a certain farm. How much should he pay for it? Looking at each ear of corn involves much time (and hence money) spent on determining the quality. Therefore, why not select a small, random sample and study its quality in hopes of inferring, to a reasonable extent, the quality of the carload. Obviously such a procedure is not restricted to corn. In fact, there are cases where one *cannot* study the quality of the entire lot. For example, if one tests the quality of bullets by firing them, one must use a sampling technique. One would not destroy his entire production to ascertain its quality. A number of the exercises at the end of this section are devoted to studying ideas associated with Example 6.1.6. If the reader has not yet guessed, $X(s)$ represents the price which a buyer might be willing to pay for the lot of 6000 ears of corn according to the quality of the *sample*.

We will have occasion to envision slot machines which select one or more electrons out of many electrons and even those which select points from inside an ellipse or which select curves from certain families of curves. Moreover, the payoff $X(s)$ need not be thought of simply as a money payoff, it could be an average income, a percentage of carbon, a value of total energy, or the area under a certain curve from 0 to $\pi$.

Similar to Example 6.1.6 but somewhat more abstract and on a somewhat reduced scale is Example 6.1.7.

## Example 6.1.7

A biased coin is tossed five times independently. The probability of heads on any particular toss is $p$ with $0 \le p \le 1$. The payoff is the number of heads which show on the five tosses. Give two mathematical versions.

Version 1: Let $S = \{(x_1, x_2, x_3, x_4, x_5); x_i \in \{H, T\}\}$. Let $u(s) =$ number of H's in $s = (x_1, x_2, x_3, x_4, x_5)$ and hence $P(s) = p^{u(s)}(1 - p)^{5 - u(s)}$ since each H among $(x_1, x_2, x_3, x_4, x_5)$ "receives probability $p$" and each T "receives probability $1 - p$" which are multiplied in view of independence. We have $X(s) = u(s)$.

Version 2: Let $S_1 = \{0, 1, 2, 3, 4, 5\}$ and let $X(n) = n$ for each $n \in S_1$. However, here we must have $P(0) = (1 - p)^5 = \binom{5}{0} p^0 (1 - p)^5$, $P(1) = \binom{5}{1} p^1 (1 - p)^4$,

$P(2) = \binom{5}{2} p^2 (1 - p)^3$, $\quad P(3) = \binom{5}{3} p^3 (1 - p)^2$, $\quad P(4) = \binom{5}{4} p^4 (1 - p)$, and

$P(5) = \binom{5}{5} p^5 (1 - p)^0$ if we are to maintain the convention that, e.g., $4 \in S_1$

signifies that four heads and hence one tail occurred in some order.

We see in the two versions of Example 6.1.7 that ambiguity remains with us just as it has in previous chapters. There just *are* many sample spaces that are usable for one problem. Which we choose depends on experience, convenience, and sometimes on taste.

Example 6.1.7 can be thought of as similar to another industrial problem. This is taken up in exercise 6.1.6.

We now give the formal definition of a random variable.

**Definition**   A *random variable* $X$ (on a sample space $S$, with its admissible system of events $\mathscr{A}$ and probability measure $P$) is a function from $S$ to $R$ with the property that for every real number $u$, $\{s \in S; X(s) \le u\}$ is an event in the admissible system of events, i.e., $\{s \in S; X(s) \le u\} \in \mathscr{A}$.

Random variables are generally denoted by letters from the latter portion of the alphabet such as $X$, $Y$, $Z$, $U$, $V$, $W$. A number of other functions which are related to random variables will be defined. It is wise to anticipate this situation and to distinguish the various functions.

The condition that for all real $u$, $\{s \in S; X(s) \le u\}$ is an event can easily be verified in the above examples in view of our conventions about events in countable and in Euclidean sample spaces. However, in the present book we do not concern ourselves with this condition. In a countable sample space such that *all* sets are events $\{s \in S; X(s) \le u\}$ is a set and therefore an event. In Euclidean sample spaces where the events are the Borel sets, any piecewise continuous function is a random variable. Exercise 6.1.13 gives an example of a situation in which a certain function on a certain $S$ is not a random variable but it will seem unnatural and perhaps silly. In more advanced treatments the difficulties of which functions are random variables are discussed more thoroughly. For most purposes the reader is advised not to concern himself with this issue. There are enough random variables to be found among the piecewise continuous functions on Euclidean spaces and among the functions on countable sample spaces to keep us busy.

Another matter which we will not dwell upon too much is the fact that if $X$ is a random variable, that is, $\{s \in S; X(s) \le u\}$ is an event for *all* real $u$, one can *deduce* that $\{s \in S; X(s) > v\}$, $\{s \in S; X(s) < v\}$, and $\{s \in S; X(s) \ge v\}$ are also events.

It should be noted that $\{s \in S; X(s) > u\}$ is analogous to the statement "the outcomes whose payoffs are greater than $u$." It is reasonable to expect that we will want to know the probabilities of events such as $\{s \in S; X(s) > 0\}$, $\{s \in S; X(s) \le -10\}$, or $\{s \in S; X(s) = 5\}$. For that reason it is necessary that these sets be events, that is, they should have probability attached to them under $P$.

Since expressions of the form $\{s \in S; X(s) \le u\}$ occur so frequently we suppress the symbols $s$ and $S$ and write $\{X \le u\}$, $\{X > u\}$, $\{X \ge u\}$ for $\{s \in S; X(s) \le u\}$, $\{s \in S; X(s) > u\}$, $\{s \in S; X(s) \ge u\}$, and so on. Similarly we write $P(X \le u)$, $P(X = u)$, for $P(\{s \in S; X(s) \le u\})$, $P(\{s \in S; X(s) = u\})$.

**EXERCISES 6.1** _____

1  A slot machine selects, at random, a whole number between 1 and 10 inclusive. The payoff is 0 if the number is not a power of a single prime number. The payoff is the exponent if the number is a power of a single prime. Express this in a mathematical version.

2  A slot machine selects a whole number between 1 and 10 inclusive with probability proportional to the square of the number. The payoff is equal to 3 minus the number selected. Express in a mathematical version.

3  A slot machine rolls a pair of fair dice—one red and one green. The payoff is the product of the number of dots showing. Describe this situation mathematically.

4  Suppose we have the same dice rolling as in exercise 3 but the payoff is the number showing on the red die minus the number showing on the green die. Describe this mathematically.

5  Describe the most general payoff book which would apply when the slot machine tosses a coin once.

6  A product such as a transistor is made by a complicated industrial process. It is convenient to assume that each transistor has probability $p$ of being defective and the production process is of such a nature that the defectiveness or acceptability of different transistors is independent. Five transistors are produced. The payoff is the number of acceptable transistors. Compare this problem to Example 6.1.7. Is there something special about the number five? Could $m$ transistors be produced? Set up the analog of version 2 of Example 6.1.7 for the case of $m$ transistors.

7  In a problem in which the various possible payoffs are $-1$, $0$, $\sqrt{2}$, $\sqrt{7}$, $9$, $10\pi$, the answer involves the sample space $S = \{1, 2, 3, 4\}$. Discuss.

8  A non-negative integer $n$ is chosen in such a way that $P(n)$ is proportional to $\lambda^n/n!$, where $\lambda > 0$ is a real number. The payoff corresponding to the choice of $n$ is exactly $n$. Set up a mathematical version.

9  Under the conditions of Example 6.1.6, is it possible that all 60 ears selected at random are of medium quality? poor quality? What are the probabilities

of those events? How much is the payoff if all of the ears in the random sample are of medium quality? poor quality?

**10**  Simulate a problem parallel to Example 6.1.6 as follows: Select four cards (at random) from a standard deck. Consider the spade suit as signifying top quality, the clubs from two to ten as signifying poor quality, and the remaining cards as signifying medium quality. From the four cards selected let $j$ be the number of spades, $l$ be the number of clubs between two and ten inclusive, and let $k$ be $4 - j - l$. Let the payoff be $6j + 4k + 2l$. Make a total of ten trials being careful to mix the cards thoroughly and to return the chosen four back to the deck. What is the maximum *possible* payoff? What is the minimum *possible* payoff? What are the maximum and minimum payoffs which you get in your ten trials?

**11**  Simulate a problem parallel to Example 6.1.6 as follows: Have a friend mark $a$ cards out of 100 cards with the letter t, $b$ cards with the letter p where $a > 0$, $b > 0$, and $a + b < 100$. The unmarked cards can be thought of as being marked m. Your friend should not inform you of the exact values of $a$ and $b$. Choose ten cards at random. If $k$ and $l$ are the number of the chosen ten marked with t and p, respectively, imagine a payoff of $5(10 - k - l) + 7k + l$. Choose the sample of 10 five times but always reinsert and mix the cards. Now ask your friend the values of $a$ and $b$. Compare the payoffs which are obtained from the five samples with the quantity $\frac{1}{10}[7a + b + 5(100 - a - b)]$.

**12**  In reference to the standard deck arranged as in exercise 10, it seems reasonable to take $\frac{4}{52}[6 \cdot 13 + 4 \cdot 30 + 2 \cdot 9] = 16 \cdot \frac{8}{13}$ as a fair price to pay for the lot if one extrapolates from the given payoff of $6j + 4k + 2l$. How many times in your ten draws of four cards was your payoff greater than 19—a price which you may consider high if a fair price is $16 \cdot \frac{8}{13}$? How many times was your payoff less than 14—a price below which the owner might not sell?

Now draw eight cards from the deck of exercise 10. Let the payoff be $3j' + 2k' + l'$ if $j'$ denotes the number of spades, $k'$ is $8 - j' - l'$, and $l'$ is the number of clubs from two to ten in the eight cards you have drawn. Do this (draw eight cards) a total of ten times being careful to replace and reshuffle after each draw. How many of the ten payoffs exceed 19? How many are less than 14?

**13**  Let $S = \{1, 2, 3\}$. Let the admissible events be only $\varnothing$ and $S$. Let $X(1) = 2$, $X(2) = 4$, $X(3) = 6$. Find $L = \{s \in S; X(s) \leq 5\}$. Is $L$ one of the admissible events? Is $X$ a function? Is $X$ a random variable? Have we adhered to our convention that all subsets of a countable sample space (which $S$ is) are admissible events? Was this convention chosen for convenience? Describe a function on $S$ which is a random variable. Describe another function which is real-valued but is not a random variable.

**14**  With reference to version 2 of Example 6.1.7, find $\{n \in S_1; X(n) \leq 3\} = \{X \leq 3\}$. Also find $\{X \leq 1\frac{1}{2}\}$, $\{X \leq 0\}$, and $\{X \leq -\pi\}$.

**15**  With reference to Example 6.1.6, find $\{s \in S; X(s) \leq 60\} = \{X \leq 60\}$. Find $P(X \leq 60)$. Find $\{X \leq 64\}$ and $\{X = 64\}$.

16    With reference to Example 6.1.5, find $\{s \in S; X(s) \le \frac{1}{2}\}$.

♦ 17    Show that if $\{X \le u\} \in \mathscr{A}$, then $\{X > u\} \in \mathscr{A}$. Also show that if $\{X \le u + 1/n\}$ $\in \mathscr{A}$ for $n = 1, 2, 3, \ldots$, then $\{X < u\} \in \mathscr{A}$.

18    Consider the random variables $u(s)$ and $v(s)$ of Example 6.1.6. Are the events $\{s \in S; u(s) \le 5\}$ and $\{s \in S; v(s) \le 3\}$ independent?

19    A pair of fair dice are rolled. Suppose one die is red and the other is green. Let $X(s)$ and $Y(s)$ be the scores on the red and green die, respectively. Are the events $\{X \le 3\}$ and $\{Y \le 2\}$ independent? Are the events $\{X \le u_1\}$ and $\{Y \le u_2\}$ independent for all $u_1, u_2$?

20    A point $s = (x, y)$ is chosen at random from the disk $S = \{(x, y); x^2 + y^2 \le 1\}$. Let $X(s) = x$ and $Y(s) = y$. Are the events $\{X \le 0\}$, $\{Y \le 0\}$ independent? Are the events $\{X \le u_1\}$, $\{Y \le u_2\}$ independent for all $u_1, u_2$?

# 6.2    Probability functions and distribution functions

It was noted above that quantities such as $P(X > u)$ or $P(X \le u)$ are of considerable importance in many probability problems. For example, in reference to Example 6.1.7 in which 250 seems to be a fair price to pay for the 6000 ears of corn, one would certainly want to know $P(X > 300)$ which is the probability that the sampling procedure would lead to a payment in excess of 300 for corn of the quality mentioned in the example. In other problems we might want to know $P(a \le X \le b)$ which is the probability that a payoff will fall between limits. In still other situations we might want to know the probability that a coin tossed 25 times will show exactly 18 heads and 7 tails. In the context of Exercise 6.1.6 this calls for $P(X = 18)$. Such quantities can be computed by the methods of Chapters 3 and 5. In this section we take up the theory associated with the quantities $P(X \le u)$, $P(X = u)$, and $P(a < X \le b)$.

**Definition**    Let $X$ be a random variable (with $S$, $\mathscr{A}$, and $P$) for which there are countably many distinct numbers $u_1, u_2, u_3, \ldots$ such that $\sum_{n=1}^{\infty} P(X = u_n) = 1$. The associated *probability function* from the reals to the reals is defined by

$$f(u) = P(X = u)$$

A random variable of this kind is called *discrete*.

If there are a number of different random variables being studied simultaneously we write $f_X$ for the probability function associated with $X$ if $X$ has such a function. If $Y$ has such a function we denote it by $f_Y$. We are thus using a subscript on a standard letter rather than using $g$, $h$, etc. for the probability functions for the other random variables.

Probability functions are called *frequency functions* or *probability point functions* by various writers.

It should be noted that if $u \notin \{u_1, u_2, u_3, \ldots\}$, $f(u) = 0$ when $f$ is given by the above definition. This follows because $\sum_{n=1}^{\infty} P(X = u_n) = 1$ and $\{X = u\} \cap \{X = u_n\} = \varnothing$ for $n = 1, 2, 3, \ldots$ under the present assumptions. Thus, if $P(X = u) > 0$ we would have $P(\{X = u\} \cup \{X = u_1\} \cup \{X = u_2\} \cup \cdots) > 1$.

In Example 6.1.1 we have $f(u) = \frac{1}{6}$ for $u = 1, 2, 3, 4, 5, 6$, since $P(X = u) = f(u) = \frac{1}{6}$ for a standard die.

In Example 6.1.2 we have $f(0) = P(X = 0) = \frac{38}{52}$, $f(1) = P(X = 1) = \frac{13}{52}$, and $f(-13) = P(X = -13) = \frac{1}{52}$.

In Examples 6.1.3–6.1.5 we have $f_X(0) = f_Y(0) = f_Z(0) = \frac{1}{2}$ and $f_X(1) = f_Y(1) = f_Z(1) = \frac{1}{2}$.

### Example 6.2.1

A random variable $X$ has its probability function given by $f(-1) = \frac{1}{2}$, $f(1) = \frac{1}{4}$, $f(2) = \frac{1}{4}$. Find $P(X \leq -3)$, $P(X \leq 1)$, and $P(0 < X \leq 1\frac{1}{2})$.

We first note that $f(-1) + f(1) + f(2) = 1$. The events $\{X = -1\}, \{X = 1\}$, and $\{X = 2\}$ are disjoint since if $s_0 \in \{X = -1\} \cap \{X = 2\}$ we would have $X(s_0) = -1$ and $X(s_0) = 2$. But a function cannot give two different values at the same sample point. Although this is a special case it so completely suggests the general case that we will not make a general argument. Therefore, if $M = \{X = -1\} \cup \{X = 1\} \cup \{X = 2\}$, $P(M) = 1$ and hence $P(M') = 0$. Technically, it is possible that $S - M \neq \varnothing$ but since $P(M') = 0$, there is nothing interesting going on in the event $M'$ from a probabilistic point of view. Now $P(X \leq -3) = 0$ since $X(s) \leq -3$ only if $s \notin M$. Similarly, $P(X \leq 1) = \frac{1}{2} + \frac{1}{4} = \frac{3}{4}$ and $P(0 < X \leq 1\frac{1}{2}) = \frac{1}{4}$.

### Example 6.2.2

A random variable $X$ has its probability function given by $f(n) = 2^{-n}$ for $n = 1, 2, 3, \ldots$. Find $P(X \leq -3), P(X \leq 3)$, and $P(X \text{"pays"}$ an even number). Again we have

$$\sum_{n=1}^{\infty} f(n) = \sum_{n=1}^{\infty} 2^{-n} = \frac{1}{2} + \frac{1}{4} + \frac{1}{8} + \frac{1}{16} + \cdots = 1$$

As in Example 6.2.1, $P(S - \bigcup_{n=1}^{\infty} \{X = n\}) = 0$ so $P(X \leq -3) = 0$, $P(X \leq 3)$ $= \frac{1}{2} + \frac{1}{4} + \frac{1}{8} = \frac{7}{8}$, and $\{X \text{ pays an even number}\} = \bigcup_{k=1}^{\infty} \{X = 2k\}$. Therefore

$$P(X \text{ pays an even number}) = \sum_{k=1}^{\infty} \frac{1}{2^{2k}} = \sum_{k=1}^{\infty} \frac{1}{4^k} = \frac{\frac{1}{4}}{1 - \frac{1}{4}} = \frac{1}{3}$$

since $\sum_{k=1}^{\infty} \frac{1}{4^k}$ is a geometric series. (See also Example 3.4.1.)

The foregoing examples illustrate the following theorem.

**Theorem 6.2.1**    Let $X$ be a random variable which has a probability function $f$. Let $\{u; f(u) \neq 0\} = \{u_1, u_2, u_3, \ldots\}$. Then

$$P(X \leq v) = \sum_{u_j \leq v} f(u_j)$$

**PROOF**    We assume that the numbers $u_1, u_2, \ldots$ are so chosen that $u_j \neq u_k$ for $j \neq k$. Therefore if $\{X = u_j\} \cap \{X = u_k\} \neq \varnothing$ for $j \neq k$ there would be some $s_0 \in \{X = u_j\} \cap \{X = u_k\}$. This says that $u_j = X(s_0) = u_k$. But a function has only one value at $s_0$. Thus $\{X = u_j\} \cap \{X = u_k\} = \varnothing$.

Let $M = \bigcup_{j=1}^{\infty} \{X = u_j\}$. Now

$$P(M) = P(\bigcup_{j=1}^{\infty} \{X = u_j\}) = \sum_{j=1}^{\infty} P\{X = u_j\} = \sum_{j=1}^{\infty} f(u_j) = 1$$

by disjointedness and since $X$ has a probability function. Therefore, $P(M') = 0$. Now $\{X \leq v\} = \{s; X(s) \leq v \text{ and } s \in M'\} \cup \{s; X(s) \leq v \text{ and } s \in M\}$. But $\{s; X(s) \leq v \text{ and } s \in M\} = \bigcup_{u_j \leq v} \{X = u_j\}$ and $P(\{s; X(s) \leq v \text{ and } s \in M'\}) \leq P(M') = 0$. Thus,

$$P(X \leq v) = 0 + \sum_{u_j \leq v} P(X = u_j) = \sum_{u_j \leq v} f(X = u_j)$$

The importance of the probability function in describing a random variable is evident. We will shortly see that many important random variables have no probability function in the sense of the above definition. Both for a study of these random variables and purposes of data description, another function is valuable.

**Definition**    Let $X$ be a random variable (with reference to a probability system). The function $F$ from the real numbers to the real numbers defined by $F(u) = P(X \leq u)$ is called the (*cumulative*) *distribution function* of $X$.

In case it is necessary to distinguish the distribution functions for different random variables we will use the notation $F_X$, $F_Y$, etc. The modifier "cumulative" is still very frequently used but we will normally refer to $F$ as the distribution function. If $X$ has a probability function, Theorem 6.2.1 describes how to find the distribution function $F$ in terms of the probability function $f$. Some books define the distribution function by $F(u) = P(X < u)$. This gives a different but equally usable function. Naturally, we must not mix the two definitions.

The distribution function may seem unnatural and contrived when compared to a probability function. But frequently one cannot give a natural answer to a complicated problem—after all, the Bohr model of the atom is also contrived. Similarly, many pieces of evidence point to the distribution function as a good concept to formulate.

## *Example 6.2.3*

A group of 1000 auto fenders has the following characteristics: 450 fenders have no noticeable imperfections, 240 fenders have one noticeable imperfection, 190 have two, 70 have three, 30 have four, and 20 have five such imperfections. Let $s$ be a fender chosen at random and let $X(s)$ = number of imperfections on fender $s$. Find the probability function of $X$ and graph its distribution function.

If $f$ is the probability function it is easy to see that $f(0) = 450/1000$, $f(1) = 240/1000$, $f(2) = 190/1000$, $f(3) = 70/1000$, $f(4) = 30/1000$, $f(5) = 20/1000$. Since the distribution function $F$ is defined by $F(u) = P(X \le u)$, we certainly have $P(X \le -1) = P(X \le -\frac{1}{2}) = P(X \le -\frac{1}{4}) = 0$. For that matter, $P(X \le u) = 0$ if $u$ is any negative real number. However, by Theorem 6.2.1, $P(X \le 0) = P(X = 0) = 0.45$. Similarly, $P(X \le \frac{1}{3}) = P(X \le \frac{1}{2}) = P(X \le \frac{3}{4}) = P(X = 0) = 0.45$. It follows that $P(X \le u) = 0.45$ whenever $0 \le u < 1$. However, $P(X \le 1) = f(0) + f(1) = 0.45 + 0.24 = 0.69$ in view of Theorem 6.2.1. We can further show that $P(X \le u) = 0.69$ whenever $1 \le u < 2$ and we leave the remaining computations to the reader.

For the graph of $F$ see Figure 2. Note that $F(u) = 1$ if $u > 5$. Note also that $F$ is a step function.

## *Example 6.2.4*

Let $X$ be a random variable with $f(-3) = \frac{1}{4}$, $f(-1) = \frac{1}{4}$, $f(2) = \frac{1}{3}$, and $f(7) = \frac{1}{6}$. Describe and graph the distribution function $F$.

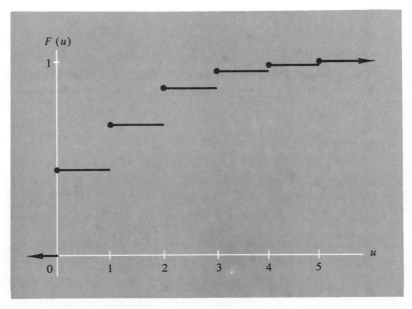

**Figure 2**

It can be verified that

$$F(u) = \begin{cases} 0, & u < -3 \\ \frac{1}{4}, & -3 \leq u < -1 \\ \frac{1}{2}, & -1 \leq u < 2 \\ \frac{5}{6}, & 2 \leq u < 7 \\ 1, & 7 \leq u \end{cases}$$

by using Theorem 6.2.1 on $u$ in the various ranges. The graph of $F$ is indicated in Figure 3. Notice that this is again a step function.

It would be profitable for the student to graph the distribution functions of a number of the random variables defined in Examples 6.1.1–6.1.7 and 6.2.1. For the moment, we leave distribution functions.

### Example 6.2.5

Let $\{u_1, u_2, u_3, \ldots\} = U$ be a countable set of real numbers and let $g: R \to R$ be such that $g(u) \geq 0$ for all $u \in R$, $g(u) \neq 0$ only if $u \in U$ and $\sum\limits_{u_j \in U} g(u_j) = 1$.

Find a probability system $\mathscr{S}$ and a random variable $X$ such that the probability function for $X$ is the given function $g$ if this is possible.

The given function $g$ has the appearances of a probability function. We must arrange for

$$g(u_j) = P(X = u_j) = P(\{s \in S; X(s) = u_j\})$$

for our eventual sample space $S$. Now we could conceivably have $\{s \in S; X(s) = u_j\}$ as a very complicated event in $S$ but why make things hard? We could just as well have a simple event, that is, an event containing only one sample point. What sample point? We are at liberty to choose such a sample point. Since the real number $u_j$ appears in the notation $\{s \in S; X(s) = u_j\}$ suppose we let $\{s \in S; X(s) = u_j\} = \{u_j\}$, i.e., $X(s) = u_j$ if and only if $s = u_j$.

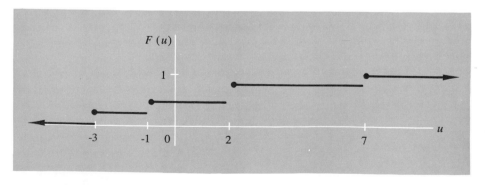

**Figure 3**

In other words, $X(u_j) = u_j$. Therefore, if $S = U \subset R$ and if we let $X(u_j) = u_j$ we have $X: U \to R$. Since $U$ is countable, we take as the admissible system of events the system of all subsets of $U$. As for the probability measure $P$, we need $P(\{s \in S; X(s) = u_j\}) = g(u_j)$ or $P(\{u_j\}) = g(u_j)$ or $P(u_j) = g(u_j)$ by our notational convention. Since $g(u_j) \geq 0$ and since $\sum_{u_j \in U} g(u_j) = 1$, Theorem 3.4.1 guarantees that $P$ is a probability measure. Since $X: U \to R$, $X$ is a random variable because any real-valued function is a random variable if the admissible system of events is the system of *all* subsets of the sample space. As for whether we have $g(u_j) = P(X = u_j)$, we need only check back to see this is true.

Example 6.2.5 shows us how to "construct" a random variable and probability system whose probability function is a given function which has the appearance of the probability function of a random variable. In view of Examples 6.1.3–6.1.5, there may be more than one way of constructing such a random variable. The construction in Example 6.2.5 is called the *standard representation* for a given probability function. One can handle all probabilistic questions about a given random variable by looking at the standard representation for the probability function of the given random variable. However, changing to the standard representation is not always an advantage.

**EXERCISES 6.2** ───────────────────────────────────

**1**   From a collection of 20 perfect shirts and ten imperfect but usable shirts a group of four shirts is chosen at random. The payoff is the number of perfect shirts in the group of four plus one-half the number of imperfect shirts in that group. Find the probability function for the resulting random variable.

**2**   Plot the distribution functions for the random variables in Examples 6.1.1–6.1.5, 6.2.1, 6.2.2.

**3**   Our standard stick is broken (at random) and the ratio of the length of the longer piece to the shorter piece is the payoff. What is the probability function for the resulting random variable? What is the distribution function?

**4**   A random variable $X$ is known to have payoffs which are integers. Is it possible that the probability function $f$ satisfies $f(3) = \frac{1}{2}$, $f(-2) = \frac{1}{4}$, $f(-5) = \frac{1}{3}$? Justify your conclusion.

**5**   Describe the standard random variables for the probability functions of Exercises 6.1.3 and 6.1.4. Describe the probability system for these standard random variables.

**6**   A random variable $X$ has its probability function given by $f(u) = 0$ if $u$ is not a positive integer and $f(u) = 1/[u(u + 1)]$ if $u$ is a positive integer. If $F$ is the distribution function, find $F(u)$ for $u \in \{1, 2, 3, 4, \ldots\}$. Also find $F(\frac{3}{2})$, $F(\frac{5}{2})$, $F(\frac{7}{2})$, ....

♦ **7** With reference to the distribution functions of Examples 6.2.3 and 6.2.4, find $\lim\limits_{\substack{u \to u_0 \\ u > u_0}} F(u)$, $\lim\limits_{\substack{u \to u_0 \\ u < u_0}} F(u)$, for $u_0 \in \{-1, 0, 1, 2\}$.

♦ **8** Prove that $P(a < X \le b) = F_X(b) - F_X(a)$.

**9** A random variable has distribution function

$$F(u) = \begin{cases} 0, & u < -1 \\ \frac{1}{3}, & -1 \le u < 0 \\ \frac{1}{2}, & 0 \le u < 4 \\ \frac{3}{4}, & 4 \le u < 7.5 \\ 1, & 7.5 \le u \end{cases}$$

Find the probability function and the standard representation of the random variable.

**10** Which of the following functions look like random variables on $S = R$? Which look like distribution functions? Which look like probability functions? Which do not look like any of these?

(a) $g(x) = \sin x$

(b) $g(x) = \tan x$

(c) $g(x) = \begin{cases} 1, & x = 0 \\ 0, & \text{otherwise} \end{cases}$

(d) $g(x) = \frac{1}{2} + \frac{1}{\pi} \arctan x$

(e) $g(x) = \sqrt{\dfrac{x}{1 + x}}$

(f) $g(x) = \begin{cases} 0, & x < 0 \\ \sqrt{\dfrac{x}{1 + x}} & x \ge 0 \end{cases}$

**11** Find the probability function of the random variable described in exercise 6.1.10. Assume equal likelihood.

**12** Can the graph of a probability function be symmetric with respect to the vertical axis? Can a distribution function have such a graph?

♦ **13** Suppose that a random variable $X$ has a distribution function $F$ which is a step function. Show that $X$ has a probability function $f$ and describe it in terms of $F$. Recall that a step function is *constant* on disjoint intervals.

# 6.3 The continuous case

In the preceding section, we studied random variables which take only countably many values with nonzero probability. It would appear that Euclidean sample spaces would allow for more complicated random variables.

### Example 6.3.1

Let $S = \{x; x$ is a real number and $0 < x \le 1\}$. Let $P(a < x \le b) = b - a$ $(0 < a < b \le 1)$ and assume the usual system of events. Let $X(x) = \cos \pi x$ for $x \in S$. Find $P(X=3)$, $P(X=\frac{1}{2})$, $P(X=-\sqrt{2}/2)$, $P(X \le 2)$, and $P(X \le -\frac{1}{2})$.

For $0 < x \le 1$, we have $1 > \cos \pi x \ge -1$. Furthermore, for those $x$ we have $\cos \pi x = \frac{1}{2}$ if and only if $x = \frac{1}{3}$ and $\cos \pi x = -\sqrt{2}/2$ if and only if $x = \frac{3}{4}$. Therefore $\{X = 3\} = \varnothing$, $\{X \le 2\} = S$, $\{X = \frac{1}{2}\} = \{\frac{1}{3}\}$, $\{X = -\sqrt{2}/2\}$ $= \{\frac{3}{4}\}$. Finally, $\{X \le -\frac{1}{2}\} = \{x \in S; \frac{2}{3} \le x \le 1\}$. The nature of this probability measure was extensively discussed in section 3.5. Example 3.5.1 guarantees that $P(X = \frac{1}{2}) = P(X = -\sqrt{2}/2) = 0 = P(X = 3)$. Naturally $P(X \le 2) = P(S) = 1$. Finally, $P(\{x; \frac{2}{3} \le x \le 1\}) = P(\{x; \frac{2}{3} < x \le 1\}) = 1 - \frac{2}{3} = \frac{1}{3}$.

Example 6.3.1 shows a typical random variable on a Euclidean sample space. A probability function is useless—in fact it does not exist by our definition since for no real numbers $u$ do we have $P(X = u) > 0$. It is easy to see that $X$ in Example 6.3.1 has a distribution function. Namely,

$$F_X(u) = P(X \le u) = \begin{cases} 0, & u < -1 \\ 1 - (1/\pi) \arccos u, & -1 \le u < 1 \\ 1, & u \ge 1 \end{cases}$$

whose graph is shown in Figure 4.

### Example 6.3.2

A needlelike spinner (see Figure 5) spins about a pivot in the center of a disk of radius 1. The pivot has a small amount of friction and various torques are applied to the needle so that it is assumed that the probability that the spinner comes to rest in any sector is proportional to the size of the central angle of the sector. A line is drawn tangent to the disk and is coordinatized in the standard units from the point of tangency. The random variable $Y$ is determined as follows: $Y(\theta) = 0$ if the spinner is parallel to the tangent line, $Y(\theta) =$ the real

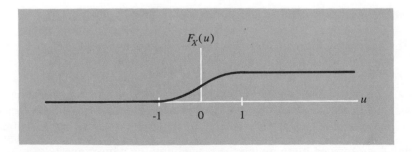

**Figure 4**

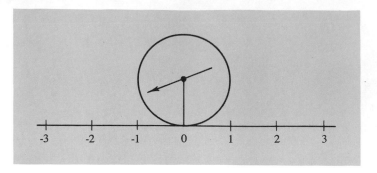

**Figure 5**

number coordinatizing the point on the tangent line at which the extension of the spinner hits the tangent line. What is the distribution function of $Y$?

A few pictures should convince the reader that there is no loss of generality in assuming that the point of the spinner points toward the line rather than away from the line. Moreover if $\theta$ is the angle between the radius to the point of tangency and the spinner we see immediately that

$$\tan \theta = \frac{\text{coordinate on the line}}{\text{radius of circle}} = \frac{Y(\theta)}{1} = Y(\theta)$$

As for $P(Y \le u)$ this is just $P(\tan \theta \le u) = P(\theta \le \arctan u)$ if $u > 0$. The size of the sector for which $\theta \le \arctan u$ is $\arctan u + \frac{1}{2}\pi$. But since probability is proportional to the sector size we want the proper constant of proportionality. So far we have $F_Y(u) = k(\frac{1}{2}\pi + \arctan u)$ for $u > 0$. But as $u \to \infty$ we must have $F(u) \to 1$. Since $\arctan u \to \frac{1}{2}\pi$ as $u \to \infty$ we find that $k = 1/\pi$ and so

$$F_Y(u) = \frac{1}{2} + \frac{1}{\pi} \arctan u$$

for $u > 0$.

A separate argument can be used to show that $F(u) = \frac{1}{2} + \frac{1}{\pi} \arctan u$ for all real $u$. The graph of $F_Y$ is given in Figure 6.

In Examples 6.3.1 and 6.3.2 it was claimed that certain random variables have certain distribution functions. It should be noted that both random variables satisfy the condition $P(X = u) = 0$ for any real $u$. Thus neither random variable has a probability function. However, it is possible to introduce an analog to a probability function. According to Theorem 6.2.1, if $Z$ has a probability function $f_Z$ we have $F_Z(u) = \sum_{u_j \le u} f_Z(u_j)$, where $\sum_{u_j} f_Z(u_j) = 1$; that is, the distribution function is obtained by summing the probability function properly. The reader should be aware that integrals are analogous to sums. Is it possible that we can

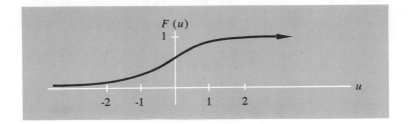

**Figure 6**

find something to integrate which will give us the distribution functions of Examples 6.3.1 and 6.3.2? Let us examine Example 6.3.2, first.

We can use $\int_{-\infty}^{u} g(x)\, dx$ as analogous to $\sum_{u_j \leq u} f_Z(u_j)$. We would seek $g$ so that

**(6.3.1)**   $\dfrac{1}{2} + \dfrac{1}{\pi}\arctan u = \int_{-\infty}^{u} g(x)\, dx$

If $g$ is a continuous function

$$\frac{d}{du}\int_{a}^{u} g(x)\, dx = g(u)$$

according to a standard result of calculus. It can be shown that the same result holds for $\int_{-\infty}^{u} g(x)\, dx$. Therefore if we differentiate both sides of (6.3.1) with respect to $u$, the right-hand side yields $g(u)$ and the left-hand side yields

$$\frac{d}{du}\left(\frac{1}{2} + \frac{1}{\pi}\arctan u\right) = \frac{1}{\pi}(1 + u^2)^{-1}$$

So $g(u) = 1/\pi(1 + u^2)$. But what if $g(u)$ is not continuous? We shall soon see what happens in this case.

Let us apply the same reasoning to Example 6.3.1. Suppose we let

$$\int_{-\infty}^{u} g(x)\, dx = \begin{cases} 0, & u < -1 \\ 1 - (1/\pi)\arccos u, & -1 \leq u < 1 \\ 1, & u \geq 1 \end{cases}$$

Providing $u \neq \pm 1$ the right-hand side can be differentiated and we get [if $g(x)$ is continuous at $u$]

$$g(u) = \begin{cases} 0, & u < -1 \\ \dfrac{1}{\pi}\dfrac{1}{\sqrt{1 - u^2}}, & -1 < u < 1 \\ 0, & u > 1 \end{cases}$$

What are $g(1)$ and $g(-1)$? We shall answer that question, shortly.

### *Example 6.3.3*

Suppose

$$F_Z(u) = \begin{cases} 0, & u < 0 \\ 1 - e^{-\lambda u}, & u \geq 0 \end{cases}$$

Find $g(x)$ so that

$$\int_{-\infty}^{u} g(x)\,dx = F_Z(u)$$

Let us try to differentiate both sides (with respect to $u$). If $u \neq 0$,

$$F_Z(u) = \begin{cases} 0, & u < 0 \\ \lambda e^{-\lambda u}, & u > 0 \end{cases}$$

If $g$ is continuous at $u$, $\dfrac{d}{du}\displaystyle\int_{-\infty}^{u} g(x)\,dx = g(u)$, so it is reasonable to want $g$ to satisfy

$$g(u) = \begin{cases} 0, & u < 0 \\ \lambda e^{-\lambda u}, & u > 0 \end{cases}$$

Then what is $g(0)$? The exact value of $g(0)$ is actually irrelevant if we are only interested in a function $g$ such that $\displaystyle\int_{-\infty}^{u} g(x)\,dx = F_Z(u)$. The properties of the integral guarantee that if $g_1$ and $g_2$ are integrable on $a \leq x \leq b$ and $g_1(x) \neq g_2(x)$ for only finitely many values of $x$, then

$$\int_a^b g_1(x)\,dx = \int_a^b g_2(x)\,dx$$

It is sometimes said that if one changes an integrable function at a finite number of places, the resulting function has the same integral. One can even change a function on more complicated sets and not change its integral. Thus there will always be many different functions $g$ satisfying $\displaystyle\int_{-\infty}^{u} g(x)\,dx = F_Z(u)$ if there is any such function.

If $F_Z(u) = \displaystyle\int_{-\infty}^{u} g(x)\,dx$ for all real $u$ then

$$F_Z(b) - F_Z(a) = \int_{-\infty}^{b} g(x)\,dx - \int_{-\infty}^{a} g(x)\,dx = \int_a^b g(x)\,dx$$

But, by Exercise 6.2.8, $P(a < Z \leq b) = F_Z(b) - F_Z(a)$. Thus $\displaystyle\int_a^b g(x)\,dx = P(a < Z \leq b)$ for any numbers $a$ and $b$.

In the above example, we were concerned about $g(1)$ and $g(-1)$. Now we see that the function can be defined arbitrarily at these points. We could choose $g(1) = g(-1) = 0$ so as to obtain one-sided continuity. However, it really does not matter.

A function $f(x)$ defined on the reals and taking non-negative values is called a *density function* for a random variable $X$ if, for $F$, the distribution function of $X$,

**(6.3.2)**     $F(u) = \int_{-\infty}^{u} f(x)\, dx,$     for all real $u$

It is tacitly assumed that the operations of integration are legitimate. We sometimes say that $f$ is a density function for $F$.

As above, there are always infinitely many different density functions if there is one density function. However, there is no more than one continuous density function. Generally speaking, we choose density functions with as few discontinuities as possible. A density function for a random variable is intimately related to the "probability mass density" of Chapter 3. We will see more of this later.

A random variable has a density function $f$ if (6.3.2) holds. Suppose that $F(u)$ is known for all $u$. How can we find $f$? As noted above, if $f$ is continuous at $u_0$ then $F'(u_0) = f(u_0)$. Thus if $F'(u)$ is a continuous function we can use $f(u) = F'(u)$. However, what if $F'(u)$ is not defined for some $u$ or if $F'(u)$ is not continuous even if it is defined for all $u$? We point out that if $g$ is a piecewise continuous function such that $F(u) = \int_{-\infty}^{u} g(x)\, dx$ for all $u$, then for all $u_0$ at which $g$ is continuous $F'(u_0) = g(u_0)$. We have, therefore, a rule of procedure which is always worth trying. Namely, compute $F'(u)$ wherever it exists. Call $F'(u) = g(u)$. Fill in the other $u$ by defining $g$ in an arbitrary way there. Then see if $F(u) = \int_{-\infty}^{u} g(x)\, dx$ for all $u$.

### Example 6.3.4
Find density functions for the following distribution functions or show that density functions do not exist.

(a)   $F(u) = \begin{cases} 0, & u < 0 \\ 1 - 1/n, & n - 1 \le u < n, \quad n = 1, 2, 3, \ldots \end{cases}$

(b)   $F(u) = \begin{cases} 0, & u < -1 \\ \frac{1}{2} - \frac{1}{2}u^2, & -1 \le u < 0 \\ \frac{1}{2} + \frac{1}{2}u^2, & 0 \le u < 1 \\ 1, & u \ge 1 \end{cases}$

(c)   $F(u) = \begin{cases} \frac{1}{2}e^u, & u < 0 \\ 1, & u \ge 0 \end{cases}$

We use our rule of procedure. In (a) $F$ is discontinuous at $u = 1, 2, 3, \ldots$. In the intervals between these discontinuities $F$ is constant so $g(u) = F'(u) = 0$

if $u \notin \{1, 2, 3, \ldots\}$. Any definition of $g$ at this discrete set of points will have no effect, i.e., $\int_{-\infty}^{u} g(x) \, dx = 0 \neq F(u)$ if $u > 1$. Hence, $F$ has no density function in case (a).

In case (b), $F$ is continuous at all $u$ since $\frac{1}{2} - \frac{1}{2}u^2 = 0$ if $u = -1$, $\frac{1}{2} - \frac{1}{2}u^2 = \frac{1}{2} + \frac{1}{2}u^2$ if $u = 0$, and $\frac{1}{2} + \frac{1}{2}u^2 = 1$ if $u = 1$. Thus $F'(u) = g(u) = 0$ if $u < -1$, $g(u) = -u$ if $-1 < u < 0$, $g(u) = u$ if $0 < u < 1$, $g(u) = 0$ if $u > 1$. Now

$$\int_{-1}^{u} (-x) \, dx = -\tfrac{1}{2}x^2 \Big|_{-1}^{u} = \tfrac{1}{2} - \tfrac{1}{2}u^2 = F(u) \qquad \text{if } -1 \leq u < 0$$

Similarly, if $0 \leq u < 1$,

$$\int_{-\infty}^{u} g(x) \, dx = \int_{-1}^{0} (-x) \, dx + \int_{0}^{u} x \, dx = -\tfrac{1}{2}x^2 \Big|_{-1}^{0} + \tfrac{1}{2}x^2 \Big|_{0}^{u}$$

$$= \tfrac{1}{2} - 0 + \tfrac{1}{2}u^2 = F(u)$$

We leave the remainder to the reader.

As for (c) if $u < 0$,

$$\int_{-\infty}^{u} \tfrac{1}{2}e^x \, dx = \tfrac{1}{2}e^x \Big|_{-\infty}^{u} = \tfrac{1}{2}e^u - \lim_{b \to -\infty} \tfrac{1}{2}e^b = \tfrac{1}{2}e^u - 0 = F(u)$$

However, $F'(u) = 0$ if $u > 0$ so

$$\int_{-\infty}^{1} g(x) \, dx = \int_{-\infty}^{0} g(x) \, dx + \int_{0}^{1} s(x) \, dx = \tfrac{1}{2}e^0 + 0 = \tfrac{1}{2} < 1 = F(1)$$

Thus case (c) has no density function.

The results of Example 6.3.4 suggest that distribution functions with jumps, i.e.,

$$\lim_{\substack{u \to u_0 \\ u < u_0}} F(u) < \lim_{\substack{u \to u_0 \\ u > u_0}} F(u)$$

do not have density functions. This is true. The proof will follow from later material. Thus it can be shown that only continuous distribution functions have density functions. Unfortunately, some continuous distribution functions do not have density functions, but these will not concern us. A more precise term for distribution functions with density functions is *absolutely continuous*. However, we shorten that to *continuous*; hence any distribution function or random variable having a density function will be called *continuous*. A distribution function or random variable having a probability function is called *discrete* as before.

### Example 6.3.5

Let $X$ be a continuous random variable on $S$, $\mathscr{A}$, $P$. Suppose $h: R \to R$ is a monotonically increasing continuous function whose derivative is continuous on $R$. Suppose $Y$ is a random variable with $h(Y(s)) = X(s)$ for all $s \in S$. Discuss the distribution function of $Y$ in terms of that of $X$.

Since $h$ is monotonically increasing $\{s \in S; Y(s) \le u\} = \{s \in S; X(s) \le h(u)\}$. Thus $F_Y(u) = F_X(h(u))$. Let $f_X$ be a density function of $X$. One exists since $X$ is a continuous random variable. Thus $F_Y(u) = \int_{-\infty}^{h(u)} f_X(x)\, dx$. If $f_X$ were continuous we would have $\int_{-\infty}^{h(u)} f_X(x)\, dx = \int_{-\infty}^{u} h'(y) f_X(h(y))\, dy$ by the substitution $x = h(y)$. This is true when $f_X$ is piecewise continuous and $h$ and $h'$ are continuous. Thus $F_Y(u) = \int_{-\infty}^{u} h'(y) f_X(h(y))\, dx$. Thus $Y$ has a density function $f_Y(y) = f_X(h(y)) h'(y)$ or $f_Y(x) = f_X(h(x)) h'(x)$.

The reader should consult exercises 6.3.2–6.3.5 for further information.

### Example 6.3.6

Consider the function

$$G(u) = \begin{cases} 0, & u < 1 \\ 1 - \dfrac{1}{u}, & u \ge 1 \end{cases}$$

Describe a sample space $S$, a system of events $\mathscr{A}$, a probability measure $P$, and a function $X: S \to R$ such that $F_X(u) = G(u)$ or state that this is impossible.

This problem is related to Example 6.2.5. Note that $\lim\limits_{u \to -\infty} G(u) = 0$, $\lim\limits_{u \to +\infty} G(u) = 1$, and $G(u)$ is monotonically increasing. Note also that if $u \ne 1$, $G'(u) = 0$ if $u < 1$ and $G'(u) = 1/u^2$ if $u > 1$. Furthermore, if $u > 1$,

$$\int_{1}^{u} \frac{1}{x^2}\, dx = -\frac{1}{x}\Big|_{1}^{u} = 1 - \frac{1}{u}$$

Thus in the region of importance $G$ is the integral of some function. We wish eventually that $P(\{x \in S; X(x) \le u\}) = G(u)$. But we have not chosen $S$ nor $X$. Let us try to accomplish this with $S = R$, the real numbers. The problem would be easy if $X(x) = x$. But then $\{x \in S; X(x) \le u\} = \{x \in R; x \le u\}$; or, in other words, the set of real numbers to the left of or at $u$. How much probability do we wish this set to have? Exactly $G(u)$. But a brief reexamination shows that if

$$f(x) = \begin{cases} 0, & x \le 1 \\ \dfrac{1}{x^2}, & x > 1 \end{cases}$$

then $G(u) = \int_{-\infty}^{u} f(x)\, dx$. Thus $P(\{x \in R; x \le u\})$ is obtained by integrating the mass density $f$.

We summarize as follows: Let $S = R$ and $\mathscr{A} = $ Borel sets of $R$. Let $P$ be determined by the probability mass density

$$p(x) = f(x) = \begin{cases} 0, & x \le 1 \\ \dfrac{1}{x^2}, & x > 1 \end{cases}$$

Let $X(x) = x$. Then $S, \mathscr{A}, P$ is a probability system, $X$ is a random variable, and $G(u) = F_X(u)$. We have not proved this assertion, however, everything is set up so that it can be proved. Since the present example is special, we do not prove this assertion. It is essentially contained in the next theorem.

**Theorem 6.3.1**     Let $g: R \to R$ be a function which can be integrated such that $g(x) \ge 0$ and $\displaystyle\int_{-\infty}^{\infty} g(x)\,dx = 1$. Then $R, \mathscr{B}, P$ is a probability system if $P$ is given by the mass density $g(x)$. Moreover, if $X(x) = x$ then

$$F_X(u) = \int_{-\infty}^{u} g(x)\,dx$$

PROOF     That $R, \mathscr{B}, P$ is a probability system follows from our general assumption as indicated in Chapter 3. Now $\{x \in R;\ X(x) \le u\} = \{x;\ x \le u\} \in \mathscr{B}$ so $X$ is a random variable. By definition of $P$, $F_X(u) = P(X \le u) = P(\{x \in R;\ x \le u\})$
$= \displaystyle\int_{-\infty}^{u} g(x)\,dx$ by definition. This is all that is required.

If Theorem 6.3.1 seems too easy, keep in mind that the difficulties are hidden in assumptions such as $P$ is countably additive on $\mathscr{B}$. This theorem is easy because a good deal of work has been done beforehand. Furthermore, the most difficult work is beyond the level of this book and has only been cited.

*Remark:* If a function, such as $G$ in Example 6.3.6 were given instead of $g$ we need only try to find a $g$ from it.

Let us return to the distribution function $F$ of a random variable $X$. For any real $u$

**(6.3.3a)**     $0 \le F(u) \le 1$ since $F(u) = P(X \le u)$,

**(6.3.3b)**     if $u < v$, $F(u) \le F(v)$ since $\{s \in S;\ X(s) \le u\} \subset \{s \in S;\ X(s) \le v\}$,

**(6.3.3c)**     $\displaystyle\lim_{n \to \infty} F(-n) = 0$  because  $\displaystyle\lim_{n \to \infty} P(x \le -n) = P(\bigcap_{n=1}^{\infty} \{X \le -n\}) = P(\varnothing)$,

**(6.3.3d)**     $\displaystyle\lim_{n \to \infty} F(n) = 1$  because  $\displaystyle\lim_{n \to \infty} P(X \le n) = P(\bigcup_{n=1}^{\infty} \{X \le n\}) = P(S)$

where (6.3.3c) and (6.3.3d) follow from Theorem 3.4.4 and its corollary. By (6.3.3b) $F$ is monotonically increasing. Therefore at each $u_0$

**(6.3.3e)**     $\lim_{\substack{u \to u_0 \\ u < u_0}} F(u)$   and   $\lim_{\substack{u \to u_0 \\ u > u_0}} F(u)$   exist

**(6.3.3f)**     $\lim_{n \to \infty} F(u_0 + 1/n) = P(\bigcap_{n=1}^{\infty} \{X \le u_0 + 1/n\}) = P(\{X \le u_0\}) = F(u_0)$

Since $F$ is monotonic, we can get the limiting behavior as general real number limits and state

**(6.3.3g)**     $\lim_{u \to -\infty} F(u) = 0$,   $\lim_{u \to +\infty} F(u) = 1$,   and   $F(u)$ is continuous from

the right in the sense that $\lim_{\substack{u \to u_0 \\ u > u_0}} F(u) = F(u_0)$

A function from $R$ to $R$ which satisfies (6.3.3a), (6.3.3b), and (6.3.3g) is called a *distribution function*. Next we show that such a distribution function is the distribution function of a random variable on a probability system. We cannot give a rigorous proof at this level, so we handle the result as an example.

### Example 6.3.7

Let $G: R \to R$ satisfy (6.3.3a), (6.3.3b), and (6.3.3g). Show that $R, \mathscr{B}, P$ is a probability system and $X(x) = x$ as a random variable has $F_X(u) = G(u)$ for all $u$ if $P(\{a < x \le b\}) = G(b) - G(a)$.

The crux of the matter is whether a countably additive probability measure can be defined on $\mathscr{B}$ when $P(\{a < x \le b\}) = G(b) - G(a)$. This is shown in more advanced books for a $G$ satisfying our conditions. The countable additivity is the difficult part. Other aspects are easy. For example,

$$P(S) = P(\bigcup_{n=1}^{\infty} \{-n < x \le n\}) = \lim_{n \to \infty} P(\{-n < x \le n\})$$

$$= \lim_{n \to \infty} [G(n) - G(-n)] = 1 - 0 = 1$$

by the definitions and (6.3.3g). Finally,

$$F_X(u) = P(X \le u) = P(\bigcup_{n=1}^{\infty} \{-n + u < X \le u\})$$

$$= \lim_{n \to \infty} P(\{-n + u < X \le u\}) = \lim_{n \to \infty} [G(u) - G(-n + u)] = G(u)$$

In view of Example 6.3.4 and Theorem 6.3.1, certain distribution functions $G$ are distribution functions of random variables which have neither a probability function nor a density function. However, we remarked earlier that we are not studying all sample spaces and all random variables when we restrict ourselves to countable sample spaces and Euclidean sample spaces with mass densities.

The representations of Theorem 6.3.1 and Example 6.3.7 are called standard representations of a density function or distribution function. Example 6.3.7 which should include Example 6.2.5 gives a somewhat different result. However, in Example 6.2.5 we are viewing a problem as definitely discrete. In Example 6.3.7 we are looking from a broader perspective.

Note that in all standard representations the same random variable $X(x) = x$ is used. This does not mean that all random variables are the same. It means that the probability measure $P$ can be adjusted from case to case.

We started the discussion of probability functions, etc. by noting that two different random variables, technically speaking, were indistinguishable from the point of view of payoffs and their probabilities. We can now make this precise. Two random variables $X$ and $Y$ are said to be *identically distributed* if $F_X(u) = F_Y(u)$ for all real $u$. In some problems we can treat them as though they are identical.

We have seen how $F$ can be constructed from a probability function $f$ or from a density function $f$. If $F$ is given we can recapture a probability function if one exists as in exercise 6.3.13 or a density function if one exists as in Examples 6.3.4 and 6.3.6. Thus, $F$ and $f$ carry the same information. We will use the term *distribution* of a random variable to mean the distribution function, probability function, or density function according to which exist and which is most convenient.

Due to the standard representation, some authors define a random variable as the coordinate variable $x$ when $R = S$ in a probability system. This simplifies some things and complicates others.

**EXERCISES 6.3**

1   Let $S$ be the solid sphere $\{(x, y, z); x^2 + y^2 + z^2 \leq 1\}$. A point $s \in S$ is chosen at random (uniform mass density on $S$). Let $X(s) =$ distance of $s$ to $(0, 0, 0)$. Find the distribution function of $X$. Find the probability or density function if either exist.

2   Let $S, \mathscr{A}, P$ be as in exercise 1. Let $Y^3(s) = X(s)$. Find the distribution of $Y$ directly and by use of Example 6.3.5.

3   Let $a < 0$. Suppose that $X$ has a density function $f$ and $(1/a)[Y(s) - b] = X(s)$. Find a density function for $Y$.

♦ 4   Example 6.3.5 treated matters as though $X$ was defined in terms of $Y$ even though it was $X$ whose density we assumed known. Use the inverse function $h^{-1}$ to restate the ideas of Example 6.3.5.

♦ 5   Show that if $h$ is monotonically decreasing, $h(Y(s)) = X(s)$, $h$ and $h'$ are continuous, and $X$ has a density function, then a density function $f_Y(x) = |h'(x)| f_X(h(x))$.

**6**  A random variable has density function

$$f(x) = \begin{cases} \frac{1}{4}, & 0 \le x \le 2 \\ \frac{1}{2}, & 2 < x \le 3 \\ 0, & \text{otherwise} \end{cases}$$

Find its distribution function $F$. Graph $F$; graph $f$.

**7**  A random variable has density function

$$f(x) = \begin{cases} 0, & |x| > 1 \\ 1 - |x|, & |x| \le 1 \end{cases}$$

Find its distribution function $F$. Graph $F$; graph $f$.

**8**  A random variable has distribution function $F(x) = 1 - 1/(e^x + 1)$ for all $x$. Do you believe it? If so find its density function or probability function. If neither exists state that.

♦ **9**  Can a density function take on values greater than 1? Explain. Can a density function be unbounded? Explain.

**10**  Deccribe the standard representation of the distribution function of exercise 1.

**11**  Let

$$G(u) = \begin{cases} 0, & u < 1 \\ 1 - 1/u^2, & u \ge 1 \end{cases}$$

Is $G(u)$ a distribution function?

**12**  Let $F(u) = Q_1(u)/Q_2(u)$ for all $u$ where the $Q_i$ are polynomials in $Q$. Show that $F$ cannot be a distribution function.

♦ **13**  Let $X$ be any random variable with distribution function $F$. Show that

$$F(u_0) = \lim_{\substack{u \to u_0 \\ u < u_0}} F(u) + P(X = u_0)$$

i.e., the size of the jump in the graph of $F$ at $u_0$ is $P(X = u_0)$. [*Hint:* Use $\{X < u_0\} = \bigcup_{n=1}^{\infty} \{X \le u_0 - 1/n\}$ and proceed as in (6.3.3).]

♦ **14**  Let $G: R \to R$ be monotonically increasing with $\lim_{u \to \infty} G(u) = 1$ and $\lim_{u \to -\infty} G(u) = 0$. Suppose $G$ is continuous from the right and there are points $u_1, u_2, \ldots$ (finite or infinite) such that if $q_j = G(u_j) - \lim_{\substack{u \to u_j \\ u < u_j}} G(u)$ then $\sum_{j=1}^{\infty} g_j = 1$. What can be said about $G$? Prove your claim.

♦ **15**  Recall that the definition of $\int_{-\infty}^{u_0} g(x)\,dx$ is $\lim_{\substack{u \to u_0 \\ u < u_0}} \int_{-\infty}^{u} g(x)\,dx$ if $g$ is not continuous at $u_0$. In view of exercise 13 what can be said about $P(X = u_0)$ if $X$ has a density function? What can be said about $F_X$ in that case? Note also that $\int_{-\infty}^{u_0} g(x)\,dx = \lim_{u \to u_0} \int_{-\infty}^{u} g(x)\,dx$ if $g$ is continuous at $u_0$, too.

◆ **16**    Prove (6.3.3c), (6.3.3d), (6.3.3f), and (6.3.3g) carefully.

◆ **17**    Show that a bounded, monotonically increasing function from $R$ to $R$ has only countably many discontinuities by showing

    (a)    any discontinuity must be a jump discontinuity, i.e., right- and left-hand limits exist and are different,

    (b)    for any integer $n > 0$ there are only finitely many points at which the jump is larger than $1/n$,

    (c)    the positive integers are countable so the discontinuities comprise a countable union of finite sets.

# 6.4  The famous distributions

In this section we point out and discuss some of the more famous distributions which have come to the fore from the applications of probability and statistics.

When a distribution is used repeatedly, it acquires a name. We will first coordinate the names with the appropriate density or probability function and then discuss the applications of some of them. More details on the applications follow in other chapters. It should be noted that the names are frequently assigned to *families* of distributions.

## The Discrete Distributions

**Binomial**

If $0 \le p \le 1$ and $n$ is a positive integer,

$$f(k) = \binom{n}{k} p^k (1 - p)^{n-k}, \qquad k = 0, 1, 2, 3, \ldots, n$$

**Poisson**

If $\lambda > 0$,

$$f(k) = e^{-\lambda} \lambda^k / k!, \qquad k = 0, 1, 2, 3, \ldots$$

**Discrete Uniform**

If $m$, $n$ are positive integers with $m < n$,

$$f(k) = \frac{1}{n - m + 1} \qquad k = m, m + 1, \ldots, n$$

**Hypergeometric**

If $m$, $n$, and $r$ are positive integers with $m < n$ and $r < n$, then

$$f(k) = \frac{\binom{m}{k}\binom{n-m}{r-k}}{\binom{n}{r}}, \qquad k = 0, 1, 2, \ldots, \min(r, m)$$

## The Continuous Distributions

**Continuous Uniform**

If $a < b$ are real numbers,

$$f(x) = \begin{cases} \dfrac{1}{b - a}, & a \le x \le b \\ 0, & \text{either } a > x \text{ or } b < x \end{cases}$$

**Normal**

If $m$ is a real number and $\sigma > 0$ is a real number,

$$f(x) = \frac{1}{\sqrt{2\pi\sigma^2}} \, e^{-(x - m)^2/2\sigma^2}, \qquad \text{for all real } x$$

**Negative Exponential**

If $\lambda > 0$,

$$f(x) = \begin{cases} 0, & x < 0 \\ \lambda e^{-\lambda x}, & x \ge 0 \end{cases}$$

**Cauchy**

$$f(x) = \frac{1}{\pi(x^2 + 1)}, \qquad \text{for all real } x$$

**Gamma**

If $\alpha > 0$,

$$f(x) = \begin{cases} 0, & x \le 0 \\ \dfrac{1}{\Gamma(\alpha)} \, x^{\alpha - 1} e^{-x}, & x > 0 \end{cases}$$

where

$$\Gamma(\alpha) = \int_0^\infty x^{\alpha - 1} e^{-x} \, dx$$

**Student's t-distribution**

If $n$ is a positive integer,

$$f(x) = (n\pi)^{-1/2} \frac{\Gamma((n + 1)/2)}{\Gamma(n/2)} \left( 1 + \frac{x^2}{n} \right)^{-(n + 1)/2}$$

**F-distribution**

If $m$, $n$ are positive integers,

$$f(x) = \begin{cases} 0, & x < 0 \\ \dfrac{m^{m/2} n^{n/2} \Gamma((m + n)/2)}{\Gamma(m/2)\Gamma(n/2)} \dfrac{x^{(m - 1)/2}}{(mx + n)^{(m + n)/2}}, & x \ge 0 \end{cases}$$

### Example 6.4.1

Suppose that $X$ has the binomial distribution with $n = 3$ and $p = \frac{1}{2}$. Find $P(X = \frac{3}{2})$, $P(X = 1)$, $P(X = 3)$, $P(X = -2)$, $P(X = 7)$.

By the definitions, for $n = 3$ and $p = \frac{1}{2}$, $f(k) = \binom{3}{k}\left(\frac{1}{2}\right)^k \left(\frac{1}{2}\right)^{3 - k}$ for $k = 0, 1,$

2, 3. That is, a random variable with a binomial distribution takes only integer values (has integer payoffs) where the integers are 0, 1, 2, ..., n. Therefore,

$$P(X = \tfrac{3}{2}) = f(\tfrac{3}{2}) = 0, \quad P(X = 1) = f(1) = \binom{3}{1}\left(\tfrac{1}{2}\right)^1\left(\tfrac{1}{2}\right)^2 = \tfrac{3}{8}, \quad P(X = 3) =$$

$$f(3) = \binom{3}{3}\left(\tfrac{1}{2}\right)^3\left(\tfrac{1}{2}\right)^0 = \tfrac{1}{8}, \quad \text{and } P(X = -2) = P(X = 7) = 0.$$ Let there be no

more speculation as to whether $f_X(u) \neq 0$ if $X$ has a binomial distribution and if $u$ is not one of the integers 0, 1, 2, ..., n.

### Example 6.4.2

Suppose $X$ has the Poisson distribution with $\lambda = 2$. Find $P(Z = 4), P(X = 53)$ $P(X = \tfrac{3}{2})$, and $P(X = -2)$.

According to the definition of the Poisson distribution, $f(k) = P(X = k) = e^{-\lambda}\lambda^k/k!$ for $k = 0, 1, 2, 3, \ldots$. Thus $P(X = \tfrac{3}{2}) = P(X = -2) = 0, \quad P(X = 4)$ $= e^{-2}2^4/4!, \quad \text{and } P(X = 53) = e^{-2}2^{53}/(53)!$

### Example 6.4.3

Let $X$ have a binomial distribution with $n = 6$ and $p = \tfrac{1}{5}$. Find $P(X \le 2)$.

By the definition of a binomial distribution, a discrete distribution

$$P(X \le 2) = F_X(2) = \sum_{k \le 2} f_X(k) = f(0) + f(1) + f(2)$$

$$= \binom{6}{0}\left(\tfrac{1}{5}\right)^0\left(\tfrac{4}{5}\right)^6 + \binom{6}{1}\left(\tfrac{1}{5}\right)^1\left(\tfrac{4}{5}\right)^5 + \binom{6}{2}\left(\tfrac{1}{5}\right)^2\left(\tfrac{4}{5}\right)^4$$

### Example 6.4.4

Let $X$ have a negative exponential distribution with $\lambda = \tfrac{1}{2}$. Find $P(X = -1)$, $P(X = 2), P(X = 0), P(X \le 3)$, and $P(1 \le X \le 4)$.

The negative exponential is a continuous distribution—it has a density function. Therefore $P(X = a) = 0$ for any real number $a$ by exercise 6.3.15.

$$P(X \le 3) = F_X(3) = \int_{-\infty}^{3} f_X(x)\, dx = \int_{0}^{3} \tfrac{1}{2}e^{-x/2}\, dx = -e^{-x/2}\Big|_0^3 = 1 - e^{-3/2}$$

As we have seen in exercises 6.2.8 and 6.3.15

$$P(1 \le X \le 4) = P(X = 1) + P(1 < X \le 4) = 0 + [F_X(4) - F_X(1)]$$

$$= -e^{-x/2}\Big|_0^4 + e^{-x/2}\Big|_0^1$$

$$= -e^{-4/2} + e^{-1/2} = e^{-1/2} - e^{-2}$$

### Example 6.4.5

Let $X$ have the Cauchy distribution. Find $P(X = -1)$ and $P(-1 \le X \le 1)$.

Since the Cauchy distribution is continuous $P(X = -1) = 0$. Therefore,

$$P(-1 \le X \le 1) = F_X(1) - F_X(-1) = \int_{-\infty}^{1} \frac{dx}{\pi(1 + x^2)} - \int_{-\infty}^{-1} \frac{dx}{\pi(1 + x^2)}$$

$$= \int_{-1}^{1} \frac{dx}{\pi(1 + x^2)} = \frac{1}{\pi} \arctan x \Big|_{-1}^{1}$$

$$= (1/\pi)[\tfrac{1}{4}\pi - (-\tfrac{1}{4}\pi)] = 2\pi/4\pi = \tfrac{1}{2}$$

## Example 6.4.6

Show that if $X$ has the normal distribution with $m = 0$ and $\sigma^2 = 1$ then for any real number $b > 0$, $P(X \ge b) \le (1/\sqrt{2\pi}b)e^{-b^2/2}$.

Since $X$ is a continuous random variable $P(X = b) = 0$ so $P(X \ge b) =$

$$P(X > b) = 1 - P(X \le b) = \int_{-\infty}^{\infty} f(x)\, dx - \int_{-\infty}^{b} f(x)\, dx = \int_{b}^{\infty} f(x)\, dx, \text{ where}$$

$f$ is the density function. Now

$$\int_{b}^{\infty} f(x)\, dx = \frac{1}{\sqrt{2\pi}} \int_{b}^{\infty} e^{-x^2/2}\, dx \le \frac{1}{\sqrt{2\pi}} \int_{b}^{\infty} \frac{x}{b} e^{-x^2/2}\, dx$$

because the curve $y = (x/b)e^{-x^2/2}$ lies above the curve $y = 1 \cdot e^{-x^2/2}$ if $x \ge b$

and $\int_{b}^{\infty} \cdots$ represents the area under the curve if the curve is above the $x$ axis.

That is certainly true here. Now $\left(\dfrac{d}{dx}\right)e^{-x^2/2} = -(2x/2)e^{-x^2/2}$ so $\int xe^{-x^2/2}\, dx = -e^{-x^2/2} + c$. Therefore

$$\int_{b}^{\infty} xe^{-x^2/2}\, dx = \lim_{q \to \infty} -e^{-x^2/2}\Big|_{b}^{q} = \lim_{q \to \infty} (-e^{-q^2/2} + e^{-b^2/2}) = e^{-b^2/2}$$

If we follow the entire chain of inequalities we have

$$P(X \ge b) \le \frac{1}{\sqrt{2\pi}\, b} e^{-b^2/2}$$

The foregoing examples ask questions which can be asked about many distributions. The interesting aspects of the famous distributions are their uses in natural problems and their interrelationships, rather than specific values.

The binomial distribution was seen before in section 6.2. There it represented the probability of having exactly $k$ defective transistors in a lot of $n$ transistors which are manufactured independently and where the probability of manufacturing a defective transistor is $p$. Equally well, the binomial distribution gives the probability of $k$ heads in $n$ tosses of a (possibly) biased coin whose probability of heads on one toss is $p$ and which is tossed $n$ times, independently. The student should be able to formulate other problems for which the binomial distribution is natural.

The Poisson distribution is related to the binomial distribution but it does not refer to $k$ occurrences out of $n$ tries. The Poisson distribution is used in problems of the following sort: "What is the probability of $k$ cosmic-ray particles hitting a region when the average cosmic-ray intensity is $\lambda$?" or "What is the probability that $k$ telephone calls will be dialed in a particular second when the average calling rate is $\lambda$?" Naturally we assume infinitely many cosmic rays and telephone users in these examples.

The relationship of the binomial and Poisson distributions is shown in Chapter 10. There the intuitive notion that the Poisson distribution is a binomial distribution with infinite $n$ is clarified.

The hypergeometric distribution arises in "sampling without replacement." Suppose a box contains $n$ balls, $m$ of which are red and the remainder white. They are thoroughly mixed. Balls are removed one at a time without being returned to the box until $r$ balls have been removed. What is the probability that $k$ of them are red? It is quite clear that we could just as well have chosen a set of $r$ balls in one fell swoop. In either case we get the hypergeometric $f(k)$.

If we sampled "with replacement" (any ball which is removed has its color noted and then is returned to the box) we would obtain the binomial probability

$$\binom{r}{k} p^k (1 - p)^{r-k}$$

where $p = m/n$ for the probability of $k$ red balls being drawn in $r$ trials.

We leave it to the reader to verify that if $m$ and $n$ are much larger than $r$, then the hypergeometric $f(k)$ is approximately equal to the above binomial probability. This is a mathematical version of the statement that for large "populations" sampling with or without replacement is essentially the same.

The continuous uniform distribution represents a choice of a real number between two fixed numbers $a$ and $b$ in about as random a fashion as is possible.

The normal distribution is one of the most amazing probability distributions. It occurs so frequently that we say $X$ is $N(m, \sigma^2)$ if its density function is

$$f(x) = \frac{1}{\sqrt{2\pi}\,\sigma}\, e^{-(x-m)^2/2\sigma^2}$$

We shall see that the normal distribution is a limiting form of binomial distributions. However, it is a limiting form of other distributions, also. More specifically, if $X$ has the normal distribution with $m = 0$ and $\sigma = 1$, and if $X_n$ has the binomial distribution with parameters $n$ and $p$ and if $np$ is very large then

**(6.4.1)**    $P(X_n - np \le u\sqrt{np(1-p)}\,) \approx P(X \le u)$

### Example 6.4.7

A fair coin is tossed 1600 times (independent tosses). Find, approximately, the probability that heads shows on no more than 780 tosses.

This problem is a binomial problem where we want to approximate $P(X_{1600} \le 780)$ when $X_{1600}$ has a binomial distribution with $n = 1600$ and $p = \frac{1}{2}$ (fair coin). According to (6.4.1) we can approximate $P(X_n - np \le u\sqrt{np(1-p)})$ which is

$$P(X_{1600} - 1600 \cdot \tfrac{1}{2} \le u\sqrt{1600 \cdot \tfrac{1}{2} \cdot \tfrac{1}{2}})) = P(X_{1600} - 800 \le u \cdot 40 \cdot \tfrac{1}{2})$$
$$= P(X_{1600} \le 800 + 20u)$$

Here we are merely using the fact that $\{X_{1600} - 800 \le 20u\} = \{X_{1600} \le 800 + 20u\}$. But we want $P(X_{1600} \le 780)$. What value of $u$ makes $800 + 20u = 780$? Obviously $u = -1$. Therefore, by (6.4.1),

$$P(X_{1600} \le 780) \approx P(X \le -1) = \int_{-\infty}^{-1} \frac{1}{\sqrt{2\pi}} e^{-x^2/2}\, dx$$

since the normal density function in question is $f(x) = (1/\sqrt{2\pi})e^{-x^2/2}$.

In a later section we discuss the use of tables which give, for various values of $u > 0$,

$$\int_0^u \frac{1}{\sqrt{2\pi}} e^{-x^2/2} = \int_{-u}^0 \frac{1}{\sqrt{2\pi}} e^{-x^2/2}\, dx$$

These, together with

$$\int_0^\infty \frac{1}{\sqrt{2\pi}} e^{-x^2/2}\, dx = \tfrac{1}{2}$$

so that

$$\int_{-\infty}^\infty \frac{1}{\sqrt{2\pi}} e^{-x^2/2}\, dx = 1$$

give numerical approximations which are useful for many problems.

The normal distribution is also valuable in describing Brownian motion and other physical phenomena.

The negative exponential distribution is frequently used as a mathematical model for the duration of telephone calls (usually where there is no time limit) or of the lifetime of simple devices, such as lightbulbs, transistors, etc.

### Example 6.4.8
It is found that the duration of telephone calls emanating from a certain central office follows the exponential distribution with $\lambda = 0.08$. Find the probability that a call will last more than 20 units.

This calls for

$$\int_{20}^\infty (0.08)e^{-0.08x}\, dx = -e^{-0.08x}\Big|_{20}^\infty = e^{(-0.08)(20)} = e^{-1.6} \approx 0.202$$

Nothing more is needed.

We reserve discussion of the other famous distributions for future chapters. It should be apparent that probability theory has a number of applications other than gambling. We will also derive additional properties of the famous distributions in other chapters. Familiarity with the properties of these distributions is helpful in deciding which (if any) famous distribution is appropriate in a model for a particular phenomenon. If no famous distribution has the correct properties the researcher may use some other distribution. Naturally, the remarks of Chapter 4 apply.

**EXERCISES 6.4**

1   Let $X$ have a binominal distribution with $n = 2m$ and $p = \frac{1}{2}$. Use Stirling's formula to estimate $P(X = m)$ for large $m$.

2   Let $f$ be the binomial probability function with $n = 18$ and $p = \frac{1}{6}$. How would you simulate this? Compare $f(0)$, $f(1)$, and $f(2)$. Compare $f(18)$, $f(17)$, and $f(16)$. Compare these to $f(\frac{2}{3})$.

3   One could conjecture that a particular major league batter bats binomially. That is (providing he is not walked) his batting chances are independent with a certain probability $p$ of getting a hit. Suppose one selects many cases of eight batting chances (without walks, etc.) and gathers data about the number of hits which he gets during these chances. One may compare the data to a binomial probability function with $n = 8$ and $p = $ the batter's season batting average. What explanations could be given for rather sharp divergence of the data from the binomial?

4   A factory makes bolts in such a fashion that the probability of a bolt being defective is $10^{-4}$. Furthermore the effectiveness or defectiveness of the various bolts is independent. What is the probability that more than two bolts out of 5000 bolts made by this factory are defective?

5   It is common to assume that a typist has a certain probability of making an error and that different strokes are independent. If a certain typist has probability 0.001 of making an error on one stroke, what is the probability of the typist making more than three errors on a page which involves 1800 typing strokes? What objections might be raised regarding the assumptions?

6   Let $X$ have a Poisson distribution with density function $f$. Show that the numbers $f(0)$, $f(1)$, $f(2)$, ..., $f(k)$, ... are monotonically nondecreasing for $k \le \lambda$.

7   Let $X$ be a negative exponential random variable and let $Y$ be a normal random variable. Show that for sufficiently large $x$, $f_X(x) > f_Y(x)$.

8   Let $np = \lambda$ where we assume $n$ is very large. Show then that if $k$ is rather small $\binom{n}{k} p^k (1 - p)^{n-k}$ is approximately $e^{-\lambda}(\lambda^k/k!)$. [*Hint:* Prove that

$$\lim_{x \to \infty} \left(1 - \frac{q}{x}\right)^{rx} = e^{-qr}$$

and that

$$\lim_{x \to \infty} \left(1 - \frac{l}{x}\right)^r = 1$$

(you can use l'Hospital's rule.)]

**9**   Let $X$ be a binomial random variable with $n = 2m + 1$ and $p = \frac{1}{2}$. Find $P(X \le m)$.

**10**   Show that a random variable $X$ which is $N(m, \sigma^2)$ has $P(X \le m) = \frac{1}{2}$.

**11**   Find the value of $\lambda > 0$ which maximizes the function $g(\lambda) = e^{-\lambda}(\lambda^n/n!)$ for fixed $n > 0$.

**12**   Find the value of $\sigma > 0$ which maximizes the function $g(\sigma) = (1/\sqrt{2\pi} \, \sigma)e^{-x^2/2\sigma^2}$ for fixed $x \ne 0$.

Exercises 11 and 12 are examples of *maximum-likelihood estimation* of parameters—an important technique in statistics.

**13**   Let $X$ have a uniform distribution (continuous). Find the distribution function of $X$.

**14**   Let $X$ have the gamma distribution with $\alpha = 1$. Find the distribution function of $X$ as explicitly as possible. For which values of $\alpha$ can the distribution function of a gamma random variable be as explicitly found as for $\alpha = 1$?

**15**   Show that for a binomial probability function $f$ the probabilities $f(k)$ are monotonically increasing up to $k \le np$.

◆ **16**   Show that if $X$ is the standard representation of a normal density function with $m = 0$ and $\sigma^2 = 1$ then $Y$ with $Y(x) = aX(x) + b$ is normal with $m = b$ and $\sigma^2 = a^2$.

**17**   What is the relationship between the Student's $t$-distribution with $n = 1$ and any other of the famous distributions?

**18**   Find the regions of increase and decrease of the density functions of the normal and gamma distributions.

**19**   State, on an intuitive basis only, which of the famous distributions may be useful as models for the following phenomena. This may be more a matter of exclusion of other famous distributions than inclusion of some famous distributions.

(a)   Lumps of coal range from 1 to 100 gm in weight. Those below 40 gm and above 60 gm are culled out. We are interested in the weight of one of the remaining lumps selected at random.

(b)   The number of bad ears of corn in bushels of 100 ears which are selected at random.

(c)   The amount of impurities in 100-gallon drums of oil coming from a certain field.

(d)   The number of calls to a police station during a half-hour period.

(e)   The number of blue-eyed people in a sample of 20 people chosen at random from a group of 100 blue-eyed and 300 non-blue-eyed people.

(f)   The cholesterol content in a blood sample of a randomly selected person.

(g)   The length of time between successive cosmic-ray particles which hit a Geiger counter.

# 6.5   Independent random variables

There are always many random variables which can be given on the sample space of any probability system and there is no reason why one cannot study a number of these random variables simultaneously. This is not done purely for amusement and generality. Most important problems in probability and statistics involve the simultaneous study of two or more random variables. One of the most famous problems is that of the mathematical model for smoking and lung cancer. In effect, the sample space is the collection of people studied. One random variable relates to the smoking of person $s$. The other random variable relates to the cancer of person $s$. Similar problems could involve (a) reading habits and income, (b) income, education, and political conservatism, (c) the impurity content of the steel, the impurity content of the aluminum, and the length of service of a product made from steel and aluminum.

Exercises 6.1.18–6.1.20 were intended to suggest a certain phenomenon which may occur when dealing with $n$ random variables $X_1, X_2, \ldots, X_n$ which are given on the sample space of a probability system. It may be true that for each $n$-tuple $(u_1, u_2, \ldots, u_n)$ of real numbers, the events

**(6.5.1)**    $\{X_1 \leq u_1\}, \{X_2 \leq u_2\}, \ldots, \{X_n \leq u_n\}$

are independent events.

In a given mathematical problem it may or may not be difficult to determine whether the independence of (6.5.1) holds for all $n$-tuples $(u_1, u_2, \ldots, u_n)$. In any event, it is this situation which corresponds to many practical problems and it also admits a very extensive mathematical theory.

**Definition**    If $X_1, X_2, \ldots, X_n$ are $n$ random variables on the sample space of a probability system, we say that they are *independent* random variables if for *each* $n$-tuple $(u_1, u_2, \ldots, u_n)$ the events $\{X_1 \leq u_1\}, \{X_2 \leq u_2\}, \ldots, \{X_n \leq u_n\}$ are independent events.

Thus the reader should convince himself that cigarette companies would be satisfied with a study which shows that the random variables relating to cancer incidence and cigarette smoking are independent.

We pass to some examples which suggest the way in which independent random variables may occur.

### Example 6.5.1

A pair of fair dice, one red and one green, are rolled with $(r, g)$ representing the numbers showing on the red and green die, respectively. Let $X(r, g) = r$ and $Y(r, g) = g$. Show that $X$ and $Y$ are independent.

The values of $u_1, u_2$ which are most interesting satisfy $0 \le u_j \le 6$. With $[u_j]$ = greatest integer less than or equal to $u_j$, we have $\{X \le u_1 \text{ and } Y \le u_2\} = \{r \le u_1 \text{ and } g \le u_2\}$ and its probability is $[u_1][u_2]/36$. Since $P(X \le u_1) = P(r \le u_1) = [u_1] \cdot 6/36 = [u_1]/6$, and similarly $\{Y \le u_2\} = [u_2]/6$, therefore

$$P(X \le u_1 \text{ and } Y \le u_2) = [u_1][u_2]/36 = P(X \le u_1)P(Y \le u_2)$$

We leave the cases where $u_j > 6$ or $u_j < 0$ to the reader.

### Example 6.5.2

A box contains ten balls which are indistinguishable except for being numbered $1, 2, 3, \ldots, 10$. One ball is drawn at random; suppose it is numbered $m$. It is replaced in the box. The balls are thoroughly mixed and a second drawing is made. This results in a ball showing $n$. Let $X(m, n) = m$ and $Y(m, n) = n$. Show that $X$ and $Y$ are independent in the conventional model for the experiment.

Since the first ball drawn is replaced and the balls are thoroughly mixed it seems reasonable to take $S = \{1, 2, 3, \ldots, 10\} \times \{1, 2, 3, \ldots, 10\}$ with equal likelihood. This is also conventional. If, for example, we consider $\{X \le 3.3, Y \le 8.2\}$ we find that $P(X \le 3.3, Y \le 8.2) = 3 \cdot 8/100$ since $\{X \le 3.3, Y \le 8.2\} = \{(m, n); m \le 3.3, n \le 8.2\} = \{(m, n); m \le 3, n \le 8, (m, n) \in S\}$. There are $3 \cdot 8$ sample points in that set. Similarly if $0 \le u_1 \le 10$, $0 \le u_2 \le 10$, and $[\alpha]$ is the greatest integer less than or equal to $\alpha$ we find

$$P(X \le u_1, \quad Y \le u_2) = \frac{[u_1][u_2]}{100}$$

However

$$P(X \le u_1) = P((m, n) \in S; m \le u_1) = \frac{[u_1] \cdot 10}{100} = \frac{[u_1]}{10}$$

A similar result holds for $P(Y \le u_2)$ and we thus see that

$$P(X \le u_1, \quad Y \le u_2) = \frac{[u_1][u_2]}{100} = \frac{[u_1]}{10}\frac{[u_2]}{10} = P(X \le u_1)P(Y \le u_2)$$

This is equivalent to the required independence. If $u_j < 0$ or $u_j > 10$ a similar analysis shows independence.

Example 6.4.2 suggests that it is our model (or interpretation) which insures that certain payoffs lead to independent random variables. However, although it may be convenient to assume independence, one should expect to pay the price of bad predictions in practical problems in which independence is a bad assumption.

Examples 6.5.1 and 6.5.2 are indicative of a general theorem concerning independent random variables. We prove this theorem in a special case which suggests the general case.

**Theorem 6.5.1**    Suppose $S_1$, $S_2$ are countable sample spaces with the usual system of events and probability measures $P_1$ and $P_2$ and suppose $U$, $V$ are random variables on $S_1$ and $S_2$, respectively. Let $S = S_1 \times S_2$ and let $P$ be defined for $(s, t) \in S$ by $P(s, t) = P_1(s)P_2(t)$. Let $X(s, t) = U(s)$ and $Y(s, t) = V(t)$. Then $X$ and $Y$ are independent with respect to $S$ (countable), the usual system of events and $P$.

PROOF    If $u$ is a real number

$$\{X \le u\} = \{(s, t); X(s, t) \le u\} = \{(s, t); U(s) \le u\}$$
$$= \{s \in S_1; U(s) \le u\} \times S_2$$

by the various set theory definitions. Similarly

$$\{Y \le v\} = S_1 \times \{t \in S_2; V(t) \le v\}$$

However, Theorem 5.2.2 proves that events of the form $M_1 \times S_2$, $S_1 \times M_2$ $(M_1 \subset S_1,\ M_2 \subset S_2)$ are independent in a probability system formed according to the hypotheses of the current theorem.

Extensions of Theorem 6.5.1 apply to cases of more than two components and the assumption of countable $S_1, S_2, \ldots$ can be dropped. To illustrate the extensions of Theorem 6.5.1 we have the following example.

*Example 6.5.3*
A point $(x, y, z)$ is chosen at random from the cube $\{(x, y, z); 0 \le x \le 1, 0 \le y \le 1,\ 0 \le z \le 1\}$. Show that the random variables $X(x, y, z) = 3x$, $Y(x, y, z) = \cos y$, $Z(x, y, z) = e^z$ are independent.

Our interpretation of the choice at random of a point from a bounded geometric region is that probability is proportional to volume in this case. So

$$P(\{a_1 < x \le b_1,\ a_2 < y \le b_2,\ a_3 < z \le b_3\}) = (b_1 - a_1)(b_2 - a_2)(b_3 - a_3)$$
$$\text{if } 0 \le a_j \le b_j \le 1$$

However we may think of $U, V, W$, respectively, as $U(x) = 3x$, $V(y) = \cos y$, $W(z) = e^z$ on $\{x; 0 \le x \le 1\}, \{y; 0 \le y \le 1\}, \{z; 0 \le z \le 1\}$ with probability assigned by $P_1(\{a_1 < x \le b_1\}) = b_1 - a$, $P_2(\{a_2 < y \le b_2\}) = b_2 - a_2$, etc. This is precisely the setup of the theorem except that the sample spaces are not countable.

**EXERCISES 6.5**

**1**   Let $X$ be a constant random variable and let $Y$ be any random variable on the same system. Show that $X$ and $Y$ are independent.

**2**   Show that if $M$ and $N$ are events and

$$X(s) = \begin{cases} 1, & s \in M \\ 0, & s \notin M \end{cases}, \qquad Y(s) = \begin{cases} 1, & s \in N \\ 0, & s \notin N \end{cases}$$

then $X$ and $Y$ are independent if and only if $M$ and $N$ are independent events.

**3**   Generalize the statement of exercise 2 to the case of three events. Look at some cases but do not go through all of the details.

**4**   Let $S = \{(x, y); x^2 + y^2 \leq 1\}$, the disk with center at $(0, 0)$ and radius 1. Let $P$ be determined by a constant mass density. Let $X(x, y) = (x^2 + y^2)^{1/2} = r$, the distance of $(x, y)$ from $(0, 0)$ and let $Y(x, y) = \theta$, the angular polar coordinate taken as the smallest positive angle from the positive $x$ axis to the line from $(0, 0)$ to $(x, y)$. Show that $X$ and $Y$ are independent.

**5**   Let $S = \{1, 2, 3, \ldots, 24\}$ with equal likelihood, $X(n) = $ the remainder after dividing $n$ by 6, i.e., $n = 6m + X(n)$, $Y(n) = $ the remainder after dividing $n$ by 8, i.e., $n = 6l + Y(n)$ where $0 \leq X(n) \leq 5$ and $0 \leq Y(n) \leq 7$. Determine whether $X$ and $Y$ are independent.

**6**   Let $S = \{1, 2, 3, \ldots, 35\}$ with equal likelihood, $X(n) = $ remainder after dividing $n$ by 5, i.e., $n = 5m + X(n)$, $Y(n) = $ remainder after dividing $n$ by 7, i.e., $n = 7l + Y(n)$ where $0 \leq X(n) \leq 4$ and $0 \leq Y(n) \leq 6$. Decide whether $X$ and $Y$ are independent.

In exercises 7–10 we assume that $\{X \leq u\}$ and $\{Y \leq v\}$ are independent events for all real numbers $u$ and $v$.

**7**   Show that $\{X < w\}$ and $\{Y \leq v\}$ are independent events for any $w$ and $v$.

[*Hint*: $\{X < w$ and $Y \leq v\} = \bigcup\limits_{n=1}^{\infty} \{X \leq w - 1/n$ and $Y \leq v\}$.]

**8**   Show that $\{X = w\}$ and $\{Y \leq v\}$ are independent for any $w$ and $v$.

**9**   Show that $\{X = w\}$ and $\{Y < z\}$ are independent for any $w$ and $z$ by changing the hint of exercise 7 to suit.

**10**   Show that $\{X = w\}$ and $\{Y = z\}$ are independent for any $w$ and $z$.

**11**   Suppose that $X$ and $Y$ have probability functions $f_X$ and $f_Y$. Suppose that for all real $u$ and $v$, $\{X = u\}$ and $\{Y = v\}$ are independent events. Show that $X$ and $Y$ are independent random variables in the sense of the definition of this section. Where is the assumption concerning probability functions used?

**12**   Two hundred balls are in a box. One hundred of them are red and the rest are white. Balls are then selected successively at random without replacement. Let $X_j = 1$ if a red ball is selected on the $j$th draw. Let $X_j = 0$ if a white ball is selected on the $j$th draw. Are $X_1$ and $X_2$ independent?

# *Mathematical Expectation*

## 7.1 Measuring the center

Although the distribution function $F$ of a random variable $X$ gives the whole story of a random variable from a probabilistic point of view, this may be very complicated and a more simplified picture may be more useful for certain purposes.

Suppose four towns A, B, C, D each with 1000 families have family income data as given in Table 1. The numbers in the body of the table are the number of families having the indicated income in the indicated town. It is best to

**Table 1**

| ANNUAL INCOME | NUMBER OF FAMILIES | | | |
| --- | --- | --- | --- | --- |
| | A | B | C | D |
| $2000 | 300 | 115 | 100 | 0 |
| $3000 | 450 | 390 | 350 | 400 |
| $4000 | 200 | 410 | 400 | 500 |
| $8000 | 50 | 50 | 125 | 100 |
| $12 000 | 40 | 35 | 25 | 0 |
| $20 000 | 10 | 0 | 0 | 0 |
| TOTAL INCOME IN MILLIONS | 3.83 | 3.86 | 4.15 | 4.00 |

assume that all figures are exact. The total incomes in each town are essentially equal. However, from certain points of view, one may consider town A to be poor. Seven hundred fifty families have incomes of $3000 or less, and 300 families have annual incomes of $2000, whereas no more than 115 families have so low an income in the other towns.

If we consider a particular town we may ask for a *single* number which would represent the "typical" income per family in that town, that is, a number which would represent the "center" of the income distribution of the town.

Three methods of obtaining such a typical number come quickly to mind and are traditional. Each may lead to a different number for the same income distribution. The numbers which they determine are named the mean, the median, and the mode. We give rough definitions for these and point out that the mean, median, and mode of the incomes of Table 1 are tabulated in Table 2.

The *mean* $\bar{m}$ = (total family income) ÷ (number of families). Roughly speaking, the *median* is the middle income when incomes are arranged in increasing order and when the income of every family is listed. The *mode* is the income which occurs most often.

### Table 2

| TOWN | MEAN INCOME | MEDIAN INCOME | MODAL INCOME |
|------|-------------|---------------|--------------|
| A | 3830 | 3000 | 3000 |
| B | 3860 | 3000 | 4000 |
| C | 4150 | 4000 | 4000 |
| D | 4000 | 4000 | 4000 |

The descriptions of the median and mode are somewhat imprecise as the following example shows.

### Example 7.1.1

Suppose the data on incomes reads 100 families have incomes of $2000 and 100 families have incomes of $8000 in a town of 200 families.

There is not a single modal income. There are two—$2000 and $8000. We might say that this distribution is bimodal (two modes). It is possible to have more than two modes. As for the median, we have

$$\underbrace{2000 \quad \cdots \quad 2000}_{100 \text{ times}} \qquad \underbrace{8000 \quad \cdots \quad 8000}_{100 \text{ times}}$$

There is no middle value. However by convention we take $\frac{1}{2}(2000 + 8000)$, the average of the values adjacent to the middle to be the median.

As Table 2 shows, the mean, median, and mode may be different from one another or they may be the same or, although not all cases are shown, any two may be equal and the third different.

**EXERCISES 7.1** _____

1   Find the mean, median, and mode of the following income distribution

| ANNUAL INCOME IN $1000 | 2 | 3 | 4 | 8 | 12 | 20 |
|---|---|---|---|---|---|---|
| NUMBER OF FAMILIES WITH THAT INCOME | | | 300 | 150 | 300 | 150 | 65 | 35 |

2   Give a hypothetical income distribution in which the mean and mode are equal but different from the median.

3   Give a hypothetical income distribution in which the mean and median are equal but different from the mode.

# 7.2  The expectation

Although the mean, median, and mode are all useful in describing features of a random variable by a single number, we will focus our attention on the mean (actually a generalization of the mean) as the basic measure of the center.

The choice of the mean as the primary object of study should not be construed as implying the inferiority of the median and mode. The mean has the most convenient arithmetical properties and so it is chosen for convenience at this stage. The student would do well to test whether the median and the mode satisfy the same algebraic relationships as the mean.

For technical reasons we view the number $\bar{m}$ from the viewpoint of random variables. Suppose that the families in town A are numbered from 1 to 1000 to form a sample space with equal likelihood. Let $X(n)$ be the income of the family numbered $n$ and consider

$$(7.2.1) \quad \sum_{n=1}^{1000} X(n) \frac{1}{1000} = \frac{1}{1000} \sum_{n=1}^{1000} X(n) = \sum_{n=1}^{1000} X(n) \div 1000$$

$$= \text{(total family income)} \div \text{(number of families)}$$

By the definition of $X(n)$, $\sum_{n=1}^{1000} X(n) = 3\,830\,000$ in the case of town A, so (7.2.1) is 3830. Our assumption of equal likelihood in the sample space of families numbered $1, 2, \ldots, 1000$ gives $p_n = \frac{1}{1000}$ where $p_n = P(n)$. We may then interpret (7.2.1) as $\sum_{n=1}^{1000} X(n)p_n$ in this case. However, whenever one has a countable sample space $S$ with $P(s_n) = p_n$ and a random variable $X$ one may study $\sum_{s_n \in S} X(s_n)p_n$. There may be convergence problems if $S$ is infinite but except for these (which we discuss later) $\sum\limits_{s_n \in S} X(s_n)p_n$ usually exists. In the case $S =$

$\{1, 2, \ldots, N\}$ with equal likelihood,

$$\sum_{s_n \in S} X(s_n)p_n = \sum_{n=1}^{N} X(n)\frac{1}{N} = \frac{1}{N}\sum_{n=1}^{N} X(n)$$

Therefore $\sum\limits_{s_n \in S} X(n)p_n$ is a generalization of $\dfrac{1}{N}\sum\limits_{n=1}^{N} X(n)$ which is the arithmetic

mean or average of the numbers $X(1)$, $X(2)$, $\ldots$, $X(N)$. The quantity $\sum\limits_{s_n \in S} X(n)p_n$

is sometimes called a weighted mean and we shall see some reasons for this.

**Example 7.2.1**
Let $S = \{1, 2, 3\}$ with $p_1 = \frac{1}{2}$, $p_2 = \frac{1}{3}$, $p_3 = \frac{1}{6}$. Let $X(n) = n - 1$. Find

$\sum\limits_{n=1}^{3} X(n)p_n$. Find the probability function of $X$.

$$\sum_{n=1}^{3} X(n)p_n = 0\cdot\tfrac{1}{2} + 1\cdot\tfrac{1}{3} + 2\cdot\tfrac{1}{6} = \tfrac{2}{3}$$

and

$$f_X(0) = \tfrac{1}{2}, \qquad f_X(1) = \tfrac{1}{3}, \qquad f_X(2) = \tfrac{1}{6}$$

**Example 7.2.2**
Let $S = \{1, 2, 3, 4, 5, 6\}$ with $p_n = \frac{1}{6}$ for each $n \in S$. Let $Y(1) = Y(2) = Y(3) = 0$, $Y(4) = Y(5) = 1$, and $Y(6) = 2$. Find $\sum\limits_{n \in S} Y(n)p_n$ and the probability function of $Y$.

$$\sum_{n \in S} Y(n)p_n = (0 + 0 + 0 + 1 + 1 + 2)\tfrac{1}{6} = \tfrac{2}{3}$$

$$f_Y(0) = \tfrac{3}{6} = \tfrac{1}{2}, \qquad f_Y(1) = \tfrac{2}{6} = \tfrac{1}{3}, \qquad f_Y(2) = \tfrac{1}{6}$$

In Examples 7.2.1 and 7.2.2 $X$ and $Y$ are identically distributed. Furthermore, $\sum\limits_{n \in S} X(n)p_n$ and its analog in Example 7.2.2 yield the same result. This will

always happen to identically distributed random variables. Finally, we observe that $Y$ is a version of $X$ within an equal-likelihood model.

After a few finite sums of the form $\sum\limits_{n=1}^{N} X(n)p_n$ one may ask about

$\sum\limits_{n=1}^{\infty} X(n)p_n$. This would be the denumerable analog of the mean and would be

relevant in the case $S = \{1, 2, 3, \ldots\}$.

**Example 7.2.3**
Let $S = \{1, 2, 3, \ldots\}$, $p_n = 1/2^n$, and $X(n) = n$. Discuss $\sum\limits_{n=1}^{\infty} X(n)p_n$.

Here $\sum\limits_{n=1}^{\infty} X(n)p_n = \sum\limits_{n=1}^{\infty} \dfrac{n}{2^n}$. This last series converges since the ratio of the

$(n+1)$st and the $n$th term is $\dfrac{n+1}{2^{n+1}} \Big/ \dfrac{n}{2^n} = \dfrac{n+1}{n} \cdot \dfrac{1}{2} \to \dfrac{1}{2}$ as $n \to \infty$. Hence, the

ratio test shows convergence. It will be shown later that $\sum\limits_{n=1}^{\infty} \dfrac{n}{2^n} = 2$, but that is not the issue at present.

### Example 7.2.4

Let $S = \{1, 2, 3, \ldots\}$ and $p_n = 1/n(n+1)$ but $X(n) = (-1)^{n+1}(n+1)$. Discuss $\sum\limits_{n=1}^{\infty} X(n)p_n$.

Here

$$\sum_{n=1}^{\infty} X(n)p_n = \sum_{n=1}^{\infty} (-1)^{n+1} \frac{1}{n} = 1 - \tfrac{1}{2} + \tfrac{1}{3} - \tfrac{1}{4} + \tfrac{1}{5} - \tfrac{1}{6} \cdots$$

This last series is conditionally convergent since the terms alternate in sign, are monotonically decreasing, and converge to zero. Conditional convergence means that although $\sum\limits_{n=1}^{\infty} X(n)p_n$ converges, $\sum\limits_{n=1}^{\infty} |X(n)p_n|$ diverges.

Since it is known that a conditionally convergent series can be rearranged to converge to a different limit we might ask if this phenomenon has any relevance to our problem. In a sense it has. Although 1, 2, 3, ... is the natural order for positive integers, why should that be the order for elements of a sample space? If we were to enumerate the elements of $S$ as 1, 2, 3, 5, 4, 7, 9, 6, 11, 13, 8, 15, 17, 10, 19, 21, 12, ... we would list every positive integer once and only once. If we computed

$$1 - \tfrac{1}{2} + \tfrac{1}{3} + \tfrac{1}{5} - \tfrac{1}{4} + \tfrac{1}{7} + \tfrac{1}{9} - \tfrac{1}{6} + \cdots = \sum_{n \in S} X(n)p_n$$

we would find that this gives us a different limit since

$$1 - \tfrac{1}{2} + \tfrac{1}{3} + \tfrac{1}{5} - \tfrac{1}{4} + \tfrac{1}{7} + \tfrac{1}{9} - \tfrac{1}{6} + \cdots > 1 - \tfrac{1}{2} + \tfrac{1}{3} + \tfrac{1}{5} - \tfrac{1}{4} + \tfrac{1}{7} + \tfrac{1}{9} - \tfrac{1}{6}$$

$$> 1 - \tfrac{1}{2} + \tfrac{1}{3} - \tfrac{1}{4} + \tfrac{1}{5} - \tfrac{1}{6} + \tfrac{1}{7}$$

$$> \sum_{n=1}^{\infty} (-1)^{n+1} \frac{1}{n}$$

This peculiarity of infinite series could lead to confusion. We therefore make a narrow definition to prevent certain complications. That is, we make some definitions which remove the conditionally convergent case to a different file, which the student may explore at his leisure.

**Definitions**     Let $S$ be a countable sample space with the usual system of events and probability measure $P$ given by $P(s_n) = p_n$ for $s_n \in S$. Let $X$ be a random variable on $S$. We say that $X$ *has an expectation* (or *the expectation of X exists*) if $\sum_{s_n \in S} |X(s_n)| p_n$ converges (to a real number). If $X$ has an expectation, we set

$$E(X) = \sum_{s_n \in S} X(s_n) p_n$$

and call $E(X)$ the *expectation* of $X$. If $\sum_{s_n \in S} |X(s_n)| p_n$ diverges we say that $X$ *has no expectation*, $E(X)$ *is not defined*, or $E(X)$ *does not exist*.

The number $E(X)$ is also called the *expected value* of $X$ and its connection with the mean should be clear from the foregoing discussion.

Notice that if $S$ has a finite number of sample points, $E(X)$ must exist because $\sum_{s_n \in S} X(s_n) p_n$ is a finite sum. Furthermore, $E(X)$ does not exist for the $X$ in Example 7.2.4, even though the series $\sum_{n=1}^{\infty} (-1)^{n+1} \dfrac{n+1}{n(n+1)}$ converges conditionally.

Finally we point out that even though $X$ has an expectation only if $\sum_{s_n \in S} |X(s_n)| p_n$ converges (to a finite limit) the value of the expectation is $\sum_{s_n \in S} X(s_n) p_n$. This last series converges because absolute convergence of a series implies ordinary convergence of that series.

Examples 7.2.1 and 7.2.2 show random variables with the same distribution function but with different sample spaces. Nevertheless, both have the same expectation. The following theorem shows that such is always the case if the expectation exists. That is, one does not need to know the underlying probability system to find the expectation—the distribution function suffices. The theorem to be proved here is conclusive only in the case of countable sample spaces but other theorems show how to handle other cases.

**Theorem 7.2.1**     Let $X$ be a random variable on a probability system with a countable sample space $S$. If $\{x_j\} = \{X(s_n); s_n \in S\}$, if $f(x_j) = P(X = x_j)$, and if $X$ has an expectation then

$$E(X) = \sum x_j f(x_j)$$

PROOF     We may assume that the $x_j$ are distinct. There are countably many $x_j$ since $S$ is countable and hence $\{X(s_n); s_n \in S\}$ is countable. Let $M_j = \{s_n \in S; X(s_n) = x_j\}$. Suppose $M_j \cap M_k \neq \varnothing$. Then there is an $s_n \in M_j \cap M_k$ in which case $x_j = X(s_n) = x_k$. Since the $x_j$ are distinct we deduce that $j = k$. Thus $M_j \cap M_k \neq \varnothing$ only if $j = k$ and we deduce that $M_j \cap M_k = \varnothing$ for $j \neq k$. Since $\{x_j\} = \{X(s_n); s_n \in S\}$, every $s_n \in S$ must belong to $M_j$ for some $j$ so $S = M_1 \cup M_2 \cup M_3 \cup \cdots$.

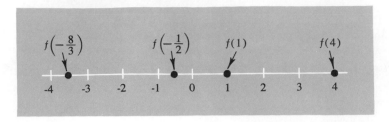

**Figure 1**

Since the expectation of $X$ exists we know that the series $\sum X(s_n)p_n$ converges absolutely. By known theorems on absolutely convergent series, any series of groups of terms from $\sum X(s_n)p_n$ where all terms are used once and only once converges to $\sum X(s_n)p_n$. Now since each $s_n \in S$ appears in one and only one of the sets $M_1, M_2, M_3, \ldots$

$$\sum_j \sum_{s_n \in M_j} X(s_n)p_n = \sum_{s_n \in S} X(s_n)p_n$$

But if $s_n \in M_j$, $X(s_n) = x_j$ and $P(M_j) = P(X = x_j) = f(x_j)$. So

$$E(X) = \sum_j \sum_{s_n \in M_j} X(s_n)p_n = \sum_j \sum_{s_n \in M_j} x_j p_n$$

$$= \sum_j x_j \sum_{s_n \in M} p_n = \sum_j x_j P(M_j) = \sum_j x_j f(x_j)$$

Although we will not show it here, it is easy to see that if $\sum x_j f(x_j)$ converges absolutely then so does $\sum_{s_n \in S} X(s_n)p_n$. Thus even the existence of the expectation can be determined from the distribution function. Nevertheless, there are occasions in which $\sum X(x_n)p_n$ is easier to calculate and to understand than $\sum x_j f(x_j)$.

A mechanical picture of the expected value of a random variable $X$ of the above sort can be obtained if one imagines a "point particle" of mass $f(x_j)$ at the point whose coordinate is $x_j$ on some coordinatized line (see Figure 1). The numbers below the line are coordinates and those above are masses. Since $\sum f(x_j) = 1$,

$$E(X) = \sum x_j f(x_j) = \frac{\sum x_j f(x_j)}{\sum f(x_j)}$$

But the last quotient is that used in determining the center of mass of a system of point particles. Thus $E(X)$ is the coordinate of the center of mass.

**EXERCISES 7.2**

**1** Let $S = \{1, 2, 3, 4\}$. Let $X(s) = s^2$ for $s \in S$. Find $E(X)$ if $S$ has the equal-likelihood measure. Find $E(X)$ if $P(1) = \frac{1}{2}$, $P(2) = \frac{1}{4}$, $P(3) = \frac{1}{8}$, $P(4) = \frac{1}{8}$.

♦ **2** Let $X$ be a random variable with $X(s) = c$, a constant, for all $s \in S$. Find $E(X)$.

3    Let $X$ have distribution function

$$F(u) = \begin{cases} 0, & u < -5 \\ \frac{1}{3}, & -5 \le u \le -2 \\ \frac{1}{2}, & -2 \le u < 0 \\ 1, & 0 \le u \end{cases}$$

Find $E(X)$. Find $E(Y)$ if $S = \{1, 2, 3, 4, 5, 6\}$ with equal likelihood and $Y(1) = -5$, $Y(2) = -5$, $Y(3) = 0$, $Y(4) = -2$, $Y(5) = 0$, $Y(6) = 0$.

4    Two fair dice are rolled independently. Find the expected value of (a) $X = $ sum of numbers showing, (b) $Y = $ product of the numbers showing, and (c) $X + Y$.

♦ 5    Let $X$ be any random variable with distribution function $F$. Let $z_l$ be the least upper bound of $\{u; F(u) < \frac{1}{2}\}$ and $z_u$ be the greatest lower bound of $\{u; F(u) > \frac{1}{2}\}$. Let $z = \frac{1}{2}(z_l + z_u)$. Find $z$ when $S = \{1, 2, 3, \ldots, n\}$ with equal likelihood and $X(n) = n$. Note the difference between the case of $n$ even and $n$ odd. Find $z$ for $X$ of Examples 7.1.1 and 7.2.2. We call $z$ the *median* of $X$.

6    In reference to $X$ and $Y$ of exercise 4 find the median of $X$, the median of $Y$, and the median of $X + Y$.

7    Let $S = \{1, 2, 3, 4, 5\}$ with equal likelihood. Let $X(s) = s$ and $Y(1) = 4$, $Y(2) = 7$, $Y(3) = 11$, $Y(4) = 9$, $Y(5) = 8$. Compute $E(X)$, $E(Y)$, $E(X + Y)$, median of $X$, median of $Y$, median of $(X + Y)$.

8    Find $E(X)$ where $X$ is the general Poisson random variable with $P(X = n) = e^{-\lambda}(\lambda^n/n!)$ for $\lambda > 0$ and $n = 0, 1, 2, 3, \ldots$.

9    Let $X$, $Y$, $Z$ represent the numbers showing on three independent distinguishable fair dice. Let $W = $ maximum $(X, Y, Z)$, i.e., $W(s) = $ largest of $(X(s), Y(s), Z(s))$ for each $s$ in a proper sample space. Find $E(W)$.

10    Let $X$ have the binomial distribution with parameters $n$ and $p$, i.e., $P(X = k) = \binom{n}{k} p^k (1 - p)^{n-k}$ if $n$ is a positive integer and $k \in \{0, 1, 2, \ldots, n\}$. Find $E(X)$.

11    An urn has 100 balls with ten black and 90 white. An experiment consists of removing five balls at random without replacement. What is the expected number of black balls removed?

12    Let $S = \{(m, n); m, n$ integers with $0 \le |n| \le m \le 10\}$, in other words, the points with integral coordinates in the triangle with vertices $(0, 0)$, $(10, 10)$, and $(10, -10)$. Suppose equal-likelihood measure is put on $S$ and let $X(s) = m$ if $s = (m, n)$. Find $E(X)$.

♦ 13    Let $S$ be a denumerable sample space with probabilities given by $P(s_n) = p_n$ for $S = \{s_1, s_2, s_3, \ldots\}$. Let $X$ be a random variable on $S$ and let $f$ be its probability function. Show that if $\sum |x_j| f(x_j)$ converges then $\sum |X(s_n)| p_n$ converges where $\{x_1, x_2, \ldots\} = \{x; f(x) \ne 0\}$.

14    Let $S = \{1, 2, 3, 4, \ldots\}$. Let $P(n) > 0$ for each $n \in S$, and $\sum_{n=1}^{\infty} P(n) = 1$. Show that there is a random variable $X$ on $S$ such that $E(X)$ does not exist.

**15** Let $X$ have the probability function $f$. Suppose that there is a number $c$ such that $f(c - x) = f(c + x)$ for all $x$. Show that if $E(X)$ exists then $E(X) = c$. Draw a diagram such as Figure 1 to illustrate the situation.

★ **16** Let $X$ have the probability function $f$. Suppose that $f(n) > 0$ only for $n = 0, 1, 2, 3, \ldots$. Let $\gamma(z) = \sum\limits_{n=0}^{\infty} f(n)z^n$ be the generating function of the sequence $f(n)$. Show, intuitively, that $E(X) = \lim\limits_{\substack{z \to 1 \\ z < 1}} \gamma'(z)$ if and only if both exist.

★ **17** Find the expected value of the binomial, Poisson, and negative binomial distributions by means of the technique in exercise 16.

◆ **18** Let $S$ be a countable sample space. Let $X$ be a random variable such that for some $A$, $|X(s)| \leq A$. What can be said about $E(X)$? Justify.

# 7.3 Euclidean sample spaces

If $S$ is a subset of $m$-dimensional Euclidean space and if the probability measure $P$ is given by an integral, namely, for an event, $M$, $P(M) = \int_M p(s)\, ds$ we might seek the analog of

$$\sum_{s_n \in S} X(s_n) p_n = \sum_{s_n \in S} X(s_n) P(s_n)$$

Before proceeding further it should be kept in mind that $s$ is a point (vector) in $m$-dimensional space and that $ds$ can stand for $dx$, $dx\, dy$, $dx\, dy\, dz$, or even an element of surface area according to the context.

The expression $\sum\limits_{s_n \in S} X(s_n) P(s_n)$ is a sum of payoffs times probabilities of the payoffs. In the case of a Euclidean sample space we would have $P(s) = 0$ for each sample point $s$.

Suppose we have $S = C_1 \cup C_2 \cup C_3 \cup \cdots$, where the $C_k$ are disjoint events on which $X(s)$ varies only slightly as $s$ ranges over $C_k$. That is, if $s_k$ is a fixed point in $C_k$, $X(s)$ is close to $X(s_k)$ whenever $s \in C_k$. It should be possible to choose such $C_k$ whenever $X$ is piecewise continuous. Since $P(C_k) = \int_{C_k} p(s)\, ds$, we have

$$\sum X(s_k) P(C_k) = \sum X(s_k) \int_{C_k} p(s)\, ds = \sum \int_{C_k} X(s_k) p(s)\, ds$$

However, since $X(s)$ is close to $X(s_k)$ when $s \in C_k$, $\int_{C_k} X(s) p(s)\, ds$ should be close to $\int_{C_k} X(s_k) p(s)\, ds$. This motivates our use of $\sum \int_{C_k} X(s) p(s)\, ds$ as analogous to the expectation in the countable case. Since the $C_k$ are disjoint and their union is $S$ it should not be surprising that (under certain restrictions on $X$)

$$\sum_{k=1}^{\infty} \int_{C_k} X(s)p(s)\, ds = \int_S X(s)p(s)\, ds$$

We therefore make the following definitions.

**Definition**   Let $X$ be a random variable on a probability system with a Euclidean sample space $S$, the usual system of events and a probability mass density. We say that $X$ *has an expectation* if $\int_S |X(s)|p(s)\, ds$ converges (to a finite value). In that case the *expectation* $E(X) = \int_S X(s)p(s)\, ds$.

Almost every remark following the definitions in the case of a countable sample space has an analog here. Hence, we do repeat them.

*Example 7.3.1*
Let $S$ be the positive real axis with density given by $p(x) = 2/(1 + x)^3$. Let $X(x) = x - 4$. Does $E(X)$ exist? If it does, find it.

We test for the convergence of $\int_0^\infty |x - 4| \dfrac{2}{(1 + x)^3}\, dx$. First of all, if $0 \le x \le 4$, $|x - 4| \le 4$. If $x \ge 4$, $|x - 4| = x - 4 \le x + 1$. Thus

$$\int_0^\infty |x - 4| \frac{2}{(1 + x)^3}\, dx \le \int_0^4 4 \frac{2}{(1 + x)^3}\, dx + \int_4^\infty (x + 1) \frac{2}{(1 + x)^3}\, dx$$

The first integral on the right is the integral of a bounded function on a finite interval, therefore it converges. The second integral can be evaluated by ordinary means. In fact $\int_4^\infty \dfrac{2}{(1 + x)^2}\, dx = \dfrac{-2}{(1 + x)}\Big|_4^\infty = \dfrac{2}{5}$. Thus $E(X)$ exists. We now can write

$$E(X) = \int_0^\infty (x - 4) \frac{2}{(1 + x)^3}\, dx = 2 \int_0^\infty (x + 1 - 5) \frac{1}{(1 + x)^3}\, dx$$

$$= 2 \int_0^\infty \left[ \frac{x + 1}{(x + 1)^3} - \frac{5}{(x + 1)^3} \right] dx = 2 \int_0^\infty \frac{dx}{(x + 1)^2} - 2 \int_0^\infty \frac{5}{(x + 1)^3}\, dx$$

$$= -\frac{2}{(x + 1)}\Big|_0^\infty - 10 \left[ -\frac{1}{2(x + 1)^2} \right]_0^\infty = 2 - 5 = -3$$

*Example 7.3.2*
A point is chosen from the sample space $S = \{x; 0 < x < 1\}$ in such a way that $p(x) = 2x$. Let $X(x) = \dfrac{\cos 4\pi x}{x}$. Does $E(X)$ exist? If so, find $E(x)$.

$E(X)$ exists if $\int_0^1 \left| \dfrac{\cos 4\pi x}{x} \right| 2x \, dx$ converges. Since $x > 0$, $\left| \dfrac{\cos 4\pi x}{x} \right| =$ $\dfrac{1}{x} |\cos 4\pi x|$ so we test for convergence of $\int_0^1 \dfrac{1}{x} |\cos 4\pi x| 2x \, dx = \int_0^1 2 |\cos 4\pi x| \, dx$. But $|\cos 4\pi x|$ defines a bounded continuous function on $0 < x < 1$ so $E(X)$ exists. Now

$$E(X) = \int_0^1 \frac{\cos 4\pi x}{x} 2x \, dx = 2 \int_0^1 \cos 4\pi x \, dx = \frac{1}{2\pi} \sin 4\pi x \Big|_0^1 = 0$$

## Example 7.3.3

Suppose that the above example were changed so that $p(x) = 1$ with the same $S$ and the same random variable. Does $E(X)$ still exist?

In testing for the convergence of $\int_0^1 \left| \dfrac{\cos 4\pi x}{x} \right| dx$ we note that $\left| \dfrac{\cos 4\pi x}{x} \right| \geq \dfrac{\frac{1}{2}}{x}$ if $0 < x < \frac{1}{12}$. This is simply because $\cos y \geq \frac{1}{2}$ if $0 < y < \frac{1}{3}\pi$ and thus $\cos 4\pi x \geq \frac{1}{2}$ if $0 < 4\pi x < \frac{1}{3}\pi$ which is $0 < x < \frac{1}{12}$. Now it is easy to verify that $\int_0^{1/12} \dfrac{\frac{1}{2}}{x} \, dx$ does not converge because $\int_\epsilon^{1/12} \dfrac{\frac{1}{2}}{x} \, dx = \frac{1}{2} \ln x \Big|_\epsilon^{1/12}$ and $-\ln \epsilon \to \infty$ as $\epsilon \to 0$.

It is interesting to note that had $S = \{x; 0 < x < 1\}$ and $p(x) = 1$ with $Y(x) = \cos (4\pi/x)/x$, $E(Y)$ does not exist although $\int_0^1 \dfrac{1}{x} \cos \dfrac{4\pi}{x} \cdot 1 \, dx$ does converge. This again involves conditional convergence versus absolute convergence. We now turn to some higher-dimensional examples.

## Example 7.3.4

A point is chosen from the rectangle $R = \{(x, y); 0 \leq x \leq 2 \text{ and } 0 \leq y \leq 10\}$ in such a way that the probability of an event is proportional to its area. Let $X(x, y)$ be the kinetic energy of a particle of mass $x$ and speed $y$. Find $E(X)$.

We first observe that the area of $R$ is $2 \cdot 10 = 20$. Thus $p(x, y) = \frac{1}{20}$ and $P(M) = \int_M \frac{1}{20} \, ds = \iint_M \frac{1}{20} \, dx \, dy$. The kinetic energy of a particle of mass $x$ and speed $y$ is $\frac{1}{2} x y^2$. Thus

$$E(X) = \int_0^2 \int_0^{10} \tfrac{1}{2} x y^2 \tfrac{1}{20} \, dy \, dx$$

if the integral converges since $\frac{1}{2} x y^2 \geq 0$ and hence $|\frac{1}{2} x y^2| = \frac{1}{2} x y^2$. It is immediate that

$$\int_0^2 \int_0^{10} \tfrac{1}{2} x y^2 \tfrac{1}{20} \, dy \, dx = \frac{1}{40} \frac{2^2}{2} \frac{10^3}{3} = \frac{50}{3}$$

Since $X(x, y) \geq 0$ for all $s = (x, y)$ in $S$, the existence of and the calculation of $E(X)$ could be done simultaneously.

### Example 7.3.5

A point $(x, y, z)$ is selected at random on $S$, the surface of the hemisphere $\{(x, y, z); x^2 + y^2 + z^2 = r^2 \text{ and } z > 0\}$. "At random" is assumed to mean that the probability of an event is proportional to its area. Let $X(x, y, z) = z$. Find $E(X)$.

From geometry, the surface area of $S$ is $2\pi r^2$. From calculus, the element of surface area of a surface $z = h(x, y)$ is $\sqrt{1 + (\partial z/\partial x)^2 + (\partial z/\partial y)^2} \, dx \, dy$. For our hemisphere $2x + 2z(\partial z/\partial x) = 0$ after taking the partial derivative of both sides of the equation $x^2 + y^2 + z^2 = r^2$ with respect to $x$. We obtain $\partial z/\partial x = -x/z$. Similarly $\partial z/\partial y = -y/z$ and the element of surface area is $\sqrt{1 + x^2/z^2 + y^2/z^2} \, dx \, dy$ $= (1/z)\sqrt{x^2 + y^2 + z^2} \, dx \, dy = (r/z) \, dx \, dy$.

Now $p(s) \, ds = \dfrac{r \, dx \, dy}{2\pi r^2 z} = \dfrac{dx \, dy}{2\pi r z}$ since $2\pi r^2$ is the surface area of $S$. Thus

$$E(X) = \iint\limits_{x^2 + y^2 \leq r^2} z \, \frac{dx \, dy}{2\pi r z} = \frac{1}{2\pi r} \iint\limits_{x^2 + y^2 \leq r^2} dx \, dy$$

The last expression is just $(1/2\pi r)(\pi r^2) = r/2$, since the region $x^2 + y^2 \leq r^2$ is just a disk of radius $r$ which has area $\pi r^2$.

### Example 7.3.6

Let $X$ have the negative exponential distribution. Find $E(X)$.

In the standard representation $S = R$, $p(x) = f(x)$, and $X(x) = x$. For the negative exponential distribution

$$f(x) = \begin{cases} 0, & x \leq 0 \\ \lambda e^{-\lambda x}, & x > 0 \end{cases}$$

Therefore $E(X) = \displaystyle\int_R X(x)p(x) \, dx = \int_0^\infty x\lambda e^{-\lambda x} \, dx$, where the second integral extends from 0 to $\infty$ since $f(x) = 0$ for $x \leq 0$. We use integration by parts with $u = x$, $du = dx$ and $dv = \lambda e^{-\lambda x} \, dx$, $v = -e^{-\lambda x}$. Thus

$$\int_0^\infty x\lambda e^{-\lambda x} \, dx = -xe^{-\lambda x} \Big|_0^\infty + \int_0^\infty e^{-\lambda x} \, dx$$

$$= -0 + 0 + (-1/\lambda)e^{-\lambda x} \Big|_0^\infty = 1/\lambda$$

The reader may ask whether we are justified in choosing the standard representation to find $E(X)$ in Example 7.3.6. Just as in the case of discrete random variables it can be shown that the distribution of a random variable completely determines its expectation. Therefore it is not necessary to use the integral over the sample space if the density function is available. Naturally, if a sample space integral is convenient it should be used. No matter which is used, the result is the same. As in Example 7.3.6 the standard representation will always give

$$(7.3.1) \qquad E(X) = \int_R xf(x)\, dx$$

if the integral is absolutely convergent; that is $\int_R |x| f(x)\, dx$ is a real number.

### Example 7.3.7

Suppose that the fraction of impurities in random samples of a particular ore has the distribution function

$$F(u) = \begin{cases} 0, & u < 0 \\ u^n, & 0 \le u \le 1 \\ 1, & u > 1 \end{cases} \qquad \text{for } n > 0$$

Find the expected value of the fraction of impurities.

We follow (7.3.1). We differentiate $F(u)$. If $u < 0$ or $u > 1$, $F'(u) = 0$. If $0 < u < 1$, $F'(u) = nu^{n-1}$. It can also be shown that if $n > 1$, $F'(0) = 0$. However it is not essential to know $F'(0)$ or $F'(1)$. It can easily be shown that

$$f(x) = \begin{cases} 0, & x \le 0 \\ nx^{n-1}, & 0 < x < 1 \\ 0, & x \ge 1 \end{cases}$$

is a density function for $F$. Therefore

$$E(X) = \int_R xf(x)\, dx = \int_0^1 xnx^{n-1}\, dx = \int_0^1 nx^n\, dx = \frac{n}{n+1} x^{n+1}\Big|_0^1 = \frac{n}{n+1}$$

We do not take up absolute convergence because $|x| = x$ for $0 \le x \le 1$ and $\int_R |x| f(x)\, dx$ is the same as $\int_R xf(x)\, dx$ in this case.

The next example shows how the expectation may be used in choosing between two random variables. It should be kept in mind that many things are being left unsaid.

### Example 7.3.8

For an admission charge of $3.25 one may choose one of two payoff books to be used in connection with the random choice of a point $\theta$ on the line segment

$\{\theta; 0 < \theta < 2\pi\}$. One payoff is $X(\theta) = \theta$ and the other payoff is $Y(\theta) = \frac{2}{9}\theta^2$. Which payoff book is preferable?

Preferability depends on one's purposes. Let us first give some probabilistic information; $E(X) = \int_0^{2\pi} \theta \frac{1}{2\pi} \, d\theta$ where $(1/2\pi) \, d\theta$ is used because of the assumption of the random choice of $\theta$. Therefore,

$$E(X) = \frac{1}{2\pi} \frac{\theta^2}{2} \Big|_0^{2\pi} = \frac{4\pi^2}{4\pi} = \pi$$

Similarly,

$$E(Y) = \int_0^{2\pi} \frac{2}{9}\theta^2 \frac{1}{2\pi} \, d\theta = \frac{1}{27\pi} \theta^3 \Big|_0^{2\pi} = \frac{8\pi^2}{27}$$

It can be seen that $\frac{27}{8} > \pi$ so $1 > \frac{8}{27}\pi$ and $\pi > \frac{8}{27}\pi^2$. Therefore, $E(X) > E(Y)$. On the basis of which has the larger expectation (a kind of average), $X$ is preferable to $Y$. The expectation of both are less than the admission price but that is not unusual.

Looking at the situation in another way, we see that $P(X < 5) = 5/2\pi$ while $P(Y < 5) = P(\frac{2}{9}\theta^2 < 5) = P(\theta \leq \frac{3}{2}\sqrt{10}) = (1/2\pi)\frac{3}{2}\sqrt{10}$. Since $5 > \frac{3}{2}\sqrt{10}$, $P(X < 5) > P(Y < 5)$ and $P(X \geq 5) = 1 - P(X < 5) < 1 - P(Y < 5) = P(Y \geq 5)$. Thus, despite the higher expectation of $X$, $Y$ is better at providing higher payoffs than \$5.

Example 7.3.8 suggests that which of two random variables is preferable is a complicated question. Nevertheless, a common convention is the choice of the one with the higher expectation.

**EXERCISES 7.3** _____

1   Let $S = \{x; 0 < x < 1\}$ and let $p(x) = 1$. Find the expectations of the random variables $X$ and $Y$ where $X(x) = 3x^2 - 7x + 2$, $Y(x) = 2e^x$.

2   Let $S = \{x; 0 < x < 1\}$ with $p(x) = (2/\pi) \sin \frac{1}{2}\pi x$. Find the expectation of $X$ given by $X(x) = x$.

3   Discuss the expectation of $X$ where $S = \{x; 0 < x < \pi\}$, $p(x) = \sin x$, and $X(x) = x$.

4   Discuss the expectation of $X$ when $X(x) = x$, $S = \{x; 1 < x\}$, and $p(x) = 1/x^2$.

5   Let $F(x) = \begin{cases} 0, & x < 0 \\ 1 - e^{-x}, & x \geq 0 \end{cases}$ be the distribution function for the random variable $X$. Find $E(X)$. Find the median of $X$.

**6**   Discuss the remarks about $Y(x)$ following Example 7.3.3.

**7**   A point $(x, y)$ is chosen at random (uniformly) from the square $\{(x, y);$ $0 \leq x \leq 2, \quad 0 \leq y \leq 2\}$. Find the expected area of an ellipse with semi-axes of length $x$ and $y$.

**8**   Consider the portion of the sphere $x^2 + y^2 + z^2 = r^2$ which lies above the plane $z = \frac{1}{2}r$. A point $(x, y, z)$ is chosen from this sample space in such a fashion that the probability of a set is proportional to its surface area. Let $X(x, y, z) = z$. Find $E(X)$.

**9**   Let $S = \{x; x \text{ real}\}$ and let $p(x) = (1/\sqrt{2\pi})e^{-x^2/2}$. Let $X(x) = x^2$. Find $E(X)$. Let $Y_m(x) = x^{2m}$ for positive integers $m$. Find $E(Y_{m+1})$ in terms of $E(Y_m)$.

**10**   Suppose the random variable $X$ has a continuous density function $f$ such that $f(x) = 0$ if $x > 0$. Let $F$ be the distribution function for $X$. Show that $E(X)$ exists if $xF(x) \to 0$ as $x \to -\infty$ and $\int_{-\infty}^{0} F(x)\, dx$ converges.

**11**   Use (7.3.1) to find the expected value of the uniform distribution.

**12**   A point $(x, y)$ is chosen as in exercise 7. Find $E(X)$ if $X(x, y) = \sqrt{x^2 + y^2}$, the distance from the origin to $(x, y)$.

**13**   Find the expectation of the random variable of exercise 12 by finding its density function and using (7.3.1).

**14**   Find an analog to the result in exercise 10 which would hold if $f(x) = 0$ for $x < 0$ when $f$ is a density function.

**15**   Use integration by parts to show that $\Gamma(\alpha + 1) = \alpha\Gamma(\alpha)$ when $\Gamma(\alpha) = \int_0^\infty x^{\alpha-1}e^{-x}\, dx$, $\alpha > 0$. Show that $\Gamma(n + 1) = n!$ for $n = 0, 1, 2, 3, \ldots$.

**16**   Find the expectation of $X$ if

$$f_{X(x)} = \begin{cases} 0, & x \leq 0 \\ \dfrac{b^\alpha x^{\alpha-1}}{\Gamma(\alpha)}\, e^{-bx}, & x > 0 \end{cases}$$

[*Hint:* Note exercise 15.]

**17**   Two points $x$ and $y$ are chosen uniformly and independently on our stick of unit length. Find $E(|x - y|)$.

**18**   Show that Student's $t$-distribution has an expectation if and only if $n > 1$.

**19**   For what values of $m$ and $n$ does the expected value of the $F$-distribution exist. Recall the $F$-distribution from section 6.4.

◆ **20**   Let $X$ have the density function $f(x)$. Suppose for some $c$, $f(c - x) = f(c + x)$ for all $x$. Show that if $E(X)$ exists, $E(X) = c$. Draw a picture of such a density function.

# 7.4   Algebraic properties of the expectation

After having established the existence of the expectation of certain random variables, it is natural to ask how the expectation relates to the arithmetic or algebraic operations on random variables. We define $(cX)(s) = cX(s)$, $(X + Y)(s) = X(s) + Y(s)$, and $(XY)(s) = X(s)Y(s)$, whenever $X$ and $Y$ are random variables on $S$.

**Theorem 7.4.1**     Let $X$ and $Y$ be random variables given with respect to a single probability system. If $E(X)$ and $E(Y)$ exist then $E(X + Y)$ exists and $E(X + Y) = E(X) + E(Y)$.

PROOF      We prove the theorem in the two cases of interest to us. First, suppose $S$ is countable and $P(s_n) = p_n$ for $s_n \in S$. Since $|a + b| \le |a| + |b|$ we have $|X(s_n) + Y(s_n)| \le |X(s_n)| + |Y(s_n)|$ so all partial sums of the series $\sum |X(s_n) + Y(s_n)|p_n$ are bounded by $\sum |X(s_n)|p_n + \sum |Y(s_n)|p_n$. Therefore the series $\sum [X(s_n) + Y(s_n)]p_n$ is absolutely convergent. But since $\sum X(s_n)p_n$ and $\sum Y(s_n)p_n$ converge, $E(X+Y)=\sum [X(s_n)+ Y(s_n)]p_n=\sum [X(s_n)p_n+ Y(s_n)p_n]= \sum X(s_n)p_n + \sum Y(s_n)p_n = E(X) + E(Y)$. If $S$ is a Euclidean probability space we must test for the convergence of $\int_S |X(s)+ Y(s)|p(s)\, ds$. Since $|X(s)+ Y(s)|p(s)$ $\le |X(s)|p(s) + |Y(s)|p(s)$ we may use theorems from the theory of integration which state that if $l(s)$ and $m(s)$ are non-negative functions which can be integrated and if $l(s) \ge m(s)$ then convergence of $\int_S l(s)\, ds$ implies that $\int_S m(s)\, ds$ converges. Therefore if we take $m(s) = |X(s) + Y(s)|p(s)$ and $l(s) = |X(s)|p(s) + |Y(s)|p(s)$ we have

$$\int_S l(s)\, ds = \int_S |X(s)|p(s)\, ds + \int_S |Y(s)|p(s)\, ds$$

which exist by the assumed existence of $E(X)$ and $E(Y)$. That $E(X + Y) = E(X) + E(Y)$ follows from the additivity of the integral.

**Theorem 7.4.2**     If $X$ is a random variable given with respect to a probability system and if $c \ne 0$ is a real number, then $E(cX) = cE(X)$ and both expectations exist or do not exist together.

PROOF      The proof follows from elementary theorems for series and integrals.

It should be noted that $E(cX)$ always exists if $c = 0$. It is not necessarily the case that $E(X)$ exists if $E(cX)$ exists and $c = 0$. Before proving another theorem we give an example which is a particular but hopefully illuminating case of the theorem to be proved.

### Example 7.4.1

Let $S = \{1, 2, \ldots, 16\}$ with $P(n) = \frac{1}{16}$ for all $n \in S$. Let

$$X(n) = \begin{cases} 5, & n = 1, 2, 3, 16 \\ -2, & n = 4, 5, 6, 11, 12, 15 \\ 0, & n = 7, 8, 13, 14 \\ 1, & n = 9, 10 \end{cases}$$

$$Y(n) = \begin{cases} 3, & n = 1, 3, 4, 6, 7, 8, 10, 15 \\ -4, & n = 2, 5, 9, 11, 12, 13, 14, 16 \end{cases}$$

Show that $X$ and $Y$ are independent and study $E(XY)$, $E(X)$, $E(Y)$.

Let us first study $E(XY)$, $E(X)$, $E(Y)$. $E(Y) = 3 \cdot \frac{8}{16} + (-4) \cdot \frac{8}{16} = -\frac{1}{2}$, $E(X) = 5 \cdot \frac{4}{16} + (-2) \cdot \frac{6}{16} + 0 \cdot \frac{4}{16} + 1 \frac{2}{16} = \frac{10}{16}$, $E(XY) = \frac{1}{16}[15 - 20 + 15 - 6 + 8 - 6 + 0 + 0 - 4 + 3 + 8 + 8 + 0 + 0 - 6 - 20] = -\frac{5}{16} = (-\frac{1}{2}) \cdot \frac{10}{16}$, so $E(XY) = E(X)E(Y)$. As for independence, we look at certain cases and leave the other cases to the reader.

$$P(X = 5 \text{ and } Y = 3) = P(\{1, 3\}) = \frac{2}{16} = \frac{4}{16} \cdot \frac{8}{16} = P(X = 5)P(X = 3)$$

$$P(X = 0 \text{ and } Y = -4) = P(\{13, 14\}) = \frac{2}{16} = \frac{4}{16} \cdot \frac{8}{16} = P(X = 0)P(Y = -4)$$

$$P(X = -2 \text{ and } Y = 3) = P(\{4, 6, 15\}) = \frac{3}{16} = \frac{6}{16} \cdot \frac{8}{16} = P(X = -2)P(Y = 3)$$

Note that if $A_1 = \{X = 5\}$, $A_2 = \{X = -2\}$, $A_3 = \{X = 0\}$, $A_4 = \{X = 1\}$, $B_1 = \{Y = 3\}$, and $B_2 = \{Y = -4\}$, then $A_1 \cap B_1$, $A_1 \cap B_2$, $A_2 \cap B_1$, $A_2 \cap B_2$, $A_3 \cap B_1$, $A_3 \cap B_2$, $A_4 \cap B_1$, $A_4 \cap B_2$ are disjoint, their union is $S$, and

$$\begin{aligned} E(XY) = {} & 5 \cdot (3)P(A_1 \cap B_1) + 5 \cdot (-4)P(A_1 \cap B_2) + (-2)(3)P(A_2 \cap B_1) \\ & + (-2)(-4)P(A_2 \cap B_2) + 0 \cdot (3)P(A_3 \cap B_1) + 0 \cdot (-4)P(A_3 \cap B_2) \\ & + 1 \cdot (3)P(A_4 \cap B_1) + 1 \cdot (-4)P(A_4 \cap B_2) \end{aligned}$$

The above example suggests a certain theorem. We prove this theorem only in the case of a countable sample space and we give a proof which is not the simplest. However, hopefully, the ideas in the proof will lead to more insight into probability theory.

**Theorem 7.4.3** If $X$ and $Y$ are independent random variables in a probability system with sample space $S$ and if $E(X)$ and $E(Y)$ exist, then $E(XY)$ exists and $E(XY) = E(X)E(Y)$.

PROOF As mentioned above, we assume that $S$ is countable. Let $\{x_1, x_2, \ldots\} = \{X(s); s \in S\}$ and let $\{y_1, y_2, \ldots\} = \{Y(s); s \in S\}$. It is conceivable that there are only finitely many $x_j$ or $y_k$ but we use the notation of a denumerable set. There cannot be uncountably many $x_j$ since $S$ is countable and $X$ is a function on $S$. We further assume that $x_j \neq x_k$ and $y_j \neq y_k$ for $j \neq k$. Let $A_j = \{s \in S; X(s) = x_j\}$ and $B_k = \{s \in S; Y(s) = y_k\}$. By the observation that $x_j \neq x_k$ for $j \neq k$ and the definitions of $X$, $Y$, $A_j$, and $B_k$ we have $A_j \cap A_{j'} = \varnothing$ for $j \neq j'$, $B_k \cap B_{k'} = \varnothing$ for $k \neq k'$, and $S = \bigcup A_j = \bigcup B_k$.

Let $M_{jk} = A_j \cap B_k$. It is immediate that $M_{jk} \cap M_{lm} = \varnothing$ if either $j \neq l$ or $k \neq m$. Furthermore,

$$\bigcup_{j,k} M_{jk} = \bigcup (A_j \cap B_k) = (\bigcup A_j) \cap (\bigcup B_k) = S \cap S = S$$

Thus each $s \in S$ is in one and only one of the events $M_{jk}$. The series $\sum_{s \in S} |X(s)Y(s)|P(s)$ must be tested for convergence. Since $|X(s)Y(s)| \geq 0$, theorems on series show that

$$\sum_{s \in S} |X(s)Y(s)|P(s) = \sum_{j,k} \sum_{s \in M_{jk}} |X(s)Y(s)|P(s)$$

in the sense that both converge if either converges and if they converge, their sums are equal. But if $s \in M_{jk}$, $s \in A_j \cap B_k$ and $X(s) = x_j$, $Y(s) = y_k$. Furthermore since $X$ and $Y$ are independent, $P(M_j) = P(A_j \cap B_k) = P(A_j)P(B_k)$. Thus

$$\sum_{j,k} \sum_{s \in M_{jk}} |X(s)Y(s)|P(s) = \sum_{j,k} |x_j| \, |y_k|P(A_j)P(B_k)$$

But by theorems on absolutely convergent series

$$\sum_{j,k} |x_j| \, |y_k|P(A_j)P(B_k) = [\sum_j |x_j|P(s_j)][\sum_k |y_k|P(B_k)]$$

Now the existence of $E(X)$ and $E(Y)$ imply the convergence of $\sum |x_j|P(A_j)$ and $\sum |y_k|P(B_k)$. Thus $E(XY)$ exists. Since all series at issue converge absolutely, the same manipulations apply without the absolute value symbols and we obtain

$$E(XY) = \sum_{s \in S} X(s)Y(s)P(s) = \sum_{j,k} \sum_{s \in M_{jk}} X(s)Y(s)P(s)$$

$$= \sum_{j,k} x_j y_k P(A_j)P(B_k) = [\sum_j x_j P(A_j)][\sum_k y_k P(B_k)] = E(X)E(Y)$$

### Example 7.4.2
If $X$ and $Y$ are not independent show it is possible that $E(XY) \neq E(X)E(Y)$ but it is also possible that $E(XY) = E(X)E(Y)$.

For the first part, let $X(1) = \frac{1}{2}$, $X(-1) = \frac{1}{2}$, and let $Y = 2X$. For the second part, let $S = \{1, 2, 3, 4\}$. Let $X(1) = X(2) = 1$, let $X(3) = X(4) = 0$. Let $Y(1) = 1$, $Y(2) = -1$ and let $Y(3) = Y(4) = 0$. We leave verification to the student.

**176** Mathematical Expectation

EXERCISES 7.4

1  Verify that which is called for in Example 7.4.2.

♦ 2  Prove that if $X$, $Y$, and $Z$ are independent random variables with expectation, then $XYZ$ has expectation and $E(XYZ) = E(X)E(Y)E(Z)$.

♦ 3  Give an example of two random variables $X$ and $Y$, neither of which has an expectation but such that $E(X + Y)$ exists.

4  Show that if $X$ and $Y$ are independent, $Y(s) \neq 0$ and $E(Y) \neq 0$, then $E(X/Y) = E(X)/E(Y)$.

5  Let $X$ and $Y$ be as in exercise 6.5.6. Compute $E(XY)$ and compare it to $E(X)E(Y)$.

6  Prove that the expected number of balls in box No. 1 is $n/k$ when $n$ indistinguishable balls are placed at random in $k$ distinguishable boxes by working with $X_j$, the number of balls in the $j$th box. Specifically, show that the $X_j$ are identically distributed and $X_1 + X_2 + \cdots + X_k = n$.

7  Let $X_1, X_2, \ldots, X_{200}$ be as in exercise 6.5.12. Find $Y = X_1 + X_2 + X_3 + \cdots + X_{200}$. Are the $X_j$ identically distributed? Calculate $E(X_j)$ based on your knowledge of $E(Y)$. Calculate $E(X_1 + X_2 + \cdots + X_{10})$. What is the probability distribution of $X_1 + X_2 + \cdots + X_{10}$? What would be involved in calculating $E(X_1 + \cdots + X_{10})$ directly from that distribution?

# 7.5  The variance

There can be many random variables with the same expectation. Table 2, below, specifies the probability functions of four random variables whose expectations are zero.

It is true that these random variables have symmetric probability functions, i.e., $f(k) = f(-k)$ and are somewhat unrepresentative on that account, but they

**Table 2**

| $n$ | $f_X(n)$ | $f_Y(n)$ | $f_Z(n)$ | $f_W(n)$ |
|---|---|---|---|---|
| 100 | 0 | $\frac{1}{2}$ | $\frac{1}{6}$ | $\frac{1}{100}$ |
| 10 | 0 | 0 | $\frac{1}{6}$ | 0 |
| 1 | $\frac{1}{2}$ | 0 | $\frac{1}{6}$ | $\frac{49}{100}$ |
| $-1$ | $\frac{1}{2}$ | 0 | $\frac{1}{6}$ | $\frac{49}{100}$ |
| $-10$ | 0 | 0 | $\frac{1}{6}$ | 0 |
| $-100$ | 0 | $\frac{1}{2}$ | $\frac{1}{6}$ | $\frac{1}{100}$ |

illustrate certain behavior of random variables with the same expected value. One might observe that $X$ and $W$ are somewhat similar. For both, one has *approximately* a probability of $\frac{1}{2}$ that the payoff will be 1 or $-1$. If a person is faced with a situation where he must win a total of 100 in three plays or fewer, a game corresponding to the random variable $X$ is useless, $W$ is probably almost as bad, $Y$ gives you probability of more than $\frac{1}{2}$ of success, and $Z$ is reasonable.

Although the probability function or distribution function contains all of the information which is probabilistically interesting in the study of a single random variable, it is sometimes desirable to describe (in part) the spread of the values of the random variable from the center with a single number—much as we specified the center with a single number. Naturally, a single number describing the spread constitutes a simplified picture of the probabilistic situation but experience has shown that this simplification may be advantageous.

There are many ways of prescribing a single number to describe the spread of values. We indicate a number of these ways in reference to the random variables in Table 2, but eventually we will focus on one way.

Consider the range $r =$ (largest value which occurs with positive probability) $-$ (smallest value which occurs with positive probability).

Thus

$$r(X) = 1 - (-1) = 2$$
$$r(Y) = 100 - (-100) = 200$$
$$r(Z) = 100 - (-100) = 200$$
$$r(W) = 100 - (-100) = 200$$

One may feel that $Y$, $Z$, and $W$ are quite different and should not have the same number to indicate their spread of values. Also, the range may not be defined since there may be no largest or smallest value with the given requirements. However, the range is a useful concept in many problems and although we will not emphasize it, it is worth much additional study.

The absolute deviation $\alpha$ can be obtained as $\alpha(U) = \sum |u_j - E(U)| f_U(u_j)$ for a random variable $U$ with probability function $fU$. In our cases

$$\alpha(X) = \frac{1}{2} \cdot 1 + \frac{1}{2} \cdot 1 = 1$$
$$\alpha(Y) = \frac{1}{2} \cdot 100 + \frac{1}{2} \cdot 100 = 100$$
$$\alpha(Z) = \frac{1}{6}(100 + 10 + 1 + 1 + 10 + 100) = \frac{1}{6}(222)$$
$$\alpha(W) = \frac{1}{110} \cdot 100 + \frac{49}{100} \cdot 1 + \frac{49}{100} \cdot 1 + \frac{1}{100} \cdot 100 = 2.98$$

This quantity seems rather useful and it distinguishes nicely among $X$, $Y$, $Z$, and $W$. However, there are technical disadvantages to its use. It is somewhat tedious to integrate absolute values of functions with varying algebraic sign. If one had avoided absolute values by using $\sum [u_j - E(U)] f_U(u_j)$ one would get nothing because $\sum [u_j - E(U)] f_U(u_j) = 0$.

Finally, we consider

$$\sigma^2(U) = \sum [u_j - E(U)]^2 f_U(u_j)$$

We have

$$\sigma^2(X) = (1)^2 \cdot \tfrac{1}{2} + (-1)^2 \cdot \tfrac{1}{2} = 1$$

$$\sigma^2(Y) = (100)^2 \cdot \tfrac{1}{2} + (-100)^2 \cdot \tfrac{1}{2} = 10\,000$$

$$\sigma^2(Z) = (100)^2 \cdot \tfrac{1}{6} + 10^2 \cdot \tfrac{1}{6} + 1^2 \cdot \tfrac{1}{6} + (-1)^2 \cdot \tfrac{1}{6} + (-10)^2 \cdot \tfrac{1}{6} + (-100)^2 \cdot \tfrac{1}{6}$$

$$= \tfrac{1}{6}(20\,202) = 3367$$

$$\sigma^2(W) = (100)^2 \cdot \tfrac{1}{100} + (1)^2 \cdot \tfrac{49}{100} + (-1)^2 \cdot \tfrac{49}{100} + (-100)^2 \cdot \tfrac{1}{100} = 200.98$$

The quantity $\sigma^2(U)$ is called the *variance* (or *dispersion*) of $U$ and although there is a squaring, experience has shown it to be technically more manageable than $\alpha(U)$. We will focus our attention on $\sigma^2$ and $\sigma = \sqrt{\sigma^2}$ which is called the *standard deviation*. It is important to observe that

$$\sigma^2(U) = \sum [u_j - E(U)]^2 f_U(u_j) = E((U - E(U))^2)$$

i.e., the expectation of the random variable $V$ given by $V(s) = [U(s) - E(u)]^2$.

Before we make a formal definition, however, it is desirable to make some observations which help to put this whole body of ideas in good perspective.

**Theorem 7.5.1**    Let $X$ be a random variable defined on the sample space $S$ of a probability system. Let $a$ and $b$ be real numbers and let $p > q > 0$. Then $E(|X - a|^p)$ exists if and only if $E(|X - b|^p)$ exists and $E(|X - a|^q)$ exists if $E(|X - a|^p$ exists.

PROOF    (When $S$ is countable, only.) We first show that the existence of $E(|X - a|^p)$ implies that $E(|X - a|^q)$ exists. Let $M = \{s_n \in S; |X - a| > 1\}$ then $S - M = \{s_n \in S; |X - a| \le 1\}$. Since $|X - a|^q \ge 0$, $\big|\,|X - a|^q\big| = |X - a|^q$ so $E(|X - a|^q)$ exists if the series $\sum_{s_n \in S} |X(s_n) - a|^q P(s_n)$ converges. Now

$$\sum_{s_n \in S} |X(s_n) - a|^q P(s_n) = \sum_{s_n \in M} |X(s_n) - a|^q P(s_n) + \sum_{s_n \in S - M} |X(s_n) - a|^q P(s_n)$$

But since for $1 < u$, $u^q \le u^p$ and we have

$$\sum_{s_n \in S - M} |X(s_n) - a|^q P(s_n) \le \sum_{s_n \in S - M} |X(s_n) - a|^p P(s_n)$$

$$\le \sum_{s_n \in S} |X(s_n) - a|^p P(s_n)$$

But the last expression converges so $\sum_{s_n \in S - M} |X(s_n) - a|^q P(s_n)$ is a definite real number. As for $\sum_{s_n \in M} |X(s_n) - a|^q P(s_n)$ we need only observe that for $s_n \in M$,

$$|X(s_n) - a|^q P(s_n) \le 1 \cdot P(s_n) = P(s_n)$$ so

$$\sum_{s_n \in M} |X(s_n) - a|^q P(s_n) \le \sum_{s_n \in M} P(s_n) = P(M) \le 1$$

Thus $\sum\limits_{s_n \in S} |X(s_n) - a|^q P(s_n)$ is the sum of two non-negative numbers, hence $E(|X - a|^q)$ exists.

We now observe that

$$|u - b|^p = |u - a + (a - b)|^p \le (|u - a| + |a - b|)^p$$

If we let $M$ (in this part of the theorem) be $\{s_n \in S;\ |X(s_n) - a| \ge |a - b|\}$ we have for $s_n \in M$, $|X(s_n) - a| + |a - b| \le 2|X(s_n) - a|$ and for $s_n \notin M$, $|X(s_n) - a| + |a - b| \le 2|a - b|$. Hence, if we let $u = X(s_n)$, we have

$$\sum_{s_n \in S} |X(s_n) - b|^p = \sum_{s_n \in M} |X(s_n) - b|^p + \sum_{s_n \in S - M} |X(s_n) - b|^p$$

$$\le \sum_{s_n \in M} (2|X(s_n) - a|)^p + \sum_{s_n \in S - M} (2|a - b|)^p$$

$$= 2^p \sum_{s_n \in M} |X(s_n) - a|^p + 2|a - b|^p P(S - M)$$

$$\le 2^p \sum_{s_n \in S} |X(s_n) - a|^p + 2|a - b|^p P(S - M)$$

So if $E(|X(s_n) - a|^p)$ exists, $E(|X(s_n) - b|^p)$ exists. Since $a$ and $b$ are arbitrary real numbers, we may interchange $a$ and $b$ and get the theorem.

Theorem 7.5.1 can be proved in general but we have not given such a proof in order to keep the Pandora's box of measure theory closed.

If we return to the mechanical picture in which point particles of mass $P(X = u_k) = f_X(u_k)$ are located at the coordinates $u_k$ on an axis we can see that $E(|X - a|^2) = \sum (u_k - a)^2 f_X(u_k)$ is the moment of inertia about an axis perpendicular to the axis of the point particles and intersecting the latter axis at coordinate $a$. We thus make a number of definitions.

**Definitions**    With the notation of Theorem 7.5.1 we say that *X has a pth absolute moment* if $E(|X|^p)$ exists. Moreover in that case the $p$th absolute moment is $E(|X|^p)$. If $p$ is an integer we say that *X has a pth moment* if $X$ has a $p$th absolute moment and the $p$th *moment* is $E(X^p)$. If $X$ has a second absolute moment [and by Theorem 7.5.1 $E(X)$ exists] we write $\sigma^2(X) = E((X - E(X))^2)$ $= E(|X - E(X)|^2)$ and we call $\sigma^2(X)$ the *variance* or *dispersion* of $X$. The *standard deviation* of $X$ is $\sigma(X) = \sqrt{\sigma^2(X)}$.

We shall be primarily concerned with the expected value and the variance of a random variable, although higher moments are valuable for describing certain properties which are of interest in more detailed studies.

**Theorem 7.5.2**    If the second moment of the random variable $X$ exists and if $c$ is a real number, then $E((X - c)^2) = E(X^2) - 2cE(X) + c^2$. If $c = E(X)$ (which exists by Theorem 7.5.1) we obtain $\sigma^2(X) = E(X^2) - [E(X)]^2$.

PROOF   We know that $E((X - c)^2) = E(X^2 - 2cX + c^2) = E(X^2) - 2cE(X) + E(c^2)$ by Theorems 7.4.1 and 7.4.2 since $E(X^2)$, $E(X)$ exist. Finally, by exercise 7.2.2, $E(c^2) = c^2$ so $E((X - c)^2) = E(X^2) - 2cE(X) + c^2$. The remaining assertion of the theorem is obvious.

**Corollary**   If $E(X^2)$ exist and if $c$ is a real number then $\sigma^2(cX) = c^2\sigma^2(X)$.

PROOF   The existence of $E((cX)^2)$ follows immediately from Theorem 7.4.2. The corollary can be proved easily from Theorems 7.5.2 and 7.4.2.

### *Example 7.5.1*

Let $S = \{x: -1 \le x \le 1\}$ and suppose $P(a < x \le b) = \frac{1}{2}(b - a)$. Let $X(x) = x^2$. Find $F_X$, $E(X)$, and $\sigma^2(X)$.

Since $P(a < x \le b) = \frac{1}{2}(b - a)$ (the probability of an interval is proportional to the length of the interval), we observe that $p(x) = \frac{1}{2}$ leads to $\int_a^b p(x)\, dx = \int_a^b \frac{1}{2}\, dx = \frac{1}{2}(b - a)$. To find $F(u) = P(X \le u) = P(\{x \in S; x^2 \le u\})$, we first observe that $F(u) = 0$ if $u < 0$. Furthermore, since $S = \{-1 \le x \le 1\}$, $P(x^2 \le 1) = P(S) = 1$. Finally, if $0 \le u \le 1$, $\{x^2 \le u\} = \{-\sqrt{u} \le x \le \sqrt{u}\}$ so $P(x^2 \le u) = \frac{1}{2}[\sqrt{u} - (-\sqrt{u})] = \sqrt{u}$. Recapitulating, we have

$$F(u) = \begin{cases} 0, & u < 0 \\ \sqrt{u}, & 0 \le u \le 1 \\ 1, & u > 1 \end{cases}$$

Now

$$E(X) = \int_{-1}^1 x^2 p(x)\, dx = \int_{-1}^1 (x^2)\tfrac{1}{2}\, dx$$

and

$$E(X^2) = \int_{-1}^1 (x^2)^2 p(x)\, dx = \int_{-1}^1 (x^4)\tfrac{1}{2}\, dx$$

We leave it to the student to evaluate these integrals, apply Theorem 7.5.2, and obtain $E(X) = \frac{1}{3}$, $\sigma^2(X) = \frac{4}{45}$.

### *Example 7.5.2*

With reference to the random variable $X$ of Example 7.5.1, calculate $E(X)$ and $E(X^2)$ from $F_X$ and $F_{X^2}$.

By the results of Example 7.5.1,

$$F_X(u) = \begin{cases} 0, & u < 0 \\ \sqrt{u}, & 0 \le u \le 1 \\ 1, & u > 1 \end{cases}$$

Therefore, $E(X) = \int_{-\infty}^{\infty} u f_X(u)\, du$ where $f_X$ is a density function for $X$. It is easy to see that

$$f_X(u) = \begin{cases} \frac{1}{2} u^{-1/2}, & 0 < u < 1 \\ 0, & \text{otherwise} \end{cases}$$

defines a density function. Thus

$$\int_{-\infty}^{\infty} u f_X(u)\, du = \int_0^1 (u)\tfrac{1}{2}u^{-1/2}\, du = \int_0^1 \tfrac{1}{2} u^{1/2}\, du$$

$$= \frac{1}{2} \frac{u^{3/2}}{\frac{3}{2}}\Big|_0^1 = \tfrac{1}{3}$$

Now $P(X^2 \le u) = P(x \le \sqrt{u}) = F_X(\sqrt{u})$ when $0 \le u \le 1$. Therefore, $F_{X^2}(u) = (\sqrt{u})^{1/2} = u^{1/4}$ for $0 \le u \le 1$. It is easy to see that

$$f_{X^2}(u) = \begin{cases} \frac{1}{4} u^{-3/4}, & 0 < u < 1 \\ 0, & \text{otherwise} \end{cases}$$

defines a density function for $X^2$. Therefore,

$$E(X^2) = \int_0^1 (u)\tfrac{1}{4} u^{-3/4}\, du = \int_0^1 \tfrac{1}{4} u^{1/4}\, du = \frac{\frac{1}{4} u^{5/4}}{\frac{5}{4}}\Big|_0^1 = \tfrac{1}{5}$$

It should be no surprise that these values agree with the results of Example 7.5.1.

### Example 7.5.3

With reference to the random variable $X$ of Examples 7.5.1 and 7.5.2, calculate $\int_{-\infty}^{\infty} u^2 f_X(u)\, du$. Why must this be $E(X^2)$?

First of all

$$\int_{-\infty}^{\infty} u^2 f_X(u)\, du = \int_0^1 (u^2)\tfrac{1}{2} u^{-1/2}\, du = \int_0^1 \tfrac{1}{2} u^{3/2}\, du = \frac{1}{2}\frac{u^{5/2}}{\frac{5}{2}}\Big|_0^1 = \tfrac{1}{5}$$

A simple reason why the integral on the left is $E(X^2)$ is that both are $\tfrac{1}{5}$. However, this is too narrow a viewpoint. A general discussion may be more useful than a specific discussion and we defer the former to the ensuing paragraphs.

The question raised in Example 7.5.3 can be broadened as follows: If $g(X)$ is a function of the random variable $X$ in the sense that $g(X)(s) = g(X(s))$ and if $X$ has density function $f$, why is $\int_{-\infty}^{\infty} g(u)f(u)\, du = E(g(X))$ assuming that the requisite quantities exist?

Under normal circumstances we should be able to approximate $\int_{-\infty}^{\infty} g(u)f(u)\, du$ by sums of the form $\sum g(u_j)f(u_j)(u_j - u_{j-1})$ where $\cdots < u_{-3} <$

$u_{-2} < u_{-1} < u_0 < u_1 < u_2 < \cdots$. Since $\int_{u_{j-1}}^{u_j} f(u)\, du = F(u_j) - F(u_{j-1})$ we would hope that $f(u_j)(u_j - u_{j-1})$ is approximately $F(u_j) - F(u_{j-1})$ if $u_j - u_{j-1}$ is small and if $f$ is smooth. Therefore, $\sum g(u_j) f(u_j)(u_j - u_{j-1})$ is approximately $\sum g(u_j)[F(u_j) - F(u_{j-1})]$ which is $\sum g(u_j) P(u_{j-1} < X \le u_j)$ and $g(u_j)$ is approximately $g(X(s))$ for $s \in \{s; u_{j-1} < X(s) \le u_j\}$ if $g$ is a smooth function. Thus, after some approximations, $\sum g(u_j) f(u_j)(u_j - u_{j-1})$ is reminiscent of $\int g(X(s)) p(s)\, ds$ since $\sum g(u_j) P(u_{j-1} < X < u_j)$ is approximately

$$\sum_j \int_{\{s; u_{j-1} < X(s) \le u_j\}} g(X(s)) p(s)\, ds$$

if $g(X(s))$ is close to $g(u_j)$ where $u_{j-1} < X(s) \le u_j$. This is not far from requiring that $g$ be continuous or at least piecewise continuous; however, we will not concern ourselves with this issue. At any rate $\int_{-\infty}^{\infty} g(u) f(u)\, du$ is approximately equal to $E(g(X))$. Hopefully as $u_j - u_{j-1} \to 0$ the approximations are better and it becomes clear that this can occur only if

(7.5.1) $\qquad E(g(X)) = \int_{-\infty}^{\infty} g(u) f(u)\, du$

In some respects, the fact that $E(g(X)) = \int_{-\infty}^{\infty} g(u) f_X(u)\, du$ if $X$ has a density function or that $E(g(X)) = \sum g(u_k) f_k(u_k)$ if $X$ has a probability function suggests that the sample space is irrelevant. This is largely true if one is studying only one random variable. However, the availability of both the sample space and the distribution function gives flexibility in making computations. Moreover, many of the remaining topics can be discussed without reference to the sample space. In other words, they treat questions about distribution functions.

**EXERCISES 7.5** _____

1  Find the variance of the random variables in exercises 7.2.1–7.2.4.

2  Let $S = \{1, 2, 3, \ldots\}$ with $P(n) > 0$ for each $n \in S$. Show that there is a random variable $X$ such that $E(X)$ exists but $\sigma^2(X)$ does not exist. Is it possible that $\sigma^2(Y)$ exists and $E(Y)$ does not exist for some random variable $Y$?

3  Discuss the variance of the random variable in Example 7.3.1.

4  Discuss the variance of the random variable in Example 7.3.2.

5  Find the variance of the random variable in exercises 7.3.3, 7.3.5, and 7.3.7.

♦ 6  Find the variance of the random variable $X$ whose density function is given by $f(x) = (1/\sqrt{2\pi}) e^{-x^2/2}$.

7  Find the expectation and variance of a random variable whose density function $f$ is given by

$$f(x) = \begin{cases} 0, & x \leq 0 \\ [1/\Gamma(\alpha)]x^{\alpha-1}e^{-x}, & x > 0 \end{cases}$$

where $\alpha > 0$ and $\Gamma(\alpha) = \int_0^\infty x^{\alpha-1}e^{-x}\,dx$. Note that $\Gamma(\alpha + 1) = \alpha\Gamma(\alpha)$.

◆ 8  Let $X$ be a random variable with second moment and let $Y = X + c$ for $c$ a constant. Find $\sigma^2(Y)$.

9  Which moments of the Student's $t$-distribution exist?

10  Which moments of the $F$-distribution exist?

11  Let $X_1, X_2, \ldots, X_n$ be random variables with the following properties: $X_1 + X_2 + \cdots + X_n = r$;  $E(X_j^2) = d^2$, which does not depend on $j$; $E(X_j X_k) = K$ which does not depend on $j$ and $k$ if $j \neq k$. Show that $nd^2 + (n^2 - n)K = r^2$ by working with $E((X_1 + X_2 + \cdots + X_n)^2)$.

12  Find $E(X_j X_k)$ for exercise 7.4.7 using the results of exercise 11.

★ 13  Compute

$$\frac{d^2}{dz^2} \sum_{n=0}^{N} f(n)z^n$$

Compute

$$\lim_{\substack{z \to 1 \\ z < 1}} \left[ \frac{d^2}{dz^2} \sum_{n=0}^{N} f(n)z^n + \frac{d}{dz} \sum_{n=0}^{N} f(n)z^n \right]$$

What does this suggest about $E(X^2)$ when $X$ has a probability function $f$ with $f(n) > 0$ only if $n = 0, 1, 2, \ldots$? Compare with exercise 7.2.16.

★ 14  Use exercise 13 to obtain $E(X^2)$ when $X$ has the Poisson distribution.

★ 15  Use exercise 13 to obtain $E(X^2)$ when $X$ has the negative binominal distribution.

★ 16  Find the variance of the Poisson distribution and the negative binominal distribution. Check Example 7.2.17.


# 7.6  Chebyshev's inequality

We now move toward the first of a number of approximation and estimation theorems.

### Example 7.6.1
A random variable $X$ with probability function $f$ has $E(X) = 3$ and $\sigma^2(X) = 9$. Is it possible that $P(X \geq 10) = \frac{1}{5}$?

From the definitions and theorems of section 7.5, $\sigma^2(X) = \sum (u_k - 3)^2 f(u_k)$ which may be written as

$$\sum_{u_k < 10} (u_k - 3)^2 f(u_k) + \sum_{u_k \geq 10} (u_k - 3)^2 f(u_k)$$

Since $f(u_k) \geq 0$ and $(u_k - 3)^2 \geq 0$ then $\sum_{u_k < 10} (u_k - 3)^2 f(u_k) \geq 0$ and $\sigma^2(X) \geq$

$\sum_{u_k \geq 10} (u_k - 3)^2 f(u_k)$. But if $u_k \geq 10$, $(u_k - 3)^2 \geq (10 - 3)^2 = 49$ so

$$9 = \sigma^2(X) \geq \sum_{u_k \geq 10} 49 f(u_k) = 49 \sum_{u_k \geq 10} f(u_k) = 49 P(X \geq 10)$$

Thus $\frac{9}{49} \geq P(X \geq 10)$. Now $\frac{1}{5} > \frac{9}{49}$ so $\frac{1}{5} > P(X \geq 10)$.

The technique of Example 7.6.1 is quite useful in other situations.

**Theorem 7.6.1** (Chebyshev's inequality.)    If $E(X^2)$ exists and if $c > 0$ is a real number then

$$P(|X - E(X)| \geq c) \leq \frac{\sigma^2(X)}{c^2}$$

PROOF    We prove this theorem in two cases although it is true for all random variables having a second moment. In the case $X$ has a probability function $f$ we have

$$\sigma^2(X) = \sum [u_k - E(X)]^2 f(u_k)$$

$$= \sum_{|u_k - E(X)| < c} [u_k - E(X)]^2 f(u_k) + \sum_{|u_k - E(X)| \geq c} [u_k - E(X)]^2 f(u_k)$$

As in Example 7.6.1,

$$\sum_{|u_k - E(X)| < c} [u_k - E(X)]^2 f(u_k) \geq 0$$

and if $|u_k - E(X)| \geq c$

$$[u_k - E(X)]^2 = |u_k - E(X)|^2 \geq c^2$$

Therefore

$$\sigma^2(X) \geq \sum_{|u_k - E(X)| \geq c} c^2 f(u_k) = c^2 P(|u_k - E(X)| \geq c)$$

and if we divide by $c^2$ we obtain $\sigma^2(X)/c^2 \geq P(|u_k - E(X)| \geq c)$. In the case $X$ has a density function (also) $f$ we have

$$\sigma^2(X) = \int_{-\infty}^{\infty} [u - E(X)]^2 f(u)\, du = \int_{E(X)-c}^{E(X)+c} [u - E(X)]^2 f(u)\, du$$

$$+ \int_{-\infty}^{E(X)-c} [u - E(X)]^2 f(u)\, du + \int_{E(X)+c}^{\infty} [u - E(X)]^2 f(u)\, du$$

Again $\int_{E(X)-c}^{E(X)+c} [u - E(X)]^2 f(u)\, du \geq 0$ since $[u - E(X)]^2 f(u) \geq 0$ and $E(X) - c$

$< E(X)+c$. Furthermore, if $u \leq E(X) - c$, $u - E(X) \leq -c$ so $|u - E(X)|^2 \geq c^2$ and if $u \geq E(X) + c$, $[u - E(X)]^2 \geq c^2$. Therefore

$$\sigma^2(X) \geq \int_{E(X)+c}^{\infty} c^2 f(u) \, du + \int_{-\infty}^{E(X)-c} c^2 f(u) \, du$$

$$= c^2 P(X - E(X) \leq -c) + c^2 P(X - E(X) \geq c)$$

$$= c^2 P(|X - E(X)| \geq c)$$

It is obvious that Chebyshev's inequality may be crude [that is, $\sigma^2(X)/c^2$ is much greater than $P(|X - E(X)| \geq c)$] since the term $\sum_{|u_k - E(X)| < c} [u_k - E(X)]^2 f(u_k)$ may be quite large. Nevertheless, there are situations in which that inequality gives a useful estimate of $P(|X - E(X)| \geq c)$. Naturally, the important values of $c$ are those where $c > \sigma(X)$.

### Example 7.6.2
Let $X$ have the uniform distribution on the interval $\{-1 \leq x \leq 1\}$. Compare Chebyshev's inequality to the true values.

We have $E(X) = 0$ by exercise 7.3.11, and $\sigma^2(X) = E(X^2) = \int_{-1}^{1} (u^2)\frac{1}{2} \, du = \frac{1}{6}u^3 \Big|_{-1}^{1} = \frac{1}{3}$. Hence $(\frac{1}{3})/c^2 = \sigma^2(X)/c^2$. Of course, $P(|X - E(X)| \geq c) = P(|X| \geq c) = P(X \leq -c \text{ or } X \geq c) = 1 - c$ if $0 < c \leq 1$. The fraction error is

$$e(c) = \frac{\frac{1}{3}/c^2 - (1 - c)}{1 - c} = \frac{1/3c^2 - 1 + c}{1 - c}$$

which tends to $+\infty$ as $c \to 1$ or $c \to 0$. Its minimum occurs at $c = \frac{2}{3}$ and $e(\frac{2}{3}) = \frac{5}{4}$ which means the *error* is $\frac{5}{4}$ of the true value so the estimate $\sigma^2(X)/c^2$ is $2\frac{1}{4}$ times the correct value.

A nice link between Chebyshev's inequality and independent random variables is provided by the consequences of the following theorem.

### Theorem 7.6.2
If $X$ and $Y$ are independent random variables and if $E(X^2)$ and $E(Y^2)$ exist, then $E((X+Y)^2)$ exists and $\sigma^2(X+Y) = \sigma^2(X) + \sigma^2(Y)$.

PROOF    If $a$ and $b$ are real numbers we have $(a - b)^2 \geq 0$ so $a^2 - 2ba + b^2 \geq 0$ and $2ab \leq a^2 + b^2$. Thus

$$|X(s) + Y(s)|^2 \leq |X(s)|^2 + 2|X(s)| \, |Y(s)| + |Y(s)|^2$$
$$\leq 2|X(s)|^2 + 2|Y(s)|^2$$

Since the convergence of a series of larger terms implies the convergence of a series of smaller terms, and similarly with the integration of sufficiently smooth functions, we deduce that $E(|X + Y|^2)$ exists because $E(2X^2 + 2Y^2)$ exists.

Now by Theorems 7.4.1–3, 7.5.1, and the hypotheses of the problem we have

$$E((X + Y)^2) = E(X^2 + 2XY + Y^2) = E(X^2) + 2E(XY) + E(Y^2)$$

By independence $E(XY) = E(X)E(Y)$ so

$$
\begin{aligned}
\sigma^2(X + Y) &= E((X + Y)^2) - (E(X + Y))^2 \\
&= E(X^2) + 2E(X)E(Y) + E(Y^2) - [E(X) + E(Y)]^2 \\
&= E(X^2) - [E(X)]^2 + E(Y^2) - [E(Y)]^2 = \sigma^2(X) + \sigma^2(Y)
\end{aligned}
$$

after a bit of computation.

We state without proof the fact that if $X_1, X_2, \ldots, X_n$ are independent and if $E(X_k^2)$ exists, then

**(7.6.1)**    $\sigma^2(X_1 + X_2 + \cdots + X_n) = \sigma^2(X_1) + \sigma^2(X_2) + \cdots + \sigma^2(X_n)$

### Example 7.6.3

A standard die is rolled 100 times. Let $X_k$ be the number of dots showing on the $k$th roll. Is it reasonable to accept a bet of \$1 against your \$10 that $X_1 + X_2 + \cdots + X_{100} < 450$?

If we assume that the $X_k$ are independent and use $E(X_k) = \frac{1}{6}(1 + 2 + \cdots + 6) = \frac{21}{6}$, $E(X_k^2) = \frac{1}{6}(1^2 + 2^2 + \cdots + 6^2) = \frac{91}{6}$, we have $E(X_k) = \frac{7}{2}$ and $\sigma^2(X_k) = \frac{91}{6} - (\frac{7}{2})^2 = \frac{35}{12}$. Thus $E(X_1 + X_2 + \cdots + X_{100}) = 350$ and

$$\sigma^2(X_1 + X_2 + \cdots + X_{100}) = \sigma^2(X_1) + \sigma^2(X_2) + \cdots + \sigma^2(X_n) = 100 \cdot \tfrac{35}{12}$$

Suppose we estimate $P(|X_1 + X_2 + \cdots + X_{100} - 350| \geq 100)$. This quantity is chosen because we finally want to deal with

$$P(X_1 + X_2 + \cdots + X_{100} \geq 450) = P(X_1 + X_2 + \cdots + X_{100} - 350 \geq 100)$$

Now the event

$$
\begin{aligned}
&\{|X_1 + X_2 + \cdots + X_{100} - 350| \geq 100\} \\
&\quad = \{(X_1 + X_2 + \cdots + X_{100} - 350) \geq 100\} \cup \{(X_1 + X_2 + \cdots + X_{100} - 350) \leq -100\}
\end{aligned}
$$

and these last events are disjoint since a number cannot be both greater than 99 and less than $-99$. Furthermore since a die is constructed so that 1 and 6, 2 and 5, and 3 and 4 are on opposite faces, any outcome in which the top faces add up to $q \geq 450$ has a corresponding outcome in which the top faces add up to $700 - q \leq 250$, i.e., $P(X_1 + \cdots + X_{100} \leq 250) = P(X_1 + X_2 + \cdots + X_{100} \geq 450)$. We now use Chebyshev's inequality to assert that

$$P(|X_1 + \cdots + X_{100} - 350| \geq 100) \leq \frac{\sigma^2(X_1 + \cdots + X_{100})}{(100)^2}$$

But $\sigma^2(X_1 + \cdots + X_{100}) = 100 \cdot \frac{35}{12}$ so

$$
\begin{aligned}
P(X_1 + \cdots + X_{100} \geq 450) &= P(X_1 + \cdots + X_{100} - 350 \geq 100) \\
&= \tfrac{1}{2} P(|X_1 + \cdots + X_{100} - 350| \geq 100) \\
&\leq \tfrac{1}{2} \cdot 100 \cdot \tfrac{35}{12} \cdot 100^{-2} = \tfrac{35}{24} \cdot \tfrac{1}{100}
\end{aligned}
$$

or about 0.015. Hence the odds seem very favorable; your $10 payment would be due less than once in 50 plays of the game.

A slight variant of Example 7.6.3 illustrates a useful technique.

### Example 7.6.4

An honest die is rolled. Estimate (too large rather than too small) the probability that the top face shows one dot or two dots fewer than 70 times in 300 rolls.

We define a new random variable $Y_k$ where

$$Y_k = \begin{cases} 1, & \text{one dot or two dots show on the } k\text{th roll} \\ 0, & \text{more than two dots show on the } k\text{th roll} \end{cases}$$

It is clear that $P(Y_k = 1) = \frac{1}{3}$, $P(Y_k = 0) = \frac{2}{3}$, and $E(Y_k) = \frac{1}{3}$, $\sigma^2(Y_k) = \frac{2}{9}$. If we let $S = Y_1 + Y_2 + \cdots + Y_{300}$ we see that $S$ is a random variable which gives the number of times that two dots or one dot shows. We assume that the $Y_1, Y_2, \ldots, Y_{300}$ are independent and note that $E(Y_1 + Y_2 + \cdots + Y_{300}) = 300 \cdot \frac{1}{3} = 100$ and $\sigma^2(Y_1 + \cdots + Y_{300}) = 300 \cdot \frac{2}{9} = \frac{200}{3}$. Chebyshev's inequality gives

$$\begin{aligned} P(Y_1 + Y_2 + \cdots + Y_{300} < 70) &= P(Y_1 + \cdots + Y_{300} - 100 < -30) \\ &= P(Y_1 + \cdots + Y_{300} - 100 \le -31) \\ &\le P(|Y_1 + Y_2 + \cdots + Y_{300} - 100| \ge 31) \\ &\le \frac{200}{3} \cdot \frac{1}{(31)^2} < \frac{5}{72} \end{aligned}$$

Better estimates are possible. It is not correct to take half of

$$P(|Y_1 + \cdots + Y_{300} - 100| \ge 31)$$

since we do not have as much symmetry as in Example 7.6.3.

The useful technique which this example illustrates is that of using a sum of random variables $Y_k$ to give a random variable which counts the number of times a certain event occurs. The $Y_k$ take 1 or 0 according as the event occurs or does not occur at the $k$th trial.

Example 7.6.4 can be examined from another viewpoint. Suppose a (possibly) biased coin is tossed $n$ times independently. Let $X_k$ be 1 or 0 according as heads occurs on the $k$th toss or not. We assume that $P(X_k = 1) = p$ and hence $P(X_k = 0) = 1 - p$, where $0 \le p \le 1$. Let $Y = X_1 + X_2 + \cdots + X_n$. Now $Y$ counts the number of times that heads shows in $n$ tosses, so $P(Y = j) =$

$P(j$ tosses show heads and $n - j$ tosses show tails$) = \binom{n}{j} p^j (1 - p)^{n-j}$ by previous arguments. Thus $Y$ has a binomial distribution with parameters $p$ and $n$. Since $Y = X_1 + \cdots + X_n$ where the $X_k$ are independent and since $E(X_k) = p$, $\sigma^2(X_k) = (1 - p)^2 p + (0 - p)^2 (1 - p) = (1 - p)p$, $E(Y) = np$, and $\sigma^2(Y) = n(1 - p)p$. Thus we have the expectation and variance of a random variable with binomial distribution.

We can deduce that if $Y$ has a binomial distribution then

$$(7.6.2) \qquad P(|Y - np| \geq c) = \sum_{|j - np| \geq c} \binom{n}{j} p^j (1 - p)^{n - j}$$

$$\leq \frac{\sigma^2(Y)}{c^2} = \frac{np(1 - p)}{c^2}$$

Here, $\displaystyle\sum_{|j - np| \geq c}$ signifies that the sum is taken over those $j$ for which $|j - np| \geq c$.

**EXERCISES 7.6** _____

**1**  An honest coin is tossed independently six times. Let $Y$ denote the number of heads on those tosses. Find $P(|Y - 3| \geq 2)$ and $P(|Y - 3| \geq 3)$ exactly. Compare the exact results with the estimate given by Chebyshev's inequality.

**2**  Let $Y$ be a random variable with $E(Y) = 0$, $\sigma^2(Y) = 1$. Show that $P(Y \geq c) \leq 1/c^2$ for any $c > 0$. Is it true that $P(Y \leq -c) \leq 1/c^2$? Is it necessarily the case that $P(Y \geq c) \leq 1/2c^2$?

**3**  Find a random variable $X$ such that $P(|X - E(X)| \geq c) = \sigma^2(X)/c^2$ for some $c > 0$. Does this hold for many values of $c$ for a given $X$?

**4**  A fair die is rolled independently 600 times. Estimate the probability that a single dot shows on fewer than 60 rolls.

**5**  Let $X$ have the density function $f(x) = \frac{1}{2} e^{-|x|}$. Calculate $E(X)$ and $\sigma^2(X)$. Find $P(|X - E(X)| \geq 3)$ exactly and by Chebyshev's inequality. Find the value of $c$ such that Chebyshev's inequality gives the smallest percentage error for $P(|X - E(X)| \geq c)$.

♦ **6**  Let $X$ be a random variable with second moment. Show that $P(|X - E(X)| < c) \geq 1 - \sigma^2(X)/c^2$.

**7**  Let $X$ have a probability function or density function $f$ such that $f(x) = 0$ if $x < 0$. Show that if $E(X)$ exists then $c^{-1} E(X) \geq P(X \geq c)$, where $c > 0$.

**8**  Let $X$ have the negative exponential distribution. Find, as a function of $c$, the fractional error between $P(X \geq c)$ and the crude estimate of it given by $c^{-1} E(X)$.

**9**  A fair die is rolled independently 900 times. Let $T$ be the total number of dots showing on all of the rolls. Obtain a lower bound on $P(2900 < T < 3400)$ by means of exercise 6.

# 7.7   The weak law of large numbers

The mathematical model that we are using implies that the probability that the ratio of the number of H's to the number of T's in $n$ tosses of a fair coin exceeds $a > 1$ is small for large $n$. To be specific, the probability of {three H's and one T or four H's and no T's} is $\frac{5}{16}$. The probability of {four H's and one T or five H's and no T's} is $\frac{6}{32}$. Presumably, experience suggests that there will be about as many H's as there are T's in $n$ tosses when $n$ is large. It is conceivable that one could toss 1000 H's in a row in independent tosses of a fair coin, but it just seems very unlikely. Similarly, in repeated independent tosses of a fair die we might expect the 1-dot face to appear on about $\frac{1}{6}$ of the tosses and we would be very surprised if it did not show at all on 1000 tosses.

We shall formulate and prove a theorem (in the mathematical model) which reflects the above situation. Although this theorem tells nothing about the real world it is reassuring that the model contains such a result.

From the discussion following Example 7.6.4 and from (7.6.2) we find that if $Y$ has a binomial distribution with parameters $n$ and $p$ then

$$P(|Y - np| \geq c) \leq np(1 - p)/c^2$$

or

$$P(|Y - np| \geq bn) \leq np(1 - p)/n^2 b^2 = p(1 - p)/nb^2$$

or

$$P(|Y/n - p| \geq b) \leq p(1 - p)/nb^2$$

We can imagine that the random variable $Y$ counts the number of heads in $n$ independent tosses of a coin whose probability of showing H is $p$. Therefore $Y/n$ gives the *average* number of H's in $n$ such tosses. Since $\{|Y/n - p| \geq b\}$ is the event that $Y/n \geq p + b$ or $Y/n \leq p - b$ we have an inequality concerning the probability that the average number of H's differs from the underlying probability of throwing a H by a specified amount $b$. Now for any fixed $b > 0$ we have $\lim_{n \to \infty} [p(1 - p)/nb^2] = 0$. We may therefore state the following theorem.

**Theorem 7.7.1**    Let $N_n$ be the random variable giving the number of H's in $n$ independent tosses of a coin whose probability of showing H is $p$. Then if $b > 0$,

$$\lim_{n \to \infty} P(|N_n/n - p| \geq b) = 0$$

Although the formulation of the theorem is in terms of a biased coin, the above theorem applies to any situation of independent repetitions of an experiment where one seeks the average number of occurrences of a certain event at each stage.

The proof of Theorem 7.7.1 follows directly from Chebyshev's inequality and a proper choice of random variable $Y$. But if $X_1, X_2, X_3, \ldots$ is any family of independent, identically distributed random variables with $E(X_k) = m$ and $\sigma^2(X_k) = s^2$ then

$$P\left(\left|\frac{1}{n}(X_1 + \cdots + X_n) - m\right| \geq b\right) \leq \frac{s^2}{n}$$

since

$$\sigma^2\left(\frac{1}{n}(X_1 + \cdots + X_n)\right) = \frac{ns^2}{n^2}$$

and

$$E\left(\frac{1}{n}(X_1 + \cdots + X_n)\right) = m$$

We therefore have the next theorem.

**Theorem 7.7.2**    If $X_1, X_2, X_3, \ldots$ are independent, identically distributed random variables with $E(X_n) = m$ and with $\sigma^2(X_n) = s^2$ then

$$\lim_{n \to \infty} P\left(\left|\frac{1}{n}(X_1 + \cdots + X_n) - m\right| \geq b\right) = 0$$

Theorems 7.7.1 and 7.7.2 are both called *the weak law of large numbers*. The name may be more appropriate for Theorem 7.7.1 but it is a corollary to Theorem 7.7.2 so the name is applied to both.

*Example 7.7.1*
A random variable $X$ has density function

$$f(x) = \begin{cases} 2/x^3, & x > 1 \\ 0, & x \leq 1 \end{cases}$$

Let $X_1, X_2, X_3, \ldots$ be independent and identically distributed with $X$. Does Theorem 7.7.2 guarantee that $\lim_{n \to \infty} P\left(\left|\frac{1}{n}(X_1 + X_2 + \cdots + X_n) - 2\right| \geq 0.001\right) = 0$?

According to Theorem 7.7.2, we must know about $E(X)$ and $\sigma^2(X)$. Therefore we compute $\int_1^\infty xf(x)\,dx$ and $\int_1^\infty x^2 f(x)\,dx$ or $\int_1^\infty x(2/x^3)\,dx$ and $\int_1^\infty x^2(2/x^3)\,dx$.

It is easy to see that $\int_1^\infty x(2/x^3)\,dx = \int_1^\infty (2/x^2)\,dx = 2$. Therefore

$$P\left(\left|\frac{1}{n}(X_1 + X_2 + \cdots + X_n) - 2\right| \geq 0.001\right)$$

is the expression in the conclusion of the theorem with $b = 0.001$ and $m = 1$.

However, $\int_1^\infty x^2(2/x^3)\,dx = \int_1^\infty (2/x)\,dx$ diverges. This implies that $\sigma^2(X)$ does not exist so Theorem 7.7.2 does not give any information on

$$\lim_{n\to\infty} P\left(\left|\frac{1}{n}(X_1 + \cdots + X_n) - 2\right| \geq 0.001\right)$$

### Example 7.7.2

A coin is biased so that the probability of H is $\frac{1}{3}$. How many times should one toss the coin independently if he wants to give himself the best chance of getting H on more than half of the tosses?

The probability of one H in one toss is $\frac{1}{3}$. The probability of two H's in two tosses is $\frac{1}{9}$. The probability of three or two H's in three tosses is $(\frac{1}{3})^3 + 3(\frac{1}{3})^2(\frac{2}{3}) = (\frac{1}{27})(1 + 6) = \frac{7}{27}$. The probability of three or four H's in four tosses is $4(\frac{2}{3})(\frac{1}{3})^3 + (\frac{1}{3})^4 = \frac{9}{81} = \frac{1}{9}$. The probability of three, four, or five heads in five tosses is $6(\frac{2}{3})^2(\frac{1}{3})^3 + 3(\frac{2}{3})(\frac{1}{3})^4 + (\frac{1}{3})^5 = (\frac{1}{243})(33) = \frac{11}{81}$. We further observe that $P(|(1/n)(X_1 + \cdots + X_n) - \frac{1}{3}| \geq b) \to 0$ as $n \to \infty$ by Theorem 7.7.1 if

$$X_k = \begin{cases} 1, & k\text{th toss shows H} \\ 0, & k\text{th toss shows T} \end{cases}$$

Therefore if $b = \frac{1}{6}$ we find

$$P\left(\left|\frac{1}{n}(X_1 + \cdots + X_n) - \frac{1}{3}\right| \geq \frac{1}{6}\right)$$

$$= P\left(\frac{1}{n}(X_1 + \cdots + X_n) \geq \frac{1}{2} \quad\text{or}\quad \frac{1}{n}(X_1 + \cdots + X_n) < \frac{1}{6}\right) \to 0$$

so $P((1/n)(X_1 + \cdots + X_n) \geq \frac{1}{2}) \to 0$ as $n \to \infty$. But $(1/n)(X_1 + \cdots + X_n) > \frac{1}{2}$ means that more than half of the tosses are H since this is precisely the way for $X_1 + X_2 + \cdots + X_n > \frac{1}{2}n$; that is, more than $\frac{1}{2}n$ of the $X_k$ are equal to 1.

Although the proof is not complete, it is quite plausible that one toss gives one the best chance of more than half of the tosses showing H.

The point of the above remarks is that when one conducts independent, identically distributed trials, then the probability that the average $(1/n)(X_1 + \cdots + X_n)$ differs perceptibly from the expectation $E(X_1)$ tends to zero as $n \to \infty$. Alternatively, the probability is high that $(1/n)(X_1 + \cdots + X_n)$ is close to $E(X_1)$ when $n$ is large.

It should be kept in mind that the above observations are statements about the mathematical model. The sequence of probabilities which converges to zero is that induced in a Cartesian product sample space associated with $(X_1, X_2, \ldots, X_n)$.

By more advanced techniques, it is possible to prove the conclusion of Theorem 7.7.2 without the hypothesis that $\sigma^2(X_n)$ exists. Using this stronger result, which we do not prove here, one may assert that $P(|(1/n)(X_1 + \cdots + X_n) - 2| > 0.001) \to 0$ in Example 7.7.1.

## EXERCISES 7.7

♦ **1** Show, carefully, that Theorem 7.7.2 implies Theorem 7.7.1.

In exercises 2–6 we assume that $X_1, X_2, X_3, \ldots$ are independent, identically distributed random variables. We let $Y_n = (1/n)(X_1 + X_2 + \cdots + X_n)$.

**2** Suppose the density function of the $X_j$ is $\begin{cases} \frac{1}{2}\sin x, & 0 < x < \pi \\ 0, & \text{otherwise} \end{cases}$ Is there a number $c$ such that for each $b > 0$, $\lim\limits_{n \to \infty} P(|Y_n - c| \geq b) = 0$? If so, find $c$. Justify your conclusion.

**3** Suppose that the $X_j$ have the Poisson distribution. Is there a number $c$ such that for each $b > 0$, $\lim\limits_{n \to \infty} P(|Y_n - c| \geq b) = 0$? If so, find $c$. Justify your conclusion.

**4** Suppose that the $X_j$ have the Cauchy distribution. Proceed as in exercises 2 and 3.

**5** Suppose that $E(X_j) = m$ and $\sigma^2(X_k) = s^2$. Show that for each $b > 0$, $\lim\limits_{n \to \infty} P(|Y_n - m| < b) = 1$.

**6** Let $X_j$ be as in exercise 5. Let $0 < \alpha < \frac{1}{2}$ for fixed $\alpha$. Show that $\lim\limits_{n \to \infty} P(|Y_n - m| \geq 1/n^\alpha) = 0$. Why is this a stronger result than Theorem 7.7.2?

**7** Suppose that $X_1, X_2, X_3, \ldots$ are independent but not necessarily identically distributed. However, suppose that $E(X_j) = m$ for all $j$, that each $\sigma^2(X_j)$ exists and there is an $A > 0$ and $\beta < 2$ such that $\sum\limits_{j=1}^{n} \sigma^2(X_j) \leq An^\beta$. Show that for each $b > 0$ $\lim\limits_{n \to \infty} P(|(X_1 + X_2 + \cdots + X_n)/n - m| \geq b) = 0$.

CHAPTER **8**

# *Random Vectors*

## 8.1 Introduction

It was previously mentioned that in some problems we must study two or more random variables simultaneously. Generally speaking, the more interesting problems involve more than one random variable on the same sample space.

**Example 8.1.1**
A study of 500 men could yield data about the annual income and annual alcohol consumption of those men. Set up a probabilistic model for the selection of a man at random and the inquiry about the above-mentioned quantities.

Let $S = \{1, 2, \ldots, 500\}$ be an enumeration of the men. Let $P(n) = \frac{1}{500}$. Let $X(n)$ be annual income of the $n$th person, and $Y(n)$ be the annual alcohol consumption of the $n$th person.

**Example 8.1.2**
A study yields information about the age and weight of mothers before giving birth and about the health of the newborn child. Set up a probabilistic model assuming random sampling of the data.

Let $S = \{1, 2, \ldots, m\}$ where $m$ is the number of mothers (assume single births). Let $X(n)$ be the age of the $n$th mother, $Y(n)$ be the weight of $n$th mother at a fixed time prior to giving birth, and $Z(n)$ be the health of the newborn child of the $n$th mother, as computed on some scale.

These examples should suggest a number of similar problems in which two or more "chance" components enter. The relationships, if any, between these quantities are apt to be sought. However it should be clear that rich and poor alike may drink a great deal, although one cannot spend more than one's total income on alcohol. Also, presumably, a man with an income of $100 000 does not use it all on alcohol for his own consumption. However, except for these extreme limitations there is a wide range of variation for both quantities.

The objectives of analysis of this data might be justification of some hypothesis or formulation of policy in addition to curiosity. For example, if one finds that heavier women have less healthy babies one might want to convince potential mothers to reduce their weight (unless one is in favor of less healthy newborn babies). Regardless of the hypotheses or policies, we will study a mathematical model which will treat such data. This model involves ideas analogous to those used in the development of a model involving one random variable.

For the present let us examine what can be done by exploiting the probability measure on the underlying sample space.

### Example 8.1.3

A point is chosen at random from the cube $\{(x, y, z); 0 \leq x \leq 1, \quad 0 \leq y \leq 1, \quad 0 \leq z \leq 1\}$. The random variables $X$, $Y$, and $Z$ are defined by $X(x, y, z) = \sqrt{x^2 + y^2 + z^2}$, $Y(x, y, z) = y$, $Z(x, y, z) = z$.
Find $P(X > 2 \text{ and } Y \leq \frac{1}{4})$. Find $P(Y \leq \frac{1}{3} \text{ and } Z \leq \frac{1}{4})$. Find $P(\max(Y, Z) \leq \frac{1}{2})$.

We first find $P(Y \leq \frac{1}{3} \text{ and } Z \leq \frac{1}{4})$ since it is easiest. The part of the cube which has $y = Y(x, y, z) \leq \frac{1}{3}$ and $z = Z(x, y, z) \leq \frac{1}{4}$ has $0 \leq x \leq 1$. It is a rectangular parallelepiped of dimensions $1 \times \frac{1}{3} \times \frac{1}{4}$. Since "at random" in such problems means that probability is proportional to volume, and since the entire sample space has unit volume, the probability in question is $\frac{1}{12}$. Secondly, $\max(Y, Z) \leq \frac{1}{2}$ means $0 \leq y \leq \frac{1}{2}$ and $0 \leq z \leq \frac{1}{2}$. Then $P(\max(Y, Z)) = 1 \cdot \frac{1}{2} \cdot \frac{1}{2} = \frac{1}{4}$. As for $P(X > 2 \text{ and } Y \leq \frac{1}{4})$ we need only note that in the cube, $x^2 + y^2 + z^2 \leq 1^2 + y^2 + 1^2$. If $y$ is restricted to lie between 0 and $\frac{1}{4}$, $x^2 + y^2 + z^2 \leq 2\frac{1}{16} \leq \frac{9}{4}$ and $\sqrt{x^2 + y^2 + z^2} \leq \frac{3}{2}$ if $y \leq \frac{1}{4}$. Therefore $P(X > 2 \text{ and } Y \leq \frac{1}{4}) = 0$.

### Example 8.1.4

Two fair dice, one red and one green, are rolled. They show $m$ and $n$, respectively. Let $X(m, n) = 3m - 5n$ and $Y(m, n) = 4m + n$. Find $P(X = 4, \quad Y = 13)$, $P(X = -2, \quad Y = 6)$, and $P(X \leq -7, Y \leq 8)$.

We must know which $(m, n)$ in our sample space satisfy

(a) $3m - 5n = 4$, $\quad 4m + n = 13$

(b) $3m - 5n = -2$, $\quad 4m + n = 6$

(c) $3m - 5n \leq -7$, $\quad 4m + n \leq 8$

The equations in (a) have the one simultaneous solution $m = 3$, $n = 1$ and only that solution. The assumption of fair dice means equal likelihood so $P(X = 4, \quad Y = 13) = \frac{1}{36}$. The equations in (b) yield $m = \frac{28}{23}$, $n = 6 - 4 \cdot \frac{28}{23}$ if solved over the real numbers. Therefore no arrangement on the dice yield $3m - 5n = -2$ and $4m + n = 6$ and hence $P(X = -2, \quad Y = 6) = 0$.

Finally, since $m, n \in \{1, 2, 3, 4, 5, 6\}$ the only way to have $4m + n \le 8$ is for $m = 1$. It is easy to verify that $n = 1, 2, 3$, or $4$ satisfy $4m + n \le 8$ if $m = 1$ but what about $3m - 5n \le -7$? We note that

$$3 \cdot 1 - 5 \cdot 1 = -2 \nleq -7, \qquad 3 \cdot 1 - 5 \cdot 2 = -7 \le -7$$
$$3 \cdot 1 - 5 \cdot 3 = -12 \le -7, \qquad 3 \cdot 1 - 5 \cdot 4 = -17 \le -7$$

Therefore $(1, 2)$, $(1, 3)$, and $(1, 4)$ are the sample points which satisfy the conditions of (c) and $P(X \le -7, \quad Y \le 8) = \frac{3}{36}$.

### Example 8.1.5

A point $(x, y)$ is chosen from the square $\{(x, y); 0 \le x \le 1 \text{ and } 0 \le y \le 1\}$ in such a way that the probability of an event is equal to its area. If $X(x, y) = 12x^2 - y$ and $Y(x, y) = 2x - y$, find $P(X \le \frac{1}{4} \text{ and } Y \le 0)$.

This is mainly a problem of identifying the event at issue. A diagram (Figure 1) is convenient.

The region where $2x - y \le 0$ is the region above the line $2x = y$. The region where $12x^2 - y \le \frac{1}{4}$ is the region inside the parabola $12x^2 - y = \frac{1}{4}$. Since our sample points are inside the square $0 \le x \le 1$, $0 \le y \le 1$, the

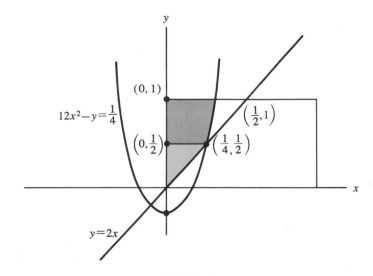

**Figure 1**

event whose probability we seek is the shaded region of Figure 1. Its area is the area of the triangle $(0, \frac{1}{2})$, $(0, 0)$, $(\frac{1}{4}, \frac{1}{2})$ which is $\frac{1}{2}(\frac{1}{2} \cdot \frac{1}{4}) = \frac{1}{16}$ plus the area given by $\int_{1/2}^{1} \sqrt{\dfrac{y + \frac{1}{4}}{12}} \, dy$ which is the area of the doubly shaded region in Figure 1, namely the region enclosed by $y = 1$, $x = 0$, $y = \frac{1}{2}$ and $12x^2 - y = \frac{1}{4}$ and to the right of the $x$ axis.

By evaluating the integral we can see that

$$P(X \le \tfrac{1}{4} \text{ and } Y \le 0) = \tfrac{1}{16} + \frac{1}{\sqrt{12}} [(\tfrac{5}{4})^{3/2} - (\tfrac{3}{4})^{3/2}]\tfrac{2}{3} \approx 0.206$$

### Example 8.1.6

Suppose $S = \{(x, y); x \text{ and } y \text{ are real numbers}\}$ and that the mass density function for the probability mass of events in $S$ is given by $p(x, y)$. Suppose that $X(x, y) = x$ and $Y(x, y) = y$. Find $P(X \le 6, \quad Y \le -2)$ and $P(a \le X \le b, \quad c \le Y \le d)$.

In Chapter 3, the concept of probability as mass had the probability of an event as the mass of that event. In this case $\{(x, y); X \le 6, \quad Y \le -2\} = \{(x, y); x \le 6, \quad y \le -2\} = \{-\infty < x \le 6, \quad -\infty < y \le -2\}$. Therefore

$$P(X \le 6, \quad Y \le -2) = \int_{-\infty}^{6} dx \int_{-\infty}^{-2} p(x, y) \, dy$$

since $p$ gives the mass density and integration of the mass density gives the mass. Similarly,

$$P(a \le X \le b, \quad c \le Y \le d) = P(\{(x, y); a \le X(x, y) \le b, \quad c \le Y(x, y) \le d\})$$

$$= P(\{(x, y); a \le x \le b, \quad c \le y \le d\})$$

$$= \int_{a}^{b} dx \int_{c}^{d} p(x, y) \, dy$$

It is important to note the simplification which results from $X(x, y) = x$ and $Y(x, y) = y$. Let us look at two more examples in which the sample space and probability measures are the same as that of Example 8.1.5 but where the random variables $X$ and $Y$ are different.

### Example 8.1.7

Let the probability system be that of Example 8.1.6. Suppose $X(x, y) = x^2 + x$ and $Y(x, y) = 3y + 7$. Find $P(X \le 6 \text{ and } Y \le -2)$.

To say $X \le 6$ and $Y \le -2$ is to say $x^2 + x \le 6$ and $3y + 7 \le -2$. This is just $3y \le -9$ or $y \le -3$ as far as the second condition is concerned. But $x^2 + x \le 6$ occurs if and only if $x^2 + x - 6 \le 0$ or $(x + 3)(x - 2) =$

$x^2 + x - 6 \le 0$. A product of two numbers is nonpositive if and only if one number is non-negative and the other is nonpositive. Therefore either

$$x + 3 \le 0 \quad \text{and} \quad x - 2 \ge 0$$

or

$$x + 3 \ge 0 \quad \text{and} \quad x - 2 \le 0$$

The first situation cannot occur because $x + 3$ is greater than $x - 2$ so if $x - 2 \ge 0$, $x + 3 > x - 2 \ge 0$. All that remains is $x + 3 \ge 0$ and $x - 2 \le 0$ or $x \ge -3$ and $x \le 2$. Therefore

$$P(X \le 6 \text{ and } Y \le -2) = P(x^2 + x \le 6 \text{ and } 3y + 7 \le -2)$$
$$= P(-3 \le x \le 2 \text{ and } y \le -2)$$
$$= \int_{-3}^{2} dx \int_{-\infty}^{-3} p(x, y)\, dy$$

It may be noted that the answer to the corresponding problem in Example 8.1.6 involved $\int_{-\infty}^{6} dx \int_{-\infty}^{-2} p(x, y)\, dy$ in which the limits on the payoff entered rather neatly as limits of integration. Here they do not.

### Example 8.1.8
Let the probability system be that of Example 8.1.6. Let $X(x, y) = 3x - y$ and $Y(x, y) = 4x + 3y$. Find $P(X \le 6 \text{ and } Y \le -2)$.

We need a description of the region in which $3x - y \le 6$ and $4x + 3y \le -2$. A diagram (Figure 2) is convenient.

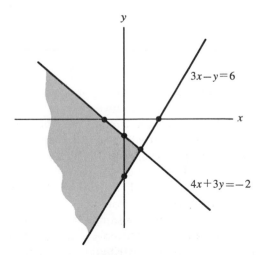

**Figure 2**

The region in which $3x - y \leq 6$ is the half plane above the line $3x - y = 6$. The region in which $4x + 3y \leq -2$ is the region below the line $4x + 3y = -2$. The region in which both occur is the shaded region. The mass attached to this region is

$$\int_{-\infty}^{16/13} \int_{3x-6}^{(-2-4x)/3} p(x, y)\, dy$$

The number $\frac{16}{13}$ is the horizontal coordinate of the point of intersection of the lines.

In this example we have *nonconstant* limits of integration even though the random variables have constant constraints $X \leq 6$ and $Y \leq -2$.

It would appear that life is getting more complicated. Generally speaking, it is. If we had three random variables it would be still more complicated.

### Example 8.1.9

Using the probability system and the random variables $X$ and $Y$ of Example 8.1.8 find $P(X \leq u$ and $Y \leq v)$.

It is not hard to see that $3x - y \leq u$ for points $x, y$ above the line $3x - y = u$. Similarly $4x + 3y \leq v$ if $(x, y)$ is below $4x + 3y = v$. Moreover these lines intersect at the point

$$\left( \frac{\begin{vmatrix} u & -1 \\ v & 3 \end{vmatrix}}{\begin{vmatrix} 3 & -1 \\ 4 & 3 \end{vmatrix}}, \frac{\begin{vmatrix} 3 & u \\ 4 & v \end{vmatrix}}{\begin{vmatrix} 3 & -1 \\ 4 & 3 \end{vmatrix}} \right) = \left( \frac{3u + v}{13}, \frac{3v - 4u}{13} \right)$$

by results of algebra. Therefore, the mass attached to the region where both inequalities hold is

$$\int_{-\infty}^{(3u+v)/13} dx \int_{3x-u}^{(v-4x)/3} p(x, y)\, dy$$

The examples of this section serve to suggest some of the questions which can be asked about two random variables on a given probability system. Similar ones can be asked about three or more random variables. So far, we have not given general theories but we have tried to answer each question as it came up with whatever technique seemed reasonable. In subsequent sections we prepare the way for a general theory. There we do for many random variables what we did in Chapter 6 for one random variable (at a time). We introduce distribution functions, probability, and density functions. To some extent these functions assume that a good deal of work with the sample space is already done, but that is true for one random variable also. Moreover, in many practical problems it is fairly realistic to assume that the distribution function or its equivalent is known.

**EXERCISES 8.1**

1   Finish Example 8.1.9 in the special case that

$$p(x, y) = \begin{cases} e^{-x}e^{-y}, & \text{both } x > 0, \quad y > 0 \\ 0, & \text{otherwise} \end{cases}$$

2   Suppose that $S$ is a countable sample space with the usual probability measure. Suppose that $X$, $Y$, and $Z$ are random variables in $S$. Describe a method for determining $P(X \leq u, \quad Y \leq v, \quad Z \leq w)$.

3   A point is chosen at random from the disk $\{(x, y); x^2 + y^2 \leq 1\}$. Let $X(x, y) = x$ and $Y(x, y) = y$. Find $P(X \leq 0, \quad Y \leq 0)$ and $P(X \leq \frac{1}{2}, \quad Y \leq \frac{1}{4})$.

◆ 4   Show that for $a < b$ and any $u$, $P(a < X \leq b, \quad Y \leq u) = P(X \leq b, \quad Y \leq u) - P(X \leq a, \quad Y \leq u)$.

◆ 5   Use exercise 4 to show that if $a < b$ and $c < d$, then $P(a < X \leq b, \quad c < Y \leq d) = P(X \leq b, \quad Y \leq d) - P(X \leq a, \quad Y \leq d) - P(X \leq b, \quad Y \leq c) + P(X \leq a, \quad Y \leq c)$. Draw a figure indicating the various sets in the $X$, $Y$ plane as though $X$ and $Y$ are numbers.

# 8.2   Distribution functions

In section 8.1 we suggested that it is important to study two or more random variables on a single probability system simultaneously. Experience has led to a number of concepts which are valuable for this study.

**Definition**   A *random vector* is an $n$-tuple $(X_1, X_2, \ldots, X_n)$ of random variables on a single probability system.

A random vector is called an $n$-dimensional random variable by some authors. We may think of $(X_1(s), X_2(s), \ldots, X_n(s))$ as a point in $n$-dimensional Euclidean space, that is, an $n$-dimensional vector. The random vector defines a function from the sample space $S$ to $n$-dimensional Euclidean space by letting $(X_1(s), \ldots, X_n(s))$ be the image of $s \in S$ under the function. Frequently we use random vectors $(X, Y)$ and $(X, Y, Z)$ since we rarely work with $n > 3$ in the present chapter.

A number of definitions suggest themselves quite readily. We make them for 3-dimensional random vectors; the analogs for other $n$ should be clear.

**Definition**   The (joint) *distribution function F* of a random vector $(X, Y, Z)$ is the function given by

$$F(u, v, w) = P(X \leq u, \quad Y \leq v, \quad Z \leq w)$$

The word "joint" is frequently omitted except for emphasis in certain contexts. For example, in section 8.6 we define conditional and marginal distribution functions indirectly. However, if no modifier is applied, one should assume that "joint" is at issue. If the joint distribution functions of a number of random vectors are being studied we write $F_{(X,Y,Z)}$ to emphasize the connection with $(X, Y, Z)$ just as we used $F_X$ to emphasize the connection with $X$.

The following facts about a distribution function $F$ follow immediately from the definitions

**(8.2.1)**    $0 \le F(u, v, w) \le 1$    for all real $u, v, w$

**(8.2.2)**    If $u \le u'$,  $v \le v'$,  and $w \le w'$,

then $F(u, v, w) \le F(u', v', w')$

**(8.2.3)**    $\lim\limits_{u \to -\infty} F(u, v, w) = \lim\limits_{v \to -\infty} F(u, v, w) = \lim\limits_{w \to -\infty} F(u, v, w) = 0$

**(8.2.4)**    $\lim\limits_{\substack{u \to +\infty \\ v \to +\infty \\ w \to +\infty}} F(u, v, w) = 1$

**(8.2.5)**    $\lim\limits_{\substack{u \to a^+ \\ v \to b^+ \\ w \to c^+}} F(u, v, w) = F(a, b, c)$

where $u \to a^+$ means $u \to a$ and $u > a$.

In exercise 8.1.5 we saw that

**(8.2.6)**    $P(a < X \le b,\; c < Y \le d)$

$= P(X \le b,\; Y \le d) - P(X \le a,\; Y \le d)$

$- P(X \le b,\; Y \le c) + P(X \le a,\; Y \le c)$

We now prove the 3-random-variable analog which should clarify the general case.

**Theorem 8.2.1**    If $u_i \le v_i$ for $i = 1, 2, 3$ then

$P(u_1 < X \le v_1,\; u_2 < Y \le v_2,\; u_3 < Z \le v_3)$

$= F(v_1, v_2, v_3) - F(u_1, v_2, v_3) - F(v_1, u_2, v_3) - F(v_1, v_2, u_3)$

$+ F(u_1, u_2, v_3) + F(u_1, v_2, u_3) + F(v_1, u_2, u_3) - F(u_1, u_2, u_3)$

PROOF    Using the argument of exercise 8.1.4, we have

$P(u_1 < X \le v_1,\; u_2 < Y \le v_2,\; Z \le w)$

**(8.2.7)**    $= P(X \le v_1,\; Y \le v_2,\; Z \le w) - P(X \le u_1,\; Y \le v_2,\; Z \le w)$

$- P(X \le v_1,\; Y \le u_2,\; Z \le w) + P(X \le u_1,\; Y \le u_2,\; Z \le w)$

But if $u_3 \le v_3$, $P(u_1 < X \le v_1,\; u_2 < Y \le v_2,\; u_3 < Z \le v_3) = P(u_1 < X \le v_1,\; u_2 < Y \le v_2,\; Z \le v_3) - P(u_1 < X \le v_1,\; u_2 < Y \le v_2,\; Z \le u_3)$. The application of (8.2.7) with $w$ replaced by $v_3$ and $u_3$, successively, gives the conclusion of the theorem.

### Example 8.2.1
Let $(X, Y)$ be a 2-dimensional random vector and let $F$ be its distribution function. Show that if $u_1 < v_1$ and $u_2 < v_2$, then $F(v_1, v_2) - F(u_1, v_2) - F(v_1, u_2) + F(u_1, u_2) \geq 0$.

By the 2-dimensional version of Theorem 8.2.1

$$P(u_1 < X \leq v_1, \quad u_2 < Y \leq v_2) = F(v_1, v_2) - F(u_1, v_2) - F(v_1, u_2) + F(u_1, u_2)$$

But $P(u_1 < X \leq v_1, \quad u_2 < Y \leq v_2) \geq 0$.

As in the case of a single random variable we have the following.

**Definition**    A random vector $(X, Y, Z)$ has a (*joint*) *probability function* if there are countably many triples $(u_1, v_1, w_1), (u_2, v_2, w_2), (u_3, v_3, w_3), \ldots$ such that $P(X = u_k, \quad Y = v_k, \quad Z = w_k) = 1$. In that case the *probability function $f$* is given by

$$f(u, v, w) = P(X = u, \quad Y = v, \quad Z = w)$$

### Example 8.2.2
Certain animals may be of type 1, 2, 3, or 4. Assume independence of different animals and that $P(\text{an animal is of type } 1) = \frac{2}{3}$, $P(\text{an animal is of type } 2) = \frac{1}{4}$, $P(\text{animal is of type } 3) = \frac{1}{16}$, and $P(\text{animal is of type } 4) = \frac{1}{48}$. Set up a random vector for the outcomes in which ten such animals are chosen and the number of each type is determined. Find the probability function.

Let $s$ be the sample of ten such animals. Let $X_k(s)$ be the number of animals of type $k$ in $s$. The probability function $f$ is given by

$$f(u_1, u_2, u_3, u_4) = \frac{10!}{u_1! u_2! u_3! u_4!} \left(\tfrac{2}{3}\right)^{u_1} \left(\tfrac{1}{4}\right)^{u_2} \left(\tfrac{1}{16}\right)^{u_3} \left(\tfrac{1}{48}\right)^{u_4}$$

where $u_1, u_2, u_3, u_4$ are non-negative integers with $\sum u_k = 10$. The quantity $10!/u_1! u_2! u_3! u_4!$ is the number of different 10-position symbols which can be made with $u_1$ ones, $u_2$ twos, $u_3$ threes, and $u_4$ fours. The numbers $\left(\tfrac{2}{3}\right)^{u_1}\left(\tfrac{1}{4}\right)^{u_2}\left(\tfrac{1}{16}\right)^{u_3} \cdot \left(\tfrac{1}{48}\right)^{u_4}$ arise from the assumption of independence.

The distribution whose probability function is given by

$$\frac{n!}{k_1! k_2! \cdots k_r!} p_1^{k_1} p_2^{k_2} \cdots p_r^{k_r}$$

where $k_1, k_2, \ldots, k_r$ are non-negative integers with $k_1 + k_2 + \cdots + k_r = n$ and $p_1 + p_2 + \cdots + p_r = 1$ is called the *multinomial* distribution.

### Example 8.2.3

An urn contains six red balls, nine white balls, 14 green balls, and eight black balls. They are identical except for color. A collection of seven balls is chosen without replacement at random from the urn. Set up a random vector for the number of balls of each color chosen among the seven. Find the probability function.

Let $R$, $W$, $G$, and $B$, respectively, represent the number of red, white, green, and black balls in the sample of seven.

$$f(r, w, g, b) \doteq P(R = r, \quad W = w, \quad G = g, \quad B = b) = \frac{1}{\binom{37}{7}} \binom{6}{r}\binom{9}{w}\binom{14}{g}\binom{8}{b}$$

where $0 \le r \le 6$, $0 \le w \le 7$, $0 \le g \le 7$, $0 \le b \le 7$, and $r + w + g + b = 7$. These probabilities follow from ideas of Chapters 1 and 3.

**Definition**     A random vector $(X, Y, Z)$ *has a (joint) density function* if there is a function $f$ from 3-dimensional Euclidean space to the reals such that $f$ can be integrated, $f(x, y, z) \ge 0$ and

(8.2.8)     $$\int_{-\infty}^{u} dx \int_{-\infty}^{v} dy \int_{-\infty}^{w} f(x, y, z)\, dz = F(u, v, w)$$

for all triples $(u, v, w)$. Any function $f$ satisfying this requirement is called a *(joint) density function* for $(X, Y, Z)$.

As in the 1-dimensional case, a random vector may have many density functions if it has any density function. This is true because two functions which agree on all but a limited set of points can both satisfy (8.2.8) for all $(u, v, w)$. We restrict ourselves to random vectors which have a density function which is piecewise continuous just as we restricted ourselves to random variables which had a piecewise continuous density function.

The same subscripting convention and the same convention about the modifier "joint" applies to probability and density functions as it does to distribution functions.

In addition to being part of the definition, equation (8.2.8) displays how the distribution function is determined by density function. Similarly, if $f$ is a probability function the definitions imply that

(8.2.9)     $$\sum_{x \le u, y \le v, z \le w} f(x, y, z) = F(u, v, w)$$

On the other hand, the distribution function determines the probability function or the density function as follows: If $(u, v, w)$ is a point of continuity of a density function $f$, theorems from calculus guarantee that

(8.2.10)     $$\frac{\partial^3 F}{\partial u\, \partial v\, \partial w}(u, v, w) = f(u, v, w)$$

For random vectors with piecewise continuous density functions, (8.2.10) guarantees that the essential character of $f$ can be gotten by operating on $F$. If $f$ is a probability function we have

$$\lim_{n \to \infty} P(u - 1/n < X \le u, \quad v - 1/n < Y \le v, \quad w - 1/n < Z \le w)$$

(8.2.11)

$$= f(u, v, w)$$

by the corollary to Theorem 3.4.4 and the definitions. By Theorem 8.2.1 the probability on the left-hand side of (8.2.11) is completely determined by $F$. Thus $f$ is completely determined (in principle) by operating on $F$.

### Example 8.2.4

Let $G$ be a function given by

$$G(u, v) = \begin{cases} (1 - e^{-u})(1 - e^{-v}), & u \ge 0 \text{ and } v \ge 0 \\ 0, & \text{otherwise} \end{cases}$$

Is $G$ the distribution function of any random vector?

Note that

$$\frac{\partial^2 G}{\partial u\, \partial v}(u, v) = \begin{cases} e^{-u}e^{-v}, & u > 0 \text{ and } v > 0 \\ 0, & u < 0 \text{ or } v < 0 \end{cases}$$

On the lines $u = 0$, $v = 0$, $\partial^2 G/\partial u\, \partial v$ may not exist.

Now $e^{-u}e^{-v} \ge 0$ and $\displaystyle\int_0^\infty \int_0^\infty e^{-u}e^{-v}\, du\, dv = 1$ so

$$p(x, y) = \begin{cases} e^{-x}e^{-y}, & x > 0 \text{ and } y > 0 \\ 0, & \text{otherwise} \end{cases}$$

gives a mass density of total mass 1. If we let $S = \{(x, y); x, y \text{ real numbers}\}$ with probability determined by the density $p(x, y)$ and we let $X(x, y) = x$, $Y(x, y) = y$ we find that if $u > 0$, $v > 0$,

$$P(X \le u, \quad Y \le v) = P(\{(x, y); x = X(x, y) \le u, \quad y = Y(x, y) \le v\})$$

$$= \int_{-\infty}^{u} dx \int_{\infty}^{v} p(x, y)\, dy = \int_0^u dx \int_0^v e^{-x}e^{-y}\, dy$$

since $p(x, y) = 0$ if either $x$ or $y$ is not positive. Now

$$\int_0^u dx \int_0^v e^{-x}e^{-y}\, dy = \int_0^u e^{-x}\, dx \int_0^v e^{-y}\, dy = \left(-e^{-x}\Big|_0^u\right)\left(-e^{-y}\Big|_0^v\right)$$

$$= (1 - e^{-u})(1 - e^{-v}) = G(u, v)$$

It can be shown that if either $u$ or $v$ is not positive $P(X \le u, \quad Y \le v) = 0$.

Example 8.2.4 shows how to construct a probability system and a random vector so as to have a certain function as the distribution function of the random vector. The key technique was to try the partial derivative of the given function as a mass density. Furthermore, the random vector had $X(x, y) = x$ and $Y(x, y) = y$. Similarly if the given function was such as to allow us to obtain a probability function by the application of (8.2.11) we could use the masses given by that probability function at the points at which the limit is not zero. The appropriate random vector will have $X(x, y) = x$ and $Y(x, y) = y$ or its higher-dimensional analog. Such a representation of a given function is called the *standard representation*. A given function may not be the distribution function of a random vector. However, if, in the proper dimensionality, it satisfies (8.2.1)–(8.2.5) and the inequality in Theorem 8.2.1, it can be shown to be the distribution function of a random vector.

**EXERCISES 8.2** ─────────────────────────────────────

◆ 1   Let $X = (X_1, X_2, X_3, X_4)$ be a random vector. Express $P(a_j \leq X_j \leq b_j$ for $j = 1, 2, 3, 4)$ in terms of $F_X(u_1, u_2, u_3, u_4)$. Prove your result.

2   Let the density function $f(x, y) = xe^{-x-y}$ if $x, y$ are both positive and $f(x, y) = 0$ otherwise. Find the distribution function.

3   Let the density function $f(x, y) = x^{-1}(1 + x)^{-2}$ if $0 < y < x$. Let $f(x, y) = 0$ otherwise. Sketch the region in which $f(x, y) > 0$. Find the distribution function.

4   Find a density function for the distribution function

$$F(u, v) = \begin{cases} e^{u+v}, & u < 0 \text{ and } v < 0 \\ e^{u}, & u < 0 \text{ and } v \geq 0 \\ e^{v}, & u \geq 0 \text{ and } v < 0 \\ 1, & u \geq 0 \text{ and } v \geq 0 \end{cases}$$

5   Find the standard representation for $(X, Y)$ if $(x, y)$ is chosen at random from $\{(x, y); 0 \leq x \leq 1, \ 0 \leq y \leq 1\}$ and $X(x, y) = x, \ Y(x, y) = \cos \pi y$.

6   What must $K$ be for $g(x, y, z) = Kx^2y \cos z$ when $0 \leq x \leq 2, \ 0 \leq y \leq 1, \ 0 \leq z \leq \frac{1}{2}\pi$ and $g(x, y, z) = 0$ otherwise, to be the density function of a random vector?

7   Let $(x, y, z)$ be chosen at random from $\{0 \leq x \leq 1, \ 0 \leq y \leq 1, \ 0 \leq z \leq 1\}$. Let $X(x, y, z) = x^2, \ Y(x, y, z) = 3y, \ Z(x, y, z) = -z$. Describe the standard representation for $(X, Y, Z)$.

8   Let $(x, y)$ be chosen as in exercise 5. Let $X(x, y) = x + y, \ Y(x, y) = y$. Describe the standard representation for $(X, Y)$. Does it seem as hard to work with the original representation as it is to find the standard representation and work with it?

◆ 9   Let $f$ be a function on 3-dimensional Euclidean space with $f(x, y, z) \geq 0$. Suppose that

$$\int_{-\infty}^{\infty} \int_{-\infty}^{\infty} \int_{-\infty}^{\infty} f(x, y, z) \, dx \, dy \, dz = 1$$

Show that $f$ is a density function for some random vector.

# 8.3   More about independent random variables

In section 6.5 the random variables $X_1, X_2, \ldots, X_n$ were said to be independent if the events $\{X_1 \le u_1\}, \ldots, \{X_n \le u_n\}$ are independent for any choice of $u_1, \ldots, u_n$. The definition of independence implies that

$$P(X_1 \le u_1, \quad X_2 \le u_2, \ldots, X_n \le u_n) = P(X_1 \le u_1)P(X_2 \le u_2) \cdots P(X_n \le u_n)$$

If this is rephrased in terms of distribution functions we have

**(8.3.1)**     $F_{(X_1, X_2, \ldots, X_n)}(u_1, u_2, \ldots, u_n) = F_{X_1}(u_1)F_{X_2}(u_2) \cdots F_{X_n}(u_n)$

That is, the (joint) distribution function of the random vector $(X_1, X_2, \ldots, X_n)$ evaluated at $(u_1, \ldots, u_n)$ is the product of the distribution functions of the various random variables $X_k$ with $F_{X_k}$ evaluated at $u_k$ for $k = 1, 2, 3, \ldots, n$.

We now show the converse relationship in a special case. The idea of this special case should make clear how the general case goes.

### Example 8.3.1
Let $(X_1, X_2, \ldots, X_6)$ be such that for all $(u_1, u_2, \ldots, u_6)$,

$$F_{(X_1, \ldots, X_6)}(u_1, u_2, \ldots, u_6) = F_{X_1}(u_1)F_{X_2}(u_2) \cdots F_{X_6}(u_6)$$

Show that

$$P(X_1 \le u_1, \quad X_3 \le u_3, \quad X_4 \le u_4, \quad X_5 \le u_5)$$
$$= P(X_1 \le u_1)P(X_3 \le u_3)P(X_4 \le u_4)P(X_5 \le u_5)$$

By the various definitions

$$P(X_1 \le u_1, \quad X_2 \le n, \quad X_3 \le u_3, \quad X_4 \le u_4, \quad X_5 \le u_5, \quad X_6 \le n)$$
$$= F_{(X_1, X_2, \ldots, X_6)}(u_1, n, u_3, u_4, u_5, n)$$
$$= F_{X_1}(u_1)F_{X_2}(n)F_{X_3}(u_3)F_{X_4}(u_4)F_{X_5}(u_5)F_{X_6}(n)$$

Now as $n \to \infty$

$$P(X_1 \le u_1, \quad X_2 \le n, \quad X_3 \le u_3, \quad X_4 \le u_4, \quad X_5 \le u_5, \quad X_6 \le n)$$
$$\to P(X_1 \le u_1, \quad X_3 \le u_3, \quad X_4 \le u_4, \quad X_5 \le u_5)$$

by Theorem 3.4.4
Furthermore as $n \to \infty$, $F_{X_2}(n) \to 1$ and $F_{X_6}(n) \to 1$. Therefore,

$$F_{X_1}(u_1)F_{X_2}(n)F_{X_3}(u_3)F_{X_4}(u_4)F_{X_5}(u_5)F_{X_6}(n)$$
$$\to F_{X_1}(u_1)F_{X_3}(u_3)F_{X_4}(u_4)F_{X_5}(u_5) = P(X_1 \le u_1)P(X_5 \le u_3)P(X_4 \le u_4)P(X_5 \le u_5)$$

However, a sequence can have only one limit so

$$P(X_1 \le u_1, \quad X_3 \le u_3, \quad X_4 \le u_4, \quad X_5 \le u_5)$$
$$= P(X_1 \le u_1)P(X_3 \le u_3)P(X_4 \le u_4)P(X_5 \le u_5)$$

It should be clear from the above example that if for all $(u_1, \ldots, u_n)$ we have equation (8.3.1) then any collection of two or more events from $\{X_1 \le u_1\}$, $\{X_2 \le u_2\}, \ldots, \{X_n \le u_n\}$ satisfies the condition that the probability of the intersection is the product of the probabilities of the events which are intersected. We thus state the next theorem without proof.

**Theorem 8.3.1**   The random variables $X_1, X_2, \ldots, X_n$ are independent if and only if for all $(u_1, u_2, \ldots, u_n)$

$$F_{(X_1, \ldots, X_n)}(u_1, \ldots, u_n) = F_{X_1}(u_1)F_{X_2}(u_2) \cdots F_{X_n}(u_n)$$

The multiplicative relationship of (8.3.1) applies to probability functions and density functions as well as to distribution functions. Regarding probability functions we consider $(X, Y)$. By exercise 8.1.5

$$P(u - 1/n < X \le u, \quad v - 1/n < Y \le v)$$

**(8.3.2)**
$$= F_{(X,Y)}(u, v) - F_{(X,Y)}(u - 1/n, v)$$
$$- F_{(X,Y)}(u, v - 1/n) + F_{(X,Y)}(u - 1/n, v - 1/n)$$

If $X$ and $Y$ are independent, the right-hand side of (8.3.2) is

**(8.3.3)**   $F_X(u)F_Y(v) - F_X(u - 1/n)F_Y(v) - F_X(u)F_Y(v - 1/n) + F_X(u - 1/n)F_Y(v - 1/n)$
$$= [F_X(u) - F_X(u - 1/n)][F_Y(v) - F_Y(v - 1/n)]$$

As $n \to \infty$ the left-hand side of (8.3.2) converges to $P(X = u$ and $Y = v)$. The right-hand side of (8.3.3) converges to $P(X = u)P(Y = v)$. Therefore if $X$ and $Y$ are independent, for any $(u, v)$, $f_{(X,Y)}(u, v) = f_X(u)f_Y(v)$ in the case of probability functions.

Conversely, if for all $(u, v)$ we have $f_{(X,Y)}(u, v) = f_X(u)f_Y(v)$, then for all $a$ and $b$

$$F_{(X,Y)}(a, b) = \sum_{\substack{u \le a \\ v \le b}} f_{(X,Y)}(u, v) = \sum_{\substack{u \le a \\ v \le b}} f_X(u)f_Y(v)$$

$$= (\sum_{u \le a} f_X(u))(\sum_{v \le b} f_Y(v)) = F_X(a)F_Y(b)$$

since all of the terms in $\sum_{\substack{u \le a \\ v \le b}} f_X(u)f_Y(v)$ appear once and only once in the formal

expansion of $(\sum_{u \le a} f_X(u))(\sum_{v \le b} f_Y(v))$ and for series of non-negative terms formal

expansions tell the whole story. We state the general result.

**Theorem 8.3.2**   If $(X_1, X_2, \ldots, X_n)$ has a probability function then the $X_j$ are independent if and only if for all $(u_1, \ldots, u_n)$

**(8.3.4)**   $f_{(X_1, \ldots, X_n)}(u_1, \ldots, u_n) = f_{X_1}(u_1)f_{X_2}(u_2) \cdots f_{X_n}(u_n)$

As for density functions, suppose that $(X, Y)$ has a density function. It can then be shown that $X$ and $Y$ have density functions. Let us make some formal calculations. First if $f_{(X,Y)}(u, v) = f_X(u)f_Y(v)$ for all $u$ and $v$ then

$$F_{(X,Y)}(u, v) = \int_{-\infty}^{u} \int_{-\infty}^{v} f_{(X,Y)}(x, y) \, dx \, dy = \int_{-\infty}^{u} \int_{-\infty}^{v} f_X(x)f_Y(y) \, dx \, dy$$

$$= \int_{-\infty}^{u} f_X(x) \, dx \int_{-\infty}^{v} f_Y(y) \, dy = F_X(u)F_Y(v)$$

Secondly, if $F_{(X,Y)}(u, v) = F_X(u)F_Y(v)$ we get, formally,

$$f_{(X,Y)}(u, v) = \frac{\partial^2}{\partial u \, \partial v} F_{(X,Y)}(u, v) = \frac{\partial}{\partial u} F_X(u) \frac{\partial}{\partial v} F_Y(v) = f_X(u)f_Y(v)$$

Although the partial differentiation cannot be justified without some restrictions on $f_{(X,Y)}$ it is true that if $X$ and $Y$ are independent there are suitable density functions such that for all $u$ and $v$, $f_{(X,Y)}(u, v) = f_X(u)f_Y(v)$. Naturally, the obvious generalization to higher-dimensional random vectors is also true.

## EXERCISES 8.3

In exercises 1–5 let $X_1, X_2, \ldots, X_n$ be independent random variables with the same distribution function $F$.

◆ 1    Let $U(s) = \max (X_1(s), X_2(s), \ldots, X_n(s))$. Show that $F_U(u) = [F(u)]^n$.

◆ 2    Let $V(s) = \min (X_1(s), X_2(s), \ldots, X_n(s))$. Show that $P(V > u) = [1 - F(u)]^n$ and $F_V(u) = 1 - [1 - F(u)]^n$.

◆ 3    Suppose that $F$ has a density function $f$. Show that $f_U(u) = n[F(u)]^{n-1}f(u)$ and $f_V(u) = n[1 - F(u)]^{n-1}f(u)$ are density functions for $U$ and $V$, respectively.

4    Suppose that $X_j$ has the density function $f(x) = (1 + x)^{-2}$ for $x > 0$, $f(x) = 0$ otherwise; i.e., $F(u) = 1 - (u + 1)^{-1}$ if $u > 0$. Show that $E(X)$ does not exist but that $E(V)$ does exist if $n \geq 2$.

◆ 5    Suppose that $F$ has a density function $f$. Show that $P(\text{two of the } X_j \text{ are equal}) = 0$. Let $n = 2m + 1$. Suppose $W(s)$ is the middle number of $X_1(s), X_2(s), \ldots, X_n(s)$ when these numbers are put in increasing order. (We need not concern ourselves with the case that two or more $X_j(s)$ are equal by the previous remark.) Show that a density function for $W$ is

$$\frac{(2m + 1)!}{m! m!} [F(u)]^m [1 - F(u)]^m f(u)$$

# 8.4 The 2-dimensional normal distribution

The most important 2-dimensional distribution is determined as follows. Let $A$, $B$, and $C$ be real constants with $A > 0$, $C > 0$, and $D = AC - B^2 > 0$. Let

**(8.4.1)**    $$f(x, y) = \frac{1}{2\pi\sqrt{D}} \exp\left[-\frac{1}{2D}(Cx^2 - 2Bxy + Ay^2)\right]$$

Clearly $f(x, y) > 0$ for all $x$, $y$ and it is continuous. By the closing remark of section 8.2 we can show that $f$ is a density function if

**(8.4.2)**    $$\int_{-\infty}^{\infty}\int_{-\infty}^{\infty} f(x, y)\, dx\, dy = 1$$

For fixed $x$ let $z = \sqrt{A/D}\, y - Bx/\sqrt{DA}$. Then $z^2 = (A/D)y^2 - 2Bxy/D + B^2x^2/DA$ since $\sqrt{A/D}\, y(Bx/\sqrt{DA}) = (B/D)xy$. Now $(1/D)(Cx^2 - 2Bxy + Ay^2) = (1/D)(C - B^2/A)x^2 + z^2$ since

$$\frac{1}{D}\left(C - \frac{B^2}{A}\right)x^2 + \frac{B^2x^2}{DA} = \frac{1}{D}\left(C - \frac{B^2}{A} + \frac{B^2}{A}\right)x^2 = \frac{C}{D}x^2$$

Furthermore $(1/D)(C - B^2/A) = (1/D)[(AC - B^2)/A] = (1/D)(D/A) = 1/A$ by the definition of $D$. For fixed $x$, $dz = \sqrt{A/D}\, dy$ or $dy = \sqrt{D/A}\, dz$. Furthermore $z \to +\infty(-\infty)$ if and only if $y \to +\infty(-\infty)$ since $\sqrt{A/D} > 0$. Taking advantage of all this we find that the left-hand side of (8.4.2) is

**(8.4.3)**    $$\frac{1}{2\pi\sqrt{D}}\int_{-\infty}^{\infty} dx \int_{-\infty}^{\infty} \exp\left(-\frac{x^2}{2A} - \frac{z^2}{2}\right)\sqrt{\frac{D}{A}}\, dz$$

$$= \frac{1}{2\pi\sqrt{A}}\int_{-\infty}^{\infty} dx \int_{-\infty}^{\infty} \exp\left(-\frac{x^2}{2A} - \frac{z^2}{2}\right) dz$$

Now we let $x = \sqrt{A}\, w$ and $dx = \sqrt{A}\, dw$. Since $A > 0$, $w \to +\infty(-\infty)$ together with $x$. Thus (8.4.3) is

**(8.4.4)**    $$\frac{1}{2\pi\sqrt{A}}\int_{-\infty}^{\infty} \sqrt{A}\, dw \int_{-\infty}^{\infty} \exp\left(-\frac{Aw^2}{2A} - \frac{z^2}{2}\right) dz$$

$$= \frac{1}{2\pi}\int_{-\infty}^{\infty} dw \int_{-\infty}^{\infty} \exp\left[-\tfrac{1}{2}(w^2 + z^2)\right] dz$$

A clever trick is known which will show that (8.3.4) is exactly equal to 1. We consider the volume of the region under the surface

$$g(w, z) = \exp\left[-\tfrac{1}{2}(w^2 + z^2)\right]$$

and above the $w$, $z$ plane. This is the right-hand side of (8.4.4) except for the factor $1/2\pi$. However we make our evaluation using the polar coordinates $w = r\cos\theta$ and $z = r\sin\theta$ for the $w$, $z$ plane. In these coordinates the top surface is given by $\exp\left[-\frac{1}{2}(r^2\cos^2\theta + r^2\sin^2\theta)\right] = e^{-r^2/2}$. It is known from calculus that the element of area in polar coordinates is $r\,dr\,d\theta$. Thus the volume we seek is

$$(8.4.5) \qquad \int_0^{2\pi}\int_0^\infty e^{-r^2/2}r\,dr\,d\theta = \int_0^{2\pi} d\theta\left(\int_0^\infty e^{-r^2/2}r\,dr\right)$$

where $0 \le \theta < 2\pi$ and $0 \le r < \infty$ covers the plane. Now $d(-e^{-r^2/2}) = -e^{-r^2/2}(-\frac{1}{2}2r\,dr) = e^{-r^2/2}r\,dr$. Thus the $r$-integration in (8.4.5) gives $-e^{-r^2/2}\big|_0^\infty$ $= 0 + 1$. The $\theta$-integration is just $\int_0^{2\pi} d\theta = 2\pi$. Combining these observations we find that the right-hand side of (8.4.4) is just $(1/2\pi)2\pi = 1$ which was to be proved.

We can now show that $\int_{-\infty}^\infty (e^{-x^2/2}/\sqrt{2\pi})\,dx = 1$, a fact which we assumed in section 6.4 and following sections. Let $J = \int_{-\infty}^\infty (e^{-x^2/2}/\sqrt{2\pi})\,dx$. By theorems about double integrals

$$J^2 = \int_{-\infty}^\infty \frac{e^{-w^2/2}}{\sqrt{2\pi}}\,dw \int_{-\infty}^\infty \frac{e^{-z^2/2}}{\sqrt{2\pi}}\,dz = \int_{-\infty}^\infty dw \int_{-\infty}^\infty \frac{e^{-w^2/2}e^{-z^2/2}}{\sqrt{2\pi}\,\sqrt{2\pi}}\,dz$$

$$= \frac{1}{2\pi}\int_{-\infty}^\infty dw \int_{-\infty}^\infty e^{-(w^2+z^2)/2}\,dz$$

However, the last integral is (8.4.4) which is 1. Therefore $J^2 = 1$. Since $J$ is obviously positive $J = 1$. This fact was used in connection with the 1-dimensional normal distribution.

A slight modification of the original density (8.4.1) gives the general 2-dimensional normal density function

$$
\begin{aligned}
& f(x-m_1, y-m_2) \\
(8.4.6) \quad &= \frac{1}{2\pi\sqrt{D}} \exp\left\{-\frac{1}{2D}\left[C(x-m_1)^2 - 2B(x-m_1)(y-m_2) + A(y-m_2)^2\right]\right\}
\end{aligned}
$$

Equation (8.4.1) gives the distribution of the general *nonsingular centered* 2-dimensional *normal* random vector and (8.4.6) gives the distribution of the general *nonsingular* 2-dimensional *normal* random vector.

## EXERCISES 8.4

♦ 1  Suppose that $(X, Y)$ has a 2-dimensional normal distribution given by (8.4.1). Show that $X$ and $Y$ are independent if and only if $B = 0$.

2  Let $f(x, y) = K/(1 + x^2 + y^2)^\alpha$ for all $x$, $y$ with $\alpha > 1$. What must $K$ be in order that $f$ is a density function? [*Hint:* Use the polar coordinate trick.]

3  Let $f$ be as in (8.4.1). Describe the equidensity curves $f(x, y) = k > 0$.

**4**   Suppose that $(X, Y)$ has the density function (8.4.1) with $B = 0$ and $A = C$. Find $P(X^2 + Y^2 \le u)$. [*Hint*: Use polar coordinates.]

**5**   Suppose that $(X, Y)$ has the density function (8.4.1). Find $P(X \ge 0, \ Y \ge 0)$.

♦ **6**   Let $f$ be as in (8.4.1). Compute $\int_{-\infty}^{\infty} dx \int_{-\infty}^{\infty} x^2 f(x, y) \, dy$.

# 8.5 Calculations with random variables and random vectors

Let $(X, Y)$ be a random vector with probability function $f$. In addition to the basic quantities such as $P(a < X \le b, \ c < Y \le d)$ we might want to calculate $P(X + Y \le 3)$, $P(X^2 + 3XY + Y^2 \ge 7)$, $P([\max(X, X + Y)] = 2)$, and $P(\cos X \le 0)$. These probabilities can be found, in principle, by finding the sums of finite or infinite series in view of the definition of a probability function. Thus, for example,

**(8.5.1)**    $P(X^2 + 3XY + Y^2 \ge 7) = \sum_{u^2 + 3uv + v^2 \ge 7} f(u, v)$

where the sigma notation means "sum over all pairs $(u, v)$ where $u^2 + 3uv + v^2 \ge 7$." Since $f(u, v) = P(X = u, \ Y = v)$ the right-hand side is the sum of the probabilities of disjoint events into which $\{s; [X(s)]^2 + 3X(s)Y(s) + [Y(s)]^2 \ge 7\}$ can be decomposed. Since $f(u, v) > 0$ for only countably many pairs $(u, v)$, the sum on the right-hand side of (8.5.1) is a finite or infinite series. Similarly

$$P([\max(X, X + Y)] = 2) = \sum_{[\max(u, u + v)] = 2} f(u, v)$$

It may not be easy to evaluate the sums in neat form but the infinite series are convergent and one can obtain arbitrary accuracy.

## Example 8.5.1
A fair coin is tossed independently six times and a payoff is made according to

$$X_j = \begin{cases} 1, & j\text{th toss is H} \\ 0, & j\text{th toss is T} \end{cases}$$

Find $P(X_1 + X_2 + \cdots + X_6 \le 3)$.

Since $X_j(s) \ge 0$ for all $s \in S$ we must have three or fewer of the random variables $X_1, \ldots, X_6$ equalling 1 and hence the rest equalling 0. This is the same as requiring the coin to show H on three or fewer tosses. The probability of this event is

$$\left(\frac{1}{2}\right)^6 + \binom{6}{1}\left(\frac{1}{2}\right)^6 + \binom{6}{2}\left(\frac{1}{2}\right)^6 + \binom{6}{3}\left(\frac{1}{2}\right)^6$$

corresponding to six T's, one H and five T's, two H's and four T's, three H's and three T's. Adding, we obtain $42/64 = 21/32$.

The discussion in Example 8.5.1 is different from our previous methods because we did not compute

$$\sum_{u_1 + u_2 + \ldots + u_6 \leq 3} f(u_1, u_2, \ldots, u_6)$$

This was mainly because $f(u_1, u_2, \ldots, u_6) = 1/2^6$ for $u_i \in \{0, 1\}$ and the critical issue is how many sextuples $(u_1, \ldots, u_6)$ satisfy $u_1 + u_2 + \cdots + u_6 \leq 3$. In a sense we retreated to the underlying sample space to solve this problem. It is frequently preferable to do so.

### *Example 8.5.2*

Let $X_1$, $X_2$, $X_3$ be independent, identically distributed random variables whose common density function $f$ is given by $f(1) = f(-1) = \frac{1}{4}$ and $f(0) = \frac{1}{2}$. Find $P(\max (X_1, X_1 + X_2, X_1 + X_2 + X_3)) > 0$.

The payoffs of $X_1$, $X_2$, $X_3$ are essentially $-1$, $0$, or $1$. The values of $(X_1, X_2, X_3)$ which lead to the maximum of $(X_1, X_1 + X_2, X_1 + X_2 + X_3) > 0$ are as follows: $(1, a, b)$ where $a, b \in \{-1, 0, 1\}$, $(0, 1, a)$ where $a \in \{-1, 0, 1\}$, $(0, 0, 1)$, and $(-1, 1, 1)$. Since the $X_j$ are independent and identically distributed, $f_{(X_1, X_2, X_3)}(u_1, u_2, u_3) = f(u_1)f(u_2)f(u_3)$ by (8.3.3). Therefore, $f_{(X_1, X_2, X_3)}(0, 0, 1) = \frac{1}{2} \cdot \frac{1}{2} \cdot \frac{1}{4}$, $f_{(X_1, X_2, X_3)}(-1, 1, 1) = \frac{1}{4} \cdot \frac{1}{4} \cdot \frac{1}{4}$,

$$\sum_{a, b \in \{-1, 0, 1\}} f_{(X_1, X_2, X_3,)}(1, a, b) = \frac{1}{4}, \qquad \sum_a f_{(X_1, X_2, X_3)}(0, 1, a) = \frac{1}{2} \cdot \frac{1}{4}$$

Therefore our answer is $\frac{1}{16} + \frac{1}{64} + \frac{1}{4} + \frac{1}{8} = \frac{29}{64}$.

There is no reason that we should restrict our interest in $P(X + Y \leq 3)$, $P(X^2 + 3XY + Y^2 \geq 7)$ etc. to the case where $(X, Y)$ has a probability function. What if $(X, Y)$ has a density function? The analog to a sum of the form $\sum_{(u,v) \in M} f(u, v)$ when $f$ is a probability function is an integral of the form

$$\iint_M f(u, v) \, du \, dv$$ where $f$ is a density function. Does this analogy lead to a correct

conclusion or are we fooling ourselves?

The standard representation for the random vector $(X, Y)$ has $S = R \times R$ and $P$ given by

$$P(\{(x, y) \in S; a < x \leq b, \quad c < y \leq d\}) = \int_c^d \left[ \int_a^b f(x, y) \, dx \right] dy$$

that is, the probability mass density at the point $(x, y)$ is $f(x, y)$. The random vector $(X, Y)$ in the standard representation is given by $X(x, y) = x$, $Y(x, y) = y$ for $(x, y) \in S$. Thus $\{X + Y \leq 3\} = \{(x, y) \in S; X(x, y) + Y(x, y) \leq 3\} =$

$\{(x, y); x + y \leq 3\}$ and by the conventions in sections 3.5 and 3.6, if the probability mass density is $f(x, y)$ then

$$P(\{(x, y); x + y \leq 3\}) = \iint\limits_{x+y\leq 3} f(x, y)\, dx\, dy = \iint\limits_{u+v\leq 3} f(u, v)\, du\, dv$$

Is it possible that we could have different answers according to whether we use the standard representation or another representation? This can be shown to be impossible under our assumptions.

It is worthwhile observing that we are asking for $P((X_1, X_2, \ldots, X_n) \in M)$ where $M$ is a set (a Borel set) in Euclidean $n$-dimensional space. The set $M$ is not always determined by equalities and inequalities but sets of that sort are common. Notice that $n = 1$ might be interesting, too.

### Example 8.5.3

Let $X$ be a random variable with density function $f$. Let $Y(s) = |X(s)|$ for all $s \in S$, the background sample space. Find a density function for $Y$.

We ignore the issue of whether $Y$ actually is a random variable at this point. We even handle the method of finding the density function in an intuitive fashion, however, intuition is often more important than rigor. For $u < 0$, $P(Y \leq u) = 0$ because $\{Y \leq u\} = \{s \in S; |X(s)| \leq u < 0\} = \varnothing$ because the absolute value of a number cannot be negative. If $u \geq 0$, $\{Y \leq u\} = \{|X| \leq u\} = \{-u \leq X \leq u\}$. As indicated above, $P(-u \leq X \leq u) = \int_{-u}^{u} f(x)\,dx$. Thus we have shown that

$$F_Y(u) = \begin{cases} 0, & u < 0 \\ \int_{-u}^{u} f(x)\, dx, & u \geq 0 \end{cases}$$

In most ordinary cases, we can find a density function by differentiating the distribution function. Therefore

$$f_Y(u) = F_Y'(u) = \begin{cases} 0, & u < 0 \\ \dfrac{d}{du}\displaystyle\int_{-u}^{u} f(x)\, dx, & u \geq 0 \end{cases}$$

From theorems of calculus

$$\frac{d}{du}\int_{-u}^{u} f(x)\, dx = \frac{d}{du}\left[\int_{0}^{u} f(x)\, dx - \int_{0}^{-u} f(x)\, dx\right]$$

$$= f(u) - (-1)f(-u) = f(u) + f(-u)$$

for $u > 0$. There may be a little trouble at $u = 0$ but one point never matters. We can, therefore, take

$$f_Y(u) = \begin{cases} 0, & u \leq 0 \\ f(u) + f(-u), & u > 0 \end{cases}$$

when differentiation is permissible. A subsequent exercise will show that $f_Y$ as given works for all $X$ with density functions $f$.

### Example 8.5.4

Let $X$ be a random variable with a density function $f$. Find the distribution function for the random variable $X^2$ defined by $X^2(s) = [X(s)]^2$ for each $s \in S$.

Again we have $F_{X^2}(u) = P(X^2 \le u) = P(\{s; [X(s)]^2 \le u\})$. Now if $u < 0$, $\{s; [X(s)]^2 \le u < 0\} = \varnothing$. If $u \ge 0$, $\{X^2 \le u\} = \{-\sqrt{u} \le X \le \sqrt{u}\}$. Now

$$P(-\sqrt{u} \le X \le \sqrt{u}) = \int_{-\sqrt{u}}^{\sqrt{u}} f(x)\, dx \text{ and}$$

$$F_{X^2}(u) = \begin{cases} 0, & u < 0 \\ \int_{-\sqrt{u}}^{\sqrt{u}} f(x)\, dx, & u \ge 0 \end{cases}$$

We now pass to an example in which a genuine random vector enters.

### Example 8.5.5

Let $(X, Y)$ be a random vector where $P(X \le 0) = 0$, and where $X$ and $Y$ are independent and have density functions. Let $Z(s) = Y(s)/X(s)$ if $X(s) > 0$. We may set $Z(s) = \pi$ if $X(s) \le 0$. Find the distribution function for $Z$.

Since $X$ and $Y$ are independent, $f_{(X,Y)}(u, v) = f_X(u) f_Y(v)$. Since $P(X \le 0) = 0$ we may assume $f_X(u) = 0$ for $u \le 0$. For those $s \in S$ where $X(s) > 0$, $Z(s) \le u$ if and only if $Y(s) \le uX(s)$. The region $y \le ux$ is indicated in Figure 3, for

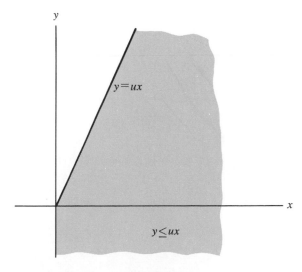

$y = ux$

$y \le ux$

**Figure 3**

$u > 0$. We can ignore cases where $x \leq 0$ for this problem. The situation for $u < 0$ is similar. Therefore,

$$P(Z \leq u) = P(Y \leq ux) = \int_0^\infty dx \int_{-\infty}^{ux} f_X(x) f_Y(y) \, dy$$

where we integrate $x$ from 0 to $\infty$ instead of from $-\infty$ to $\infty$ because $P(X \leq 0)$ $= 0$. By definition $\int_{-\infty}^a f_Y(y) \, dy = F_Y(a)$ so $P(Z \leq u) = \int_0^\infty f_X(x) F_Y(ux) \, dx$ after setting $a = ux$.

We have been dealing indirectly with functions of a random variable. To facilitate our studies we make a formal definition.

**Definition**   Let $g(x_1, x_2, \ldots, x_n)$ define a function from $n$-dimensional Euclidean space to the real numbers. Let $(X_1, X_2, \ldots, X_n)$ be a random vector. We define the function $g(X_1, \ldots, X_n): S \to R$ by $g(X_1, X_2, \ldots, X_n)(s) = g(X_1(s), X_2(s), \ldots, X_n(s))$. If $g$ has an algebraic or standard mathematical notation (for example, $e^x$, $\sin x$, $\cos x$, $\log x$, etc.) we also denote $g(X_1, \ldots, X_n)$ by the substitution of $X_1, \ldots, X_n$ into the notation for $g$.

Under mild smoothness requirements on $g$, $g(X_1, \ldots, X_n)$ is a random variable, i.e., $\{s \in S; g(X_1(s), X_2(s), \ldots, X_n(s)) \leq u\} \in \mathscr{A}$ for all real $u$. We will not discuss these smoothness requirements since it is hard for us to write a function $g$ which does not satisfy them. Thus we will have random variables such as $|X|$ (see Example 8.5.3), $X^2$ (Example 8.5.4), $\sin X$, $X^2 + XY + 3Y^2$, $e^{X+Y}$ whenever we have random variables $X$ and $Y$.

It is not always necessary that $g(x_1, \ldots, x_n)$ have its domain equal to all of $n$-dimensional Euclidean space. If there are proper restrictions on the random variables $X_1, X_2, \ldots, X_n$, it is possible that $g(X_1(s), X_2(s), \ldots, X_n(s))$ is a genuine function from $S$ to $R$ even though $g$ is not defined on all of $n$-dimensional Euclidean space. Thus we have $Y/X$ if $\{X = 0\} = \varnothing$ or $\sqrt{X}$ if $\{X < 0\} = \varnothing$.

**Example 8.5.6**
Ten different, fair coins are tossed independently. Let $X_1, X_2, \ldots, X_{10}$ be random variables defined by

$$X_n = \begin{cases} 1, & n\text{th coin shows H} \\ 0, & n\text{th coin shows T} \end{cases}$$

for $n = 1, 2, \ldots, 10$. Let

$$g(x_1, x_2, \ldots, x_{10}) = \frac{1}{10} \sum_{n=1}^{10} \left( x_n - \frac{x_1 + x_2 + \cdots + x_{10}}{10} \right)^2$$

Find the probability function for $Y = g(X_1, X_2, \ldots, X_{10})$.

The function $g$ is symmetric in the sense that if we permute the symbols $x_1, x_2, \ldots, x_{10}$ to the symbols $x_1', x_2', \ldots, x_{10}'$, $g(x_1', x_2', \ldots, x_{10}') = g(x_1, x_2, \ldots, x_{10})$.

Let $s$ be the sample point for which the first $k$ coins show H and the remaining $10 - k$ show T. The probability of this event is $(\frac{1}{2})^k(\frac{1}{2})^{10-k} = (\frac{1}{2})^{10}$. Furthermore $X_1(s) = 1, \ldots, X_k(s) = 1, \quad X_{k+1}(s) = 0, \ldots, X_{10}(s) = 0$ if $k < 10$. Therefore,

$$
\begin{aligned}
Y(s) &= \frac{1}{10} \sum_{n=1}^{10} \{X_n(s) - \tfrac{1}{10}[X_1(s) + \cdots + X_{10}(s)]\}^2 \\
&= \tfrac{1}{10}[k(1 - \tfrac{1}{10}k)^2 + (10 - k)(0 - \tfrac{1}{10}k)^2] \\
&= \tfrac{1}{10}[\tfrac{1}{100}k(10 - k)^2 + (10 - k)\tfrac{1}{100}k^2] = \tfrac{1}{1000}[10 - k + k]k(10 - k) \\
&= \tfrac{1}{100}k(10 - k)
\end{aligned}
$$

We get this value of $g(s)$ for all $s$ which have $k$ coins showing H and $10 - k$ coins showing T by the observed symmetry. Moreover the probability of exactly $k$ H's and $(10 - k)$ T's is $\binom{10}{k}\left(\frac{1}{2}\right)^k\left(\frac{1}{2}\right)^{10-k} = \binom{10}{k}\left(\frac{1}{2}\right)^{10}$. Notice that $k(10 - k)$ is the same for the following pairs of values of $k$: $k = 0, \quad k = 10$; $k = 1, \quad k = 9$; $k = 2, \quad k = 8$; $\ldots$; $k = 4, \quad k = 6$. Therefore $f_Y\left(\dfrac{5 \cdot 5}{100}\right) = \binom{10}{5}\left(\frac{1}{2}\right)^{10}$ and $f_Y\left(\dfrac{k(10 - k)}{100}\right) = 2\binom{10}{k}\left(\frac{1}{2}\right)^{10}, \quad k = 0, 1, 2, \ldots, 4$.

In Example 6.3.5 we saw that if $X$ has a density function $f_X$ and $Y = g(X)$ where $g$ is monotone, then

**(8.5.2)**     $f_Y(g(x))|g'(x)| = f_X(x)$

This is a change of variables theorem. We can change variables in $n$ dimensions, too. We work in two dimensions.

Suppose that $(X, Y)$ has a density function $f$ and $U = g(X, Y)$, $V = h(X, Y)$ where $g$ and $h$ are continuous and have continuous first partial derivatives. We may assume that $(X, Y)$ is in the standard representation. Then $s = (x, y)$, $X(x, y) = x, \quad Y(x, y) = y, \quad U(x, y) = g(x, y), \quad$ and $V(x, y) = h(x, y)$. Let

$$
J(x, y) = \begin{vmatrix} \dfrac{\partial g(x, y)}{\partial x} & \dfrac{\partial g(x, y)}{\partial y} \\[2mm] \dfrac{\partial h(x, y)}{\partial x} & \dfrac{\partial h(x, y)}{\partial y} \end{vmatrix}
$$

The determinant $J$ is called the *Jacobian*. Other notation for the Jacobian is $\partial(g, h)/\partial(x, y)$.

One significant aspect of $J$ is as follows. Suppose that $M$ is a set in the $xy$ plane which has area. Frequently $M$ will be a rectangle. Let $T(M) = \{(g(x, y),$

$h(x, y))$; $(x, y) \in M$}. Let $\|M\| = $ l.u.b.$_{s_1, s_2 \in M}$ (distance between $s_1$ and $s_2$). Suppose that we restrict our attention to such sets $M$ with $(x_0, y_0) \in M$. Let $A(L)$ be the area of the set $L$. Then

(8.5.3)      as $\|M\| \to 0$,      $\dfrac{A(T(M))}{A(M)} \to |J(x_0, y_0)|$

We will find a density function for the random vector $(U, V)$ in a heuristic way. Let $u = g(x, y)$ and $v = h(x, y)$. Let $N$ be a rectangle centered at $(u_0, v_0)$ as in Figure 4. Suppose that $(x_1, y_1), (x_2, y_2), \ldots$ are the points such that $(u_0, v_0) = (g(x_j, y_j), h(x_j, y_j))$. Normally there are only finitely many such points. Suppose that there are disjoint sets $M_j$ with $(x_j, y_j) \in M_j$ and $T(M_j) = N$. We have

(8.5.4)      $P(\{(x, y) \in S; (g(x, y), h(x, y)) \in N\}) = \sum_j P(M_j) = \sum_j \int_M f\, ds$

If $\|N\|$ and the $\|M_j\|$ are small, $\int_{M_j} f\, ds \approx f(x_j, y_j) A(M_j)$ by definition of a density and a standard representation. Furthermore,

$$P((u, v) \in N) \approx f_{(U,V)}(u_0, v_0) A(N)$$

By (8.5.4),

$$f_{(U,V)}(u_0, v_0) A(N) \approx \sum_j f(x_j, y_j) A(M_j)$$

or

$$f_{(U,V)}(u_0, v_0) \approx \sum_j f(x_j, y_j) \frac{A(M_j)}{A(N)}$$

If $J(x_j, y_j) \neq 0$, by the reciprocation of (8.5.3), $A(M_j)/A(N) \approx 1/|J(x_j, y_j)|$. Under our present assumptions and with the additional assumption that $J(x, y) = 0$ at most on a family of curves whose total area is zero, we can assert that

(8.6.5)      $f_{(U,V)}(u_0, v_0) = \sum_j f(x_j, y_j)/|J(x_j, y_j)|$

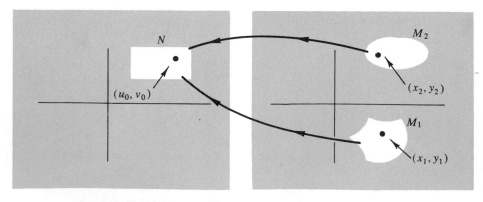

**Figure 4**

This technique of computing $f_{(U,V)}$ is called the "change of variables" or "Jacobian" technique.

### Example 8.5.7

In the current notation let $g(x, y) = ax + by$ and $h(x, y) = cx + dy$ where $a$, $b$, $c$, and $d$ are constants. Discuss $f_{(U,V)}$.

In our current notation $u = ax + by$ and $v = cx + dy$. Also,

$$J(x, y) = \begin{vmatrix} a & b \\ c & d \end{vmatrix}$$

by an easy computation of the appropriate partial derivatives. Note that $J$ is the determinant of the coefficients of the system of linear equations $ax + by = u$, $cx + dy = v$. If $J(x, y) = ad - bc \neq 0$ we have $f_{(U,V)}(u, v) = f(x, y)/|ad - bc|$ since $u$ and $v$ can be solved for uniquely in terms of $x$ and $y$, when the determinant of the coefficient is not zero. With that solution

$$f_{(U,V)}(u, v) = f\left( \frac{\begin{vmatrix} u & b \\ v & d \end{vmatrix}}{\begin{vmatrix} a & b \\ c & d \end{vmatrix}}, \frac{\begin{vmatrix} a & u \\ c & v \end{vmatrix}}{\begin{vmatrix} a & b \\ c & d \end{vmatrix}} \right) \Big/ |ad - bc|$$

The change of variable technique is valid for all random vectors regardless of dimension. It is just more difficult to carry out the calculations in higher dimensions.

### EXERCISES 8.5

1  Calculate $f_{X^2}$ if $X$ is $N(0, 1)$.

♦ 2  Let $(X, Y)$ have the density (8.4.1). What is the density of $(X + m_1, Y + m_2)$?

♦ 3  Let $X$ and $Y$ be independent and $N(0, 1)$. Find the density of $(aX + bY, cX + dY)$ where $ad - bc \neq 0$.

4  Show that $f_Z(u) = \int_0^\infty x f_X(x) f_Y(ux)\, dx$ is a density function for $Z$ of Example 8.5.5 by showing that $\int_0^a f_Z(u)\, du = P(Z \leq a)$. [*Hint:* Interchange order of integration.]

♦ 5  Let $(X, Y)$ have the density (8.4.1). Find the density of $(U, V) = (aX + bY, cX + dY)$ where $ad - bc \neq 0$.

♦ 6  Show that $U$ and $V$ of exercise 5 are independent if and only if $acA + bdC + (ad + bc)B = 0$.

◆ 7  Show that if $(X, Y)$ is as in exercise 5 there are two independent normal random variables $U$ and $V$ such that $X = qU + rV$, $Y = sU + tV$.

8  Apply the change of variables technique in the 1-dimensional form to get the density functions for $f_{|X|}$ and $f_{X^2}$. See Examples 8.5.3 and 8.5.4.

◆ 9  Prove that $|ad - bc|$ is the area of the parallelogram with adjacent edges $(0, 0)$, $(a, b)$ and $(0, 0)$, $(c, d)$. Draw a diagram.

10  Let $X$, $Y$, and $Z$ be independent and be $N(0, t_1)$, $N(0, t_2 - t_1)$, and $N(0, t_3 - t_2)$, respectively. Find the joint density function of $(X, X + Y, X + Y + Z)$.

11  Let $X_1$, $X_2$, $X_3$ be independent, identically distributed negative exponential random variables. Find the joint density function of $(X_1, X_2, X_1 + X_2 + X_3)$.

12  Let $g(x, y) = x + \sqrt{y - 1}$. Let $(X, Y)$ be independent and suppose $X$ is $N(0, 1)$ and $Y$ has the uniform distribution with $a = 1$, $b = 5$. Why does $g(X, Y)$ make perfectly good sense even though $g$ is not defined for all $(x, y)$?

13  Let $X$ have the density function $f(x) = 3x^2$ for $0 \le x \le 1$ and $f(x) = 0$ for other $x$. Find a density function for $X^3$. What is the distribution function for $X$?

14  Let $X$ have the negative exponential distribution. Find a density function for $1 - e^{-\lambda X}$. What is the distribution function for $X$?

15  Let $X$ have the Cauchy distribution. Find a density function for $\frac{1}{2} + (1/\pi) \arctan X$. What is the distribution function for $X$? Do exercises 13–15 have a pattern? See exercise 16.

16  Let $X$ have the standard representation with density function $f$ and distribution function $F$. Suppose that $F(u) = 0$ for all $u \le a$, $F(u) = 1$ for all $u \ge b$, and $F(u)$ is strictly monotonically increasing on $a < u < b$. Show that $F(X)$ has the uniform distribution on the interval $\{y; 0 < y < 1\}$ by showing that if $0 < v < 1$, $P(F(X) \le v) = v$.

# 8.6  Conditional and marginal densities

In some problems a density function for a multidimensional random vector is known and density functions for the component random variables are desired. For example, we may have $(X, Y, Z)$ with density function $f_{(X,Y,Z)}$. We can consider $F_{(X,Z)}(u, w) = P[X \le u \text{ and } Z \le w]$. This is just

$$\int_{-\infty}^{u} dx \int_{-\infty}^{w} dz \int_{-\infty}^{\infty} f_{(X,Y,Z)}(x, y, z) \, dy$$

By the definition of a density function as one whose integral is the distribution function we have $f_{(X,Z)}(x, z) = \int_{-\infty}^{\infty} f_{(X,Y,Z)}(x, y, z) \, dy$. Similarly, $f_{(Y,Z)}(y, z) = \int_{-\infty}^{\infty} f_{(X,Y,Z)}(x, y, z) \, dx$.

We might want $f_X(x)$. Since

$$f_X(u) = P(X \le u) = \int_{-\infty}^{u} dx \int_{-\infty}^{\infty} dy \int_{-\infty}^{\infty} f_{(X,Y,Z)}(x, y, z) \, dz$$

then

$$f_X(x) = \int_{-\infty}^{\infty} dy \int_{-\infty}^{\infty} f_{(X,Y,Z)}(x, y, z) \, dz$$

The density functions arising this way are called *marginal density functions* associated with the (joint) density function $f_{(X,Y,Z)}$. A density function for a 2-dimensional random vector $(X, Y)$ has only the marginal density functions

$$f_X(x) = \int_{-\infty}^{\infty} f_{(X,Y)}(x, y) \, dy \quad \text{and} \quad f_Y(y) = \int_{-\infty}^{\infty} f_{(X,Y)}(x, y) \, dx$$

## Example 8.6.1

The random vector $(X, Y)$ has the density function $f(x, y) = \dfrac{K}{(1 + x^2 + y^2)^2}$ where $K$ is a constant chosen to insure that $\int_{-\infty}^{\infty} \int_{-\infty}^{\infty} \dfrac{K}{(1 + x^2 + y^2)^2} \, dx \, dy = 1$. Find $f_X(x)$.

We need only compute $\int_{-\infty}^{\infty} \dfrac{K}{(1 + x^2 + y^2)^2} \, dy$. If we set $y = \sqrt{1 + x^2} \tan \theta$, it is clear that $\theta \to \pm \frac{1}{2}\pi$ according as $y \to \pm\infty$ for any fixed $x$. Therefore our integral is

$$K \int_{-\pi/2}^{\pi/2} \frac{\sqrt{1 + x^2} \sec^2 \theta}{(1 + x^2)^2 (1 + \tan^2 \theta)^2} \, d\theta = \frac{K}{(1 + x^2)^{3/2}} \int_{-\pi/2}^{\pi/2} \frac{\sec^2 \theta}{\sec^4 \theta} \, d\theta$$

$$= \frac{K}{(1 + x^2)^{3/2}} \int_{-\pi/2}^{\pi/2} \cos^2 \theta \, d\theta$$

It is easy to see that $\int_{-\pi/2}^{\pi/2} \cos^2 \theta \, d\theta = \frac{1}{2}\pi$. Thus $f_X(x) = \dfrac{K\pi}{2(1 + x^2)^{3/2}}$. It can also be shown that $K = 1/\pi$. Thus $f_X(x) = \frac{1}{2}(1 + x^2)^{-3/2}$.

The analog for probability functions in the case that $f_{(X,Y)}$ is a 2-dimensional probability function is defined by

$$f_X(u) = \sum_y f_{(X,Y)}(u, y)$$

where we may take

$$y \in \{v_j; f_{(X,Y)}(u_j, v_j) > 0 \text{ and } \sum f_{(X,Y)}(u_j, v_j) = 1\}$$

If $(X, Y)$ is a random vector with a probability function $f_{(X,Y)}$ and if $f_{(X,Y)}(u_k, v_k) > 0$, we can ask for $P(X = u \mid Y = v_k)$, a conditional probability. According to the definitions this is

$$\frac{P(X = u \text{ and } Y = v_k)}{P(Y = v_k)} = \frac{f_{(X,Y)}(u, v_k)}{f_Y(v_k)}$$

**Definition**    If $(X, Y)$ is a random vector with a probability function $f_{(X,Y)}$ we define the *conditional probability functions* $f_{X|Y}$ and $f_{Y|X}$ by

$$f_{X|Y}(u \mid v_k) = \frac{f_{(X,Y)}(u, v_k)}{f_Y(v_k)}, \qquad f_Y(v_k) \neq 0$$

and

$$f_{Y|X}(v \mid u_k) = \frac{f_{(X,Y)}(u_k, v)}{f_X(u_k)}, \qquad f_X(u_k) \neq 0$$

So, for example, $f_{X|Y}$ is called the conditional probability function of $X$ given $Y$ or "$f$ sub $X$ given $Y$."

## Example 8.6.2

Six fair coins are tossed independently. Let

$$X(s) = \begin{cases} 1, & \text{if the sixth coin is H} \\ 0, & \text{if the sixth coin is T} \end{cases}$$

Let $Y(s)$ be the number of H's connected with the sample point $s$, that is, the sample point $s$ specifies the top face of each of the six coins. Find $f_{Y|X}$.

We need only find $f_{Y|X}(n \mid 0)$ for $n = 0, 1, \ldots, 6$ and $f_{Y|X}(n \mid 1)$ for $n = 0, 1, \ldots, 6$. For other values of $u$, $f_{Y|X}(u \mid 1) = f_{Y|X}(u \mid 0) = 0$. We will do $f_{Y|X}(n \mid 0)$ only. By definition

$$f_{Y|X}(n \mid 0) = \frac{P(Y = n \text{ and } X = 0)}{P(X = 0)} = \frac{P(Y = n \text{ and } X = 0)}{\frac{1}{2}}$$

$$= 2P(Y = n \text{ and } X = 0)$$

Now $P(Y = 6 \text{ and } X = 0) = 0$ since we cannot have the sixth coin showing T and a total of six H's. Otherwise $P(Y = n \text{ and } X = 0) = \binom{5}{n}\left(\frac{1}{2}\right)^n \left(\frac{1}{2}\right)^{5-n}\frac{1}{2}$ where $\binom{5}{n}$ is the number of ways of choosing $n$ coins from the first five. These coins will show heads. The terms $(\frac{1}{2})^n$, $(\frac{1}{2})^{5-n}$, $\frac{1}{2}$ represent the probabilities that the $n$ coins chosen do show H, the other $5 - n$ of the first five show T, and the last coin shows $T$, respectively. Therefore

$$f_{Y|X}(n \mid 0) = \begin{cases} 0, & n = 6 \\ \binom{5}{n}\left(\frac{1}{2}\right)^n \left(\frac{1}{2}\right)^{5-n}, & n = 0, 1, \ldots, 5 \end{cases}$$

Similarly

$$f_{Y|X}(n \mid 1) = \begin{cases} 0, & n = 0 \\ \binom{5}{n-1}\left(\frac{1}{2}\right)^{n-1} \left(\frac{1}{2}\right)^{5-n+1}, & n = 1, 2, \ldots, 6 \end{cases}$$

If we turn from probability functions to density functions we may ask, "what is the significance of $f_{(X,Y)}(x, y)/f_X(x)$ if $f_X(x) > 0$?" We see that

$$\frac{\int_x^{x+h} du \int_y^{y+k} f_{(X,Y)}(u, v)\, dv}{\int_x^{x+h} f_X(u)\, du} = \frac{P(x \le X \le x + h, \ y \le Y \le y + k)}{P(x \le X \le x + h)}$$

But if $(x, y)$ is a continuity point of $f_{(X,Y)}$ we have

$$\lim_{\substack{h \to 0 \\ k \to 0}} \frac{\int_x^{x+h} du \int_y^{y+k} f_{(X,Y)}(u, v)\, dv}{k \int_x^{x+h} f_X(u)\, du} = \lim_{\substack{h \to 0 \\ k \to 0}} \frac{hk f_{(X,Y)}(x, y)}{kh f_X(x)}$$

$$= \frac{f_{(X,Y)}(x, y)}{f_X(x)}$$

Therefore, if $(x, y)$ is a continuity point of $f_{(X,Y)}$

$$\lim_{\substack{h \to 0 \\ k \to 0}} \frac{P(x \le X \le x + h, \ y \le Y \le y + k)}{kP(x \le X \le x + h)}$$

$$= \lim_{\substack{h \to 0 \\ k \to 0}} \frac{1}{k} P(y \le Y \le y + k \mid x \le X \le x + h)$$

We set $f_{(X,Y)}(x, y)/f_X(x) = f_{Y|X}(y \mid x)$. The function $f_{Y|X}$ is called the conditional density function of $Y$ given $X$.

### Example 8.6.3
Suppose that a particle of mass $x$ will have a random displacement $y$ having a density function $(2\pi)^{-1/2}xe^{-x^2y^2/2}$ one unit of time after it has fallen into a fluid. Particles whose masses are random with a density function

$$f(x) = \frac{2}{\Gamma(\alpha)} x^{2\alpha-1}e^{-x^2}, \qquad x > 0$$

fall into this fluid at a fixed point. Find the distribution of the resulting displacements.

Let $X$ be a random variable giving the mass of these particles and $Y$ a random variable giving the resulting displacements after unit time. By our assumptions $f_X(x) = \frac{2}{\Gamma(\alpha)} x^{2\alpha-1} e^{-x^2}$ for $x > 0$ and $f_{Y|X}(y \mid x) = \frac{x}{\sqrt{2\pi}} e^{-x^2 y^2/2}$.

Therefore, $f_{(X,Y)} = \frac{2}{\sqrt{2\pi}\,\Gamma(\alpha)} x^{2\alpha} e^{-x^2(1+y^2/2)}$ for $x > 0$ and

$$f_Y(y) = \int_0^\infty \frac{2}{\sqrt{2\pi}\,\Gamma(\alpha)} x^{2\alpha} e^{-x^2(1+y^2/2)}\, dx$$

We make the substitution $x^2(1 + y^2/2) = z$ with $2x(1 + \frac{1}{2}y^2)\, dx = dz$. Now

$$f_Y(y) = \int_0^\infty \frac{1}{\sqrt{2\pi}\,\Gamma(\alpha)} (x^2)^{\alpha-1/2} e^{-x^2(1+y^2/2)} 2x\, dx$$

$$= \int_0^\infty \frac{1}{\sqrt{2\pi}\,\Gamma(\alpha)} \left[\frac{z}{1 + \frac{1}{2}y^2}\right]^{\alpha-1/2} e^{-z} \frac{dz}{(1 + \frac{1}{2}y^2)}$$

$$= \frac{1}{\sqrt{2\pi}\,\Gamma(\alpha)} (1 + \tfrac{1}{2}y^2)^{-\alpha-1/2} \int_0^\infty z^{\alpha-1/2} e^{-z}\, dz$$

The $z$ integration is from $0$ to $\infty$ because $z = (1 + \frac{1}{2}y^2)x$ and $1 + \frac{1}{2}y^2$ is positive while $x$ goes from $0$ to $\infty$. Now $\Gamma(\beta) = \int_0^\infty z^{\beta-1} e^{-z}\, dz$. With $\beta = \alpha + \frac{1}{2}$, we see that $\int_0^\infty z^{\alpha-1/2} e^{-z}\, dz = \Gamma(\alpha + \frac{1}{2})$. Therefore,

$$f_Y(y) = \frac{\Gamma(\alpha + \frac{1}{2})}{\sqrt{2\pi}\,\Gamma(\alpha)} (1 + \tfrac{1}{2}y^2)^{-\alpha-1/2}$$

### Example 8.6.4

Let $X$ have the density function $f_X(x) = (1/\sqrt{2\pi t})e^{-(1/2t)x^2}$ where $t > 0$. Let $f_{Y|X}(y \mid x) = [1/\sqrt{2\pi(s - t)}] \exp\{-[1/2(s - t)](y - x)^2\}$ where $s > t$. Find $f_Y(y)$.
   By definition $f_{(X,Y)}(x, y) = f_{Y|X}(y \mid x)f_X(x)$. Therefore

$$f_{(X,Y)}(x, y) = \frac{1}{2\pi\sqrt{t(s - t)}} \exp\left[-\frac{1}{2(s - t)} (y - x)^2\right] \exp\left(-\frac{1}{2t} x^2\right)$$

Also

$$f_Y(y) = \int_{-\infty}^\infty f_{(X,Y)}(x, y)\, dx$$

$$= \int_{-\infty}^\infty \frac{1}{2\pi\sqrt{t(s - t)}} \exp\left[-\frac{1}{2(s - t)} (y - x)^2 - \frac{1}{2t} x^2\right] dx$$

We observe that

$$\frac{1}{(s-t)}(y^2 - 2xy + x^2) + \frac{1}{t}x^2$$

$$= \left(\frac{1}{t} + \frac{1}{s-t}\right)x^2 - \frac{2}{s-t}xy + \frac{y^2}{s-t}$$

$$- \frac{s-t+t}{t(s-t)}x^2 - \frac{2txy}{(s-t)t} + \frac{ty^2}{t(s-t)}$$

$$= \frac{s}{t(s-t)}\left(x^2 + 2\frac{t}{s}xy + \frac{t^2}{s^2}y^2\right) + \frac{ty^2}{t(s-t)} - \frac{ty^2}{s(s-t)}$$

If we let $z = \sqrt{\dfrac{s}{t(s-t)}}\left(x - \dfrac{t}{s}y\right)$, $dz = \sqrt{\dfrac{s}{t(s-t)}}\,dx$. Since $\sqrt{\dfrac{s}{t(s-t)}} > 0$,

$z \to \pm\infty$ as $x \to \pm\infty$. We have

$$\int_{-\infty}^{\infty} \frac{1}{2\pi\sqrt{t(s-t)}} \exp\left\{-\tfrac{1}{2}z^2 - \tfrac{1}{2}y^2\left[\frac{1}{s-t} - \frac{t}{s(s-t)}\right]\right\}\sqrt{\frac{t(s-t)}{s}}\,dz$$

But $\dfrac{1}{s-t} - \dfrac{t}{s(s-t)} = \dfrac{1}{s-t}\left(1 - \dfrac{t}{s}\right) = \dfrac{1}{s}$. Therefore the last integral is

$$\int_{-\infty}^{\infty} \frac{1}{2\pi\sqrt{s}} \exp\left(-\tfrac{1}{2}z^2 - \frac{1}{2}\frac{y^2}{s}\right) dz = \frac{e^{-y^2/2s}}{\sqrt{2\pi s}}\int_{-\infty}^{\infty} \frac{e^{-z^2/2}}{\sqrt{2\pi}}\,dz$$

$$= \frac{1}{\sqrt{2\pi s}}e^{-y^2/2s} = f_Y(y)$$

since $\displaystyle\int_{-\infty}^{\infty} (e^{-z^2/2}/\sqrt{2\pi})\,dz = 1$.

In the case of a 3-dimensional random vector $(X, Y, Z)$ we would have two types of conditional densities, namely,

$$f_{X|(Y,Z)}(x \mid y, z) = \frac{f_{(X,Y,Z)}(x, y, z)}{f_{(Y,Z)}(y, z)}$$

and

$$f_{(X,Y)|Z}(x, y \mid z) = \frac{f_{(X,Y,Z)}(x, y, z)}{f_Z(z)}$$

where the definitions are for those $(y, z)$ and $z$ where $f_{(Y,Z)}(y, z) \neq 0$ and $f_Z(z) \neq 0$, respectively.

**EXERCISES 8.6**

**1**   Let

$$f_{(X,Y)}(x, y) = \begin{cases} \dfrac{1}{\pi}, & x^2 + y^2 \leq 1 \\ 0, & \text{otherwise} \end{cases}$$

Compute $f_X(x)$. Give a geometric description for the behavior of $(X, Y)$.

**2**  Find the marginal and conditional densities of (8.4.1).

**3**  Let

$$f_{(X,Y)}(x, y) = \begin{cases} 2e^{-x-y}, & 0 < y < x \\ 0, & \text{otherwise} \end{cases}$$

Find $f_Y(y)$.

**4**  Let

$$f_X(m) = e^{-\lambda}\lambda^m/m!, \qquad m = 0, 1, 2, 3, \ldots$$

and

$$f_{Y|X}(n \mid m) = \begin{cases} \binom{m}{n} p^n (1 - p)^{m-n}, & n = 0, 1, 2, \ldots, m \\ 0, & \text{otherwise} \end{cases}$$

Find $f_Y(n)$. Sketch the points where $f_{(X,Y)}(m, n) > 0$. Compare with exercise 8.1.6.

**5**  Let $g(x, y) = (1/2\pi)[e^{-(x^2+y^2)/2} - xe^{-4(x^2+y^2)}]$ for all $x$, $y$. Show that $g(x, y)$ is the density function for some 2-dimensional random vector, say $g(x, y) = f_{(X,Y)}(x, y)$. Find $f_X(x)$ and $f_Y(y)$. Is $(X, Y)$ a 2-dimensional normal random vector? Compare with exercise 2.

**6**  Suppose that two points are chosen on our standard stick as follows: The first point, $x$, is chosen with a uniform distribution. The second point, $y$, is chosen on the longer piece with uniform conditional density. Let $X(x, y) = x$ and $Y(x, y) = y$. Find $f_{(X,Y)}$.

**7**  Let $X$ and $Y$ be two independent $N(0, 1)$ random variables. Let $h(x, y) = x$ and $k(x, y) = y/x$. Find the joint density function for $(X, Y/X)$ by the change of variable method. Then find the marginal density function giving the density of $Y/X$.

**8**  Let $f(x, y, z)$ be the joint density function of $(X, Y, Z)$ of exercise 8.5.10. Find the marginal density of $(Y, Z)$ directly from $f$.

**9**  A number $x$ is chosen from the positive real axis with the negative exponential density $\lambda e^{-\lambda x}$, $\lambda > 0$; conditional on that choice the number $y$ is chosen from the real line with a $N(0, x)$ density. Let $X(x, y) = x$ and $Y(x, y) = y$. Find the joint density function $f_{(X,Y)}$. Set up an expression for $f_Y(y)$.

**10**  Let $X$ and $Y$ be independent random variables with density function $f_X$ and $f_Y$, respectively. Show that if $U = 2X + 3Y$,

$$f_U(u) = \int_{-\infty}^{\infty} \tfrac{1}{3} f_X(v) f_Y[\tfrac{1}{3}(u - 2v)] \, dv$$

[*Hint:* Consider the random vector $(U, V)$ where $U = 2X + 3Y$, $V = X$ and use change of variables.]

**11**  Work exercise 10 as follows: Consider the random vector $(U, W)$ where

$U = 2X + 3Y$, $\quad W = Y$. Then $f_U(u) = \int_{-\infty}^{\infty} \frac{1}{2} f_X(\frac{1}{2}(u - 3w)) f_Y(w) \, dw$. Get the same result from exercise 10 by setting $w = \frac{1}{3}(u - 2v)$.

**12**  Let $X_1$, $X_2$, $X_3$, $X_4$ be independent random variables with density functions $f_{X_j}$. By using the substitution $U_1 = X_1$, $\quad U_2 = X_2$, $\quad U_3 = X_3$, and $U = X_1 + X_2 + X_3 + X_4$ find a density function for $U$ in terms of the density functions for the $X_j$.

**13**  Suppose that $X$ and $Y$ are independent with density functions $f_X$ and $f_Y$. Find a density function for $U = X^2 - 3Y$.

◆ **14**  Let $(X, Y)$ have the density function (8.4.1). Show that $\sigma^2(X) = A$ and $\sigma^2(Y) = C$ by using the marginal densities of $X$ and $Y$.

# 8.7  Covariance and correlation

For random vectors $(X, Y, Z)$ it is common to define moments of the sort $E(X^l Y^m Z^n)$. It is important to note that

$$(8.7.1) \qquad E(X^l Y^m Z^n) = \int_{-\infty}^{\infty} \int_{-\infty}^{\infty} \int_{-\infty}^{\infty} x^l y^m z^n f_{(X,Y,Z)}(x, y, z) \, dx \, dy \, dz$$

when the integral converges absolutely.

**Example 8.7.1**
If $(X, Y)$ is a random vector with $E(X^2)$, $E(Y^2)$ existing, show that $E(XY)$ exists.

Since $0 \le (a \pm b)^2$ for any real numbers $a$, $b$ then $0 \le a^2 \pm 2ab + b^2$ so $\mp ab \le \frac{1}{2}(a^2 + b^2)$ and $|X(s) Y(s)| \le \frac{1}{2}|X(s)|^2 + \frac{1}{2}|Y(s)|^2$. Therefore

$$E(|XY|) \le \tfrac{1}{2} E(X^2) + \tfrac{1}{2} E(Y^2)$$

This assures the existence of $E(XY)$.

**Definitions**    If $(X, Y)$ is a random vector such that $E(X^2)$ and $E(Y^2)$ exist then $E((X - E(X))(Y - E(Y)))$ is called the *covariance* of $X$ and $Y$. It is abbreviated as cov $(X, Y)$. If $\sigma^2(X) > 0$ and $\sigma^2(Y) > 0$, then $\dfrac{\text{cov}(X, Y)}{\sigma(X)\sigma(Y)} =$ $r(X, Y)$ is called the *correlation coefficient*.

We know that $E((X - E(X))(Y - E(Y)))$ exists because if $E(X^2)$ exists $E(X)$ exists by Theorem 7.5.1 and $E((X - E(X))^2)$ exists by Theorem 7.5.2. It is immediate that

$$(8.7.2) \qquad E((X - E(X))(Y - E(Y))) = E(XY) - E(X)E(Y)$$

### Example 8.7.2

If $(X, Y)$ is as in Example 8.7.1 with $\sigma^2(X) > 0$, $\sigma^2(Y) > 0$ show that $-1 \leq r(X, Y) \leq 1$.

Let $X_1 = X - E(X)$, $Y_1 = Y - E(Y)$. $E(X_1^2) = \sigma^2(X)$, $E(Y_1^2) = \sigma^2(Y)$, and $E(X_1 Y_1)$ is the covariance of $X$ and $Y$. For any real numbers $\alpha, \beta$ we have $E((\alpha X_1 \pm \beta Y_1)^2) \geq 0$. So $0 \leq \alpha^2 E(X_1^2) \pm 2\alpha\beta E(X_1 Y_1) + \beta^2 E(Y_1^2)$. Therefore,

$$\mp 2\sigma(X_1)\sigma(Y_1)E(X_1 Y_1) \leq \sigma^2(Y_1)\sigma^2(X_1) + \sigma^2(X_1)\sigma^2(Y_1)$$
$$= 2\sigma^2(X_1)\sigma^2(Y_1)$$

Therefore, $\mp E(X_1 Y_1) \leq \sigma(X_1)\sigma(Y_1)$ or $\mp r(X, Y) = \mp E(X, Y)/\sigma(X_1)\sigma(Y_1) \leq 1$. This is equivalent to the $-1 \leq r(X, Y) \leq 1$.

### Example 8.7.3

Let $(X, Y)$ have the density function

$$f(x, y) = \frac{1}{2\pi\sqrt{D}} \exp\left[-\frac{1}{2D}(Cx^2 - 2Bxy + Ay^2)\right]$$

where $D = AC - B^2$. Show that $B = \text{cov}(X, Y)$.

By equation (8.7.1) we calculate $\int_{-\infty}^{\infty}\int_{-\infty}^{\infty} xyf(x, y)\, dx\, dy$ and note its existence because $\sigma^2(X) = A$ and $\sigma^2(Y) = C$ by exercise 8.6.14. As in section 8.4 we let $z = \sqrt{A/D}\, y - Bx/\sqrt{DA}$ for fixed $x$. We then let $x = \sqrt{A}\, w$. Our integral becomes

$$\int_{-\infty}^{\infty}\int_{-\infty}^{\infty} \sqrt{A}\, w\left(\sqrt{\frac{D}{A}}\, z + \sqrt{\frac{D}{A}}\frac{B}{\sqrt{DA}}\sqrt{A}\, w\right)$$

$$\times \frac{1}{2\pi\sqrt{D}} \exp\left[-\tfrac{1}{2}(z^2 + w^2)\right]\sqrt{\frac{D}{A}}\sqrt{A}\, dw\, dz$$

$$= \int_{-\infty}^{\infty}\int_{-\infty}^{\infty} (\sqrt{D}\, wz + Bw^2)\frac{1}{2\pi} \exp\left[-\tfrac{1}{2}(z^2 + w^2)\right] dw\, dz$$

$$= \int_{-\infty}^{\infty} ze^{-z^2/2}\, dz \int_{-\infty}^{\infty} \frac{\sqrt{D}\, we^{-w^2/2}}{2\pi}\, dw + \int_{-\infty}^{\infty} \frac{Bw^2 e^{-w^2/2}}{\sqrt{2\pi}}\, dw \int_{-\infty}^{\infty} \frac{e^{-z^2/2}}{\sqrt{2\pi}}\, dz$$

$$= 0 + B \cdot 1 = B$$

It is easy to show that $E(X) = E(Y) = 0$. Therefore by equations (8.7.1) and (8.7.2) the covariance of $X$ and $Y$ is $B$.

### Example 8.7.4

Let $Y = bX$. Find $r(X, Y)$.

We assume $b \neq 0$ for otherwise $r(X, Y) = 0$. We also assume $\sigma^2(X) > 0$ to make the problem interesting. Then $E(Y) = bE(X)$ and $\sigma^2(Y) = b^2\sigma^2(X)$ so $\sigma(Y) = |b|\sigma(X)$. Finally, $E(XY) = E(XbX) = bE(X^2)$. Therefore,

$$\text{cov}(X, Y) = E(XY) - E(X)E(Y) = bE(X^2) - E(X)bE(X)$$
$$= b\{E(X^2) - [E(X)]^2\} = b\sigma^2(X)$$

Therefore,

$$r(X, Y) = \frac{\text{cov}(X, Y)}{\sigma(X)\sigma(Y)} = \frac{b\sigma^2(X)}{\sigma(X)|b|\sigma(X)} = \frac{b}{|b|} = \begin{cases} 1, & b > 0 \\ -1, & b < 0 \end{cases}$$

As a converse we prove the following theorem.

**Theorem 8.7.1** If $r(X, Y) = \pm 1$ [and $\sigma(X) > 0$ is assumed] then $P(Y = aX + c) = 1$ for some real number $a$ which is positive if $r = 1$ and negative if $r = -1$.

PROOF    Let $Y_1 = Y - E(Y)$ and $X_1 = X - E(X)$. Then $\sigma^2(X) = E(X_1^2)$, $\sigma^2(Y) = E(Y_1^2)$, and cov $(X, Y) = E(X_1 Y_1)$. Since $Z = [\sigma(Y)X_1 - \sigma(X)Y_1]^2 \geq 0$ we have

$$0 \leq E([\sigma(Y)X_1 - \sigma(X)Y_1]^2) = \sigma^2(Y)E(X_1^2) - 2\sigma(Y)\sigma(X)E(X_1 Y_1) + \sigma^2(X)E(Y_1^2)$$
$$= \sigma^2(Y)\sigma^2(X) - 2\sigma(Y)\sigma(X)E(X_1 Y_1) + \sigma^2(X)\sigma^2(Y)$$

If $r(X, Y) = 1$, $\quad 1 = \dfrac{\text{cov}(X, Y)}{\sigma(X)\sigma(Y)} = \dfrac{E(X_1 Y_1)}{\sigma(X)\sigma(Y)}$ and $E(X_1 Y_1) = \sigma(X)\sigma(Y)$ so

$$0 \leq E(Z) = 2\sigma^2(X)\sigma^2(Y) - 2\sigma(Y)\sigma(X)\sigma(X)\sigma(Y) = 0$$

However, since $Z$ is a non-negative random variable, its expectation is zero if and only if $P(Z = 0) = 1$. Therefore, $P([\sigma(Y)X_1 - \sigma(X)Y_1]^2 = 0) = 1$ which is equivalent to

$$1 = P(\sigma(Y)X_1 - \sigma(X)Y_1 = 0) = P(\sigma(Y)[X - E(X)] - \sigma(X)[Y - E(Y)] = 0)$$
$$= P(Y = [\sigma(Y)/\sigma(X)][X - E(X)] + E(Y))$$

so $P(Y = aX + c) = 1$ where $a = \sigma(Y)/\sigma(X) > 0$ and $c = [\sigma(Y)/\sigma(X)]E(X) + E(Y)$. If $r = -1$, the proof can be repeated with $Z = [\sigma(Y)X_1 + \sigma(X)Y_1]^2$.

**EXERCISES 8.7** _____

1  Find cov $(X, Y)$ for $X$ and $Y$ of exercise 8.6.2.

2  Find cov $(X, Y)$ for $X$ and $Y$ of exercise 8.6.6.

3  Find cov $(X, Y)$ for $X$ and $Y$ of exercise 8.6.3.

4  Find $r(X, Y)$ for $X$ and $Y$ of exercise 8.6.5.

In exercises 5–10 suppose that data consisting of a number of ordered pairs $(x_j, y_j)$ of numbers are treated as sample points in a sample space. Equal likelihood is given except repetitions are counted. A graph of such data is called a *scatter diagram*. It is not damaging to assume that there are no repetitions of $(x_j, y_j)$ in the ensuing discussion. All sums are $\sum_{j=1}^{n}$.

**5**  Suppose that there are $n$ points. What does the quantity $a = \sum (x_j y_j / n)$ suggest? What does the quantity $b_x = \sum x_j / n$ suggest? What does the quantity $c_x = \sum (1/n)(x_j - b_x)^2$ suggest?

**6**  What would be true about $q = (a - b_x b_y)/\sqrt{c_x c_y}$ if $b_x = \sum (x_j / n) \approx 0$, $b_y = \sum (y_j / n) \approx 0$ and the scatter diagram looks like Figure 5(a)? Figure 5(b)?

**7**  Draw a scatter diagram which would result in $(a - b_x b_y)/\sqrt{c_x c_y}$ being approximately $-0.8$.

**8**  Given points $(x_j, y_j)$ of a scatter diagram with $n \geq 2$, show that numbers $\hat{m}$ and $\hat{B}$ which minimize $Q(m, B) = \sum (y_j - mx_j - B)^2$ satisfy $\sum (y_j - \hat{m}x_j - \hat{B}) = 0$ and $\sum x_j (y_j - \hat{m}x_j - \hat{B}) = 0$.

**9**  Show that $\hat{m}$ and $\hat{B}$ of exercise 8 are $\hat{m} = (a - b_x b_y)/c_x$ and $\hat{B} = b_y - mb_x$ if $c_x \neq 0$.

**10**  Discuss how minimizing $\sum (y_j - \hat{m}x_j - \hat{B})^2$ can be thought of as finding the line $y = \hat{m}x + \hat{B}$ which best fits the data of the scatter diagram. What would it mean if $c_x = 0$?

In exercises 11–17 let $(X, Y)$ have the probability function $f_{(X,Y)}$. Assuming that the series is absolutely convergent, the function (of $y$) $E(X \mid y) = \sum_{x} x f_{X|Y}(x \mid y)$ is called the *conditional expectation* of $X$ at $y$. Naturally, it only is defined for $y$ for which $P(Y = y) > 0$. We define $E(X \mid y) = 0$ for $y$ with $P(Y = y) = 0$.

**11**  Suppose $X$ and $Y$ are independent. Find $E(X \mid y)$.

**12**  Show that if $E(X)$ exists, $E(X \mid y)$ exists for all $y$ with $P(Y = y) > 0$.

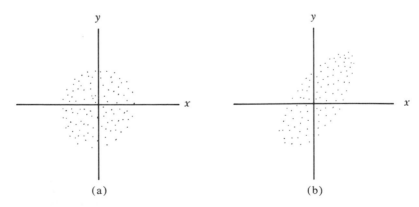

(a)                                      (b)

**Figure 5**

**13** Let $(X, Y)$ be as in exercise 8.6.3. Find $E(X \mid y)$.

**14** Compute $\sum E(X \mid y)f_Y(y)$.

In exercises 15–17 we work with density functions.

**15** What would be a reasonable analog to $E(X \mid y) = \sum_x x f_{X \mid Y}(x \mid y)$ when $(X, Y)$ has a density function? Check your definition to see if the analog to exercises 11 and 14 is true for your definition of $E(X \mid y)$ in the density function case.

**16** Find $E(X \mid y)$ when $(X, Y)$ has the density (8.4.6).

**17** Find $E(Y \mid x)$ for $(X, Y)$ of exercise 8.6.3.

# CHAPTER 9

# *Characteristic Functions*

## 9.1 Introduction

In this chapter we do a good deal of work with complex numbers. For our purposes a complex number $z$ is a symbol $a + bi$ where $a$ and $b$ are real numbers. The arithmetic of complex numbers is the usual arithmetic subject to the condition that $i^2 = -1$. That is,

**(9.1.1)** $\quad (a + bi) \pm (c + di) = (a \pm c) + (b \pm d)i$

**(9.1.2)** $\quad (a + bi)(c + di) = (ac - bd) + (bc + ad)i$

It is frequently convenient to think of $z$ as the point $(a, b)$ in the usual Cartesian coordinate system. Therefore the distance from $(0, 0)$ to $(a, b)$ is $\sqrt{a^2 + b^2}$. We define $|z| = |a + bi| = \sqrt{a^2 + b^2}$, and call $|z|$ the *modulus* of $z$. We call $a$ the *real part* of $z$ and $b$ (not $bi$) the *imaginary part* of $z$. We write $a = \operatorname{Re} z$ and $b = \operatorname{Im} z$. It is easy to verify that if $w = c + di$ then

**(9.1.3)** $\quad |zw| = |z|\,|w|$

**(9.1.4)** $\quad |z + w| \le |z| + |w|$

**(9.1.5)** $\quad |\operatorname{Re} z| \le |z|, \qquad |\operatorname{Im} z| \le |z|$

**(9.1.6)** $\quad |z| \le |\operatorname{Re} z| + |\operatorname{Im} z|$

We say that an infinite sequence $z_1, z_2, z_3, \ldots$ of complex numbers *converges* to $z$ and write $\lim_{n \to \infty} z_n = z$ if $\lim_{n \to \infty} |z_n - z| = 0$. Note that $\lim_{n \to \infty} |z_n - z| = 0$ is a statement about real limits so we can use all we know about real limits. We say that the sequence $z_1, z_2, z_3, \ldots$ *converges*, if there is a complex number $z$ such that $\lim\limits_{n \to \infty} z_n = z$.

### Example 9.1.1

If $z_n = a_n + b_n i$ and $z = a + bi$ show that $\lim\limits_{n \to \infty} z_n = z$ if and only if both $\lim\limits_{n \to \infty} a_n = a$ and $\lim\limits_{n \to \infty} b_n = b$.

First suppose $\lim\limits_{n \to \infty} z_n = z$. By (9.1.1) and (9.1.5) $0 \le |a_n - a| = |\mathrm{Re}\,(z_n - z)|$ $\le |z_n - z|$. But since $\lim\limits_{n \to \infty} z_n = z$, $\lim\limits_{n \to \infty} |z_n - z| = 0$, which implies $\lim\limits_{n \to \infty} a_n = a$. Similarly $\lim\limits_{n \to \infty} b_n = b$. If $\lim\limits_{n \to \infty} a_n = a$ and $\lim\limits_{n \to \infty} b_n = b$ we have $\lim\limits_{n \to \infty} |a_n - a| = 0$ and $\lim\limits_{n \to \infty} |b_n - b| = 0$. But by (9.1.1) and (9.1.6)

$$0 \le |z_n - z| \le |\mathrm{Re}\,(z_n - z)| + |\mathrm{Im}\,(z_n - z)| = |a_n - a| + |b_n - b|$$

Here, known results show that $\lim\limits_{n \to \infty} |z_n - z| = 0$. So $\lim\limits_{n \to \infty} z_n = z$.

These results can be applied to series of complex numbers $\sum\limits_{n=1}^{\infty} w_n$ where we write $w = \sum\limits_{n=1}^{\infty} w_n$ if $w = \lim\limits_{N \to \infty} \sum\limits_{n=1}^{N} w_n$. It is immediate that if $w_n = c_n + d_n i$ then $w = c + di = \sum\limits_{n=1}^{\infty} w_n$ if and only if $c = \sum\limits_{n=1}^{\infty} c_n$ and $d = \sum\limits_{n=1}^{\infty} d_n$ where the last two equations are in terms of real numbers only.

If a series $\sum\limits_{n=1}^{\infty} w_n$ has $\sum\limits_{n=1}^{\infty} |w_n|$ a convergent (real, non-negative) series we find that $\sum\limits_{n=1}^{N} |c_n| \le \sum\limits_{n=1}^{N} |w_n| \le \sum\limits_{n=1}^{\infty} |w_n|$. By theorems on real series $\sum\limits_{n=1}^{\infty} |c_n|$ converges. Therefore $\sum\limits_{n-1}^{\infty} c_n$ converges since absolute convergence of a series implies ordinary convergence. Similarly $\sum\limits_{n=1}^{\infty} d_n$ converges so

$$\sum_{n=1}^{\infty} w_n = \left( \sum_{n=1}^{\infty} c_n \right) + \left( \sum_{n=1}^{\infty} d_n \right) i$$

We say that $\sum\limits_{n=1}^{\infty} w_n$ is *absolutely convergent* if $\sum\limits_{n=1}^{\infty} |w_n|$ converges. We have just shown that if $\sum\limits_{n=1}^{\infty} w_n$ is absolutely convergent, then it is convergent.

Since $\sum\limits_{n=1}^{\infty} |w_n|$ is a real series, one can use certain results on real series to establish absolute convergence of $\sum\limits_{n=1}^{\infty} w_n$, if it is true, and thus frequently deduce convergence of $\sum\limits_{n=1}^{\infty} w_n$ this way.

### Example 9.1.2

Show that $\sum\limits_{n=0}^{\infty} z^n/n!$ converges.

We study $|z^n/n!| = |z^n|/n! = |z|^n/n!$ by an extension of (9.1.3.). The series $\sum\limits_{n=0}^{\infty} |z^n/n!| = \sum\limits_{n=0}^{\infty} |z|^n/n!$ is a series of positive real numbers to which we apply the ratio test

$$\frac{|z|^{n+1}}{(n+1)!} \div \frac{|z|^n}{n!} = \frac{|z|^{n+1}}{|z|^n} \frac{n!}{(n+1)!} = \frac{|z|}{n+1} \to 0$$

as $n \to \infty$. Therefore the series $\sum\limits_{n=0}^{\infty} |z|^n/n!$ converges. The series $\sum\limits_{n=0}^{\infty} z^n/n!$ is absolutely convergent and, hence, convergent.

It can be shown that any series obtained by rearranging the terms of an absolutely convergent series has the same sum. Other manipulations of absolutely convergent real series can be performed on absolutely convergent complex series with the same results.

It is customary to write $e^z = \sum\limits_{n=0}^{\infty} z^n/n!$ for any complex number $z$. Some of the exercises show that $e^z$ behaves as one would expect an exponential to behave. Thus if $t$ is real

$$e^{it} = \left(1 + \frac{(it)^2}{2!} + \frac{(it)^4}{4!} + \frac{(it)^6}{6!} + \frac{(it)^8}{8!} + \cdots\right)$$

$$+ \left(\frac{it}{1!} + \frac{(it)^3}{3!} + \frac{(it)^5}{5!} + \frac{(it)^7}{7!} + \cdots\right)$$

Since $i^{4n} = 1$, $i^{4n+1} = i$, $i^{4n+2} = -1$, and $i^{4n+3} = -i$ for $n = 0, 1, 2, 3,$ $\ldots$, we have

$$e^{it} = \left(1 - \frac{t^2}{2!} + \frac{t^4}{4!} - \frac{t^6}{6!} + \frac{t^8}{8!} + \cdots\right) + \left(t - \frac{t^3}{3!} + \frac{t^5}{5!} - \frac{t^7}{7!} + \cdots\right)i$$

But the parenthesized terms are the MacLaurin series for $\cos t$ and $\sin t$, respectively. Thus we have the equation

**(9.1.7)**     $e^{it} = \cos t + (\sin t)i = \cos t + i \sin t$

If $I$ is an interval of real numbers, we may have for each $t \in I$ a complex number $g(t) = h(t) + ik(t)$ associated with $t$. We call such an assignment or association a *complex-valued function* of the real variable $t$. The functions $h$ and $k$ are real-valued functions of $t$.

If $t_0 \in I$ we define $\lim_{t \to t_0} g(t) = z$ if and only if $\lim_{t \to t_0} |g(t) - z| = 0$. If $t_0$ is an interior point of $I$ the above limits are understood to be 2-sided limits. If $t_0$ is an end point of $I$ the limits are the obvious 1-sided limits. It is easy to show that $\lim_{t \to t_0} g(t) = z = u + iv$ if and only if we have $\lim_{t \to t_0} h(t) = u$ and $\lim_{t \to t_0} k(t) = v$.

If $I$ is a suitable interval we can define $\lim_{t \to \infty} g(t) = z$ and $\lim_{t \to +\infty} g(t) = z$, $\lim_{t \to -\infty} g(t) = z$ in the obvious ways. Finally, we say that $g$ is *continuous* at $t_0 \in I$ if $\lim_{t \to t_0} g(t) = g(t_0)$. We say that $g$ is *differentiable* at $t_0$ if $\lim_{t \to t_0} \dfrac{g(t) - g(t_0)}{(t - t_0)} = w$ and we write $g'(t_0) = w$. It should be noted that

$$\frac{g(t) - g(t_0)}{t - t_0} = \frac{h(t) - h(t_0)}{t - t_0} + i\,\frac{k(t) - k(t_0)}{t - t_0}$$

since $t - t_0$ is real. Thus $g'(t_0)$ exists if and only if $h'(t_0)$ and $k'(t_0)$ exist. In this case $g'(t_0) = h'(t_0) + ik'(t_0)$. This type of differentiation is not the complex differentiation which one finds in the theory of functions of a complex variable. There, one has $\lim_{z \to z_0} \dfrac{f(z) - f(z_0)}{(z - z_0)}$ where $z$ and $z_0$ are complex numbers and where $f$ is defined in an open region of the plane containing $z_0$.

Integration of $g$ may be defined in terms of $h$ and $k$. Thus we define

$$\int_{t_1}^{t_2} g(t)\, dt = \int_{t_1}^{t_2} h(t)\, dt + i \int_{t_1}^{t_2} g(t)\, dt$$

if and only if the two integrals on the right exist. It can be shown that

$$\int_{t_1}^{t_2} g(t)\, dt = \lim_{\|\pi\| \to 0} \sum_{j=1}^{n} f(\xi_j')(\xi_j - \xi_{j-1})$$

where $\pi$ is the partition $t_1 = \xi_0 < \xi_1 < \cdots < \xi_n = t_2$ and $\|\pi\|$ is the $\max_{1 \le j \le n} (\xi_j - \xi_{j-1})$ and $\xi_{j-1} \le \xi_j' \le \xi_j$.

Finally if $g$ is continuous on $\{t; t_1 \le t \le t_2\}$ and if there is a function $G$ which is differentiable for all $t$ with $t_1 \le t \le t_2$ and $G'(t) = g(t)$ it is easy to verify that

(9.1.8)   $$\int_{t_1}^{t_2} g(t)\, dt = G(t_2) - G(t_1)$$

This follows because $G(t) = H(t) + iK(t)$ has $H'(t) + iK'(t) = G'(t) = g(t) = h(t) + ik(t)$. Therefore by the fundamental theorem of calculus

$$\int_{t_1}^{t_2} g(t)\, dt = \int_{t_1}^{t_2} h(t)\, dt + i \int_{t_1}^{t_2} k(t)\, dt$$
$$= H(t_2) - H(t_1) + i[K(t_2) - K(t_1)] = G(t_2) - G(t_1)$$

We now point out a useful numerical result. An intuitive proof which can be made rigorous is given in exercises 9.1.10–9.1.12. This result states that

**(9.1.9)** $$\int_0^\infty \frac{\sin t}{t}\, dt = \lim_{M \to \infty} \int_0^M \frac{\sin t}{t}\, dt = \frac{\pi}{2}$$

### Example 9.1.3

Show that $\lim_{M \to \infty} \int_0^M \frac{\sin t}{t}\, dt$ exists.

Note that $|(\sin t)/t| \le 1$ and $\lim_{t \to 0} (\sin t)/t = 1$. Therefore there is no question of convergence of the integral at $t = 0$. As for the situation as $M \to \infty$

$$\int_0^M \frac{\sin t}{t}\, dt = \int_0^{\pi/2} \frac{\sin t}{t}\, dt + \int_{\pi/2}^M \frac{\sin t}{t}\, dt$$

Clearly, the first integral on the right is convergent. For the second integral, in order to do integration by parts, let $u = 1/t$ and $dv = \sin t\, dt$. Then $du = -t^{-2}\, dt$ and $v = -\cos t$. Thus

$$\int_{\pi/2}^M \frac{\sin t}{t}\, dt = -\frac{\cos t}{t}\Big|_{\pi/2}^M - \int_{\pi/2}^M \frac{\cos t}{t^2}\, dt$$

$$= -\frac{\cos M}{M} + \frac{\cos \pi/2}{\pi/2} - \int_{\pi/2}^M \frac{\cos t}{t^2}\, dt$$

As $M \to \infty$, $-\dfrac{\cos M}{M} \to 0$ and $\displaystyle\int_{\pi/2}^M \frac{\cos t}{t^2}\, dt$ converges since $\left|\dfrac{\cos t}{t^2}\right| \le \dfrac{1}{t^2}$ and

$\lim_{M \to \infty} \displaystyle\int_{\pi/2}^M \frac{1}{t^2}\, dt = \dfrac{2}{\pi}$. Therefore $\displaystyle\int_{\pi/2}^M \frac{\sin t}{t}\, dt$ is the difference of two functions which have limits as $M \to \infty$. Therefore, it, too, has a limit as $M \to \infty$.

### EXERCISES 9.1 ───────────────────────

1  Show that $e^z e^w = e^{z+w}$ for all complex numbers $z$ and $w$. Show that $(e^z)^n = e^{nz}$ for any positive integer $n$ and complex number $z$.

2  Show that if $g$ is a continuous complex-valued function on $t_1 \le t \le t_2$ and if $|g(t)| \le M$ then

$$\left| \int_{t_1}^{t_2} g(t)\, dt \right| \le M(t_2 - t_1)$$

3  Let $g(t) = e^{it}$ for $-\infty < t < \infty$. Show that $g'(t) = ie^{it}$.

4  If we let $\left(\dfrac{d}{dt}g\right)(t) = g'(t)$ show that $\dfrac{d}{dt}(g_1 + g_2)(t) = g_1'(t) + g_2'(t)$ and

$\dfrac{d}{dt}(g_1 g_2)(t) = g_1'(t)g_2(t) + g_2'(t)g_1(t)$ if $g_1$ and $g_2$ are differentiable at $t$ and $g_1$ and $g_2$ are complex-valued functions of the real variable $t$.

**5** Let $f$ be a real-valued differentiable function and let $g$ be a complex-valued differentiable function such that $G(t) = g(f(t))$ is defined. Show that $G'(t) = g'(f(t))f'(t)$.

**6** For some purposes it is desirable to set $(1 - it)^{-\alpha} = (1 + t^2)^{-\alpha/2} e^{i\alpha \arctan t}$ for $-\infty < t < +\infty$ and real $\alpha > 0$. Show that if $\alpha$ is a positive integer,

$$(1 - it)^{-\alpha} = [(1 - it)^{-1}]^\alpha = \left[ \frac{(1 + it)}{(1 + t^2)} \right]^\alpha \text{ if}$$

$$(1 + it)^\alpha = (1 + t^2)^{\alpha/2}(\cos \arctan t + i \sin \arctan t)^\alpha$$

Show that if $\alpha, \beta > 0$, $(1 - it)^{-\alpha}(1 - it)^{-\beta} = (1 - it)^{-(\alpha + \beta)}$.

**7** If $z = u + iv$ and $t$ is real, let $g(t) = e^{tz}$. Show that $g'(t) = ze^{tz}$. Show that

$$\int_{t_1}^{t_2} g(t)\, dt = z^{-1}(e^{t_2 z} - e^{t_1 z}), \qquad z \neq 0$$

**8** Prove that if $z = a + ib$, $|a| \leq |z|$, $|b| \leq |z|$, and $|z| \leq |a| + |b|$.

**9** Evaluate

$$S(a, b, c, d) = \int_0^1 [t(a + ib) + (1 - t)(c + id)]^n [a - c + i(b - d)]\, dt$$

and $S(x_1, y_1, x_2, y_2) + S(x_2, y_2, x_3, y_3) + S(x_3, y_3, x_1, y_2)$ for $n = 0, 1, 2, 3, \ldots$.

**10** In view of exercise 7, show that

$$\int_{t_1}^{t_2} e^{-at} \sin bt\, dt = \text{Im} \int_{t_1}^{t_2} e^{-at + bit}\, dt$$

$$= \text{Im} \left\{ \frac{1}{-a + bi} (e^{[-a + bi]t_2} - e^{[-a + bi]t_1}) \right\}$$

and if $a > 0$, show that

$$\int_0^\infty e^{-at} \sin t\, dt = \frac{1}{1 + a^2}$$

**11** Let $r > 0$. Compute $\displaystyle\int_r^\infty \frac{1}{1 + a^2}\, da$. Compute $\displaystyle\int_0^\infty \left( \int_r^\infty e^{-at} \sin t\, da \right) dt$ if one can change order of integration.

**12** Assuming that

$$\lim_{r \to 0} \int_0^\infty dt \int_r^\infty e^{-at} \sin t\, da = \int_0^\infty \left( \lim_{r \to 0} \int_r^\infty e^{-at}\, da \right) \sin t\, dt$$

show that

$$\int_0^\infty \frac{\sin t}{t}\, dt = \frac{\pi}{2}$$

**13** Show, by mathematical induction, that the binomial formula is valid for complex $z$ and $w$; i.e., if $n$ is a positive integer $(w + z)^n = \sum_{k=0}^\infty \binom{n}{k} w^k z^{n-k}$.

**14** Show that the series $\sum_{n=1}^{\infty} z^n/n$ converges for all $z$ with $|z| < 1$ by using the absolute convergence of $\sum_{n=1}^{\infty} |z|^n/n$ when $|z| < 1$.

**15** If $w$ is a complex number and $w \neq 1$ show that $w + w^2 + w^3 + \cdots + w^N = (w - w^{N+1})/(1 - w)$.

**16** Show that

$$\int_0^1 (z + tz^2 + t^2z^3 + t^3z^4 + \cdots + t^{N-1}z^N)\, dt = \int_0^1 \frac{1}{t}\left[\frac{zt - (tz)^{N+1}}{1 - tz}\right] dt$$

If $|z| = 1$ but $z \neq 1$ show that for $0 < t < 1$, $\;|1 - tz| \geq \delta > 0$ for some $\delta$ depending on $z$. Now $\int_0^1 \dfrac{zt\, dt}{t(1 - tz)}$ must be something. Why? Show that

$$\left| \int_0^1 \frac{(tz)^{N+1}}{t(1 - tz)}\, dt \right| \to 0 \qquad \text{as } N \to \infty$$

How does this argument show that $\sum_{n=1}^{\infty} z^n/n$ converges to $\int_0^1 \dfrac{z\, dt}{1 - tz}$ if $|z| = 1$, $z \neq 1$?

**17** Using exercise 12 show that

$$\int_{-\infty}^{\infty} \frac{\sin \alpha t}{t}\, dt = \begin{cases} \pi, & \alpha > 0 \\ 0, & \alpha = 0 \\ -\pi, & \alpha < 0 \end{cases}$$

# 9.2 The characteristic function

Let $S = \{s_1, s_2, s_3, \ldots\}$ be a discrete sample space with $P(s_k) = p_k$. Let $X$ be a random variable defined on $S$. Since $\sin tX$ and $\cos tX$ define bounded random variables on $S$ for each fixed real $t$ then $E(\cos tX)$ and $E(\sin tX)$ exist by the results of section 7.2. We define $E(e^{itX}) = E(\cos tX) + iE(\sin tX)$ since $e^{ix} = \cos x + i \sin x$ for any real $x$. In essence $e^{itX}$ is a complex-valued random variable defined on $S$. However, we will not have occasion to consider $E(e^{itZ})$ where $Z$ is a complex random variable.

By our conventions

$$E(e^{itX}) = \sum \cos [tX(s_k)]p_k + i \sum \sin [tX(s_k)]p_k$$
$$= \sum e^{itX(s_k)}p_k$$

Now

$$|E(e^{itX})| = \left| \lim_{n \to \infty} \sum_{k=1}^{n} e^{itX(s_k)}p_k \right|$$

if $S$ is denumerable. By the generalized version of (9.1.4),

$$\left| \sum_{k=1}^{n} e^{itX(s_k)}p_k \right| \leq \sum_{k=1}^{n} \left| e^{itX(s_k)}p_k \right| = \sum_{k=1}^{n} p_k \leq 1$$

since $|e^{itX(s_k)}| = 1$ and since both $X(s_k)$ and $t$ are real numbers. Thus all partial sums $\sum_{k=1}^{n} e^{itX(s_k)}p_k$ lie inside (or on) the circle of radius 1 center at the origin in the Cartesian representation of the complex numbers. Thus $\lim_{n \to \infty} \sum_{k=1}^{n} e^{itX(s_k)}p_k$ cannot be outside of that circle. Hence $|E(e^{itX})| \leq 1$ for any real number $t$.

**Definition**    If $X$ is a (real-valued) random variable, the function $\phi$ defined by $\phi(t) = E(e^{itX})$ is called the *characteristic function* of $X$.

The characteristic function is a complex-valued function whose domain is the set of real numbers. Since we may be involved with characteristic functions for many random variables we will normally use $\phi_X$, $\phi_Y$ for the characteristic functions of $X$ and $Y$, respectively.

So far we have defined the characteristic function clearly when $S$ is denumerable and in that case we have $|\phi(t)| \leq 1$. It is easy to see that $\phi(0) = 1$.

### Example 9.2.1
Let $S = \{H, T\}$ with $P(H) = p$ where $0 \leq p \leq 1$. Let $X(H) = 1$, $X(T) = 0$. Find the characteristic function of $X$.

$$\phi(t) = e^{itX(H)}p + e^{itX(T)}(1 - p) = pe^{it} + (1 - p) = 1 - p + pe^{it}$$

(It should be noted that this is the generating function for $X$ with $e^{it}$ substituted for $z$).

### Example 9.2.2
Let $S = \{0, 1, 2, 3, \ldots\}$, $P(n) = e^{-\lambda}(\lambda^n/n!)$. Let $X(n) = n$ so $X$ is a Poisson random variable. Compute the characteristic function of $X$.

$$\phi(t) = \sum_{n=0}^{\infty} e^{itX(n)}e^{-\lambda}\frac{\lambda^n}{n!} = \sum_{n=0}^{\infty} e^{int}e^{-\lambda}\frac{\lambda^n}{n!}$$

$$= e^{-\lambda}\sum_{n=0}^{\infty}\frac{(e^{it}\lambda)^n}{n!} = e^{-\lambda}e^{\lambda e^{it}} = e^{-\lambda(1 - e^{it})}$$

This, too, is the generating function at $e^{it}$.

### Example 9.2.3
Let $S = \{-2, -1, 0, 1\}$. Let $P(-2) = (1 - p)/3$, $P(-1) = (1 - p)/3$, $P(0) = (1 - p)/3$, where $0 \leq p \leq 1$. Let

$$X(n) = \max(n, 0) = \begin{cases} 1, & n = 1 \\ 0, & \text{otherwise} \end{cases}$$

Find the characteristic function of $X$.

$$\phi(t) = e^{itX(-2)}\frac{1-p}{3} + e^{itX(-1)}\frac{1-p}{3} + e^{itX(0)}\frac{1-p}{3} + e^{itX(1)}p$$

$$= 1 - p + pe^{it}$$

The characteristic functions of Examples 9.2.1 and 9.2.3 are identical. This should not be surprising since the random variables of those examples are identically distributed.

### Example 9.2.4

Let $S = \{(\delta_1, \delta_2, \ldots, \delta_n): \delta_i \in \{H, T\}\}$, that is, $S$ consists of $2^n$ sample points. The probability measure is equal likelihood. Let $Y(\delta_1, \delta_2, \ldots, \delta_n) = $ number of H's in $\delta_1, \delta_2, \ldots, \delta_n$. The characteristic function of $Y$ can be obtained as follows:

$$P(Y = k) = \frac{1}{2^n}\binom{n}{k} = \frac{1}{2^n}\frac{n!}{k!(n-k)!}, \qquad k = 0, 1, 2, \ldots, n$$

Therefore,

$$\phi_Y(t) = \sum_{k=0}^{n}\frac{e^{itk}}{2^n}\frac{n!}{k!(n-k)!} = \sum_{k=0}^{n}\binom{n}{k}\left(\frac{e^{it}}{2}\right)^k\frac{1}{2^{n-k}}$$

where $\binom{n}{k}$ is the coefficient of $a^k b^{n-k}$ in the binomial expansion of $(a + b)^n$.

Thus $\phi_Y(t) = (\frac{1}{2}e^{it} + \frac{1}{2})^n$ if we use $a = \frac{1}{2}e^{it}$, $b = \frac{1}{2}$.

Notice that if $p = \frac{1}{2}$ in Example 9.2.1, $\phi_X(t) = \frac{1}{2} + \frac{1}{2}e^{it}$. Thus $\phi_Y(t) = [\phi_X(t)]^n$ for the $X$ of Example 9.2.1. We will see that this is not a coincidence.

Let $Y = aX + b$ where $X$ is a random variable and $a$ and $b$ are real constants. Since $E(cZ) = cE(Z)$ even if $c$ and $Z$ are complex,

$$E(e^{itY}) = E(\exp(it(aX + b))) = E(\exp(itaX + itb))$$
$$= E(\exp itb \exp itaX) = e^{itb}E(\exp i(at)X) = e^{itb}\phi_X(at)$$

Thus

**(9.2.1)**    $\phi_Y(t) = e^{itb}\phi_X(at)$

This result is not restricted to the discrete case. It holds for all random variables $X$.

If $X$ and $Y$ are independent random variables and $g_1$ and $g_2$ are suitable real-valued functions we are assuming that $g_1(X)$ and $g_2(Y)$ are independent random variables by Theorem 6.5.1 in generalized form. By Theorem 7.4.3 $E(UV) = E(U)E(V)$ if $U$ and $V$ are independent random variables. Thus

$$\phi_{X+Y}(t) = E(\exp it(X + Y)) = E(\exp itX \exp itY)$$
$$= E(\exp itX)E(\exp itY) = \phi_X(t)\phi_Y(t)$$

One might object that $U$, $V$, $g_1(X)$, $g_2(Y)$ are real-valued whereas $\exp itX$ is not. However, if we take advantage of the fact that $e^{iu} = \cos u + i \sin u$ and $E(W + iZ) = E(W) + iE(Z)$ all details can be checked. Thus if $X$ and $Y$ are independent random variables

**(9.2.2)** $\qquad \phi_{X+Y}(t) = \phi_X(t)\phi_Y(t)$

and, in fact, if $X_1, X_2, \ldots, X_n$ are independent random variables $\phi_{(X_1 + \cdots + X_n)}(t) = \phi_{X_1}(t)\phi_{X_2}(t) \cdots \phi_{X_n}(t)$. The results of Example 9.2.3 should be understood in this context. There the random variable $Y = X_1 + X_2 + \cdots + X_n$ where

$$X_k = \begin{cases} 0, & \delta_k = T \\ 1, & \delta_k = H \end{cases}$$

Suppose $E(X)$ exists. In the discrete case this means that $\sum X(s_k)p_k$ is absolutely convergent. The series for $\phi_X(t)$ is $\sum e^{itX(s_k)}p_k$ and the term-by-term derivative with respect to $t$ is $\sum iX(s_k)e^{itX(s_k)}p_k$. Theorems on term-by-term differentiation guarantee that if $\sum |iX(s_k)e^{itX(s_k)}p_k|$ converges then $\dfrac{d}{dt}\phi_X(t) = \sum iX(s_k)e^{itX(s_k)}p_k$. But $\sum |iX(s_k)e^{itX(s_k)}p_k| = \sum |X(s_k)|p_k$ since $|ie^{itX(s_k)}| = 1$. Thus existence of $E(X)$ guarantees that $\dfrac{d}{dt}\phi_X(t) = \sum iX(s_k)e^{itX(s_k)}p_k$. Now if we set $t = 0$ and recall that $e^{i0} = 1$, we have

**(9.2.3)** $\qquad \phi_X'(0) = i \sum X(s_k)e^{i0X(s_k)}p_k = i \sum X(s_k)p_k = iE(X)$

The same theorems cited above guarantee that

$$\phi_X''(t) = \frac{d^2}{dt^2}\phi_X(t) = \sum i^2[X(s_k)]^2 e^{itX(s_k)}p_k$$

if this last series is absolutely convergent, i.e., if $\sum [X(s_k)]^2 p_k$ converges since $|i^2 e^{itX(s_k)}| = 1$. But $\sum [X(s_k)]^2 p_k = E(X^2)$. Thus if we set $t = 0$ we obtain

**(9.2.4)** $\qquad \phi_X''(0) = i^2 \sum [X(s_k)]^2 e^{i0X(s_k)}p_k = i^2 E(X^2)$

### Example 9.2.5
If $X$ has a Poisson distribution, find $E(X)$ and $\sigma^2(X)$ from the characteristic function.

By Example 9.2.2, $\phi(t) = \sum_{n=0}^{\infty} e^{-\lambda}e^{int}(\lambda^n/n!) = e^{-\lambda(1-e^{it})} = e^{-\lambda}e^{\lambda e^{it}}$. The ratio test shows that $\sum_{n=0}^{\infty} n^r e^{-\lambda}(\lambda^n/n!)$ converges for each $r \geq 0$ so all moments of $X$ exist. Now, $\phi'(t) = e^{-\lambda}e^{\lambda e^{it}}\lambda ie^{it}$ and $\phi''(t) = \lambda ie^{-\lambda}(e^{\lambda e^{it}}\lambda ie^{it}e^{it} + ie^{\lambda e^{it}}e^{it})$ so $\phi'(0) = e^{-\lambda}e^{\lambda}\lambda i = \lambda i$ and $\phi''(0) = \lambda ie^{-\lambda}(e^{\lambda}\lambda i + ie^{\lambda}) = i^2(\lambda^2 + \lambda)$. By (9.2.3) and (9.2.4) $E(X) = \lambda$ and $\sigma^2(X) = E(X^2) - [E(X)]^2 = \lambda^2 + \lambda - \lambda^2$. Thus $\sigma^2(X) = \lambda$.

So far we have assumed that the sample space $S$ is countable. However, for any random variable $X$, $\cos tX$ and $\sin tX$ are both bounded random variables and hence $E(\cos tX)$ and $E(\sin tX)$ exist for any real number $t$. We can define $E(e^{itX})$ as $E(\cos tX) + iE(\sin tX)$ and see that this definition is compatible with all theorems on integration, summation, etc., of complex numbers. In particular if $X$ has density function $f$ we have

$$\phi_X(t) = E(e^{itX}) = \int_{-\infty}^{\infty} \cos tx\, f(x)\, dx + i \int_{-\infty}^{\infty} \sin tx\, f(x)\, dx$$

$$= \int_{-\infty}^{\infty} e^{itx} f(x)\, dx$$

It can be verified that equations (9.2.1)–(9.2.4) hold for all random variables.

### Example 9.2.6
Let $X$ have density function

$$f(x) = \begin{cases} e^{-x}, & x > 0 \\ 0, & x \leq 0 \end{cases}$$

Find the characteristic function of $X$.

$$\phi(t) = \int_0^{\infty} e^{itx} e^{-x}\, dx = \lim_{M \to \infty} \int_0^M e^{(it-1)x}\, dx$$

By exercise 9.1.9,

$$\int_0^M e^{(it-1)x}\, dx = \frac{1}{it-1}(e^{(it-1)M} - 1)$$

But

$$\lim_{M \to \infty} \frac{1}{it-1}(e^{(it-1)M} - 1) = \frac{-1}{it-1} = \frac{1}{1-it}$$

since $\lim_{M \to \infty} e^{itM-M} = 0$. Therefore $\phi(t) = 1/(1 - it)$.

It is easy to verify that $\int_0^{\infty} x^r e^{-x}\, dx$ converges so $E(X^r)$ exists for $r > 0$. Therefore, since $\phi'(t) = i/(1 - it)^2$ and $\phi''(t) = 2i^2/(1 - it)^3$, $E(X) = 1$, $\sigma^2(X) = E(X^2) - [E(X)]^2 = 2 - 1^2 = 1$.

### Example 9.2.7
Let $\lambda$ be a nonzero real number and let $X$ be the random variable of Example 9.2.6. Find the characteristic function of $Y = \lambda^{-1}X$.

By equation (9.2.1)

$$\phi_{\lambda^{-1}X}(t) = \phi_X(\lambda^{-1}t) = \frac{1}{1 - i(t/\lambda)} = \frac{\lambda}{\lambda - it}$$

### Example 9.2.8

Let $X$ have density function $f(x) = (1/\sqrt{2\pi})e^{-x^2/2}$. Find the characteristic function, $E(X)$ and $\sigma^2(X)$.

From the standard representation of $X$,

$$\phi(t) = \int_{-\infty}^{\infty} e^{itx}\frac{1}{\sqrt{2\pi}} e^{-x^2/2}\, dx = \int_{-\infty}^{\infty} \frac{1}{\sqrt{2\pi}}(\cos tx + i \sin tx)e^{-x^2/2}\, dx$$

It is easy to see that $\int_{-\infty}^{\infty} \sin tx e^{-x^2/2}\, dx = 0$; thus

$$\phi(t) = \frac{1}{\sqrt{2\pi}}\int_{-\infty}^{\infty} \cos tx e^{-x^2/2}\, dx$$

We will evaluate $\phi(t)$ by a trick. Note that $\phi'(t)=(1/\sqrt{2\pi})\int_{-\infty}^{\infty} -x\sin tx\, e^{-x^2/2}\, dx$. In the notation of integration by parts we take $u = \sin tx/\sqrt{2\pi}$ and $dv = -xe^{-x^2/2}\, dx$. Thus

$$\phi'(t) = \frac{1}{\sqrt{2\pi}} \sin tx e^{-x^2/2}\Big|_{-\infty}^{+\infty} - \int_{-\infty}^{\infty} \frac{t\cos tx}{\sqrt{2\pi}} e^{-x^2/2}\, dx$$

Since $\lim_{M \to \infty} \sin tM\, e^{-M^2/2} = 0$ we obtain

$$\phi'(t) = -\int_{-\infty}^{\infty} t\frac{\cos tx}{\sqrt{2\pi}} e^{-x^2/2}\, dx = -t\phi(t)$$

Thus we have the differential equation $\phi'(t) = -t\phi(t)$ for $\phi$. If we write $\phi'(t)/\phi(t) = -t$, we get $\ln \phi(t) = -t^2/2 + c$. For any random variable we have $\phi(0) = 1$. Thus $0 = \ln 1 = -0 + c$ and $c = 0$. We see that $\phi(t) = e^{-t^2/2}$. Theorems on differential equations show that this is the only solution to the differential equation which satisfies $\phi(0) = 1$. We see also that $\phi''(t) = -\phi(t) - t\phi'(t)$ and $\phi''(0) = -\phi(0)$. Since $E(X) = 0$, $\sigma^2(X) = E(X^2) = -\phi''(0) = \phi(0) = 1$, which agrees with previous results.

### EXERCISES 9.2

1  Let $P(X = n) = 1/2^n$ for $n = 1, 2, 3, \ldots$. Find the characteristic function, $E(X)$, and $\sigma^2(X)$.

♦ 2  Let $X$ have density function

$$f(x) = \begin{cases} 1, & 0 \le x \le 1 \\ 0, & \text{otherwise} \end{cases}$$

Find the characteristic function of $X$.

♦ 3  Let $Y = aX + b$ where $X$ is that of Example 9.2.8. Find $\phi_Y(t)$. Find the density function of $Y$.

4 Let $X_1$ and $X_2$ be independent and identically distributed with $X$ of exercise 2. Find $\phi_{X_1 + X_2}(t)$.

5 Show that if $X$ has density function $f$ and if $f(x) = f(-x)$, then $\phi_X(t)$ is a real number for each $t$.

6 Show that if $E(X^n)$ exists then $E(X^n) = i^n \phi^{(n)}(0)$ if $X$ is a discrete random variable.

7 Let $X$ have density function $f(x) = \frac{1}{2}e^{-|x|}$ for all real $x$. Find $\phi(t)$, $E(X)$, $E(X^2)$.

♦ 8 Show that if $X$ has the generating function $\gamma$, then the characteristic function $\phi(t) = \gamma(e^{it})$. Find the characteristic function of $X$ if its generating function is given by $\gamma(z) = 2^{-\alpha}(1 - \frac{1}{2}z)^{-\alpha}$ for $\alpha > 0$.

9 Find the characteristic function of a random variable $X$ whose density function is given by

$$f(x) = \begin{cases} 1 - |x|, & -1 \le x \le 1 \\ 0, & |x| > 1 \end{cases}$$

Hint:

$$f(x) = \begin{cases} 1 - x, & 0 \le x \le 1 \\ 1 + x, & -1 \le x \le 0 \\ 0, & |x| > 1 \end{cases}$$

10 Let $X$ be the random variable of Example 9.2.6. Find the characteristic function of $Y = aX + b$.

11 Let $X$ have the density function

$$f(x) = \begin{cases} 0, & x < -\frac{1}{2}\pi \\ 0, & x > \frac{1}{2}\pi \\ \frac{1}{2}\cos x, & -\frac{1}{2}\pi \le x \le \frac{1}{2}\pi \end{cases}$$

Find the characteristic function of $X$.

12 Let $p \ge 0$ be an integer. Let $X_p$ have the density function

$$f(x) = \begin{cases} (p + 1)x^p, & 0 \le x \le 1 \\ 0, & \text{otherwise} \end{cases}$$

Show that

$$it\phi_{X_{p+1}}(t) = [e^{it} - \phi_{X_p}(t)](p + 2)$$

13 Check the details leading up to (9.2.2). That is, show that $E(\exp itX \exp itY) = \phi_X(t)\phi_Y(t)$ using Theorem 7.4.3 for real-valued functions of $X$ and $Y$.

14 Let $X_1$, $X_2$, and $X_3$ be independent and identically distributed with the distribution of Example 9.2.6. Find the characteristic function of $X_1 + X_2 + X_3$.

15 Find the characteristic function of $Y$ where $f_Y(x) = \frac{1}{2}x^2 e^{-x}$ for $x > 0$ and $f_Y(x) = 0$ for $x < 0$. Compare with exercise 14.

♦ **16**   Let $X$, $Y$ have binomial distributions with parameters $n, p$ and $m, p$, respectively. Find $\phi_{X+Y}(t)$.

♦ **17**   Let $X$ have a characteristic function $\phi$ such that $\phi(t) = \sum_{n=0}^{\infty} a_n t^n$ for $0 \le t < \epsilon$ where $\epsilon$ is some positive number. Show that $E(X^n) = i^{-n} n! a_n$.

**18**   Using exercise 17 find all of the moments for the random variable in Example 9.2.7.

**19**   Find all of the moments for the random variable of Example 9.2.8.

♦ **20**   Suppose that $f_X(x) = f_X(-x)$ for all $x$ where $f_X$ is the density function or the probability function. Show that $\phi_X(t) = E(\cos tX)$.

♦ **21**   Find the expectation and variance of the binomial, Poisson, exponential, normal, and gamma distributions by means of characteristic functions.

# 9.3   Uniqueness of the characteristic function

Let $X$ be a random variable and let $F$ be its distribution function. If $u_1$ and $u_2$ are continuity points of $F$, i.e., $\lim_{u \to u_1} F(u) = F(u_1)$ and $\lim_{u \to u_2} F(u) = F(u_2)$, then it can be shown that

**(9.3.1)**   $$F(u_2) - F(u_1) = \lim_{T \to \infty} \frac{1}{2\pi} \int_{-T}^{T} \frac{e^{-itu_2} - e^{-itu_1}}{-it} \phi(t)\, dt$$

if $\phi$ is the characteristic function of $X$. We do not propose to prove the result in this book nor even to use the result to calculate $F(u_2) - F(u_1)$ for many special random variables. We will use (9.3.1) to prove a fundamental uniqueness theorem.

**Theorem 9.3.1**   If two random variables $X$ and $Y$ have the same characteristic function, they have the same distribution function.

PROOF   By exercise 6.3.17, the distribution functions $F_X$ and $F_Y$ can have at most countably many discontinuity points. Let $C = \{u; u$ is a discontinuity point of $F_X$ or of $F_Y\}$. The set $C$ is countable.

If $a < b$, the interval $\{u; a < u < b\}$ can easily be seen to contain uncountably many points and hence it must contain (uncountably many) points in the complement of $C$. Now by (9.3.1) for every $u_2 \notin C$ and $u_1 \notin C$ we have

$$F_X(u_2) - F_X(u_1) = \lim_{T \to \infty} \frac{1}{2\pi} \int_{-T}^{T} K(u_2, u_1, t)\phi_X(t)\, dt$$

$$= \lim_{T \to \infty} \frac{1}{2\pi} \int_{-T}^{T} K(u_2, u_1, t)\phi_Y(t)\, dt = F_Y(u_2) - F_Y(u_1)$$

since $\phi_X(t) = \phi_Y(t)$ for all real numbers $t$. The quantity $K(u_2, u_1, t)$ is the obvious quantity in (9.3.1). Thus $F_X(u_2) - F_X(u_1) = F_Y(u_2) - F_Y(u_1)$. We may let $u_1 \to -\infty$ (always using $u_1 \notin C$) and since $F_X$ and $F_Y$ are distribution functions, $F_X(u_1) \to 0$ and $F_Y(u_1) \to 0$. We thus have $F_X(u_2) = F_Y(u_2)$ for each $u_2 \notin C$.

By the definition of limit and the above observation we have for any real number a

$$F_X(a) = \lim_{\substack{u_2 \notin C,\, u_2 > a \\ u_2 \to a}} F_X(u_2) = \lim_{\substack{u_2 \notin C,\, u_2 > a \\ u_2 \to a}} F_Y(u_2) = F_Y(a)$$

since distribution functions are continuous from the right. Thus $F_X$ and $F_Y$ are the same function.

The theorem proves that the characteristic function determines the distribution function (in principle). Previous results indicate that the distribution function determines the characteristic function.

Furthermore, if $X$ has a density function $f$, at many points $u$ we have

**(9.3.2)** $\quad f(u) = \lim_{T \to \infty} \frac{1}{2\pi} \int_{-T}^{T} e^{-itu} \phi(t)\, dt$

Nonetheless, we will not normally use (9.3.1) or (9.3.2) to find density or distribution functions. Instead we concentrate on assembling a reasonable collection of basic characteristic functions and hope that we can use the methods of equations (9.2.1) and (9.2.2) to extend our knowledge of characteristic functions.

### Example 9.3.1
A random variable $X$ has its characteristic function $\phi$ given by $\phi(t) = \frac{1}{3}\cos 2t + \frac{2}{3}\cos t$. Describe the random variable in another way.

We observe that

$$\tfrac{1}{3}\cos 2t + \tfrac{2}{3}\cos t = \tfrac{1}{3}[\tfrac{1}{2}(e^{2it} + e^{-2it})] + \tfrac{2}{3}[\tfrac{1}{2}(e^{it} + e^{-it})]$$
$$= \tfrac{1}{6}e^{2it} + \tfrac{1}{6}e^{-2it} + \tfrac{1}{3}e^{it} + \tfrac{1}{3}e^{-it}$$

This is exactly the characteristic function of a random variable $Y$ which has $P(Y = 2) = \frac{1}{6}$, $P(Y = -2) = \frac{1}{6}$, $P(Y = 1) = \frac{1}{3}$, and $P(Y = -1) = \frac{1}{3}$. But the probability function determines and is determined by the distribution function. Thus $X$ and $Y$ are identically distributed by Theorem 9.3.1. We therefore have $P(X = 2) = P(X = -2) = \frac{1}{6}$ and $P(X = 1) = P(X = -1) = \frac{1}{3}$.

### Example 9.3.2
A random variable $X$ has its characteristic function $\phi$ given by $\phi(t) = e^{-3it}e^{-t^2/4}$. Describe $X$ in another way.

By Example 9.2.7 a random variable $Y$ with density function given by $f_Y(x) = (1/\sqrt{2\pi})e^{-x^2/2}$ has $\phi_Y(t) = e^{-t^2/2}$. Furthermore $\phi_{aY+b}(t) = e^{itb}\phi_Y(at)$. Now $e^{-3it}e^{-t^2/4} = e^{-it\cdot 3}e^{-(t/\sqrt{2})^2/2}$. If we set $b = -3$ and $a = 1/\sqrt{2}$ we can cite

Theorem 9.3.1 and argue that $X$ and $(1/\sqrt{2})Y - 3$ are identically distributed. Thus, the density function of $X$ is given by

$$f_X(x) = \sqrt{2}\,\frac{1}{\sqrt{2\pi}}\,e^{-[\sqrt{2}(x+3)]^2/2} = \frac{1}{\sqrt{\pi}}\,e^{-(x+3)^2}$$

### Example 9.3.3

A random variable $X$ has characteristic function $\phi$ given by $\phi(t) = e^{-|t|}$ for all $t$. Find the density function of $X$.

Here we will use (9.3.2) because of our failure to find random variables whose characteristic function looks like $e^{-|t|}$. Now

$$\frac{1}{2\pi}\int_{-T}^{T} e^{-iut}e^{-|t|}\,dt = \frac{1}{2\pi}\int_{0}^{T} e^{-iut}e^{-t}\,dt + \frac{1}{2\pi}\int_{-T}^{0} e^{-iut}e^{t}\,dt$$

$$= \frac{1}{2\pi}\left(\frac{e^{(-iu-1)t}}{-iu-1}\right)\Big|_{0}^{T} + \frac{1}{2\pi}\left(\frac{e^{(-iu+1)t}}{-iu+1}\right)\Big|_{-T}^{0}$$

$$= \frac{1}{2\pi}\left(\frac{1}{iu+1} + \frac{1}{-iu+1}\right) + \frac{1}{2\pi}\left(\frac{e^{(-iu-1)T}}{-iu-1} - \frac{e^{(-iu+1)(-T)}}{-iu+1}\right)$$

Now

$$\lim_{T\to\infty}\left(ke^{(-iu-1)T} + le^{(-iu+1)(-T)}\right) = 0$$

since $|e^{iuT}| = 1$ and $|e^{-iuT}| = 1$ and $e^{-T}\to 0$ as $T\to\infty$. Thus

$$f_X(u) = \frac{1}{2\pi}\left(\frac{1}{iu+1} + \frac{1}{-iu+1}\right) = \frac{1}{2\pi}\left(\frac{1}{1-iu} + \frac{1}{1+iu}\right)$$

$$= \frac{1}{2\pi}\frac{2}{(1-iu)(1+iu)} = \frac{1}{\pi(1+u^2)}$$

so $X$ has the Cauchy distribution.

This is the only example in which we will use (9.3.1) or (9.3.2). However, it is desirable to point out that the key to (9.3.1) is the fact that

$$(9.3.3) \qquad \lim_{T\to\infty}\int_{-T}^{T}\frac{\sin\alpha x}{x}\,dx = \begin{cases} \pi, & \alpha > 0 \\ 0, & \alpha = 0 \\ -\pi, & \alpha < 0 \end{cases}$$

In Chapter 6 we defined the standard normal distribution as that with density function $f(x) = (2\pi)^{-1/2}e^{-x^2/2}$ and, in general, a random variable $X$ is $N(m, \sigma^2)$ if its density function is given by $(2\pi\sigma^2)^{-1/2}\exp\left[-(1/2\sigma^2)(x-m)^2\right]$. It is easy to see that if $Y$ is $N(0, 1)$ then $X = \sigma Y + m$ is $N(m, \sigma^2)$. Thus

$$\phi_X(t) = e^{imt}e^{-(\sigma t)^2/2} = e^{imt-\sigma^2 t^2/2}$$

### Example 9.3.4

Show that the sum of two independent normal random variables is a normal random variable.

Let $Y_1$ be $N(m_1, \sigma_1^2)$ and $Y_2$ be $N(m_2, \sigma_2^2)$. By what we have just observed $\phi_{Y_j}(t) = e^{im_j t - \sigma_j^2 t^2/2}$ for $j = 1, 2$. Also, $\phi_{Y_1+Y_2}(t) = \phi_{Y_1}(t)\phi_{Y_2}(t)$ since $Y_1$ and $Y_2$ are independent. Thus

$$\phi_{Y_1+Y_2}(t) = e^{im_1 t - \sigma_1^2 t^2/2} e^{im_2 t - \sigma_2^2 t^2/2}$$
$$= \exp\left[i(m_1 + m_2)t - \tfrac{1}{2}(\sigma_1^2 + \sigma_2^2)t^2\right]$$
$$= \exp\left(imt - \tfrac{1}{2}\sigma^2 t^2\right)$$

where $m = m_1 + m_2$ and $\sigma^2 = \sigma_1^2 + \sigma_2^2$. Thus by Theorem 9.3.1, $Y_1 + Y_2$ is $N(m, \sigma^2)$.

The crux of Example 9.3.4 is that $Y_1 + Y_2$ is normal not that $E(Y_1 + Y_2) = E(Y_1) + E(Y_2)$ and $\sigma^2(Y_1 + Y_2) = \sigma^2(Y_1) + \sigma^2(Y_2)$. These last expressions are valid for the sum of any independent random variables with second moments. However, it is not the case that the sum of any two independent random variables in our families is in the same family as the next example shows.

### Example 9.3.5

Prove that the sum of two independent exponential random variables with positive parameter is not an exponential random variable.

By Example 9.2.7, the characteristic function of an exponential random variable with parameter $\lambda > 0$ is $\phi(t) = \lambda/(\lambda - it)$. Now if $\phi_{X_1}(t) = \lambda_1/(\lambda_1 - it)$ and $\phi_{X_2}(t) = \lambda_2/(\lambda_2 - it)$,

$$\phi_{X_1+X_2}(t) = \frac{\lambda_1}{\lambda_1 - it}\frac{\lambda_2}{\lambda_2 - it} = \frac{\lambda_1\lambda_2}{\lambda_1\lambda_2 - it(\lambda_1 + \lambda_2) - t^2}$$

Now it is easy to see that

$$\frac{\lambda_1\lambda_2}{\lambda_1\lambda_2 - it(\lambda_1 + \lambda_2) - t^2} = \frac{\lambda}{\lambda - it}$$

cannot hold for any choice of $\lambda > 0$ and all real $t$. Thus, by Theorem 9.3.1, $X_1 + X_2$ is not an exponential random variable.

Let us move on to a discussion of the characteristic function of a gamma random variable.

### Example 9.3.6

Suppose $X$ has density function

$$f(x) = \begin{cases} \dfrac{1}{\Gamma(\alpha)} x^{\alpha-1} e^{-x}, & x > 0 \\ 0, & x \leq 0 \end{cases}$$

where $\alpha > 0$ is the parameter. Here of course, $\Gamma(\alpha) = \int_0^\infty x^{\alpha-1}e^{-x}\,dx$. Find the characteristic function.

Since $\alpha$ is fixed

**(9.3.4)**    $\Gamma(\alpha)\phi_X(t) = \int_0^\infty e^{itx}x^{\alpha-1}e^{-x}\,dx$

$$= \int_0^\infty x^{\alpha-1}e^{-x(1-it)}\,dx$$

It would be ideal if we could let $x(1 - it) = y$. This would leave

$$\Gamma(\alpha)\phi_X(t) = \lim_{M\to\infty}\int_0^{M(1-it)}\left(\frac{y}{1-it}\right)^{\alpha-1}e^{-y}\frac{1}{1-it}\,dy$$

But what is the significance of $\int_0^{M(1-it)} g(y)\,dy$? The theory of functions of a complex variable shows that

**(9.3.5)**    $\displaystyle\lim_{M\to\infty}\int_0^{M(1-it)}\left(\frac{y}{1-it}\right)^{\alpha-1}e^{-y}\frac{1}{1-it}\,dy = \int_0^\infty y^{\alpha-1}e^{-y}(1-it)^{-\alpha}\,dy$

so

$$\Gamma(\alpha)\phi_X(t) = (1-it)^{-\alpha}\Gamma(\alpha)$$

and

$$\phi_X(t) = (1-it)^{-\alpha}$$

The proof of (9.3.5) is not easy. Instead we will approach the problem from the viewpoint of differential equations. We differentiate (9.3.4) with respect to $t$ and get

$$\Gamma(\alpha)\phi_X'(t) = \int_0^\infty ixe^{itx}x^{\alpha-1}e^{-x}\,dx$$

$$= i\int_0^\infty x^\alpha e^{itx}e^{-x}\,dx$$

In order to integrate by parts we let $dv = ie^{itx}\,dx$ and $u = x^\alpha e^{-x}$. Then $v = t^{-1}e^{itx}$ and $du = (-x^\alpha e^{-x} + \alpha x^{\alpha-1}e^{-x})\,dx$. Thus

$$\Gamma(\alpha)\phi_X'(t) = t^{-1}e^{itx}x^\alpha e^{-x}\Big|_0^\infty - \int_0^\infty t^{-1}e^{itx}(-x^\alpha + \alpha x^{\alpha-1})e^{-x}\,dx$$

if $t \neq 0$. Since $\alpha > 0$, $\displaystyle\lim_{x\to 0} x^\alpha = 0$ and $\displaystyle\lim_{x\to\infty} x^\alpha e^{-x} = 0$. We can therefore write

$$\Gamma(\alpha)\phi_X'(t) = t^{-1}\int_0^\infty e^{itx}(x^\alpha - \alpha x^{\alpha-1})e^{-x}\,dx$$

or

$$t\Gamma(\alpha)\phi_X'(t) - \int_0^\infty e^{itx}x^\alpha e^{-x}\,dx = -\alpha\int_0^\infty e^{itx}x^{\alpha-1}e^{-x}\,dx$$

## Table 1

| DISTRIBUTION OF $X$ | PROBABILITY OR DENSITY FUNCTION | $E(X)$ | $\sigma^2(X)$ | CENTER OF SYMMETRY | CHARACTERISTIC FUNCTION |
|---|---|---|---|---|---|
| BINOMIAL | $\binom{n}{k}p^k(1-p)^{n-k}$ <br> $k=0, 1, 2, \ldots, n$ | $np$ | $np(1-p)$ | $n/2$ if $p=\frac{1}{2}$ | $(1-p+pe^{it})^n$ |
| POISSON | $e^{-\lambda}\lambda^k/k!$ <br> $k=0, 1, 2, \ldots$ | $\lambda$ | $\lambda$ | none | $e^{-\lambda(1-e^{it})}$ |
| DISCRETE UNIFORM | $1/(n-m+1)$ <br> $k=m, m+1, \ldots, n$ | $\dfrac{n+m}{2}$ | $\dfrac{(n-m)(n-m+2)}{12}$ | $\dfrac{n+m}{2}$ | $\displaystyle\sum_{k=m}^{n} e^{ikt}/(n-m+1)$ |
| HYPER-GEOMETRIC | $\binom{m}{k}\binom{n-m}{r-k} \div \binom{n}{r}$ <br> $k=0, 1, 2\ldots, \min(n,m)$ | $r\dfrac{m}{n}$ | $\dfrac{mr(n-r)(n-m)}{n^2(n-1)}$ | $\frac{1}{2}r$ <br> if $n=2m$ | — |
| NEGATIVE BINOMIAL | $(-1)^k\binom{-r}{k}p^k(1-p)^r$ <br> $k=0, 1, 2, \ldots$ | $\dfrac{rp}{1-p}$ | $\dfrac{rp^2}{(1-p)^2}$ | none | $(1-p)^r(1-pe^{it})^{-r}$ |
| CONTINUOUS UNIFORM | $1/(b-a)$ <br> $a\le x\le b$ | $\dfrac{a+b}{2}$ | $\dfrac{(b-a)^2}{12}$ | $\dfrac{a+b}{2}$ | $\dfrac{e^{ibt}-e^{iat}}{it}$ |
| NORMAL | $\dfrac{e^{-(x-m)^2/2\sigma^2}}{\sqrt{2\pi}\,\sigma}$ <br> $-\infty<x<\infty$ | $m$ | $\sigma^2$ | $m$ | $e^{imt}e^{-\sigma^2t^2/2}$ |
| NEGATIVE EXPONENTIAL | $\lambda e^{-\lambda x}$ <br> $x>0$ | $1/\lambda$ | $1/\lambda^2$ | none | $\dfrac{\lambda}{\lambda-it}$ |
| CAUCHY | $(\pi(1+x^2))^{-1}$ <br> $-\infty<x<\infty$ | does not exist | does not exist | $0$ | $e^{-\lvert t\rvert}$ |
| GAMMA | $x^{\alpha-1}e^{-x}/\Gamma(\alpha)$ <br> $x>0$ | $\alpha$ | $\alpha$ | none | $(1-it)^{-\alpha}$ |
| STUDENT'S $t$ | $\dfrac{\Gamma\left(n+\frac{1}{2}\right)}{\sqrt{n\pi}\,\Gamma\left(\frac{n}{2}\right)}\left(1+\dfrac{x^2}{n}\right)^{-n+1/2}$ <br> $-\infty<x<\infty$ | $0$ <br> if $n>1$ | $\dfrac{n}{n-2}$ <br> if $n>2$ | $0$ | — |
| $F$ | $\dfrac{m^{m/2}n^{n/2}\Gamma\left(\dfrac{m+n}{2}\right)x^{m/2-1}}{\Gamma\left(\dfrac{m}{2}\right)\Gamma\left(\dfrac{n}{2}\right)(mx+n)^{(m+n)/2}}$ <br> $x>0$ | $\dfrac{n}{n-2}$ <br> if $n>2$ | $\dfrac{2n^2(n+m-2)}{(n-2)^2(n-4)m}$ <br> if $n>4$ | none | |

Characteristic functions for the hypergeometric, Student's $t$ and $F$ distribution are omitted since they are beyond the scope of this book.

and

$$t\Gamma(\alpha)\phi'_X(t) - (1/i)\Gamma(\alpha)\phi'_X(t) = -\alpha\Gamma(\alpha)\phi_X(t)$$

and

$$(1 - it)\phi'_X(t) = i\alpha\phi_X(t)$$

It is easy to see that $(1 - it)^\alpha = \phi_X(t)$ satisfies this last equation and that $(1 - it)^\alpha = 1$ if $t = 0$. Of course, one may raise the objection that there may be other solutions, but we do not take that matter up here.

We have seen that the characteristic function is extremely useful in the study of random variables. In Table 1 we list the characteristic function for some of the famous distributions, together with their probability or density function, expectation, variance and center of symmetry.

**EXERCISES 9.3** _____

1   Find the distribution of the random variable $X$ for each of the following characteristic functions.

(a)   $\phi_X(t) = \frac{1}{3}e^{-2it} + \frac{1}{2}e^{it} + \frac{1}{6}$

(b)   $\phi_X(t) = \frac{1}{4}\cos 3t + \frac{1}{2} + \frac{1}{8}a^{it} + \frac{1}{8}e^{-6it}$

(c)   $\phi_X(t) = \frac{1}{2}\cos 2t + \frac{1}{2}\sin 3t$

(d)   $\phi_X(t) = (1 - p + pe^{2it})^{18}$

(e)   $\phi_X(t) = e^{-4it}(1 - p + pe^{it})^{18}$

(f)   $\phi_X(t) = e^{-it/2}\dfrac{(e^{it} - 1)}{it} = \dfrac{2\sin\frac{1}{2}t}{t}$ [*Hint*: See exercise 9.2.2.]

(g)   $\phi_X(t) = \dfrac{\sin 2t}{2t}$ [*Hint:* See part (f).]

(h)   $\phi_X(t) = (\frac{1}{3} + \frac{2}{3}e^{it})^6 e^{-3(1 - e^{it})}$

◆ 2   Let $X$ be $N(0, 1)$. Find $\phi_{X^2}(t)$ in two ways. First, use $F_{X^2}(u)$ from exercise 8.5.1 and proceed as in Example 9.3.6. Secondly, find $\displaystyle\int_{-\infty}^{\infty} e^{itx^2}\,\frac{1}{\sqrt{2\pi}}\,e^{-x^2/2}\,dx$

by assuming that the limits of integration with the substitution $y = \sqrt{1 - 2it}\,x$ are still $-\infty$ and $\infty$.

◆ 3   Let $X_1, X_2, \ldots, X_n$ be independent and $N(0, 1)$. Let $Y = X_1^2 + X_2^2 + \cdots + X_n^2$. Find $\phi_Y$. Compare $\phi_Y$ to $\phi_{2Z}$ where $Z$ is a gamma random variable with $\alpha = \frac{1}{2}n$.

4   Is there a pattern which relates the results of exercise 9.2.7 and Example 9.3.3? Note how (9.3.2) is used in the example.

5   Let $\phi_X(t) = e^{ibt}$. Use (9.3.1) to show that if $u_1 < b < u_2$, $F(u_2) - F(u_1) = 1$.

**6**   Show that the random variable of exercise 9.2.9 is the sum of two independent, identically distributed random variables with a uniform distribution on $\{x; -\frac{1}{2} \le x \le \frac{1}{2}\}$.

**7**   Suppose that $X$ and $Y$ have Poisson distributions with parameters $\lambda$ and $\mu$, respectively. What is the distribution of $X + Y$?

# 9.4 Characteristic functions of random vectors

An extremely satisfactory definition of the characteristic function of a random vector is as follows:

**(9.4.1)**   $\phi_{(X_1, X_2, \ldots, X_n)}(t_1, t_2, \ldots, t_n) = E(\exp(i \sum_{j=1}^{n} t_j X_j))$

This makes the characteristic function of an $n$-dimensional random vector a function of $n$ real variables $t_1, t_2, \ldots, t_n$. If $X = (X_1, X_2, \ldots, X_n)$ we also use the notation $\phi_X$. It is clear that $\phi_X(0, 0, \ldots, 0) = 1$ and $|\phi_X(t_1, t_2, \ldots, t_n)| \le 1$. Furthermore,

**(9.4.2)**   $\phi_X(t_1, 0, 0, \ldots, 0) = E(\exp(it_1 X_1 + 0)) = \phi_{X_1}(t_1)$

$\phi_X(0, t_2, 0, 0, \ldots, 0) = \phi_{X_2}(t_2)$

$\vdots$

Similarly, the characteristic function of groups of components of $X$ can be obtained by substituting $t_j = 0$ for the other $j$ into $\phi_X$.

It is not always easy to calculate characteristic functions of complicated random vectors. However, it is not easy to calculate the characteristic function of a complicated random variable. The theory of characteristic functions of random vectors completely parallels that for characteristic functions of random variables. We have the following theorem.

**Theorem 9.4.1**   If $X$ and $Y$ are two $n$-dimensional random vectors and for all $t_1, t_2, \ldots, t_n$

$$\phi_X(t_1, t_2, \ldots, t_n) = \phi_Y(t_1, t_2, \ldots, t_n)$$

then for all real $b_1, b_2, \ldots, b_n$

$$F_X(b_1, b_2, \ldots, b_n) = F_Y(b_1, b_2, \ldots, b_n)$$

PROOF   We will not prove this theorem. It follows from a multidimensional inversion formula analogous to equation (9.3.1).

As a result of Theorem 9.4.1, it follows that the characteristic function of a random vector uniquely determines the distribution of the random vector. Thus we have the following corollary.

**Corollary**

**(9.4.3)**     $\phi_X(t_1, t_2, \ldots, t_n) = \phi_{X_1}(t_1)\phi_{X_2}(t_2) \cdots \phi_{X_n}(t_n)$

if and only if $X_1, X_2, \ldots, X_n$ are independent.

PROOF     If $X_1, X_2, \ldots, X_n$ are independent, equation (9.4.3) follows from $\exp(iY_1 + iY_2) = \exp(iY_1)\exp(iY_2)$, the independence of $\exp(iY_1)$ and $\exp(iY_2)$ when $Y_1$ and $Y_2$ are independent, and Theorem 7.4.3.

Conversely, suppose that $X_1, X_2, \ldots, X_n$ are random vectors satisfying (9.4.3). By the use of Cartesian products and the standard representation we can form $n$ independent random variables $Z_1, Z_2, \ldots, Z_n$ such that, individually, $X_j$ and $Z_j$ are identically distributed and therefore,

**(9.4.4)**     $\phi_{X_j}(t_j) = \phi_{Z_j}(t_j), \qquad j = 1, 2, \ldots, n$

By the above argument, since the $Z_j$ are independent, $\phi_{(Z_1, \ldots, Z_n)}(t_1, \ldots, t_n) = \phi_{Z_1}(t_1) \cdots \phi_{Z_n}(t_n) = \phi_{X_1}(t_1) \cdots \phi_{X_n}(t_n) = \phi_X(t_1, t_2, \ldots, t_n)$ by assumption. By Theorem 9.4.1, $X$ and $Z = (Z_1, \ldots, Z_n)$ have the same distribution function. Thus

**(9.4.5)**     $F_X(b_1, b_2, \ldots, b_n) = F_Z(b_1, b_2, \ldots, b_n)$

$\qquad\qquad\qquad\quad = F_{Z_1}(b_1)F_{Z_2}(b_2) \cdots F_{Z_n}(b_n)$

$\qquad\qquad\qquad\quad = F_{X_1}(b_1)F_{X_2}(b_2) \cdots F_{X_n}(b_n)$

by the independence of the $Z_j$ and Theorem 9.3.1 applied to equation (9.4.4). However (9.4.5) is the definition of independence for the $X_j$.

*Example 9.4.1*
Let $X_1$ and $X_2$ be independent $N(0, 1)$ random variables. Show that $Y_1 = X_1 - X_2$ and $Y_2 = X_1 + X_2$ are independent.

By the independence of the $X_j$'s

$$\phi_{(Y_1, Y_2)}(t_1, t_2) = E(\exp(it_1(X_1 - X_2) + it_2(X_1 + X_2)))$$

$$= E(\exp(i(t_1 + t_2)X_1 + i(-t_1 + t_2)X_2))$$

$$= \phi_{X_1}(t_1 + t_2)\phi_{X_2}(-t_1 + t_2)$$

However, by our assumptions about the $X_j$'s, $\phi_{X_1}(t) = \phi_{X_2}(t) = e^{-t^2/2}$. Thus

$$\phi_{(Y_1, Y_2)}(t_1, t_2) = \exp\left[-\tfrac{1}{2}(t_1 + t_2)^2\right]\exp\left[-\tfrac{1}{2}(-t_1 + t_2)^2\right]$$

$$= \exp\left[-\tfrac{1}{2}(t_1^2 + 2t_1t_2 + t_2^2 + t_1^2 - 2t_1t_2 + t_2^2)\right]$$

$$= \exp\left[-\tfrac{1}{2}(2t_1^2 + 2t_2^2)\right] = \exp\left[-\tfrac{1}{2}(2t_1^2)\right]\exp\left[-\tfrac{1}{2}(2t_2^2)\right]$$

By Theorem 9.4.1 and exercise 9.3.1, $(Y_1, Y_2)$ has the same distribution as the random vector whose components are independent and $N(0, 2)$.

Example 9.4.1 is indicative of the "practical" problems which can be handled by characteristic functions. This use is limited. The characteristic function is more valuable as a theoretical device. Not withstanding this disclaimer we compute the characteristic function of another random vector.

### Example 9.4.2

Let $S = \{(x, y); 0 \le y \le x\}$ with probability mass density $p(x, y) = 2e^{-x}e^{-y}$. Let $X(x, y) = x$ and $Y(x, y) = y$. Find $\phi_{(X,Y)}$.

$$E(\exp{(it_1 X + it_2 Y)}) = \int_0^\infty dx \int_0^x e^{it_1 x + it_2 y} 2e^{-x}e^{-y}\, dy$$

This is $2 \int_0^\infty e^{it_1 x - x} \left( \int_0^x e^{it_2 y - y}\, dy \right) dx$. By exercise 9.1.7 this equals

$$2 \int_0^\infty e^{it_1 x - x} \left( \frac{e^{it_2 y - y}}{it_2 - 1} \right) \Big|_0^x dx = \frac{2}{it_2 - 1} \int_0^\infty e^{it_1 x - x}(e^{it_2 x - x} - 1)\, dx$$

$$= \frac{2}{it_2 - 1} \left( \frac{\exp{[(it_1 + it_2 - 2)x]}}{it_1 + it_2 - 2} - \frac{\exp{[(it_1 - 1)x]}}{it_1 - 1} \right) \Big|_0^\infty$$

$$= \frac{2}{it_2 - 1} \left( \frac{1}{it_1 - 1} - \frac{1}{it_1 + it_2 - 2} \right)$$

since $\lim_{b \to \infty} \exp{(iA - 1)b} = 0$.

### EXERCISES 9.4

1   Find the characteristic function for the random vector $(X, Y)$ of exercise 8.6.4.

2   Find the characteristic function for the random vector $(X, Y)$ of exercise 8.6.3.

◆ 3   Find the characteristic function for the random vector $(aX + bY, cX + dY)$ when $X$ and $Y$ are independent $N(0, 1)$ random variables. Show that $aX + bY$ and $cX + dY$ are independent if and only if $ac + db = 0$.

◆ 4   Let $X_1, X_2, X_3$ be independent $N(0, 1)$ random variables. Show that $a_{11} X_1 + a_{12} X_2 + a_{13} X_3$, $a_{21} X_1 + a_{22} X_2 + a_{23} X_3$, $a_{31} X_1 + a_{32} X_2 + a_{33} X_3$ are independent if and only if $a_{j1}a_{k1} + a_{j2}a_{k2} + a_{j3}a_{k3} = 0$ for $j \ne k$ and $j, k \in \{1, 2, 3\}$.

◆ 5   Consider the random vectors $(U, V) = (aX + bY, cX + dY)$ where $X, Y$ are independent and $N(0, 1)$. Suppose that $ad - bc = 0$. Show that for some $\alpha, \beta, \alpha U + \beta V = 0$.

6   Let $(N_1, N_2, \ldots, N_r)$ have the multinomial distribution. Find its characteristic function.

# 9.5 Summary

It is reasonable to ask, "What is the probabilistic significance of the characteristic function of a random variable?" We ignore random vectors. Before giving a direct, formal answer let us recall the properties of the characteristic function.

**(9.5.1)**    $\phi_{aX+b}(t) = e^{itb}\phi_X(t)$

**(9.5.2)**    If $X$ and $Y$ are independent, $\phi_{X+Y}(t) = \phi_X(t)\phi_Y(t)$.

**(9.5.3)**    If $E(X)$ exists, then $iE(X) = \phi'(0)$.

**(9.5.4)**    If $u$ and $v$ are continuity points of $F_X$, then

$$F_X(v) - F_X(u) = \lim_{T \to \infty} \frac{1}{2\pi} \int_{-T}^{T} \frac{e^{-itv} - e^{-itu}}{-it} \phi(t)\, dt$$

Finally, we prove in the next chapter that if $X_1, X_2, X_3, \ldots$ is a sequence of random variables and if $X$ is a random variable, then

**(9.5.5)**    $X_1, X_2, X_3, \ldots$ converge to $X$ in a certain sense if, and only if, for each $t$, $\phi_{X_n}(t) \to \phi_X(t)$ as $n \to \infty$.

Therefore, regardless of the significance of $\phi_X$ it carries a great deal of information about $X$. In fact, (9.5.4) shows that the distribution function $F$ can be recaptured from $\phi_X$ by conducting certain integral and limiting operations. These operations are frequently difficult to conduct explicitly but they are very standard mathematical operations. Furthermore, one can imagine that it is easier to deduce and/or prove certain results by means of characteristic functions.

Therefore, why is it necessary that the characteristic function have some easily described significance? It is a tool. It has advantages and disadvantages. One learns to use it by seeing others use it, by experimenting oneself, and by thinking about its relationship to other things.

# 10

# *Limit Theorems*

## 10.1  Convergence in distribution

Although the exactness of mathematics seems to set it apart from other disciplines, much of the subject matter of mathematics deals with approximations. Approximations are extremely important in probability theory as well as in other branches of mathematics and they are the main focus of this chapter.

First we recall that $\lim_{n \to \infty} (1 + 1/n)^n = e$. Furthermore if $k > 0$, $\lim_{n \to \infty} (1 + k/n)^n$

$= \lim_{n \to \infty} [(1 + k/n)^{n/k}]^k$ so if $u = n/k$, $\lim_{n \to \infty} (1 + k/n)^n = \lim_{u \to \infty} [(1 + 1/u)^u]^k = e^k$.

It can be shown in other ways that $\lim_{n \to \infty} (1 + z/n)^n = e^z$ for any complex

number $z$.

### Example 10.1.1

A biased coin is tossed $n$ times independently. Suppose $P(\text{H}) = p$ is very close to zero. Suppose $k$ is relatively small and $n$ is quite large. Estimate $P$ (exactly $k$ heads show in $n$ tosses).

By the assumed independence

$$P(\text{exactly } k \text{ H's in } n \text{ tosses}) = \binom{n}{k} p^k (1 - p)^{n-k}$$

Let $p = n\lambda$ or $\lambda = p/n$. Then

$$\binom{n}{k} p^k (1-p)^{n-k} = \frac{n(n-1)\cdots(n-k+1)}{k!} \left(\frac{\lambda}{n}\right)^k \left(1-\frac{\lambda}{n}\right)^{n-k}$$

$$= \frac{n}{n}\frac{n-1}{n}\cdots\frac{n-k+1}{n}\frac{\lambda^k}{k!}\left(1-\frac{\lambda}{n}\right)^n\left(1-\frac{\lambda}{n}\right)^{-k}$$

It is assumed that $n$ is large, $k$ is relatively small, and $p = \lambda/n$ is small. Therefore $1 - \lambda/n$ and $(1 - \lambda/n)^{-k}$ are close to 1 as are $n/n$, $(n-1)/n, \ldots, (n-k+1)/n$, and the product

$$\frac{n}{n}\frac{n-1}{n}\cdots\frac{n-k+1}{n}$$

Therefore, $\binom{n}{k} p^k (1-p)^{n-k}$ is approximately equal to $(\lambda^k/k!)(1 - \lambda/n)^n$. But since $n$ is large $(1 - \lambda/n)^n$ is approximately $e^{-\lambda}$. Thus

$$\binom{n}{k} p^k (1-p)^{n-k} \approx e^{-\lambda}\lambda^k/k!$$

The quantity $e^{-\lambda}\lambda^k/k!$ is exactly the probability that a random variable $Y$ having a Poisson distribution with parameter $\lambda$ takes on the value $k$. In other words, if $X$ has a binomial distribution with parameters $p$ and $n$, then $P(X = k)$ is approximately equal to $P(Y = k)$ if $p$ is small, $\lambda = np$, and $k$ is considerably smaller than $n$.

Furthermore, if we can permit a larger error (a cruder approximation)

$$P(X \le j) = \sum_{k=0}^{j} P(X = k) \approx \sum_{k=0}^{j} P(Y = k) = P(Y \le j)$$

### Example 10.1.2

A fair coin is tossed $n$ times where $n$ is very large. Approximate $P$(average number of H's $\ge 0.501$).

If $N_n$ is the random variable giving the actual number of H's in $n$ tosses, we can rephrase the quantity as $P(N_n/n \ge 0.501) = P(N_n/n - \frac{1}{2} \ge 0.001) \le P(N_n/n - \frac{1}{2} \ge 0.001$ or $N_n/n - \frac{1}{2} \le -0.001) = P(|N_n/n - \frac{1}{2}| \ge 0.001)$. But Chebyshev's inequality shows that

$$P(|N_n/n - \tfrac{1}{2}| \ge 0.001) \le \frac{(0.001)^2}{\frac{1}{2}n}$$

which is very small for very large $n$. We therefore say that $P$(average number of H's $\ge 0.501$) is approximately zero.

### Example 10.1.3

A large collection of identical balls are numbered $1, 2, 3, \ldots, n$ and placed in a box. They are stirred up well and one is chosen. Accordingly, a payoff $X = k/n$ is made if the ball numbered $k$ is chosen. Find, approximately, $P(X \le u)$.

If $u \leq 0$,   $P(X \leq u) = 0$ since $k/n > 0$ and cannot satisfy $k/n \leq u \leq 0$. Since $k \leq n$,   $k/n \leq 1$ for all $k = 1, 2, \ldots, n$ so $P(X \leq 1) = P(k/n \leq 1) = 1$. Thus $P(X \leq u) = 1$ for all $u > 1$. The interesting cases of $0 < u < 1$ remain. For $k/n \leq u$ it is necessary and sufficient that $k \leq nu$. Since $k$ is an integer it is necessary and sufficient that $0 < k \leq [nu]$, where $[x]$ is the greatest integer less than or equal to $x$. Thus $P(X \leq u) = [nu]/n$. But by definition $nu \geq [nu] > nu - 1$ since there must be some integer between $nu$ and $nu - 1$. Thus $u \geq [nu]/n = P(X \leq u) \geq (nu - 1)/n = u - 1/n$. Since we have assumed $n$ to be large we argue $P(X \leq u) \approx u$ for $0 < u < 1$.

The above examples are designed to illustrate a certain viewpoint. However, it is better to view them in a different context. Suppose $X_1, X_2, X_3, \ldots$ is an infinite sequence of random variables. It is unimportant as to whether they have the same or different sample spaces.

**Definition**     The sequence of random variables $X_1, X_2, X_3, \ldots$ is said to *converge in distribution to the random variable* $Y$ if for each real $u$ which is a continuity point for the distribution function $F_Y$ we have

$$\lim_{n \to \infty} F_{X_n}(u) = F_Y(u)$$

We say that the sequence $X_1, X_2, X_3, \ldots$ *converges in distribution* if there is a random variable $Y$ such that $X_1, X_2, X_3, \ldots$ converges in distribution to $Y$.

In view of this definition, Example 10.1.3 may be reformulated as follows: Let $f_n(k/n) = P(X_n = k/n) = 1/n$ for $k = 1, 2, 3, \ldots, n$, i.e., $P(X_n = 1/n) = P(X_n = 2/n) = P(X_n = 3/n) = \cdots = P(X_n = (n-1)/n) = P(X_n = n/n) = 1/n$. Let the random variable $Y$ have uniform distribution on $0 \leq u \leq 1$, i.e., the density function $f_Y$ satisfies

$$f_Y(u) = \begin{cases} 1, & 0 \leq u \leq 1 \\ 0, & \text{otherwise} \end{cases}$$

Then the sequence $X_1, X_2, X_3, \ldots$ converges in distribution to $Y$ since

$$\lim_{n \to \infty} F_{X_n}(u) = \lim_{n \to \infty} P(X_n \leq u) = \begin{cases} 0, & u \leq 0 \\ u, & 0 \leq u \leq 1 = \int_{-\infty}^{u} f_Y(x)\,dx = F_Y(u) \\ 1, & u > 0 \end{cases}$$

If we recast Example 10.1.1 in this fashion, we can let $\lambda > 0$ be fixed and let $X_n$ be a binomial random variable with parameters $n$, $\lambda/n = p$; i.e.,

$$P(X_n = k) = \binom{n}{k}\left(\frac{\lambda}{n}\right)^k \left(1 - \frac{\lambda}{n}\right)^{n-k}$$

for $k = 0, 1, 2, \ldots, n$. Let $Y$ be a Poisson random variable with parameter $\lambda$. We saw after Example 10.1.1 that $P(X_n \leq j) \approx P(Y \leq j)$. A closer inspection

of the calculations in Example 10.1.1 shows that (for fixed $j$) $\lim\limits_{n \to \infty} P(X_n \le j) =$
$P(Y \le j)$ for $j = 0, 1, 2, 3, \ldots$. It is easy to see that if $u$ is any real number, we
have $\lim\limits_{n \to \infty} P(X_n \le u) = P(Y \le u)$; i.e., $\lim\limits_{n \to \infty} F_{X_n}(u) = F_Y(u)$. Thus the sequence
$X_1, X_2, X_3, \ldots$ converges in distribution to $Y$. It is worth observing that
$\lim\limits_{n \to \infty} F_{X_n}(u) = F_Y(u)$ even at discontinuity points of $F_Y$. This is an extra advantage
for some purposes. It is not necessary for convergence in distribution since it
may or may not occur.

As for Example 10.1.2, we could let $X_n = N_n/n$. It would have been just as
easy to conclude that $\lim\limits_{n \to \infty} P(X_n \ge \frac{1}{2} + \delta) = \lim\limits_{n \to \infty} P(N_n/n \ge \frac{1}{2} + \delta) = 0$ for any
$\delta > 0$ as it was when $\delta = 0.001$. Similarly it can be shown that $\lim\limits_{n \to \infty} P(X_n \le \frac{1}{2} - \delta)$
$= 0$ for any $\delta > 0$. Now $P(X_n \le u) = 1 - P(X_n > u)$. If $u > \frac{1}{2}$, $u = \frac{1}{2} + \delta$
where $\delta > 0$. Therefore, $\lim\limits_{n \to \infty} P(X_n \le u) = 1 - \lim\limits_{n \to \infty} P(X_n > \frac{1}{2} + \delta) = 1$. If
$u < \frac{1}{2}$, $u = \frac{1}{2} - \delta$ for some $\delta > 0$ so $\lim\limits_{n \to \infty} P(X_n \le \frac{1}{2} - \delta) = 0$. Therefore

$$\lim_{n \to \infty} F_{X_n}(u) = \begin{cases} 1, & u > \frac{1}{2} \\ 0, & u < \frac{1}{2} \end{cases}$$

However if $P(Y = \frac{1}{2}) = 1$, that is, $Y$ is the constant random variable $\frac{1}{2}$,

$$F_Y(u) = \begin{cases} 1, & u > \frac{1}{2} \\ 0, & u < \frac{1}{2} \\ 1, & u = \frac{1}{2} \end{cases}$$

In any case, $u = \frac{1}{2}$ is the only discontinuity and we have shown that $\lim\limits_{n \to \infty} F_{X_n}(u)$
$= F_Y(u)$ if $u \ne \frac{1}{2}$. Thus the sequence $N_1/1, N_2/2, N_3/3, N_4/4, \ldots$ converges in
distribution to the constant random variable $Y = \frac{1}{2}$.

**EXERCISES 10.1** _____

**1** The production characteristics of a certain bolt are of such a nature that the
probability of producing a defective is 0.001. Assume that bolts are produced
independently and use the Poisson approximation to estimate $P$(fewer than
five defectives in 5000 bolts) and $P$(fewer than seven defectives in 5000 bolts).

**2** Let $X_{40}$ be a binomial random variable with $p = \frac{1}{4}$ and $n = 40$. Find
$P(X_{40} \le 3)$. What is $P(Y \le 3)$ where $Y$ is a Poisson random variable with
$\lambda = np$? Estimate $P(X_{40} \le 3) \div P(Y \le 3)$ using logarithms. Why is the
quotient so far from unity?

**3** Suppose a hypothetical organism is of such a nature that the probability of a
mutant of a certain kind is 0.001. Find, approximately, $P$(fewer than eight
mutants in a population of 5000). Discuss the approximations and any other
assumption which you make.

**4**  In the context of Example 10.1.2 with $n = 1000$, estimate $P(X \le \sqrt{2}/2)$. Estimate $P(X \le \sqrt{2}/2)$ with $n = 5000$ and $10\,000$. If $n = 1000$, find $P(X \le \sqrt{2}/2)$ exactly.

**5**  Let $X_n$ be a random variable with $P(X_n = 1) = \frac{1}{2} + (1/2n)$, $P(X_n = 0) = \frac{1}{2} - (1/2n)$ for $n = 1, 2, 3, \ldots$. Show that the sequence $X_1, X_2, X_3, \ldots$ converges in distribution to the random variable $Y$ where $P(Y=1)=\frac{1}{2}$, $P(Y=0)=\frac{1}{2}$.

**6**  Let $X_n$ be a random variable with $P(X_n = 1 + 1/n) = \frac{1}{2}$, $P(X_n = 1/n) = \frac{1}{2}$. Discuss convergence in distribution of the sequence $X_1, X_2, X_3, \ldots$.

**7**  (Continuation of 6.) Let $X$ be any random variable. Let $a_1, a_2, a_3, \ldots$ be a sequence of real numbers with $\lim_{n \to \infty} a_n = 0$. Discuss convergence in distribution of the sequence $X + a_1, X + a_2, X + a_3, \ldots$.

**8**  Random variables $X_n$ have distribution functions

$$F_{X_n}(u) = \begin{cases} 0, & u < 0 \\ u^n, & 0 \le u \le 1 \\ 1, & u > 1 \end{cases}$$

Discuss convergence in distribution of $X_1, X_2, X_3, \ldots$.

**9**  For $n = 1, 2, 3, \ldots$ random variables $X_n$ have probability functions $f_{X_n}((k/n)^2) = 1/n$, where $k = 1, 2, 3, \ldots, n$. Find $P(X_6 = \frac{1}{4})$, $P(X_3 = \frac{1}{9})$, $P(X_6 \le \frac{1}{4})$, $P(X_6 \le \frac{1}{3})$, and $P(X_{10} \le \frac{2}{3})$. Describe $P(X_n \le u)$ in terms of greatest-integer notation. Find $\lim_{n \to \infty} P(X_n \le u) = \lim_{n \to \infty} F_{X_n}(u)$. What does your conclusion say about convergence in distribution of $X_1, X_2, X_3, \ldots$?

**10**  The random variable $X_n$ has density function

$$f_{X_n}(x) = \begin{cases} 0, & x \le 0 \\ ne^{-nx}, & x > 0 \end{cases}$$

which means $X_n$ is an exponentially distributed random variable with parameter $\lambda = n$. Discuss convergence in distribution of $X_1, X_2, X_3, \ldots$. Do the same if $X_n$ is exponentially distributed with parameter $\lambda = 1/n$.

**♦ 11**  Suppose $f_1, f_2, f_3, \ldots$ are non-negative functions continuous on $\{x; 0 \le x \le 1\}$ with $\int_0^1 f_n(x)\, dx = 1$ for $n = 1, 2, 3, \ldots$. Suppose the sequence $f_1, f_2, f_3, \ldots$ converges uniformly to $f$. Discuss convergence in distribution of the sequence of random variables $X_1, X_2, X_3, \ldots$ with

$$f_{X_n}(x) = \begin{cases} 0, & x < 0 \\ f_n(x), & 0 \le x \le 1 \\ 0, & x > 1 \end{cases}$$

Is uniform convergence essential to convergence in distribution?

**12**  Let $X_n$ be $N(m, 1/n)$. Discuss convergence in distribution of the sequence $X_1, X_2, X_3, \ldots$.

# 10.2 The central limit theorem

The examples of section 10.1 are basically simple. We now state a theorem which deals with convergence in distribution and which has dominated the field of probability for many years.

**Theorem 10.2.1** (Central limit theorem.)    Let $X_1, X_2, X_3, \ldots$ be a sequence of independent, identically distributed random variables where $E(X_n^2)$ exists. Let $E(X_n) = \mu$ and $\sigma^2(X_n) = \sigma^2$. Let

$$Y_n = (X_1 + X_2 + X_3 + \cdots + X_n - n\mu)/\sqrt{n\sigma^2}.$$

Let $Y$ have density function $f_Y(u) = e^{-u^2/2}/\sqrt{2\pi}$ for all real $u$. Then the sequence $Y_1, Y_2, Y_3, \ldots$ converges in distribution to $Y$.

The proof of this theorem is deferred until the next section. At this stage we emphasize its generality. Nothing need be known about the distribution function common to the $X_n$ except that its second moment exists.

### Example 10.2.1
A fair coin is tossed 2500 times. Find, approximately, $P$(number of H's in 2500 tosses $\leq 1300$).

Let

$$X_k = \begin{cases} 1, & \text{if the } k\text{th toss shows H} \\ 0, & \text{if the } k\text{th toss shows T} \end{cases}$$

Let $N_n = X_1 + X_2 + \cdots + X_n$ (the number of H's in $n$ tosses). It is easy to see that $E(X_k) = \frac{1}{2}$, $\sigma^2(X_k) = \frac{1}{4}$. Let

$$Y_n = \frac{1}{\sqrt{n\sigma^2}} (X_1 + X_2 + \cdots + X_n - n\mu)$$

$$= \frac{1}{\sqrt{n \cdot \frac{1}{4}}} (N_n - \tfrac{1}{2}n) = \frac{2}{\sqrt{n}} (N_n - \tfrac{1}{2}n)$$

According to the above theorem $\lim_{n \to \infty} F_{Y_n}(u) = F_Y(u) = \int_{-\infty}^{u} (1/\sqrt{2\pi})e^{-x^2/2} \, dx$.

Since 2500 is quite large we would hope that $F_{Y_{2500}}(u) \approx \int_{-\infty}^{u} (1/\sqrt{2\pi})e^{-x^2/2} \, dx$. Now

$$F_{Y_{2500}}(u) = P(Y_{2500} \leq u) = P\left(\frac{2}{\sqrt{2500}} (N_{2500} - 1250) \leq u\right)$$

We are eventually interested in $P(N_{2500} \leq 1300)$ so we must choose a value of $u$ such that $P\left(\dfrac{2}{\sqrt{2500}} (N_{2500} - 1250) \leq u\right) = P(N_{2500} \leq 1300)$. But

$$P(\tfrac{2}{50}(N_{2500} - 1250) \leq u) = P(N_{2500} - 1250 \leq 25u) = P(N_{2500} \leq 1250 + 25u)$$

Therefore, $u = 2$ will do and we have $P(N_{2500} \leq 1300) \approx \int_{-\infty}^{2} (1/\sqrt{2\pi})e^{-x^2/2} \, dx$
$= 0.977$.

Although $\int (1/\sqrt{2\pi})e^{-x^2/2} \, dx$ cannot be computed in terms of elementary functions there are extensive tables of $\int_{-\infty}^{u} (1/\sqrt{2\pi})e^{-x^2/2} \, dx$ for values of $u$ between 0 and 4. These suffice for most purposes. One such table is Table 1 which appears in the appendix.

### Example 10.2.2

Pears which grow in a certain region and which ripen at a certain time have varying weight; e.g., the weight of a typical pear is given by the random variable $X$ where $E(X) = \frac{1}{2}$ and $\sigma^2(X) = \frac{4}{25}$. What is the probability that a batch of 10 000 pears selected at random weighs less than 4800?

We assume that the weight of any group of pears is independent of that of another group of pears. Let $X_k$ denote the weight of the $k$th pear. According to the central limit theorem

$$\lim_{n \to \infty} P\left(\frac{1}{\sqrt{n \cdot \frac{4}{25}}} (X_1 + X_2 + \cdots + X_n - \tfrac{1}{2}n) \leq u\right) = \int_{-\infty}^{u} \frac{1}{\sqrt{2\pi}} e^{-x^2/2} \, dx$$

Thus we would expect that

$$P\left(\frac{1}{\sqrt{10\,000 \cdot \frac{4}{25}}} (\text{weight of } 10\,000 \text{ pears} - 5000) \leq u\right) \approx \int_{-\infty}^{u} \frac{1}{\sqrt{2\pi}} e^{-x^2/2} \, dx$$

We must find $u$. If we let $W =$ weight of 10 000 pears, we have

$$\frac{1}{\sqrt{10\,000 \cdot \frac{4}{25}}} (W - 5000) \leq u$$

or $W - 5000 \leq \frac{2}{5}100u = 40u$ so $W \leq 5000 + 40u$. Since we eventually want $P(W \leq 4800)$, we want to choose $u$ so that $4800 = 5000 + 40u$; i.e., $-200 = 40u$ and $u = -5$. It can be shown that $\int_{-\infty}^{-5} (1/\sqrt{2\pi})e^{-x^2/2} \, dx < 10^{-5}$ although it should be allowed that the error which one makes in approximating $P(W \leq 4800)$ by $\int_{-\infty}^{-5} (1/\sqrt{2\pi})e^{-x^2/2} \, dx$ is not so small. We will not take up the matter of the error in the approximation here. Suffice it to say that for $n = 10\,000$ one usually obtains very good accuracy.

One should be able to formulate many examples of the above type. Moreover, one can organize the statement of the theorem so as to make computation very easy. First of all, we recall that

$$P(a < Y_n \leq b) = P(Y_n \leq b) - P(Y_n \leq a) = F_{Y_n}(b) - F_{Y_n}(a)$$

Therefore

$$\lim_{n \to \infty} P(a < Y_n \le b) = \lim_{n \to \infty} [F_{Y_n}(b) - F_{Y_n}(a)] = F_Y(b) - F_Y(a)$$

$$= P(a < Y \le b)$$

where the density function of $Y$ is $(1/\sqrt{2\pi})e^{-x^2/2}$. Thus we have

**(10.2.1)** 
$$\lim_{n \to \infty} P\left(a < \frac{1}{\sqrt{n\sigma^2}}(X_1 + X_2 + \cdots + X_n - n\mu) \le b\right)$$

$$= \int_a^b \frac{1}{\sqrt{2\pi}} e^{-x^2/2} \, dx$$

where the notation is that of Theorem 10.2.1. Now

$$P\left(a < \frac{1}{\sqrt{n\sigma^2}}(X_1 + X_2 + \cdots + X_n - n\mu) \le b\right)$$

$$= P(a\sqrt{n\sigma^2} + n\mu < X_1 + X_2 + \cdots + X_n \le b\sqrt{n\sigma^2} + n\mu)$$

So, in problems in which we want to estimate $P(q < X_1 + X_2 + \cdots + X_n \le r)$ for definite $n, q, r$ we set $q = a\sqrt{n\sigma^2} + n\mu$ and $r = b\sqrt{n\sigma^2} + n\mu$. The solutions are

**(10.2.2)** 
$$a = \frac{q - n\mu}{\sqrt{n}\sigma}, \qquad b = \frac{r - n\mu}{\sqrt{n}\sigma}$$

and $\int_a^b (1/\sqrt{2\pi})e^{-x^2/2} \, dx$ gives the approximate value which we seek.

### Example 10.2.3

Four hundred and twenty fair dice (assumed independent) are rolled. Find, approximately, the probability that the total number of dots showing is between 1435 and 1498.

If $X_k$ is the number of dots showing on the $k$th die we have $E(X_k) = \frac{7}{2}$ and $\sigma^2(X_k) = \frac{91}{6} - \frac{49}{4} = \frac{1}{12}(182 - 147) = \frac{35}{12}$ so $\sqrt{n\sigma^2} = \sqrt{\frac{35}{12} \cdot 420} = \sqrt{35 \cdot 35} = 35$. Therefore, $a = \frac{1}{35}(1435 - 1470) = -1$ and $b = \frac{1}{35}(1498 - 1470) = \frac{28}{35} = 0.8$. Accordingly our approximation is

$$\int_{-1}^{0.8} \frac{1}{\sqrt{2\pi}} e^{-x^2/2} \, dx = \int_0^{0.8} \frac{1}{\sqrt{2\pi}} e^{-x^2/2} \, dx + \int_{-1}^0 \frac{1}{\sqrt{2\pi}} e^{-x^2/2} \, dx$$

$$= \int_0^{0.8} \frac{1}{\sqrt{2\pi}} e^{-x^2/2} \, dx + \int_0^1 \frac{1}{\sqrt{2\pi}} e^{-x^2/2} \, dx$$

$$= \int_{-\infty}^{0.8} \frac{1}{\sqrt{2\pi}} e^{-x^2/2} \, dx - \left(1 - \int_{-\infty}^1 \frac{1}{\sqrt{2\pi}} e^{-x^2/2} \, dx\right) \approx 0.63$$

In Example 10.2.3 the words "between 1435 and 1498" were interpreted as $1435 <$ number of dots showing $\leq 1498$. It is legitimate to ask about the approximation if the question asked for $1435 <$ number of dots showing $<$ 1498. Should this be interpreted as $1435 <$ number of dots showing $\leq 1497$? There are various conventions made to handle such questions. They have little to do with the limit theorem in general.

### Example 10.2.4

Let $X_1, X_2, X_3, \ldots$ be independent, identically distributed random variables whose common probability function is

$$f(k) = (-1)^k \binom{-\tfrac{1}{3}}{k}(\tfrac{1}{2})^k(\tfrac{1}{2})^{1/3} \qquad k = 0, 1, 2, 3, \ldots$$

Find, approximately, $P(210 < X_1 + X_2 + \cdots + X_{600} \leq 240)$.

By exercise 1.5.4

$$\phi_{X_n}(t) = \sum_{k=0}^{\infty} (-1)^k \binom{-\tfrac{1}{3}}{k}(\tfrac{1}{2})^k(\tfrac{1}{2})^{1/3}e^{ikt} = (\tfrac{1}{2})^{1/3}(1 - \tfrac{1}{2}e^{it})^{-1/3} = (2 - e^{it})^{-1/3}$$

$$\phi'_{X_n}(t) = -\tfrac{1}{3}(-ie^{it})(2 - e^{it})^{-4/3}$$

and

$$\phi''_{X_n}(t) = \tfrac{1}{3}[ie^{it}(2 - e^{it})^{-4/3} + e^{it}(-\tfrac{4}{3})(-ie^{it})(2 - e^{it})^{-7/3}]$$

Therefore $\phi'_{X_n}(0) = \tfrac{1}{3}i$, $E(X_n) = \tfrac{1}{3}$, $n\mu = \tfrac{1}{3} \cdot 600 = 200$, and $\sigma^2(X_n) = -\phi''_{X_n}(0) - (\tfrac{1}{3})^2 = \tfrac{2}{3}$. Since $n = 600$ and $n\sigma^2 = 400$ we want $a = (210 - 200)/20 = 0.5$ and $b = (240 - 200)/20 = 2$. From the tables we see that

$$\int_{0.5}^{2} \frac{1}{\sqrt{2\pi}} e^{-x^2/2}\,dx = \int_{0}^{2} \frac{1}{\sqrt{2\pi}} e^{-x^2/2}\,dx - \int_{0}^{0.5} \frac{1}{\sqrt{2\pi}} e^{-x^2/2}\,dx \approx 0.28$$

where we are using the central limit theorem since $n = 600$ is rather sizeable.

It is natural to ask how large $n$ should be in order that the central limit theorem provides a good approximation to $P(a < X_1 + X_2 + \cdots + X_n \leq b)$? There are two meaningful answers. One, the conventional and empirical, includes the commonly accepted rule of thumb that $n \geq 25$ is good providing that the $X_j$ have a fairly smooth and symmetrical density function. We will not comment on the discrete case because a correction should be made to our method of handling end points before accuracy is discussed. We take up the second meaningful answer in section 10.3.

**EXERCISES 10.2**

1  The experience of a hypothetical airline is that the distribution of the weight in pounds of a random passenger with baggage has an expected value of 200 and a variance of 400. Estimate the probability that 144 passengers and their baggage will exceed 30 000 lbs. Assume independence.

**2**   A fair die is rolled 1800 times. Estimate the probability that the faces 5 or 6 will show more than 500 times but no more than 650 times.

**3**   Suppose that the probability of $X$ cosmic ray particles of a certain type hitting a region in 1 sec is given by a Poisson distribution with parameter $\lambda$, i.e., $P(X = n) = e^{-\lambda}\lambda^n/n!$. If the region is suitably chosen we could arrange for $\lambda = 1$. Assuming independence, estimate the probability that more than 3780 of those particles hit the region in 1 hour. What is the probability of fewer than 3480 of the particles hitting the region in an hour?

**4**   Suppose that the energy $E$ of one of the above particles is a random variable such that (in certain units) $P(E = 1) = 0.05$, $P(E = 2) = 0.15$, $P(E = 3) = 0.40$, $P(E = 4) = 0.30$, and $P(E = 5) = 0.10$. Estimate the probability that the total energy of $10^4$ of such particles is between $(3.1) \times 10^4$ and $(3.3) \times 10^4$ in the same units if the energies of different particles are independent.

**5**   Let $X_1, X_2, X_3, \ldots$ be independent random variables each having density function $f(x) = (1/\sqrt{2\pi})e^{-x^2/2}$. Estimate $P(X_1^2 + X_2^2 + \cdots + X_{800}^2 \geq 880)$.

**6**   Let $X_1, X_2, X_3, \ldots$ be independent random variables each having the uniform density on the interval $\{x; 0 < y < 1\}$. Let $Y_n = X_1 X_2 X_3 \cdots X_n$. Estimate $P(e^{-420} < Y_{400} < e^{-360})$. [*Hint:* Let $Z_k = \ln X_k$.]

Exercises 7–10 take up a routine for rapid calculation of the approximation given by the central limit theorem. We call it *counting standard deviations*.

**7**   Let $X_1, X_2, \ldots, X_n$ be independent, identically distributed random variables. Suppose that $\sigma^2(X_j) = \sigma^2$. What is the standard deviation of $Z = X_1 + X_2 + \cdots + X_n$? Let $E(X_j) = \mu$. What is the expectation of $Z$?

**8**   Consider the random variable $Z$ of exercise 7. Suppose that we wish to estimate $P(q < Z \leq r)$. Show that we need only let $a$ = number of standard deviations of $Z$ that $q$ is from $E(Z)$ and $b$ = number of standard deviations of $Z$ that $r$ is from $E(Z)$, where $a$ and $b$ may be negative. Then show that $P(q < Z \leq r)$ is approximately the probability that an $N(0, 1)$ random variable gives a value between $a$ and $b$ standard deviation from its expected value.

**9**   Let $Z$ be an exponential random variable with $f_Z(x) = 0.01e^{-x/100}$ for $x > 0$. We have seen that $E(Z) = 100$ and $\sigma^2(Z) = 10\,000$ so $\sigma(Z) = 100$. Suppose that we wish to approximate $P(50 < Z \leq 200)$. By counting standard deviations we get $a = -\frac{1}{2}$ and $b = 1$. By Table 1 we get $0.34134 + 0.19146 = 0.53280$. However, if we do not approximate we get

$$\int_{50}^{200} 0.01e^{-x/100}\,dx = e^{-1/2} - e^{-2} = 0.47099$$

to five decimal places. This is an error of about 13%. What, if anything, is wrong?

**10**   Count the number of standard deviations in the following:

(a)   $E(Z) = 750$, $\sigma^2(Z) = 400$, $q = 660$, $r = 790$

(b)   $E(Z) = -750$, $\sigma^2(Z) = 400$, $q = -780$, $r = -700$

(c) $E(Z) = 750, \quad \sigma^2(Z) = 400, \quad q = 660, \quad r = 720$

(d) $E(Z) = -750, \quad \sigma^2(Z) = 400, \quad q = -780, \quad r = -760$

**11** What would you want to know about the $Z$ in exercise 10 before using Table 1 with the number of standard deviations which you found if you want a good approximation?

◆ **12** Let $X_1, X_2, X_3, \ldots$ be a sequence of independent, identically distributed random variables with $E(X_k) = \mu$ and $\sigma^2(X_k) = \sigma^2$. Show that the sequence of $(1/n)(X_1 + X_2 + \cdots + X_n - n\mu)$ converges in distribution to a random variable which is $N(0, \sigma^2)$.

# 10.3 Proof of the central limit theorem

A rigorous proof of the central limit theorem is beyond the scope of this book. One must either introduce (and prove) many advanced results or go into many detailed calculations. However, we can take up certain ideas related to a proof of Theorem 10.2.1 so that the reader gains insight into other limit theorems as well as Theorem 10.2.1.

We first state an important theorem without proof. It is sometimes called the *continuity theorem* for characteristic functions.

**Theorem 10.3.1** Let $Y_1, .Y_2, Y_3, \ldots$ be a sequence of random variables and let $Y$ be a random variable. Then the sequence $Y_1, Y_2, Y_3, \ldots$ converges in distribution to $Y$ if and only if for each real $t$ we have

$$\lim_{n \to \infty} \phi_{Y_n}(t) = \phi_Y(t)$$

that is, the sequence of characteristic functions $\phi_{Y_1}, \phi_{Y_2}, \ldots$ converges pointwise to $\phi_Y$.

*Example 10.3.1*

Let $Y_n$ be binomially distributed with parameters $n$ and $\lambda/n$, where $\lambda > 0$ is fixed and $n > \lambda$. Let $Y$ have a Poisson distribution with parameter $\lambda$. Show that the sequence $Y_{[\lambda]+1}, Y_{[\lambda]+2}, Y_{[\lambda]+3}, \ldots$ converges to $Y$ in distribution by using Theorem 10.3.1.

We have seen before that $\phi_Y(t) = \exp(-\lambda + \lambda e^{it})$ and that $\phi_{Y_n}(t) = \left(1 - \dfrac{\lambda}{n} + \dfrac{\lambda}{n} e^{it}\right)^n$. Now it was pointed out that $(1 + z/n)^n \to e^z$ for any complex $z$ as $n \to \infty$. Therefore, for each real $t$

$$\lim_{n \to \infty} \phi_{Y_n}(t) = \lim_{n \to \infty} \left(1 + \frac{\lambda e^{it} - \lambda}{n}\right)^n = \exp(\lambda e^{it} - \lambda) = \exp(-\lambda + \lambda e^{it})$$

By Theorem 10.3.1, $Y_{[\lambda]+1}, Y_{[\lambda]+2}, Y_{[\lambda]+3}, \ldots$ converges to $Y$ in distribution.

It is true that this is another proof of a previous result but it does show how Theorem 10.3.1 is used.

### Example 10.3.2

Let $P(Y_n = k/n) = 1/n$ for $k = 1, 2, 3, \ldots, n$ and for $n = 1, 2, 3, 4, \ldots$. Let $Y$ have the density function

$$f_Y(x) = \begin{cases} 1, & 0 \le x \le 1 \\ 0, & \text{otherwise} \end{cases}$$

Discuss in the light of Theorem 10.3.1.

It was shown in Example 10.1.3 that $Y_1, Y_2, Y_3, \ldots$ converges in distribution to $Y$. We should, therefore, be able to show that for each real $t$, $\lim_{n \to \infty} \phi_{Y_n}(t) = \phi_Y(t)$. By the definition,

$$\phi_{Y_n}(t) = [e^{it(1/n)} + e^{it(2/n)} + e^{it(3/n)} + \cdots + e^{it(n/n)}] \frac{1}{n}$$

Now $e^{it(k/n)} = (e^{it/n})^k$ so we have the sum of a geometric progression $e^{it/n}$, $(e^{it/n})^2$, $(e^{it/n})^3, \ldots, (e^{it/n})^n$ in the brackets. If $t/n = 2m\pi$ for some integer $m$, $e^{it/n} \ne 1$. Now unless $t = 0$, $t/n \ne 2m\pi$ for sufficiently large $n$, which is our primary interest. By elementary algebra

$$y + y^2 + y^3 + \cdots + y^n = \frac{y^{n+1} - y}{y - 1}$$

if $y \ne 1$. Therefore if $t \ne 0$,

$$\phi_{Y_n}(t) = \frac{(e^{it/n})^{n+1} - e^{it/n}}{n(e^{it/n} - 1)}$$

for sufficiently large $n$. Now $(e^{it/n})^{n+1} - e^{it/n} = e^{it/n}[(e^{it/n})^n - 1] = e^{it/n}(e^{it} - 1)$ and $e^{it/n} - 1 = it/n + i^2t^2/2!n^2 + i^3t^3/3!n^3 + \cdots$. Thus

$$\phi_{Y_n}(t) = \frac{e^{it/n}(e^{it} - 1)}{n(e^{it/n} - 1)} = \frac{e^{it/n}(e^{it} - 1)}{n\left(\dfrac{it}{n} + \dfrac{i^2t^2}{2!n^2} + \dfrac{i^3t^3}{3!n^3} + \cdots\right)}$$

$$= \frac{e^{it/n}(e^{it} - 1)}{it + \dfrac{i^2t^2}{2!n} + \dfrac{i^3t^3}{3!n^2} + \dfrac{i^4t^4}{4!n^3} + \cdots}$$

Since $\lim_{n \to \infty} e^{it/n} = e^0 = 1$, we have that for $t \ne 0$, $\lim_{n \to \infty} \phi_{Y_n}(t) = (e^{it} - 1)/it$. As for $t = 0$, $\phi(0) = 1$ for any characteristic function $\phi$. Thus $\lim_{n \to \infty} \phi_{Y_n}(0) = 1$.

We now look at $\phi_Y(t) = \int_0^1 e^{itx} \cdot 1 \, dx$. If $t \ne 0$,

$$\frac{d}{dx} \frac{e^{itx}}{it} = e^{it}$$

so

$$\int_0^1 e^{itx}\,dx = \frac{e^{itx}}{it}\Big|_0^1 = \frac{e^{it} - e^0}{it} = \frac{e^{it} - 1}{it}$$

If $t = 0$, $\int_0^1 e^{itx}\,dx = \int_0^1 1\,dx = 1$. Thus $\lim_{n\to\infty} \phi_{Y_n}(t) = \phi(t)$ for each real $t$ which is exactly what we would expect.

Another approach to Example 10.3.2 goes as follows:

$$\phi_{Y_n}(t) = \sum_{k=1}^n e^{it(k/n)}\frac{1}{n} = \sum_{k=1}^n \left(\cos\frac{tk}{n} + i\sin\frac{tk}{n}\right)\frac{1}{n}$$

$$= \sum_{k=1}^n \left(\cos\frac{tk}{n}\right)\left(\frac{k}{n} - \frac{k-1}{n}\right) + i\sum_{k=1}^n \left(\sin\frac{tk}{n}\right)\left(\frac{k}{n} - \frac{k-1}{n}\right)$$

A sum of the type

$$\sum_{k=1}^n h\left(\frac{k}{n}\right)\left(\frac{k}{n} - \frac{k-1}{n}\right)$$

is the kind of sum which one gets when he is approximating the integral $\int_0^1 h(x)\,dx$ by dividing the interval $0 \le x \le 1$ into $n$ equal subdivisions and evaluating $h$ at the right-hand end points of each subdivision. If $h$ is continuous, it is known that such sums converge to $\int_0^1 h(x)\,dx$ as $n \to \infty$. Thus

$$\lim_{n\to\infty} \phi_{Y_n}(t) = \int_0^1 \cos tx\,dx + i\int_0^1 \sin tx\,dx = \int_0^1 (\cos tx + i\sin tx)\,dx$$

$$= \int_0^1 e^{itx}\,dx = \phi_Y(t)$$

### Example 10.3.3

Ten thousand identical slips of paper are marked with numbers and put in a box. All but four are marked with 0 and those four are marked 1, 2, 3, 4. A slip is drawn at random and the player is paid the amount marked on the slip. That slip is returned to the box and all slips are stirred very well. This procedure is repeated for a total of 10 000 plays. Estimate, from the viewpoint of Theorem 10.3.1, $P(\text{total winnings} \le 10)$.

Let $P(X_k = 0) = 1 - 4/n$, $P(X_k = 1) = P(X_k = 2) = P(X_k = 3) = P(X_k = 4) = 1/n$ for $k = 1, 2, 3, \ldots, n$. It is possible to use the central limit theorem. We leave it to the reader to estimate $P(\text{total winnings} \le 10)$ in that fashion. Let $Y_n = X_1 + X_2 + \cdots + X_n$. Then $\phi_{Y_n}(t) = [\phi_{X_1}(t)]^n$ by the independence of the $X_i$ and their identical distribution. By remarks in Chapter 9,

$$\phi_{Y_k}(t) = \left(1 - \frac{4}{n}\right) + e^{it}\frac{1}{n} + e^{2it}\frac{1}{n} + e^{3it}\frac{1}{n} + e^{4it}\frac{1}{n}$$

$$= 1 + \frac{1}{n}(-4 + e^{it} + e^{2it} + e^{3it} + e^{4it})$$

and

$$\phi_{Y_n}(t) = [\phi_{X_1}(t)]^n = \left[1 + \frac{1}{n}(-4 + e^{it} + e^{2it} + e^{3it} + e^{4it})\right]^n$$

Since $\lim_{n\to\infty} (1 + a/n)^n = e^a$ we have

$$\lim_{n\to\infty} \phi_{Y_n}(t) = \exp\left(-4 + e^{it} + e^{2it} + e^{3it} + e^{4it}\right)$$

$$= e^{-1+e^{it}}e^{-1+e^{2it}}e^{-1+e^{3it}}e^{-1+e^{4it}}$$

By Chapter 9, $e^{-\lambda(1-e^{it})}$ is the characteristic function of a Poisson random variable with parameter $\lambda > 0$. Furthermore $e^{-\lambda(1-e^{ikt})}$ is the characteristic function of $k$ times a Poisson random variable with parameter $\lambda > 0$. Thus if $Z_1, Z_2, Z_3, Z_4$ are independent Poisson random variables with parameter $\lambda = 1$, we have

$$\phi_{Z_1 + 2Z_2 + 3Z_3 + 4Z_4}(t) = e^{-1+e^{it}}e^{-1+e^{2it}}e^{-1+e^{3it}}e^{-1+e^{4it}}$$

Therefore $Y_1, Y_2, Y_3, \ldots$ converges in distribution to $Y = Z_1 + 2Z_2 + 3Z_3 + 4Z_4$ as given. It is probably most feasible to estimate $P(Y_{10\,000} \leq 10)$, which is really what the problem calls for, by approximating it by $P(Y \leq 10) = P(Z_1 + 2Z_2 + 3Z_3 + 4Z_4 \leq 10)$. This last quantity can be found by adding the coefficients of $e^{ijt}$ for $j \leq 10$ in the characteristic function

$$e^{-4}\left(1 + e^{it} + \frac{e^{2it}}{2!} + \frac{e^{3it}}{3!} + \cdots\right)\left(1 + e^{2it} + \frac{e^{4it}}{2!} + \frac{e^{6it}}{3!} + \cdots\right)$$

$$\times \left(1 + e^{3it} + \frac{e^{6it}}{2!} + \frac{e^{9it}}{3!} + \cdots\right)\left(1 + e^{4it} + \frac{e^{8it}}{2!} + \frac{e^{12it}}{3!} + \cdots\right)$$

or in

$$e^{-4}\left[1 + e^{it} + e^{2it} + e^{3it} + e^{4it} + \frac{(e^{it} + e^{2it} + e^{3it} + e^{4it})^2}{2!}\right.$$

$$\left. + \frac{(e^{it} + e^{2it} + e^{3it} + e^{4it})^3}{3!} + \cdots\right]$$

We get

$$e^{-4}\left(5 + \frac{16}{2!} + \frac{60}{3!} + \cdots\right) \approx 0.567$$

as the approximation.

### Example 10.3.4

Let $X_1, X_2, X_3, \ldots$ be independent, identically distributed random variables with $E(X_k)$ existing. Does $Y_n = (1/n)(X_1 + X_2 + \cdots + X_n)$ converge in distribution?

If $E(X_k^2)$ exists we would know that $P(|Y_n - E(X_1)| \geq \epsilon) \to 0$ for each $\epsilon > 0$ as $n \to \infty$ and that this implies convergence in distribution to the constant random variable $Y = E(X_1)$. Let $\phi(t) = E(e^{itX_k}) = E(e^{itX_1})$ where we are using the identical distribution of the $X_k$. Therefore,

$$\phi_{Y_n}(t) = \phi_{(1/n)(X_1 + X_2 + \cdots + X_n)}(t) = [\phi(t/n)]^n$$

by the results of Chapter 9. By definition,

$$
\begin{aligned}
E(e^{iX_1 t}) &= E(1 + itX_1 + e^{itX_1} - 1 - itX_1) \\
&= E(1) + itE(X_1) + E(e^{itX_1} - 1 - itX_1) \\
&= 1 + itE(X_1) + E(\cos tX_1 - 1 + i(\sin tX_1 - tX_1))
\end{aligned}
$$

Let $E(X_1) = m$. Since $\phi(t) = E(e^{iX_1 t})$,

$$\phi\left(\frac{t}{n}\right) = 1 + i\frac{tm}{n} + E\left(\cos\frac{t}{n}X_1 - 1 + \left(\sin\frac{t}{n}X_1 - \frac{t}{n}X_1\right)\right)$$

It is easily verified by trigonometry that $|\sin u - u| \leq 2u$ and $|\cos u - 1| \leq 2u$ for all $u$. Furthermore $\lim_{u \to 0} [(\sin u - u)/u] = 0$ and $\lim_{u \to 0} [(\cos u - 1)/u] = 0$. Therefore we may write

$$E\left(\cos\frac{t}{n}X_1 - 1\right) = E\left(\frac{\cos\frac{t}{n}X_1 - 1}{\frac{t}{n}X_1}\frac{t}{n}X_1\right) = \frac{t}{n}E\left(\frac{\cos\frac{t}{n}X_1 - 1}{\frac{t}{n}X_1}X_1\right)$$

and

$$
\begin{aligned}
E\left(\sin\frac{t}{n}X_1 - \frac{t}{n}X_1\right) &= E\left(\frac{\sin\frac{t}{n}X_1 - \frac{t}{n}X_1}{\frac{t}{n}X_1}\frac{t}{n}X_1\right) \\
&= \frac{t}{n}E\left(\frac{\sin\frac{t}{n}X_1 - \frac{t}{n}X_1}{\frac{t}{n}X_1}X_1\right)
\end{aligned}
$$

Theorems in the theory of measure and integration which we will shortly cite imply that for each $t$,

$$\lim_{n \to \infty} E\left(\frac{\cos\frac{t}{n}X_1 - 1}{\frac{t}{n}X_1}X_1\right) = 0$$

and

$$\lim_{n \to \infty} E\left( \frac{\sin \frac{t}{n} X_1 - \frac{t}{n} X_1}{\frac{t}{n} X_1} X_1 \right) = 0$$

Therefore, we can write, for each $t$

$$\phi\left(\frac{t}{n}\right) = 1 + \frac{itm}{n} + \frac{t}{n}[q_1(n) + iq_2(n)]$$

where

$$q_1(n) = E\left( \frac{\cos \frac{t}{n} X_1 - 1}{\frac{t}{n} X_1} X_1 \right), \qquad q_2(n) = E\left( \frac{\sin \frac{t}{n} X_1 - \frac{t}{n} X_1}{\frac{t}{n} X_1} X_1 \right)$$

so $\lim_{n \to \infty} q_j(n) = 0$ for $j = 1, 2$. Therefore,

$$\phi_{Y_n}(t) = \left\{ 1 + \frac{itm}{n} + \frac{t}{n}[q_1(n) + iq_2(n)] \right\}^n$$

Since $\lim_{n \to \infty} (1 + a/n)^n = e^a$, and since $\lim_{n \to \infty} t[q_1(n) + iq_2(n)] = 0$, it is easy to show that

$$\lim_{n \to \infty} \left\{ 1 + \frac{itm}{n} + \frac{t}{n}[q_1(n) + iq_2(n)] \right\}^n = e^{itm}$$

However, $e^{itm}$ is the characteristic function of the constant random variable $Y = m = E(X_1)$. Thus $Y_1, Y_2, Y_3, \ldots$ converges in distribution to the constant random variable $Y$. This is a form of the weak law of large numbers without the assumption that $\sigma^2$ exists.

In Examples 10.3.3 and 10.3.4, we have used the following approach: if $Y_1, Y_2, Y_3, \ldots$ form a sequence of random variables and if $\lim_{n \to \infty} \phi_{Y_n}(t) = \phi(t)$ can be recognized as the characteristic function of a random variable $Y$, then $Y_1, Y_2, Y_3, \ldots$ converges to $Y$ in distribution. There are cases in which it is easy to find $\lim_{n \to \infty} \phi_{Y_n}(t)$ and not so easy to recognize $\phi$ as the characteristic function of a random variable. There are also cases in which it is hard to find $\lim_{n \to \infty} \phi_{Y_n}(t) = \phi(t)$, but once it is found, it is easy to find $Y$. There are cases where everything is difficult and there are cases in which the use of characteristic functions is disadvantageous.

Before we proceed we state two theorems which we do not prove. These theorems were used in the discussion of Example 10.3.4. Theorem 10.3.3 is relatively easy to prove but Theorem 10.3.2 is more difficult.

**Theorem 10.3.2**    If $V$ is a random variable on a probability system $S$, $\mathscr{A}$, $P$ such that $E(V)$ exists and if $U_1$, $U_2$, $U_2$, ... are random variables on the same system which are uniformly bounded, say $|U_n(s)| \leq b$ for each positive integer $n$ and each $s \in S$ ($b$ is independent of $n$ and $s$) and if for each $s \in S$, $\lim_{n \to \infty} U_n(s) = 0$, then $\lim_{n \to \infty} E(U_n V) = 0$.

**Theorem 10.3.3**    If $a$ is a fixed complex number and $c_1$, $c_2$, $c_3$, ... is a sequence of complex numbers with $\lim_{n \to \infty} c_n = 0$, then

$$\lim_{n \to \infty} \left[ 1 + \frac{1}{n}(a + c_n) \right]^n = e^a$$

We now sketch a proof of the central limit theorem. Let $Z_1, Z_2, Z_3, \ldots$ be independent, identically distributed random variables. Let $Z = Z_1$ and suppose $E(Z) = 0$ and $E(Z^2) = \sigma^2(Z) = 1$. Let $Y_n = (1/\sqrt{n})(Z_1 + Z_2 + \cdots + Z_n)$ and let $\phi(t) = \phi_Z(t)$. Then, in analogy to Example 10.3.3, we have $\phi_{Y_n}(t) = [\phi(t/\sqrt{n})]^n$ since $Y_n = Z_1/\sqrt{n} + Z_2/\sqrt{n} + \cdots + Z_n/\sqrt{n}$ and $\phi_{Z_k/\sqrt{n}}(t) = \phi(t/\sqrt{n})$. Furthermore

$$\phi(t) = E(e^{itZ}) = E\left( 1 + itZ + \frac{i^2 t^2}{2} Z^2 + e^{itZ} - 1 - itZ - \frac{i^2 t^2}{2} Z^2 \right)$$

$$= E(1) + it E(Z) + \frac{i^2 t^2}{2} E(Z^2) + E\left( e^{itZ} - 1 - itZ - \frac{i^2 t^2}{2} Z^2 \right)$$

$$= 1 + it \cdot 0 - \frac{t^2}{2} + E\left( \cos tZ - 1 + \frac{t^2 Z^2}{2} + i(\sin tZ - tZ) \right)$$

So

(10.3.1)    $\phi\left( \dfrac{t}{\sqrt{n}} \right) = 1 - \dfrac{t^2}{2n} + E\left( \cos \dfrac{tZ}{\sqrt{n}} - 1 + \dfrac{t^2 Z^2}{2n} + i\left( \sin \dfrac{tZ}{\sqrt{n}} - \dfrac{tZ}{\sqrt{n}} \right) \right)$

It is easy to verify that for any real $u$, $|(\sin u - u)/u^2| \leq 2$, and

$$\lim_{u \to 0} \frac{\sin u - u}{u^2} = 0$$

as well as to verify that

$$\left| \frac{\cos u - 1 + \frac{1}{2} u}{u^2} \right| = \left| \frac{-2 \cdot \sin^2 \frac{1}{2} u + \frac{1}{2} u^2}{u^2} \right| \leq \frac{2|\sin^2 \frac{1}{2} u| + \frac{1}{2}|u^2|}{u^2} \leq 1$$

since $|\sin v/v| \leq 1$. Moreover $\lim_{u \to 0} [(\cos u - 1 + \frac{1}{2} u^2)/u^2] = 0$. Therefore

$$E\left( \cos \frac{tZ}{\sqrt{n}} - 1 + \frac{t^2 Z^2}{2n} \right) = E\left( \frac{\cos \dfrac{tZ}{\sqrt{n}} - 1 + \dfrac{t^2 Z^2}{2n}}{(tZ/\sqrt{n})^2} \left( \frac{tZ}{\sqrt{n}} \right)^2 \right)$$

$$= \frac{t^2}{n} E\left( \frac{\cos \dfrac{tZ}{\sqrt{n}} - 1 + \dfrac{t^2 Z^2}{2n}}{(tZ/\sqrt{n})^2} Z^2 \right)$$

We let

$$p_1(n) = E\left(\frac{\cos\dfrac{tZ}{\sqrt{n}} - 1 + \dfrac{t^2 Z^2}{2n}}{(tZ/\sqrt{n})^2}\, Z^2\right)$$

and apply Theorem 10.3.2 with $V = Z^2$ [recalling $E(Z^2) = 1$] and

$$U_n = \frac{\cos\dfrac{tZ}{\sqrt{n}} - 1 + \dfrac{t^2 Z^2}{2n}}{(tZ/\sqrt{n})^2}$$

As we saw above, $|U_n(s)| \le 1$ since

$$U_n(s) = \frac{\cos\dfrac{tZ(s)}{\sqrt{n}} - 1 + \dfrac{t^2 [Z(s)]^2}{2n}}{(tZ(s)/\sqrt{n})^2}$$

We take $(\cos u - 1 + \tfrac{1}{2}u^2)/u^2$ to be 0 when $u = 0$ since

$$\lim_{u\to 0} \frac{\cos u - 1 + \tfrac{1}{2}u^2}{u^2} = 0$$

This gives a continuous extension to $u = 0$. Since $\lim_{u\to 0} [(\cos u - 1 + \tfrac{1}{2}u^2)/u^2] = 0$ we have that for each $s \in S$, $\lim_{n\to\infty} U_n(s) = 0$. Thus, by Theorem 10.3.2, $\lim_{n\to\infty} p_1(n)$ $= \lim_{n\to\infty} E(U_n V) = 0$. Similarly, we may now let

$$U_n = \frac{\sin\dfrac{tZ}{\sqrt{n}} - \dfrac{tZ}{\sqrt{n}}}{(tZ/\sqrt{n})^2}$$

and $V = Z^2$ and let

$$p_2(n) = E\left(\frac{\sin\dfrac{tZ}{\sqrt{n}} - t\dfrac{Z}{\sqrt{n}}}{(tZ/\sqrt{n})^2}\, Z^2\right)$$

But Theorem 10.3.2 and the above remarks guarantee that $\lim_{n\to\infty} p_2(n) = 0$. Thus (10.3.1) can be rewritten as

(**10.3.2**)   $$\phi\left(\frac{t}{\sqrt{n}}\right) = 1 - \frac{t^2}{2n} + \frac{t^2}{n} p_1(n) + \frac{it^2}{n} p_2(n)$$

Since $\lim_{n\to\infty} p_1(n) = 0$ and $\lim_{n\to\infty} p_2(n) = 0$ then $\lim_{n\to\infty} [t^2 p_1(n) + it^2 p_2(n)] = 0$ and by Theorem 10.3.3, for each real $t$,

$$\lim_{n\to\infty} \phi\left(\frac{t}{\sqrt{n}}\right) = \lim_{n\to\infty} \left(1 + \frac{1}{n}\left\{-\frac{t^2}{2} + t^2[p_1(n) + ip_2(n)]\right\}\right)^n = e^{-t^2/2}$$

Now $e^{-t^2/2}$ gives the characteristic function of a random variable $Z_0$ whose density function $f(x) = (1/\sqrt{2\pi})e^{-x^2/2}$. Since $Z_0$ has a density function its distribution function is continuous for all reals and we have by Theorem 10.3.1, that if $Y_n = (1/\sqrt{n})(Z_1 + Z_2 + \cdots + Z_n)$,

$$\lim_{n\to\infty} P(Y_n \le u) = \int_{-\infty}^{u} \frac{1}{\sqrt{2\pi}} e^{-x^2/2}\, dx$$

for any real $u$.

Now if $X_1, X_2, X_3, \ldots$ are independent, identically distributed random variables with $E(X_n) = \mu$ and $\sigma^2(E_n) = \sigma^2 > 0$ for all $n$ we let $Z_n = (1/\sigma)(X_n - \mu)$. It has been shown before that $E(Z_n) = 0$ and $\sigma^2(Z_n) = 1$ and it is clear that the $Z_n$ are independent and identically distributed. Therefore

$$\lim_{n\to\infty} P\left(\frac{1}{\sqrt{n}}\left(\frac{X_1 - \mu}{\sigma} + \frac{X_2 - \mu}{\sigma} + \cdots + \frac{X_n - \mu}{\sigma}\right) \le u\right) = \int_{-\infty}^{u} \frac{1}{\sqrt{2\pi}} e^{-x^2/2}\, dx$$

or

$$\lim_{n\to\infty} P\left(\frac{1}{\sqrt{n}\,\sigma}(X_1 + X_2 + \cdots + X_n - n\mu) \le u\right) = \int_{-\infty}^{u} \frac{1}{\sqrt{2\pi}} e^{-x^2/2}\, dx$$

and that is the central limit theorem.

If $E(X^3)$ exists it is a theorem, proved independently by Berry and Esseen, that

**(10.3.3)** $$\left| P(X_1 + X_2 + \cdots + X_n - n\mu \le b\sqrt{n}\,\sigma) - \int_{-\infty}^{b} (2\pi)^{-1/2} e^{-1/2x^2}\, dx \right|$$
$$\le c\tau\sigma^{-3}n^{-1/2}$$

where $\tau = E(|X^3|)$, $c = 7.5$ in the Esseen version and $\tau = E(|X - \mu|^3)$, $c = 1.88$ in the Berry version. These authors actually obtain results for the case when the $X_j$ are not identically distributed.

Since (10.3.3) is so general we must realize that $c\tau\sigma^{-3}n^{-1/2}$ is not a good estimate when the $X_j$ have a smooth distribution.

**EXERCISES 10.3** _____

♦ 1  Use L'Hospital's rule or the MacLaurin series to show that

$$\lim_{u\to 0} \frac{\sin u - u}{u^2} = \lim_{u\to 0} \frac{\cos u - 1 + \frac{1}{2}u^2}{u^2} = 0$$

♦ 2  Let $k(u) = 2u^2 - \sin u + u$. Show that $k(0) = 0$ and that $k'(u) \ge 0$ for $u \ge 0$. Thus $2u^2 \ge \sin u - u$ for $u \ge 0$. Deduce that $|(\sin u - u)/u^2| \le 2$ for all $u$.

3  Let $S = \{s; 0 < s < 1, s \text{ real}\}$. Let $P$ be given by a uniform mass density on $S$. Let $V(s) = 1$ for each $s \in S$ and $U_n(s) = n^2 e^{-ns}$. Show that for each $s \in S$ $\lim_{n\to\infty} U_n(s) = 0$. Show that $\lim_{n\to\infty} E(U_n V)$ does not exist. Why does this not contradict Theorem 10.3.2?

**4**  Using the same $S$ and $P$ as in the above exercise, let $V(s) = 1$ for each $s \in S$. Let $U_n(s) = (s - \frac{1}{2})^{2n}$. Show that $|U_n(s)| \le 1$ for each $n$ and $s \in S$ and that $\lim_{n \to \infty} U_n(s) = 0$ for each $s \in S$. Verify that $\lim_{n \to \infty} E(U_n V) = 0$. Does the sequence of functions $U_1, U_2, U_3, \ldots$ converge uniformly on $S$ to the zero function?

**5**  Discuss the convergence of $\phi_{X_n}(t)$ for the two types of $X_n$ in exercise 10.1.10.

**6**  Let $P(X_n = n) = \frac{1}{2}$ and $P(X_n = -n) = \frac{1}{2}$. Discuss the convergence of $\phi_{X_n}(t)$.

**7**  Let $X_1, X_2, X_3, \ldots$ be independent, identically distributed random variables with $E(X_k) = 0$ and $\sigma^2(X_j) = \sigma^2 > 0$. Discuss convergence in distribution of $n^{-c}(X_1 + X_2 + \cdots + X_n)$ for as many values of $c$ as seems appropriate.

♦ **8**  Let $X_1, X_2, X_3, \ldots$ be independent random variables with the Cauchy density function $f(x) = [\pi(1 + x^2)]^{-1}$. What is the distribution of $(1/n)(X_1 + X_2 + \cdots + X_n)$? Discuss convergence in distribution of $(1/n)(X_1 + X_2 + \cdots + X_n)$.

♦ **9**  Prove Theorem 10.3.3 by showing that for any $\epsilon > 0$ there is an $N$ such that for sufficiently large $n$

$$\left| \sum_{k=N+1}^{\infty} a^k/k! \right| < \tfrac{1}{3}\epsilon$$

$$\left| \sum_{k=N+1}^{n} \binom{n}{k}\left(\frac{a + c_n}{n}\right)^k \right| < \tfrac{1}{3}\epsilon$$

$$\left| \sum_{k=0}^{N} \left[ \binom{n}{k}\left(\frac{a + c_n}{n}\right)^k - a^k/k! \right] \right| < \tfrac{1}{3}\epsilon$$

**10**  Let $X_1, X_2, X_3, \ldots$ be independent, identically distributed random variables with $E(X_k) = \mu$ and $\sigma^2(X_k) = \sigma^2$. Let $Y_n = (1/\sqrt{n}\,\sigma)[X_1 + X_2 + \cdots + X_n - n\mu]$. Verify that $\phi_{Y_n}(t) \to e^{-t^2/2}$ when $X_k$ has the following distributions: (a) Poisson; (b) negative exponential; (c) gamma; (d) binomial.

**11**  A random variable $X_n$ in a sequence of random variable has

$$P(X = n^{-1}(k - \tfrac{1}{2}n)) = 2k/n(n + 1)$$

for $k = 0, 1, 2, \ldots, n$. Discuss convergence in probability of $X_1, X_2, X_3, \ldots$.

# 10.4  Sequences of random vectors

We previously mentioned that Theorem 10.3.2 is called the continuity theorem for characteristic functions. It is very valuable in cases where the characteristic functions are easier to deal with than the distribution functions. There is a continuity theorem for characteristic functions of sequences of random vectors which we shall shortly state. First of all we must formulate a definition of convergence in distribution of a sequence of random vectors. Although it can be shown that the specialization to 1-dimensional random vectors is equivalent to the original definition, there is a somewhat different viewpoint in this definition.

**Definition**    Let $Z_1, Z_2, Z_3, \ldots$ be a sequence of random vectors of fixed dimension $k$. Let $Z$ be a $k$-dimensional random vector. The sequence $Z_1, Z_2, Z_3, \ldots$ is said to *converge in distribution to Z* if for any Borel set $B \subset E^k$, $k$-dimensional space, such that $P(Z \in$ boundary of $B) = 0$ we have

$$\lim_{n \to \infty} P(Z_n \in B) = P(Z \in B)$$

There is little loss if one restricts the sets $B$ to fairly simple sets such as $k$-dimensional cubes, cylinders, Cartesian products of lower-dimensional figures, and sets defined by algebraic and transcendental inequalities.

**Theorem 10.4.1** (Continuity theorem.)    The sequence of $k$-dimensional random vectors $Z_1, Z_2, Z_3, \ldots$ converges in distribution to the $k$-dimensional random vector $Z$ if and only if for each $(t_1, t_2, \ldots, t_k)$

$$\phi_{Z_n}(t_1, t_2, \ldots, t_k) \to \phi_Z(t_1, t_2, \ldots, t_k) \text{ as } n \to \infty$$

In order to point out the advantages of Theorem 10.4.1 we work an example using Theorem 10.2.1 but avoiding the former.

### Example 10.4.1
Let $N_n$ be the random variable which gives the number of H's in $n$ independent tosses of a fair coin. Discuss the convergence in distribution of the sequence whose $n$th term is

$$Y_n = \frac{2}{n} [(N_n - \tfrac{1}{2}n)^2 + (\text{number of T's in } n \text{ tosses} - \tfrac{1}{2}n)^2]$$

Let

$$X_k = \begin{cases} 1, & \text{if the } k\text{th toss is H} \\ 0, & \text{if the } k\text{th toss is T} \end{cases}$$

Then $N_n = X_1 + X_2 + \cdots + X_n$ and $E(X_k) = \tfrac{1}{2}$, $\sigma^2(X_k) = \tfrac{1}{4}$. Since the number of T's in $n$ tosses is $n - N_n$, we have

$$Y_n = \frac{2}{n}[(N_n - \tfrac{1}{2}n)^2 + (n - N_n - \tfrac{1}{2}n)^2] = \frac{2}{n}[(N_n - \tfrac{1}{2}n)^2 + (\tfrac{1}{2}n - N_n)^2]$$

$$= \frac{4}{n}(N_n - \tfrac{1}{2}n)^2 = \frac{4}{n}(X_1 + X_2 + \cdots + X_n - \tfrac{1}{2}n)^2$$

We have just seen that

$$\lim_{n \to \infty} P\left(\frac{1}{\sqrt{\tfrac{1}{4}n}}(X_1 + X_2 + \cdots + X_n) \leq u\right) = \int_{-\infty}^{u} \frac{1}{\sqrt{2\pi}} e^{-x^2/2} \, dx$$

and

$$Y_n = \left( \frac{X_1 + X_2 + \cdots + X_n - \frac{1}{2}n}{\sqrt{\frac{1}{4}n}} \right)^2$$

Can these be put together? First of all $P(Y_n < 0) = 0$ so $\lim_{n \to \infty} P(Y_n \le u) = 0$ if $u < 0$. Secondly, if $u \ge 0$,

$$P(Y_n \le u) = P\left( \left( \frac{X_1 + \cdots + X_n - \frac{1}{2}n}{\sqrt{\frac{1}{4}n}} \right)^2 \le u \right)$$

$$= P\left( -\sqrt{u} \le \frac{X_1 + X_2 + \cdots + X_n - \frac{1}{2}n}{\sqrt{\frac{1}{4}n}} \le \sqrt{u} \right)$$

Now

$$\lim_{n \to \infty} P\left( -\sqrt{u} < \frac{X_1 + X_2 + \cdots + X_n - \frac{1}{2}n}{\sqrt{\frac{1}{4}n}} \le \sqrt{u} \right) = \int_{-\sqrt{u}}^{\sqrt{u}} \frac{1}{\sqrt{2\pi}} e^{-x^2/2} \, dx$$

It is not difficult to show that

$$\lim_{n \to \infty} P\left( -\sqrt{u} = \frac{X_1 + X_2 + \cdots + X_n - \frac{1}{2}n}{\sqrt{\frac{1}{4}n}} \right) = 0$$

Thus

**(10.4.1)**   $\lim_{n \to \infty} P(Y_n \le u) = \begin{cases} \int_{-\sqrt{u}}^{\sqrt{u}} \frac{1}{\sqrt{2\pi}} e^{-x^2/2} \, dx, & u \ge 0 \\ 0, & u < 0 \end{cases}$

Theorem 10.4.1 with $k = 1$, $B = \{x; x^2 \le u\}$, $X_k = Z_k$, $Z$ an $N(0, 1)$ random variable and exercise 10.3.10(d) would bring us to (10.4.1) directly.

The right hand side of (10.4.1) is the distribution function of a random variable $Y$ whose density function

$$f_Y(u) = \begin{cases} \frac{1}{\sqrt{2\pi u}} e^{-u/2}, & u > 0 \\ 0, & u \le 0 \end{cases}$$

Example 10.4.1 is a special case of a limit theorem which we will state but not prove until Chapter 12. This limit theorem is the basis for the statistical test known as the $\chi^2$ test.

The theorem deals with independent, identically distributed random variables $X_1, X_2, X_3, \ldots$ and Borel sets on the real line, $C_1, C_2, \ldots, C_m$ $(m > 1)$

which are disjoint and whose union is the entire real line. Furthermore we assume that $P(X_1 \in C_j) = \pi_j > 0$ and let

$$g_j(x) = \begin{cases} 1, & x \in C_j \\ 0, & x \notin C_j \end{cases}$$

Let $N_j^{(n)} = \sum_{k=1}^{n} g_j(X_k)$. That is, the number of the first $n$ random variables $X_1, X_2, \ldots, X_n$ which are in $C_j$. Let $Z_n = \sum_{j=1}^{m} (1/n\pi_j)(N_j^{(n)} - n\pi_j)^2$.

The limit theorem in question states that for $r > 0$

$$\lim_{n \to \infty} P(Z_n \le r) = \int_0^r \frac{1}{2^{(m-1)/2}\Gamma((m-1)/2)} x^{(m-2)/2} e^{-x/2} \, dx$$

where $\Gamma(\alpha) = \int_0^\infty x^{\alpha-1} e^{-x} \, dx$. Therefore the sequence $Z_1, Z_2, Z_3, \ldots$ converges in distribution to a certain multiple of a gamma random variable of section 6.4. Random variables with density function

$$f(x) = \begin{cases} 0, & x \le 0 \\ 2^{-m/2}[\Gamma(m/2)]^{-1} x^{(m-1)/2} e^{-x/2} & x > 0 \end{cases}$$

are said to have the $\chi^2$ distribution with $m$ degrees of freedom. Thus the limit theorem states that $Z_1, Z_2, Z_3, \ldots$ converges to a $\chi^2$ random variable with $m - 1$ degrees of freedom.

The limit theorems of this chapter are essentially classical and elementary in comparison to the many recent limit theorems. An account of the current techniques used in finding and proving limit theorems could fill a number of books.

**EXERCISES 10.4**

1  Let $(X, Y)$ be a random vector. Suppose that $a_n \to 0$ and $b_n \to 0$ as $n \to \infty$. Show that $(X + a_n, Y + b_n)$ converges to $(X, Y)$ in distribution.

2  Let $X_1, X_2, X_3, \ldots$ be independent random variables with each of the $X_k$ being normal $N(0, \sigma^2)$. Let $Y_n = n^{-1/2}(X_1 + X_2 + \cdots + X_n)$ and $Z_n = n^{-3/2}[nX_1 + (n-1)X_2 + (n-2)X_3 + \cdots + 2X_{n-1} + X_n]$. Discuss the convergence in distribution of the sequence of random vectors $(Y_n, Z_n)$.

[*Hint:* Use the continuity theorem. Recall that $\sum_{k=1}^{n} k = n(n+1)/2$ and $\sum_{k=1}^{n} k^2 = n(n+1)(2n+1)/6$. Use the independence of the $X_j$.]

3  Let $X_1, X_2, X_3, \ldots$ be independent, 2-dimensional random vectors with probability function $f$ satisfying $f(0, 1) = f(0, -1) = f(1, 0) = f(-1, 0) = f(1, 1) = f(-1, -1) = \frac{1}{6}$. Find the characteristic function of $X_j$. Recall that for a vector $(a, b)$, $c(a, b) = (ca, cb)$. Find the characteristic function of

$n^{-1/2}X_j$. Assume that the characteristic function of a sum of independent random vectors is the product of their characteristic functions. Discuss convergence in distribution of the sequence of random vectors

$$Y_n = n^{-1/2}(X_1 + X_2 + \cdots + X_n)$$

[*Hint:* $e^{ix} + e^{-ix} = 2 \cos x$; $\cos x = 1 - x^2/2! + x^4/4! + \cdots$.]

4  Let $X_1, X_2, X_3, \ldots$ be independent and uniformly distributed on the interval $0 \le x \le 1$; hence the density function of each of the $x_j$ is

$$f(x) = \begin{cases} 1, & 0 \le x \le 1 \\ 0, & \text{otherwise} \end{cases}$$

Let $C_1 = \{x; x < \frac{1}{4}\}$, $C_2 = \{x; \frac{1}{4} \le x \le \frac{1}{2}\}$, $C_3 = \{x; \frac{1}{2} \le x < \frac{3}{4}\}$, $C_4 = \{x; x \ge \frac{3}{4}\}$. Using the notation of the closing material of section 10.4 estimate $\alpha$ such that $P(Z_{100} \ge \alpha) = 0.05$.

5  Let $Y_n$ have characteristic function

$$\phi_{Y_n}(t_1, t_2) = (\sum_{j=1}^{k} p_j e^{i(a_j t_1 + b_j t_2)/\sqrt{n}})^n$$

where $\sum p_j = 1$, $\sum p_j a_j = 0$, and $\sum p_j b_j = 0$. Discuss convergence in distribution of $Y_n$.

# 10.5 The *n*-dimensional normal distribution

The following result follows from section 9.3 and exercise 8.5.7. Let $(X_1, X_2)$ be a 2-dimensional normal random vector with $E(X_1) = E(X_2) = 0$. Then there are independent random variables $Z_1$ and $Z_2$ which are $N(0, 1)$ and there are real numbers $a_{11}, a_{12}, a_{21}, a_{22}$ such that $X_1 = a_{11}Z_1 + a_{12}Z_2$ and $X_2 = a_{21}Z_1 + a_{22}Z_2$. Moreover, we can have $a_{12} = 0$.

We could define the *n*-dimensional normal density function in analogy to the 2-dimensional density function. However, there are advantages to the indirect route we take. We will obtain certain interesting degenerate cases in addition to the nondegenerate *n*-dimensional normal distribution.

Henceforth let $a_{jk}$, $j = 1, 2, \ldots, n$ and $k = 1, 2, \ldots, n$ be $n^2$ real numbers. Let $Z_1, Z_2, \ldots, Z_n$ be independent random variables which are $N(0, 1)$. Let

(10.5.1)     $X_j = \sum_{k=1}^{n} a_{jk}Z_k, \quad j = 1, 2, \ldots, n$

Let us consider an example with these $X_j$'s.

**Example 10.5.1**
If $n \ge 3$, compute $E(X_1 X_3)$.

By definition $E(X_1 X_3) = E((\sum_{k=1}^{n} a_{1k} Z_k)(\sum_{k=1}^{n} a_{3k} Z_k))$ which equals

**(10.5.2)** $\quad E(a_{11} a_{31} Z_1 Z_1 + a_{12} a_{32} Z_2 Z_2 + \cdots + a_{1n} a_{3n} Z_n Z_n + a_{11} a_{32} Z_1 Z_2$

$\qquad\qquad\qquad\qquad\qquad + \text{terms involving } Z_l Z_m \text{ for } l \neq m)$

Now the $Z_k$'s are independent and $N(0, 1)$. If $l \neq m$ the independence of the $Z_k$'s gives $E(Z_l Z_m) = 0$. If $l = m$ we have $E(Z_l Z_l) = \sigma^2(Z_l) + [E(Z_l)]^2 = 1 + 0 = 1$. Therefore,

$$E(X_1 X_3) = a_{11} a_{31} \cdot 1 + a_{12} a_{32} \cdot 1 + \cdots + a_{1n} a_{3n} \cdot 1 + 0 + \cdots + 0$$

$$= a_{11} a_{31} + a_{12} a_{32} + \cdots + a_{1n} a_{3n}$$

It is clear that this same approach would give

**(10.5.3)** $\quad E(X_j X_l) = a_{j1} a_{l1} + a_{j2} a_{l2} + \cdots + a_{jn} a_{ln}$

Next we point out that if $Z_1, Z_2, \ldots, Z_n$ are independent and $N(0, 1)$ then

**(10.5.4)** $\quad \phi_{(Z_1, Z_2, \ldots, Z_n)}(t_1, t_2, \ldots, t_n) = e^{-t_1^2/2} e^{-t_2^2/2} \cdots e^{-t_n^2/2}$

$$= \exp\left[-\tfrac{1}{2}(t_1^2 + t_2^2 + \cdots + t_n^2)\right]$$

Let us now find the characteristic function of the random vector $(X_1, X_2, \ldots, X_n)$, namely, $E[\exp(it_1 X_1 + it_2 X_2 + \cdots + it_n X_n)]$. For the present we compute $t_1 X_1 + t_2 X_2 + \cdots + t_n X_n = \sum_{j=1}^{n} t_j X_j$. By (10.5.1),

$$\sum_{j=1}^{n} t_j X_j = \sum_{j=1}^{n} t_j \sum_{k=1}^{n} a_{jk} Z_k = \sum_{k=1}^{n} \sum_{j=1}^{n} t_j a_{jk} Z_k = \sum_{k=1}^{n} (\sum_{j=1}^{n} t_j a_{jk}) Z_k$$

We let $u_k = \sum_{j=1}^{n} t_j a_{jk} = t_1 a_{1k} + t_2 a_{2k} + \cdots + t_n a_{nk}$ for $k = 1, 2, \ldots, n$. Equation (10.5.4) is equivalent to $\phi_{(Z_1, Z_2, \ldots, Z_n)}(v_1, v_2, \ldots, v_n) = \exp\left[-\tfrac{1}{2}(v_1^2 + v_2^2 + \cdots + v_n^2)\right]$ for all $(v_1, v_2, \ldots, v_n)$. Now

$$E(\exp(it_1 X_1 + \cdots + it_n X_n)) = E(\exp(i(\sum_{j=1}^{n} t_j X_j)))$$

$$= E(\exp(i \sum_{k=1}^{n} (\sum_{j=1}^{n} t_j a_{jk}) Z_k))$$

$$= E(\exp(i \sum_{k=1}^{n} u_k Z_k))$$

$$= \phi_{(Z_1, Z_2, \ldots, Z_n)}(u_1, u_2, \ldots, u_n)$$

$$= \exp\left[-\tfrac{1}{2}(u_1^2 + u_2^2 + \cdots + u_n^2)\right]$$

$$= \exp\left(-\tfrac{1}{2} \sum_{k=1}^{n} u_k^2\right) = \exp\left[-\tfrac{1}{2} \sum_{k=1}^{n} (\sum_{j=1}^{n} t_j a_{jk})^2\right]$$

Note that for all $(t_1, \ldots, t_n)$, $\sum\limits_{k=1}^{n} (\sum\limits_{j=1}^{n} t_j a_{jk})^2 \geq 0$ since it is a sum of squares of real numbers.

Expressions such as $\sum\limits_{k=1}^{n} (\sum\limits_{j=1}^{n} t_j a_{jk})^2$ are more conveniently described in another way. Note that $\sum\limits_{j=1}^{n} t_j a_{jk} = \sum\limits_{l=1}^{n} t_l a_{lk}$. Therefore,

$$\sum_{k=1}^{n} \left(\sum_{j=1}^{n} t_j a_{jk}\right)^2 = \sum_{k=1}^{n} \left(\sum_{j=1}^{n} t_j a_{jk}\right)\left(\sum_{l=1}^{n} t_l a_{lk}\right)$$

$$= \sum_{j=1}^{n} \sum_{l=1}^{n} \left(\sum_{k=1}^{n} a_{jk} a_{lk}\right) t_j t_l$$

This maneuver displays $\sum\limits_{k=1}^{n} a_{jk} a_{lk}$ as the coefficient of $t_j t_l$ in a certain "expansion" which we now discuss.

Let $b_{jl} = \sum\limits_{k=1}^{n} a_{jk} a_{lk}$. Note that $b_{1j} = \sum\limits_{k=1}^{n} a_{lk} a_{jk} = \sum\limits_{k=1}^{n} a_{jk} a_{lk} = b_{jl}$. So

$$E(\exp(i(t_1 X_1 + \cdots + t_n X_n))) = \exp\left(-\tfrac{1}{2} \sum_{j=1}^{n} \sum_{l=1}^{n} b_{jl} t_j t_l\right)$$

Note, from (10.5.3), that $b_{jl} = E(X_j X_l)$. Now if $E(X_j) = 0$, $\sigma^2(X_j) = E(X_j^2)$, and cov $(X_j X_l) = E(X_j X_l)$ so $\sigma^2(X_j) = b_{jj}$ and cov $(X_j X_l) = b_{jl}$ for $j \neq l$.

**Theorem 10.5.1**   With our current conventions suppose that for each $j$, $l$ with $j \neq l$

$$a_{j1} a_{l1} + a_{j2} a_{l2} + \cdots + a_{jn} a_{ln} = 0$$

Then the random variable $X_1, X_2, \ldots, X_n$ are independent.

PROOF   We have

(10.5.5)     $\phi_{(X_1, X_2, \ldots, X_n)}(t_1, t_2, \ldots, t_n) = \exp\left(-\tfrac{1}{2} \sum\limits_{j=1}^{n} \sum\limits_{l=1}^{n} b_{jl} t_j t_l\right)$

where $b_{jl} = \sum\limits_{k=1}^{n} a_{jk} a_{lk}$. Under our assumptions $b_{jl} = 0$ if $j \neq l$. Therefore

$$\sum_{j=1}^{n} \sum_{l=1}^{n} b_{jl} t_j t_l = b_{11} t_1^2 + b_{22} t_2^2 + \cdots + b_{nn} t_n^2$$

and

$$\phi_{(X_1, X_2, \ldots, X_n)}(t_1, t_2, \ldots, t_n) = \exp\left[-\tfrac{1}{2}(b_{11} t_1^2 + b_{22} t_2^2 + \cdots + b_{nn} t_n^2)\right]$$
$$= e^{-b_{11} t_1^2/2} e^{-b_{22} t_2^2/2} \cdots e^{-b_{nn} t_n^2/2}$$

The last is the product of characteristic functions of independent random variables which are $N(0, \sqrt{b_{kk}})$, respectively. By the uniqueness theorem for characteristic functions of random vectors the components of $(X_1, X_2, \ldots, X_n)$ are independent.

Suppose a random vector $(Y_1, Y_2, \ldots, Y_n) = Y$ has a characteristic function

$$\phi_Y(t_1, t_2, \ldots, t_n) = \exp\left(-\tfrac{1}{2} \sum_{j=1}^{n} \sum_{l=1}^{n} c_{jl} t_j t_l\right), \qquad c_{1j} = c_{j1}$$

What can we say about $Y$? Note the relationship between $\phi_Y$ and our random vector $(X_1, X_2, \ldots, X_n)$ defined by (10.5.1). Since $|\phi_Y(t_1, \ldots, t_n)| \leq 1$ for all $(t_1, t_2, \ldots, t_n)$ we must have $\sum_{j=1}^{n} \sum_{l=1}^{n} c_{jl} t_j t_l \geq 0$.

In linear algebra it is shown that if $\sum_{j=1}^{n} \sum_{l=1}^{n} c_{jl} t_j t_l \geq 0$ for all $(t_1, \ldots, t_n)$ and $c_{jl} = c_{lj}$ then there are numbers $\alpha_{jk}$ such that with $t_j = \sum_{k=1}^{n} \alpha_{jk} u_k$ then $\sum_{j=1}^{n} \sum_{l=1}^{n} c_{jl} t_j t_l = \sum_{k=1}^{n} \beta_k u_k^2$ with $\beta_k \geq 0$. But then

$$E\left(\exp\left(i \sum_{j=1}^{n} t_j Y_j\right)\right) = \phi_Y(t_1, \ldots, t_n) = \exp\left(-\tfrac{1}{2} \sum \sum c_{jl} t_j t_l\right)$$

$$= \exp\left(-\tfrac{1}{2} \sum_{k=1}^{n} \beta_k u_k^2\right)$$

or

$$E\left(\exp\left(i \sum_{j=1}^{n} \sum_{k=1}^{n} \alpha_{jk} u_k Y_j\right)\right) = \exp\left(-\tfrac{1}{2} \sum_{k=1}^{n} \beta_k u_k^2\right)$$

or

$$E\left(\exp\left(i \sum_{k=1}^{n} u_k\left(\sum_{j=1}^{n} \alpha_{jk} Y_j\right)\right)\right) = \exp\left(-\tfrac{1}{2} \sum_{k=1}^{n} \beta_k u_k^2\right)$$

This says that $\sum_{j=1}^{n} \alpha_{jk} Y_j = \sqrt{\beta_k}\, U_k$ where $U_k$ is $N(0, 1)$ and the $U_k$ are independent. Furthermore, it can be shown that $Y_j = \sum_{k=1}^{n} \alpha_{kj} \sqrt{\beta_k}\, U_k$. Thus a random vector $Y$ whose characteristic function has the form $\exp\left(-\tfrac{1}{2} \sum \sum c_{jl} t_j t_l\right)$ with $c_{jl} = c_{lj}$ has $Y_j = \sum_{k=1}^{n} a_{jk} Z_k$, where $a_{jk} = \alpha_{kj} \sqrt{\beta_k}$ and $Z_k = U_k$. Thus we have come full circle on (10.5.1).

**Definition**    A random vector $X = (X_1, X_2, \ldots, X_n)$ is called a *centered n*-dimensional normal random vector if $\phi_X(t_1, t_2, \ldots, t_n) = \exp\left(-\tfrac{1}{2} \sum_{j=1}^{n} \sum_{l=1}^{n} b_{jl} t_j t_l\right)$, where $b_{jl} = b_{lj}$ and $\sum_{j=1}^{n} \sum_{l=1}^{n} b_{jl} t_j t_l \geq 0$.

If $\sum_{j=1}^{n} \sum_{l=1}^{n} b_{jl}t_jt_l = 0$ *only if* $t_j = 0$, $j = 1, 2, \ldots, n$, $X$ is said to be *non-singular*. A random vector of the form $(X_1 + m_1, X_2 + m_2, \ldots, X_n + m_n)$ where $(X_1, X_2, \ldots, X_n)$ is a centered *n*-dimensional normal random vector is called an *n*-dimensional normal random vector. Naturally, it is centered if $m_1 = m_2 = \cdots = m_n = 0$. It is nonsingular if $X$ is nonsingular.

The discussion has shown that centered *n*-dimensional normal random vectors are the same as random vectors defined by equation (10.5.1) with independent, $N(0, 1)$ $Z_1, Z_2, \ldots, Z_n$.

Suppose we have $X_j = \sum_{k=1}^{n} a_{jk}Z_k$ as in (10.5.1) but suppose that the $a_{jk}$ satisfy certain relationships. We would then expect the $X_j$ to satisfy certain relationships. An important result of this sort arises when for each $j$, $l$,

$$\sum_{k=1}^{n} a_{jk}a_{lk} = \sum_{k=1}^{n} a_{kj}a_{kl} = \begin{cases} 1, & j = l \\ 0, & j \neq l \end{cases}$$

**Theorem 10.5.2**    If the $a_{jk}$ in $X_j = \sum a_{jk}Z_k$ satisfy

$$\sum_{j=1}^{n} a_{jk}a_{jl} = \begin{cases} 1, & k = l \\ 0, & k \neq l \end{cases}$$

then $X_1^2 + X_2^2 + \cdots + X_n^2 = Z_1^2 + Z_2^2 + \cdots + Z_n^2$.

PROOF

$$\sum_{j=1}^{n} X_j^2 = \sum_{j=1}^{n} \left( \sum_{k=1}^{n} a_{jk}Z_k \right)^2 = \sum_{j=1}^{n} \left( \sum_{k=1}^{n} a_{jk}Z_k \right) \left( \sum_{l=1}^{n} a_{jl}Z_l \right)$$

$$= \sum_{k=1}^{n} \sum_{l=1}^{n} \left( \sum_{j=1}^{n} a_{jk}a_{jl}Z_kZ_l \right)$$

But

$$\sum_{j=1}^{n} a_{jk}a_{jl} = \begin{cases} 1, & k = l \\ 0, & k \neq l \end{cases}$$

Thus only if $k = l$ do we get a nonzero contribution. But then $\sum_{j=1}^{n} a_{jk}a_{jl}Z_kZ_l = Z_kZ_k = Z_k^2$. Therefore, $\sum_{j=1}^{n} X_j^2 = \sum_{k=1}^{n} Z_k^2$ or $X_1^2 + X_2^2 + \cdots + X_n^2 = Z_1^2 + Z_2^2 + \cdots + Z_n^2$.

Note that the result of Theorem 10.5.2 states that for each $s$ in the underlying sample space $S$

$$\sum_{j=1}^{n} X_j^2(s) = \sum_{j=1}^{n} Z_j^2(s)$$

Furthermore, linear algebra theorems state that

**(10.5.6)**   $\sum\limits_{j=1}^{n} a_{jk}a_{jl} = \begin{cases} 1, & k = l \\ 0, & k \neq l \end{cases}$ is equivalent to $\sum\limits_{j=1}^{n} a_{kj}a_{lj} = \begin{cases} 1, & k = l \\ 0, & k \neq l \end{cases}$

in our present case where $k = 1, 2, \ldots, n,$   $l = 1, 2, \ldots, n.$

### Example 10.5.2

Let $Z_1, Z_2, Z_3, Z_4$ be independent and $N(0, 1)$. Let $X_1 = \frac{1}{2}(Z_1 + Z_2 + Z_3 + Z_4)$. Find $X_2, X_3, X_4$ which are $N(0, 1)$ and are such that $X_1, X_2, X_3, X_4$ are independent.

In reference to Theorem 10.5.1 we can consider $a_{11} = \frac{1}{2},$   $a_{12} = \frac{1}{2},$   $a_{13} = \frac{1}{2},$ $a_{14} = \frac{1}{2}$. We want the other $a_{jk}$ so that

$$\sum_{j=1}^{4} a_{jk}a_{jl} = \begin{cases} 1, & k = l \\ 0, & k \neq l \end{cases}$$

Note that by Theorem 10.5.1 the $X_j$ will be independent. We can take $a_{21} = 1/\sqrt{2},$   $a_{22} = -1/\sqrt{2},$   $a_{23} = 0,$   $a_{24} = 0;$   $a_{31} = 1/\sqrt{6},$   $a_{32} = 1/\sqrt{6},$ $a_{33} = -2/\sqrt{6},$   $a_{34} = 0;$   $a_{41} = 1/\sqrt{12},$   $a_{42} = 1/\sqrt{12},$   $a_{43} = 1/\sqrt{12},$ $a_{44} = -3/\sqrt{12}$. It can be verified that

$$\sum_{j=1}^{4} a_{kj}a_{lj} = \begin{cases} 1, & k = l \\ 0, & k \neq l \end{cases}$$

which by (10.5.6), settles matters.

**EXERCISES 10.5** _____

1   Let $Z_1, Z_2,$ and $Z_3$ be independent and $N(0, 1)$. Let $X_1 = 3Z_1 - Z_2 + Z_3,$ $X_2 = Z_1 - Z_2 - Z_3,$   $X_3 = Z_1 + 6Z_2 - 2Z_3$. Find the characteristic function of $(X_1, X_2, X_3)$ and $E(X_j X_k)$.

2   With $Z_j$ as in exercise 1 let $X_1 = Z_1 + Z_2 + Z_3,$   $X_2 = Z_1 - 2Z_2 - 2Z_3,$ and $X_3 = -2Z_1 + Z_2 + Z_3$. Show that $(X_1, X_2, X_3)$ is a degenerate 3-dimensional normal random vector. What is $X_1 + X_2 + X_3$?

♦ 3   Let $(X_1, X_2, \ldots, X_n)$ be an $n$-dimensional normal random vector. Let $Y_k = \sum\limits_{j=1}^{n} c_{kj}X_j$ for $k = 1, 2, \ldots, n$. Show that $(Y_1, Y_2, \ldots, Y_n)$ is an $n$-dimensional normal random vector.

♦ 4   Let $(X_1, X_2, \ldots, X_n)$ be a centered, $n$-dimensional, normal random vector. Show that two components $X_j$ and $X_k$ are independent if and only if $E(X_j X_k) = 0$. [*Hint:* Use characteristic functions. Let $t_l = 0$ if $l \neq j,$   $l \neq k$.]

♦ 5   With the conventions of Theorem 10.5.1, show that if

$$\sum_{k=1}^{n} a_{jk}a_{lk} = \begin{cases} 0, & j \neq l \\ 1, & j = l \end{cases} = \delta_{jl}$$

then the $X_j$ are $N(0, 1)$ as well as independent. The function $\delta_{jl}$ is called the *Kronecker delta*.

◆ **6**   Let $\sum_{k=1}^{n} c_{jk}c_{lk} = 0$ for all $j \neq l$ and $1 \leq j$, $l \leq n$ and $c_j = \sum_{k=1}^{n} c_{jk}^2 > 0$ for $j = 1, 2, \ldots, n$. Let $a_{jk} = c^{-1/2}c_{jk}$. Show that $\sum_{k=1}^{n} a_{jk}a_{kl} = \delta_{jl}$ as given above.

◆ **7**   Let $\pi_1 + \pi_2 + \pi_3 + \pi_4 = 1$ with $\pi_j > 0$. Let $c_{11} = \sqrt{\pi_1}$, $c_{12} = \sqrt{\pi_2}$, $c_{13} = \sqrt{\pi_3}$, $c_{14} = \sqrt{\pi_4}$; $c_{21} = (1/\sqrt{\pi_1})(\pi_1 - 1)$, $c_{22} = \sqrt{\pi_2}$, $c_{23} = \sqrt{\pi_3}$, $c_{24} = \sqrt{\pi_4}$; $c_{31} = 0$, $c_{32} = (1/\sqrt{\pi_2})(\pi_1 + \pi_2 - 1)$, $c_{33} = \sqrt{\pi_3}$, $c_{34} = \sqrt{\pi_4}$; $c_{41} = 0$, $c_{42} = 0$, $c_{43} = (1/\sqrt{\pi_3})(\pi_1 + \pi_2 + \pi_3 - 1)$, $c_{44} = \sqrt{\pi_4}$. Show that $\sum_{k=1}^{4} c_{jk}c_{lk} = 0$ if $j \neq l$.

◆ **8**   Suppose that $c_{3k} = 0$ for $1 \leq k < m$, $c_{2k} = c_{1k}$ for $m \leq k \leq n$ and $\sum_{k=1}^{n} c_{1k}c_{3k} = 0$. Show that $\sum_{k=1}^{n} c_{2k}c_{3k} = 0$. How does this relate to exercise 7?

**9**   Let $\pi_k > 0$ for $k = 1, 2, \ldots, 5$ and $\sum_{k=1}^{5} \pi_k = 1$. Show that there are $a_{jk}$ for $j, k = 1, 2, 3, 4, 5$ with $a_{1k} = \sqrt{\pi_k}$ and $\sum_{k=1}^{5} a_{jk}a_{lk} = \delta_{jl}$. [*Hint:* Extend exercise 7 and use exercises 6 and 8.]

◆ **10**   Let $\pi_k > 0$ for $k = 1, 2, \ldots, n$ and $\sum_{k=1}^{n} \pi_k = 1$. Show that there are $a_{jk}$ for $j, k = 1, 2, \ldots, n$ with $a_{1k} = \sqrt{\pi_k}$ and $\sum_{k=1}^{n} a_{jk}a_{lk} = \delta_{jl}$. See exercise 9.

**11**   Let $\pi_k = \frac{1}{4}$ for $k = 1, 2, 3, 4$. Compare Example 10.5.2 with the results of exercise 7.

# CHAPTER 7

# *Random Processes*

## 11.1  Random walks

It is now convenient to view the $x$ axis as a long, coordinatized street. Suppose we start to walk at $x = x_0$ at time $t = 0$. We flip a possibly biased coin. If H appears we move to $x = x_0 + 1$ while if T appears we move to $x = x_0 - 1$. We flip the coin again independently or otherwise as compared to the first flip. If H appears on this toss we move one unit in the positive sense of the coordinatization. If T appears on the second toss we move one unit in the negative sense. We continue the flipping and movement. For purposes of idealization we assume that the coin is flipped at one-second intervals and that the flip is instantaneous. We assume that the movement is uniform and of speed one unit per second.

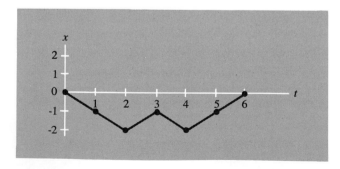

**Figure 1**

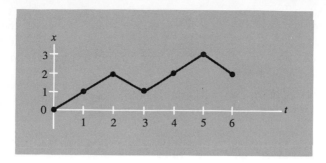

**Figure 2**

The nature of the coin toss sequence completely determines the pattern of the walk after $x_0$ is known. For example, if the coin toss sequence is TTHTHH up to $t = 6$ seconds and $x_0 = 0$ the graph of the motion is that in Figure 1. If the coin toss sequence is HHTHHT up to $t = 6$ and $x_0 = 0$ the graph of the motion is that in Figure 2.

Normal plotting of a graph for six tosses may not give enough insight into the possibilities. Consider the graph in Figure 3 corresponding to the coin toss sequence THHHHHTHHTHHTHTTTTTH with $x_0 > 0$. Coordinates have been intentionally omitted from Figure 3 so as not to prejudice the viewer by too much scale. Such a graph could suggest a performance of a stock over a period of time, a graph of static or "noise" in an electronic device over a period of time, or a graph of a component of the position of a particle undergoing Brownian motion during a period of time. One may feel that graphs of such phenomena should have somewhat different characteristics, but the graph in Figure 3 could be a plausible approximation. Thus random walks can be used as mathematical models for other more interesting phenomena.

Let us replace H by 1 and T by $-1$ in our descriptions of coin tossing sequences. Thus instead of writing $s = $ HTHHHT we write $s = (1, -1, 1, 1, 1, -1)$. With this convention and with $s = (s_1, s_2, \ldots, s_n)$ where $s_j = 1$ or $-1$ we have

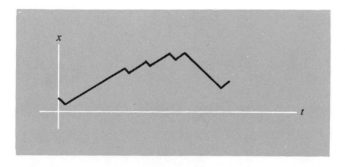

**Figure 3**

a most convenient device for describing random walks. Note that the movement of the walker from time $t = j - 1$ to time $t = j$ is exactly $s_j$. The reason that the time period is $j - 1$ to $j$ is that the first coin toss occurs at time $t = 0$. The reason that the movement is $s_j$ is that $s_j = 1$ if the toss is H and $s_j = -1$ if the toss is T. Next we note that if $k = 1, 2, \ldots, n$ the coordinate of the walker at time $t = k$ is $x_0 + s_1 + s_2 + \cdots + s_k$; i.e., the starting position plus the sum of the movements which have occurred up to time $t = k$.

Let $s = (s_1, s_2, \ldots, s_n)$ be one of our coin toss sequences. Let $x_s(t)$ be the position of the walker walking according to $s$ at time $t$ for $0 \leq t \leq n$. We have seen that $x_s(0) = x_0$ and

$$x_s(k) = x_0 + s_1 + s_2 + \cdots + s_k \qquad k = 1, 2, \ldots, n$$

Suppose that $k < t < k + 1$. At time $k$ the coordinate of the walker is $x_s(k)$. His speed is one unit per second. Therefore from time $k$ to time $t$ he moves a distance of $t - k$ units positively or negatively according to the sign of $s_{k+1}$. Therefore his position at time $t$ is $x_s(k) + (t - k)s_{k+1}$. If we set $0 \cdot s_{n+1} = 0$ and let $[t]$ be the greatest integer less than or equal to $t$ we have

$$(11.1.1) \qquad x_s(t) = x_s([t]) + (t - [t])s_{[t]+1}$$

Note that this is true for all $t$ with $0 \leq t \leq n$.

It may seem to the reader that we are replacing a simple idea, the coin tossing sequence, by a complicated idea, the function describing the position of the walker, when the latter is completely determined by the former and contains nothing new. This is true but there is good reason for it. In many applications we see only the "movements" but not that which causes the "movements." In fact we may know only the probabilistic aspects of the movements. Also, the random walk functions $x_s$ can be used to approximate other *functions*. Finally, from a pictorial viewpoint only, the eye seems to extract information more readily from continuous patterns rather than from isolated points on a graph. Therefore we do not hesitate to connect points and that is essentially what $x_s$ is.

So far we have not done any probability. Let

$$S = \{s = (s_1, s_2, \ldots, s_n); \ s_j \in \{1, -1\}\}$$

Under normal conditions we assume that $S$ has a probability measure. The probability measure used is determined by our assumptions about the coin tossing; i.e., independent tosses or not, fair coin or not, etc. Let

$$\Omega_S = \{x_s; s \in S\}$$

The set $\Omega_S$ is a set of functions which are defined on the interval $0 \leq t \leq n$. We define a probability measure on $\Omega_S$ as follows:

$$(11.1.2) \qquad P(x_s) = P(s)$$

That is, the probability of the function $x_s$ is the same as the probability of the coin toss sequence $s$. The alert reader may notice that we are doing little else except renaming a number of mathematical entities. Our reasons will soon be clear.

It is desirable to allow for unending sequences of coin tosses. Under the usual conditions we cannot give positive measure to any single such sequence. However, useful measures on $S = \{s = (s_1, s_2, s_3, \ldots); s_j \in \{1, -1\}\}$ are completely determined by the measure of events of the form

$$M_{u_1, u_2, \ldots, u_k} = \{(a_1, a_2, \ldots, a_k, s_{k+1}, s_{k+2}, \ldots); s_j \in \{1, -1\}, \quad j > k\}$$

Such an event should be regarded as one in which the outcomes of the first $k$ tosses are prescribed and the other tosses are unrestricted. The index $k$ ranges over all positive integers and the $a_j$ are 1 or $-1$ but are specified in the case of a particular event of the form $M_{a_1, a_2, \ldots, a_k}$. We then let $\Omega_S = \{x_s; s \in S\}$, i.e., the set of functions describing random walks corresponding to unending sequences of coin tosses. We define

$$P(\{x_s; s \in M_{a_1, a_2, \ldots, a_k}\}) = P(M_{a_1, a_2, \ldots, a_k})$$

With these definitions of probability in certain sets of functions we can ask for the probabilities of other events whose descriptions involve properties of functions.

### Example 11.1.1

Suppose the coin is fair, tosses are independent, and $x_0 = 0$. Let $n = 6$. Calculate $P(x_s(3) \leq 2)$, $P(x_s(4) = 0$ but $x_s(t) > 0$ for $0 < t < 4)$, and $P(x_s(2) \leq 0$ and $x_s(6) \leq 0)$.

We can easily compute $P(x_s(3) \leq 2)$. Note that with $x_0 = 0$, $x_s(3) = s_1 + s_2 + s_3$. The only way in which $s_1 + s_2 + s_3$ can exceed 2 subject to $s_j = 1$ or $-1$ is for $s_1 = s_2 = s_3 = 1$. Now $P$(sequences in which the first three tosses are H) $= (\frac{1}{2})^3$. We want

$$P(x_s(3) \leq 2) = 1 - P(x_s(3) > 2) = 1 - (\tfrac{1}{2})^3 = \tfrac{7}{8}$$

For $x_s(4) = 0$ and $x_s(t) > 0$ during $0 < t < 4$ we must have $s_1 = 1$ and $s_2 = 1$. Otherwise $x_s(2) = 0$. But if $s_1 = s_2 = 1$ the only way for $x_s(4) = 0$ is for $s_3 = s_4 = -1$. Therefore

$$P(x_s(4) = 0 \text{ and } x_s(t) > 0 \text{ during } 0 < t < 4) = (\tfrac{1}{2})^4$$

the probability of a coin tossing sequence having H in the first two positions, T in the next two positions, and anything in the last two positions.

Finally, we cannot find $P(x_s(2) \leq 0$ and $x_s(6) \leq 0)$ by multiplying the probabilities of the events $\{x_s(2) \leq 0\}$ and $\{x_s(6) \leq 0\}$ since these events are not independent. However

$$\{x_s(2) \leq 0 \text{ and } x_s(6) \leq 0\} = \bigcup_{k=0}^{1} \{x_s(2) = -2k \text{ and } x_s(6) \leq 0\}$$

since $x_s(2)$ can be only $-2$, $0$ or $2$ and only $-2$ and $0$ are less than or equal to 0. Furthermore the events in the union are disjoint. Now given $x_s(2)=0$, $x_s(6)\le0$ if and only if two or more of the tosses $s_3$, $s_4$, $s_5$, $s_6$ are equal to $-1$. Since those tosses are independent of any event determined by $s_1$ and $s_2$ we have

$$P(x_s(6) \le 0 \mid x_s(2) = 0)$$
$$= P(\text{two or more of the tosses } s_3, s_4, s_5, s_6 \text{ equal } -1 \mid x_s(2))$$
$$= (\tfrac{1}{2})^4 + \binom{4}{3}(\tfrac{1}{2})^3(\tfrac{1}{2}) + \binom{4}{2}(\tfrac{1}{2})^2(\tfrac{1}{2})^2 = \tfrac{11}{16}$$

By a similar argument we see that

$$P(x_s(6) \le 0 \mid x_s(2) = -2)$$
$$= P(\text{one or more of the tosses } s_3, s_4, s_5, s_6 \text{ equal } -1 \mid x_s(2) = -2)$$
$$= 1 - (\tfrac{1}{2})^4 = \tfrac{15}{16}$$

Since $P(A \cap B) = P(A \mid B)P(B)$ and $P(x_s(2) = -2) = \tfrac{1}{4}$ while $P(x_s(2) = 0) = \tfrac{1}{2}$ we have

$$P(x_s(2) \le 0 \text{ and } x_s(6) \le 0) = \tfrac{1}{2}\cdot\tfrac{11}{16} + \tfrac{1}{4}\cdot\tfrac{15}{16} = \tfrac{37}{64}$$

The technique used in the last part of Example 11.1.1 is used extensively in this chapter. Let us see it again in another example. Another useful technique appears here, too.

### Example 11.1.2

Assume independent tossing of a fair coin with $n = 10$ and $x_0 = 0$. Find $P(x_s(k) < 0$ for some $k \mid x_s(10) = 4)$.

We use the formula $P(A \mid B) = P(A \cap B) \div P(B)$. We note that

$$P(x_s(10) = 4) = P(s_1 + s_2 + \cdots + s_{10} = 4)$$
$$= P(\text{seven H's and three T's appearing in any way in ten tosses})$$
$$= \binom{10}{7}(\tfrac{1}{2})^7(\tfrac{1}{2})^3$$

We next compute $P(x_s(k) < 0$ for some $k$ and $x_s(10) = 4)$. Since the steps of the random walk are of length one unit the random walk must reach the co-ordinate $-1$ if it is $< 0$ at any time. Let $A_j = \{x_s(j) = -1$ for the first time at $t = j$ and $x_s(10) = 4\}$. The events $A_j$ are disjoint because a walk cannot reach $-1$ for the *first* time at two *different* times. Also the union of the events $A_j$ is $\{x_s(k) < 0$ for some $k$ and $x_s(10) = 4\}$. A random walk starting at 0 cannot reach $-1$ at $t = 0, 2, 4, \ldots$ because there must have been one more T than H thrown up to the time it reaches $-1$. Also, to get from $x = -1$ to $x = 4$ requires at least five steps so $P(A_j) = 0$ if $j > 5$.

Now for $A_1$ we need $s_1 = -1$ and $s_2, s_3, \ldots, s_{10}$ must signify seven H's and two T's in any order. Since the tosses are independent, $P(A_1) = \frac{1}{2} \binom{9}{7} (\frac{1}{2})^7 (\frac{1}{2})^2$.

For $A_3$ we must have $s_1 = 1$ and $s_2 = s_3 = -1$ in order for the walk to reach $-1$ for the first time at $t = 3$. After that $s_4, s_5, \ldots, s_{10}$ must signify six H's and one T in any order. Using independence we have $P(A_3) = (\frac{1}{2})^3 \binom{7}{6} (\frac{1}{2})^6 (\frac{1}{2})^1$.

Finally for $A_5$ we need $s_1 = 1$ but then we can have $s_2 = -1$ and $s_3 = 1$ or $s_2 = 1$ and $s_3 = -1$ but both $s_4 = -1$ and $s_5 = -1$ in order for the walk to reach $-1$ for the first time at $t = 5$. Then to get to $x = 4$ at $t = 10$, $s_5 = s_6 = \cdots = s_{10} = 1$. Therefore $P(A_5) = 2(\frac{1}{2})^5(\frac{1}{2})^5$. Thus

$$P(x_s(k) < 0 \text{ for some } k \mid x_s(10) = 4)$$
$$= [P(A_1) + P(A_3) + P(A_5)] \div P(x_s(10) = 4) = \tfrac{3}{8}$$

Note how the breaking up of an event in terms of the first occasion when some key subsidiary event occurs gives us disjoint component events. Note also that we do not have to concern ourselves with the way that the walk reaches $x = 4$ once it has reached $-1$ for the first time. It can go farther to the left or oscillate just as long as it reaches $x = 4$ at time $t = 10$.

We allowed for the possibility that the coin tosses may not be independent. This is contrary to the usual definition of a random walk. Most writers assume independent tosses but do not restrict themselves to one-step movements. Despite this deviation from standard terminology we will allow for dependence of the tosses but we will move in steps of equal length in our random walks.

**EXERCISES 11.1** ———————————————————————

1   Plot the graphs of the random walks controlled by coin tossing and starting at $x_0 = 0$ for the coin tossing sequences
   (a)   HHTHHHTTHHTTHTHTTTTTH
   (b)   HTTTTHTHHTTHHHTTHTHTT
   (c)   HHHHTTTHTHTTTHTTTHHTH
   (d)   HHHTTHHTHTTHHHHTHTTTH

2   Find the time, if it exists, that the random walks in exercise 1 first return to $x = 0$. Naturally all $t \le 21$.

3   Suppose that the coin tosses controlling the random walk are independent and $P(H) = p$. Find the probability of the set of random walks whose first 21 moves are determined by the sequences in exercise 1.

4   Find the probability that the random walk described in exercise 3 returns to its starting point for the first time at $t = 6$.

5   Find the amount of time which the random walks described in exercise 1 spend in the positive part of their coordinate axis.

**6**    Assume the standard coin tossing random walk with a fair coin starting at $x_0 = 0$. Find the probability that it spends four of the first six time units in the positive coordinate region.

**7**    Consider the following random walk with dependence and starting at $x_0 = 0$. On the first toss H and T are equally likely. On any toss after the first

$P(H$ on $n$ + 1st toss $\mid$ H on $n$th toss and anything on previous tosses$) = \frac{1}{3}$

$P(H$ on $n$ + 1st toss $\mid$ T on $n$th toss and anything on previous tosses$) = \frac{2}{3}$

Find the probability distribution of the position of the walk after the third toss.

In exercises 8–11 we restrict our attention to the random process most authors call the random walk; namely that in which the position of the walker at time $t = n$ is $x(n) = x_0 + X_1 + X_2 + \cdots + X_n$, where the $X_j$ are independent, identically distributed random variables.

**8**    Describe how to specialize the above defined process to obtain the coin tossing random walk with independent tosses.

**9**    Suppose that the $X_j$ mentioned above are $N(0, 1)$. Find the distribution of $x(n)$.

**10**    Still supposing that the $X_j$ are $N(0, 1)$ find $P(x(n) \leq a$ and $x(n + m) \leq b)$.

**11**    Suppose that the $X_j$ mentioned above have probabilities $P(X_j = a + b) = p$ and $P(X_j = a - b) = 1 - p$ where $a$, $b$, and $p$ are fixed. Let $x_0 = 0$. Find the characteristic function of $x(n)$.

Exercises 12–14 deal with random walks on the $xy$ plane.

**12**    Let $X_1, X_2, X_3, \ldots$ and $Y_1, Y_2, Y_3, \ldots$ be independent random variables with $P(X_j = 1) = P(Y_j = 1) = P(X_j = -1) = P(Y_j = -1) = \frac{1}{2}$. Let $W_n = (X_1 + X_2 + \cdots + X_n, Y_1 + Y_2 + \cdots + Y_n)$ for $n = 1, 2, 3, \ldots$. This can be thought of as the position of a random walk in the $xy$ plane at time $t = n$. Find (a) $P(W_6 = (4, -2))$,   (b) $P(W_9 = (-1, 5))$,   (c) the distribution of $W_n$,   (d) $P(W_4 = (0, 0)$ but $W_2 \neq (0, 0))$.

**13**    Suppose that $W_n = (X_1 + \cdots + X_n, Y_1 + \cdots + Y_n)$ where the $X_j$ and $Y_k$ are independent and $N(0, 1)$.

(a)    Find the distribution of $W_n$.

(b)    Find the probability that the distance from $W_n$ to $(0, 0)$ is less than $a$.

**14**    Let $V_j$ be a 2-dimensional random vector with $P(V_j = (k, l)) = \frac{1}{6}$ when $(k, l)$ is any one of $(0, 1)$, $(0, -1)$, $(1, 0)$, $(-1, 0)$, $(1, 1)$, or $(-1, -1)$. Suppose that $V_1, V_2, V_3, \ldots$ satisfy

$$\phi_{V_1 + \cdots + V_n}(u_1, u_2) = \prod_{j=1}^{n} \phi_{V_j}(u_1, u_2)$$

where $\phi$ denotes the characteristic function. Find $\phi_{V_1 + V_2 + \cdots + V_n}(u_1, u_2)$.

# ★ 11.2 First passage probabilities

In the discussion of Example 11.1.2 we used the events

$$A_j = \{x_s(j) = -1 \text{ for the first time at } t = j \text{ and } x_s(10) = 4\}, \qquad j \le 5$$

It is natural to ask about events in which something happens for the first time at $t = j$. In a study of probability we seek the probabilities of such events. Problems of this kind are called *first-passage* problems.

We assume $x_0 = 0$ and independent losses. We will find the probability that

(a)  the random walk returns to $x = 0$ for the first time at $t = 2n$,

(b)  the random walk first reaches $x = 1$ at $t = 2n + 1$,

(c)  the random walk first reaches $x = b$ at $t = n$ and never reaches $x = -a$ before $t = n$.

To be more specific, we will find generating functions for $p_{2n}$, $v_{2n+1}$, and $r_n(a, b)$ where

**(11.2.1)**    $p_{2n} = P(x_s(2n) = 0 \text{ and } x_s(t) \ne 0 \text{ for } 0 < t < 2n), \quad n = 1, 2, 3, \dots$

**(11.2.2)**    $v_{2n+1} = P(x_s(2n + 1) = 1 \text{ and } x_s(t) < 1 \text{ for } 0 \le t < 2n + 1),$
$$n = 0, 1, 2, \dots$$

and for integers $a$, $b$ with $-a < 0 < b$

**(11.2.3)**    $r_n(a, b) = P(x_s(n) = b \text{ and } x_s(t) > -1 \text{ for } 0 \le t \le n),$
$$n = 1, 2, 3, \dots$$

In the case of $p_{2n}$ and $v_{2n+1}$ we can find more explicit expressions than generating functions. It should be clear that

$$p_{2n} = P\big(x_s(j + 2n) = c, \quad x_s(t) \ne c \text{ for } j < t < j + 2n \mid x_s(j) = c\big)$$

and

$$v_{2n+1} = P\big(x_s(j + 2n + 1) = c + 1, \quad x_s(t) < c + 1$$
$$\text{for } j < t < j + 2n + 1 \mid x_s(j) = c\big)$$

These are the conditional probabilities of first return to $x = c$ after $t = j$ at $t = j + 2n$ given that $x_s(j) = c$ and of first reaching $x = c + 1$ after $t = j$ at $t = j + 2n + 1$ given that $x_s(j) = c$ but the formulas may speak better for themselves. Such an interpretation can be made for $r_n(a, b)$.

As a reminder to the reader the generating functions for these probabilities are the series

**(11.2.4)**    $\displaystyle\sum_{n=1}^{\infty} p_{2n} z^{2n}, \qquad \sum_{n=0}^{\infty} v_{2n+1} z^{2n+1}, \quad \text{and} \quad \sum_{n=0}^{\infty} r_n(a, b) z^n$

Note that $x_s(t) = 0$ only if $t$ is an even integer and $x_s(t) = 1$ only if $t$ is an odd integer.

***Theorem 11.2.1***   Let a random walk be controlled by independent tossing of a coin with $P(H) = p$, $P(T) = q = 1 - p$, and $x_0 = 0$. Then $p_2 = 2pq$ and

$$p_{2n} = \frac{(\frac{1}{2})(\frac{3}{2})(\frac{5}{2}) \cdots (\frac{1}{2} + n - \frac{1}{2})}{2(n!)} (4pq)^n, \qquad n = 2, 3, 4, \ldots$$

We will first prove some nonprobabilistic lemmas.

***Lemma 11.2.1***   If $n \geq 1$

$$\underbrace{(-\tfrac{1}{2})(-\tfrac{3}{2})(-\tfrac{5}{2}) \cdots [-\tfrac{1}{2} - (n - 1)]}_{n \text{ factors}}(-4)^n = \frac{(2n)!}{n!}$$

PROOF   A precise proof should use mathematical induction. However, such a proof is apt to hide the reason for the truth of the result. We therefore give an informal proof.
Now,

$$\underbrace{(-\tfrac{1}{2})(-\tfrac{3}{2})(-\tfrac{5}{2}) \cdots [-\tfrac{1}{2} - (n - 2)][-\tfrac{1}{2} - (n - 1)]}_{n \text{ factors}}(-4)^n$$

$$= 4^n(\tfrac{1}{2})(\tfrac{3}{2})(\tfrac{5}{2}) \cdots (\tfrac{1}{2} + n - 2)(\tfrac{1}{2} + n - 1)$$
$$= 2^n(2 \cdot \tfrac{1}{2})(2 \cdot \tfrac{3}{2})(2 \cdot \tfrac{5}{2}) \cdots [2 \cdot (\tfrac{1}{2} + n - 2)][2 \cdot (\tfrac{1}{2} + n - 1)]$$
$$= 2^n[1 \cdot 3 \cdot 5 \cdots (2n - 3)(2n - 1)]$$

$$\frac{(2n)!}{n!} = \frac{(2n)(2n - 1)(2n - 2)(2n - 3)(2n - 4) \cdots 6 \cdot 5 \cdot 4 \cdot 3 \cdot 2 \cdot 1}{n(n - 1)(n - 2) \cdots 3 \cdot 2 \cdot 1}$$

$$= \underbrace{\frac{2n}{n} \frac{2n - 2}{n - 1} \frac{2n - 4}{n - 2} \cdots \frac{6}{3} \frac{4}{2} \frac{2}{1}}_{n \text{ factors}} (2n - 1)(2n - 3) \cdots 5 \cdot 3 \cdot 1$$

where certain rearrangements have been made to suit our purpose. But this is exactly

$$2^n(2n - 1)(2n - 3) \cdots 5 \cdot 3 \cdot 1 = 2^n[1 \cdot 3 \cdot 5 \cdots (2n - 3)(2n - 1)]$$

This completes the proof.

***Lemma 11.2.2***   If $w$ is a real number and $|w| < \tfrac{1}{4}$ then

$$(1 - 4w)^{-1/2} = \sum_{n=0}^{\infty} \binom{2n}{n} w^n$$

PROOF     Let $f(w) = (1 - 4w)^{-1/2}$ for $|w| < \frac{1}{4}$. The MacLaurin series for $f(w)$ is given by $\sum\limits_{n=0}^{\infty} \dfrac{f^{(n)}(0)}{n!} w^n$. Now

$$f^{(0)}(0) = f(0) = (1 - 4 \cdot 0)^{-1/2} = 1$$
$$f'(w) = (-\tfrac{1}{2})(1 - 4w)^{-3/2}(-4)$$
$$f''(w) = (-\tfrac{1}{2})(-\tfrac{3}{2})(1 - 4w)^{-5/2}(-4)(-4)$$
$$f'''(w) = (-\tfrac{1}{2})(-\tfrac{3}{2})(-\tfrac{5}{2})(1 - 4w)^{-7/2}(-4)(-4)(-4)$$
$$\vdots$$

It is fairly clear that

$$f^{(n)}(0) = (-\tfrac{1}{2})(-\tfrac{3}{2})(-\tfrac{5}{2}) \cdots [-\tfrac{1}{2} - (n - 1)](-4)^n(1 - 4 \cdot 0)^{-1/2 - n}$$

But from Lemma 11.2.1

$$f^{(n)}(0) = \frac{(2n)!}{n!}$$

so

$$\frac{f^{(n)}(0)}{n!} = \frac{(2n)!}{n!n!} = \binom{2n}{n}$$

We leave it to calculus courses to show that the MacLaurin series converges and represents $f(w)$ for $|w| < \frac{1}{4}$. Therefore,

$$(1 - 4w)^{-1/2} = \sum_{n=0}^{\infty} \frac{f^n(0)}{n!} w^n = \sum_{n=0}^{\infty} \binom{2n}{n} w^n$$

We now proceed to the proof of Theorem 11.2.1.

PROOF     For $n = 0, 1, 2, 3, \ldots$

$$P(x_s(2n) = 0) = \binom{2n}{n} p^n q^n$$

since that is the probability of $n$ H's and $n$ T's in $2n$ tosses. This is a necessary and sufficient condition for the walk to be at $x = 0$ at $t = 2n$. Naturally, the walk may have been at $x = 0$ before $t = 2n$, also.

Define $p_0 = P(x_s(t)$ *returns* to $x = 0$ at $t = 0) = 0$. We would not say that the walk has *returned* to zero at $t = 0$.

In order to be at $x = 0$ at time $t = 2n \geq 2$ the random walk must have returned to $x = 0$ for the first time at $t = 2k$ for $k = 1, 2, \ldots, n$ and then it must have moved as many steps to the left as to the right. That is, for $n \geq 1$,

(11.2.5)     $\{x_s(2n) = 0\} = \bigcup\limits_{k=1}^{n} \{x_s(2k) = 0, \quad x_s(t) \neq 0$

for $0 < t < 2k$ and $x_s(2n) = 0\}$

where the events on the right are disjoint because a random walk cannot return to $x = 0$ *for the first time* at two different times. Since tosses preceding $t = 2k$ are independent of tosses at and after $t = 2k$

$$P(x_s(2k) = 0, \ x_s(t) \neq 0 \text{ for } 0 < t < 2k \text{ and } x(2n) = 0) = p_{2k}\binom{2n - 2k}{n - k}p^{n-k}q^{n-k}$$

since $\binom{2n - 2k}{n - k}p^{n-k}q^{n-k}$ is the probability of taking as many steps to the left

as to the right between $t = 2k$ and $t = 2n$ and this must happen if $x_s(2k) = 0$ and $x_s(2n) = 0$.

We take the probability of both sides of (11.2.5) and use $p_0 = 0$ to obtain, for $n \geq 1$,

**(11.2.6)** $\qquad \binom{2n}{n}p^n q^n = \sum_{k=1}^{n} p_{2n}\binom{2n - 2k}{n - k}p^{n-k}q^{n-k}$

Let $a_n = \binom{2n}{n}p^n q^n$ and $b_n = p_{2n}$ for $n = 0, 1, 2, \ldots$.

Note that equation (11.2.6) is

$$a_n = \sum_{k=0}^{n} b_k a_{n-k}$$

since $b_k = p_{2k}$ and

$$a_{n-k} = \binom{2(n - k)}{n - k}p^{n-k}q^{n-k} = \binom{2n - 2k}{n - k}p^{n-k}q^{n-k}$$

By exercises 1.5.11 and 1.5.14 we have, with $x$ replaced by $y$,

**(11.2.7)** $\qquad \sum_{n=1}^{\infty} a_n y^n = \left(\sum_{n=1}^{\infty} b_n y^n\right)\left(\sum_{n=0}^{\infty} a_n y^n\right)$ if $a_n = \sum_{k=1}^{\infty} b_k a_{n-k}$ for all $n \geq 1$

Now each of these series converges for $|y| < 1$ because $0 \leq a_n \leq 1$ and $0 \leq b_n \leq 1$ as $a_n$ and $b_n$ are probabilities of events. By lemma 11.2.2 if $|w| < \frac{1}{4}$

$$\sum_{n=0}^{\infty}\binom{2n}{n}w^n = (1 - 4w)^{-1/2}$$

Let $w = pqy = p(1 - p)y$. It can be seen that $p(1 - p) \leq \frac{1}{4}$ so if $|y| < 1$, $|pqy| < \frac{1}{4}$. Therefore,

$$\sum_{n=0}^{\infty} a_n y^n = \sum_{n=0}^{\infty}\binom{2n}{n}p^n q^n y^n = \sum_{n=0}^{\infty}\binom{2n}{n}(pqy)^n = (1 - 4pqy)^{-1/2}$$

and

$$\sum_{n=1}^{\infty} a_n y^n = (1 - 4pqy)^{-1/2} - a_0 y^0 = (1 - 4pqy)^{-1/2} - 1$$

Therefore, (11.2.7) can be rewritten as

$$(1 - 4pqy)^{-1/2} - 1 = \left( \sum_{n=1}^{\infty} b_n y^n \right) (1 - 4pqy)^{-1/2}$$

We multiply both sides of this equation by $(1 - 4pqy)^{1/2}$ and obtain

**(11.2.8)** $\quad 1 - (1 - 4pqy)^{1/2} = \sum_{n=1}^{\infty} b_n y^n = \sum_{n=1}^{\infty} p_{2n} y^n$

The MacLaurin series for $(1 - 4w)^{1/2}$ is

$$1 + \frac{1}{2} \frac{(-4)}{1!} w + \frac{1}{2} \left( -\frac{1}{2} \right) \frac{(-4w)^2}{2!} + \frac{(\frac{1}{2})(-\frac{1}{2})(-\frac{3}{2})(-4w)^3}{3!} + \cdots$$

so

$$1 - (1 - 4pqy)^{1/2} = -\frac{1}{2} \frac{4pqy}{1!} + \frac{1}{2} \left( \frac{1}{2} \right) \frac{(4pqy)^2}{2!} + \frac{1}{2} \left( \frac{1}{2} \right) \left( \frac{3}{2} \right) \frac{(4pqy)^3}{3!}$$

$$+ \frac{1}{2} \left( \frac{1}{2} \right) \left( \frac{3}{2} \right) \left( \frac{5}{2} \right) \frac{(4pqy)^4}{4!} + \cdots$$

after proper allowance for algebraic signs has been made. Since coefficients of equal powers of $y$ must agree we have

$$p_2 = \frac{1}{2} \frac{4pq}{1!} = 2pq$$

and

$$p_{2n} = \frac{1}{2} \left[ \left( \frac{1}{2} \right) \left( \frac{3}{2} \right) \cdots \left( \frac{2n - 3}{2} \right) \right] \frac{(4pq)^n}{n!}, \qquad n > 1$$

This is equivalent to the statement of the theorem.

This is a long proof but the result we seek is complicated. The second part of Example 11.1.1 is related but $n = 2$ is small. Example 11.1.2 is also related. In order to appreciate Theorem 11.2.1 the reader should draw the graphs of all random walks satisfying $x_s(10) = 0$ and $x_s(t) \neq 0$ for $0 < t < 10$ in the interval $0 \leq t \leq 10$.

### Example 11.2.1
What is the probability that the random walker never returns to $x = 0$?

The probability that he returns sometimes is

$$p_2 + p_4 + p_6 + p_8 + \cdots = \sum_{n=1}^{\infty} p_{2n} y^n, \qquad y = 1$$

But from (11.2.8), $\sum_{n=1}^{\infty} p_{2n} y^n = 1 - (1 - 4pqy)^{1/2}$. Therefore $p_2 + p_4 + p_6 + \cdots$
$= 1 - (1 - 4pq)^{1/2}$.

The probability that he never returns is

$$1 - (p_2 + p_4 + p_6 + \cdots) = 1 - [1 - (1 - 4pq)^{1/2}]$$
$$= (1 - 4pq)^{1/2} = [1 - 4p(1 - p)]^{1/2}$$

Note that this is zero if and only if $p = \frac{1}{2}$. Therefore only random walks of our type with fair coins have probability 1 of returning to $x = 0$.

Note that if we let $y = z^2$ in (11.2.8) we obtain

**(11.2.8a)**    $\displaystyle\sum_{n=1}^{\infty} p_{2n} z^{2n} = 1 - (1 - 4pqz^2)^{1/2}$

This is more convenient than (11.2.8) because the exponent of $z$ is the time of first return.

### Example 11.2.2
In how many ways can one perform a random walk between $t = 0$ and $t = 16$ and return to the starting point for the first time at $t = 16$?

Since probabilities are not at issue, only the ways of walking, we can choose probabilities for our convenience. Suppose $p = \frac{1}{2}$. Then all $2^{16}$ ways of taking 16 steps are equally likely. But then

$P$(returning for first time to zero at $t = 16$)

$$= \frac{\text{number of ways of returning in this way}}{2^{16}}$$

or

the number of ways of returning for first time at $t = 16$

$$= 2^{16} \cdot \frac{1}{2} \frac{\frac{1}{2} \cdot \frac{3}{2} \cdot \frac{5}{2} \cdot \frac{7}{2} \cdot \frac{9}{2} \cdot \frac{11}{2} \cdot \frac{13}{2}}{8!} (4 \cdot \tfrac{1}{2} \cdot \tfrac{1}{2})^8$$

$$= 2^8 \frac{1 \cdot 3 \cdot 5 \cdot 7 \cdot 9 \cdot 11 \cdot 13}{8 \cdot 7 \cdot 6 \cdot 5 \cdot 4 \cdot 3 \cdot 2 \cdot 1} = 2 \cdot 3 \cdot 11 \cdot 13 = 858$$

### Example 11.2.3
Suppose $p = \frac{2}{3}$. Find the probability that the first return to zero occurs after $t = 4$.

If $p = \frac{2}{3}$,

$P$(returning to zero for first time after $t = 4$)

$$= p_6 + p_8 + p_{10} + \cdots = 1 - (1 - 4pq)^{1/2} - p_2 - p_4$$

$$= 1 - (1 - \tfrac{8}{9})^{1/2} - 2 \cdot \tfrac{2}{3} \cdot \tfrac{1}{3} - \frac{1}{2} \cdot \frac{\frac{1}{2}}{2!} (4 \cdot \tfrac{2}{3} \cdot \tfrac{1}{3})^2$$

$$= 1 - \tfrac{1}{3} - \tfrac{4}{9} - \tfrac{8}{81} = \tfrac{10}{81}$$

We now turn to $v_{2n+1}$. Naturally, $v_1 = p = (1/2q)2pq = (1/2q)p_2$.

**Theorem 11.2.2**     Under our current hypotheses with $P(H) = p$, $\quad P(T) = q$ for $n = 1, 2, 3, \ldots$

$$v_{2n+1} = P(x_s(2n+1) = 1 \text{ and } x_s(t) < 1 \text{ for } 0 < t < 2n + 1)$$

$$= \frac{1}{2q} p_{2n+2} = \frac{1 \cdot \frac{1}{2} \cdot \frac{3}{2} \cdots (\frac{1}{2} + n - 1)}{4 \quad (n+1)!} 4^{n+1} p^{n+1} q^n$$

PROOF     Let $s = (b_1, b_2, \ldots, b_{2n+1})$ be a sequence such that $b_j \in \{1, -1\}$ and $b_1 + b_2 + \cdots + b_{2n+1} = 1$ while

(11.2.9)     $b_1 + b_2 + \cdots + b_k < 1, \qquad k = 1, 2, \ldots, 2n$

A coin toss sequence such as $s$ is the typical sequence for which $x_s(2n+1) = 1$ and $x_s(t) < 1$ for $0 \le t < 2n + 1$.

Let $s^* = (-1, b_1, b_2, \ldots, b_{2n+1})$. The sequence $s^*$ is of length $2n + 2$. Since

$$-1 + b_1 + b_2 + \cdots + b_{2n+1} = -1 + 1 = 0$$

and by (11.2.9)

$$-1 + b_1 + b_2 + \cdots + b_k < 0, \qquad k = 1, 2, \ldots, 2n$$

this says that

$$x_{s^*}(2n+2) = 0 \quad \text{and} \quad x_{s^*}(t) < 0 \quad \text{for } 0 < t < 2n + 2$$

Conversely, if $x_{s^*}(2n+2) = 0$ and $x_{s^*}(t) < 0$ for $0 < t < 2n + 2$ then $s^* = (-1, a_2, a_3, \ldots, a_{2n+2})$. But this states that

$$-1 + a_2 + a_3 + \cdots + a_{2n+2} = 0 \quad \text{or} \quad a_2 + a_3 + \cdots + a_{2n+2} = 1$$

and

$$-1 + a_2 + a_3 + \cdots + a_k < 0 \quad \text{or} \quad a_2 + a_3 + \cdots + a_k < 1,$$
$$k = 2, 3, \ldots, 2n + 1$$

Thus the random walks which reach $x = 1$ for the first time at $t = 2n + 1$ are in one-to-one correspondence with the random walks which return to $x = 0$ for the first time at $t = 2n + 2$ but are always to the left of $x = 0$ for $0 < t < 2n + 2$. By the ideas of Example 11.2.2 there are

$$\frac{1 \cdot \frac{1}{2} \cdot \frac{3}{2} \cdot \frac{5}{2} \cdots (\frac{1}{2} + n - 1)}{2 \quad (n+1)!} 4^{n+1}$$

random walks which return to $x = 0$ for the first time at $t = 2n + 2$. Half of them are always to the left of $x = 0$ for $0 < t < 2n + 2$. Each random walk which reaches $x = 1$ for the first time at $t = 2n + 1$ has probability $p^{n+1} q^n$ so

$$v_{2n+1} = \frac{1}{2} \cdot \frac{1}{2} \frac{1 \cdot \frac{1}{2} \cdot \frac{3}{2} \cdot \frac{5}{2} \cdots (\frac{1}{2} + n - 1)}{(n+1)!} 4^{n+1} p^{n+1} q^n$$

$$= \frac{1}{2q} \left[ \frac{1 \cdot \frac{1}{2} \cdot \frac{3}{2} \cdot \frac{5}{2} \cdots (\frac{1}{2} + n - 1)}{2 \quad (n+1)!} (4pq)^{n+1} \right] = \frac{1}{2q} p_{2n+2}$$

**Corollary**     If $z \neq 0$ and $|z| < 1$,

$$\sum_{n=0}^{\infty} v_{2n+1} z^{2n+1} = \frac{1}{2qz} [1 - (1 - 4pqz^2)^{1/2}]$$

PROOF

$$v_1 = p = \frac{1}{2q} 2pq = \frac{1}{2q} p_2$$

The other coefficients are covered by the theorem. Therefore,

$$\sum_{n=0}^{\infty} v_{2n+1} z^{2n+1} = \sum_{n=0}^{\infty} \frac{1}{2qz} p_{2n+2} z^{2n+2} = \frac{1}{2qz} \sum_{n=0}^{\infty} p_{2(n+1)} z^{2(n+1)}$$

$$= \frac{1}{2qz} \sum_{m=1}^{\infty} p_{2m} z^{2m} = \frac{1}{2qz} [1 - (1 - 4pqz^2)^{1/2}]$$

by (11.2.8a).

### Example 11.2.4
Find the probability that our random walk never reaches $x = 1$.

As in Example 11.2.1, the probability of ever reaching $x = 1$ is

$$v_1 + v_3 + v_5 + \cdots = \sum_{n=0}^{\infty} v_{2n+1} 1^{2n+1} = \frac{1}{2q \cdot 1} [1 - (1 - 4pq \cdot 1^2)^{1/2}]$$

$$= \frac{1}{2q} [1 - (1 - 4pq)^{1/2}]$$

Therefore the probability of never reaching $x = 1$ is $1 - (1/2q)[1 - (1 - 4pq)^{1/2}]$.

### Example 11.2.5
Let $v_{2n}^{(2)} = P(x_s(2n) = 2$ and $x_s(t) < 2$ for $t < 2n)$. This is the probability of reaching $x = 2$ for the first time at $t = 2n$, for $n = 1, 2, 3, \ldots$. Show that

$$\sum_{n=1}^{\infty} v_{2n}^{(2)} z^{2n} = \left\{ \frac{1}{2qz} [1 - (1 - 4pqz^2)^{1/2}] \right\}^2$$

In order to reach $x = 2$ for the first time at $t = 2n$ the walker must have reached $x = 1$ for the first time at $t = 2k + 1$ and from there have reached $x = 2$ for the first time at $t = 2n$ (see Figure 4) where $2n = 12$ and $2k + 1 = 7$.

Now reaching $x = 2$ for the first time at $t = 2n$ given that we are starting from $x = 1$ at time $t = 2k + 1$ has probability $v_{2n-(2k+1)}$. Therefore

$P(x_s(2k + 1)) = 1$ for first time starting from $x = 0$ and $x_s(2n)$

$\qquad = 2$ for first time after starting from $x = 1$ at time $t = 2k + 1)$

$\qquad = v_{2k+1} v_{2n-2k-1}$

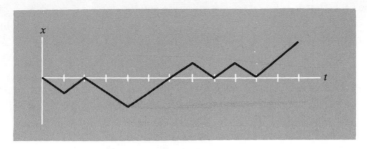

**Figure 4**

Since a walk may reach $x = 1$ for the first time for $t = 2k + 1$ where $k = 0$, $1, 2, \ldots, n - 1$ and since it cannot reach $x = 1$ for the *first time* at two different time instants, we have

$$v_{2n}^{(2)} = \sum_{k=0}^{n-1} v_{2k+1} v_{2n-2k-1}$$

We multiply both sides by $z^{2n}$ and sum on $n = 1, 2, 3, \ldots$ to get

$$\sum_{n=1}^{\infty} v_{2n}^{(2)} z^{2n} = \sum_{n=1}^{\infty} \sum_{k=0}^{n-1} v_{2k+1} z^{2k+1} v_{2n-2k-1} z^{2n-2k-1}$$

$$= \left( \sum_{k=0}^{\infty} v_{2k+1} z^{2k+1} \right) \left( \sum_{r=0}^{\infty} v_{2r+1} z^{2r+1} \right)$$

since $2k + 1$ and $2n - 2k - 1$ represent any two odd numbers greater than zero whose sum is $2n$. By the corollary to Theorem 11.2.2

$$\left( \sum_{k=0}^{\infty} v_{2k+1} z^{2k+1} \right) \left( \sum_{r=0}^{\infty} v_{2r+1} z^{2r+1} \right) = \left( \frac{1}{2qz} [1 - (1 - 4pqz^2)^{1/2}] \right)^2$$

It seems quite clear, on the basis of Example 11.2.5, that if $v_m^{(r)} = P(x_s(r) = m$ and $x_s(t) < m$ for $t < r)$ where $r > 1$ is some positive integer, then

**(11.2.10)** $$\sum_{m=0}^{\infty} v_m^{(r)} z^m = \left\{ \frac{1}{2qz} [1 - (1 - 4pqz^2)^{1/2}] \right\}^r$$

Needless to say, $v_m^{(r)} = 0$ unless a random walk can reach $x = r$ in $m$ steps. Note that $r$ and $m$ must both be odd or both be even, i.e., $r + m$ is even if $v_m^{(r)} > 0$.

Before we deal with $r_n(a, b)$ we give some intuition on this problem in a slightly different form. Suppose a gambling house has a coin with bias $P(H) = p$, $P(T) = q$. A gambler starts out with $c \geq 0$ dollars and plays so that he wins \$1 on H and loses \$1 on T. The fortune of the gambling house is $l \geq c$ dollars. Find the probability

$$u_n(c) = P\begin{pmatrix} \text{gambler's fortune reaches } \$l \text{ for the first time} \\ \text{at } n\text{th turn and he does not go broke before} \end{pmatrix}$$

It can be seen that $u_n(c) = r_n(c, l - c)$ since starting at $x = c$ and being banned from $x = 0$ before reaching $x = l > c > 0$ is the same as starting from $x = 0$ and being banned from $x = -a$ before reaching $x = b > 0 > -a$ if $l - c = b - 0 = b$ and $c - 0 = 0 - (-a) = a$.

**Theorem 11.2.3**    A random walk is governed by independent tosses with $P(H) = p$, $P(T) = q = 1 - p$ and our current definitions. Then, if $d(z) = \sqrt{1 - 4pqz^2}$ and $l \geq c \geq 0$,

$$U(z, c) = \sum_{n=0}^{\infty} u_n(c)z^n = (2pz)^{l-c}\frac{[1 + d(z)]^c - [1 - d(z)]^c}{[1 + d(z)]^l - [1 - d(z)]^l}$$

PROOF    Certain cases have easy answers. For $n = 0, 1, 2, 3, \ldots$, $u_n(0) = 0$; that is, the gambler starts out broke so he goes broke before reaching $l$. Next,

$$u_n(l) = \begin{cases} 1, & n = 0 \\ 0, & n = 1, 2, 3, \ldots \end{cases}$$

i.e., if he starts with $l$ he reaches $l$ for the first time at $n = 0$. If $0 < c < l$ and $n = 0, 1, 2, 3, \ldots$

(11.2.11)    $u_{n+1}(c) = pu_n(c + 1) + qu_n(c - 1)$

This follows because he can get H or T on the first toss. If he gets H he need only reach $l$ from $c + 1$ in $n$ steps. If he gets T he must reach $l$ from $c - 1$ in $n$ steps. In both cases he must reach $l$ for the first time in $n$ steps and he cannot go broke first. We also are using the independence of the first toss from the next $n$ tosses. Note how $u_n(0) = 0$ and $u_n(l)$ fit into equation (11.2.11) when $c = 1$ or $c = l - 1$.

We multiply both sides of equation (11.2.11) by $z^{n+1}$ and sum over $n = 0, 1, 2, 3, \ldots$. For $0 < c < l$,

$$\sum_{n=0}^{\infty} u_{n+1}(c)z^{n+1} = U(z, c) - u_0(c) = U(z, c)$$

since $u_0(c) = 0$ in that case. This settles the left-hand side. We note that $U(z, l) = 1$ and $U(z, 0) = 0$. This procedure applied to the right-hand side of (11.2.11) gives

$$p \sum_{n=0}^{\infty} u_n(c + 1)z^{n+1} + q \sum_{n=0}^{\infty} u_n(c - 1)z^{n-1}$$

$$= zp \sum_{n=0}^{\infty} u_n(c + 1)z^n + zq \sum_{n=0}^{\infty} u_n(c - 1)z^n = zpU(z, c + 1) + zqU(z, c - 1)$$

We equate those results and manipulate to get

(11.2.12)    $zpU(z, c + 1) = U(z, c) - zqU(z, c - 1)$   for $0 < c < l$
          and $U(z, l) = 1$,     $U(z, 0) = 0$.

By successively setting $c = 1, 2, 3, \ldots, l - 1$ in (11.2.12) we can find $U(z, c + 1)$ in terms of $U(z, 1)$ and other expressions. But since $U(z, l) = 1$ we then determine $U(z, 1)$ and all $U(z, c)$ for $2 \leq c \leq l - 1$. Thus there is one and only one solution of equation (11.2.12) with $U(z, 0) = 0$ and $U(z, l) = 1$. The reader can verify that

$$U(z, c) = (2pz)^{l-c} \frac{[1 + d(z)]^c - [1 - d(z)]^c}{[1 + d(z)]^l - [1 - d(z)]^l}$$

satisfies the conditions. Admittedly, we are not providing insight on why one should try such an expression.

### Example 11.2.6

What is the probability that the gambler will break the gambling house before going broke if $p < \frac{1}{2}$? What if $p = \frac{1}{2}$?

This problem calls for

$$\sum_{n=0}^{\infty} u_n(c) = \sum_{n=0}^{\infty} u_n(c) 1^n = U(1, c)$$

$$= (2p)^{l-c} \frac{(1 + \sqrt{1 - 4pq})^c - (1 - \sqrt{1 - 4pq})^c}{(1 + \sqrt{1 - 4pq})^l - (1 - \sqrt{1 - 4pq})^l}$$

As we can see, if $p < \frac{1}{2}$, $\sqrt{1 - 4pq} = \sqrt{1 - 4p + 4p^2} = 1 - 2p$. In that case our answer is

$$(2p)^{l-c} \frac{(2 - 2p)^c - (2p)^c}{(2 - 2p)^l - (2p)^l} = 2^{l-c} p^{l-c} \frac{2^c[(1 - p)^c - p^c]}{2^l[(1 - p)^l - p^l]}$$

$$= \frac{p^{l-c}(q^c - p^c)}{q^l - p^l} = \frac{\left(\dfrac{q}{p}\right)^c - 1}{\left(\dfrac{q}{p}\right)^l - 1}$$

If $p = \frac{1}{2}$, this expression for $\sum_{n=0}^{\infty} u_n(c)$ breaks down. However, it seems plausible that we can take

$$\lim_{\substack{p \to 1/2 \\ p < 1/2}} \frac{\left(\dfrac{q}{p}\right)^c - 1}{\left(\dfrac{q}{p}\right)^l - 1} = \lim_{\substack{p \to 1/2 \\ p < 1/2}} \frac{\left(\dfrac{1-p}{p}\right)^c - 1}{\left(\dfrac{1-p}{p}\right)^l - 1}$$

$$= \lim_{w \to 1} \frac{w^c - 1}{w^l - 1} = \lim_{w \to 1} \frac{cw^{c-1}}{lw^{l-1}} = \frac{c}{l}$$

by l'Hospitals rule. Now we note that

$$\sum_{n=0}^{\infty} r_n(a, b) z^n = \sum_{n=0}^{\infty} u_n(c) z^n, \qquad c = a \text{ and } l = a + b$$

Therefore,

**(11.2.13)**  $$\sum_{n=0}^{\infty} r_n(a, b)z^n = (2pz)^b \frac{[1 + d(z)]^a - [1 - d(z)]^a}{[1 + d(z)]^{a+b} - [1 - d(z)]^{a+b}}$$

We now ask for the probability $u_n^*(c)$ of the gambler going broke for the first time at the $n$th toss when he has not previously broken the bank and starts with \$$c$. He still wins \$1 with probability $p$ and losses \$1 with probability $q$. However, if we reverse the roles of going broke and breaking the bank this is the same as first reaching $c > 0$ at the $n$th step when the probability of a positive movement is $q$ and not previously reaching $-(l - c) < 0$ given that the random walk starts at $x = 0$. By (11.2.13) we have

**(11.2.14)**  $$\sum_{n=0}^{\infty} u_n^*(c)z^n = \sum_{n=0}^{\infty} r_n(l - c, c)z^n = (2qz)^c \frac{[1 + d(z)]^{l-c} - [1 - d(z)]^{l-c}}{[1 + d(z)]^l - [1 - d(z)]^l}$$

since $d(z) = \sqrt{1 - 4pqz^2}$ does not change when $p$ and $q$ are interchanged.

In this section we have presented the more accessible first-passage probabilities. They are so named because they are related to the first time a random walk passes to some kind of "boundary." The reader may not feel that they are so accessible. However, there are many applications for these probabilities. Furthermore, first-passage probabilities for more complicated random processes or for more complicated boundaries are extremely difficult to evaluate.

**EXERCISES 11.2** _____

1   Show that if $0 < p < 1$,  $p(1 - p) \le \frac{1}{4}$ and $p(1 - p) = \frac{1}{4}$ if and only if $p = \frac{1}{2}$.

2   Show that if $p \le \frac{1}{2}$,  $\sqrt{1 - 4pq} = 1 - 2p$ and if $p \ge \frac{1}{2}$,  $\sqrt{1 - 4pq} = 1 - 2q$.

3   Find the probability that our random walk starting at $x = 0$ with $p < \frac{1}{2}$
    (a) ever reaches $x = 1$,   (b) never reaches $x = 1$,   (c) ever reaches $x = 4$,
    (d) never reaches $x = 4$.

4   In how many different ways can our random walk ($x_0 = 0$) be performed between $t = 0$ and $t = 18$ so that it returns to $x = 0$ for the first time at $t = 18$?

5   Let $\alpha_{2n+1} = P$(our random walk reaches $x = 1$ for the second time at $t = 2n + 1$) and $x_0 = 0$. Show that

$$\sum_{n=0}^{\infty} \alpha_{2n+1} z^{2n+1} = \left(\sum_{n=1}^{\infty} p_{2n} z^{2n}\right)\left(\sum_{n=0}^{\infty} v_{2n+1} z^{2n+1}\right), \qquad |z| < 1$$

[*Hint:* Find $\alpha_{2n+1}$ in terms of the $p_{2k}$ and the $v_{2l+1}$ and use exercises 1.5.10–1.5.14.]

6   Find $v_8^{(2)}$ and $v_8^{(4)}$ if $p = \frac{1}{2}$ by finding the coefficient of $z^8$ in equation (11.2.10) with proper choices of $r$.

7   If $p > \frac{1}{2}$ show that $P(x_s$ eventually reaches $x = 1) = 1$.

8   Prove Lemma 11.2.1 by mathematical induction.

9   In reference to Lemma 11.2.2 show, using mathematical induction, that

$$f^{(n)}(w) = \frac{(2n)!}{n!}(1 - 4w)^{-1/2 - n}$$

In exercises 10 and 11 let $\beta(p)$ be the probability of our gambler (who starts with $c > 0$ dollars) breaking the bank (which starts with $l > c$ dollars) before he goes broke. He still wins \$1 with probability $p$ and loses \$1 with probability $q$.

10   Show that if $p_1 < p_2$, $\beta(p_1) < \beta(p_2)$ by studying the ways in which he can break the bank and comparing the probability of each under $p_1$ and $p_2$.

11   Find $\beta(p)$ if $p > \frac{1}{2}$. Show that

$$\lim_{\substack{p \to 1/2 \\ p > 1/2}} \beta(p) = \frac{c}{l}$$

How does this prove that $\beta(\frac{1}{2}) = c/l$ when combined with Example 11.2.6 and exercise 7?

12   Let $l = 4$. Compute $U(z, 2)$, $U(z, 3)$, and $U(z, 4)$ by successive substitution in equation (11.2.12). Then determine $U(z, 1)$ from the relationship $U(z, 4) = 1$ if $l = 4$.

13   Show that

$$U(z, c) = (2pz)^{l-c} \frac{[1 + d(z)]^c - [1 - d(z)]^c}{[1 + d(z)]^l - [1 - d(z)]^l}$$

satisfies equation (11.2.12).

14   Let $c_n \geq 0$ for $n = 0, 1, 2, \ldots$. Suppose $g(z) = \sum_{n=0}^{\infty} c_n z^n$ and the series converges for $0 < z < 1$ and possibly other $z$. Show that $\sum_{n=0}^{\infty} c_n$ converges to $C$ if and only if $g(z) \to C$ as $z \to 1$, $z < 1$. Note that $\sum_{n=0}^{\infty} c_n$ diverges if and only if $g(z) \to \infty$ as $z \to 1$, $z < 1$ since $g(z)$ is monotonically increasing as $z \to 1$.

[*Hint:* A series of positive terms converges if and only if there is some $B$ which is greater than all of its partial sums. If $N$ is fixed $\lim_{z \to 1} \sum_{n=0}^{N} c_n z^n = \sum_{n=0}^{N} c_n$.]

15   If $p \neq \frac{1}{2}$ show that $\sum_{u=0}^{\infty} v_{2n+1} z^{2n+1}$ converges for $|z| < 1 + \epsilon$ for suitable $\epsilon$. Find

$$\frac{d}{dz}\left(\sum_{n=0}^{\infty} v_{2n+1} z^{2n+1}\right)$$

for those $z$.

In exercises 16 and 17 let $\Omega$ be the sample space of random walks of this type described in this chapter.

**16** Let $Y$ be defined on $\Omega$ by $Y(x_s) = 2n + 1$ if $x_s$ reaches $x = 1$ for the first time at time $2n + 1$. Assume $p > \frac{1}{2}$ so that $P(0 < Y < \infty) = 1$ and $Y$ is a genuine random variable. Find $E(Y)$.

[*Hint:*

$$E(Y) = \sum_{n=0}^{\infty} (2n + 1)v_{2n+1}$$

and

$$\frac{d}{dz} \sum_{n=0}^{\infty} v_{2n+1}z^{2n+1} = \sum_{n=0}^{\infty} (2n + 1)v_{2n+1}z^{2n}, \qquad 0 < z < 1 + \epsilon]$$

**17** Let $X$ be defined on $\Omega$ with $p = \frac{1}{2}$ by $X(x_s) = 2n$ if $x$ returns to $x = 0$ for the first time at $t = 2n$. Note that $X$ is a genuine random variable only if $p = \frac{1}{2}$. Show that $E(X)$ does not exist.

♦ **18** Let $x_0 = 0$ and $T_{2n} = $ amount of time up to $t = 2n$ during which the graph of $x_s$ is above the $t$ axis.

(a) Show that if $0 < k < n$,

$$P(T_{2n} = 2k) = \sum_{j=1}^{n} P(x_s(u) > 0 \text{ for } 0 < u < 2j, \ x_s(2j) = 0)P(T_{2n-2j} = 2k - 2j)$$

$$+ \sum_{j=1}^{n} P(x_s(u) < 0 \text{ for } 0 < u < 2j, \ x_s(2j) = 0)P(T_{2n-2j} = k)$$

(b) Show that if $n \geq 1$,

$$P(T_{2n} = 2n) = \sum_{j=1}^{n} P(x_s(u) > 0 \text{ for } 0 < u < 2j, \ x_s(2j) = 0)P(T_{2n-2j} = 2n - 2j)$$

$$+ \sum_{j=n+1}^{\infty} P(x_s(u) > 0 \text{ for } 0 < u < 2j, \ x_s(2j) = 0)$$

$$P(T_{2n} = 0) = \sum_{j=1}^{n} P(x_s(u) < 0 \text{ for } 0 < u < 2j, \ x_s(2j) = 0)P(T_{2n-2j} = 0)$$

$$+ \sum_{j=n+1}^{\infty} P(x_s(u) < 0 \text{ for } 0 < u < 2j, \ x_s(2j) = 0)$$

and

$$P(T_0 = 0) = 1$$

[*Hint:* Just analyze the different ways in which $x_s$ can spend various amounts of time above the $t$ axis in terms of its first return to the $t$ axis and whether it was above or below before its first return. Draw some pictures.]

♦ **19** Show that if the random walk is governed by a fair coin and $x_0 = 0$, then for $|\alpha| < 1$, $|\beta| < 1$

$$Q(\alpha, \beta) = \sum_{n=0}^{\infty} \sum_{k=0}^{n} P(T_{2n} = 2k)\alpha^{2n} \beta^{2k} = [(1 - \alpha^2)(1 - \alpha^2 \beta^2)]^{-1/2}$$

[*Hint:* Exercise 18, (11.2.8a) and the ideas of section 1.5.]

# 11.3  Speeding up the random walk

In section 11.1 we considered a leisurely random walk. Moves occurred every second and the standard step was one unit. Now suppose we imagine the coin to be tossed every $\Delta$ seconds where $\Delta < 1$. For example, if $\Delta = \frac{1}{5}$ there are five tosses per second.

No new ideas are needed to obtain results because a sequence of tosses does not have an intrinsic time scale. Furthermore, we need not assume that steps of length one unit are taken. However, we do assume that in any fixed case the steps are of equal magnitude. Unless otherwise mentioned we assume independent tosses of a fair coin.

## Example 11.3.1

Consider a random walk controlled by a coin tossed $n$ times per second where each step is of length $c > 0$. The initial position $x_0 = 0$. Suppose $a < b$. For large $n$, estimate

$P$(position of the walk at $t = 1$ is greater than $a$ and less than or equal to $b$)

If $s = (s_1, s_2, s_3, \ldots)$ is a coin toss sequence which controls a random walk and if the steps are of length $c$, then the change of position due to the $k$th toss is $cs_k$; i.e., $c$ if $s_k$ refers to H and $-c$ if $s_k$ refers to T. Since $x_0 = 0$ and there are $n$ tosses per second, the position of the walk at time $t = 1$ is

$$cs_1 + cs_2 + \cdots + cs_n = c(s_1 + s_2 + \cdots + s_n)$$

We may set $X_k(s) = s_k$ and use the central limit theorem.

$$P(a < c(s_1 + s_2 + \cdots + s_n) \leq b) = P\left(\frac{a}{c} < s_1 + s_2 + \cdots + s_n \leq \frac{b}{c}\right)$$

$$= P\left(\frac{a}{c\sqrt{n}} < \frac{s_1 + s_2 + \cdots + s_n}{\sqrt{n}} \leq \frac{b}{c\sqrt{n}}\right)$$

But $\sigma^2(X_k) = 1$ and $E(X_k) = 0$. Therefore, for large $n$,

$$P\left(\frac{a}{c\sqrt{n}} < \frac{s_1 + s_2 + \cdots + s_n}{\sqrt{n}\,\sigma^2(X_k)} \leq \frac{b}{c\sqrt{n}}\right) \approx \int_{a/c\sqrt{n}}^{b/c\sqrt{n}} \frac{e^{-x^2/2}}{\sqrt{2\pi}}\,dx$$

Since $0 \leq e^{-x^2/2}/\sqrt{2\pi} \leq 1/\sqrt{2\pi}$,

$$0 \leq \int_{a/c\sqrt{n}}^{b/c\sqrt{n}} \frac{e^{-x^2/2}}{\sqrt{2\pi}}\,dx \leq \frac{1}{\sqrt{2\pi}}\left(\frac{b}{c\sqrt{n}} - \frac{a}{c\sqrt{n}}\right) = \frac{1}{\sqrt{2\pi}}(b - a)\frac{1}{c\sqrt{n}}$$

as $n \to \infty$, $\quad \dfrac{1}{\sqrt{2\pi}}(b - a)\dfrac{1}{c\sqrt{n}} \to 0$.

It seems quite clear that, for example,

$P$(position of the walk at $t = 1$ is greater than $a$ and less than or equal to $b$) $\to 0$,

as $n \to \infty$

Thus, if $c$ is fixed, it is very unlikely that the random walk will be found in a fixed finite interval when $n$ is very large. In a sense, the walk shoots off to infinity before $t = 1$. However, if $c$ varies with $n$ and if $c\sqrt{n} \to l \neq 0$ the random walk will not shoot off to infinity before $t = 1$. In a sense we should say that the system of random walks will not do this. For definiteness we assume $c = 1/\sqrt{n}$ henceforth. In that case

$$\lim_{n \to \infty} P\left(a < \frac{1}{\sqrt{n}}(s_1 + s_2 + \cdots + s_n) \leq b\right) = \int_a^b \frac{e^{-x^2/2}}{\sqrt{2\pi}}\, dx$$

### Example 11.3.2

Consider a sequence of random walks in which $n = 2^r$ tosses per second are used and $c = 1/\sqrt{n}$ is the step length. As usual $x_0 = 0$. Let $x(\tfrac{7}{8})$ and $x(\tfrac{9}{4})$ be the positions of the walk at times $t = \tfrac{7}{8}$ and $\tfrac{9}{4}$, respectively. Find $E(e^{iux(7/8) + ivx(9/4)})$ for each $n$ and find $\lim_{r \to \infty} E(e^{iux(7/8) + ivx(9/4)})$.

If $r \geq 2$, $\tfrac{7}{8} \cdot 2^r = 7 \cdot 2^{r-3}$ and $\tfrac{9}{4} \cdot 2^r = 9 \cdot 2^{r-2}$ are integers.

$$x(\tfrac{7}{8}) = \frac{1}{2^{r/2}}(s_1 + s_2 + \cdots + s_{7 \cdot 2^{r-3}})$$

and

$$x(\tfrac{9}{4}) = \frac{1}{2^{r/2}}(s_1 + s_2 + \cdots + s_{9 \cdot 2^{r-2}})$$

Therefore

$$E(e^{iux(7/8) + ivx(9/4)}) = E\left(\exp\left[\frac{iu}{2^{r/2}}(s_1 + s_2 + \cdots + s_{7 \cdot 2^{r-3}})\right.\right.$$

$$\left.\left. + \frac{iv}{2^{r/2}}(s_1 + s_2 + \cdots + s_{9 \cdot 2^{r-2}})\right]\right)$$

$$= E\left(\exp\left[\frac{i(u + v)}{2^{r/2}}(s_1 + s_2 + \cdots + s_{7 \cdot 2^{r-3}})\right.\right.$$

$$\left.\left. + \frac{iv}{2^{r/2}}(s_{7 \cdot 2^{r-3} + 1} + \cdots + s_{9 \cdot 2^{r-2}})\right]\right)$$

because $7 \cdot 2^{r-3} < 9 \cdot 2^{r-2}$. Now the $s_j$ define independent, identically distributed random variables, and

$$E(e^{i\xi s_k}) = \tfrac{1}{2}e^{i\xi} + \tfrac{1}{2}e^{-i\xi} = \cos \xi$$

Therefore we have

$$E(e^{iux(7/8) + ivx(9/4)}) = \left(\cos\frac{u + v}{2^{r/2}}\right)^{7 \cdot 2^{r-3}}\left(\cos\frac{v}{2^{r/2}}\right)^{9 \cdot 2^{r-2} - 7 \cdot 2^{r-3}}$$

$$= \left[\left(\cos\frac{u + v}{2^{r/2}}\right)^{2^r}\right]^{7/8}\left[\left(\cos\frac{v}{2^{r/2}}\right)^{2^r}\right]^{9/4 - 7/8}$$

By Theorem 10.3.3 and the MacLaurin series for the cosine function

$$\lim_{n \to \infty} \left( \cos \frac{\xi}{\sqrt{n}} \right)^n = e^{-\xi^2/2}$$

Therefore, with $n = 2^r$ and $\sqrt{n} = 2^{r/2}$,

**(11.3.1)**  $\lim_{r \to \infty} E(e^{iux(7/8) + ivx(9/4)}) = (e^{-(u+v)^2/2})^{7/8}(e^{-v^2/2})^{9/4 - 7/8}$

It seems clear that if $t_1 < t_2$ and if $t_1 = k_1 2^{-j_1}$ and $t_2 = k_2 2^{-j_2}$ for positive integers $k_1, k_2, j_1, j_2$ then equation (11.3.1) can be generalized to

**(11.3.2)**  $\lim_{r \to \infty} E(e^{iux(t_1) + ivx(t_2)}) = (e^{-(u+v)^2/2})^{t_1}(e^{-v^2/2})^{t_2 - t_1}$

$$= e^{-t_1(u+v)^2/2} e^{-(t_2 - t_1)v^2/2}$$

Note that the dyadic rationals $k2^{-j}$ have the property that for any $r > j$, $k2^{-j} \cdot 2^r = k2^{r-j}$ is an integer. We use these numbers to avoid interpolation; it is just a helpful device. Note also that between any two real numbers $t_4 > t_3 \geq 0$ there is a number of the form $k2^{-j}$.

We now look at

$$e^{-t_1(u+v)^2/2} e^{-(t_2 - t_1)v^2/2} = \exp\left[ -\tfrac{1}{2}(t_1 u^2 + 2t_1 uv + t_2 v^2) \right]$$

It is not surprising that this is reminiscent of the characteristic function of a normal random vector. (Example 11.3.1 did suggest the normal distribution.)

However, what kind of normal random vector gives rise to the characteristic function of (11.3.2)? Let $X$ be $N(0, t_1)$, $Y$ be $N(0, t_2 - t_1)$, and let $X$ and $Y$ be independent. Then

$$E(\exp[iuX + iv(X + Y)]) = E(\exp[i(u + v)X + ivY])$$

$$= e^{-(u+v)^2 t_1/2} e^{-v^2(t_2 - t_1)/2}$$

since $X$ and $Y$ are independent and $\phi_X(\xi) = e^{-\xi^2 t_1/2}$ while $\phi_Y(\xi) = e^{-\xi^2(t_2 - t_1)/2}$. Therefore $(X, X + Y)$ gives rise to the characteristic function of (11.3.2).

Let us look at this result in another way which is more intuitive. First note that $(1/\sqrt{n})(s_1 + s_2 + \cdots + s_{nt_1})$ is approximately $N(0, t_1)$ and $(1/\sqrt{n})(s_{nt_1 + 1} + s_{nt_1 + 2} + \cdots + s_{nt_2})$ is approximately $N(0, t_2 - t_1)$. They are independent because they are controlled by independent tosses. But

$$x(t_1) = \frac{1}{\sqrt{n}}(s_1 + s_2 + \cdots + s_{nt_1})$$

and

$$x(t_2) = \frac{1}{\sqrt{n}}(s_1 + s_2 + \cdots + s_{nt_1}) + \frac{1}{\sqrt{n}}(s_{nt_1 + 1} + s_{nt_1 + 2} + \cdots + s_{nt_2})$$

By section 10.5 $(x(t_1), x(t_2))$ should be a normal random vector, approximately.

We have allowed our notation to degenerate. Note that the subscript "$s$" of $x_s$ has been dropped. This has been done not only for convenience but also because the random walk determined by a fixed sequence $(a_1, a_2, a_3, \ldots)$ is different for different values of $n$. We are also using $x(t_1)$ to denote the coordinate of infinitely many different things since $x(t_1) = (1/\sqrt{n})(s_1 + s_2 + \cdots + s_{nt_1})$ depends on $n$. However, if we stay alert we should have no trouble with the random walks that change as $n$ changes.

Suppose we consider equation (11.3.2) in still another way. Let $t_1 < t_2$ be numbers of the form $k \cdot 2^{-j}$. Let $x(t_1)$ and $x(t_2)$ be the positions at time $t_1$ and $t_2$, respectively, of our random walk with $n = 2^{-r}$. If $Z_1$ is $N(0, t_2)$, $Z_2$ is $N(0, t_2 - t_1)$ and $Z_1$ and $Z_2$ are independent we have seen that (11.3.2) states that

$$\phi_{(x(t_1), x(t_2))}(u, v) \to \phi_{(Z_1, Z_1 + Z_2)}(u, v)$$

By Theorem 10.4.1 we have for all $a_1 < b_1$ and $a_2 < b_2$

$$\lim_{r \to \infty} P(a_1 < x(t_1) \leq b_1 \text{ and } a_2 < x(t_2) \leq b_2)$$

$$= P(a_1 < Z_1 \leq b_1 \text{ and } a_2 < Z_1 + Z_2 \leq b_2)$$

For a picture suggesting the events related to $\{a_1 < x(t_1) \leq b_1\}$ and $\{a_2 < x(t_2) \leq b_2\}$ we have Figure 5. The event $\{a_i < x(t_i) \leq b_i\}$ may be interpreted as the set of walks whose graphs $x(t)$ pass through the slot determined by $a_i$ and $b_i$ at time $t_i$.

We now turn to a theorem which generalizes the results of Example 11.3.2.

It is convenient to define a product notation similar to the $\Sigma$ notation for sums. Henceforth, if $a_j$ are numbers with $j = m, m + 1, \ldots, n$ we let

$$\prod_{j = m}^{n} a_j = a_m a_{m+1} a_{m+2} \cdots a_n$$

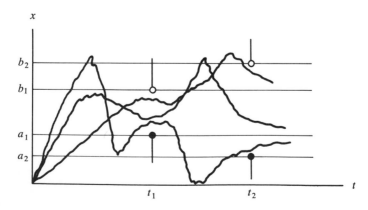

**Figure 5**

From this point on we will be concerned with times $0 = t_0 < t_1 < t_2 < \cdots < t_m$ and with random variables $Z_1, Z_2, \ldots, Z_m$ which are independent and each $Z_l$ is $N(0, t_l - t_{l-1})$.

**Theorem 11.3.1**     Let a random walk starting at $x_0$ be controlled by independent tosses of a fair coin occurring $2^r$ times per unit time. The step length is $2^{-r/2}$. Let $x(t)$ be the position of the walk at time $t$. Then

(i)   $\displaystyle\lim_{r \to \infty} P(x(t_1) \leq b_1, x(t_2) \leq b_2, \ldots, x(t_m) \leq b_m) = F_X(b_1, b_2, \ldots, b_m)$

where $F_X$ is the distribution function of

$$X = (x_0 + Z_1, x_0 + Z_1 + Z_2, x_0 + Z_1 + Z_2 + Z_3, \ldots, x_0 + Z_1 + Z_2 + \cdots + Z_m)$$

(ii)   The density function of $X$ is

$$f(x_1, x_2, \ldots, x_m) = \prod_{l=1}^{m} \frac{1}{\sqrt{2\pi(t_l - t_{l-1})}} \exp\left[-\frac{(x_l - x_{l-1})^2}{2(t_l - t_{l-1})}\right]$$

PROOF     We restrict ourselves to $m = 3$ and assume that $t_l = k \cdot 2^{-j}$, $k, j$ positive integers. Then, for sufficiently large $r$, $2^r t_l$ is an integer. We first calculate the density function of $X$. The density function of $(Z_1, Z_2, Z_3)$ is

$$\prod_{l=1}^{3} \frac{1}{\sqrt{2\pi(t_l - t_{l-1})}} \exp\left[-\frac{z_l^2}{2(t_l - t_{l-1})}\right]$$

This is so because $Z_1, Z_2, Z_3$ are independent and each is $N(0, t_l - t_{l-1})$. Let $x_1 = x_0 + z_1$, $x_2 = x_0 + z_1 + z_2$, $x_3 = x_0 + z_1 + z_2 + z_3$ be a change of coordinates from $z_1, z_2, z_3$ to $x_1, x_2, x_3$. The number $x_0$ is a constant but its notation is convenient. Note that $z_1 = x_1 - x_0$, $z_2 = x_2 - x_1$, and $z_3 = x_3 - x_2$. The Jacobian

$$\frac{\partial(x_1, x_2, x_3)}{\partial(z_1, z_2, z_3)} = \begin{vmatrix} 1 & 0 & 0 \\ 1 & 1 & 0 \\ 1 & 1 & 1 \end{vmatrix} = 1 \cdot \begin{vmatrix} 1 & 0 \\ 1 & 1 \end{vmatrix} - 0 \cdot \begin{vmatrix} 1 & 0 \\ 1 & 1 \end{vmatrix} + 0 \cdot \begin{vmatrix} 1 & 1 \\ 1 & 1 \end{vmatrix}$$

$$= \begin{vmatrix} 1 & 0 \\ 1 & 1 \end{vmatrix} = 1 \cdot 1 - 1 \cdot 0 = 1$$

Let $X_1 = x_0 + Z_1$, $X_2 = x_0 + Z_1 + Z_2$, $X_3 = x_0 + Z_1 + Z_2 + Z_3$. The 3-dimensional form of (8.5.5) corresponding to a one-to-one transformation with $X_k = g_k(Z_1, Z_2, Z_3)$ and $k = 1, 2, 3$ is

$$f_{(X_1, X_2, X_3)}(x_1, x_2, x_3) = \frac{f_{(Z_1, Z_2, Z_3)}(z_1, z_2, z_3)}{|J(z_1, z_2, z_3)|}$$

where

$$J(z_1, z_2, z_3) = \begin{vmatrix} \partial g_1/\partial z_1 & \partial g_1/\partial z_2 & \partial g_1/\partial z_3 \\ \partial g_2/\partial z_1 & \partial g_2/\partial z_2 & \partial g_2/\partial z_3 \\ \partial g_3/\partial z_1 & \partial g_3/\partial z_2 & \partial g_3/\partial z_3 \end{vmatrix}$$

In the present case our notation includes $z_1 = x_1 - x_0$, $z_2 = x_2 - x_1$, $z_3 = x_3 - x_2$, and $|J(z_1, z_2, z_3)| = 1$. Therefore

$$f_X(x_1, x_2, x_3) = \prod_{l=1}^{3} \frac{1}{\sqrt{2\pi(t_l - t_{l-1})}} \exp\left[-\frac{(x_l - x_{l-1})^2}{2(t_l - t_{l-1})}\right]$$

This demonstrates (ii) of the conclusion of the theorem. We note that by the definition of $Z_1, Z_2, Z_3$

$$\phi_{(Z_1, Z_2, Z_3)}(u_1, u_2, u_3) = \prod_{l=1}^{3} \exp\left[-\tfrac{1}{2}(t_l - t_{l-1})u_l^2\right]$$

$$= \exp\left[-\tfrac{1}{2}\sum_{l=1}^{3} (t_l - t_{l-1})u_l^2\right]$$

Therefore,

$$\phi_{(x_0 + Z_1, x_0 + Z_1 + Z_2, x_0 + Z_1 + Z_2 + Z_3)}(u_1, u_2, u_3)$$
$$= E(\exp\{i[u_1(x_0 + Z_1) + u_2(x_0 + Z_1 + Z_2) + u_3(x_0 + Z_1 + Z_2 + Z_3)]\})$$
$$= E(\exp[i(u_1 + u_2 + u_3)x_0 + i(u_1 + u_2 + u_3)Z_1 + i(u_2 + u_3)Z_2 + iu_3 Z_3])$$
$$= \exp(i(u_1 + u_2 + u_3)x_0) \exp\left(-\tfrac{1}{2}\sum_{l=1}^{3}(t_l - t_{l-1})(u_l + \cdots + u_3)^2\right)$$

For sufficiently large $r$

$$x(t_l) = x_0 + 2^{r/2}(s_1 + s_2 + \cdots + s_{t_l \cdot 2^r})$$

Therefore, using the characteristic function computations of Example 11.3.2,

$$E(\exp\{i[u_1 x(t_1) + u_2 x(t_2) + u_3 x(t_3)]\})$$
$$= \exp\{i(u_1 + u_2 + u_3)x_0 + i(u_1 + u_2 + u_3)2^{-r/2}(s_1 + \cdots + s_{2^r t_1})$$
$$+ i(u_2 + u_3)2^{-r/2}(s_{2^r t_1 + 1} + \cdots + s_{2^r t_2})$$
$$+ iu_3 2^{-r/2}(s_{2^r t_2 + 1} + \cdots + s_{2^r t_3})\}$$
$$= \exp[i(u_1 + u_2 + u_3)x_0]\left[\left(\cos\frac{(u_1 + u_2 + u_3)}{2^{r/2}}\right)^{2^r}\right]^{t_1}$$

**(11.3.4)**
$$\times \left[\left(\cos\frac{(u_2 + u_3)}{2^{r/2}}\right)^{2^r}\right]^{t_2 - t_1}\left[\left(\cos\frac{u_3}{2^{r/2}}\right)^{2^r}\right]^{t_3 - t_2}$$

The limit of expression (11.3.4) as $r \to \infty$ is

**(11.3.5)**   $\exp[i(u_1 + u_2 + u_3)x_0]\prod_{l=1}^{3} \exp\left[-\tfrac{1}{2}(t_l - t_{l-1})(u_l + \cdots + u_3)^2\right]$

But expression (11.3.5) is exactly (11.3.3). Since the random vector $X$ has a density function, Theorem 10.4.1 implies that

$$\lim_{r \to \infty} P(x(t_1) \le b_1, \ldots, x(t_m) \le b_m) = F_X(b_1, b_2, \ldots, b_m)$$

since (11.3.5) is the limit of characteristic functions of $(x(t_1), \ldots, x(t_m))$.

It is reasonable to ask the significance of such rapid changes of direction which are inherent in our "fast" random walks. Stock market fluctuations do not occur more frequently than a few times per minute. Furthermore, they usually occur in multiples of $\$\frac{1}{8}$. Our model may not apply to the stock market.

However, Brownian motion, in which a small but macroscopic particle is bombarded by molecules may be rather well described by our "fast" random walk. Another phenomenon which may also be so described is "noise" in an electronic device.

Aside from these concrete problems there are theoretical reasons to study this fast random walk. Unfortunately, in a first course it is impossible to show the connections between the fast random walk and other topics in probability let alone the connections with other branches of mathematics.

**EXERCISES 11.3** ————————————————————

1   Graph the random walks corresponding to the sequences in exercise 11.1.1 if the coin is tossed 16 times per second, the step length is $\frac{1}{4}$ and $x_0 = 0$.

2   Let $X_1, X_2, X_3, \ldots$ be independent random variables with $P(X_j = \pm 1) = \frac{1}{2}$ for all $j$. Find the limiting distribution of $x_0 + An^{-1/2}(X_1 + X_2 + \cdots + X_{nt})$ where $t > 0$ is a dyadic rational and $n = 2, 4, 8, 16, \ldots$.

3   With the notation of exercise 2 and $u \, (> t)$ another dyadic rational find the limiting (joint) distribution of

$$(x_0 + An^{-1/2}(X_1 + X_2 + \cdots + X_{nt}), \, x_0 + An^{-1/2}(X_1 + X_2 + \cdots + X_{nu}))$$

4   Let $X_1, X_2, X_3, \ldots$ be independent random variables with $P(X_j = 1) = \frac{1}{2} + Bn^{-1/2}$ and $P(X_j = -1) = \frac{1}{2} - Bn^{-1/2}$ for $|Bn^{-1/2}| < \frac{1}{2}$. Find the expectation and variance of $Y_n = An^{-1/2}(X_1 + X_2 + \cdots + X_{nt})$ when $t > 0$ is a dyadic rational, $n = 2^m$, and $nt$ is an integer. Find the limiting distribution of $Y_n$. Find the expectation and variance of the limiting distribution. (We have a kind of random walk with drift.)

5   Find the limiting distribution of $n^{-1/2}(X_1 + X_2 + \cdots + X_{nt})$ where the $X_j$ are independent, identically distributed random variables with $E(X_j) = 0$, $\sigma^2(X_j) = 1$, and $t$ as above.

6   Find the limiting distributions of the sequence of random vectors $n^{-1/2}W_{nt}$, where $n$ and $t$ are as above and (a) $W_k$ is as in exercise 11.1.12 and (b) $W_k$ is as in exercise 11.1.14. Recall that $a(X, Y) = (aX, aY)$.

# ★ 11.4  The Wiener* Brownian Motion Process

In section 11.3 we sped up a random walk. We found that if $0 = t_0 < t_1 < t_2$ and $x(t)$ is the position of the random walk at time $t$ then

$$\lim_{r \to \infty} P(a_1 < x(t_1) \le b_1, a_2 < x(t_2) \le b_2) = \int_{a_1}^{b_1} \int_{a_2}^{b_2} \left[ \prod_{j=1}^{2} \frac{e^{-(x_j - x_{j-1})^2/2(t_j - t_{j-1})}}{\sqrt{2\pi(t_j - t_{j-1})}} \right] dx_1 \, dx_2$$

if the random walk is controlled by a fair coin tossed $2^r$ times per unit time and the step length is $2^{-r/2}$. It is natural to ask if the limit operation cannot be disposed of in some way. This is analogous to a transition from difference quotients to derivatives or from binomial distributions to normal distributions.

Henceforth we let

**(11.4.1)**  $\quad g(y; m, t) = \dfrac{1}{\sqrt{2\pi t}} e^{-(y-m)^2/2t}$

because it occurs very often.

For any fixed $r$ and any fixed sequence of tosses we get a particular continuous function when we graph the resulting random walk. If we were to plot only for $0 \le t \le 1$ we would get $2^{2^r}$ different functions since there are $2^r$ tosses which govern the walk between $t = 0$ and $t = 1$ and each toss can have two outcomes.

Let $y_0$ (to be used in place of $x_0$) be a fixed real number. Let $\Omega = \{\omega; \omega$ is a real-valued continuous function on $t \ge 0$ with $\omega(0) = y_0\}$.

For each $t \ge 0$ we define a random variable (now using lower-case letters for random variables) $x_t: \Omega \to R$ by

**(11.4.2)**  $\quad x_t(\omega) = \omega(t)$

This is a different symbolism than the $x_s(t)$ of sections 11.1–11.3.

What is $P(x_t \le u)$? We cannot say at this stage because we have no probability measure on $\Omega$. We do not even have a system $\mathscr{A}$ of events on $\Omega$. We take our basic system of events $\mathscr{A}$ as the smallest admissible system containing

$$A_{t_1, a} = \{\omega \in \Omega; \omega(t_1) \le a\}$$

as $t_1$ ranges over all non-negative real numbers and $a$ ranges over all real numbers. Note that these events belonging to $\mathscr{A}$ guarantee that each $x_t$ is a random variable. An intersection of a finite number of such events is

$$\{\omega \in \Omega; \omega(t_1) \le a_1, a(t_2) \le a_2, \ldots, \omega(t_m) \le a_m\}$$

for $0 \le t_1 < t_2 < \cdots < t_m$ and $a_1, a_2, \ldots, a_m$ arbitrary real numbers. Such an event can be thought of as a system of "infinite slots" illustrated in Figure 6.

Another event in $\mathscr{A}$ is $\{\omega \in \Omega; \omega(t_1) > a\}$ since $\mathscr{A}$ is closed under complementation. Therefore if $0 < t_1 < t_2 < \cdots < t_m$ and

$$B = \{\omega \in \Omega; a_1 < \omega(t_1) \le b_1, a_2 < \omega(t_2) \le b_2, \ldots, a_m < \omega(t_n) \le b_m\}$$

* Named for Norbert Wiener (1894–1964). We will call it the Wiener process.

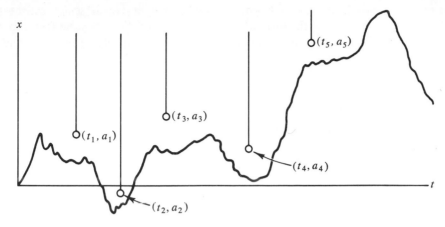

**Figure 6**

is an event in $\mathcal{A}$, what should $P(B)$ equal? We are really asking what should we use as the probability measure on $\Omega$? There are infinitely many possibilities. We mention certain preliminary measures. Let $r$ be fixed. Certain of the functions $\omega \in \Omega$ are graphs of random walks of our current type. Our preliminary measure essentially gives all *other* $\omega$ probability zero. We define

> $P_r(B) = $ probability of the graph of a random walk controlled by our usual setup with $2^r$ tosses per unit time and step length $2^{-r/2}$ passing through the slots of $B$; i.e., the probability that the walk has coordinate greater than $a_k$ and less than or equal to $b_k$ at each time $t_k$ for $k = 1, 2, 3, \ldots, m$.

The measures $P_r$ are similar to discrete probability measures. If we had restricted ourselves to a bounded $t$ interval $0 \le t \le T$ we could make the same definitions only with $0 < t_1 < t_2 < \cdots < t_m \le T$. Then a $P_r$ as defined above would be a discrete probability measure since for *fixed $r$* there are only finitely many random walks up to time $T$. The graphs of these walks would be equally likely, too.

From Theorem 11.3.1 we have

$$\lim_{r \to \infty} P_r(B) = \int_{a_1}^{b_1} \int_{a_2}^{b_2} \cdots \int_{a_m}^{b_m} [\prod_{k=1}^{m} g(y_k; y_{k-1}, t_k - t_{k-1})] \, dy_1 \, dy_2 \cdots dy_m$$

if we set $t_0 = 0$ as usual. What could be nicer than to have

**(11.4.3)**   $$P(B) = \int_{a_1}^{b_1} \int_{a_2}^{b_2} \cdots \int_{a_m}^{b_m} [\prod_{k=1}^{m} g(y_k; y_{k-1}, t_k - t_{k-1})] \, dy_1 \, dy_2 \cdots dy_m$$

The only complication is whether this can be done consistently. Recall that just because there is a (partial) measure defined on certain events it does not follow that it can be extended to all of $\mathcal{A}$ in such a way as to have all of the properties of a probability measure, especially countable additivity. Fortunately in this case there is one and only one *probability measure* on $\mathscr{A}$ which satisfies (11.4.3) on events having the form of our $B$. A proof of this is beyond the scope of this book. We will assume its validity, and state it and related results in the theorem to follow.

**Theorem 11.4.1**    Let $y_0$ be a fixed real number. Let $\Omega$, $\mathscr{A}$, $x_t$, $g$, etc. be as defined above. Then

(i)   There is one and only one probability measure $P$ on $\Omega$, $\mathscr{A}$ with

**(11.4.4)**    $P(\{\omega \in \Omega; a_k < \omega(t_k) \le b_k \text{ for } 0 < t_1 < t_2 < \cdots < t_m\})$

$$= \int_{a_1}^{b_1} \cdots \int_{a_m}^{b_m} [\prod_{k=1}^{m} g(y_k; y_{k-1}, t_k - t_{k-1})] \, dy_1 \cdots dy_m$$

(ii)   For the probability measure $P$ in (i) the random variables $x_{t_k} - x_{t_{k-1}}$, $k = 1, 2, \ldots, m$ are independent and $N(0, t_k - t_{k-1})$.

(iii)   If $P^*$ is any measure on $\Omega$, $\mathscr{A}$ such that the random variables $x_{t_k} - x_{t_{k-1}}$, $k = 1, 2, \ldots, m$ are independent and $N(0, t_k - t_{k-1})$ under $P^*$, then $P^* = P$.

PROOF    We assume the assertion in (i) since its proof is beyond the scope of this book. We note next that by definition of $x_t$

**(11.4.5)**    $\{\omega \in \Omega; a_k < x_{t_k}(\omega) \le b_k \text{ for } k = 1, 2, \ldots, m\}$

$$= \{\omega \in \Omega; a_k < \omega(t_k) \le b_k \text{ for } k = 1, 2, \ldots, m\}$$

If $Z_k = x_{t_k} - x_{t_{k-1}}$ the $Z_k$ are independent and $N(0, t_k - t_{k-1})$ if and only if the density function relation

**(11.4.6)**    $f_{(Z_1, Z_2, \ldots, Z_m)}(z_1, \ldots, z_m) = \prod_{k=1}^{m} g(z_k; 0, t_k - t_{k-1})$

holds. By the techniques used to prove Theorem 11.3.1, (11.4.6) is equivalent to

**(11.4.7)**    $f_{(x_{t_1}, x_{t_2}, \ldots, x_{t_m})}(y_1, \ldots, y_m) = \prod_{k=1}^{m} g(y_k - y_{k-1}; 0, t_k - t_{k-1})$

By definition of $g$,

**(11.4.8)**    $g(u - v; 0, t) = g(u; v, t)$

so (11.4.7) is equivalent to asserting

**(11.4.9)**    $f_{(x_{t_1}, \ldots, x_{t_m})}(y_1, \ldots, y_m) = \prod_{k=1}^{m} g(y_k; y_{k-1}, t_k - t_{k-1})$

We next recall that $f(w_1, \ldots, w_m)$ is a density function for $(W_1, W_2, \ldots, W_m)$ if and only if for all $a_k < b_k$, $k = 1, 2, \ldots, m$

**(11.4.10)**    $\int_{a_1}^{b_1} \cdots \int_{a_m}^{b_m} f(w_1, \ldots, w_m) \, dw_1 \cdots dw_m$

$$= P(a_1 < W_1 \le b_1, \ldots, a_m < W_m \le b_m)$$

Now if $P$ on $\Omega$ and $\mathscr{A}$ satisfies (11.4.4), then (11.4.5) and (11.4.10) imply (11.4.9) which implies (11.4.7) and assertion (ii) follows. Finally if assertion (iii) holds, (11.4.6), (11.4.7), (11.4.9), (11.4.10), and (11.4.5) imply that

$$P^*(\{\omega \in \Omega; a_k < \omega(t_k) \le k_k \text{ for } 0 < t_1 < \cdots < t_m\})$$

$$= \int_{a_1}^{b_1} \cdots \int_{a_m}^{b_m} [\prod_{k=1}^{m} g(y_k; y_{k-1}, t_k - t_{k-1})] \, dy_1 \cdots dy_m$$

But since there is only one measure satisfying (11.4.4)

$$P^* = P$$

Theorem 11.4.1 states that the relationship expressed by assertion (ii) characterizes the measure $P$, just as (11.4.4) characterizes $P$. Later in this section we consider some corollaries to the above result.

We now take up some problems which have beautiful heuristic solutions in the case of the Wiener process. Rigorous proofs may be found in more advanced books.

### Example 11.4.1

Consider $\Omega$, $\mathscr{A}$, and $P$ as usual. Let $y_0 < b$ and $t > 0$. Find

$$P(\omega(u) < b \text{ for all } u \text{ with } 0 \le u \le t)$$

First we find the probability of the complementary event $P(\omega(u) \ge b$ for some $u$ between 0 and $t$). Since $\omega$ is continuous and $\omega(0) = y_0 < b$, for any $\omega$ in the complementary event $u^* = \text{g.l.b.} \{u \le t, \ \omega(u) \ge b\}$ must satisfy $\omega(u^*) = b$. We leave proof of this for those interested in continuous functions.

Consider Figure 7. For every $\omega$ which hits $b$ for the first time at $u = u^*$ and continues on so that $\omega(t) > b$ there is another $\omega$ (reflection beyond $u = u^*$) for which $\omega(t) < b$. Since their behaviors from $u^*$ to $t$ are reflections of one another and since the increments $x_v - x_u$ for $v > u \ge u^*$ are independent and $N(0, v - u)$—hence symmetrical—the original $\omega$ and its reflection beyond $u^*$ have equal probability. Hence

$$P(\omega(t) > b) = P(\omega(t) < b \text{ but } \omega(u) = b \text{ for some } u < t)$$

Since $x_t$ is $N(y_0, t)$, $P(\omega(t) = b) = 0$. We therefore have

$$1 = P(\omega(t) > b) + P(\omega(t) < b \text{ but } \omega(u) = b \text{ for some } u < t)$$
$$+ P(\omega(u) < b \text{ for all } u, \quad 0 \le u \le t)$$

Now

$$P(\omega(t) > b) = P(x_t > b) = \int_b^{\infty} g(y_1; y_0; t)$$

$$= \int_b^{\infty} \frac{1}{\sqrt{2\pi t}} \exp\left[-\frac{1}{2t}(y_1 - y_0)^2\right] dy_1$$

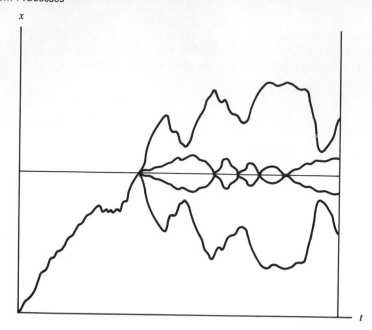

**Figure 7**

Therefore

**(11.4.11)**   $P(\omega(u) < b \text{ for all } u \le t) = 1 - 2 \int_b^\infty \dfrac{1}{\sqrt{2\pi t}} \exp\left[-\dfrac{1}{2t}(y_1 - y_0)^2\right] dy_1$

If we set $z = (y_1 - y_0)/\sqrt{t}$ and $dz = dy_1/\sqrt{t}$ we have

**(11.4.11′)**    $1 - \dfrac{2}{\sqrt{2\pi}} \int_{(b-y_0)/\sqrt{t}}^\infty e^{-z^2/2} \, dz$

Note that as $t \to \infty$,

**(11.4.12)**    $P(\omega(u) < b \text{ for all } u \le t) \to 1 - \dfrac{2}{\sqrt{2\pi}} \int_0^\infty e^{-z^2/2} \, dz$

$$= 1 - 2\int_0^\infty \frac{e^{-z^2/2}}{\sqrt{2\pi}} \, dz = 1 - 2 \cdot \tfrac{1}{2} = 0$$

Let $y_0 = 0$, $b > 0$, and $T_{b(\omega)} = \text{g.l.b. } \{u; \omega(u) \ge b\}$. We may think of $T_{b(\omega)}$ as the (random) time of first passage of $\omega$ to $b$. For $v > 0$,

$$P(T_b > v) = P(\omega(u) < b \text{ for } 0 \le u \le v)$$

Then

$$P(T_b \le v) = 1 - \left(1 - \frac{2}{\sqrt{2\pi}} \int_{b/\sqrt{v}}^\infty e^{-z^2/2} \, dz\right)$$

since $y_0 = 0$. Now

$$\frac{d}{dv} P(T_b \leq v) = \frac{d}{dv} \left( \frac{2}{\sqrt{2\pi}} \int_{b/\sqrt{v}}^{\infty} e^{-z^2/2} \, dz \right)$$

$$= \frac{2}{\sqrt{2\pi}} e^{-(b/\sqrt{v})^2/2} (-1) \frac{dv}{dv} \left( \frac{b}{\sqrt{v}} \right)$$

$$= \frac{1}{\sqrt{2\pi}} \frac{b}{v^{3/2}} e^{-b^2/2v}$$

Naturally since $P(T_b > 0) = 1$ the density is zero for $v \leq 0$.

We now pass to a theorem on conditional probabilities.

**Theorem 11.4.2**   Let $\Omega$, $P$, and $y_0$ be as usual. Let $0 < \zeta < t_1 < t_2 < \cdots$ $< t_m$. With the probabilities of (11.4.4) let

$$f(y_1, y_2, \ldots, y_m \mid x_\zeta = z_0)$$

be the conditional density of $P(x_{t_1} \leq y_1, x_{t_2} \leq y_2, \ldots, x_{t_m} \leq y_m \mid x_\zeta = z_0)$. Let $f^*$ be the density function of probabilities given by (11.4.4) in reference to $\Omega^*$, $P^*$, and $z_0$; i.e., if $\omega^* \in \Omega^*$, $\omega^*(0) = z_0$. Then

$$f(y_1, y_2, \ldots, y_m \mid x_\zeta = z_0) = f^*_{(x_{t_1 - \zeta}, x_{t_2 - \zeta}, \ldots, x_{t_m - \zeta})}(y_1, y_2, \ldots, y_m)$$

PROOF      Since $x_t(\omega) = \omega(t)$,

$f^*_{(x_{t_1 - \zeta}, x_{t_2 - \zeta}, \ldots, x_{t_m - \zeta})}(y_1, y_2, \ldots, y_m)$

$$= g(y_1; z_0, t_1 - \zeta - 0) \prod_{j=2}^{m} g(y_j; y_{j-1}, t_j - \zeta - (t_{j-1} - \zeta))$$

$$= g(y_1; z_0, t_1 - \zeta) \prod_{j=2}^{m} g(y_j; y_{j-1}, t_j - t_{j-1})$$

since $t_j - \zeta - (t_{j-1} - \zeta) = t_j - t_{j-1}$.

By section 8.6, $f(y_1, y_2, \ldots, y_m \mid x_\zeta = z_0)$ is the quotient of the density for $(x_\zeta, x_{t_1}, x_{t_2}, \ldots, x_{t_m})$ at $(z_0, y_1, y_2, \ldots, y_m)$ by the density of $x_\zeta$ at $z_0$; i.e., from (11.4.4)

$$[g(z_0; y_0, \zeta) g(y_1; z_0, t_1 - \zeta) \prod_{j=1}^{m} g(y_j; y_{j-1}, t_j - t_{j-1})] \div g(z_0; y_0, \zeta)$$

$$= g(y_1; z_0, t_1 - \zeta) \prod_{j=1}^{m} g(y_j; y_{j-1}, t_j - t_{j-1})$$

This theorem is interpreted as stating that conditional probabilities of events determined by times $t_j > \zeta$ given $\omega(\zeta) = z_0$ are the same as the probabilities of the corresponding events determined by times $t_j - \zeta$ with respect to the measure of (11.4.4) on the space of continuous functions starting at $z_0$.

Let $\omega \in \Omega$. We say that $\omega$ is *eventually less than b* if there exists a $w$ (depending on $\omega$) such that $\omega(u) < b$ if $u \geq w$.

### Example 11.4.2

Show that $P(\omega$ is eventually less than $b) = 0$.

We first note that for fixed $n > 0$,

$$P(\omega(u) < b \text{ if } u \geq n) = \lim_{m \to \infty} P(\omega(u) < b \text{ if } n \leq u \leq n + m),$$

$$m = 1, 2, 3, \ldots$$

According to Theorem 11.4.2 since $\{\omega(u) < b \text{ if } n \leq u \leq n + m\}$ is determined by "events occurring after $u = n$" we can use (11.4.11') and state for $y < b$,

$$P(\omega(u) < b \text{ for } n \leq u \leq n + m \mid \omega(n) = y) = 1 - \frac{2}{\sqrt{2\pi}} \int_{(b-y)/\sqrt{m}}^{\infty} e^{-z^2/2} \, dz$$

Now

$P(\omega(u) < b \text{ for } n \leq u \leq n + m)$

$$= \int_{-\infty}^{b} \left( 1 - \frac{2}{\sqrt{2\pi}} \int_{(b-y)/\sqrt{m}}^{\infty} e^{-z^2/2} \, dz \right) g(y; y_0, n) \, dy$$

since this is

$$\int_{-\infty}^{b} P(\omega(u) < b \text{ for } n \leq u \leq n + m \mid \omega(n) = y) g(y; y_0, n) \, dy$$

and $g(y; y_0, n)$ is the density of $x_n$. Therefore,

$$P(\omega(u) < b \text{ if } u \geq n) = \lim_{m \to \infty} \int_{-\infty}^{b} \left( 1 - \frac{2}{\sqrt{2\pi}} \int_{(b-y)/\sqrt{m}}^{\infty} e^{-z^2/2} \, dz \right) g(y; y_0, n) \, dy$$

$$= \int_{-\infty}^{b} \left[ \lim_{m \to \infty} \left( 1 - \frac{2}{\sqrt{2\pi}} \int_{(b-y)/\sqrt{m}}^{\infty} e^{-z^2/2} \, dz \right) \right] g(y; y_0, n) \, dy$$

where the interchange of limit and integral can be justified. However, as observed in (11.4.12) the limit at issue is zero. Thus

$$P(\omega(u) < b \text{ if } u \geq n) = \int_{-\infty}^{b} 0 \cdot g(y; y_0, n) \, dy = 0$$

It is well known that for any real $w$ there is a positive integer $n$ such that $n > w$ so $\{\omega(u) < b \text{ if } u \geq w\} \subset \{\omega(u) < b \text{ if } u \geq n > w\}$. Thus

$$\{\omega \text{ is eventually less than } b\} \subset \bigcup_{n=1}^{\infty} \{\omega(u) < b \text{ if } u \geq n\}$$

Since each event on the right has probability zero and there are countably many of them, their union has probability zero. Therefore $0 \geq P(\omega$ is eventually less than $0) \geq 0$.

The concept of a function being eventually greater than $b$ can be defined in the obvious way and it can be shown that $P(\omega$ is eventually greater than $b) = 0$. Thus continuous functions which are eventually greater than $+1$ or eventually less than $-1$ form an event in $\Omega$ of probability zero under the probability measure $P$.

### Example 11.4.3

Suppose $y_0 = 0$. Let $V(\omega)$ be the amount of time for which $\omega(t) \geq 0$ during $0 \leq t \leq 1$. Find the distribution function of $V$.

This problem is akin to exercise 11.2.19. We found there that if $T_{2n}$ is the amount of time up to $t = 2n$ that a standard random walk is non-negative then

**(11.4.13)**   $$\sum_{n=0}^{\infty} \sum_{k=0}^{n} P(T_{2n} = 2k)\alpha^{2n}\beta^{2k} = [(1 - \alpha^2)(1 - \alpha^2\beta^2)]^{-1/2}$$

We consider a random walk starting at $x = 0$ controlled by $2n$ tosses per unit time with the usual step length of $(2n)^{-1/2}$. The graphs of such random walks are obtained by changing the scale on graphs of a standard random walk as in Figure 8. The time scale is expanded by $2n$ and the space scale by $(2n)^{1/2}$. Thus if a standard random walk is non-negative for $2k$ units up to $t = 2n$ its rescaled "version" is non-negative for $2k/2n$ units up to $t = 1$. Since the portions above the axis stay the same, only the amount of time represented by those portions is rescaled. If things are to work out nicely, the limit as $n \to \infty$ should represent

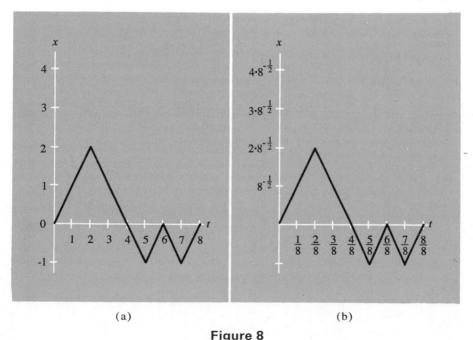

(a)                              (b)

**Figure 8**

the distribution of $V$. Let us continue heuristically, assuming that our expectations are justified.

From (11.4.13), $P(T_{2n} = 2k)$ is the coefficient of $\alpha^{2n}\beta^{2k}$ in the MacLaurin expansion of $[(1 - \alpha^2)(1 - \alpha^2\beta^2)]^{-1/2}$. Now by Lemma 11.2.2 with $w = z/4$,

$$(1 - z)^{-1/2} = \sum_{m=0}^{\infty} \binom{2m}{m} 2^{-2m} z^m$$

so we consider

$$\left(\sum_{l=0}^{\infty} \binom{2l}{l} 2^{-2l} \alpha^{2l}\right)\left(\sum_{m=0}^{\infty} \binom{2m}{m} 2^{-2m} \alpha^{2m}\beta^{2m}\right)$$

after substituting $\alpha^2$ and $\alpha^2\beta^2$ for $z$ and using different indices when these series are multiplied together. The coefficient of $\alpha^{2n}$ in the resulting series is

$$\sum_{m=0}^{n} \binom{2n - 2m}{n - m} 2^{-2(n-m)} \binom{2m}{m} 2^{-2m}\beta^{2m}$$

since $l + m = n$ if $\alpha^{2l}\alpha^{2m} = \alpha^{2n}$. The coefficient of $\beta^{2k}$ is therefore

$$\binom{2n - 2k}{n - k} 2^{-2(n-k)} \binom{2k}{k} 2^{-2k}$$

Also

$$P\left(\frac{T_{2n}}{2n} \le \frac{2j}{2n}\right) = P(T_{2n} \le 2_j) = \sum_{k=0}^{j} P(T_{2n} = 2k)$$

$$= \sum_{k=0}^{j} \binom{2n - 2k}{n - k} 2^{-2(n-k)} \binom{2k}{k} 2^{-2k}$$

By Example 1.4.2,

$$\binom{2m}{m} 2^{-2m} \sim \frac{1}{\sqrt{\pi m}}$$

We would hope (after ignoring the small term with $k = 0$) that

$$P\left(\frac{T_{2n}}{2n} \le \frac{2j}{2n}\right) \approx \sum_{k=1}^{j} \frac{1}{\sqrt{\pi(n - k)}\sqrt{\pi k}} = \frac{1}{\pi} \sum_{k=1}^{j} \frac{1}{\sqrt{\left(1 - \dfrac{k}{n}\right)\dfrac{k}{n}}} \frac{1}{n}$$

But as $n \to \infty$ if $j/n = u$,

$$\sum_{k=0}^{j} \frac{1}{\sqrt{\left(1 - \dfrac{k}{n}\right)\dfrac{k}{n}}} \frac{1}{n}$$

should approach

$$\int_0^u \frac{1}{\sqrt{v(1 - v)}} \, dv$$

since it is a sum of bases, $1/n$, times heights of the form $1/\sqrt{v(1-v)}$ where $v = k/n$. Now $P(V \le u)$ should be

$$\lim_{n \to \infty} P\left(\frac{T_{2n}}{2n} \le \frac{2j}{2n}\right), \qquad \frac{j}{n} = u$$

But this would imply

**(11.4.14)**    $P(V \le u) = \dfrac{1}{\pi} \displaystyle\int_0^u \dfrac{1}{\sqrt{v(1-v)}}\, dv = \dfrac{2}{\pi} \arcsin \sqrt{u}, \quad \text{for } 0 \le u \le 1$

All of the questioned remarks can be rigorously proved. We will assume (11.4.14).

It is sometimes said that $V$ obeys the arcsin law. A graph of the density function of the arcsin law is given in Figure 9. Thus the density is smallest when $v = \frac{1}{2}$. In a sense it is least likely that $V = \frac{1}{2}$ as compared to $V = u \ne \frac{1}{2}$ and $0 < u < 1$. For random walks, Figure 9 gives an approximate version of the truth. It is least likely that $T_{2n} = n$. In most cases $T_{2n} = 2k < n$ or $T_{2n} = 2k > n$ since once you go positive or negative it is likely that you will stay positive or negative. This plausible result seemed paradoxical to some when it was first pointed out because it is most probable that there would be equally many H's as T's.

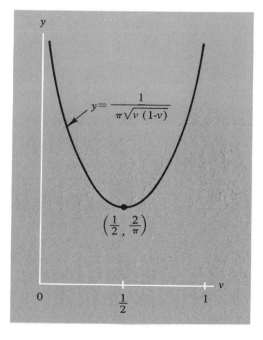

**Figure 9**

We now turn to the easiest of a number of results which show that the "well-behaved" functions in $\Omega$ form a subset of a set of measure zero under the measure $P$.

### Example 11.4.4

Let $y_0 = 0$ in reference to $\Omega$ and $P$. Then there is an event $A$ with $\{\omega \in \Omega; \omega$ is differentiable from the right at $t = 0\} \subset A$ and $P(A) = 0$.

To say that $\omega$ is differentiable from the right at $t = 0$ is to say

$$\lim_{\substack{h \to 0 \\ h > 0}} \frac{\omega(h) - \omega(0)}{h}$$

exists. Thus if $\omega$ is differentiable from the right at $t = 0$ and its derivative is $l$ there is a $\delta > 0$ such that if $0 < h < \delta$,

$$\left| \frac{\omega(h) - \omega(0)}{h} - l \right| < 1$$

and since $\omega(0) = 0$ we can say that $|\omega(h)/h| < l + 1$ for $0 < h < \delta$. This is only a consequence. It does not imply differentiability at $t = 0$. Thus

$$\{\omega; \omega \text{ is differentiable from the right at } t = 0\} \subset \bigcup_{r=1}^{\infty} \bigcup_{m=1}^{\infty} \bigcap_{n=m}^{\infty} \left\{\omega; \left| \frac{\omega(1/n)}{1/n} \right| < r \right\}$$

since this union asserts that there is some integer $r$ and some integer $m$ such that

$$\left| \frac{\omega(1/n)}{1/n} \right| < r$$

whenever $1/n \leq 1/m$. This will certainly hold because there is an $r > l + 1$ and $1/m < \delta$ in the above notation. For fixed $r$ we look at

$$P\left(\omega; \left| \frac{\omega(1/n)}{1/n} \right| < r \right) = P\left(\omega; \left| \omega\left(\frac{1}{n}\right) \right| < \frac{1}{n} r \right)$$

$$= P\left(\omega; -\frac{1}{n} r < \omega\left(\frac{1}{n}\right) < \frac{1}{n} r \right) = P\left(-\frac{1}{n} r < x_{1/n} < \frac{1}{n} r \right)$$

This is

$$\int_{(-1/n)r}^{(1/n)r} g\left(y; 0, \frac{1}{n}\right) dy = \int_{(-1/n)r}^{(1/n)r} \frac{1}{\sqrt{2\pi n^{-1}}} e^{-y^2/2n^{-1}} dy$$

where the $<$ or $\leq$ are interchangeable since $x_{1/n}$ has a continuous distribution. Let $y = z/\sqrt{n}$ so $dy = dz/\sqrt{n}$. Then

$$\int_{(-1/n)r}^{(1/n)r} \frac{1}{\sqrt{2\pi n^{-1}}} e^{-y^2/2n^{-1}} dy = \int_{(-1/n)r}^{(1/n)r} \sqrt{\frac{n}{2\pi}} e^{-ny^2/2} dy$$

$$= \int_{\sqrt{n}[(-1/n)r]}^{\sqrt{n}(1/n)r} \sqrt{\frac{n}{2\pi}} e^{-z^2/2} \frac{dz}{\sqrt{n}} = \int_{-r/\sqrt{n}}^{r/\sqrt{n}} \frac{e^{-z^2/2}}{\sqrt{2\pi}} dz$$

Now

$$\bigcap_{n=m}^{\infty} \left\{\omega; \left|\frac{\omega(1/n)}{1/n}\right| < r\right\} \subset \left\{\omega; \left|\frac{\omega(1/m)}{1/m}\right| < r\right\},$$

So

$$0 \le P\left(\bigcap_{n=m}^{\infty} \left\{\omega; \left|\frac{\omega(1/n)}{1/n}\right| < r\right\}\right) \le P\left(\omega; \left|\frac{\omega(1/m)}{1/m}\right| < r\right) = \int_{-r/\sqrt{m}}^{r/\sqrt{m}} e^{-z^2/2} \, dz$$

We are not prevented from letting $m \to \infty$. Now

$$\int_{-r/\sqrt{m}}^{r/\sqrt{m}} e^{-z^2/2} \, dz \to \int_{0}^{0} e^{-z^2/2} \, dz = 0, \quad \text{as } m \to \infty$$

since $(r/\sqrt{m}) \to 0$ as $m \to \infty$ and integrals of continuous functions are continuous functions of the upper and lower limits of integration. Thus

$$P\left(\bigcap_{n=m}^{\infty} \left\{\omega; \left|\frac{\omega(1/n)}{1/n}\right| < r\right\}\right) = 0$$

since the intersection at issue has some definite probability for each $m$. Now

$$\bigcup_{r=1}^{\infty} \bigcup_{m=1}^{\infty} \left[\bigcap_{n=m}^{\infty} \left\{\omega; \left|\frac{\omega(1/n)}{1/n}\right| < r\right\}\right]$$

is a union of countably many sets each of which has probability zero so it must have probability zero. This is a set $A$ which contains

$$\{\omega; \omega \text{ is differentiable from the right at } t = 0\}$$

The reader may ask whether the set of $\omega$ which are right differentiable at $t = 0$ is not then of probability zero. We answer by saying that if that set is not in $\mathscr{A}$ we know nothing about its probability. After all, $\mathscr{A}$ does not have all subsets of $\Omega$.

The reader may point out that we did not show that sets such as $\{\omega; \omega(u) < b$ for $0 \le u \le t\}$ are in $\mathscr{A}$. We did not. However, it is not difficult to show that this set is in $\mathscr{A}$.

Thus in some sense, the functions $\omega$ which are differentiable at $t = 0$ are unimportant to $\Omega$ under the measure $P$. How many continuous functions which are not differentiable at $t = 0$ do we know? A famous example is

$$\omega(t) = \begin{cases} 0, & t = 0 \\ t \sin \dfrac{1}{t}, & t > 0 \end{cases}$$

Once we have a measure $P$ on $\Omega$ and $\mathscr{A}$ we can obtain the distribution or moments of a myriad of random variables on $\Omega$. It is convenient to think of $\omega$

as the graph of noise in an electronic device or as the graph of a component of the position of a particle undergoing Brownian motion. Now let us prove some corollaries to Theorem 11.4.1.

**Corollary 1**    Using the notation in Theorem 11.4.1, $E(x_t) = y_0$, $\sigma^2(x_t) = t$, and $E(x_t x_u) = y_0^2 + \text{minimum } (t, u)$.

PROOF    First let $m = 1$ and $t_1 = t$. Now $x_{t_1} - x_{t_0} = x_t - y_0$ since $t_0 = 0$ by convention and $x_0(\omega) = \omega(0) = y_0$. By assertion (ii) $x_t - y_0$ is $N(0, t - 0)$. Therefore $0 = E(x_t - y_0) = E(x_t) - y_0$ and $E(x_t) = y_0$. Note also that $\sigma^2(x_t) = \sigma^2(x_t - y_0) = t$ since $x_t - y_0$ is $N(0, t)$. If $u = t$, $E(x_t x_u) = \sigma^2(x_t) + [E(x_t)]^2 = y_0^2 + t = y_0^2 + \min (u, t)$. Otherwise, if $u \neq t$ let us arrange notation so that $t < u$. We choose $m = 2$ with $t_1 = t$, $t_2 = u$. Now

(11.4.15)    $E(x_t x_u) = E(x_t(x_t + x_u - x_t)) = E(x_t^2) + E(x_t(x_u - x_t))$

But as seen above $E(x_t^2) = \sigma^2(x_t) + [E(x_t)]^2 = t + y_0^2$, so

$$E(x_t(x_u - x_t)) = E((x_t - y_0 + y_0)(x_u - x_t))$$
$$= E((x_t - y_0)(x_u - x_t)) + E(y_0(x_u - x_t))$$

Now by (ii) $x_t - y_0$ and $x_u - x_t$ are independent and $N(0, t)$ and $N(0, u - t)$, respectively. Thus $E((x_t - y_0)(x_u - x_t)) = E(x_t - y_0)E(x_u - x_t) = 0 \cdot 0$ and $E(y_0(x_u - x_t)) = y_0 E(x_u - x_t) = 0$ since $y_0$ is constant. By (11.4.11)

$$E(x_t x_u) = t + y_0^2 + 0 = y_0^2 + t = y_0^2 + \min (u, t)$$

since $t < u$.

**Corollary 2**    The covariance of $x_{t_j} - y_0$ and $x_{t_k} - y_0$ is $\min (t_j, t_k)$.

PROOF

$$\text{cov } (x_{t_j} - y_0, x_{t_k} - y_0) = E((x_{t_j} - y_0)(x_{t_k} - y_0))$$
$$= E(x_{t_j} x_{t_k}) - E(y_0 x_{t_j}) - E(y_0 x_{t_k}) + E(y_0^2)$$
$$= \min (t_j, t_k) + y_0^2 - E(y_0 x_{t_j}) - E(y_0 x_{t_k}) + y_0^2$$
$$= \min (t_j, t_k) + y_0^2 - y_0^2 - y_0^2 + y_0^2 = \min (t_j, t_k)$$

by Corollary 1 and the usual properties of expectations.

If $y_0 = 0$ this corollary states that the covariance of "noise" in a particular model of an electronic device at time $t$ and at time $u$ is $\min (t, u)$.

Let $Q_t(\omega) = \int_0^t [\omega(u)]^2 \, du$. It can be shown (but not so easily) that $Q_t$ is a random variable; $Q_t$ can be thought of as the power of the noise from 0 to $t$. Since the noise is random, the power is random.

## Example 11.4.5

Compute $E(Q_t)$ without assuming $y_0 = 0$.

We have

$$E(Q_t) = E\left(\int_0^t [\omega(u)]^2 \, du\right)$$

Now the expectation is a kind of "sum" over $\Omega$ and the integral is a "sum" over the numbers $0 \le u \le t$. Everything of interest is non-negative so it should be possible to interchange these sums. This is indeed possible and we obtain

(11.5.16)     $\displaystyle E(Q_t) = \int_0^t E([\omega(u)]^2) \, du = \int_0^t E(x_u^2) \, du$

$$= \int_0^t (u + y_0^2) \, du = \tfrac{1}{2}t^2 + ty_0^2$$

in view of Corollary 2 and in view of the definition $x_u(\omega) = \omega(u)$.

## Example 11.4.6

Let $\alpha(u)$ give a fixed function which is continuous on $0 \le u \le t$. Let $V_t(\omega) = \int_0^t \alpha(u)\omega(u) \, du$. Find $E(V_t)$ and $\sigma^2(V_t)$.

Let us first find $E(V_t)$ and $\sigma^2(V_t)$. Using the technique suggested in Example 11.4.5 (and it can be justified),

$$E(V_t) = E\int_0^t \alpha(u)\omega(u) \, du = \int_0^t E(\alpha(u)\omega(u)) \, du = \int_0^t E(\alpha(u)x_u) \, du$$

But by Corollary 1 with $u = t$,

$$E(\alpha(u)x_u) = \alpha(u)E(x_u) = \alpha(u)y_0$$

Therefore

(11.4.17)     $\displaystyle E(V_t) = \int_0^t \alpha(u)y_0 \, du = y_0 \int_0^t \alpha(u) \, du$

Let us compute $E(V_t^2)$:

$$E(V_t^2) = E\left(\left[\int_0^t \alpha(u)\omega(u) \, du\right]^2\right)$$

Now

$$\left[\int_0^t \alpha(u)\omega(u) \, du\right]^2 = \int_0^t \alpha(u)\omega(u) \, du \int_0^t \alpha(v)\omega(v) \, dv$$

$$= \int_0^t \int_0^t \alpha(u)\omega(u)\alpha(v)\omega(v) \, du \, dv$$

by results of calculus. Therefore

$$E(V_t^2) = E\left(\int_0^t \int_0^t \alpha(u)\omega(u)\alpha(v)\omega(v) \, du \, dv\right)$$

This is essentially a triple sum but we should be able to interchange the orders of summation. Therefore

**(11.4.18)**    $$E(V_t^2) = \int_0^t \int_0^t [E(\alpha(u)\omega(u)\alpha(v)\omega(v))] \, du \, dv$$

$$= \int_0^t \int_0^t [\alpha(u)\alpha(v)E(x_u x_v)] \, du \, dv$$

$$= \int_0^t \int_0^t \alpha(u)\alpha(v)[\min(u, v) + y_0^2] \, du \, dv$$

$$= \int_0^t \int_0^t \alpha(u)\alpha(v) \min(u, v) \, du \, dv + \int_0^t \int_0^t \alpha(u)\alpha(v) y_0^2 \, du \, dv$$

But

**(11.4.19)**    $$\int_0^t \int_0^t \alpha(u)\alpha(v) y_0^2 \, du \, dv = y_0^2 \left[\int_0^t \alpha(u) \, du\right]^2$$

since

$$\int_0^t \int_0^t \alpha(u)\alpha(v) \, du \, dv = \int_0^t \alpha(u) \, du \int_0^t \alpha(v) \, dv = \left[\int_0^t \alpha(u) \, du\right]^2$$

Therefore

**(11.4.20)**    $$\int_0^t \int_0^t \alpha(u)\alpha(v) \min(u, v) \, du \, dv$$

$$= \int_0^t dv \int_0^v \alpha(u)\alpha(v)u \, du + \int_0^t dv \int_v^t \alpha(u)\alpha(v)v \, du$$

since if $0 \le u \le v$, $\min(u, v) = u$ and if $v \le u \le t$, $\min(u, v) = v$.
Now

$$\sigma^2(V_t) = E(V_t^2) - [E(V_t)]^2 = \int_0^t \int_0^t \alpha(u)\alpha(v) \min(u, v) \, du \, dv$$

in view of (11.4.17), (11.4.18), and (11.4.19). Therefore $\sigma^2(V_t)$ is given by (11.4.20).

Moments of a random variable do not tell its whole probabilistic story. It is desirable to know the distribution. We now study the distribution of $V_t$ as defined in Example 11.4.6.

As an integral, $V_t$ is a limit of sums. More precisely, since $\alpha$ and each $\omega$ is a continuous function we divide the interval $0 \le u \le t$ into $n$ equal subintervals and have

**(11.4.21)**    $$V_t(\omega) = \int_0^t \alpha(u)\omega(u) \, du = \lim_{n \to \infty} \left[\sum_{j=1}^n \alpha\left(\frac{jt}{n}\right)\omega\left(\frac{jt}{n}\right)\frac{t}{n}\right]$$

where $\sum_{j=1}^{n} \alpha\left(\dfrac{jt}{n}\right)\omega\left(\dfrac{jt}{n}\right)\dfrac{t}{n}$ is a precise rendition of a sum of bases $t/n = jt/n -$ $(j-1)t/n$ times heights $\alpha(jt/n)\omega(jt/n)$. Put another way

$$V_t = \lim_{n\to\infty}\left[\frac{t}{n}\sum_{j=1}^{n}\alpha\left(\frac{jt}{n}\right)x_{jt/n}\right]$$

Heuristically speaking $V_t$ is the limit of sums of normal random variables $x_u$. Therefore, it should be a normal random variable although it may be constant, that is, $\sigma^2 = 0$.

**Lemma 11.4.1**     Let $\alpha$ be continuous on $0 \le v \le t$. Then

$$\lim_{n\to\infty}\sum_{j=1}^{n}\frac{t}{n}\left[\sum_{k=j}^{n}\alpha\left(\frac{kt}{n}\right)\frac{t}{n}\right]^2 = \int_0^t\left[\int_u^t\alpha(v)\,dv\right]^2 du$$

PROOF     We first note that, as usual,

$$\left[\sum_{k=j}^{n}\alpha\left(\frac{kt}{n}\right)\frac{t}{n}\right]^2 = \left[\sum_{k=j}^{n}\alpha\left(\frac{kt}{n}\right)\frac{t}{n}\right]\left[\sum_{l=j}^{n}\alpha\left(\frac{lt}{n}\right)\frac{t}{n}\right]$$

In equation (11.4.21) we saw a single integral as a limit of sums. The same is true for a triple integral, i.e., if $\beta$ is continuous and $u_j$, $v_k$, $w_l$ are suitable

$$\sum_{j}\sum_{k}\sum_{l}\beta(u_j, v_k, w_l)\,\Delta u_j\,\Delta v_k\,\Delta w_l$$

converges to

$$\int\int\int_M \beta(u, v, w)\,du\,dv\,dw$$

in a certain sense. Here $M$ is the region of integration. In our problem we have

**(11.4.22)**   $$\sum_{j=1}^{n}\frac{t}{n}\left[\sum_{k=j}^{n}\alpha\left(\frac{kt}{n}\right)\frac{t}{n}\right]^2 = \sum_{j=1}^{n}\sum_{k=j}^{n}\sum_{l=j}^{n}\alpha\left(\frac{kt}{n}\right)\alpha\left(\frac{lt}{n}\right)\left(\frac{t}{n}\right)^3$$

Let $u_j = jt/n$, $v_k = kt/n$, $w_l = lt/n$. Note that

$$\Delta u_j = \frac{jt}{n} - \frac{(j-1)t}{n} = \frac{t}{n}$$

Similarly $\Delta v_k = t/n$ and $\Delta w_l = t/n$. Thus if $\beta(u, v, w) = \alpha(v)\alpha(w)$ the right-hand side of equation (11.4.22) has the proper form. We must now identify $M$. In Figure 10, we have indicated a face of the cube $0 \le u, v, w \le t$. The points heavily marked have coordinates $(jt/n, kt/n, 0)$ where $k = j, j+1, \ldots, n$. These suggest where points of the form $(jt/n, kt/n, lt/n)$ for $k \ge j$ and $l \ge j$ are located. They fall in the region $M = \{(u, v, w); u \le v \le t \text{ and } u \le w \le t\}$. The small cubes whose corners are $(jt/n, kt/n, lt/n)$, $k \ge j$ and $l \ge j$, have volume $(t/n)^3$

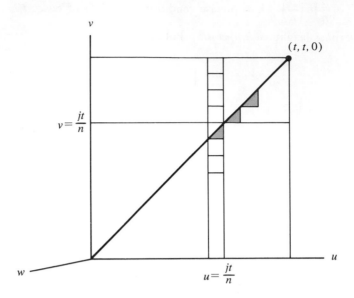

**Figure 10**

and their union is all of $M$ plus a bit extra suggested by the shaded corners. However as $n \to \infty$ the volume of the extra approaches zero. The theory of triple integrals guarantees that

$$\lim_{n \to \infty} \sum_{j=1}^{n} \frac{t}{n} \left[ \sum_{k=j}^{n} \alpha\left(\frac{kt}{n}\right) \frac{t}{n} \right]^2 = \int_0^t du \int_u^t \int_u^t \alpha(v)\alpha(w) \, dv \, dw$$

since

$$\int_u^t \int_u^t \alpha(v)\alpha(w) \, dv \, dw = \int_u^t \alpha(v) \, dv \int_u^t \alpha(w) \, dw = \left[ \int_w^t \alpha(v) \, dv \right]^2$$

Hence, the lemma is proved.

**Theorem 11.4.3**   The random variable $V_t$ defined in Example 11.4.6 is a normal random variable.

PROOF   We compute the characteristic function of $V_t$,

$$E(\exp i\xi V_t(w)) = E\left( \exp i\xi \left[ \lim_{n \to \infty} \sum_{k=1}^{n} \frac{t}{n} \alpha\left(\frac{kt}{n}\right) \omega\left(\frac{kt}{n}\right) \right] \right)$$

$$= \lim_{n \to \infty} E\left( \exp \left[ i\xi \sum_{k=1}^{n} \frac{t}{n} \alpha\left(\frac{kt}{n}\right) x_{kt/n} \right] \right)$$

by advanced theorems on interchange of $E$ and lim and the fact that $x_{kt/n}(\omega) = \omega(kt/n)$. Now

$$x_{kt/n} = x_0 + (x_{t/n} - x_0) + (x_{2t/n} - x_{t/n}) + (x_{3t/n} - x_{2t/n}) + \cdots$$
$$+ (x_{kt/n} - x_{(k-1)t/n})$$

$$= y_0 + \sum_{j=1}^{n} (x_{jt/n} - x_{(j-1)t/n})$$

Thus

$$\sum_{k=1}^{n} \alpha\left(\frac{kt}{n}\right) x_{kt/n} = \sum_{k=1}^{n} \alpha\left(\frac{kt}{n}\right) y_0 + \sum_{k=1}^{n} \alpha\left(\frac{kt}{n}\right) \sum_{j=1}^{k} (x_{jt/n} - x_{(j-1)t/n})$$

Note that

$$\sum_{k=1}^{n} \alpha\left(\frac{kt}{n}\right) \sum_{j=1}^{k} (x_{jt/n} - x_{(j-1)t/n}) = \sum_{j=1}^{n} \left[\sum_{k=j}^{n} \alpha\left(\frac{kt}{n}\right)\right](x_{jt/n} - x_{(j-1)t/n})$$

because on the left $j = 1, 2, \ldots, k$ so $x_{jt/n} - x_{(j-1)t/n}$ enters as a coefficient of $\alpha(kt/n)$ if and only if $k \geq j$. We note by Theorem 11.4.1 that the random variables $x_{jt/n} - x_{(j-1)t/n}$ are independent and $N(0, t/n)$ since

$$\frac{jt}{n} - \frac{(j-1)t}{n} = \frac{t}{n}$$

Therefore by (9.2.1) and the extension of (9.2.2)

$$E\left(\exp\left[i\xi \sum_{j=1}^{n} a_j(x_{jt/n} - x_{(j-1)t/n})\right]\right) = \prod_{j=1}^{n} \exp\left(-\tfrac{1}{2}\xi^2 a_j^2 \frac{t}{n}\right)$$

Now

(11.4.23)    $$E\left(\exp\left[i\xi \sum_{k=1}^{n} \frac{t}{n} \alpha\left(\frac{kt}{n}\right) x_{kt/n}\right]\right)$$

$$= E\left(\exp\left[i\xi \sum_{k=1}^{n} \frac{t}{n} \alpha\left(\frac{kt}{n}\right) y_0 + i\xi \sum_{k=1}^{n} \frac{t}{n} \alpha\left(\frac{kt}{n}\right) \sum_{j=1}^{n} (x_{jt/n} - x_{(j-1)t/n})\right]\right)$$

$$= \exp\left(i\xi y_0 \sum_{k=1}^{n} \frac{t}{n} \alpha\left(\frac{kt}{n}\right)\right) E\left(\exp\left\{i\xi \sum_{j=1}^{n} \left[\sum_{k=j}^{n} \frac{t}{n} \alpha\left(\frac{kt}{n}\right)\right]\right.\right.$$

$$\left.\left. \times (x_{jt/n} - x_{(j-1)t/n})\right\}\right)$$

$$= \exp\left(i\xi y_0 \sum_{k=1}^{n} \frac{t}{n} \alpha\left(\frac{kt}{n}\right)\right) \prod_{j=1}^{n} \exp\left\{-\tfrac{1}{2}\xi^2 \left[\sum_{k=j}^{n} \frac{t}{n} \alpha\left(\frac{kt}{n}\right)\right]^2 \frac{t}{n}\right\}$$

But

$$\prod_{j=1}^{n} \exp b_j = \exp \left(\sum_{j=1}^{n} b_j\right)$$

Therefore (11.4.23) is

$$\exp\left[i\xi y_0 \sum_{k=1}^{n} \frac{t}{n} \alpha\left(\frac{kt}{n}\right)\right] \exp\left\{-\tfrac{1}{2}\xi^2 \sum_{j=1}^{n}\left[\sum_{k=j}^{n} \frac{t}{n}\alpha\left(\frac{kt}{n}\right)\right]^2 \frac{t}{n}\right\}$$

Since the exponential function is continuous we can take the limit of the terms inside the "exp" and get, by Lemma 11.4.1 and (11.4.21)

$$E(\exp i\xi V_t) = \exp\left\{i\xi y_0 \int_0^t \alpha(t)\, dt - \tfrac{1}{2}\xi^2 \int_0^t \left[\int_u^t \alpha(v)\, dv\right]^2 du\right\}$$

This is the characteristic function of a normal random variable since it is of the form $\exp(i\xi a - \tfrac{1}{2}\xi^2 b)$. Actually, if $b = 0$ the random variable is a constant. We will consider this to be a degenerate normal.

Note that

$$\int_0^t \left[\int_u^t \alpha(v)\, dv\right]^2 du = 0$$

if and only if $\alpha(v) = 0$, $0 \le v \le t$.

It may seem that we have gone through a lot of agony to show the normality of $V_t$ but there are a number of techniques in the example, the lemma, and the theorem that are useful in other cases. Furthermore, electronics engineers feel that in certain devices $\int_0^t \alpha(u)\omega(u)\, du$ describes the noise at time $t$ better than $\omega(t)$.

We close this section with an example that introduces some of the ideas of modern probability research.

### ★Example 11.4.7

Let $\Omega$, $P$ be as usual with $y_0 = 0$. Define the random variables

$$I_n(\omega) = \sum_{k=1}^{n} \omega\left(\frac{k-1}{n}\right)\left[\omega\left(\frac{k}{n}\right) - \omega\left(\frac{k-1}{n}\right)\right]$$

Show that the sequence of $I_n$ converges in distribution as $n \to \infty$ to $\tfrac{1}{2}x_1^2 - \tfrac{1}{2}$ or $\tfrac{1}{2}[\omega(1)]^2 - \tfrac{1}{2}$.

Since $x_0 = y_0 = 0$

$$x_{k/n} = (x_{1/n} - x_0) + (x_{2/n} - x_{1/n}) + (x_{3/n} - x_{2/n}) + \cdots + (x_{k/n} - x_{(k-1)/n})$$

and

(11.4.24)    $$I_n = \sum_{k=1}^{n} x_{(k-1)/n}(x_{k/n} - x_{(k-1)/n})$$

$$= \sum_{k=1}^{n} \left[\sum_{j=1}^{k-1} (x_{j/n} - x_{(j-1)/n})\right](x_{k/n} - x_{(k-1)/n})$$

Now

$$x_1^2 = [(x_{1/n} - x_0) + (x_{2/n} - x_{1/n}) + (x_{3/n} - x_{2/n}) + \cdots + (x_{n/n} - x_{(n-1)/n})]^2$$

This is

$$\sum_{k=1}^{n} \sum_{j=1}^{n} (x_{j/n} - x_{(j-1)/n})(x_{k/n} - x_{(k-1)/n})$$

Since the sum on the extreme right of (11.4.24) has each product $(x_{j/n} - x_{(j-1)/n}) \cdot$

$(x_{k/n} - x_{(k-1)/n})$ for $1 \leq j < k$, $2I_n + \sum_{k=1}^{n} (x_{k/n} - x_{(k-1)/n})^2 = x_1^2$ and

$$I_n = \tfrac{1}{2}[x_1^2 - \sum_{k=1}^{n} (x_{k/n} - x_{(k-1)/n})^2]$$

We look next at $(x_{k/n} - x_{(k-1)/n})^2$. By Theorem 11.4.1 we have, for each $n$,

that $x_{k/n} - x_{(k-1)/n}$ is $N\left(0, \dfrac{k}{n} - \dfrac{k-1}{n}\right)$ or $N\left(0, \dfrac{1}{n}\right)$. Moreover these random

variables are independent. Let $Z_1, Z_2, \ldots, Z_n$ be independent $N(0, 1)$ random variables. We can arrange things so that

$$x_{k/n} - x_{(k-1)/n} = \frac{1}{\sqrt{n}} Z_k$$

since

$$E\left(\frac{1}{\sqrt{n}} Z_k\right) = 0$$

and

$$\sigma^2\left(\frac{1}{\sqrt{n}} Z_k\right) = \frac{1}{n} \cdot 1 = \frac{1}{n}$$

Thus

$$\sum_{k=1}^{n} (x_{k/n} - x_{(k-1)/n})^2 = \sum_{k=1}^{n} \frac{1}{n} Z_k^2$$

Now $E(Z_k^2) = \sigma^2(Z_k) + [E(Z_k)]^2 = 1$ and the second moment of $Z_k^2$ or $E(Z_k^4)$ exists since all moments of an $N(0, 1)$ random variable exist. By the weak law of large numbers

$$\lim_{n \to \infty} P\left(\left|\frac{Z_1^2 + Z_2^2 + \cdots + Z_n^2}{n} - 1\right| \geq \epsilon\right) = 0$$

for all $\epsilon > 0$. That is equivalent to

$$\lim_{n \to \infty} P\left(\left|\frac{Z_1^2 + Z_2^2 + \cdots + Z_n^2}{2n} - \frac{1}{2}\right| \geq \frac{\epsilon}{2}\right) = 0, \quad \text{for each } \epsilon > 0$$

or

$$0 = \lim_{n \to \infty} P\left(\left|\frac{\sum_{k=1}^{n} (x_{k/n} - x_{(k-1)/n})^2}{2} - \frac{1}{2}\right| \geq \frac{\epsilon}{2}\right) = P(|\tfrac{1}{2}x_1^2 - I_n - \tfrac{1}{2}| \geq \tfrac{1}{2}\epsilon)$$

This is equivalent to $P(\frac{1}{2}x_1^2 + I_n \geq \frac{1}{2} + \frac{1}{2}\epsilon$ or $\frac{1}{2}x_1^2 - I_n \leq \frac{1}{2} - \frac{1}{2}\epsilon)$ converges to zero as $n \to \infty$. This shows that

$$P(\tfrac{1}{2}x_1^2 - \tfrac{1}{2} \leq u - \tfrac{1}{2}\epsilon) - \delta_n < P(I_n \leq u) < P(\tfrac{1}{2}x_1^2 - \tfrac{1}{2} \leq u + \tfrac{1}{2}\epsilon) + \delta_n$$

where $\delta_n \to 0$ as $n \to \infty$. If $u$ is a continuity point of the distribution function of $\frac{1}{2}x_1^2 - \frac{1}{2}$ for any $\eta > 0$ we can choose $\frac{1}{2}\epsilon$ so that $P(\frac{1}{2}x_1^2 - \frac{1}{2} \leq u - \frac{1}{2}\epsilon)$ and $P(\frac{1}{2}x_1^2 - \frac{1}{2} \leq u + \frac{1}{2}\epsilon)$ are within $\frac{1}{2}\eta$ of $P(\frac{1}{2}x_1^2 - \frac{1}{2} \leq u)$. For that $\frac{1}{2}\epsilon$ we know that there is an $m$ such that for $n \geq m$, $\delta_n < \frac{1}{2}\eta$ and

$$|P(\tfrac{1}{2}x_1^2 - \tfrac{1}{2} \leq u) - P(I_n \leq u)| < \eta, \qquad n \geq m$$

Thus $I_n$ converges in distribution to $\frac{1}{2}x_1^2 - \frac{1}{2}$.

The limit of $I_n$ is a special case of a *stochastic integral*. Since the theory of stochastic integrals was developed by K. Ito, this limit is frequently called the *Ito integral*. It is extremely important. Exercise 11.4.20 contrasted with Example 11.4.7 gives some insight into the complexities of integrals of this type. They cannot be manipulated like ordinary integrals.

### EXERCISES 11.4

In exercises 1–3, 10 assume that $\mathscr{A}$ is an admissible system of sets of $\Omega$ and each $A_{t,a} = \{\omega; \omega(t) \leq a\}$ is in $\mathscr{A}$ for each fixed $t \geq 0$ and real $a$.

1  (a)  Show that $\{\omega; \omega(t) < b\}$ is in $\mathscr{A}$.

   (b)  If $t > 0$ show that $P(\omega(t) \leq b) = P(\omega(t) < b)$.

   (c)  Is this true if $t = 0$? Explain.

2  Show that $\{\omega \in \Omega; a_1 < \omega(t_1) \leq b_1, a_2 < \omega(t_2) \leq b_2, \ldots, a_m < \omega(t_m) \leq b_m\}$ is in $\mathscr{A}$.

3  Let $A$ be a Borel set of the real line. Show that if $t > 0$, $\{\omega; \omega(t) \in A\}$ is in $\mathscr{A}$.

4  Let $y_0 = -2$, $t_1 = 2$, $t_2 = 4$, $t_3 = 10$. Let $a_1 = -5$, $b_1 = 6$, $a_2 = 4$, $b_2 = 5.2$, $a_3 = -10$, $b_3 = -5$. Sketch some curves for $\omega$'s passing through all slots and some passing through the first and third slot but not the second slot.

5  Prove equation (11.4.8).

6  If $t > 0$ and $b \neq y$ show that

$$\frac{\partial}{\partial t} g(y; b, t) = \frac{1}{2} \frac{\partial^2}{\partial y^2} (y; b, t)$$

7  By showing that

$$\int_{-\infty}^{\infty} e^{iux} \left( \int_0^{\infty} \frac{e^{-st} e^{-x^2/2t}}{\sqrt{2\pi t}} \, dt \right) dx = \int_{-\infty}^{\infty} e^{iux} \frac{e^{-\sqrt{2s}|x|}}{\sqrt{2s}} \, dx$$

show that

$$\int_0^\infty e^{-st} \frac{e^{-x^2/2t}}{\sqrt{2\pi t}} \, dt = \frac{e^{-\sqrt{2s}\,|x|}}{\sqrt{2s}}$$

**8**  Calculate $E(x_t x_u)$ directly using (11.4.4).

**9**  If $u > t > 0$, calculate

$$\int_0^\infty \int_0^\infty \exp\left(-\frac{x^2}{2t} - \frac{(y - x)^2}{2(u - t)}\right) \frac{dx \, dy}{2\pi \sqrt{t(u - t)}}$$

by substituting $x = \sqrt{t}\, w$ and $y = \sqrt{t}\, w + \sqrt{u - t}\, z$ and using the polar coordinate substitutions $w = r \cos \theta$, $z = r \sin \theta$.

**10**  Show that $\{\omega;\ \omega(u) \le b \text{ for } 0 \le u \le t\} = \bigcap_r \{\omega;\ \omega(r) \le b\} \in \mathscr{A}$, where $r$ ranges over all rational numbers between 0 and $t$. [*Hint:* The $\omega$ are continuous.]

**11**  Show that if $y_0 < b$ and $t > 0$

$$P(\omega(u) \le b \text{ for } 0 \le u \le t) = P(\omega(u) < b \text{ for } 0 \le u \le t)$$

**12**  Let $y_0 = 0$. Using the tables for the normal distribution, find $P(\omega(u) < 2$ for $0 \le u \le 9)$.

**13**  Show that (11.4.11′) is just

$$\frac{1}{\sqrt{2\pi}} \int_{-|b - y_0|t^{-1/2}}^{|b - y_0|t^{-1/2}} e^{-z^2/2} \, dz$$

**14**  Reconcile Theorem 11.4.3 with Example 11.4.6 by showing that

$$\int_0^t \int_0^t \min\,(u,\,v)\alpha(u)\alpha(v) \, du \, dv = \int_0^t \left[\int_u^t \alpha(v) \, dv\right]^2 du$$

[*Hint:* $\left(\int_u^t \alpha(v) \, dv\right)^2 = \int_u^t \int_u^t \alpha(v)\alpha(w) \, dv \, dw$.]

**15**  Let $W_t(\omega) = \int_0^t \omega^4(u) \, du$, where $y_0 = 0$. Find $E(W_t)$. [*Hint:* Interchange expectation and integral on $u$.]

**16**  If $b < y_0$, use a heuristic argument to find $P(\omega(u) > b$ for $0 \le u \le t)$.

**17**  Show that $P(\omega$ is eventually greater than $b) = 0$.

**18**  Show that if $b > y_0$, $E(T_b)$ does not exist.

**19**  Find the distribution of $V_t$ in the case $y_0 = 0$ and

(a)  $V_t(\omega) = \int_0^t \omega(u) \, du$

(b)  $V_t(\omega) = \int_0^t (t - u)\omega(u) \, du$

**20** Show that the limiting distribution as $n \to \infty$ of

$$J_n(\omega) = \sum_{k=1}^{n} \omega\left(\frac{k}{n}\right)\left[\omega\left(\frac{k}{n}\right) - \omega\left(\frac{k-1}{n}\right)\right]$$

is that of $\frac{1}{2}\omega^2(1) + \frac{1}{2}$ in the case $y_0 = 0$.

**21** Referring to $V$ of Example 11.4.3 find $P(0 < V \le \frac{1}{4})$ and $P(\frac{1}{4} \le V \le \frac{1}{2})$.

# 11.5 Stochastic processes

A very large part of advanced probability theory is concerned with stochastic processes. In this section we introduce this concept and mention some of the more intensely studied special cases.

Let $T$ be an infinite set of real numbers. Normally $T$ is one of (a) the non-negative integers, (b) all integers, (c) an interval $a \le t \le b$, (d) the non-negative real numbers, or (e) all real numbers.

Suppose that for each finite collection of elements of $T$, say $t_1, t_2, \ldots, t_n$ with $t_1 < t_2 < \cdots < t_n$, we have an $n$-dimensional joint distribution function $F_{t_1, t_2, \ldots, t_n}$. Suppose the distribution functions are related to one another by

**(11.5.1)** $$\lim_{x_j \to \infty} F_{t_1, t_2, \ldots, t_{j-1}, t_j, t_{j+1}, \ldots, t_n}(x_1, x_2, \ldots, x_{j-1}, x_j, x_{j+1}, \ldots, x_n)$$

$$= F_{t_1, t_2, \ldots, t_{j-1}, t_{j+1}, \ldots, t_n}(x_1, x_2, \ldots, x_{j-1}, x_{j+1}, \ldots, x_n)$$

for any $j$, with the obvious interpretation when $j = 1$ or $j = n$.

It is not essential that the $t_j$ be indexed in increasing order. Thus we could have $F_{3, 1/7, \pi}(x_1, x_2, x_3)$. If we allow this then

$$F_{3, 1/7, \pi}(x_1, x_2, x_3) = F_{1/7, 3, \pi}(x_2, x_1, x_3) = F_{\pi, 3, 1/7}(x_3, x_1, x_2)$$

etc. However, under normal conditions it is most convenient to use $t_1, \ldots, t_n$ when they are in increasing order. Unless it is mentioned to the contrary, we will assume that $t_1 < t_2 < \cdots < t_n$.

The distribution functions $F_{t_1, \ldots, t_n}$ should be interpreted intuitively, as giving the probability of the phenomena being less than or equal to $x_1, x_2, \ldots, x_n$ at times $t_1, t_2, \ldots, t_n$, respectively. However, $T$ need not be a time "axis." In some applications it is a reciprocal temperature. Equation 11.5.1 is a consistency condition which states that if the restriction on the $j$th position of $F_{t_1, t_2, \ldots, t_n}$ is removed by letting $x_j \to \infty$ the resulting probability is the same as that given by

$$F_{t_1, t_2, \ldots, t_{j-1}, t_{j+1}, \ldots, t_n}(x_1, \ldots, x_{t_{j-1}}, x_{t_{j+1}}, \ldots, x_{t_n})$$

in which "time" $t_j$ is not at issue.

The reader may observe that this is essentially the basic information that we had in Theorem 11.4.1 about the Wiener Brownian motion process. However, there we had density functions and there $F_0$ was not directly specified. However, since $\omega(0) = y_0$ for all $\omega$ we could have used

$$F_{0,t_1 \ldots ,t_n}(z_0, z_1, \ldots, z_n) = \begin{cases} 0, & z_0 < y_0 \\ F_{t_1,\ldots,t_n}(z_1, \ldots, z_n), & z_0 \geq y_0 \end{cases}$$

Such a system of distribution functions will be called a *pre-stochastic process*. Naturally, a system of probability or density functions equivalent to the distribution functions $F_{t_1,t_2,\ldots,t_n}$ could be thought of as the pre-stochastic process, too.

We now point out certain types of pre-stochastic processes which have been singled out for study. We will not always use the prefix "pre."

(1) *Gaussian Processes:* For each collection $t_1, t_2, \ldots, t_n$ the distribution function is that of an $n$-dimensional normal random vector which may be degenerate.

(2) *Stationary processes:* If $t_1, t_2, \ldots, t_n \in T$ and $t_1 + u, t_2 + u, \ldots, t_n + u$ $\in T$, $F_{t_1,t_2,\ldots,t_n} = F_{t_1+u,t_2+u,\ldots,t_n+u}$.

(3) *Process with independent increments:* If $F_{t_1,t_2,\ldots,t_n}$ is the distribution function of the random vector $(Z_1, Z_2, \ldots, Z_n)$ then $Z_1, Z_2 - Z_1, Z_3 - Z_2, \ldots,$ $Z_n - Z_{n-1}$ are independent random variables.

(4) *Markov Chain:* Assuming probability functions $f_{t_1,t_2,\ldots,t_n}$ instead of distribution functions,

$$\frac{f_{t_1,t_2,\ldots,t_n+1}(x_1, x_2, \ldots, x_{n+1})}{f_{t_1,t_2,\ldots,t_n}(x_1, x_2, \ldots, x_n)} = \frac{f_{t_n,t_{n+1}}(x_n, x_{n+1})}{f_{t_n}(x_n)}$$

for all choices of $(x_1, x_2, \ldots, x_n)$ where the left-hand denominator is not zero.

We first point out that the Wiener process is Gaussian and has independent increments. The random walks controlled by independent tosses are Markov chains. They also have independent increments.

**Theorem 11.5.1** Let $S, \mathscr{A}, P$ be a probability system. Suppose that to each $t \in T$ there is a random variable $X_t$ on $S$. Let $F_{t_1,t_2,\ldots,t_n} = F_{(X_{t_1},\ldots,X_{t_n})}$. Then the system $\{F_{t_1,t_2,\ldots,t_n}\}$ is a pre-stochastic process.

PROOF We need only verify the consistency condition (11.5.1). To simplify notation we assume $n = 3$, $j = 2$, and let $t_1 = u$, $t_2 = v$, $t_3 = w$. By definition of a distribution function

$$\lim_{y \to \infty} F_{(X_u, X_v, X_w)}(x, y, z) = \lim_{m \to \infty} F_{(X_u, X_v, X_w)}(x, m, z)$$

where $m$ ranges over integers. By Theorem 3.4.4

$$\lim_{m \to \infty} P(X_u \leq x, \quad X_v \leq m, \quad X_w \leq z) = P(X_u \leq x, \quad X_w \leq z)$$

This means

$$\lim_{y \to \infty} F_{u,v,w}(x, y, z) = F_{u,w}(x, z)$$

This is the content of (11.5.1) in our case. Other cases can be treated in the same manner.

It may seem that Theorem 11.5.1 represents circular reasoning. This is not so. In the definition of a pre-stochastic process the distribution functions were assumed to exist apart from a fixed probability system. They are connected to one another by equation (11.5.1).

Theorem 11.5.1 states that a pre-stochastic process may arise from a single probability system and a collection of random variables. This was the case in section 11.4. There $S = \Omega$. However, in that case the proof of the existence of $P$ was left to more advanced books. Theorem 11.5.1 enables the reader to construct some pre-stochastic processes in cases in which $S$ and $P$ are more elementary. Furthermore we are led to a more useful mathematical entity than the pre-stochastic process.

A system consisting of a set $T \subset R$, a probability system $S$, $\mathscr{A}$, $P$, and a random variable $X_t$ for each $t \in T$ is called a *stochastic process*.

The system of distribution functions

**(11.5.2)** $\qquad F_{t_1,\ldots,t_n}(x_1,\ldots,x_n) = P(X_{t_1} \le x_1, \ldots, X_{t_n} \le x_n),$

$$\text{for all } (x_1, \ldots, x_n)$$

is called the pre-stochastic process of the given stochastic process.

In a sense stochastic processes are models or realizations of the pre-stochastic process. They are realizations in a relative way only, since they can be extremely abstract. Stochastic processes have been found to be more useful than pre-stochastic processes since they unify the system of distribution functions and refer them to a single probability system.

Every stochastic process has a pre-stochastic process which is determined by (11.5.2). Two stochastic processes with the same pre-stochastic process are said to be *equivalent*. We will apply the modifiers "Gaussian," "stationary," etc. to the corresponding stochastic processes as well as to pre-stochastic processes.

We have started with pre-stochastic processes because many of the preliminary results about stochastic processes are really results about pre-stochastic processes. Most works on the subject go immediately to stochastic processes.

The most important stochastic processes are those in which $S$ is a set of functions on $T$. These functions may be continuous or not. When we have a stochastic process in which

$$S = \{\omega; \ \omega \text{ is a particular type of function on } T\}$$

if no further mention is made we assume by convention that $\mathscr{A}$ is generated by $\{\omega; a < \omega(t_j) \le b\}$ for all $a \le b$ and all $t_j \in T$. If no special mention is made we also assume that $X_t(\omega) = \omega(t)$. We then have

$$F_{t_1,\ldots,t_n}(x_1,\ldots,x_n) = P(\omega(t_1) \le x_1, \ldots, \omega(t_n) \le x_n)$$

This will be the standard stochastic process on a set of functions on $T$. Analogous to the way in which we called elements of $S$ sample points, we frequently refer to the functions $\omega$ in such a sample space as *sample functions*.

### Example 11.5.1

Let $S = \{s; 0 \leq s \leq 1\}$, $\mathscr{A} =$ Borel subsets of $S$ and $P$ be determined by the uniform density on $S$. Let $T = \{0, 1, 2, \ldots\}$. Let $X_t(s) = s^t$. With the above definition of $F_{t_1,\ldots,t_n}$ find $F_{1,3}(\frac{1}{3}, \frac{1}{8})$ and $F_{1,3}(2, \frac{1}{8})$

By our conventions,

$$F_{1,3}(\tfrac{1}{3}, \tfrac{1}{8}) = P(X_1(s) \leq \tfrac{1}{3} \text{ and } X_3(s) \leq \tfrac{1}{8})$$
$$= P(s \in S, s \leq \tfrac{1}{3} \text{ and } s^3 \leq \tfrac{1}{8})$$

But $s^3 \leq \frac{1}{8}$ for $s \in S$ if and only if $0 \leq s \leq \frac{1}{2}$. To have $s \leq \frac{1}{3}$ and $0 \leq s \leq \frac{1}{2}$ it is necessary and sufficient that $0 \leq s \leq \frac{1}{3}$. Therefore

$$F_{1,3}(\tfrac{1}{3}, \tfrac{1}{8}) = P(s \in S; 0 \leq s \leq \tfrac{1}{3}) = \tfrac{1}{3}$$

because of uniform density. Next,

$$F_{1,3}(2, \tfrac{1}{8}) = P(s \leq 2 \text{ and } s^3 \leq \tfrac{1}{8})$$

Since $s \leq 1$ for all $s \in S$,

$$P(s \leq 2 \text{ and } s^3 \leq \tfrac{1}{8}) = P(s^3 \leq \tfrac{1}{8}) = P(0 \leq s \leq \tfrac{1}{2}) = \tfrac{1}{2}$$

We could give many more examples of stochastic processes based on the probability system of Example 11.5.1. Some of them would be less artificial than Example 11.5.1 in which all of the random variables $X_t$ are functions of one another. This makes for too much dependence. The interesting stochastic processes are those in which there is some dependence but not too much.

### Example 11.5.2

Suppose that each distribution function $F_{t_1,\ldots,t_n}$ in a pre-stochastic process is determined by a probability function $f_{t_1,\ldots,t_n}$. When $j = 1$, show that the consistency condition (11.5.1) is equivalent to

**(11.5.3)**    $\displaystyle\sum_{x_1} f_{t_1,\ldots,t_n}(x_1, \ldots, x_n) = f_{t_2,t_3,\ldots,t_n}(x_2, x_3, \ldots, x_n),$

We will consider only $n = 2$ and let $t_1 = u$, $t_2 = v$. For a distribution function to be determined by a probability function we have

$$F_{u,v}(x, y) = \sum_{\substack{a \leq x \\ b \leq y}} f_{u,v}(a, b)$$

where the sum on the right has only countable many nonzero terms.

If (11.5.1) holds we have

$$\sum_{b \le y} f_v(b) = F_v(y) = \lim_{x \to \infty} F_{u,v}(x, y)$$

$$= \lim_{x \to \infty} \sum_{\substack{a \le x \\ b \le y}} f_{u,v}(a, b) = \sum_{b \le y} \sum_{a} f_{u,v}(a, b)$$

Now $\sum_{a} f_{(u,v)}(a, b)$ is a function of $b$ only; call it $g(b)$.

We need only show that if, for all $y$,

**(11.5.4)** $$\sum_{b \le y} f_v(b) = \sum_{b \le y} g(b)$$

then $f_v(y_0) = g_v(y_0)$ for all $y_0$. However,

$$f_v(y_0) = \sum_{b \le y_0} f_v(b) - \lim_{\substack{y \to y_0 \\ y < y_0}} [\sum_{b \le y} f_v(b)]$$

Since (11.5.4) holds for all $y$ and $f_v(y_0)$ and $g(y_0)$ are determined in the same way from the left- and right-hand sides of (11.5.4), respectively, $f_v(y_0) = g(y_0)$ for all $y_0$. Thus, (11.5.3) holds in our case.

Suppose (11.5.3) holds in our case. Then $\sum_{a} f_{u,v}(a, b) = f_v(b)$ and

$$\lim_{x \to \infty} F_{u,v}(x, y) = \lim_{x \to \infty} \sum_{\substack{a \le x \\ b \le y}} f_{u,v}(a, b)$$

$$= \sum_{b \le y} \sum_{a} f_{u,v}(a, b) = \sum_{b \le y} f_v(b)$$

But

$$\sum_{b \le y} f_v(b) = F_v(y)$$

The proof of the result of Example 11.5.2 in other cases is simply a matter of notation. We will use the general result freely.

Theorem 11.5.1 has a kind of converse which is extremely important. Its proof is beyond the scope of this book, hence we just state the result. It is due to A. N. Kolmogorov.

**Theorem 11.5.2** (Kolmogorov.)   Let $T$ be a subset of the reals of the type we are using and let $\{F_{t_1, t_2, \ldots, t_n}\}$ be a pre-stochastic process. Suppose

$$\Omega = \{\omega; \omega \text{ is a function from } T \text{ to the real numbers}\}$$

Let $\mathscr{A}$ be the smallest admissible system of subsets of $\Omega$ which contains $\{\omega; \omega(t_1) \le a\}$ as $t_1$ ranges over $T$ and $a$ ranges over the real numbers. Then there is a probability measure $P$ on $\Omega$ and $\mathscr{A}$ such that for each finite set $t_1, t_2, \ldots, t_n$ in $T$ with $t_1 < t_2 < \cdots < t_n$

**(11.5.5)**     $F_{t_1, t_2, \ldots, t_n}(x_1, x_2, \ldots, x_n) = P(\omega(t_1) \leq x_1, \omega(t_2) \leq x_2, \ldots, \omega(t_n) \leq x_n)$

At first glance Theorem 11.5.2 looks like the theorem on the Wiener process. However, in that case $\Omega$ was a set of *continuous* functions. Here $\Omega$ is the set of all real-valued functions on $T$.

Kolmogorov's theorem says that given a pre-stochastic process there is always a stochastic process using the set of all functions on $T$ as the $S$ part of the realization and using $X_t(\omega) = \omega(t)$. There may be more than one stochastic process corresponding to a given pre-stochastic process. The Wiener process has a realization using continuous functions as well as the realization given by Theorem 11.5.2. Mathematicians have found it advantageous to use different realizations for a given pre-stochastic process. An advantage of the single probability measure $P$ which knits the various distribution functions together can be seen with regard to the Wiener process,

**(11.5.6)**     $\displaystyle \lim_{n \to \infty} F_{t_1, t_2, \ldots, t_n}(b, b, \ldots, b) = 1 - \frac{2}{\sqrt{2\pi}} \int_{(b - y_0)/\sqrt{t}}^{\infty} e^{-x^2/2} \, dx$

where $t_1, t_2, t_3, \ldots$ is an enumeration of the rational numbers between 0 and $t$ and $b > y_0$, $t > 0$. The expression on the right was obtained by means other than manipulation of the $F$'s on the left. Also one can use the expectation induced by the single probability measure $P$ as a "weapon" against some or all of the $X_t$ and against the $F_{t_1, \ldots, t_n}$.

In this book it will not be essential to distinguish between pre-stochastic processes and stochastic processes. We will frequently just call both stochastic processes or just processes. We take up Markov chains and Gaussian processes in later sections. At this point we consider an important process which is called a Markov chain by certain authors. It does not fit our definition of a Markov chain so we discuss it here, rather than in the section on Markov chains.

Let $T$ be the non-negative real numbers. Suppose we seek a process to describe the number of radioactive disintegrations from a mass of radioactive material after time zero. Naturally, the physical description is just an intuitive background for the problem. At any rate, it seems clear that the process should "start" at zero at time zero. Imagine that there have been $j$ "disintegrations" up to time $t$. What should happen from $t$ to $t + \Delta$? There *may* be one or more additional disintegrations in the time interval. This is a chance event. If $\Delta$ is extremely small, it is quite likely that there will be no disintegrations at all. Also when $\Delta$ is extremely small it would be incredible to have more than one disintegration. The probability that there will be one disintegration increases as $\Delta$ increases. Since it is a probability it cannot exceed 1, but we may feel that for small $\Delta$ it is *essentially* proportional to $\Delta$. We next turn to the connection between happenings before $t$ and happenings after $t$. Both physical evidence and a desire for simplicity suggest that we assume that those happenings are independent. We summarize our assumptions as follows:

(1)   Happenings after time $t$ are independent of those before time $t$

(2)   Probability of one disintegration between $t$ and $t + \Delta$ is $b\Delta + \epsilon_1(\Delta)$

**(11.5.7)**   (3)   Probability of more than one disintegration in the interval from $t$ to $t + \Delta$ is $\epsilon_2(\Delta)$

(4)   Probability of no disintegration in the interval from $t$ to $t + \Delta$ is $1 - b\Delta + \epsilon_3(\Delta)$

where $\epsilon_j(\Delta)$, $j = 1, 2, 3$, may be negative or positive and go to zero faster than $\Delta$ in the sense that $\epsilon_j(\Delta)/\Delta \to 0$.

We are now ready to work out some results. Let $f_t(j)$ signify the probability that there have been exactly $j$ disintegrations up to time $t$.

Let $g(t) = f_t(0)$. The probability that the process is still at zero at time $t + \Delta$ is $g(t)[1 - b\Delta + \epsilon_3(\Delta)]$ since we must have no disintegrations from times $t$ to $t + \Delta$ as well as none up to $t$. Parts (1) and (4) of (11.5.7) explain the multiplication as well as the term $1 - b\Delta + \epsilon_3(\Delta)$. Thus

$$g(t + \Delta) = g(t)[1 - b\Delta + \epsilon_3(\Delta)] = g(t) - b\Delta g(t) + g(t)\epsilon_3(\Delta)$$

$$g(t + \Delta) - g(t) = -b\Delta g(t) + \epsilon_3(\Delta)g(t)$$

$$\frac{g(t + \Delta) - g(t)}{\Delta} = \frac{-\Delta b g(t)}{\Delta} + \frac{\epsilon_3(\Delta)}{\Delta} g(t) = -bg(t) + \frac{\epsilon_3(\Delta)}{\Delta} g(t)$$

We let $\Delta \to 0$, and if all of these operations are justified $g'(t) = -bg(t)$ since $\epsilon_3(\Delta)/\Delta \to 0$ as $\Delta \to 0$.

It is well known that the solutions to $g'(t) = -bg(t)$ are $g(t) = c_0 e^{-bt}$. But since at $t = 0$ there definitely are no disintegrations,

$$1 = P(\text{no disintegrations at up to } t + 0) = f_0(0) = 1$$

Therefore, $1 = c_0 e^{-b \cdot 0} = c_0$ and $f_t(0) = e^{-bt}$. For $j > 0$ let

$$k_j(t) = f_t(j) = P(j \text{ disintegrations up to time } t)$$

By conditions (1)–(4) of (11.5.7)

$k_j(t + \Delta) = P(j \text{ disintegration up to } t \text{ and none between } t \text{ and } t + \Delta)$

$+ P(j - 1 \text{ disintegrations up to } t \text{ and one between } t \text{ and } t + \Delta)$

$+ P(j - 2 \text{ or fewer disintegrations up to } t \text{ and two or more between } t \text{ and } t + \Delta)$

$= k_j(t)[1 - b\Delta + \epsilon_3(\Delta)] + k_{j-1}(t)[b\Delta + \epsilon_1(\Delta)] + q\epsilon_2(\Delta)$

where $0 < q < 1$ since it is the probability of $j - 2$ or less up to $t$. Again

$$k_j(t + \Delta) = k_j(t) - b\Delta k_j(t) + \epsilon_3(\Delta)k_j(t)$$
$$+ b\Delta k_{j-1}(t) + \epsilon_1(\Delta)k_{j-1}(t) + q\epsilon_2(\Delta)$$
$$k_j(t + \Delta) - k_j(t) = -b\Delta k_j(t) + b\Delta k_{j-1}(t)$$
$$+ \epsilon_3(\Delta)k_j(t) + \epsilon_1(\Delta)k_{j-1}(t) + q\epsilon_2(\Delta)$$
$$\frac{k_j(t + \Delta) - k_j(t)}{\Delta} = -bk_j(t) + bk_{j-1}(t) + \frac{\epsilon_3(\Delta)}{\Delta}k_j(t)$$
$$+ \frac{\epsilon_1(\Delta)}{\Delta}k_{j-1}(t) + q\frac{\epsilon_2(\Delta)}{\Delta}$$

Letting $\Delta \to 0$ we have

$$k_j'(t) = -bk_j(t) + bk_{j-1}(t)$$

since $\epsilon_j(\Delta)/\Delta \to 0$.

This looks difficult because it is a differential equation with two "dependent variables" $k_j(t)$ and $k_{j-1}(t)$. However, we can work our way up, step by step. Take $j = 1$,

$$k_1'(t) = -bk_1(t) + bk_0(t) = -bk_1(t) + be^{-bt}$$

since $k_0(t) = g(t) = e^{-bt}$. This is a common linear differential equation sometimes written $y' + by = be^{-bt}$. Its solution is of the form $y = c_0e^{-bt} + u(t)$, where $u' + bu = be^{-bt}$. We try $u = ate^{-bt}$ since $u = e^{-bt}$ does not work. First, $u' = ae^{-bt} - bate^{-bt}$. Now $ae^{-bt} - bate^{-bt} + bate^{-bt} = ae^{-bt}$. In order that $ae^{-bt} = be^{-bt}$, we must have $a = b$. So far our solution is

$$k_1(t) = c_1e^{-bt} + bte^{-bt}$$

However, since $P(\text{one disintegration up to } t = 0) = 0$,

$$0 = k_1(t) = c_1e^{-b \cdot 0} + 0 \cdot e^{-b \cdot 0} = c_1$$

We have $f_1(t) = k_1(t) = bte^{-bt}$.

Next with $j = 2$,

$$k_2'(t) = -bk_2(t) + bk_1(t)$$

Here we have, essentially, $y' + by = b^2te^{-bt}$. We leave it to the reader to show that $y = c_2e^{-bt} + \frac{1}{2}(bt)^2e^{-bt} = k_2(t)$. Since $k_2(0) = 0$, $c_2 = 0$.

The next case should make the situation clear:

$$k_3'(t) + bk_3(t) = b(\tfrac{1}{2}b^2t^2e^{-bt})$$

or

$$y' + by = \tfrac{1}{2}b^2t^2e^{-bt}$$

We can use $y = c_3e^{-bt} + at^3e^{-bt}$. Therefore, we need only verify whether

$$3at^2e^{-bt} - bat^3e^{-bt} + bat^3e^{-bt} = \tfrac{1}{2}b^3t^2e^{-bt}$$

This is true precisely if $a = \dfrac{1}{3 \cdot 2} b^3$. Again $c_3 = 0$ since $k_3(0) = 0$. We have

$$k_3(t) = \frac{1}{3 \cdot 2} b^3 t^3 e^{-bt}. \text{ Since } k_2(t) = \tfrac{1}{2} b^2 t^2 e^{-bt}, \quad k_1(t) = (bt)e^{-bt} = \frac{bt}{1!} e^{-bt}, \text{ and}$$

$$k_0(t) = e^{-bt} = \frac{(bt)^0}{0!} e^{-bt} \text{ this suggests that } f_t(j) = k_j(t) = \frac{1}{j!}(bt)^j e^{-bt}. \text{ For}$$

fixed $t$ this is the Poisson distribution. This process is called the *Poisson process*. However, we do not have a pre-stochastic process. We have only candidates for probability functions $f_t$. We need probability functions $f_{t_1,t_2,\dots,t_n}$ which obey a consistency condition.

We treat the case of $n = 3$. We let $t_1 = t$, $t_2 = u$, $t_3 = v$ with $t \le u \le v$. Then we have $f_{t,u,v}(j, k, l)$ since $j$ is the number of disintegrations up to $t$, $k$ is the number of disintegrations up to $u$, and $l$ is the number of disintegrations up to $v$. Now if $j \le k \le l$, the event of $j$ disintegrations from 0 up to $t$, $k$ up to $u$, and $l$ up to $v$ means that there must be $k - j$ from $t$ to $u$ and $l - k$ from $u$ to $v$. Since we assume that happenings before $t$ are independent of those after $t$ and since we are assuming that the "rate of disintegration" $b$ is the same at all times, $P(k - j \text{ disintegrations from } t \text{ to } u) = P(k - j \text{ disintegrations from } 0 \text{ to } u - t)$

$$= \frac{[b(u - t)]^{k-j}}{(k - j)!} e^{-b(u-t)}$$

We could argue similarly about the $l - k$ disintegrations from $u$ to $v$. Therefore,

$$f_{t,u,v}(j, k, l) = \frac{(bt)^j}{j!} \frac{[b(u - t)]^{k-j}}{(k - j)!} \frac{[b(v - u)]^{l-k}}{(l - k)!} e^{-bt} e^{-b(u-t)} e^{-b(v-u)}$$

$$= \frac{t^j (u - t)^{k-j}(v - u)^{l-k}}{j!(k - j)!(l - k)!} b^l e^{-bv}, \qquad j \le k \le l$$

We will verify only one phase of condition (11.5.1). We show that

$$\sum_{k=0}^{\infty} f_{t,u,v}(j, k, l) = f_{t,v}(j, l)$$

First, $f_{t,u,v}(j, k, l) = 0$ if $k < j$ or $k > l$. Therefore,

**(11.5.8)** $$\sum_{k=0}^{\infty} f_{t,u,v}(j, k, l) = \sum_{k=j}^{l} \frac{t^j (u - t)^{k-j}(v - u)^{l-k}}{j!(k - j)!(l - k)!} b^l e^{-bv}$$

Let $m = k - j$. As $k$ goes from $j$ to $l$, $m$ goes from 0 to $l - j$, so (11.5.8) is

$$b^l e^{-bv} \frac{t^j}{j!} \sum_{m=0}^{l-j} \frac{(u - t)^m (v - u)^{l-(m+j)}}{[l - (m + j)]! m!}$$

If we multiply numerator and denominator by $(l - j)!$ we have

$$b^l e^{-bv} \frac{t^j}{j!} \frac{1}{(l - j)!} \sum_{m=0}^{l-j} \frac{(l - j)!}{m!(l - j - m)!} (u - t)^m (v - u)^{l-j-m}$$

But the sum is just the binomial expansion of $[(v - u) + (u - t)]^{l-j} = (v - t)^{l-j}$. Therefore, (11.5.8) is

$$b^l e^{-bv} \frac{t^j}{j!} \frac{1}{(l-j)!} (v-t)^{l-j} = \frac{(bt)^j}{j!} \frac{[b(v-t)]^{l-j}}{(l-j)!} e^{-bt} e^{-b(v-t)}$$

$$= f_{t,v}(j, l)$$

It seems clear that we could do the same for any $t_1 < t_2 < \cdots < t_n$.

**EXERCISES 11.5** ─────────────────────────────

1   Suppose that $T$ is a finite set (technically not allowed in the definition) with $n$ elements. Show that a pre-stochastic process on $T$ is essentially an $n$-dimensional distribution function with all of its marginal distributions.

2   Show that the distribution functions of a pre-stochastic process satisfy

$$\lim_{x_1 \to \infty} \lim_{x_2 \to \infty} F_{t_1,t_2,\ldots,t_n}(x_1, x_2, \ldots, x_n) = F_{t_3,t_4,\ldots,t_n}(x_3, x_4, \ldots, x_n)$$

3   Suppose that a pre-stochastic process is given by probability functions. Show that if $f_{t_1,\ldots,t_n}(x_1, \ldots, x_n) > 0$, then

$$f_{t_1,\ldots,t_{j-1},t_{j+1},\ldots,t_n}(x_1, \ldots, x_{j-1}, x_{j+1}, \ldots, x_n) > 0$$

4   Let $S = \{s; 0 < s \le 1\}$ with the uniform mass density and the usual system of events. Let $T = \{1, 2, 3, \ldots\}$ and, for $t \in T$, $X_t(s) = t$th digit in the decimal expansion of $s$ as in section 2.3. Clearly, this pre-stochastic process is determined by probability functions. Find $f_{2,3,5}(4, 2, 7)$ and $F_{2,3,5}(4, 2, 7)$.

5   In the context of exercise 4 find $E(\cos \alpha X_t)$.

6   Let $S, \mathscr{A}, P$ be as in exercise 4. Let $T$ be the non-negative reals. Let $d_n(s) = n$th digit in the decimal expansion of $s \in S$. [Actually $d_n(s) = X_n(s)$ for the $X_t$ of exercise 4. We will use a different $X_t$ here.] Let

$$X_t(s) = \sum_{n=1}^{100} [d_n(s) - \tfrac{9}{2}]n^{-1} \cos nt$$

Find $X_t(\tfrac{1}{3})$. Compute $E(X_t)$, $\sigma^2(X_t)$, and $E(X_t X_u)$ for $t \ne u$.

Exercises 7–12 deal with the Poisson process. We work with the realization of the process whose existence is guaranteed by Theorem 11.5.2. It is a poor realization so we must work around some of its inadequacies.

7   Find the characteristic function of the random vector $(\omega(t), \omega(u), \omega(v))$, where $0 \le t < u < v$ and $\omega$ is the typical sample function in $\Omega$.

8   Show that the process has independent increments by examining

$$E(\exp\{i\alpha[\omega(u) - \omega(t)] + i\beta[\omega(Y) - \omega(u)]\})$$
$$= E(\exp\{-i\alpha\omega(t) + i[\alpha - \beta]\omega(u) + i\beta\omega(v)\})$$

and using exercise 7. Find the distribution of $\omega(u) - \omega(t)$.

**9**   Show that for any fixed $t \in T$ and any integer $n \geq 0$, $P(n < \omega(t) < n + 1) = 0$. Let $Q$ be the non-negative rational numbers. Show that

$$P\left(\bigcup_{t \in Q} \bigcup_{n=0}^{\infty} \{\omega; n < \omega(t) < n + 1\}\right) = 0$$

**10**   Show that if $t < u$, $P(\omega(t) > \omega(u)) = 0$. Show that

$$P\left(\bigcup_{t \in Q} \bigcup_{\substack{u \in Q \\ u > t}} \{\omega; \omega(t) > \omega(u)\}\right) = 0$$

**11**   Find $P(\omega(t) = n$ and $\omega(t + 1/m) > n)$. Show that its limit as $m \to \infty$ is zero. Let

$$\Omega_1 = \Omega - \bigcup_{t \in Q}\left(\bigcup_{n=0}^{\infty} \{\omega; n < \omega(t) < n + 1\} \cup \bigcup_{\substack{u \in Q \\ u > t}} \{\omega; \omega(t) > \omega(u)\}\right)$$

By exercises 8 and 9, $P(\Omega_1) = 1$. Describe the sample functions in $\Omega_1$ in words by referring to their behavior on $Q$. Show that

$A = \{\omega \in \Omega_1, \quad \omega(t) = n$ for some $n$ and some $t \in Q$ and
$$\omega(t + 1/m) > n \text{ for all } m = 1, 2, 3, \ldots\}$$

has $P(A) = 0$. Describe the sample functions in $\Omega_1 - A$ by referring to their behavior on $Q$.

**12**   Let $\tau_n(\omega) = $ g.l.b. $\{t \in Q; \omega(t) \geq n\}$. We may think of $\tau_n(\omega)$, loosely, as the time of first passage of $\omega$ to the region $\{m; m \geq n\}$. Show that

$$\{\omega; \tau_n(\omega) \leq u\} = \{\omega; \omega(t) \geq n \text{ for all } t \in Q, \quad t \geq u\}$$

Show that $P(\tau_1 \leq u) = 1 - e^{-bu}$. Find $P(\tau_2 \leq u)$.

# 11.6   Markov chains

A very important and basic stochastic process is the Markov chain with stationary transition probabilities. It is a model for a broad class of phenomena. We define it, study some of its properties, and give some examples.

Let $T = \{0, 1, 2, \ldots\}$. Let $J$, called the *state space*, be a countable set. It is usually assumed that $J$ is a subset of the reals. The elements of $J$ are called *states*. Let $p_{jk}$ for $j \in J$, $k \in J$ be a system of non-negative numbers which satisfy the condition

**(11.6.1)**   $\sum_{k \in J} p_{jk} = 1, \quad$ for each $j \in J$

The number $p_{jk}$ is called the *transition probability* from state $j$ to state $k$. It represents the conditional probability that the system is in state $k$ at time $n + 1$ given that it was in state $j$ at time $n$. These are the basic ingredients of our Markov chain with stationary transition probabilities.

Next we define the *n*-step transition probabilities. Let

**(11.6.2a)**     $p_{jk}^{(1)} = p_{jk}$

**(11.6.2b)**     $p_{jk}^{(n+1)} = \sum\limits_{i \in J} p_{ji}^{(n)} p_{ik}, \quad n = 1, 2, \ldots$

for all pairs $j \in J, \quad k \in J$.

Those who are familiar with matrices should note that, for each *n* the system of numbers, $p_{jk}^{(n)}$, can be thought of as a matrix. However, the matrix entries are indexed by elements of *J* rather than by $1, 2, \ldots, m$. If *J* is infinite the matrix may be infinite. The system of $p_{jk}$ is often called the *transition matrix*. It would not be important that the $p_{jk}$ can be thought of as matrix elements were it not for the fact that (11.6.2b) is the definition of matrix multiplication. Indeed, (11.6.2b) implies that if *A* is the matrix of the $p_{jk}$, then $A^n$ is the matrix of the $p_{jk}^{(n)}$. Thus many problems in Markov chains are essentially matrix problems. We will not approach them in this way.

To show that the *n*-step transition probabilities behave reasonably we prove the following theorem.

**Theorem 11.6.1**     Let *m* and *n* be positive integers and let *j* and *k* belong to *J*. Then

$$p_{jk}^{(n+m)} = \sum\limits_{l \in J} p_{jl}^{(n)} p_{lk}^{(m)}$$

PROOF     We use mathematical induction on *m*. If $m = 1$ the result follows from the definition of $p_{lk}^{(1)}$ in (11.6.2a) and the definition of $p_{jk}^{(n+1)}$ in (11.6.2b).

Assume the theorem for $m \geq 1$. We prove it for $m + 1$. By (11.6.2b)

$$p_{jk}^{(n+m+1)} = \sum\limits_{l \in J} p_{jl}^{(n+m)} p_{lk}$$

But by the induction hypothesis

$$p_{jl}^{(n+m)} = \sum\limits_{i \in J} p_{ji}^{(n)} p_{il}^{(m)}$$

Therefore

$$p_{jk}^{(n+m+1)} = \sum\limits_{l \in J} \left( \sum\limits_{i \in J} p_{ji}^{(n)} p_{il}^{(m)} \right) p_{lk}$$

$$= \sum\limits_{i \in J} \sum\limits_{l \in J} p_{ji}^{(n)} p_{il}^{(m)} p_{lk}$$

since *i* and *l* range freely over *J*. However, by a change of notation in (11.6.2b)

$$\sum\limits_{l \in J} p_{ji}^{(n)} p_{il}^{(m)} p_{lk} = p_{ji}^{(n)} \sum\limits_{l \in J} p_{il}^{(m)} p_{lk} = p_{ji}^{(n)} p_{ik}^{(m+1)}$$

Therefore

$$p_{jk}^{(n+m+1)} = \sum\limits_{i \in J} p_{ji}^{(n)} p_{ik}^{(m+1)}$$

This is that which is to be proved with the index *i* instead of *l*.

Loosely speaking, this theorem states that the probability of going from state $j$ to state $k$ in $n + m$ seconds is the probability of going from state $j$ to some state in $n$ seconds and from that state to state $k$ in the next $m$ seconds. It would be a disaster were this not the case.

### Example 11.6.1

For the standard random walk with probability $p$ of moving to the right and probability $q = 1 - p$ of moving to the left, describe $J$. Find $p_{jk}$ and $p_{jk}^{(2)}$.

We can move one step in either direction so $J$ is all of the integers and $p_{jk} = 0$ unless $k = j + 1$ or $k = j - 1$. Now $p_{j,j+1} = p$, the conditional probability of moving to the right given that the walk is at $j$. Similarly $p_{j,j-1} = q$. By definition

$$p_{jk}^{(2)} = \sum_{l \in J} p_{jl}^{(1)} p_{lk} = \sum_{l = -\infty}^{\infty} p_{jl} p_{lk}$$

Therefore

$$p_{jj}^{(2)} = p_{j,j+1} p_{j+1,j} + p_{j,j-1} p_{j-1,j} = pq + qp = 2pq$$
$$p_{j,j+2}^{(2)} = p_{j,j+1} p_{j+1,j+2} = p^2$$

and

$$p_{j,j-2} = p_{j,j-1} p_{j-1,j-2} = q^2$$

Furthermore, $p_{jk}^{(2)} = 0$ unless $k = j - 2, j,$ or $j + 2$.

### Example 11.6.2

Describe the $p_{jk}$ for a "random walk with absorbing barriers" at 0 and $N$. Describe $J$.

We may require the random walk to be an ordinary random walk at $1, 2, \ldots, N - 1$. Thus $J = \{0, 1, 2, \ldots, N\}$, $p_{j,j-1} = q$, and $p_{j,j+1} = p$ if $j = 1, 2, \ldots, N - 1$. To bring in the absorbing barriers we let $p_{00} = 1$ and $p_{NN} = 1$. This requires the walker to stay at 0 or $N$ if he ever reaches 0 or $N$. All other $p_{jk}$ are zero.

### Example 11.6.3

Describe the $p_{jk}$ for a "random walk with perfect reflecting barrier at 0 and a partially absorbing barrier at $N$." Describe $J$.

As above if $1 \le j \le N - 1$, $p_{j,j-1} = q$ and $p_{j,j+1} = p$. To bring in the perfect reflecting barrier at 0 we let $p_{01} = 1$. Thus if you get to 0 you are re-instated at 1 on the next "move."

How we treat the partially absorbing barrier at $N$ depends on our interpretation of "partially absorbing." We will give one interpretation; in applications of Markov chains one would usually have more detail on the nature of the barriers. Let us assume that if the walker gets to the barrier at $N$ he is reflected to $N - 1$ with probability $r$ and he becomes permanently attached with probability $1 - r$. In view of the reflection we have $p_{N,N-1} = r$. To get the permanent attachment we use $N + 1$ as a "garbage" state and set $p_{N,N+1} = 1 - r$ and $p_{N+1,N+1} = 1$. This ensures that an absorbed particle stays put. In other problems we may wish to have some wandering after absorption. All $p_{jk}$ which we have not specified are zero, and $J = \{0, 1, 2, \ldots, N, N + 1\}$.

### Example 11.6.4 (Ehrenfest model.)

A collection of $2N$ numbered balls and a box are available. At any stage a number from 1 to $2N$ is chosen with equal likelihood. The ball with that number is put into the box if it was outside of the box or it is removed from the box if it was inside the box. The state of the chain is the number of balls in the box minus the number of balls outside of the box.

If the state at any time is $j$ then there are $\frac{1}{2}j + N$ balls in the box and $-\frac{1}{2}j + N$ balls outside the box since $(\frac{1}{2}j + N) - (-\frac{1}{2}j + N) = j$ and $(\frac{1}{2}j + N) + (-\frac{1}{2}j + N) = 2N$. Needless to say $j$ must be an even integer. It cannot be greater than $2N$ nor less than $-2N$. Thus $J = \{-2N, -2N + 2, -2N + 4, \ldots, 0, 2, \ldots, 2N\}$. Now $p_{jk} = 0$ unless $k = j + 2$ or $j - 2$, since after a ball is changed the number of balls inside minus the number outside increases or decreases by 2. The probability that a ball inside the box will be moved is (number of balls inside) $\div (2N)$ since the numbers $1, \ldots, 2N$ are selected at random. Therefore $p_{j,j-2} = (\frac{1}{2}j + N)/2N = j/4N + \frac{1}{2}$ and $p_{j,j+2} = -j/4N + \frac{1}{2}$.

Examples 11.6.2–11.6.4 constitute *finite Markov chains* in the sense that the state space $J$ is a finite set.

### Example 11.6.5

Describe a set of transition probabilities which would constitute a model for the growth of a bacteria culture or the dynamics of a population whose members are born and die off. Discuss the model.

We let $J = \{0, 1, 2, \ldots\}$ in general although we may wish to use only $J = \{0, 1, 2, \ldots, N\}$. The chain is in the $j$th state at a particular time if there are $j$ members alive at that time. We may assume that the time scale is chosen so that at any instant, at most one birth or one death may occur. As such $p_{jk} = 0$ unless $k = j - 1, j$, or $j + 1$. If $j \geq 1$ we could let $p_{jj} = a_j$ be the conditional probability of no births or deaths when there are $j$ individuals alive, $p_{j,j+1} = b_j$ be the conditional probability of a birth out of $j$ individuals, and $p_{j,j-1} = c_j$ be the conditional probability of a death out of $j$ individuals. If $j = 0$ we let $p_{00} = 1$ and $p_{0j} = 0$ for $j > 1$. Naturally $a_j + b_j + c_j = 1$. It seems reasonable for $a_j$ to decrease as $j$ increases whereas $b_j$ and $c_j$ increase as $j$ increases. If it were desirable to model a population with a limited food supply we could assume that $c_j \to 1$ as $j \to j_{\max}$ or as $j \to \infty$.

This model does not allow for multiple births but this is no problem if the time scale is chosen properly. This model does not account for the ages of the population members. This could be serious if fertility varies with age. The model does not account for the sex of the members of the population but this is probably not a serious defect. The model could account for a decrease in fertility due to food scarcity by having $b_j$ decrease properly with large $j$.

Example 11.6.5 is a crude version of the so-called birth and death process. One could envision some applications of this process but it would be naive to believe that reliable values of $a_j$, $b_j$, and $c_j$ are readily available. However, one could hope for answers to qualitative questions of population dynamics; for example, is there an equilibrium size of the population; how does it depend on the $p_{jk}$'s? Naturally, the answers would be answers in the model. The model may not be adequate for practical work.

So far we have not defined a Markov chain with stationary transition probabilities nor have we shown that anything in this section is a pre-stochastic process. We remedy this matter.

Let $f_0: J \to R$ be such that $f_0(j) \geq 0$ and $\sum_{j \in J} f_0(j) = 1$. This function is a probability function. It is called the *initial distribution*. For $j$, $k$, $l$ in $J$ and timelike integers $0 < t < u < v$, we define

**(11.6.3)**   $$f_{t,u,v}(j, k, l) = \sum_{i \in J} f_0(i) p_{ij}^{(t)} p_{jk}^{(u-t)} p_{kl}^{(v-u)}$$

$$f_{0,t,u,v}(i, j, k, l) = f_0(i) p_{ij}^{(t)} p_{jk}^{(u-t)} p_{kl}^{(v-u)}$$

We are making the definition for three timelike values to keep notation simple. Specializations to fewer than three time values or generalizations to more than three should be obvious.

**Theorem 11.6.2**   The functions defined by (11.6.3) satisfy the consistency conditions for probability functions, (11.5.3), and its obvious extension.

PROOF   We prove the theorem in two cases. First,

$$\sum_{i \in J} f_{0,t,u,v}(i, j, k, l) = \sum_{i \in J} f_0(i) p_{ij}^{(t)} p_{jk}^{(u-t)} p_{kl}^{(v-u)}$$

$$= f_{t,u,v}(j, k, l)$$

by definition (11.6.3). For the second case we have

$$\sum_{k \in J} f_{t,u,v}(j, k, l) = \sum_{k \in J} \sum_{i \in J} f_0(i) p_{ij}^{(t)} p_{jk}^{(u-t)} p_{kl}^{(v-u)}$$

$$= \sum_{i \in J} f_0(i) p_{ij}^{(t)} \sum_{k \in J} p_{jk}^{(u-t)} p_{kl}^{(v-u)}$$

By Theorem 11.6.1 with modified notation

$$\sum_{k \in J} p_{jk}^{(u-t)} p_{kl}^{(v-u)} = p_{jl}^{(u-t+v-u)}$$

But $u - t + v - u = v - t$. Therefore, with the obvious definition of $f_{t,v}(j, l)$ we have

$$\sum_{k \in J} f_{t,u,v}(j, k, l) = \sum_{i \in J} f_0(i) p_{ij}^{(t)} p_{jl}^{(v-t)} = f_{t,v}(j, l)$$

We leave the other cases to the reader.

Theorem 11.6.2 states that the probability functions of (11.6.3) define a pre-stochastic process. We call any stochastic process obtained in this fashion a *Markov chain with stationary transition probabilities*.

**Theorem 11.6.3**     A Markov chain with stationary transition probabilities satisfies the definition of Markov chain given in section 11.5.

PROOF     We consider a special case with four nonzero time values. We show that if $0 < t < u < v < w$

$$(11.6.4) \qquad \frac{f_{t,u,v,w}(j, k, l, m)}{f_{t,u,v}(j, k, l)} = \frac{f_{v,w}(l, m)}{f_v(l)}$$

when both sides are defined. By definition,

$$f_{t,u,v}(j, k, l) = \sum_{i \in J} f_0(i) p_{ij}^{(t)} p_{jk}^{(u-t)} p_{kl}^{(v-u)}$$

and

$$f_{t,u,v,w}(j, k, l, m) = \sum_{i \in J} f_0(i) p_{ij}^{(t)} p_{jk}^{(u-t)} p_{kl}^{(v-u)} p_{lm}^{(w-v)}$$

This is exactly $p_{lm}^{(w-v)} f_{t,u,v}(j, k, l)$. Therefore the left-hand side of equation (11.6.4) is $p_{lm}^{(w-v)}$. Also

$$f_{v,w}(l, m) = \sum_{i \in J} f_0(i) p_{il}^{(v)} p_{lm}^{(w-v)} = p_{lm}^{(w-v)} \sum_{i \in J} f_0(i) p_{il}^{(v)} = p_{lm}^{(w-v)} f_v(l)$$

Thus the right-hand side of equation (11.6.4) is $p_{lm}^{(w-v)}$.

By Theorem 11.5.2 there is a realization of a Markov chain on $\Omega = \{$all functions on $T\}$. We may think of these functions as graphs of the movement of the process. We may also think of first passage and first return to various states of the Markov chain. Let us restrict our attention to the first return problem. For $m \geq 1$ and $\omega \in \Omega$ let

$$v_j^{(m)} = P(\omega(m) = j \text{ but } \omega(t) \neq j \text{ for } 0 < t < m \mid \omega(0) = j)$$

i.e., the conditional probability that the chain is back in state $j$ for the first time at time $m$ given that it started in state $j$ at time 0. Using the techniques of the proof of Theorem 11.2.1 it can be shown that for $|z| < 1$,

$$(11.6.5) \qquad \sum_{m=1}^{\infty} p_{jj}^{(m)} z^m = \left( \sum_{m=1}^{\infty} v_j^{(m)} z^m \right)\left( \sum_{m=0}^{\infty} p_{jj}^{(m)} z^m \right)$$

if $p_{jj}^{(0)}$ is defined to be 1. Let

$$A_j(z) = \sum_{m=0}^{\infty} p_{jj}^{(m)} z^m$$

and

$$B_j(z) = \sum_{m=1}^{\infty} v_j^{(m)} z^m$$

Using exercises 1.5.11 and 1.5.14 as in the proof of Theorem 11.2.1 we obtain

**(11.6.6)** $\qquad A_j(z) - 1 = B_j(z) A_j(z)$

Since the $v_j^{(m)}$ are probabilities of disjoint events, $v_j = \sum_{m=1}^{\infty} v_j^{(m)} \leq 1$. The power

series $A_j(z)$ and $B_j(z)$ have non-negative coefficients. By exercise 11.2.14

$$\lim_{\substack{z \to 1 \\ z < 1}} B_j(z) = \sum_{m=1}^{\infty} v_j^{(m)} = v_j$$

We rewrite equation (11.6.6) as

**(11.6.7)** $\qquad 1 - [A_j(z)]^{-1} = B_j(z)$

Since the coefficients of the power series $A_j(z)$ are non-negative,

$$\lim_{\substack{z \to 1 \\ z < 1}} A(z) = \sum_{m=0}^{\infty} p_{jj}^{(m)} \geq p_{jj}^{(0)} = 1$$

if we include the possibility that both the $\lim_{\substack{z \to 1 \\ z < 1}} A(z)$ and the series $\sum_{m=0}^{\infty} p_{jj}^{(m)}$ may

diverge to $+\infty$. With $a = \lim_{\substack{z \to 1 \\ z < 1}} [A(z)]^{-1}$ we have, clearly, $a = (\sum_{m=0}^{\infty} p_{jj}^{(m)})^{-1}$ if the

series converges and $a = 0$ if the series diverges. Within this framework we have proved the next theorem.

**Theorem 11.6.4** Let $j$ be a state of a Markov chain with stationary transition probabilities. Then $v_j$, the conditional probability of returning to the state $j$ at some time $t > 0$ given that the chain starts in the state $j$, satisfies

**(11.6.8)** $\qquad v_j = 1 - a = 1 - (\sum_{m=0}^{\infty} p_{jj}^{(m)})^{-1}$

We think of $v_j$ as the probability of ever returning to the state $j$ if the chain starts at $j$.

### Example 11.6.6

Let $j$ be a state for which $v_j = 1$. Suppose that $P(\omega(0) = j) = 1$. Show that the expected time of first return of the chain to the state $j$ is

$$\lim_{\substack{z \to 1 \\ z < 1}} \frac{A_j'(z)}{A_j^2(z)}$$

where $A_j'(z)$ denotes $\dfrac{d}{dz} A_j(z)$.

By (11.6.7)

$$B_j(z) = \sum_{m=1}^{\infty} v_j^{(m)} z^m = 1 - [A_j(z)]^{-1}$$

The expected time of first return under consideration is

$$\sum_{m=1}^{\infty} m v_j^{(m)} = \lim_{\substack{z \to 1 \\ z < 1}} \sum_{m=1}^{\infty} m v_j^{(m)} z^{m-1}$$

by exercise 11.2.14 since $v_j^{(m)} \geq 0$. Note that although the $v_j^{(m)}$ were defined as conditional probabilities, the requirement that $P(\omega(0) = j) = 1$ essentially states that the chain definitely starts in the state $j$. Thus $v_j^{(m)} = P(\omega(0) = j$, $\omega(m) = j$ for first time after $t = 0)$. But

$$\sum_{m=1}^{\infty} m v_j^{(m)} z^{m-1} = \frac{d}{dz} B_j(z) = \frac{d}{dz} \{1 - [A_j(z)]^{-1}\} = \frac{A_j'(z)}{A_j^2(z)}$$

If we reintroduce the limiting operation we have

$$\sum_{m=1}^{\infty} m v_j^{(m)} = \lim_{\substack{z \to 1 \\ z < 1}} \frac{A_j'(z)}{A_j^2(z)}$$

Needless to say the right-hand side may not exist. This was the case in the standard random walk with $p = \frac{1}{2}$.

It is advantageous to imagine nice random walks on systems other than the line. For example, we can consider the points $(j, k)$ in 2-dimensional space in which $j$ and $k$ are integers. Such points are called *lattice points*. The basic random walk involves movements from $(j, k)$ to $(j + 1, k), (j - 1, k), (j, k + 1)$, or $(j, k - 1)$ with equal probability. Similarly, in 3-dimensional space we can imagine movements on the lattice points from $(j, k, l)$ to $(j + 1, k, l)$, $(j - 1, k, l), (j, k + 1, l), (j, k - 1, l), (j, k, l + 1)$, or $(j, k, l - 1)$ with equal probability. Naturally, $j$, $k$, and $l$ are integers and all movements that are not mentioned have probability zero. The set of lattice points in $n$-dimensional space is always a countable set. In order to minimize notation we set up a one-to-one correspondence between the set of lattice points and the integers. For definiteness, we let $(0, 0)$ or $(0, 0, 0)$ always correspond to 0 in any problem. It is not necessary to know the details of the correspondence in our problems; it is just a useful code.

*Example 11.6.7*

Show that for the random walk in 3-dimensional space just described

$P$(walk eventually returns to $(0, 0, 0)$ given that it starts there) $< 1$

By our convention we use the state space $J = \{\text{integers}\}$ and a code in which 0 corresponds to $(0, 0, 0)$. By Theorem 11.6.4 we need only show that the series

$\sum\limits_{m=0}^{\infty} p_{00}^{(m)}$ converges in the usual sense; that is that it does not tend to infinity.

For convenience we call the points $(\pm 1, 0, 0)$, $(0, \pm 1, 0)$, and $(0, 0, \pm 1)$ N, S, E, W, U, D for North, South, East, West, up, and down. These are in correspondence with six integers $k$ and $p_{0k} = \frac{1}{6}$ for each of those particular $k$. Regardless of the code we may ask, "What is $p_{00}^{(m)}$?" It is determined by the number of $m$-letter words using the letters N, S, E, W, U, D in which the number of N's and S's are the same, the number of E's and W's are the same, and the number of U's and D's are the same. This is true because there must be as many positive sense moves as negative sense moves in each component in order to start at and return to $(0, 0, 0)$. Furthermore, if this is the case, no matter what the order of the moves, the random walker would return to $(0, 0, 0)$. Furthermore, all of this must carry over under the code system.

By section 1.3 the number of words of the type at issue is

$$\sum \frac{m!}{j!\,j!\,k!\,k!\,l!\,l!}$$

where $j$ = number of N's, $k$ = number of E's, and $l$ = number of U's. Naturally $j + j + k + k + l + l = m$ and the sum is taken over all possible non-negative integers $j, k, l$ of this type. It is clear that $m$ must be an even integer. Henceforth, we will work with $2m$ rather than $m$. The probability of any *particular* arrangement of $2m$ moves is $(\frac{1}{6})^{2m}$ since the moves are independent and equally likely. Thus

$$\sum\limits_{m=0}^{\infty} p_{00}^{(m)} = \sum\limits_{m=0}^{\infty} \sum\limits_{j,k,l} \frac{(2m)!}{j!\,j!\,k!\,k!\,l!\,l!} \left(\frac{1}{6}\right)^{2m}$$

and the inner sum on the right is taken on all $j, k, l \geq 0$ with $2j + 2k + 2l = 2m$. How do we determine whether this series converges?

In order to show that the series converges, we use a trick that is hard to motivate. It has a long history and it has been found to be extremely important.

Let $j$ and $k$ be two integers. If $j = k$, $e^{ij\theta}e^{-ik\theta} = 1$ and

$$\frac{1}{2\pi} \int_{-\pi}^{\pi} e^{ij\theta}e^{-ik\theta}\, d\theta = 1$$

If $j \neq k$ and $j - k = l$,

$$\frac{1}{2\pi} \int_{-\pi}^{\pi} e^{ij\theta}e^{-ik\theta}\, d\theta = \left.\frac{e^{i(j-k)\theta}}{2\pi i(j-k)}\right|_{-\pi}^{\pi} = \frac{e^{il\pi} - e^{-il\pi}}{2\pi i l} = 0$$

since $e^{ix} - e^{-ix} = 2i \sin x$ and $l$ is an integer. From sections 1.3 and 1.5

$$(e^{i\alpha} + e^{-i\alpha} + e^{i\beta} + e^{-i\beta} + e^{i\gamma} + e^{-i\gamma})^{2m}$$

$$= \sum \frac{(2m)!}{j!\,j'!\,k!\,k'!\,l!\,l'!}\, e^{ij\alpha}e^{-ij'\alpha}e^{ik\beta}e^{-ik'\beta}e^{il\gamma}e^{-il'\gamma}$$

where the sum is taken over all sets of non-negative integers $j, j', k, k', l$, and $l'$

for which $j + j' + k + k' + l + l' = 2m$. To eliminate the contributions from those cases in which $j \neq j'$ or $k \neq k'$ or $l \neq l'$ we take

$$\left(\frac{1}{2\pi}\right)^3 \int_{-\pi}^{\pi} d\alpha \int_{-\pi}^{\pi} d\beta \int_{-\pi}^{\pi} (e^{i\alpha} + e^{-i\alpha} + e^{i\beta} + e^{-i\beta} + e^{i\gamma} + e^{-i\gamma})^{2m} \, d\gamma$$

This retains contributions precisely from $j = j'$, $k = k'$, $l = l'$ since

$$\frac{1}{2\pi} \int_{-\pi}^{\pi} e^{ij\theta} e^{-ij\theta} \, d\theta = 1$$

By section 9.1, $e^{ix} + e^{-ix} = 2 \cos x$. Thus

$$\sum_{m=0}^{\infty} \sum_{j,k,l} \frac{(2m)!}{j!j!k!k!l!l!} \left(\frac{1}{6}\right)^{2m}$$

$$= \left(\frac{1}{2\pi}\right)^3 \sum_{m=0}^{\infty} \int_{-\pi}^{\pi} d\alpha \int_{-\pi}^{\pi} d\beta \int_{-\pi}^{\pi} 2^{2m}(\cos \alpha + \cos \beta + \cos \gamma)^{2m}(\tfrac{1}{6})^{2m} \, d\gamma$$

It can be shown that interchange of the order of summation and integration is permissible since all terms are non-negative. Since $2^{2m}(\tfrac{1}{6})^{2m} = (\tfrac{1}{3})^{2m}$ we have the key quantity of

**(11.6.9)** $\quad \left(\dfrac{1}{2\pi}\right)^3 \displaystyle\int_{-\pi}^{\pi} d\alpha \int_{-\pi}^{\pi} d\beta \int_{-\pi}^{\pi} \sum_{m=0}^{\infty} \left(\dfrac{\cos \alpha + \cos \beta + \cos \gamma}{3}\right)^{2m} d\gamma$

Now if $-1 \leq a \leq 1$,

$$\sum_{m=0}^{\infty} a^{2m} = \sum_{m=0}^{\infty} (a^2)^m = \frac{1}{1 - a^2}$$

if we use the convention and consequences of $1/0 = \infty$.

We only wish to know whether (11.6.9), which is $\sum_{m=0}^{\infty} p_{00}^{(m)}$, converges or diverges. We can ignore the factor $(1/2\pi)^3$ for this purpose. We are thus left with

$$\int_{-\pi}^{\pi} d\alpha \int_{-\pi}^{\pi} d\beta \int_{-\pi}^{\pi} \frac{1}{1 - [\tfrac{1}{3}(\cos \alpha + \cos \beta + \cos \gamma)]^2} \, d\gamma$$

For all $\alpha$, $\beta$, $\gamma$, $\tfrac{1}{3}|\cos \alpha + \cos \beta + \cos \gamma| \leq 1$. In our region of integration $\tfrac{1}{3}(\cos \alpha + \cos \beta + \cos \gamma) = 1$ if and only if $\alpha = 0$, $\beta = 0$, $\gamma = 0$; it is $-1$ if and only if $\alpha = \pi$, $\beta = \pi$, $\gamma = \pi$ or $\alpha = -\pi$, $\beta = -\pi$, $\gamma = -\pi$. For small $\alpha$, $\cos \alpha = 1 - \tfrac{1}{2}\alpha^2 + \cdots$ where the unspecified terms sum to less than $\tfrac{1}{4}\alpha^2$. Therefore, if $\alpha$, $\beta$, $\gamma$ are sufficiently close to zero,

$$1 - [\tfrac{1}{3}(\cos \alpha + \cos \beta + \cos \gamma)]^2 = 1 - \left(1 - \frac{\alpha^2 + \beta^2 + \gamma^2}{6} + \text{small terms}\right)^2$$

$$= 1 - 1 + \tfrac{2}{6}(\alpha^2 + \beta^2 + \gamma^2)$$

$$- \text{(terms involving fourth powers and higher)}$$

$$\geq \tfrac{1}{6}(\alpha^2 + \beta^2 + \gamma^2)$$

after making a crude estimate.

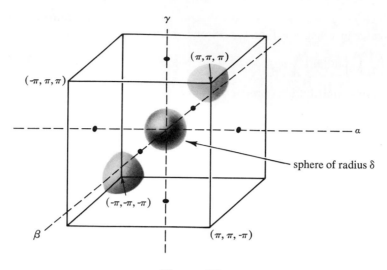

**Figure 11**

The integration in (11.6.9) is taken over a cube of side $2\pi$ as in Figure 11. If $(\alpha, \beta, \gamma)$ is sufficiently close to $(0, 0, 0)$, say in a sphere of radius $\delta$ and center $(0, 0, 0)$, the integrand is less than $6/(\alpha^2 + \beta^2 + \gamma^2)$ because, as noted previously, the denominator of the integrand is greater than $\frac{1}{6}(\alpha^2 + \beta^2 + \gamma^2)$. In the partial spheres at $(\pi, \pi, \pi)$ and $(-\pi, -\pi, -\pi)$ the integrand behaves in essentially the same way because $\cos(\pi - \theta) = \cos(-\pi - \theta) = -\cos\theta$ and $(-\cos\alpha' - \cos\beta' - \cos\gamma')^2 = (\cos\alpha' + \cos\beta' + \cos\gamma')^2$. In the portion of the cube outside of these spheres the denominator of the integrand is at least $\epsilon > 0$ because $[\frac{1}{3}(\cos\alpha + \cos\beta + \cos\gamma)]^2$ is less than $1 - \epsilon$ for some $\epsilon > 0$. In that situation the integrand is less than or equal to $\epsilon^{-1}$. If we combine these observations we have

$$\int_{-\pi}^{\pi} d\alpha \int_{-\pi}^{\pi} d\beta \int_{-\pi}^{\pi} \frac{d\gamma}{1 - [\frac{1}{3}(\cos\alpha + \cos\beta + \cos\gamma)]^2}$$

$$\leq \frac{10}{8} \iiint_{\substack{\text{sphere at} \\ (0,0,0)}} \frac{6\, d\alpha\, d\beta\, d\gamma}{\alpha^2 + \beta^2 + \gamma^2} + \iiint_{\substack{\text{rest of} \\ \text{cube}}} \epsilon^{-1}\, d\alpha\, d\beta\, d\gamma$$

where the factor $\frac{10}{8}$ accounts for the partial spheres.
    Now

$$\iiint_{\substack{\text{rest of} \\ \text{cube}}} \epsilon^{-1}\, d\alpha\, d\beta\, d\gamma \leq \iiint_{\substack{\text{whole} \\ \text{cube}}} \epsilon^{-1}\, d\alpha\, d\beta\, d\gamma = (2\pi)^3 \epsilon^{-1}$$

Integrals over spheres are usually best done in spherical coordinates. Let

$\alpha = r \sin \theta \cos \phi$, $\beta = r \sin \theta \sin \phi$, and $\gamma = r \cos \theta$. Then $\alpha^2 + \beta^2 + \gamma^2 = r^2$ and the crucial integral over the sphere is

$$\int_0^\delta dr \int_0^\pi d\theta \int_0^{2\pi} \frac{6r^2 \sin \theta \, d\phi}{r^2}$$

where $r^2 \sin \theta$ is the Jacobian of the change of coordinates or, if we please, the volume element in spherical coordinates.

The last integral is just $\int_0^\delta dr \int_0^\pi d\theta \int_0^{2\pi} 6 \sin \theta \, d\phi = 6 \cdot \delta \cdot 2 \cdot 2\pi$ which is finite.

Therefore $\sum_{m=0}^\infty p_{00}^{(m)}$ converges and $P$(walk returns to $(0, 0, 0)$ given that it starts there) $< 1$.

We have chosen this lengthy method of proof because it illustrates current ideas in probability. It is more general than it appears. It also has a number of applications to problems outside of probability theory and mathematics. Furthermore, the question in the example is not easy and any solution would have to be somewhat lengthy.

## EXERCISES 11.6 _____

1   Compute $p_{jk}^{(2)}$ for the Markov chain of Example 11.6.2.

2   What would it mean if a Markov chain has $f_0(k) = 1$ for a fixed state $k \in J$ and $f_0(j) = 0$ for $j \neq k$?

3   Find transition probabilities for a Markov chain which is an ordinary random walk on $\{1, 2, \ldots, N-1\}$ and which has reflecting barriers at both 0 and $N$.

4   Find $p_{jk}^{(3)}$ for the Markov chain of Example 11.6.1.

5   Find the $p_{jk}^{(2)}$ for the Ehrenfest model of Example 11.6.4. if $-2N + 2 < j < 2N - 2$.

6   Find transition probabilities which would describe the following phenomena. We have an ordinary random walk on the coordinates $1, 2, \ldots, N-1$. There is a reflecting barrier at 0. At $N$ there is an absorbing barrier which releases absorbed particles at coordinate $N-2$ three time units after they hit $N$. Use three artificial states to handle the particle after it hits coordinate $N$.

7   Find transition probabilities for a random walk on the coordinates $1, 2, 3, \ldots$ which has an absorbing barrier at 0. Let $p$ be the probability of a move to the right. Find $P$(the particle is absorbed at the $n$th step | the particle starts at coordinate $k > 0$).

8   Let $J$ be the integers. Suppose that the transition probabilities of a Markov chain satisfy $p_{jk} = f(k - j)$. Let $X_1, X_2, \ldots, X_n$ be independent random variables all having probability function $f$. Suppose that $X_0$ has probability function $f_0$. Show that $\omega(n) = X_0 + X_1 + \cdots + X_n$.

**9**   Show that $\sum_{k \in J} p_{jk}^{(2)} = 1$ for each $j \in J$.

**10**   Generalize (11.6.3) to $0 < t_1 < t_2 < t_3 < \cdots < t_n$.

**11**   Discuss why the proof of Theorem 11.2.1 can be used to prove the validity of (11.6.5) and (11.6.6).

**12**   Let $j$ and $k$ be two states of a Markov chain. Suppose that $P$(chain eventually reaches $k \mid$ chain starts at $j$) $= 1$ and $P$(chain eventually reaches $j \mid$ chain starts at $k$) $= 1$. For the $v_j$ of Theorem 11.6.4 show that $v_j = v_k = 1$.

Exercises 13–16 refer to the $r$-dimensional random walks with equal probability of moving to the $2r$ adjacent points.

**13**   In view of exercise 12, what can be said about $P$(walk eventually reaches $(j, k, l) \mid$ walk starts at $(0, 0, 0)$) when $(j, k, l) \neq (0, 0, 0)$.

**14**   Find a finite set, $M$, of lattice points in 3-dimensional Euclidean space such that $P$(walk is eventually at some point of $M \mid$ walk starts at $(0, 0, 0)$) $= 1$.

**15**   Show that

$$\left(\frac{1}{2\pi}\right)^3 \int_{-\pi}^{\pi} d\alpha \int_{-\pi}^{\pi} d\beta \int_{-\pi}^{\pi} (e^{i\alpha} + e^{-i\alpha} + e^{i\beta} + e^{-i\beta} + e^{i\gamma} + e^{-i\gamma})^4 \, dr$$

$$= \sum \frac{4!}{j!j!k!k!l!l!}$$

where the sum is over all non-negative integers $j, k, l$ with $2j + 2k + 2l = 4$.

**16**   Ascertain whether the basic 2-dimensional random walk starting at $(0, 0)$ eventually returns to $(0, 0)$ with probability 1.

# 11.7   Gaussian processes

In section 11.5 we defined a Gaussian process as a stochastic process such that for any $\{t_1, \ldots, t_n\} \subset T$, $F_{t_1, \ldots, t_n}$ is the distribution function of an $n$-dimensional normal random vector (which may be a degenerate normal random vector). In section 10.5 we saw that the distribution of a normal random vector $(X_1, X_2, \ldots, X_n)$ is completely determined by the system of numbers $E(X_j)$ and $E(X_j X_k)$ for $j, k = 1, 2, \ldots, n$ inasmuch as its characteristic function is completely determined by the $E(X_j)$ and the $E(X_j X_k) - E(X_j)E(X_k)$.

Suppose that we have a Gaussian process and $t \in T$, $u \in T$. In principle we can find the expectations, variances, and covariance of the components of the random vector with distribution function $F_{t,u}$. We would get the same values if we took the $t$ and $u$ components from $F_{s,t,v,u,w}$ since the latter yields $F_{t,u}$ after applying the consistency condition (11.5.1) successively to $s$, $v$, and $w$. Thus for each $t \in T$ there is a number $m(t)$, the "expected value of $F_t$" and for each pair of numbers $t, u \in T$ there is a number $r(t, u)$ which is the "covariance of $F_{t,u}$." (We are abusing language by talking about moments of $F$ instead of moments of the components of random vectors.) We regard the variance of $X_j$ as the covariance of $X_j$ and $X_j$. The function $m(t)$ is called the *mean function* and $r(t, u)$ is called the *covariance function*.

Just as the distribution of a normal random vector is determined by the $E(X_j)$ and $E(X_j X_k)$, a Gaussian process is characterized by its mean and covariance functions. Indeed, given such functions and $\{t_1, \ldots, t_n\} \subset T$ we can find the expected values, variances, and covariances of the components of the random vector whose distribution function is $F_{t_1,\ldots,t_n}$; namely,

$$E(X_{t_k}) = m(t_k), \quad \sigma^2(X_{t_k}) = r(t_k, t_k), \quad \text{cov}(X_{t_j}, X_{t_k}) = r(t_j, t_k)$$

As noted above these quantities completely determine $F_{t_1,\ldots,t_n}$.

### Example 11.7.1

Find $m(t)$ and $r(t, u)$ for the Wiener process of section 11.4.

Corollaries 1 and 2 of Theorem 11.4.1 show that

**(11.7.1)**      $m(t) = y_0$   and   $r(t, u) = \min(t, u)$

An important feature of Gaussian processes is their closure under many important mathematical operations. That is, if one performs certain operations on a Gaussian process the result is also a Gaussian process. This should not be interpreted to mean that the performance of various mathematical operations on other processes is not interesting. However, this fact yields important dividends in the realm of Gaussian processes. It is a useful fact that a linear transformation of a normal random vector is also a normal random vector, or that the limit of normal random vectors is also a normal random vector.

In the course of studying the results of operating on Gaussian processes we will take various viewpoints. In Examples 11.7.3–11.7.7 we set up standard processes. In Examples 11.7.2, 11.7.8 and 11.7.9 we have non-standard processes. Nonetheless, we do have functions of a timelike variable. To emphasize this we use $\lambda_\omega(t)$ just as we used $x_s(t)$ in section 11.1.

Suppose, now, that $\Omega, \mathscr{A}, P$ is a Gaussian process and that $\Omega$ is a set of functions on $T$ with

$$P(\omega(t_1) \leq x_1, \ldots, \omega(t_n) \leq x_n) = F_{t_1,\ldots,t_n}(x_1, \ldots, x_n)$$

We will see how to obtain other Gaussian processes from this one.

### Example 11.7.2

Consider the Wiener process with $y_0 = 0$. Let $b$ be any positive number. For $\omega \in \Omega$ let $\lambda_\omega(t) = b^{-1/2}\omega(bt)$. Show that

$$P(\lambda_\omega(t_1) \leq x_1, \ldots, \lambda_\omega(t_n) \leq x_n) = P(\omega(t_1) \leq x_1, \ldots, \omega(t_n) \leq x_n)$$

We consider the random vector $(\lambda_\omega(t_1), \ldots, \lambda_\omega(t_n)) = (b^{-1/2}\omega(bt_1), \ldots, b^{-1/2}\omega(bt_n))$. By section 11.4 the Wiener process is Gaussian so $(\omega(u_1), \ldots, \omega(u_n))$ is a normal random vector. Therefore, by exercise 10.5.3 $(b^{-1/2}\omega(bt_1), \ldots, b^{-1/2}\omega(bt_n))$ is a normal random vector. Now

$$E(\lambda_\omega(t)) = E(b^{-1/2}\omega(bt)) = b^{-1/2}E(\omega(bt)) = b^{-1/2}\cdot 0 = 0$$

Now, since $b > 0$, $\min(bt, bu) = b\min(t, u)$. Also $E(\omega(u)) = 0$ and $\mathrm{cov}(\omega(t), \omega(u)) = \min(t, u)$ so

$$E(\lambda_\omega(t)\lambda_\omega(u)) = E(b^{-1/2}\omega(bt)b^{-1/2}\omega(bu)) = b^{-1}\min(bt, bu)$$
$$= b^{-1}b\min(t, u) = \min(t, u)$$

Thus $(\lambda_\omega(t_1), \ldots, \lambda_\omega(t_n))$ and $(\omega(t_1), \ldots, \omega(t_n))$ have the same distribution since the expected values and covariances of corresponding components agree.

The result in Example 11.7.2 is not profound. It is just a matter of changing the time scale and space scale properly. It could be anticipated from the fast random walk. The next result is a bit deeper and it has a good many consequences. We are able to take up only one or two.

### Example 11.7.3
Let $\Omega$, $\mathscr{A}$, $P$ be the standard realization of the Wiener process. Let $\Lambda = \{\lambda; \lambda$ is a continuous function on $t > 0$ and $\lambda(0) = 0\}$. Define the probability measure $P_1$ on $\mathscr{B}$, the usual system of events on $\Lambda$, by
$$P_1(\lambda(0) \le x_0, \quad \lambda(t_1) \le x_1, \ldots, \lambda(t_n) \le x_n)$$

$$= \begin{cases} 0, & x_0 < 0 \\ P\left(t_1\omega\left(\dfrac{1}{t_1}\right) \le x_1, \ldots, t_n\omega\left(\dfrac{1}{t_n}\right) \le x_n\right), & x_0 \ge 0 \end{cases}$$

for $0 < t_1 < t_2 < \ldots < t_n$.
Discuss the resulting stochastic process on $T = \{t; t \ge 0$ and real$\}$.

We first take $n = 1$ and let $x_1 \to +\infty$. We obtain

$$P_1(\lambda(0) \le x_0) = \begin{cases} 0, & x_0 < 0 \\ 1, & x_0 \ge 0 \end{cases}$$

Therefore, $\lambda(0)$ has the same distribution as the constant random variable zero. If we take $x_0 > 0$ we see that $(\lambda(0), \lambda(t_1), \ldots, \lambda(t_n))$ has the same distribution as $\left(0, t_1\omega\left(\dfrac{1}{t_1}\right), \ldots, t_n\omega\left(\dfrac{1}{t_n}\right)\right)$, the last $n$ components of which form an $n$-dimensional normal random vector. A simple use of characteristic functions shows that $\left(0, t_1\omega\left(\dfrac{1}{t_1}\right), \ldots, t_n\omega\left(\dfrac{1}{t_n}\right)\right)$ is a degenerate $(n + 1)$-dimensional normal random vector. Hence $\Lambda$, $\mathcal{B}$, $P_1$ is a Gaussian process. By the definition of $P_1$, if $t > 0$, $u > 0$

$$E_{P_1}(\lambda(t)) = E_P\left(t\omega\left(\tfrac{1}{t}\right)\right) = tE_P\left(\omega\left(\tfrac{1}{t}\right)\right) = t \cdot 0 = 0$$

and

$$E_{P_1}(\lambda(t)\lambda(u)) = E_P\left(t\omega\left(\tfrac{1}{t}\right)u\omega\left(\tfrac{1}{u}\right)\right)$$

Note that we have indicated as a subscript the measure on which the expectations are based. Suppose that $0 < t \le u$. Then $1/t \ge 1/u$ so

$$E_P\left(t\omega\left(\tfrac{1}{t}\right)u\omega\left(\tfrac{1}{u}\right)\right) = tu\min\left(\tfrac{1}{t},\tfrac{1}{u}\right) = tu\tfrac{1}{u} = t$$

The corresponding result holds if $0 < u \le t$ and the case $t$ or $u = 0$ can be handled easily because, essentially, $\lambda(0) = 0$.

Therefore, for $\Lambda$, $\mathcal{B}$, $P_1$ the mean function $m(t) = 0$ and the covariance function $r(t, u) = \min(t, u)$. Thus this process is equivalent to the Wiener process.

In section 11.4 we saw that if $y_0 = 0$ and $b > 0$

$$P(\omega(u) \le b \text{ for } 0 \le u \le t) = 1 - \sqrt{\tfrac{2}{\pi t}}\int_b^\infty e^{-x^2/2t}\,dx$$

Since the functions $\omega$ are continuous

$$\{\omega(t) \le b \text{ for } 0 \le u \le t\} = \{\omega(t) \le b \text{ for } 0 < u \le t \text{ and } u \text{ rational}\}$$

$$= \bigcap_{k=1}^\infty \{\omega(t_1) \le b, \quad \omega(t_2) \le b, \ldots, \omega(t_k) \le b\}$$

where $t_1, t_2, \ldots$ is an enumeration of the rational numbers between 0 and $t$ excluding 0. By convention we set $t_j = 1/u_j$ or $u_j = 1/t_j$. By definition

$$P_1(\lambda(u_1) \le u_1 b, \quad \lambda(u_2) \le u_2 b, \ldots, \lambda(u_k) \le u_k b)$$

$$= P\left(u_1\omega\left(\tfrac{1}{u_1}\right) \le u_1 b, \quad u_2\omega\left(\tfrac{1}{u_2}\right) \le u_2 b, \ldots, u_k\omega\left(\tfrac{1}{u_k}\right) \le u_k b\right)$$

$$= P(\omega(t_1) \le b, \quad \omega(t_2) \le b, \ldots, \omega(t_k) \le b)$$

By our definitions, $u_1, u_2, u_3, \ldots$ is an enumeration of the rational numbers greater than or equal to $1/t$ since reciprocals of rationals are rational and reciprocation of positive numbers inverts order.

From Chapter 3, the fact that $P$ is a probability measure, and since

$$\{\omega(t_1) \le b, \ldots, \omega(t_k) \le b\} \subset \{\omega(t_1) \le b, \ldots, \omega(t_{k+1}) \le b\}$$

$$P(\bigcap_{k=1}^\infty \{\omega(t_1) \le b, \ldots, \omega(t_k) \le b\}) = \lim_{k\to\infty} P(\omega(t_1) \le b, \ldots, \omega(t_k) \le b) =$$

$$\lim_{k\to\infty} P_1(\lambda(u_1) \le u_1 b, \ldots, \lambda(u_k) \le u_k b) = P_1(\bigcap_{k=1}^\infty \{\lambda(u_1) \le u_1 b, \ldots, \lambda(u_k) \le u_k b\})$$

The functions $\lambda \in \Lambda$ are continuous except possibly at $t = 0$. Furthermore $\lambda(u) \le ub$ is equivalent to $\lambda(u)/u \le b$. Since $t > 0$, for $u \ge 1/t > 0$, $\lambda(u)/u$ is continuous and

$$\bigcap_{k=1}^{\infty} \{\lambda(u_1) \le bu_1, \ldots, \lambda(u_k) \le bu_k\} = \left\{\lambda(u) \le bu \text{ for rational } u \ge \frac{1}{t}\right\}$$

$$= \left\{\frac{\lambda(u)}{u} \le b \text{ for rational } u \ge \frac{1}{t}\right\}$$

$$= \left\{\lambda(u) \le ub \text{ for all } u \ge \frac{1}{t}\right\}$$

Thus, by Example 11.7.3 we have for the Weiner process with $y_0 = 0$

**(11.7.2)**  $P\left(\omega(u) \le ub \text{ for all } u \ge \frac{1}{t}\right) = 1 - \sqrt{\frac{2}{\pi t}} \int_b^{\infty} e^{-x^2/2t}\, dx$

It is not our objective to obtain results about the Wiener process in this section. We wish only to show the ways in which processes give information on one another.

Examples 11.7.2 and 11.7.3 suggest a general pattern which we state only in rough form. A change of the "time scale" and the "space scale" of a Gaussian process gives a Gaussian process. However, the time scale must be changed in a one-to-one manner. We use this procedure again in Example 11.7.4.

### *Example 11.7.4* (Ornstein–Uhlenbeck velocities.)

Let $\Omega, \mathscr{A}, P$ be the usual Wiener process with $y_0 = 0$. Let $\Lambda$ (different from the previous $\Lambda$) be the set of continuous functions on $-\infty < t < \infty$ with the usual admissible system of events. Let

$$P_1(\lambda(t_1) \le b_1, \ldots, \lambda(t_n) \le b_n) = P(e^{-t_1}\omega(e^{2t_1}) \le b_1, \ldots, e^{-t_n}\omega(e^{2t_n}) \le b_n)$$

Discuss the resulting process.

The process is Gaussian because $e^t > 0$ for any real $t$ and $(\lambda(t_1), \ldots, \lambda(t_n))$ has the same distribution as $(e^{-t_1}\omega(e^{2t_1}), \ldots, e^{-t_n}\omega(e^{2t_n}))$ which, for $\omega$ ranging over $\Omega$, is clearly normal.

We have

$$E_{P_1}(\lambda(t)) = E_P(e^{-t}\omega(e^{2t})) = e^{-t}E_P(\omega(e^{2t})) = e^{-t}y_0 = 0$$

Furthermore,

$$E_{P_1}(\lambda(t)\lambda(u)) = E_P(e^{-t}\omega(e^{2t})e^{-u}\omega(e^{2u}))$$
$$= e^{-t}e^{-u}E_P(\omega(e^{2t})\omega(e^{2u}))$$

Now $t \le u$ if and only if $e^{2t} \le e^{2u}$. Therefore if $t \le u$,

$$E_{P_1}(\lambda(t)\lambda(u)) = e^{-t}e^{-u} \min(e^{2t}, e^{2u}) = e^{-t}e^{-u}e^{2t} = e^{-u+t} = e^{-|u-t|}$$

It can be seen that $r(t, u) = e^{-|u-t|}$ for $t > u$ as well. We are finished with the basic description because the mean function and covariance function characterize a Gaussian process.

This process was intensely studied by Ornstein and Uhlenbeck as a model for the velocities of particles undergoing Brownian motion. This was done because the Wiener process model for Brownian motion implies infinite velocities at all times. We will point out an additional result about this process later. At present we mention that it is stationary according to the definition in section 11.5.

### Example 11.7.5

Let $S, \mathscr{A}, P$ be a probability system on which we have independent, $N(0, 1)$ random variables $X_1, X_2, \ldots, X_k$. Let $h_1(t), h_2(t), \ldots, h_k(t)$ be real-valued functions on $T$. Show that the set of functions $\Omega = \{\omega\}$ where

$$\omega(t) = h_1(t)X_1 + h_2(t)X_2 + \cdots + h_k(t)X_k$$

with probabilities induced by $X_1, \ldots, X_k$ is a Gaussian process. Find $m(t)$ and $r(t, u)$ for that process.

We need only show that for each set $\{t_1, t_2, \ldots, t_n\} \subset T$ the random vector $(\omega(t_1), \omega(t_2), \ldots, \omega(t_n))$ on $\Omega$ is a normal random vector. Now since the probability measure on $\Omega$ is induced by $X_1, X_2, \ldots, X_k$ and $P$ on $S$, the random variables

$$
\begin{aligned}
Y_1 &= \omega(t_1) = h_1(t_1)X_1 + h_2(t_1)X_2 + \cdots + h_k(t_1)X_k \\
Y_2 &= \omega(t_2) = h_1(t_2)X_1 + h_2(t_2)X_2 + \cdots + h_k(t_2)X_k \\
&\;\;\vdots \\
Y_n &= \omega(t_n) = h_1(t_n)X_1 + h_2(t_n)X_2 + \cdots + h_k(t_n)X_k
\end{aligned}
$$

form a normal random vector by exercise 10.5.3. We have

$$E(\omega(t)) = h_1(t)E(X_1) + h_2(t)E(X_2) + \cdots + h_k(t)E(X_k) = 0$$

since for each fixed $t \in T$ the numbers $h_j(t)$ are constants as far as $E$ is concerned and $X_j$ is $N(0, 1)$. Thus $m(t) = 0$. Similarly,

$$E(\omega(t)\omega(u)) = E\left(\sum_{j=1}^{k} \sum_{l=1}^{k} h_j(t)h_l(u)X_jX_l\right)$$

$$= \sum_{j=1}^{k} \sum_{l=1}^{k} h_j(t)h_l(u)E(X_jX_l)$$

Since the $X_j$ are independent and $N(0, 1)$, $E(X_jX_l) = \delta_{jl}$ and

**(11.7.3)**     $r(t, u) = E(\omega(t)\omega(u)) - E(\omega(t))E(\omega(u))$

$$= \sum_{j=1}^{k} h_j(t)h_j(u) - 0 = \sum_{j=1}^{k} h_j(t)h_j(u)$$

The result in Example 11.7.5 is the easiest in a sequence of techniques of constructing more complicated Gaussian processes out of simpler ones. It is usually assumed that the only constants $a_1, a_2, \ldots, a_k$ such that $\sum_{j=1}^{k} a_j h_j(t) = 0$ for all $t \in T$ are $a_j = 0$ for $j = 1, 2, \ldots, k$; i.e., the functions $h_j$ are linearly independent. The conclusion of the example holds regardless of this.

### Example 11.7.6

Let $S$ be $k$-dimensional Euclidean space with probability mass density

$$p(s_1, \ldots, s_k) = \prod_{j=1}^{k} (2\pi)^{-1/2} \exp(-\tfrac{1}{2}s_j^2)$$

The coordinate random variables $X_j(s) = s_j$ for $s \in S$ are independent and $N(0, 1)$. Let $T$ be the real numbers and let $h_j(t) = \cos jt$, $j = 1, 2, \ldots, k$. Discuss the process defined in Example 11.7.5 as specialized to this case.

For each $(s_1, s_2, \ldots, s_k) \in S$ we get the sample function $\omega(t) = \sum_{j=1}^{k} s_j \cos jt$.

Since $S$ is all of $k$-dimensional space, the $s_j$ range freely over all real numbers. As in Example 11.7.5 $m(t) = 0$ and $r(t, u) = \sum_{j=1}^{k} \cos jt \cos ju$. Since $\cos j(t + 2\pi)$ $= \cos jt$ for all $t$, $\omega(t + 2\pi) = \omega(t)$ so the typical sample function is a periodic function on $T$. We also have $\cos(-t) = \cos t$ so $\omega$ has a graph which is symmetrical with respect to the vertical axis. Some graphs are shown in Figure 12. It cannot be said that the functions shown there are typical because any fixed $\omega$ has probability 0.

### Example 11.7.7

Let $S$ be as in Example 11.7.6. Let $T = \{0 \le t \le 1\}$. Let $h_j(t)$ be alternatively $+1$ and $-1$ in the intervals $0 < t < \dfrac{1}{j}$, $\dfrac{1}{j} < t < \dfrac{2}{j}, \ldots, \dfrac{j-1}{j} < t < \dfrac{j}{j}$ starting with $+1$ on the left. Let $h_j(t) = 0$ at the jump points $0, \dfrac{1}{j}, \dfrac{2}{j}, \ldots, \dfrac{j-1}{j}, \dfrac{j}{j}$. What can be said about the continuity of the sample functions $\omega$ and of $r$ as determined in Example 11.7.5?

By definition $\omega(t) = s_1 h_1(t) + s_2 h_2(t) + \cdots + s_k h_k(t)$. Furthermore the function $h_j$ has a discontinuity at the $j + 1$ points $0, \dfrac{1}{j}, \dfrac{2}{j}, \ldots, \dfrac{j-1}{j}, \dfrac{j}{j}$. It does not follow, however, that each sample function $\omega$ has discontinuities at all of the points at which the relevant $h_j$ have discontinuities. As an extreme case if $s_j = 0$ for all $j \le k$, $\omega(t) = 0$ and it is continuous. Furthermore if $k = 4$ and $s_2 = -s_4$,

$$\lim_{\substack{t \to 1/2 \\ t < 1/2}} \omega(t) = \lim_{\substack{t \to 1/2 \\ t > 1/2}} \omega(t)$$

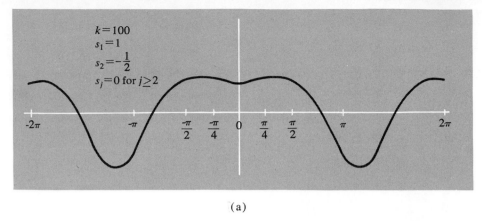

$k=100$
$s_1=1$
$s_2=-\dfrac{1}{2}$
$s_j=0$ for $j\geq2$

(a)

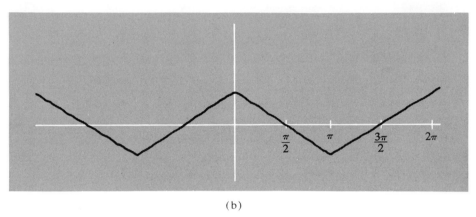

(b)

**Figure 12**

as in Figure 13. However, it seems clear that it is difficult to cancel out the discontinuities of the $h_j$'s when we form the $\omega$'s. The $s_j$'s must balance one another or be zero. This balancing or being zero can occur only on a set of $s \in S$ which has probability zero. We would expect that

$$P(\omega \text{ has discontinuities at } l/j \text{ for } 0 \leq l \leq j \leq k) = 1$$

As a matter of fact, since the $h_j$ are continuous except at those special points, each $\omega$ is continuous at the other points of $T$.

As before

$$r(t, u) = \sum_{j=1}^{k} h_j(t)h_j(u)$$

It is an elementary result of the definition of continuity at a point that if neither $t$ nor $u$ are in $Q = \{l/j; l, j \text{ integers } 0 \leq l \leq j \leq k\}$ then $r$ is continuous at the point $(t, u)$. Thus when $k = 3$, $r$ is continuous in the interiors of the subrectangles of the unit square which are indicated in Figure 14. Moreover, since the

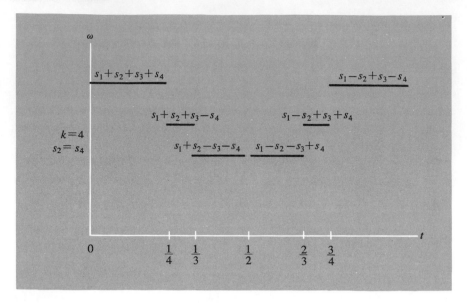

**Figure 13**

functions $h_j$ take on only the values 0, 1, or $-1$ and $h_j(t) = 0$ only if $t \in Q$ we find that $r$ is constant on the interior of the indicated rectangles. It takes on the value marked in Figure 14 in those subrectangles. It is discontinuous only on the boundaries of the indicated subrectangles and if $k = 3$ it is discontinuous at each point of those boundaries. However, if $k = 4$, $r$ may not be discontinuous

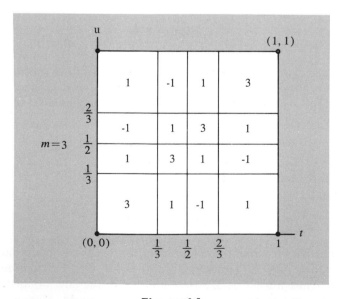

**Figure 14**

at each point of those boundaries. That case will be covered in an exercise. Finally we point out that on the interior of the diagonal rectangles $r(t, u) = 3$ since $r(t, t) = 3$ if $t \in Q$.

Examples 11.7.6 and 11.7.7 hint at a certain general situation. In the context of Example 11.7.5 in which the sample functions depend on a finite number of $X_j$ the continuity or certain other properties of the $h_j$ carry over so that the sample functions have the same properties; for example, if each $h_j$ is $l$ times differentiable each $\omega$ will be $l$ times differentiable. However, the functions $h_j$ are not intrinsic to the process. A Gaussian process is characterized by $m$ and $r$. In our case $m(t) = 0$. However,

$$r(t, u) = \sum_{j=1}^{k} h_j(t)h_j(u)$$

so continuity and differentiability properties of the $h_j$ carry over to the covariance function $r$ in some sense.

It can be shown that (and how) continuity and differentiability of $m(t)$ and $r(t, u)$ imply continuity and differentiability for all sample functions except those in a set of probability zero. This is true not only in the context of Example 11.7.5 but in general. It would take us too far afield to state, let alone prove, results of this type, hence we stop after these few remarks.

The situation in Example 11.7.5 is somewhat too simple. The more interesting case is that in which there are infinitely many $X_j$ and, of course, infinitely many functions $h_j$. Along these lines it can be shown that if $0 \le t, u \le \pi$

**(11.7.4)**    $\displaystyle \min(t, u) - \frac{1}{\pi} tu = \sum_{j=1}^{\infty} \frac{2}{\pi} \frac{\sin jt \sin ju}{j^2}$

Thus if $X_0, X_1, X_2, \ldots$ are independent and $N(0, 1)$, in some sense

**(11.7.5)**    $\displaystyle \omega(t) = \frac{1}{\sqrt{\pi}} tX_0 + \sum_{j=1}^{\infty} \sqrt{\frac{2}{\pi}} \frac{\sin jt}{j} X_j$

is the Wiener process with $y_0 = 0$ and $t$ restricted to $0 \le t \le \pi$. There are other series representations for the Wiener process. These, too, are beyond the scope of this book. However, we can point out that the first question in the study of series representations such as (11.7.5) is: "In what sense does the series on the right-hand side converge?" This is serious because the values of a normal random variable $X$ can be extremely large.

We come now to the final general topic in this section. Let $\Omega, \mathscr{A}, P$ be a Gaussian process in which $\Omega$ is a set of continuous functions on $T$. We assume $T \supset \{t; a \le t \le b\}$, $U$ is an interval of real numbers, and for each $u \in U$, $K(u, t)$ is bounded and piecewise continuous in $t$ on $a \le t \le b$. Let

**(11.7.6)**    $\displaystyle \lambda_\omega(u) = \int_a^b K(u, t)\omega(t) \, dt$

For each fixed $u$, $\lambda_\omega(u)$ is a function from $\Omega$ to the real numbers as $\omega$ varies. However, for each fixed $\omega \in \Omega$, $\lambda_\omega(u)$ defines a function from

$$U = \{u; u_0 \le u \le u'\}$$

to the real numbers. Let $\Lambda$ be the set of such functions. It is in correspondence to $\Omega$ although the correspondence may be many to one since it is just possible that for two different elements $\omega_1$ and $\omega_2$, $\lambda_{\omega_1}(u) = \lambda_{\omega_2}(u)$ for all $u \in U$. We will want to use both aspects of $\lambda_\omega(u)$. Although we do not prove it, it is true that for each fixed $u \in U$, $\lambda_\omega(u)$ is a random variable on $\Omega$.

### Example 11.7.8

Let $u$ and $v$ be fixed elements of $U$. In view of definition (11.7.6) and considering $\lambda_\omega(u)$ and $\lambda_\omega(v)$ as random variables on $\Omega$, calculate $E(\lambda_\omega(u))$ and

$$\text{cov}(\lambda_\omega(u), \lambda_\omega(v)).$$

Without worrying about convergence we have

$$E(\lambda_\omega(u)) = E\left(\int_a^b K(u, t)\omega(t)\, dt\right)$$

The operator $E$ is a kind of sum over $\Omega$. The integral $\int_a^b \ldots dt$ is a kind of sum over $T$. These sums are independent of one another so we should be able to take them in reverse order. Thus

$$E\left(\int_a^b K(u, t)\omega(t)\, dt\right) = \int_a^b E(K(u, t)\omega(t))\, dt = \int_a^b K(u, t)E(\omega(t)\, dt)$$

since $K(u, t)$ is constant as far as $E$ is concerned. However the $\omega$ are sample functions for a Gaussian process so

$$E(\omega(t)) = m(t)$$

and

$$E(\lambda_\omega(u)) = \int_a^b K(u, t)m(t)\, dt$$

By the same reasoning $E(\omega(t)\omega(w)) = r(t, w) + m(t)m(w)$ and

$$E(\lambda_\omega(u)\lambda_\omega(v)) = E\left(\left[\int_a^b K(u, t)\omega(t)\, dt\right]\left[\int_a^b K(v, w)\omega(w)\, dw\right]\right)$$

$$= E\left(\int_a^b dt \int_a^b K(u, t)K(v, w)\omega(t)\omega(w)\, dw\right)$$

$$= \int_a^b dt \int_a^b K(u, t)K(v, w)E(\omega(t)\omega(w))\, dw$$

$$= \int_a^b dt \int_a^b K(u, t)K(v, w)[r(t, w) + m(t)m(w)]\, dw$$

But

$$\int_a^b dt \int_a^b K(u, t)K(v, w)m(t)m(w)\, dw$$

$$= \left[\int_a^b K(u, t)m(t)\, dt\right]\left[\int_a^b K(v, w)m(w)\, dw\right]$$

$$= E(\lambda_\omega(u))E(\lambda_\omega(v))$$

Therefore,

**(11.7.7)**     $\operatorname{cov}(\lambda_\omega(u), \lambda_\omega(v)) = \int_a^b \int_a^b K(u, t)K(v, w)r(t, w)\, dtdw$

and

$$E(\lambda_\omega(u)) = \int_a^b K(u, t)m(t)\, dt$$

The interchanges of operations can be justified under the hypotheses of continuity of $K$ and of the sample functions $\omega$.

The alert reader may feel that we would not be calculating a mean function and covariance function if there were not a Gaussian process present. This is not strictly true. A mean function and a covariance function can exist for many processes. However, it is true that there is a Gaussian process of interest here. Let us state a theorem and sketch its proof.

**Theorem 11.7.1**     Let $\Omega, \mathscr{A}, P$ be a Gaussian process and let $K$ be continuous as in the hypotheses preceding (11.7.6). Then $\Omega, \mathscr{A}, P$ and the random variables $\lambda_\omega(u)$ on $\Omega$ form a (nonstandard) realization for a Gaussian process.

OUTLINE OF PROOF     We need only show that for $u_1, u_2, \ldots, u_n$ in $U$, the random vector $(\lambda_\omega(u_1), \ldots, \lambda_\omega(u_n))$ is normal. If $a = t_0 < t_1 < t_2 < \cdots < t_k = b$ and $\Delta_j = t_j - t_{j-1}$,

$$\lambda_\omega^{(k)}(u_l) = \sum_{j=1}^k K(u_l, t_j)\omega(t_j)\Delta_j$$

is an approximation to $\lambda_\omega(u_l)$ for each $l$ because it is the kind of sum of bases times heights which converges to an integral. But $(\lambda_\omega^{(k)}(u_1), \lambda_\omega^{(k)}(u_2), \ldots, \lambda_\omega^{(k)}(u_n))$ is a normal random vector for each $k$ because $(\omega(t_1), \omega(t_2), \ldots, \omega(t_k))$ is a normal random vector and the $\lambda_\omega^{(k)}(u_l)$ are sums of constants times the $\omega(t_j)$. Under suitable conditions, limits of normal random vectors are normal random vectors. We take the limit as $k \to \infty$ and the largest $\Delta_j \to 0$. Then $\lambda_\omega^{(k)}(u_l) \to \lambda_\omega(u_l)$.

Note that the boundedness and piecewise continuity of $K$ guarantees that for each $u \in U$ and $\omega \in \Omega$, $\int_a^b K(u, t)\omega(t)\, dt$ exists. Any other hypotheses on $K$ which allows this and allows the steps of the theorem are equally acceptable.

### Example 11.7.9

Let $\Omega, \mathscr{A}, P$ be the Wiener process with $y_0 = 0$, $a = 0$, $b = \infty$, $U = \{u; u \geq 0, u \text{ real}\}$. Let

$$K(u, t) = \begin{cases} u - t, & 0 \leq t \leq u \\ 0, & t \geq u \end{cases}$$

Discuss the process $\lambda_\omega(u) = \int_0^\infty K(u, t)\omega(t)\, dt$.

For each fixed $u$, $K(u, t)$ is piecewise continuous because 0 and $u - t$ are continuous. Furthermore, $K(u, t)$ is bounded on and zero outside of $0 \leq t \leq u$. Thus $\lambda_\omega(u) = \int_0^u (u - t)\omega(t)\, dt$. Since $E(\omega(t)) = y_0 = 0$, Example 11.7.8 gives $E(\lambda_\omega(u)) = 0$. If $v \geq u \geq 0$,

$$\mathrm{cov}(\lambda_\omega(v), \lambda_\omega(u)) = \int_0^v dw \int_0^u (u - t)(v - w) \min(t, w)\, dt$$

Now $\min(t, w)$ is sometimes $t$ and sometimes $w$. Hence, it is necessary to split up the last integral in order to evaluate it. Now $v \geq u$ so whenever $v \geq w \geq u \geq t$, $\min(t, w) = t$. If $u \geq t \geq w \geq 0$, $\min(t, w) = w$. If $u \geq w \geq t \geq 0$, $\min(t, w) = t$. Exactly one of these three cases must apply to any pair $(t, w)$ with $0 \leq t \leq u$, $0 \leq w \leq v$ except if $w = t$. More than one case applies there but that occurs on a line and makes no contribution to a double integral. Thus our integral for the covariance is equal to

$$\int_u^v dw \int_0^u (u - t)(v - w)t\, dt + \int_0^u dt \int_0^t (u - t)(v - w)w\, dw$$

$$+ \int_0^u dw \int_0^w (u - t)(v - w)t\, dt$$

The usual techniques give

$$\mathrm{cov}(\lambda_\omega(v), \lambda_\omega(u)) = \frac{u^3}{12}\left(v^2 - \frac{uv}{2} + \frac{u^2}{10}\right)$$

if $v \geq u$. If $u \geq v$, interchange the roles of $u$ and $v$. By Theorem 11.7.1 the $\lambda$-process is Gaussian.

Processes such as that in Example 11.7.9 arise in physics and engineering. Specifically $\lambda_\omega(u)$ is the solution of the differential equation $\dfrac{d^2 y}{du^2} = \omega(u)$ with $y = 0$ and $\dfrac{dy}{du} = 0$ when $u = 0$. It can be interpreted as the motion of a particle starting from rest but subject to random accelerations whose nature is that of the sample functions of the Wiener process. In other problems we may have

$$L\frac{d^2y}{du^2} + R\frac{dy}{du} + \frac{1}{C}y = \omega(u)$$

This would occur if a series circuit consisting of a resistor, a capacitor, and an inductance were subject to random voltages coming from a stochastic process whose sample functions were the $\omega$. This is a very simple case of "noise" in an electrical device. Naturally, if you have nothing but the probabilistic nature of the noise voltages, you cannot expect anything better than the probabilistic nature of the resulting currents in the circuit.

We now return to a previous idea for an additional result.

### Example 11.7.10

Let $\Omega, \mathscr{A}, P$ be the standard realization of the Wiener process with $y_0 = 0$. Show that

$P(\text{for each integer } n > 0 \text{ there are } t, u < \dfrac{1}{n} \text{ with } \omega(t) < 0, \quad \omega(u) > 0) = 1$

We make use of $\Lambda, \mathscr{B}, P_1$ of Example 11.7.3. We define a number of sets which can be shown to be events. Let

$A_{-1} = \{\omega \in \Omega; \omega \text{ is eventually greater than } -1\}$

$A_{+1} = \{\omega \in \Omega; \omega \text{ is eventually less than } +1\}$

$A = \{\omega \in \Omega; \text{ for each integer } n > 0 \text{ there are } t, u > n \text{ with } \omega(t) \le -1, \\ \omega(u) \ge 1\}$

$A^* = \{\omega \in \Omega; \text{ for each integer } n > 0 \text{ there are rational numbers } t^*, u^* > n \\ \text{ with } \omega(t^*) \le -\frac{1}{2} \text{ and } \omega(u^*) \ge \frac{1}{2}\}$

$B^* = \{\lambda \in \Lambda; \text{ for each integer } n > 0 \text{ there are rational numbers } 0 < v^*, \\ w^* < \frac{1}{n} \text{ with } \lambda(v^*) \le -\frac{1}{2}v^*, \quad \lambda(w^*) \ge \frac{1}{2}w^*\}$

$B = \{\lambda \in \Lambda; \text{ for each integer } n > 0 \text{ there are } v, w < \frac{1}{n} \text{ with } \lambda(v) < 0, \\ \lambda(w) > 0\}$

Now by the definitions and the continuity of the $\omega$, $A = (A_{-1} \cup A_{+1})'$, $A^* \supset A$, and $B \supset B^*$. By Example 11.4.2, $P(A_{+1}) = 0$. By exercise 11.4.17, $P(A_{-1}) = 0$ so $P(A_{-1} \cup A_{+1}) = 0$ and $P(A) = 1$. But then $P(A^*) = 1$. By Example 11.7.3 and the fact that the rationals are countable, $P_1(B^*) = P(A^*) = 1$. Thus $P_1(B) = 1$. But since $\Lambda, \mathscr{B}, P_1$ is a realization of the Wiener process we obtain our result with it.

As a result of Example 11.7.10 we see that the sample functions of the Wiener process oscillate a good bit above and below their starting value.

## EXERCISES 11.7

**1** Let $\Omega$, $\mathscr{A}$, $P$ be the standard Wiener process with $\omega(0) = y_0$. Define the random variables $\lambda_t(\omega) = -\omega(t)$ for $\omega \in \Omega$ and $t \geq 0$. Show that the $\lambda_t$ on $\Omega$, $\mathscr{A}$, $P$ give the Wiener process "starting" at $-y_0$.

**2** Show that the Ornstein–Uhlenbeck velocity process does not have independent increments. [*Hint:* Calculate $E((\lambda(t_3) - \lambda(t_2))(\lambda(t_2) - \lambda(t_1)))$ for $t_1 < t_2 < t_3$.]

**3** Show that the Ornstein–Uhlenbeck velocity process has $r(t + v, u + v) = r(t, u)$. Show that its distribution functions satisfy $F_{t_1 + v, t_2 + v, \dots, t_n + v} = F_{t_1, t_2, \dots, t_n}$ for all $v$. That is, the process is stationary.

**4** Let $r(t, u)$ be the covariance function of a Gaussian process. Show that for any $t_1, t_2, \dots, t_n$ in $T$ and any $n$ real numbers $a_1, a_2, \dots, a_n$,

$$\sum_{j,k=1}^{n} r(t_j, t_k)a_j a_k \geq 0$$

[*Hint:* $E((a_1 Z_1 + a_2 Z_2 + \cdots + a_n Z_n)^2) \geq 0$ for any $(Z_1, Z_2, \dots, Z_n)$.]

**5** Use the consequences of exercise 1 to show that if $c < y_0 = \omega(0)$ then

$$P(\omega(u) > c \text{ for } 0 \leq u \leq t) = \int_{-|c-y_0|t^{-1/2}}^{|c-y_0|t^{-1/2}} \frac{1}{\sqrt{2\pi}} e^{-z^2/2} \, dz$$

[*Hint:* $x < y$ if and only if $-x > -y$.]

**6** Plot a diagram such as Figure 14 for Example 11.7.7 with $k = 4$.

**7** Consider the differential equation $\dfrac{d^2 y}{dt^2} - 25y = \omega(t)$ subject to $y = 0$ and

$\dfrac{dy}{dt} = 0$ when $t = 0$. Show that

$$y(u) = \int_0^u \tfrac{1}{5}(-\cos 5u \sin 5t + \sin 5u \cos 5t)\omega(t) \, dt$$

by solving the differential equation by means of variation of parameters. Fit the result into the context of Example 11.7.9.

**8** Let $\lambda$ denote the Ornstein–Uhlenbeck process. Show that for $t > 0$ the conditional density of $P(\lambda(t) \leq x \mid \lambda(0) = a)$ is

$$[2\pi(1 - e^{-2t})]^{-1/2} \exp\left[-\tfrac{1}{2}(x - ae^{-t})^2(1 - e^{-2t})^{-1}\right]$$

[*Hint:* Sections 8.4 and 8.6.]

**9** Let $\Omega$, $\mathscr{A}$, $P$ be a Gaussian process in which $\Omega$ is a set of functions on $T$. Let $U$ be a set of real numbers and $q: U \to T$ a one-to-one function. Let $a(u)$ and $b(u)$ be real-valued functions on $U$. Show that the system of random variables $\lambda_u(\omega) = a(u) + b(u)\omega(q(u))$ forms a Gaussian process on $\Omega$, $\mathscr{A}$, $P$. Find its mean function and covariance function in terms of the first process.

**10** Suppose that the functions $h_1, \dots, h_{2n}$ of Example 11.7.5 are $a_1 \cos t$, $a_1 \sin t$, $a_2 \cos 2t$, $a_2 \sin 2t, \dots, a_n \cos nt$, $a_n \sin nt$. Show that the resulting process has $r(t, u) = q(t - u)$ where $q$ is some function. Show that the process is stationary. [*Hint:* See exercise 3.]

# CHAPTER 12

# Preliminaries of Statistics

## 12.1 Data

In order to study statistics effectively one must have the proper frame of mind. One should be keenly interested in examining data in order to apply the results of probability theory.

For example, the following data were obtained in actual experiments which one may regard as the independent tossing of a biased coin.

| | |
|---|---|
| **(12.1.1)** | HHHHHHHTTTHHHHHHHHTTH |
| **(12.1.2)** | HTHHTHHHHHHHTTTHHTHHH |
| **(12.1.3)** | THTHTTTHTHHTHHHTTTHH |
| **(12.1.4)** | HTHTTHHHHTHHTHTTHTTH |
| **(12.1.5)** | THHTTHHHTHHTHHHHHHHH |

The reader should feel justified (and a bit self-satisfied) if he wonders whether the same coin tossed independently produced the ten T's of (12.1.3) and the nine T's of (12.1.4) while producing five in (12.1.1) and (12.1.5) and six in (12.1.2). One should be suspicious of assertions of independence. For example, if a coin lands H, it is usually held on the thumb with H facing upwards on the next trial. Perhaps in the flipping there is a tendency to land H as a result of the way in which the experimenter flips coins. Another experimenter may have a tendency to flip so that the coin lands opposite to its previous face. However, since the reader has only tacit assurances from the author that a specific biased coin was flipped in a manner constituting independent trials, the data may be suspicious for reasons of fraud on the part of the author.

As partial reassurance, we will describe the actual experiments as much as possible. Outcomes (12.1.1) and (12.1.2) were obtained by choosing a card from a fairly well-shuffled standard deck. If the card was from the spade suit, T was recorded. If the card was from one of the other suits, H was recorded. The card was then returned to the deck which was then reshuffled. The procedure was repeated until 20 trials were made. In the experiment yielding (12.1.3)–(12.1.5) a U.S. quarter coin minted in 1941 was flipped. If in two tosses both tosses showed tails, T was recorded. If any other arrangement occurred in two tosses, H was recorded. Both experiments were done in such a way as to simulate a biased coin in which $P(H) = \frac{3}{4}$. Furthermore, the experiments were done in the sequence (12.1.1), (12.1.3), (12.1.4), (12.1.2), and (12.1.5). Finally, in (12.1.3) and (12.1.4) the coin was flipped so as to fall upon a bed. In (12.1.5) the coin fell upon a wooden floor and bounced around a good deal. It did turn over in the air many times in (12.1.3)–(12.1.5). Note that in (12.1.1) and (12.1.2) there is the appearance of three T's in a row out of five and six T's, respectively.

The reader should now flip a coin a number of times to generate some data for himself. This is especially so if he suspects the above presentation. After all, each of sequences (12.1.1) to (12.1.5) and similar such sequences to follow may have been written without having occurred in actual experiments.

Let us list some sample points which one may consider in the context of tossing a coin 20 times:

**(12.1.6)**     HHHHHHHHHHHTTTTTTTTTT

**(12.1.7)**     HTHTHTHTHTHTHTHTHTHT

**(12.1.8)**     TTTTTTTTHTHHTTHHHHHT

**(12.1.9)**     TTTHTHTTHTHHTHHTHHTT

**(12.1.10)**   HTTHHHHHHTTTHHTTTHTTH

**(12.1.11)**   TTHHHHHTTTHHTTTTTTTH

The sequences (12.1.6)–(12.1.11) could be records of actual experiments as well as sample points in an abstract mathematical system. The reader should be well aware of real-world problems for which a probabilistic model seems appropriate but for which the probability measure is not readily determinable. In such cases, no amount of thought per se seems to yield the proper probability measure, just as no thought per se would yield the value of the speed of light. Experiments must be made and data must be gathered.

For definiteness let us focus upon the problem of determining whether a coin is tossed independently and whether the coin is biased. Suppose we consider (12.1.6)–(12.1.11) as results of experiments with a coin. One could envision two positions.

*Pessimistic:*   How can you ascertain the bias of a coin with only a finite amount of data? Does the occurrence of (12.1.6) or (12.1.7) suggest that the tosses are not independent? If so, how can the others suggest independence? After all, for a fair coin the sample points (12.1.6)–(12.1.11) have probability $2^{-20}$ according to the theory. How can one contain more information than another? Furthermore, unless the coin is quite biased (12.1.6)–(12.1.11) have approximately the same probability (in independent tosses). If the coin is quite biased, each of (12.1.6)–(12.1.11) is rather unlikely. Note also, that a coin with only a slight bias toward tails has the property that in independent tosses

**(12.1.12)**     $P$(TTTTTTTTTTTTTTTTTTTT)

$$> P(\text{HHHTTTHHHTTHTTTHTTTH})$$

Yet 20 T's in 20 tosses seems unusual for a coin with only a slight bias toward tails.

In other words, no decent conclusions can be ascertained from data. Any experiment will yield a specific piece of data which will have its own special characteristics and, in a sense, not be random. Anything which happens is just chance.

*Optimistic:*   Great! If you want to find the bias of a coin, flip it about 1000 times and divide the number of H's by 1000. Outcomes (12.1.6), (12.1.7), and (12.1.8) would never occur in independent tossing. Even if (12.1.12) is true for a coin with a slight bias toward T try throwing 20 T's in a row, or 20 H's for that matter. The outcome HHHTTTHHHTTHTTTHTTTH looks like what you get even though the T's seems to occur in blocks.

With a large, carefully designed experiment you can get results about bias, etc., which if not exactly correct, are close enough for useful deductions.

Before discussing the above viewpoints, let us discuss how sample points (12.1.6)–(12.1.11) arose. Due to lack of patience (12.1.6) and (12.1.7) were simply written down. Sequences (12.1.8), (12.1.9), and (12.1.10) were thrown with the same coin, a 1962 penny minted in Denver. Sequence (12.1.11) was determined from (12.1.10) as follows. Take the first three symbols in (12.1.10) and record what occurs most frequently, i.e., for HTT, record T. Take the second, third, and fourth symbol and record the majority symbol, i.e., TTH so record T. Take the third, fourth, and fifth symbol THH so H is recorded. Continue to the fourth, fifth, and sixth—HHH so H is recorded. Thus (12.1.11) has TTHH.... Two extra tosses were needed to get a total of 20 trials in (12.1.11). Note that we are simulating dependent random variables.

Now we can discuss the viewpoints. The pessimistic viewpoint is not without merit—it is just fearful. All of the assertions of probabilities are correct. Nonetheless, if the pessimist were to look beyond specific dangers he would see that if a fair coin is flipped 20 times independently, it is quite unusual to have fewer than 5 H's or more than 15 H's. However, if a rather biased coin is flipped independently, those events occur fairly frequently. Therefore, if one is willing to venture forth and take a few risks he is apt to be correct in most cases, providing he uses proper procedures.

The optimistic viewpoint is too glib. Real-life sampling can be extremely expensive. But, if one uses modest size samples mistakes can be fairly frequent. Furthermore, in order to get reasonable results one inevitably makes simplifying assumptions which stubborn skeptics will not grant.

In this book, we will take a position of cautious optimism. To follow the pessimists' position would be to commit the sin of noninvolvement.

Let us study some of the data mentioned earlier using methods which hint at the ideas to be expounded upon in subsequent chapters.

Suppose now that biased coins come to an experimenter. They are believed to have $P(H) = \frac{3}{4}$ but they may have another bias. The experimenter wishes to ascertain which coins have $P(H) = \frac{3}{4}$ with some degree of reliability. He uses the following decision procedure: He tosses the coin 20 times independently. If 14, 15, or 16 H's show he declares (rightly or wrongly) that the coin has $P(H) = \frac{3}{4}$. Otherwise he declares it to have *some* other bias.

Note that if sequence (12.1.1), (12.1.2), or (12.1.5) occurs he declares the coin to have bias $P(H) = \frac{3}{4}$. If (12.1.3) or (12.1.4) occurs, he declares it to have some other bias.

This decision procedure is quite arbitrary and no mathematics is directly involved. We analyze this decision procedure and a similar one in Example 12.1.1.

### Example 12.1.1

Find the probability that a coin whose bias is actually $P(H) = \frac{3}{4}$ will be declared to have that bias under the above decision procedure. Find the same for coins of bias $P(H) = \frac{1}{2}$, $P(H) = \frac{2}{3}$, and $P(H) = \frac{4}{5}$. Contrast these results with those obtained by using a procedure whereby $P(H) = \frac{3}{4}$ is declared if 13, 14, 15, 16, or 17 H's show in independent tosses.

If $P(H) = p$ we have frequently seen that in independent tosses

$$P(k \leq \text{number of H's in 20 tosses} \leq l) = \sum_{j=k}^{l} \binom{20}{j} p^j (1 - p)^{20-j}$$

We summarize the computation in Table 1.

### Table 1

| $p$ | $\frac{1}{2}$ | $\frac{2}{3}$ | $\frac{3}{4}$ | $\frac{4}{5}$ |
|---|---|---|---|---|
| FIRST PROCEDURE | 0.056 | 0.419 | 0.561 | 0.526 |
| SECOND PROCEDURE | 0.131 | 0.644 | 0.808 | 0.793 |

The probabilities in the second row are greater than those in the first row as they must be. This reflects how one pays for recognizing $P(H) = \frac{3}{4}$ more surely by failing to recognize the other biases when the size of the sample is fixed. To be specific, under the first procedure a coin of bias $P(H) = \frac{3}{4}$ is erroneously declared in about four cases in nine. Under the second procedure a coin of bias $P(H) = \frac{1}{2}$ is declared to have $P(H) = \frac{3}{4}$ in about one case in eight compared to less than half that rate under the first procedure.

One of the objectives of statistics is the study of decision procedures as in Example 12.1.1 in order to determine good, if not optimal, decision procedures for certain problems. The main tool for this study is probability theory. In the next example we use more advanced methods of probability theory to indicate the flavor of what is to come.

**Example 12.1.2**

Use the Central Limit Theorem to find $a$ and $b$ so that if $P(H) = \frac{3}{4}$,

$$P(a \leq \text{number of H's in 300 independent tosses} \leq b) \approx 0.90$$

Let $N_n = $ number of H's in $n$ tosses. When $P(H) = p$ we have

$$\lim_{n \to \infty} P\left(\alpha < \frac{N_n - np}{\sqrt{np(1-p)}} \leq \beta\right) = \int_\alpha^\beta \frac{1}{\sqrt{2\pi}} e^{-x^2/2} \, dx$$

or

$$P\left(\alpha < \frac{N_n - np}{\sqrt{np(1-p)}} \leq \beta\right) \approx \int_\alpha^\beta \frac{1}{\sqrt{2\pi}} e^{-x^2/2} \, dx$$

for large $n$. From Table 1 (appendix)

$$\int_{-1.65}^{1.65} \frac{1}{\sqrt{2\pi}} e^{-x^2/2} \, dx \approx 0.90$$

Also for $n = 300$ and $p = \frac{3}{4}$ we have $np = 225$ and $\sqrt{np(1-p)} = 7.5$. Now

$$P\left(\alpha < \frac{N_n - np}{\sqrt{np(1-p)}} \leq \beta\right)$$

$$= P(np - \alpha\sqrt{np(1-p)} < N_n \leq np + \beta\sqrt{np(1-p)})$$

Therefore

$$P(225 - 12.375 < N_{300} \leq 225 + 12.375) \approx 0.90$$

Since $N_n$ is a whole number, $225 - 12.375 < N_n \leq 225 + 12.375$ is equivalent to $213 \leq N_n \leq 237$. We have $P(213 \leq N_{300} \leq 237) \approx 0.90$ so we take $a = 213$ and $b = 237$.

Notice that the computations of Example 12.1.2 furnish decision procedures for our problem of biased coins. These procedures, for sufficiently large sample size $n$, are correct approximately 90% of the time for coins whose $P(H) = \frac{3}{4}$. In some of the exercises to follow the reader is asked how well these procedures work for coins of other biases.

The decision procedures that we have discussed so far can be regarded as procedures for making judgments about a proper or preferable probability measure for the probabilistic model for a real-world system. To be more specific, given a particular real coin of the type under consideration, we flip it 20 times. If the number of H's which appear lies in a certain range we judge that the proper probability measure is that determined by $P(\text{H}) = \frac{3}{4}$. Furthermore, the judgment is made by the examination of data. Generally speaking, statistics* is the study of the use of data to make judgments about probability measures for the models of real-world systems.

In the material to come we propose methods that are intuitively plausible. These methods will seem to be good on intuitive grounds alone. However, in order to more fully appreciate the ideas, the reader should try to imagine that there may be better methods which are not so intuitive.

## EXERCISES 12.1

**1**  How can a coin of bias $P(\text{H}) = \frac{3}{8}$ be simulated?

**2**  Verify the probabilistic contentions in the "pessimistic position."

**3**  Let $X_1$, $X_2$, $X_3$, $X_4$ be independent random variables with $P(X_k = -1) = \frac{1}{2}$, $P(X_k = 1) = \frac{1}{2}$ for each $k$. Compute $E(X_1 + X_2 + X_3)$, $E(X_2 + X_3 + X_4)$, and $E((X_1 + X_2 + X_3)(X_2 + X_3 + X_4))$. Are $X_1 + X_2 + X_3 = Y$ and $X_2 + X_3 + X_4 = Z$ independent?

**4**  For sequence (12.1.11) compute

$$\frac{\text{Number of H's} - 10}{\sqrt{20 \cdot \frac{1}{2} \cdot \frac{1}{2}}}$$

Compare with $-1.65$ and $1.65$.

**5**  Is it proper to make the comparison indicated in exercise 4 if one is testing for a fair coin and tosses are independent? Explain.

**6**  Let the random variable $V$ give the number of H's which occur in 400 independent tosses of a coin of bias $P(\text{H}) = 0.55$. Find, approximately, the distribution of

$$Y = \frac{V - 200}{\sqrt{400 \cdot \frac{1}{2} \cdot \frac{1}{2}}}$$

with the help of the Central Limit Theorem.

Exercises 7–12 deal with runs in coin tossing. A run of heads in a coin toss sequence is a subsequence consisting of one or more H's which is preceded and succeeded by T's or nothing. We suppose that a fair coin is tossed 20 times.

---

* Here it is worthwhile to point out that the term "statistics" functions in two different ways. We study a certain subject matter called statistics. One of the concepts in this subject matter is a *statistic* (plural: statistics). The reader should be able to make this distinction in the ensuing material.

◆ **7**   Let $r_T$ and $r_H$ be the number of tail runs and head runs, respectively, in a sequence of heads and tails. Find $r_T$ and $r_H$ for (12.1.1), (12.1.2), and (12.1.3). Show that in any such sequence $|r_H - r_T| \le 1$.

**8**   What is the probability that the runs consist of two head runs of four and six H's, respectively, and three tail runs of two, three, and five T's, respectively, when tosses are independent? The runs may occur in any legitimate order.

**9**   What is the probability that in 20 independent tosses there are ten head runs and ten tail runs, each of length one?

**10**   What is the probability that in 20 independent tosses there are four head runs of length 1, 2, 3, 5 and three tail runs of length 2, 3, 4 and these runs occur in some order?

**11**   Certain tests for independence of trials are keyed to the number of runs. What do exercises 8–10 suggest about how evidence concerning the number of runs which confirm independence should look?

**12**   Could one get the same number of runs in dependent tosses as in independent tosses? What would this suggest about tests for independence which are based only on the number of runs in a sequence of tosses?

**13**   Add a column for $P(H) = \frac{7}{8}$ to Table 1; that is, find the probabilities that the experimenter will declare a coin of bias $P(H) = \frac{7}{8}$ to have bias $\frac{3}{4}$ under the two procedures.

Exercises 14–18 suggest the so-called *Bayesian* approach. We do not emphasize this type of decision making. However, when it can be applied, it has many advantages. The exercises follow Example 12.1.1. We assume that the probability that a coin of a particular bias appears is known.

**14**   Suppose that 90% of the coins have bias $\frac{3}{4}$, 6% have bias $\frac{1}{2}$, and 4% have bias $\frac{7}{8}$ (see exercise 13) and that the coins appear at random subject to these probabilities. Show that the probability that a coin will be incorrectly declared under the first procedure is $0.90(0.439) + 0.06(0.056) + 0.04(0.226)$.

**15**   Assume the probabilities of exercise 14. Suppose that the user of these coins loses \$10 if a bias $\frac{3}{4}$ coin is discarded, loses \$100 if a bias $\frac{1}{2}$ coin is accepted, and loses \$50 if a bias $\frac{7}{8}$ coin is accepted. Naturally, he accepts a coin if and only if the experimenter declares it to have $P(H) = \frac{3}{4}$. Find his expected loss if the first decision procedure is used by the experimenter.

**16**   Under the assumptions of exercises 14 and 15 find the expected loss if the second procedure is used.

**17**   Find the expected loss under the two decision procedures if 80% of the coins have bias $\frac{3}{4}$, 10% have bias $\frac{1}{2}$, and 10% have bias $\frac{7}{8}$.

**18**   Work exercise 17 if the respective percentages are 70, 20, 10.

**19**   Discuss the assumptions in exercises 14–18. Instead of coins imagine a food processor who uses potatoes. Think of $P(H) = p$ as signifying the fraction of good potatoes in a shipment; $\frac{3}{4}$ being standard. The loss incurred when $p = \frac{7}{8}$ is accepted as standard could represent the processor's missed chance at a higher price. The loss incurred when $p = \frac{1}{2}$ is accepted as standard could represent purchasers who will shift to another brand because of this brand's low quality.

# 12.2 Sampling and statistics

Statistics seeks to make judgements about probability measures by studying data rather than by other means. It is desirable to set up a mathematical model for the procedure of gathering data. However, we will not attempt to formulate the most general procedure for gathering data.

In the background of any application of the theory of statistics is a *population*. It is from this population that one obtains the data. The method of obtaining data is called *sampling* and a list of pieces of data is called a *sample*. Frequently the data occurs in the form of numbers in which case a sample will be a list of numbers. If the data occur in some other form one may code it by numbers if one chooses.

### Example 12.2.1

The Schmutz soap company wishes to try out a new package for its soap. It sends some employees to various stores. These employees are instructed to ask various shoppers whether they like the new package compared to the old package. The shoppers are asked whether the new package is much more attractive, more attractive, just about as attractive, less attractive, much less atttactive, or no opinion with reference to the old package. Discuss this procedure from our present point of view.

We would use the collection of all shoppers as our population. A sample would be a set of responses to the question of whether the new package is much more, more, equally, less, much less attractive than the old package. We may code these responses as 3, 1, 0, $-1$, $-3$, respectively, if desirable.

### Example 12.2.2

The Bureau of Agriculture wishes to make a study of the yield per acre of wheat in the state of Minnesota. Formulate a sampling procedure and state the population and the sample.

The Bureau could make a map of the wheat growing areas, select, for example, 100 plots, pay the owner to record carefully the yield on those plots, and gather the information. The population can be taken as the set of acres of wheatgrowing land. The sample is the list of 100 numbers giving the yield per acre of the 100 plots.

As one can see, the population is not unique. For example, one might question whether the collection of all shoppers is the population being sampled in Example 12.2.1. For instance, if certain geographical, social, or temporal restrictions occurred in the supermarket study one might be deceiving himself about the population to which his results apply. We will not dwell on such matters in this book. These are important questions but we restrict ourselves to the more mathematical aspects of statistics. As a crude rule of thumb the population consists of all "individuals" who might have been sampled for data under the sampling procedure. Other populations may be the set of subatomic particles which hit a certain instrument in a one-hour period, the set of numbers of defective bulbs in each box of 1000 boxes of light bulbs, the set of lifetimes of 10 000 transistors operating under certain conditions, or the set of incomes of families in a town.

Associated with a population is a sample space which we call $S_0$. This sample space is to mimic the population in a mathematically useful way. It is a mathematical entity which is one phase of a probabilistic model just as before. Just how much it mimics the population is a complicated matter since there can be more than one sample space. It need not be in one-to-one correspondence with the population. For example, although there are only 10 000 lifetimes of 10 000 transistors we might still wish to use the sample space

$$S_0 = \{x; x \text{ is a non-negative real number}\}$$

Similarly, even though I.Q. scores of people are non-negative we may wish to use as sample space for such a population,

$$S_0 = \{x; x \text{ is any real number}\}$$

In gathering data we must allow for more than one piece of data. We would want to toss a coin more than once to ascertain its bias, or we would want to test more than one transistor.

**Definition**   The *sample space of a sample (of size n)* from a population whose sample space is $S_0$ is

$$S = \underbrace{S_0 \times S_0 \times \cdots \times S_0}_{n}$$

Notice the clumsy phrase "sample space of a sample." Actually since $n$ is unspecified there are many such spaces. There is no standard terminology for such spaces. Perhaps it could be called a data space. However, we wish to reserve the word *data* for actual observations of the population.

Since most data can be coded into numbers or lists of numbers we should expect to have a random variable $X$ on $S_0$. For some purposes we will have a random vector $X$ on $S_0$. The latter could occur if we were doing medical research and wanted the weight, age, and height of the patient as well as the dosage of the drug for each case. Frequently $S_0$ can be chosen to be a set of real numbers and $X(x) = x$ for each $x \in S_0$.

With any random variable $X$ on $S_0$ we inherit $n$ random variables on $S$ defined as follows: If $s = (s_1, s_2, \ldots, s_n) \in S_0 \times S_0 \times, \cdots, \times S_0$,

**(12.2.1)** $\qquad X_1(s) = X(s_1),\ X_2(s) = X(s_2), \ldots, X_n(s) = X(s_n)$

Henceforth we take (12.2.1) as our official definition of the $(X_1, \ldots, X_n)$.

### Example 12.2.3

A coin is flipped three times in order to gather data. Describe the population, the data, the sample space of the sample, and, if one scores 1 for H and $-1$ for T, describe $X_1$, $X_2$, and $X_3$.

The population can be taken as all tosses of the coin for ever and ever. The data are precisely the pattern of H and T which you get. We can take

$$S = \{H, T\} \times \{H, T\} \times \{H, T\}$$

Finally,

$$X_1(H, T, T) = X_1(H, T, H) = X_1(H, H, T) = X_1(H, H, H) = \quad 1$$
$$X_1(T, T, T) = X_1(T, T, H) = X_1(T, H, T) = X_1(T, H, H) = -1$$
$$X_2(T, H, T) = X_2(T, H, H) = X_2(H, H, T) = X_2(H, H, H) = \quad 1$$
$$X_2(T, T, T) = X_2(T, T, H) = X_2(H, T, T) = X_2(H, T, H) = -1$$
$$X_3(T, T, H) = X_3(T, H, H) = X_3(H, T, H) = X_3(H, H, H) = \quad 1$$
$$X_3(T, T, T) = X_3(T, H, T) = X_3(H, T, T) = X_3(H, H, T) = -1$$

We now come to the key concept in our mathematical theory of statistics.

**Definition**    A random variable on the sample space of a sample is called a *statistic*.

This is a fairly innocuous definition. Nevertheless, a good deal of the remaining material is devoted to discussing the properties of various statistics.

We now describe a number of important statistics. We assume that $S = S_0 \times S_0 \times \cdots \times S_0$, a typical point $s \in S$ is $s = (s_1, s_2, \ldots, s_n)$, and $X$ and $Y$ are random variables on $S_0$.

**(12.2.2)** $\qquad \bar{X}(s) = \dfrac{X(s_1) + X(s_2) + \cdots + X(s_n)}{n}$

**(12.2.3)** $\qquad d^2(s) = \dfrac{1}{n-1} \sum_{k=1}^{n} [X(s_k) - \bar{X}(s)]^2, \qquad n > 1$

**(12.2.4)** $\qquad k(s) = \dfrac{1}{n} \sum_{k=1}^{n} X(s_k) Y(s_k) - \bar{X}(s)\bar{Y}(s)$

**(12.2.5)**    $r(s) = \max(X(s_1), X(s_2), \ldots, X(s_n)) - \min(X(s_1), \ldots, X(s_n))$

**(12.2.6)**    $\chi^2(s) = \sum\limits_{j=1}^{k} \dfrac{[N_j(s) - n\pi_j]^2}{n\pi_j}$,   where $\sum\limits_{j=1}^{k} \pi_j = 1$,   $\pi_j \geq 0$, and $N_j(s)$ is

the number of numbers from $\{X(s_1), \ldots, X(s_n)\}$ which satisfy the $j$th out of $k$ conditions.

**(12.2.7)**    $t(s) = \dfrac{\bar{X}(s)\sqrt{n}}{\sqrt{d^2(s)}}$,   where $\bar{X}$ and $d^2$ are defined above

**(12.2.8)**    $c(s) = \begin{cases} k/10, & X(s_1), X(s_2), \ldots, X(s_k) \leq a \\ N + l, & X(s_l) \geq b, \quad 1 \leq l/10 \leq n \\ n/10, & \text{one of } X(s_1), \ldots, X(s_k) > a \\ & \text{but all of } X(s_1), \ldots, X(s_n) < b \end{cases}$

where $a < b$ and $1 \leq k \leq n$.

In many cases $S_0$ is the set of all real numbers and $X(x) = x$ for $x \in S_0$. In such a situation we would see a formula such as

$$\bar{X}(s) = \frac{x_1 + x_2 + \cdots + x_n}{n}, \qquad s = (x_1, x_2, \ldots, x_n) \in S$$

All of the statistics (12.2.2)–(12.2.7) are symmetric in the sense that if $X(s_j)$ and $X(s_k)$ are interchanged the value of the statistic is not changed. This is not the case for (12.2.8).

Note also that we are suspending our convention of using capital letters near the end of the alphabet for random variables. Many classical statistics acquired symbols, and standard tables use these symbols. Therefore we suspend our convention in order to use the notation of such tables.

The statistics (12.2.2)–(12.2.7) were not established arbitrarily. The significance of some of them is suggested by their names; $\bar{X}$ is called the sample mean, $r$ the sample range, and $d^2$ can be called a modified sample variance. Note how these correspond to the mean, range, and (modified) variance of the numbers $X(s_1), X(s_2), \ldots, X(s_n)$ if those numbers have probability $n^{-1}$ each. The statistic $k$ can be thought of as a sample covariance. Furthermore, it seems intuitively clear that for large $n$, $\bar{X}(s)$ will be close to $E(X)$ for "most" sample points $s$ since the numbers $X(s_1), X(s_2), \ldots, X(s_n)$ will take on many of the possible values of $X$ with the proper frequency. We will make this intuitive statement more precise shortly.

The statistics (12.2.2)–(12.2.7) are mathematical analogs of calculations which occur in practical problems. For example, $X$ may refer to rust resistance and $Y$ to tensile strength of samples of steel made by a new process. Or, $X$ may refer to the I.Q. of a child and $Y$ to his protein intake in a survey of children. For practice, we formulate a purely mathematical example.

## Example 12.2.4

Let $S_0 = \{1, 2, \ldots, 6\}$, $X(m) = m$, and $Y(m) = (-1)^m$ for $m \in S_0$. Let $n = 8$ and $s' = (2, 5, 6, 1, 2, 1, 3, 5)$. Compute $\bar{X}(s')$, $d^2(s')$, and $k(s')$.

$$\bar{X}(s') = \tfrac{1}{8}(2 + 5 + 6 + 1 + 2 + 1 + 3 + 5) = \tfrac{25}{8}$$

$$d^2(s') = \tfrac{1}{7}(2 - \tfrac{25}{8})^2 + (5 - \tfrac{25}{8})^2 + (6 - \tfrac{25}{8})^2 + \cdots + (5 - \tfrac{25}{8})^2$$

$$= \tfrac{1}{7}(\tfrac{215}{8}) = 4.02$$

$$k(s') = \tfrac{1}{8}(2 - 5 + 6 - 1 + 2 - 1 - 3 - 5) - \tfrac{25}{8}(-\tfrac{2}{8})$$

$$= \tfrac{5}{32}$$

We may also think of Example 12.2.4 as relating to the rolling of a die eight times. Note how the evaluation of a statistic at a sample point $s'$ models a computation made from data which arises in an $n$-fold repetition of an experiment.

## Example 12.2.5

A buyer of apples wishes to purchase shipments of high quality apples. He ranks apples for taste as 0, 1, 2, 3, 4 with 0 worst and 4 best. He is only interested in shipments whose average rank is between 3 and 4. Within that range he wishes to pay less for poorer quality shipments. How can he decide what to bid for various shipments?

The buyer may taste 20 apples chosen randomly from a shipment. Let $q_1, q_2, \ldots, q_{20}$ be their taste rankings. Let $\bar{q} = \tfrac{1}{20}(q_1 + q_2 + \cdots + q_{20})$. He may bid as follows:

$$\begin{array}{lll} \text{No more than \$100} & \text{if} & 4.0 \geq \bar{q} > 3.7 \\ \text{No more than \$90} & \text{if} & 3.7 \geq \bar{q} > 3.5 \\ \text{No more than \$75} & \text{if} & 3.5 \geq \bar{q} > 3.2 \\ \text{No more than \$50} & \text{if} & 3.2 \geq \bar{q} \geq 3.0 \end{array}$$

Of course, we assume that these are reasonable bids for such apples. He may also wish to pay less if a shipment has excessively many apples of rank 0 or 1 even if the rest are of rank 4. To this end he can choose a cutoff value $a$ and bid less if

$$\tilde{q} = \frac{1}{19} \sum_{j=1}^{20} (q_j - \bar{q})^2 \geq a$$

since this quantity is large when $3.0 \leq \bar{q} \leq 4.0$ and quite a few of the $q_j$ are 0 or 1.

The problem of this example may be artificial. However, it does suggest how statistics are used in certain types of decisions. Actually, $\bar{q}$ is not a statistic. It is intended to be a single number obtained from real-world ratings of real apples. On the other hand $\bar{X}$ refers to a mathematical function. The similarity is intentional as is that between $\tilde{q}$ and $d^2$.

Rather than setting up special letters we will use a "^" mark over a statistic to denote the number obtained when real-world data are substituted into a statistic which is the model for that data. Thus, instead of $\bar{q}$ and $\tilde{q}$ we use $\hat{\bar{X}}$ and $\hat{d}^2$ respectively. The reader should note the logical difference between $\bar{X}(s')$ and $\hat{\bar{X}}$ although for a specific point $s'$ and specific data, $\bar{X}(s')$ and $\hat{\bar{X}}$ may be the same number.

The discussion of Example 12.2.5 only gives a decision procedure for limiting bids. The techniques of statistics are not in use. These techniques are concerned with certain risks; e.g., suppose that a shipment is generally bad but by chance a sample of 20 apples contains mostly good apples. Our apple buyer might pay an unusually high price for the shipment. It would be nice to sample all of the apples in the shipment but this is out of the question. He must risk some exceptional samples. Statistics studies decision procedures in the context of risks. Normally it seeks to find the best procedure; i.e., the one with the least risk. There may be no best procedure in certain problems.

Example 12.2.5 suggests that decision procedures may be related to statistics. We now study some properties of the statistics (12.2.2)–(12.2.7). We will not take up the statistic (12.2.8); it was included in order to suggest that there can be unsymmetrical and curious statistics as well as symmetrical statistics.

We have not yet discussed the choice of probability measure on $S = S_0 \times S_0 \times \cdots \times S_0$. In principle, $S$ can have all kinds of probability measures; however, certain of them are germane to statistics. We will always want symmetry, for example in the case $n = 3$

(12.2.9)     $P(A \times B \times C) = P(B \times A \times C) = P(C \times A \times B)$ and so on

However, normally we will require more than just symmetry.

For purposes of mathematical analysis we usually study independent sampling. It will be seen that this allows us to utilize our results on independent random variables. The real-world analog to independent sampling is sampling with replacement. This could result in the same person being questioned twice in a survey and this is inefficient because the information is redundant. Sampling without replacement, which we study briefly in section 15.2, is more practical but is more difficult mathematically. However, as we pointed out in section 6.4, for large populations, sampling with and without replacement give essentially the same results. Thus we study independent sampling with the objective of getting results more easily and getting results which approximate the corresponding results for more practical modes of sampling.

The mathematical definition of independent sampling is determined by the choice of probability measure $P$ on $S$. We use the usual probability measure on $S$ which arises from a measure $P_0$ on $S_0$ as follows: If $A_1, A_2, \ldots, A_n$ are events contained in $S_0$

(12.2.10)     $P(A_1 \times A_2 \cdots \times A_n) = P_0(A_1)P_0(A_2)\ldots P_0(A_n)$

This equation also includes (12.2.9).

### Example 12.2.6

A biased coin is tossed three times independently and data are obtained. Formulate the sample space of the sample and describe its probability measure. Find the probability distribution of $X_1 + X_2 + X_3$ if

$$X_j = \begin{cases} 1, & j\text{th toss is H} \\ 0, & j\text{th toss is T} \end{cases}$$

We may take $S_0 = \{H, T\}$ and $P_0(H) = p$, $P_0(T) = 1 - p$. Then $S = S_0 \times S_0 \times S_0 = \{H, T\} \times \{H, T\} \times \{H, T\}$, and

$$P(H, H, H) = p^3, \quad P(H, H, T) = p^2(1 - p), \quad P(H, T, H) = p^2(1 - p)$$
$$P(T, H, H) = p^2(1 - p), \quad P(H, T, T) = p(1 - p)^2, \quad P(T, H, T) = p(1 - p)^2$$
$$P(T, T, H) = p(1 - p)^2, \quad P(T, T, T) = (1 - p)^3$$

It is obvious that

$$P(X_1 + X_2 + X_3 = 0) = (1 - p)^3$$
$$P(X_1 + X_2 + X_3 = 1) = 3p(1 - p)^2$$
$$P(X_1 + X_2 + X_3 = 2) = 3p^2(1 - p)$$
$$P(X_1 + X_2 + X_3 = 3) = p^3$$

that is, $X_1 + X_2 + X_3$ has a binomial distribution.

The above example can easily be generalized to the case of $n$ tosses. We will freely use that generalization.

When the sampling is independent, the random variables $X_1, X_2, X_3, \ldots, X_n$ as defined in (12.2.1) are independent. They are identically distributed by (12.2.10). Hence, we should expect to bring to bear our previous results on independent, identically distributed random variables. With this in mind we proceed to some theorems about the statistics (12.2.2)–(12.2.7).

**Theorem 12.2.1**   Assume independent sampling. Suppose that $\sigma^2(X) = D^2$ and $E(X) = m$ exist. Suppose that $\sigma^2(Y)$ exists. Then

(12.2.11)   $E(\bar{X}) = m$

(12.2.12)   $E(d^2) = D^2$

(12.2.13)   $E(k) = \dfrac{n - 1}{n} \operatorname{cov}(X, Y)$

PROOF

$$E(\bar{X}) = E\left(\frac{1}{n}(X_1 + X_2 + \cdots + X_n)\right) = \frac{1}{n}\underbrace{(m + m + \cdots + m)}_{n \text{ times}} = m$$

We next compute $E((X_1 - \bar{X})^2)$. It is

$$E(X_1^2 - 2X_1\bar{X} + \bar{X}^2) = E(X_1^2) - 2E(X_1\bar{X}) + E(\bar{X}^2)$$

Now $E(X_1^2) = \sigma^2(X_1) + [E(X_1)]^2 = D^2 + m^2$.

$$E(X_1\bar{X}) = E\left(X_1 \frac{X_1 + X_2 + \cdots + X_n}{n}\right)$$

$$= \frac{1}{n} E(X_1^2 + X_1X_2 + X_1X_3 + \cdots + X_1X_n)$$

$$= \frac{1}{n} [E(X_1^2) + E(X_1X_2) + E(X_1X_3) + \cdots + E(X_1X_n)]$$

Now $E(X_1^2) = D^2 + m^2$ and $E(X_1X_k) = E(X_1)E(X_k) = m^2$ if $k \neq 1$ by the independence of the sampling. Therefore,

$$E(X_1\bar{X}) = \frac{1}{n}\left(D^2 + m^2 + \underbrace{m^2 + \cdots + m^2}_{n - 1 \text{ times}}\right) = \frac{D^2}{n} + \frac{nm^2}{n}$$

$$E(\bar{X}^2) = \frac{1}{n^2} E((X_1 + \cdots + X_n)(X_1 + \cdots + X_n))$$

$$= \frac{1}{n^2} E(X_1^2 + X_2^2 + \cdots + X_n^2 + \sum_{j \neq k} X_j X_k)$$

Again we use independence and identical distribution and obtain

$$E(\bar{X}^2) = \frac{1}{n^2} [n(D^2 + m^2) + (n^2 - n)m^2] = \frac{D^2}{n} + m^2$$

since there are $n^2 - n$ terms $X_j X_k$ for $j \neq k$. We have

$$E((X_1 - \bar{X})^2) = D^2 + m^2 - 2\left(\frac{D^2}{n} + m^2\right) + \left(\frac{D^2}{n} + m^2\right) = D^2 - \frac{D^2}{n}$$

By identical distribution

$$E((X_j - \bar{X})^2) = D^2 - \frac{D^2}{n}, \qquad j = 1, 2, \ldots, n$$

So

$$E(d^2) = E\left(\frac{1}{n-1} \sum_{j=1}^{n} (X_j - \bar{X})^2\right) = \frac{1}{n-1} n\left(D^2 - \frac{D^2}{n}\right) = \frac{n-1}{n-1} D^2 = D^2$$

Finally,

$$E(k) = E\left(\frac{1}{n} \sum_{j=1}^{n} X_j Y_j - \bar{X}\bar{Y}\right)$$

$$= \frac{1}{n} \sum_{j=1}^{n} E(X_j Y_j) - \frac{1}{n^2} E((X_1 + \cdots + X_n)(Y_1 + \cdots + Y_n))$$

$$= \frac{1}{n} \sum_{j=1}^{n} E(X_j Y_j) - \frac{1}{n^2} \sum_{j=1}^{n} E(X_j Y_j) - \frac{1}{n^2} \sum_{j \neq l} E(X_j Y_l)$$

By identical distribution $E(X_j Y_j) = E(XY)$. By the independence of the sampling, $X_j$ and $Y_l$ are independent if $j \neq l$. Therefore for $j \neq l$, $E(X_j Y_l) = E(X)E(Y)$ and

$$E(k) = E(XY) - \frac{n}{n^2} E(XY) - \frac{1}{n^2} (n^2 - n)E(X)E(Y)$$

$$= E(XY)\left(1 - \frac{1}{n}\right) - E(X)E(Y)\left(1 - \frac{1}{n}\right) = \left(1 - \frac{1}{n}\right) \text{cov}(X, Y)$$

$$= \frac{n-1}{n} \text{cov}(X, Y)$$

The proof of Theorem 12.2.1 contains false reasons in some places. Normally, identical distribution of random variables $X_1, X_2, \ldots, X_n$ implies nothing about the identical distribution of functions of more than one of the random variables. Thus, one cannot infer that $X_1 - \bar{X}$ and $X_2 - \bar{X}$ are identically distributed from that property of $X_1, \ldots, X_n$. However, equation (12.2.10) actually guarantees more than the identical distribution of $X_1, \ldots, X_n$. It guarantees that any function of $X_1, \ldots, X_n$ has the same distribution as the same function of the $X_j$ rearranged in some other order. We leave it to the reader to verify that this is what is needed.

Theorem 12.2.1 describes the expected value of some of our statistics in terms of the moments of $X$ and $Y$. We may wish to know the variance of these statistics. We retain the assumptions $m = E(X)$, $\sigma^2(X) = D^2$, and independent sampling.

A previous result about independent, identically distributed random variables is

**(12.2.14)**  $\sigma^2(\bar{X}) = \frac{1}{n} \sigma^2(X) = \frac{D^2}{n}$

Next we look at $\sigma^2(d^2)$.

$$\sigma^2(d^2) = E(d^4) - [E(d^2)]^2 = E(d^4) - (D^2)^2$$

To find $E(d^4)$ we first let $Y_j = X_j - E(X_j) = X_j - m$. Note that $E(Y_j) = 0$ and $E(Y_j^2) = E((X_j - m)^2) = \sigma^2(X_j) = D^2$. We set $q = E(Y_j^4)$. It is easy to verify that $Y_j - \bar{Y} = X_j - \bar{X}$. Therefore

$$\frac{1}{n-1} \sum_{j=1}^{n} (X_j - \bar{X})^2 = \frac{1}{n-1} \sum_{j=1}^{n} (Y_j - \bar{Y})^2$$

A computation parallel to that of Theorem 7.5.2 gives

$$\sum_{j=1}^{n} \frac{1}{n} (Y_j - \bar{Y})^2 = \sum_{j=1}^{n} \frac{1}{n} Y_j^2 - \bar{Y}^2$$

or

$$\sum_{j=1}^{n} (Y_j - \bar{Y})^2 = \sum_{j=1}^{n} Y_j^2 - n\bar{Y}^2$$

after multiplying by $n$. Now

$$(\sum_{j=1}^{n} Y_j^2 - n\bar{Y}^2)^2 = (\sum_{j=1}^{n} Y_j^2)^2 - 2n\bar{Y}^2 \sum_{j=1}^{n} Y_j^2 + n^2\bar{Y}^4$$

It will be shown in the exercises that

$$E((\sum_{j=1}^{n} Y_j^2)^2) = nq + n(n-1)D^4$$

$$E(\bar{Y}^2 \sum_{j=1}^{n} Y_j^2) = n^{-2}[nq + n(n-1)D^4]$$

and

$$E(\bar{Y}^4) = n^{-4}[nq + 3n(n-1)D^4]$$

Therefore,

$$E\left(\left(\frac{1}{n-1} \sum_{j=1}^{n} (X_j - \bar{X})^2\right)^2\right)$$

$$= E\left(\left(\frac{1}{n-1} \sum_{j=1}^{n} Y_j^2 - n\bar{Y}^2\right)^2\right)$$

$$= \left(\frac{1}{n-1}\right)^2 \{nq + n(n-1)D^4 - 2nn^{-2}[nq + n(n-1)D^4]$$

$$+ n^2 n^{-4}[nq + 3n(n-1)D^4]\}$$

$$= \frac{1}{(n-1)^2}\left\{\left(n - 2 + \frac{1}{n}\right)q + \left[n(n-1) - 2(n-1) + \frac{3}{n}(n-1)\right]D^4\right\}$$

$$= \frac{(n^2 - 2n + 1)}{n(n-1)^2} q + \frac{n-1}{(n-1)^2}\left(n - 2 + \frac{3}{n} D^4\right)$$

and

(12.2.15)    $$\sigma^2(d^2) = E(d^4) - D^4 = \frac{1}{n}q + \frac{1}{n-1}\left(n - 2 + \frac{3}{n}\right)D^4 - D^4$$

$$= \frac{1}{n}q + \frac{3-n}{n(n-1)} D^4$$

We restate these results as a theorem.

**Theorem 12.2.2**    Assume independent sampling and the usual $X_j$. Suppose that $E(X^4)$ exists, $m = E(X)$, $D^2 = \sigma^2(X)$, and $q = E((X - m)^4)$. Then

(12.2.15)    $$\sigma^2\left(\frac{1}{n-1} \sum_{j=1}^{n} (X_j - \bar{X})^2\right) = \frac{1}{n}q + \frac{3-n}{n(n-1)} D^4$$

This rather lengthy proof indicates the difficulties inherent in finding the moments of the other statistics in the general case. We therefore focus on aspects other than moments.

We continue to assume independent sampling. Let $M_1, M_2, \ldots, M_k$ be disjoint sets of real numbers whose union is the set of all real numbers. Let $\pi_j = P(X \in M_j)$ for $j = 1, 2, \ldots, k$. We assume that $\pi_j > 0$. By the disjointness of the $M_j$ and the fact that their union is the entire real line, we have $\pi_1 + \pi_2 + \cdots + \pi_k = 1$. For any integer $n > 0$ we define $k$ random variables $N_1, N_2, \ldots, N_k$ by the relationship

$N_j(s) = \{$number of the numbers $X_1(s), X_2(s), \ldots, X_n(s)$ which are in the set $M_j$

Since the $M_j$ are disjoint and cover the real line, each number $X_m(s)$ belongs to one and only one set $M_j$ if $m$ and $s$ are fixed. Therefore, for each $s \in S$

$$N_1(s) + N_2(s) + \cdots + N_k(s) = n.$$

At this point we are in a position to study ideas related to the distribution of the statistic (12.2.6).

### Example 12.2.7

With the above notation, independent sampling, $n = 8$ and $k = 3$, find $P(N_2 = 0)$, $P(N_2 = 3)$, and $P(N_1 = 4, \ N_2 = 2, \ N_3 = 2)$.

$$P(N_2 = 0) = P(X_1 \notin M_2, \ X_2 \notin M_2, \ldots, X_8 \notin M_2) = (1 - \pi_2)^8$$

by independence.

$P(N_2 = 3) = P($for three indexes $l$, $X_l \in M_2$ and for the remaining five indexes $l'$,
$$X_{l'} \notin M_2)$$

$$= \binom{8}{3} \pi_2^3 (1 - \pi_2)^5$$

$P(N_1 = 4, \ N_2 = 2, \ N_3 = 2) = P($for four indexes $l$, $X_l \in M_1$, for two other indexes $l'$, $X_{l'} \in M_2$ and for two other indexes $l''$, $X_{l''} \in M_3)$

$$= \frac{8!}{4!2!2!} \, \pi_1^4 \pi_2^2 \pi_3^2$$

where $\pi_1^4 \pi_2^2 \pi_3^2$ follows from the assumed independence and $8!/4!2!2!$ arises because we can write $M_1$, $M_2$, or $M_3$ over the integers $1, 2, 3, \ldots, 8$ to tell which random variable falls where in $8!/4!2!2!$ ways.

It should be quite clear in general that if $M_1, M_2, \ldots, M_k$ are disjoint events as above and if $P(X(s_0) \in M_j) = \pi_j > 0$ and we are studying *independent* sampling, then if $N_j(s)$ is the number of indexes $l$ from $\{1, 2, 3, \ldots, n\}$ for which $X_l(s) \in M_j$ we have

$$(12.2.16) \qquad P(N_1 = r_1, \ N_2 = r_2, \ldots, N_k = r_k) = \frac{n!}{r_1! r_2! \cdots r_k!} \pi_1^{r_1} \pi_2^{r_2} \cdots \pi_k^{r_k}$$

for any integers $r_j \geq 0$ with $r_1 + r_2 + \cdots + r_k = n$

With the joint distribution of the random vector $(N_1, N_2, \ldots, N_k)$ provided by (12.2.16) we can, in principle, obtain the distribution of the statistic

$$\frac{(N_1 - n\pi_1)^2}{n\pi_1} + \frac{(N_2 - n\pi_2)^2}{n\pi_2} + \cdots + \frac{(N_k - n\pi_k)^2}{n\pi_k}$$

of (12.2.6). It is not easy to give an explicit description of this distribution except in special cases.

If $X$ is $N(m, D^2)$, that is, $E(X) = m$, $\sigma^2(X) = D^2$, then $\bar{X}$ has a normal distribution by Chapter 9. (We assume independent sampling.) But by (12.2.11) and (12.2.14)

**(12.2.17)**      $\bar{X}$ is $N(m, D^2/n)$

If $X$ is binomial with parameters $r$ and $p$ then by Chapter 9, $X_1 + X_2 + \cdots + X_n$ is binomial with parameters $nr$ and $p$. Thus, in this case, we have

**(12.2.18)**      $P\left(\bar{X} = \dfrac{k}{n}\right) = P(X_1 + \cdots + X_n = k) = \dbinom{nr}{k}p^k(1 - p)^{nr-k},$

$$k = 0, 1, 2, \ldots, nr$$

Similarly, if $X$ is Poisson with parameter $\lambda$

**(12.2.19)**      $P\left(\bar{X} = \dfrac{k}{n}\right) = P(X_1 + \cdots + X_n = k) = e^{-n\lambda}(n\lambda)^k/k!,$

$$k = 0, 1, 2, 3, \ldots$$

We now find the distribution of the statistic $r$ of (12.2.5) when $X$ has a density function. We assume independent sampling. The distribution and density functions of $X$ are $F$ and $f$, respectively. By the definitions and section 8.3, $U = \max(X_1, X_2, \ldots, X_n) \leq b$ and $V = \min(X_1, X_2, \ldots, X_n) > a$ if and only if $a < X_j \leq b$ for $j = 1, 2, \ldots, n$. This event has probability $[F(b) - F(a)]^n$ by the independence of the $X_j$. We take $\partial^2/\partial b\partial a$ and find that $(U, V)$ has the joint density function

$$n(n - 1)[F(u) - F(v)]^{n-2}f(u)f(v)$$

for $u > v$. For $c > 0$,

$$P(U - V \leq c) = P(U \leq V + c)$$

$$= \int_{-\infty}^{\infty} dv \int_{v}^{v+c} n(n - 1)[F(u) - F(v)]^{n-2}f(u)f(v)\,du$$

where the range of the inner integration starts at $v$ because $u > v$. This is just

**(12.2.20)**      $\displaystyle\int_{-\infty}^{\infty} n[F(v + c) - F(v)]^{n-1}f(v)\,dv = P(r \leq c)$

since $F$ is the integral of $f$.

### Example 12.2.8

If $X$ has the uniform distribution on $[-1, 3]$ find the distribution of $r$ when $n = 8$.

According to our calculations, for $c > 0$ and $c \leq 4$ we have

$$P(r \leq c) = \int_{-\infty}^{\infty} 8[F(v + c) - F(v)]^7 f(v) \, dv$$

Now

$$f(v) = \begin{cases} \frac{1}{4}, & -1 < v < 3 \\ 0, & \text{otherwise} \end{cases}$$

and

$$F(v) = \begin{cases} 0, & v \leq -1 \\ \frac{1}{4}(v + 1), & -1 < v < 3 \\ 1, & v \geq 3 \end{cases}$$

Our integral is

$$\int_{-1}^{3} 8\{[\tfrac{1}{4}(v + c + 1)] \wedge 1 - \tfrac{1}{4}(v + 1)\}^7 \tfrac{1}{4} \, dv$$

where $a \wedge 1 = $ the smaller of $a$ and 1. This is just

$$\int_{-1}^{3-c} 2[\tfrac{1}{4}(v + c + 1) - \tfrac{1}{4}(v + 1)]^7 \, dv + \int_{3-c}^{3} 2[1 - \tfrac{1}{4}(v + 1)]^7 \, dv$$

or

$$\int_{-1}^{3-c} 2(\tfrac{1}{4})^7 c^7 \, dv + 2 \int_{3-c}^{3} (\tfrac{1}{4})^7 (3 - v)^7 \, dv$$

$$= \frac{2}{4^7} [c^7(4 - c) + \tfrac{1}{8}c^8] = \frac{2c^7}{4^7} (4 - \tfrac{7}{8}c) = P(r \leq c)$$

### EXERCISES 12.2

◆ **1**  Show that

$$\frac{1}{n} \sum_{j=1}^{n} (X_j - \bar{X})^2 = \frac{1}{n} \sum_{j=1}^{n} X_j^2 - (\bar{X})^2$$

**2**  How should the surveyors of Example 12.2.1 proceed in order to be sure that the population which they sample is the entire soap buying population?

**3**  Work out Example 12.2.4 in the case $n = 10$ and $s' = (4, 3, 4, 1, 5, 2, 4, 3, 2, 4)$.

◆ **4**  Let $S_0 = \{H, T\}$ with equal likelihood. Let $X(H) = 1$ and $X(T) = -1$. Compute $E(X)$. If $n = 3$ show that $\bar{X}(s) \neq E(X)$ for all $s \in S$. Assuming independent sampling, when is it true that $\bar{X}(s) = E(X)$ for all $s \in S$? Let this be a warning to those who would confuse $\bar{X}$ and $E(X)$.

5   Describe a procedure of simulating decision procedures such as that in Example 12.2.5 with a deck of cards if clubs are poorest apples, diamonds next, hearts next, and spades are best apples.

6   Construct an example of identically distributed random variables $X_1$, $X_2$, $X_3$ with the property that $X_1 - \bar{X}$ and $X_2 - \bar{X}$ are not identically distributed. Naturally, the $X_j$ cannot be independent.

♦ 7   In reference to the notation of Theorem 12.2.2 following (12.2.14) show that

$$E(Y_j Y_k Y_l Y_m) = 0, \qquad j, k, l, m \text{ are different}$$
$$E(Y_j Y_k Y_l Y_m) = q, \qquad j = k = l = m$$
$$E(Y_j Y_k Y_l Y_m) = D^4, \qquad j = k, \; l = m, \; \text{but } j \neq l$$

In exercises 8–12 we continue using the notation that is developed prior to Theorem 12.2.2.

♦ 8   Show that $(\sum_{j=1}^{n} Y_j^2)^2$ is a sum of $n$ terms of the form $Y_j^4$ plus $n^2 - n$ terms of the form $Y_j^2 Y_k^2$ for $j \neq k$.

♦ 9   Show that

$$\bar{Y}^2 (\sum_{j=1}^{n} Y_j^2) = n^{-2}[\text{sum of } n \text{ terms of the form } Y_j^4 \text{ plus } n(n-1) \text{ terms of the}$$
form $Y_j^2 Y_k^2$ for $j \neq k$ plus terms of the form $Y_j^2 Y_k Y_l$ for $k \neq l]$

From the viewpoint of the proof of Theorem 12.2.2, why are the terms $Y_j^2 Y_k Y_l$ with $k \neq l$ irrelevant?

♦ 10   How many 4-letter words can be made with two $j$'s and two $k$'s?

♦ 11   Show that

$$\bar{Y}^4 = n^{-4}[\text{sum of } n \text{ terms of the form } Y_j^4 \text{ plus } \binom{n}{2} \frac{4!}{2!2!} \text{ terms of the form}$$

$Y_j^2 Y_k^2$ for $j \neq k$ plus terms of the form $Y_j^2 Y_k Y_l$ for $k \neq l$ plus terms of the form $Y_j Y_k Y_l Y_m$ where $j, k, l, m$ are all different]

♦ 12   How do exercises 7–11 fill in the missing details in the proof of (12.2.15)?

13   Consider the statistic $c(s)$ of (12.2.8). Suppose that $P(X \geq b) = 0.01$, $P(X \leq a) = 0.80$, $P(a < X < b) = 0.19$, and the random variables $X(s_j)$ are independent. Compute $E(c)$ when $n = 40$ and $k = 16$. Such a statistic could arise in the attribution of research costs; e.g., if each test costs $\$\frac{1}{10}$, the whole research cost is $\$N$, $b$ means failure, and $a$ or less on the first $k$ tries means success.

14   Find the characteristic function of $\bar{X}$ if $X$ has the uniform distribution on $0 \leq x \leq a$ and independent sampling is used.

In exercises 15–18 we use the statistics $Y_k^{(n)}$ defined by

$$Y_k^{(n)}(s) = k\text{th number from among } X(s_1), X(s_2), \ldots, X(s_n) \text{ when these are}$$
arranged in increasing order

The statistic $Y_k^{(n)}$ is called the *order statistic* of *rank k*. Normally we will assume that $S_0$ is the reals and $X(x) = x$.

**15**   Find $Y_4^{(6)}(s)$ when $s = (-2, 3, \sqrt{\pi}, \frac{1}{4}, -1, e)$, $s = (4, -2, -2, 3, 6, 0)$, and $s = (6, -1, 2, 0, \pi, \pi)$.

**16**   Describe $Y_1^{(n)}$ and $Y_n^{(n)}$ using other terminology.

**17**   If $n = 2m + 1$, describe $Y_{m+1}^{(n)}$ with other terminology.

**18**   Assume independent sampling and $X$ uniformly distributed on $0 \le x \le 1$. What can be said, intuitively, about $P(Y_n^{(4n)} > \frac{1}{3})$ for large $n$?

**19**   Find the density function for the sample range $r$ when $X$ has the negative exponential distribution and sampling is independent.

# 12.3   Normal random variables and statistics

Generally speaking, the normal distribution is the most intensely studied of the various probability distributions. This distribution has certain properties which seem to force its consideration; we will therefore give it special attention.

Suppose that sampling is independent and that $X$ is $N(m, D^2)$. We noted in (12.2.17) that $\bar{X}$ is $N(m, D^2/n)$. Does $d^2$ have a special distribution? We shall see that it does—it is a constant multiple of one of our famous random variables. Furthermore, $\bar{X}$ and $d^2$ are independent when $X$ is normal and sampling is independent. In an exercise we ask for an example of an $X$ with a binomial distribution in which $\bar{X}$ and $d^2$ are not independent even though the $X_1, X_2, \ldots, X_n$ are independent.

**Theorem 12.3.1**   Let $X$ be $N(0, 1)$ with independent sampling. Then $\bar{X}$ and $d^2$ are independent.

PROOF   We need only show that $U = X_1 + X_2 + \cdots + X_n$ and

$$V^2 = \sum_{j=1}^{n} \left( X_j - \frac{U}{n} \right)^2$$

are independent since constant multiples of independent random variables are independent and $\bar{X} = U/n$ and $d^2 = V^2/(n - 1)$.

By exercise 12.2.1,

$$V^2 = \sum_{j=1}^{n} (X_j - \bar{X})^2 = \sum_{j=1}^{n} X_j^2 - n\bar{X}^2$$

$$= \sum_{j=1}^{n} X_j^2 - n\left(\frac{U}{n}\right)^2 = \sum_{j=1}^{n} X_j^2 - \frac{U^2}{n}$$

But $U^2/n = (U/\sqrt{n})^2$. Therefore

**(12.3.1)**      $V^2 = X_1^2 + X_2^2 + \cdots + X_n^2 - (U/\sqrt{n})^2$

Suppose we let the $X_j$ play the role of the $Z_j$ of section 10.5. Let $a_{1k} = 1/\sqrt{n}$ for $k = 1, 2, \ldots, n$. Let

$$
a_{jk} = \begin{cases}
0, & k > j \\[2mm]
-\dfrac{j-1}{\sqrt{j(j-1)}}, & k = j \\[2mm]
\dfrac{1}{\sqrt{j(j-1)}}, & k < j
\end{cases}
$$

for $j = 2, 3, \ldots, n$.

In an exercise we ask the reader to show that

$$
\sum_{k=1}^{n} a_{jk} a_{lk} = \begin{cases} 1, & j = l \\ 0, & j \neq l \end{cases}
$$

Theorem 10.5.1 states that the $W_j$ given by $W_j = \sum_{k=1}^{n} a_{jk} X_k$ are independent.

Exercise 10.5.5 shows that the $W_j$ are $N(0, 1)$. Furthermore Theorem 10.5.2 shows that

$$
W_1^2 + W_2^2 + \cdots + W_n^2 = X_1^2 + X_2^2 + \cdots + X_n^2
$$

or

$$
W_2^2 + W_3^2 + \cdots + W_n^2 = X_1^2 + X_2^2 + \cdots + X_n^2 - W_1^2
$$

Now by the definition of $a_{1k}$

$$
W_1 = \frac{1}{\sqrt{n}} (X_1 + X_2 + \cdots + X_n) = \frac{U}{\sqrt{n}}
$$

so

$$
W_2^2 + W_3^2 + \cdots + W_n^2 = X_1^2 + X_2^2 + \cdots + X_n^2 - \left(\frac{U}{\sqrt{n}}\right)^2
$$

Therefore, by equation (12.3.1)

$$
V^2 = W_2^2 + W_3^2 + \cdots + W_n^2
$$

Thus $V^2$ is a function of $W_2, W_3, \ldots, W_n$ which are independent of $W_1 = U/\sqrt{n}$ so $V^2$ and $U$ are independent.

**Corollary 1**      Under the above hypotheses, $V^2$ has a $\chi^2$ distribution with $n - 1$ degrees of freedom.

PROOF      We have $V^2 = W_2^2 + W_3^2 + \cdots + W_n^2$, a sum of $n - 1$ independent random variables each of which is the square of an $N(0, 1)$ random variable. By the definition at the end of section 8.5, $V^2$ is a $\chi^2$ random variable with $n - 1$ degrees of freedom.

**Corollary 2**    Assume independent sampling but suppose that $X$ is $N(m, D^2)$. Then $\bar{X}$ and $d^2$ are independent and $d^2 = \dfrac{D^2}{n-1} V^2$ where $V^2$ has a $\chi^2$ distribution with $n - 1$ degrees of freedom.

PROOF    Let $Y_j = D^{-1}(X_j - m)$. We have seen that the $Y_j$ are $N(0, 1)$ and independent since the $X_j$ are independent. We have $X_j = DY_j + m$ so

$$\bar{X} = \frac{1}{n}(DY_1 + m + \cdots + DY_n + m) = D\bar{Y} + m$$

Also

$$\frac{1}{n-1} \sum_{j=1}^{n} (X_j - \bar{X})^2 = \frac{1}{n-1} \sum_{j=1}^{n} [DY_j + m - (D\bar{Y} + m)]^2$$

$$= \frac{1}{n-1} \sum_{j=1}^{n} D^2(Y_j - \bar{Y})^2$$

Theorem 12.3.1 states that $\bar{Y}$ and $V^2 = \sum\limits_{j=1}^{n} (Y_j - \bar{Y})^2$ are independent so $\bar{X} = D\bar{Y} + m$ and $d^2 = \dfrac{D^2}{n-1} V^2$ are independent. Corollary 1 proves that $V^2$ is $\chi^2$ with $n - 1$ degrees of freedom.

Ideas of section 8.5 can be used to show that the density function

$$(12.3.2) \qquad f_{d^2}(q) = \frac{(cq)^{(n-3)/2} e^{-cq/2}}{2^{(n-1)/2} \Gamma((n-1)/2)} c$$

where $c = \left(\dfrac{D^2}{n-1}\right)^{-1}$.

Fortunately it is not necessary to integrate $f_{d^2}(q)$ since tables can be used to find most of the necessary values.

Continuing with the notation and assumptions of Corollary 2 we have $\bar{X} - m = D\bar{Y}$ and $d^2 = [D^2/(n-1)]V^2$ or $\sqrt{d^2} = (D/\sqrt{n-1})V$. Since the $Y_j$ are $N(0, 1)$, $\bar{Y}$ is $N(0, 1)$ and $\bar{Y} = n^{-1/2}Z$ where $Z$ is $N(0, 1)$. We have

$$(12.3.3) \qquad \frac{(\bar{X} - m)n}{\sqrt{d^2}} = \frac{D\bar{Y}\sqrt{n}}{\dfrac{D}{\sqrt{n-1}}V} = \frac{\bar{Y}\sqrt{n}}{\dfrac{1}{\sqrt{n-1}}V} = \frac{Z\sqrt{n-1}}{V}$$

The random variables $Z$ and $V$ in equation (12.3.3) are independent since $\bar{Y}$ and $V^2$ are independent. Note that $Z$ is $N(0, 1)$ and $V^2$ is $\chi^2$ with $n - 1$ degrees of freedom. Note that the distribution of $(\bar{X} - m)\sqrt{n}/\sqrt{d^2}$ does not depend on the standard deviation of $X$.

We study the distribution of $(\bar{X} - m)\sqrt{n}/\sqrt{d^2}$ by studying that of $Z\sqrt{n - 1}/V$. Let $r = n - 1$, $L = Z\sqrt{r}$, and $K = \sqrt{V^2}$. Note that $L$ is $N(0, r)$ and independent of $K$.

We recall from exercise 8.5.4 that if $K$ and $L$ are independent random variables with density functions $f_K$ and $f_L$, respectively, and if $P(K \le 0) = 0$ then the density function of $W = L/K$ is

$$f_W(w) = \int_0^\infty u f_L(uw) f_K(u) \, du$$

Now, since $L$ is $N(0, r)$, $f_L(v) = (1/\sqrt{2\pi r})e^{-v^2/2r}$. We want $K^2$ to be a $\chi^2$ random variable with $r$ degrees of freedom. Since $f_K(u) = 2u f_{K^2}(u^2)$, by exercise 8.5.8

$$f_K(u) = 2u[2^{r/2}\Gamma(r/2)]^{-1}(u^2)^{r - 1/2}e^{-u^2/2}, \qquad u > 0$$

Therefore

$$f_W(w) = \int_0^\infty 2u^2[2^{r/2}\Gamma(r/2)]^{-1}(u^2)^{(r/2) - 1}e^{-u^2/2}\frac{1}{\sqrt{2\pi r}}e^{-u^2w^2/2r} \, du$$

$$(12.3.4) \qquad = \frac{2}{\sqrt{2\pi r}}[2^{r/2}\Gamma(r/2)]^{-1}\int_0^\infty (u^2)^{r/2}e^{-[(1 + w^2/r)u^2]/2} \, du$$

Let $z = \frac{1}{2}(1 + w^2/r)u^2$ or $u^2 = 2z(1 + w^2/r)^{-1}$. Then

$$du = \sqrt{2}\,\tfrac{1}{2}z^{-1/2}(1 + w^2/r)^{-1/2} \, dz$$

and the last integral in (12.3.4) becomes

$$\int_0^\infty (2z)^{r/2}\left(1 + \frac{w^2}{r}\right)^{-r/2}e^{-z}\sqrt{2}\,\tfrac{1}{2}z^{-1/2}\left(1 + \frac{w^2}{r}\right)^{-1/2} \, dz$$

$$= 2^{(r-1)/2}\left(1 + \frac{w^2}{r}\right)^{-(r+1)/2}\int_0^\infty z^{(r-1)/2}e^{-z} \, dz$$

By definition

$$\int_0^\infty z^{(r-1)/2}e^{-z} \, dz = \int_0^\infty z^{[(r+1)/2] - 1}e^{-z} \, dz = \Gamma(r + 1)/2$$

After some simplification we have

$$(12.3.5) \qquad f_W(w) = \frac{1}{\sqrt{\pi r}}\frac{\Gamma((r + 1)/2)}{\Gamma(r/2)}\left(1 + \frac{w^2}{r}\right)^{-(r+1)/2}$$

**Definition**   The distribution given by (12.3.5) is called *Student's\* t-distribution with r degrees of freedom*. We usually denote a random variable having such a distribution by $t$.

Note that such a $t = Z\sqrt{r}/V$, where $Z$ and $V$ are independent, $Z$ is $N(0, 1)$ and $V^2$ is a $\chi^2$ random variable with $r$ degrees of freedom.

### Example 12.3.1

The mathematical model for the movements of certain particles assumes that the change of coordinate in one unit of time is given by a random variable $X$ which is $N(0, \sigma^2)$. Suppose eight independent observations of this change of position in the unit of time yield $x_1 = 5.2$, $x_2 = -1.6$, $x_3 = -2.9$, $x_4 = 0.7$, $x_5 = -1.8$, $x_6 = -3.7$, $x_7 = -0.4$, $x_8 = 0.8$. Are such data compatible with the assumptions?

First,

$$\tfrac{1}{8}(x_1 + x_2 + \cdots + x_8) = -(0.4625) = \hat{\bar{X}}$$

$$
\begin{aligned}
\hat{d}^2 &= \tfrac{1}{7} \sum_{j=1}^{n} (x_j - \bar{x})^2 \\
&= \tfrac{1}{7}[(5.6625)^2 + (-1.1375)^2 + (-2.4375)^2 + (1.1625)^2 \\
&\qquad + (-1.3375)^2 + (-3.2375)^2 + (0.0625)^2 + (1.2625)^2] \\
&\approx \frac{54.5}{7}
\end{aligned}
$$

Now $\sqrt{8}\,\bar{X} = Z$ is $N(0, 1)$ and $d^2 = \tfrac{1}{7}V^2$ where $V^2$ is a $\chi^2$ random variable with seven degrees of freedom. Therefore

$$\frac{\bar{X}\sqrt{8}}{\sqrt{d^2}} = \frac{Z\sqrt{7}}{\sqrt{V^2}}$$

has Student's $t$-distribution with seven degrees of freedom. Let

$$\hat{t} = \frac{\hat{\bar{X}}\sqrt{8}}{\sqrt{\hat{d}^2}} \approx \frac{-(0.4625)\sqrt{8}\sqrt{7}}{\sqrt{54.5}} \approx -0.47$$

Tables of the $t$-distribution give $P(t \geq 0.553) = P(t \leq -0.553) = 0.3$ when $t$ has seven degrees of freedom. That is, a normal random variable of the type assumed would frequently give $t$ values less than the value that we obtained just by chance. We would usually say that the data are compatible with the assumptions since $|\hat{t}|$ is not unusually large compared to 0.553.

A particle of the above sort would be judged not to be subject to drift on the given evidence too.

---

\* "Student" is a pseudonym for W. L. Gossett (1876–1937) who worked for Guinness Breweries and published under the above pseudonym.

## Example 12.3.2

The Atlas Wire Company makes wire which is spun into cables for bridges. Experience has shown that a good model for the force required to break a piece of wire of a standard size and of a particular sort is a normally distributed random variable. A process in current use produces wire whose expected breaking force is 200 units. A new wire-making process is being studied. Twenty randomly chosen standard size pieces of wire made by the new process break under the following forces:

200.3,  200.7,  199.2,  200.2,  200.4,  199.7,  199.8,  200.4,  201.1,

200.5,  200.6,  199.7,  198.6,  200.1,  200.0,  199.6,  200.8,  199.9,

200.9,  199.5

Does this indicate a wire of fundamentally different quality?

If we call the values listed $x_1, x_2, \ldots, x_{20}$ it can be seen that

$$\tfrac{1}{20}(x_1 + x_2 + \cdots + x_{20}) = 200.1$$

Could this be a chance fluctuation from a process whose typical breaking force is 200?

We compute $\tfrac{1}{19} \sum (x_j - 200.1)^2 = 7.50/19 = 0.3947$. These computations are taken as evaluations of the statistics $\bar{X}$ and $d^2$. Now $(\bar{X} - m)\sqrt{20}/\sqrt{d^2}$ has a $t$-distribution with 19 degrees of freedom. Call that $t$-random variable $t$.

$$P(|t| \geq 1.7291) = 0.10 \quad \text{and} \quad P(|t| \geq 1.3277) = 0.20$$

Also $P(t \geq 1.3277) = 0.10$. Now

$$\frac{(200.1 - 200.0)\sqrt{20}}{\sqrt{0.3947}} = 0.7102$$

and

$$|0.7102| < 1.7291, \qquad 0.7102 < 1.3277$$

Note that the probability is 0.80 that a $t$-random variable with 19 degrees of freedom will yield values whose absolute value is less than 1.3277. Furthermore, with probability 0.90 such a $t$-random variable will yield values less than 1.7291. Our data, whose abstraction yields such a $t$-random variable, led to 0.7102 which would be construed to be well within the range of chance fluctuations of breaking strengths which are normally distributed with expected value 200. We would say that the new process yields wire which is no better but certainly no worse than that made by the old process.

In order not to pre-empt material from subsequent chapters we stop our discussion of Student's $t$-distribution.

Many students ask, "How do you know what class of distributions should be used for the underlying random variable in a particular statistical problem?" In coin tossing the answer is clear. But why assume the normal distribution for IQ's? The question here is one of choosing the proper model for real-world phenomena. The comments of Chapter 4 apply. Considerable insight if not genius comes into play in the original studies of the problem. After that, one may use the conventions of the subject. We give some examples in the text. However, in the final analysis, the choice of a model in difficult, nonstandard problems cannot be taught by giving a few rules.

Because of the difficulties in choosing a good model in certain cases, as well as the avoidance of oversimplification, statisticians have developed the theory of "nonparametric statistics." In it they study methods which do not depend on knowing much detail about the distribution of the basic random variable. We will not emphasize nonparametric methods.

Suppose that we are dealing with a particular real-world problem of chance and we feel that a particular distribution is appropriate. However, we are not certain. How could we use experimental data to confirm our feelings? Let us consider an example.

### Example 12.3.3

A proposed ecological theory implies that the probabilities of catching fish of types 1, 2, ..., 5 under certain circumstances are 0.2, 0.4, 0.2, 0.1, 0.1, respectively. One hundred fish of these types are caught under the given circumstances. Would 13, 34, 23, 14, 16 fish of the respective types contradict the theory? How about catches of 25, 30, 26, 7, 12 or 12, 35, 27, 13, 13?

We tabulate the numbers in Table 2.

### Table 2

| TYPE | 1 | 2 | 3 | 4 | 5 |
|------|---|---|---|---|---|
| EXPECTED | 20 | 40 | 20 | 10 | 10 |
| A | 13 | 34 | 23 | 14 | 16 |
| B | 25 | 30 | 26 | 7 | 12 |
| C | 12 | 35 | 27 | 13 | 13 |

None of these catches look very favorable compared to what we would expect by the theory. Naturally there should be some chance deviation from the expected. A catch of 18, 45, 19, 8, 10 would be nice.

In rows A, B, C of the table we see a total of 13 more than the expected in some types and a total of 13 fewer than the expected in other types. Could the theory be correct but the fluctuations merely a matter of chance? Let us note the maximum deviation from the expected in each row. The maximum deviations are 7, 10, and 8, respectively. Note, though, that the deviation of 10 in row B occurs in type 2 where 40 is expected. That is not such a large percentage deviation. The percentage deviation of 6 compared to the expected 10 in row A, type 5 is much larger than 10 compared to 40. It would be nice if we could weigh all of these deviations on a common scale. One such scale requires the computation of

$$\sum_{j=1}^{r} \frac{(\theta_j - e_j)^2}{e_j} = \hat{\chi}^2$$

where $e_1, e_2, \ldots, e_r$ are the expected frequencies in the classes and $\theta_1, \theta_2, \ldots, \theta_r$ are the observed frequencies in the classes, namely the number of fish caught. In reference to Table 2 we have $\hat{\chi}^2(A) = 9$, $\hat{\chi}^2(B) = 6.85$, and $\hat{\chi}^2(C) = 8.75$. These values are compared to the distributions of certain random variables with certain additional criteria. The theory to follow will show why this is done.

We will use the Kronecker delta

$$\delta_{jl} = \begin{cases} 1, & j = l \\ 0, & j \neq l \end{cases}$$

**Lemma 12.3.1**    Let $\pi_j > 0$ and $\pi_1 + \pi_2 + \cdots + \pi_r = 1$. Let $W = (W_1, W_2, \ldots, W_r)$ have characteristic function

$$\phi_W(t_1, t_2, \ldots, t_r)$$
$$= \exp\{-\tfrac{1}{2}[(t_1^2 + t_2^2 + \cdots + t_r^2) - (\sqrt{\pi_1}t_1 + \sqrt{\pi_2}t_2 + \cdots + \sqrt{\pi_r}t_r)^2]\}$$

Then there are independent random variables $Z_1, Z_2, \ldots, Z_r$ with $Z_1 = 0$, $Z_j \ N(0, 1)$ for $2 \leq j \leq r$ and $u_{jk}$ with $\sum_{k=1}^{r} u_{jk}u_{lk} = \delta_{jl}$ such that $W$ has the same distribution as $(X_1, X_2, \ldots, X_r)$ where $X_j = \sum_{k=1}^{r} u_{jk}Z_k$. Furthermore

$$P(W_1^2 + W_2^2 + \cdots + W_r^2 \leq b) = P(Z_1^2 + Z_2^2 + \cdots + Z_r^2 \leq b)$$

**PROOF**    In exercise 10.5.10 it was noted that if $\pi_j > 0$ and $\pi_1 + \pi_2 + \cdots + \pi_r = 1$ there are numbers $a_{jk}$ such that $a_{1j} = \sqrt{\pi_j}$ for $j = 1, 2, \ldots, r$ and $\sum_{k=1}^{r} a_{jk}a_{kl} = \delta_{jl}$. In the remainder of this proof we take all unmarked summations to be from 1 to $r$. Note that by theorems in linear algebra we also have $\sum a_{kj}a_{kl} = \delta_{jl}$. Let $u_{jk} = a_{kj}$ so

$$\sum u_{jk}u_{lk} = \sum a_{kj}a_{kl} = \delta_{jl}$$

Let

(12.3.6)    $\xi_k = \sum u_{jk}t_j$

Note that

$$\xi_1 = \sum u_{j1}t_j = \sum a_{1j}t_j = \sum \sqrt{\pi_j}t_j$$

Using the technique of Theorem 10.5.2 it can be seen that $\sum \xi_j^2 = \sum t_k^2$. Therefore,

(12.3.7)
$$\xi_2^2 + \xi_3^2 + \cdots + \xi_r^2 = t_1^2 + t_2^2 + \cdots + t_r^2 - \xi_1^2$$
$$= t_1^2 + t_2^2 + \cdots + t_r^2$$
$$- (\sqrt{\pi_1}t_1 + \sqrt{\pi_2}t_2 + \cdots + \sqrt{\pi_r}t_r)^2$$

Let $Z_1, Z_2, \ldots, Z_r$ be independent random variables with $Z_1 = 0$ and $Z_j$ $N(0, 1)$ for $2 \le j \le r$. Let $X_j = \sum u_{jk}Z_k$. Now

$$\sum t_j X_j = \sum_{j=1}^{r} t_j \sum_{k=1}^{r} u_{jk}Z_k = \sum_{k=1}^{r} \left(\sum_{j=1}^{r} t_j u_{jk}\right)Z_k = \sum \xi_k Z_k$$

by (12.3.6). Therefore,

$$E(\exp[(t_1 X_1 + t_2 X_2 + \cdots + t_r X_r)]) = E(\exp(i \sum t_j X_j))$$
$$= E(\exp(i \sum \xi_k Z_k))$$

But by the definition of the $Z_k$ we have

$$E(\exp(i \sum \xi_k Z_k)) = \exp[-\tfrac{1}{2}(0 \cdot \xi_1^2 + \xi_2^2 + \cdots + \xi_r^2)]$$

Now by (12.3.7)

$$\exp[-\tfrac{1}{2}(\xi_2^2 + \xi_3^2 + \cdots + \xi_r^2)]$$
$$= \exp\{-\tfrac{1}{2}[t_1^2 + t_2^2 + \cdots + t_r^2 - (\sqrt{\pi_1}t_1 + \cdots + \sqrt{\pi_r}t_r)^2]\}$$

so

$$E(\exp[i(t_1 X_1 + t_2 X_2 + \cdots + t_r X_r)])$$
$$= \exp\{-\tfrac{1}{2}[t_1^2 + \cdots + t_r^2 - (\sqrt{\pi_1}t_1 + \cdots + \sqrt{\pi_r}t_r)^2]\}$$

By the uniqueness theorem for characteristic functions of random vectors we find that $W$ has the same distribution as $X = (X_1, X_2, \ldots, X_r)$ since their characteristic functions are identical. Now since $\sum u_{jk}u_{lk} = \delta_{jl}$, $\sum X_j^2 = \sum Z_j^2$ by Theorem 10.5.2. Therefore since $W$ and $X$ have the same distribution $P(W_1^2 + W_2^2 + \cdots + W_r^2 \le b) = P(Z_1^2 + \cdots + Z_r^2 \le b)$.

It is interesting to note that the random vector $Z = (0, Z_2, Z_3, \ldots, Z_r)$ has the character of an $(r - 1)$-dimensional, nonsingular normal random vector tucked in $r$-dimensional space. Moreover, $W$ is a rotated version of $Z$.

Henceforth we let $M_1, M_2, \ldots, M_r$ be disjoint sets whose union is the set of all real numbers. Let $X$ be a random variable on a sample space $S_0$. Assume independent sampling and for $s \in S$ let $N_j(s) = $ number of the numbers $X_1(s), X_2(s), \ldots, X_n(s)$ which are in $M_j$. Let $\pi_j = P(X \in M_j) > 0$. Let

$$Y_n^2(s) = \sum_{j=1}^{r} \frac{[N_j(s) - n\pi_j]^2}{n\pi_j}$$

We use the subscript "$n$" on $Y_n$ because next we prove a limit theorem.

**Theorem 12.3.2**      With the above notation

$$\lim_{n \to \infty} P(Y_n^2 \le b) = P(Z_{2.}^2 + Z_3^2 + \cdots + Z_r^2 \le b)$$

where $Z_2, Z_3, \ldots, Z_r$ are independent and $N(0, 1)$.

**PROOF**      Let

$$V_n = \left( \frac{N_1 - n\pi_1}{\sqrt{n\pi_1}}, \frac{N_2 - n\pi_2}{\sqrt{n\pi_2}}, \ldots, \frac{N_r - n\pi_r}{\sqrt{n\pi_r}} \right)$$

We will show that the sequence of $r$-dimensional characteristic functions $\phi_{V_n}(t_1, t_2, \ldots, t_r) = \phi_{V_n}(t)$ converges to

$$\exp\{-\tfrac{1}{2}[t_1^2 + t_2^2 + \cdots + t_r^2 - (\sqrt{\pi_1}t_1 + \sqrt{\pi_2}t_2 + \cdots + \sqrt{\pi_r}t_r)]\} = \phi_W(t)$$

The continuity theorem for multidimensional characteristic functions, Theorem 10.4.1, can then be applied and we will have

$$\lim_{n \to \infty} P(Y_n^2 \le b) = P(W_1^2 + W_2^2 + \cdots + W_r^2 \le b)$$

But by Lemma 12.3.1

$$P(W_1^2 + W_2^2 + \cdots + W_r^2 \le b) = P(Z_1^2 + Z_2^2 + \cdots + Z_r^2 \le b)$$
$$= P(0 + Z_2^2 + \cdots + Z_r^2 \le b)$$

since $Z_1 = 0$. We will then have

$$(12.3.8) \qquad \lim_{n \to \infty} P\left( Y_n^2 = \frac{(N_1 - n\pi_1)^2}{n\pi_1} + \frac{(N_2 - n\pi_2)^2}{n\pi_2} + \cdots + \frac{(N_r - n\pi_r)^2}{n\pi_r} \le b \right)$$
$$= P(Z_1^2 + Z_2^2 + \cdots + Z_r^2 \le b)$$

where the $Z_j$ are independent and $N(0, 1)$.

Let us continue with the proof. From section 1.5 we note that

$$(12.3.9) \qquad (q_1 + q_2 + \cdots + q_r)^n = \sum \frac{n!}{n_1! n_2! \cdots n_r!} q_1^{n_1} q_2^{n_2} \cdots q_r^{n_r}$$

where $n_j \ge 0$, $n_1 + n_2 + \cdots + n_r = n$, and the sum is taken over all such arrangements of integers $n_j$. From equation (12.2.16) we have

$$P(N_1 = n_1, \quad N_2 = n_2, \ldots, N_r = n_r) = \frac{n!}{n_1! n_2! \cdots n_r!} \pi_1^{n_1} \pi_2^{n_2} \cdots \pi_r^{n_r}$$

Therefore, by the definition of characteristic function, equation (12.3.9), and the relationship $(e^\lambda)^k = e^{k\lambda}$ we have

$$(12.3.10) \qquad E(\exp[i(t_1 N_1 + t_2 N_2 + \cdots + t_r N_r)])$$
$$= \sum P(N_1 = n_1, \quad N_2 = n_2, \ldots, N_r = n_r) e^{in_1 t_1} e^{in_2 t_2} \cdots e^{in_r t_r}$$
$$= \sum \frac{n!}{n_1! n_2! \cdots n_r!} (\pi_1 e^{it_1})^{n_1} \cdots (\pi_r e^{it_r})^{n_r}$$
$$= (\pi_1 e^{it_1} + \pi_2 e^{it_2} + \cdots + \pi_r e^{it_r})^n$$

Let $q = \sqrt{\pi_1}\,t_1 + \sqrt{\pi_2}\,t_2 + \cdots + \sqrt{\pi_r}t_r$, $T = t_1^2 + t_2^2 + \cdots + t_r^2$, and $\alpha_j = 1/\sqrt{n\pi_j}$. Then

$$\phi_{V_n}(t) = E\left(\exp\left[i\left(t_1\frac{N_1 - n\pi_1}{\sqrt{n\pi_1}} + t_2\frac{N_2 - n\pi_2}{\sqrt{n\pi_2}} + \cdots + t_r\frac{N_r - n\pi_r}{\sqrt{n\pi_r}}\right)\right]\right)$$

$$= E(\exp[i(t_1\,\alpha_1 N_1 + t_2\alpha_2 N_2 + \cdots + t_r\alpha_r N_r)])$$

$$\times \exp[i(-t_1\sqrt{n\pi_1} - \cdots - t_r\sqrt{n\pi_r})]$$

since $\exp(a + b) = \exp(a)\exp(b)$, $n\pi_j/\sqrt{n\pi_j} = \sqrt{n\pi_j}$, and the $n\pi_j$ do not depend on $n_1, n_2, \ldots, n_r$, the indices of summation for the expectation. Therefore by equation (12.3.10) with $\alpha_j t_j$ instead of $t_j$

$$\phi_{V_n}(t) = (\pi_1 e^{i\alpha_1 t_1} + \pi_2 e^{i\alpha_2 t_2} + \cdots + \pi_r e^{i\alpha_r t_r})^n \exp(-i\sqrt{n}\,q)$$

Now $\exp(\sqrt{n}a) = e^{\sqrt{n}a} = (e^{a/\sqrt{n}})^n$ and $a^n b^n = (ab)^n$ so

$$\phi_{V_n}(t) = (\pi_1 e^{i\alpha_1 t_1} + \pi_2 e^{i\alpha_2 t_2} + \cdots + \pi_r e^{i\alpha_r t_r})^n (e^{-iq/\sqrt{n}})^n$$

$$= [(\pi_1 e^{i\alpha_1 t_1} + \pi_2 e^{i\alpha_2 t_2} + \cdots + \pi_r e^{i\alpha_r t_r})e^{-iq/\sqrt{n}}]^n$$

The MacLaurin series for $e^x$ gives us

$$\pi_j e^{i\alpha_j t_j} = \pi_j\left(1 + \frac{it_j}{\sqrt{n\pi_j}} + \frac{1}{2!}\left(\frac{it_j}{\sqrt{n\pi_j}}\right)^2 + \frac{1}{3!}\left(\frac{it_j}{\sqrt{n\pi_j}}\right)^3 + \cdots\right)$$

Since $i^2 = -1$ and $\pi_1 + \pi_2 + \cdots + \pi_r = 1$,

$$\sum_{j=1}^{r} \pi_j e^{i\alpha_j t_j} = \pi_1 + \cdots + \pi_r + \frac{i}{\sqrt{n}}(\sqrt{\pi_1}t_1 + \cdots + \sqrt{\pi_r}t_r) - \frac{1}{2n}T + n^{-3/2}[\quad]$$

$$= 1 + i\frac{q}{\sqrt{n}} - \frac{1}{2n}T + \frac{1}{n^{3/2}}[\quad]$$

where we group all terms from the MacLaurin series whose powers of $n^{-1/2}$ are three or more in the brackets [ ].

Now

$$\left(1 + i\frac{q}{n} - \frac{1}{2n}T + n^{-3/2}[\quad]\right)e^{iq/\sqrt{n}}$$

$$= \left(1 + \frac{iq}{\sqrt{n}} - \frac{1}{2n}T + n^{-3/2}[\quad]\right)\left(1 - \frac{iq}{\sqrt{n}} - \frac{1}{2}\frac{q^2}{n} + n^{-3/2}[\quad]\right)$$

$$= 1 - \frac{iq}{\sqrt{n}} + \frac{iq}{\sqrt{n}} - \frac{i^2 q^2}{(\sqrt{n})^2} - \frac{1}{2n}T - \frac{1}{2}\frac{q^2}{n} + n^{-3/2}[\quad]$$

$$= 1 - \frac{1}{2n}T + \frac{q^2}{n} - \frac{1}{2}\frac{q^2}{n} + n^{-3/2}[\quad]$$

Therefore

$$\phi_{V_n}(t) = \left\{ 1 - \frac{1}{2n}(T - q^2) + n^{-3/2}[\quad] \right\}^n$$

By Theorem 10.3.3 we have

$$\lim_{n \to \infty} \left\{ 1 - \frac{1}{2n}(T - q^2) + n^{-3/2}[\quad] \right\}^n$$

$$= \exp[-\tfrac{1}{2}(T - q^2)]$$

$$= \exp\{-\tfrac{1}{2}[t_1^2 + t_2^2 + \cdots + t_r^2 - (\sqrt{\pi_1}\,t_1 + \sqrt{\pi_2}\,t_2 + \cdots + \sqrt{\pi_r}\,t_r)^2]\}$$

$$= \phi_W(t_1, t_2, \ldots, t_r)$$

This is what we set out to show, hence the proof is complete.

By definition $Z_2^2 + Z_3^2 + \cdots + Z_r^2$ has a $\chi^2$ distribution with $r - 1$ degrees of freedom. Its distribution is given by

**(12.3.11)**    $P(Z_2^2 + Z_3^2 + \cdots + Z_r^2 \le b)$

$$= \int_0^b 2^{-(r-1)/2} \frac{1}{\Gamma(r-1)/2} x^{(r-3)/2} e^{-x/2} \, dx$$

This theorem explains why statistics of the form

$$\frac{(N_1 - n\pi_1)^2}{n\pi_1} + \frac{(N_2 - n\pi_2)^2}{n\pi_2} + \cdots + \frac{(N_r - n\pi_r)^2}{n\pi_r}$$

are denoted by $\chi^2$. It is not true that they have exactly a $\chi^2$ distribution. But if $n$ is large they have, approximately, a $\chi^2$ distribution.

Although the proof of the theorem does not tell why one must use $r - 1$ degrees of freedom it is certainly the case that one must. It is a useful rule of thumb to feel that one loses one degree of freedom because

$$N_1 + N_2 + \cdots + N_r = n$$

That is, there is one constraint on the $N_j$.

It is not necessary to evaluate the integral on the right-hand side of (12.3.11). These values are tabulated in a useful way in Table 2 in the appendix.

### Example 12.3.4
The following data are observed under circumstances which can reasonably be assumed to be independent observations of a random variable. Does this random variable seem like a normal random which is $N(0, 4)$? The observed values are as follows:

7.62, 5.44, 4.21, 3.2, 2.7, 2.41, 2.17, 1.93, 1.71, 1.58, 1.40, 1.31, 1.17, 1.12, 0.99, 0.92, 0.86, 0.77, 0.71, 0.64, 0.59, 0.52, 0.49, 0.47, 0.43, 0.37, 0.31, 0.28,

0.22, 0.19, 0.16, 0.13, 0.09, 0.04, 0.01, −0.02, −0.03, −0.04, −0.07, −0.13, −0.18, −0.22, −0.27, −0.29, −0.36, −0.44, −0.49, −0.53, −0.60, −0.68, −0.81, −0.88, −0.95, −0.99, −1.05, −1.20, −1.43, −1.52, −1.55, −1.68, −1.88, −1.95, −2.17, −2.25, −2.36, −2.60, −2.94, −3.86, −4.72, −5.18.

We can let $S_0 = \{x; -\infty < x < \infty\}$. Let the mass density on $S_0$ be $p(x) = (1/2\sqrt{2\pi})e^{-x^2/8}$. We let

$$a_1 = -2.8, \quad a_2 = -2.0, \quad a_3 = -1.5, \quad a_4 = -1, \quad a_5 = -0.6, \quad a_6 = -0.14,$$
$$a_7 = 0.26, \quad a_8 = 0.68, \quad a_9 = 1.16, \quad a_{10} = 1.62, \quad a_{11} = 2.30$$

Let $A_1 = \{x; x \le -x_1\}$, and let $A_k = \{x; x_{k-1} < x \le x_k\}$ for $k = 2, 3, \ldots, 11$ and let $A_{12} = \{x; x > x_{12}\}$. It can be seen that

$$P(\{x; a < x \le b\}) = \int_{a/2}^{b/2} \frac{1}{\sqrt{2\pi}} e^{-x^2/2} \, dx$$

so we can use Table 1 (appendix) after dividing by 2. It can be verified that

$$\pi_1 = 0.0808, \quad \pi_2 = 0.0779, \quad \pi_3 = 0.0679, \quad \pi_4 = 0.0819, \quad \pi_5 = 0.0736,$$
$$\pi_6 = 0.0900, \quad \pi_7 = 0.0796, \quad \pi_8 = 0.0814, \quad \pi_9 = 0.0859, \quad \pi_{10} = 0.0720,$$
$$\pi_{11} = 0.0839, \quad \pi_{12} = 0.1251$$

We use the data in an evaluation of

$$Y_{70} = \sum_{j=1}^{12} \frac{(N_j - 70\pi_j)^2}{70\pi_j}$$

This gives

$$\frac{(4 - 5.656)^2}{5.656} + \frac{(4 - 5.453)^2}{5.953} + \frac{(5 - 4.753)^2}{4.753} + \frac{(3 - 5.736)^2}{5.736}$$

$$+ \frac{(6 - 5.152)^2}{5.152} + \frac{(8 - 6.300)^2}{6.300} + \frac{(12 - 5.572)^2}{5.572} + \frac{(9 - 5.698)^2}{5.698}$$

$$+ \frac{(6 - 6.013)^2}{6.013} + \frac{(4 - 5.040)^2}{5.040} + \frac{(3 - 5.873)^2}{5.873} + \frac{(6 - 8.757)^2}{8.757} \approx 14$$

We assume that 70 is large enough so that $P(Y_{70} \le u) \approx P(\chi^2$ with 11 degrees of freedom $\le u)$. Now $P(\chi^2$ with 11 degrees of freedom $\le 17.275) = 0.90$. Thus, our data do not furnish a value that is strikingly large since with probability about 0.10, 70 independent samplings of an $N(0, 4)$ random variable will yield a $Y_{70}$ value greater than 17.275. Moreover, random variables whose distributions are materially different from $N(0, 4)$ will usually yield much larger values if one should substitute data generated by them into a $Y_{70}$ expression whose $\pi_j$'s are determined by $N(0, 4)$.

**EXERCISES 12.3** _____

1   Let $U_n$ have Student's $t$-distribution with $n$ degrees of freedom and let $V_n$ have the $\chi^2$ distribution with $n$ degrees of freedom. Find $v$ such that (a) $P(V_7 \geq v) = 0.80$, (b) $P(V_{12} \leq v) = 0.70$, (c) $P(V_{18} \geq v) = 0.05$, and (d) $P(V_4 \leq v) = 0.05$. Find $u$ such that (e) $P(U_{15} \geq u) = 0.05$, (f) $P(U_{28} \leq u)$ $= 0.35$, and (g) $P(-u \leq U_4 \leq u) = 0.80$.

♦ 2   Show that if $X$ is $N(m, \sigma^2)$ and sampling is independent then $(\bar{X} - m)\sqrt{n}/\sqrt{\hat{d}^2}$ has Student's $t$-distribution with $n - 1$ degrees of freedom.

♦ 3   Show that the $a_{jk}$ defined by $a_{1k} = n^{-1/2}$ for $k = 1, 2, \ldots, n$, $a_{jk} = 0$ for $k > j$; $a_{jk} = [j(j - 1)]^{-1/2}$ if $k < j$ but $j > 1$, and $a_{jj} = -(j - 1)^{1/2}j^{-1/2}$ for $j = 2, 3, \ldots, n$ satisfy $\sum_{k=1}^{n} a_{jk}a_{lk} = \delta_{jl}$.

4   Would you judge a die to be fair if in 90 (independent) rolls it showed 10 ones, 13 twos, 19 threes, 8 fours, 18 fives, and 22 sixes? Naturally we assume that there is some reason to feel that it may not be fair.

5   It is claimed that an independent sample of size 25 yields $\hat{X} = 18.6$ and $\hat{d}^2 = 4.76$. Can this reasonably result from simulation of a normal random variable with expected value 17.5? What if $\hat{d}^2 = 10.24$?

6   Let $X_1$ and $X_2$ be independent random variables with $P(X_j = 1) = \frac{1}{2}$ and $P(X_j = 0) = \frac{1}{2}$. Show that $X_1 + X_2$ and

$$[X_1 - \tfrac{1}{2}(X_1 + X_2)]^2 + [X_2 - \tfrac{1}{2}(X_1 + X_2)]^2$$

are not independent random variables.

7   A nutritionist knows that the average weight of male adults in Anitnegra was 152.38 lbs. five years ago when the last extensive survey was taken. He weighs 16 such people selected at random and finds the average of their weights to be 148.64 with a sampled $\hat{d}^2 = 121$. Is this a significant difference? What are you assuming?

♦ 8   For the random variable $V_n$ of exercise 1 compute $E(V_n)$ and $\sigma^2(V_n)$. Why can it be said that as $n \to \infty$,

$$P\left(\frac{V_n - E(V_n)}{\sigma(V_n)} \leq x\right) \to \int_{-\infty}^{x} \frac{1}{\sqrt{2\pi}} e^{-x^2/2}\, dx$$

What can be said about $P(V_n \leq u)$ for large $n$?

♦ 9   Let $f_n(x)$ be the density function for Student's $t$-distribution with $n$ degrees of freedom. Show that for each fixed $x$,

$$\lim_{n \to \infty} f_n(x) = (2\pi)^{-1/2} e^{-x^2/2}$$

Show that there is a $K > 0$ such that for all $n \geq 1$, $f_n(x) \leq K(1 + t^2)^{-1}$. If these are linked with Theorem 10.3.2, what can be said about

$$\lim_{n \to \infty} \int_{a}^{b} f_n(x)\, dx$$

Find an approximation for $P(U_{55} \leq 0.10)$ where $U_n$ is as in exercise 1 without looking in Table 3 (appendix).

# CHAPTER 13

# *Estimation*

## 13.1 Parameters

In Chapter 12 we set up a mathematical model for statistical problems. It includes a sample space $S_0$ as a model for the population and a probability measure $P_0$ on $S_0$. Our objective is that of making judgments about the nature of $P_0$. In this chapter it is convenient to assume that $S_0$ is a subset of $R$ and that the data are such that $X(x) = x$ for each $x \in S_0$ models one piece of the data. With this setup $P_0$ determines $F_X$ via the relationship

$$F_X(u) = P_0(X \le u) = P_0(\{x \in S_0 ; x \le u\})$$

In this chapter we assume independent sampling unless otherwise noted.

In many problems we have some information about the structure of $P_0$ but we lack certain numbers which separate out the proper $P_0$ from among similar probability measures. Examples of such problems include the following: a biased coin whose bias $p$ is unknown, certain electronic devices whose lifetimes are assumed to have an exponential distribution but where $\lambda$ is unknown, or the weights of eggs which are modeled by an $N(m, \sigma^2)$ distribution with unknown $m$ and $\sigma^2$. Constants such as $\lambda$, $p$, $m$, and $\sigma^2$ are called *parameters*.

It should be clear that one would like to know the values of parameters in statistical problems. If one plans to gamble with a biased coin, it is good to know the bias. If one is responsible for keeping in operation a computer which uses certain electronic devices it would help to know the behavior of the lifetimes of the devices in order to plan the number of replacements to keep in stock.

We use a parameter instead of the distribution $P_0$ because, for example, it is easier to write "$\lambda$" than "$\lambda e^{-\lambda x}$ for $x > 0$." However, it should be recognized that a parameter does not single out specific distributions until the class of distributions of the problem have been agreed upon.

We usually denote a general parameter by $\theta$. It should be kept in mind that $\theta$ could denote $(m, \sigma^2)$ or even something more complicated as well as a single real number. As a result of the mode of sampling (independent sampling in this chapter) the probability measure on $S_0$ denoted by $\theta$ induces a measure on $S$, the sample space of the sample. We call the latter measure $P_\theta$ and use the subscript to suggest the dependence of $P_\theta$ on $\theta$. Corresponding to $P_\theta$ on $S$ are the expectation $E_\theta$ and the variance $\sigma_\theta^2$ which are determined by $P_\theta$ and can be applied to random variables on $S$. Recall that such random variables are called statistics. We use $E$ and $\sigma^2$ without subscripts to denote the expectation and variance for random variables on $S_0$ under $P_0$ (or equivalently, under $\theta$).

### Example 13.1.1

Consider independent tossing of a coin whose bias is described by $P_0(H) = \theta$ for $0 < \theta < 1$. Let the statistic $Z =$ number of H's $-$ number of T's in $n$ tosses. Compute $P_\theta(Z > 0)$, $E_\theta(Z)$, $\sigma_\theta^2(Z)$ and, in the case $n = 4$, $P_\theta(\text{TTHT})$.

We save $P_\theta(Z)$ for the last. Thus, by past results

$$E_\theta(Z) = E_\theta(\text{number of H's}) - E_\theta(\text{number of T's}) = n\theta - n(1 - \theta)$$

$$\sigma_\theta^2(Z) = n((2 - 2\theta)^2\theta + (-2\theta)^2(1 - \theta) = 4n\theta(1 - \theta))$$

and

$$P_\theta(\text{TTHT}) = (1 - \theta)^3\theta$$

Finally

$$P_\theta(Z > 0) = \sum_{j = v}^{n} \binom{n}{j}\theta^j(1 - \theta)^{n-j}$$

where $v = \tfrac{1}{2}n + 1$ if $n$ is even and $v = \tfrac{1}{2}(n + 1)$ if $n$ is odd.

### EXERCISES 13.1

1  Let $\theta > 0$ signify the Poisson distribution. That is, $P_0(k) = e^{-\theta}\theta^k/k!$ for $k = 0, 1, 2, \ldots$.
  (a)  Describe $P_\theta$ for independent sampling.
  (b)  Find $P_\theta(4, 1, 0, 2, 1, 3, 1)$ if $n = 7$.
  (c)  Find $P_5(4, 1, 0, 2, 1, 3, 1)$ if $n = 7$.

2  Using the distribution in exercise 1. Let

$$Z(x_1, x_2, \ldots, x_n) = x_1 + x_2 + \cdots + x_n$$

  (a)  Find $E_\theta(Z)$.
  (b)  Find $E_1(Z)$, $E_3(Z)$, and $E_{\sqrt{2}}(Z)$.

# 13.2 Estimators

Many of the special distributions are indexed by parameters which are related to their moments. Hence, it is reasonable to expect that we could make use of the moment or an analog to the moment to estimate the parameter.

### Example 13.2.1

A company produces bullets. It is assumed that there is a certain probability $p$ that a bullet will be defective. It is also assumed that the effectiveness or defectiveness of different bullets is independent. Fifty bullets are chosen at random and are test fired. Three are defective. Estimate $p$.

The assumptions of the model are such that the number $X$ of defectives in the 50 chosen bullets has a binomial distribution, i.e.,

$$P_p(X = k) = \binom{50}{k} p^k (1 - p)^{50-k}$$

for $k = 0, 1, 2, \ldots, 50$. Now $E_p(X) = 50p$. Furthermore, although we cannot say $50p = 3$, it seems more plausible that $50p$ is nearly 3 than that $50p = 30$ or $50p = 0.005$. Therefore, why not estimate $50p$ by 3. Equivalently we could estimate $p$ by $\frac{3}{50} = 0.06$.

Another way of analyzing Example 13.2.1 is as follows: We could imagine a standard random variable $Y$ for the typical bullet. $Y = 1$ if the bullet is defective and $Y = 0$ if the bullet is acceptable. Then $E(Y) = p$. Furthermore, the study of 50 (independent) bullets leads one to $Y_1, Y_2, \ldots, Y_{50}$ which are independent and identically distributed with $Y$. Moreover the random variable $X = Y_1 + Y_2 + \cdots + Y_{50}$ models the number of defectives. Finally we point out that $\frac{1}{50}(Y_1 + Y_2 + \cdots + Y_{50})$ is the mathematical abstraction of our computation $\frac{3}{50}$ in Example 13.2.1.

**Definition**     A statistic which is used to estimate a parameter is called an *estimator* for the parameter. A value which one obtains by substituting data in place of the appropriate random variables in an estimator is called an *estimate* of the parameter.

We write $\hat{\theta}$ for an estimated value of the parameter $\theta$. This is consistent with our use of "^". Example 13.2.1 is meant to show that $\bar{X}$ is frequently used as an estimator for the expected value of the underlying random variable $X$.

### Example 13.2.2

Certain particle motions modeled by $Z$ are assumed to be independent and normally distributed with $E(Z) = 0$. Find an estimator of $\sigma^2$ for a sample of size $n$. Find an estimate of $\sigma^2$ if eight experiments produced data as follows: $-3.28, -0.30, 2.21, 1.08, 1.62, -4.79, 0.78, 2.84$.

Our study of the normal distribution indicated that if $E(Z) = 0$, $\sigma^2(Z) = E(Z^2)$. Now, if we look at a number of copies of $Z$ we would expect them to mimic the distribution of the payoffs of $Z$. Therefore we take as our estimator $W = (1/n)(Z_1^2 + Z_2^2 + \cdots + Z_n^2)$. Accordingly our estimate is

$$\tfrac{1}{8}[(-3.28)^2 + (-0.30)^2 + (2.21)^2 + (1.08)^2 + (1.62)^2 + (-4.79)^2$$
$$+ (0.78)^2 + (2.84)^2] \approx \tfrac{1}{8}[51.142]$$

It seems intuitively desirable to use as an estimator for $E(X^r)$ the *sample* r th *moment*

(13.2.1) $$m_r = \frac{X_1^r + X_2^r + \cdots + X_n^r}{n}$$

when one has a sample of size $n$.

### Example 13.2.3
A sample of size 10 yields values 3, $-1$, 5, 0, 0, $-1$, 4, $-1$, $-2$, 2. Estimate the third moment for the random variable $X$ which models these values.

If we use the estimator $m_3$, we obtain

$$\tfrac{1}{10}[3^3 + (-1)^3 + 5^3 + 0^3 + 0^3 + (-1)^3 + 4^3 + (-1)^3$$
$$+ (-2)^3 + 2^3] = \tfrac{1}{10}[203] = 20.3$$

as the estimate.

### Example 13.2.4
A shipment has 4800 items. A random sample of 48 items shows four defectives. Find the probability of 420 or fewer defectives in the entire shipment based on the estimated value from the sample.

As in Example 13.2.1 we could estimate $p$, the probability of a defective, by $p = \tfrac{1}{48}$(number of defectives in the sample) $= \tfrac{4}{48}$. Since there are 4800 items we may use the approximation of the central limit theorem and argue $P_p$(number of defectives in shipment $\leq 420$)

$$= P_p\left\{\frac{X_1 + X_2 + \cdots + X_{4800} - 4800p}{\sqrt{4800p(1-p)}} \leq u\right\} \approx \int_{-\infty}^{u} \frac{e^{-x^2/2}}{\sqrt{2n}} \, dx$$

where

$$X_j = \begin{cases} 1, & j\text{th item is defective} \\ 0, & \text{otherwise} \end{cases}$$

From section 9.2

$$P\left\{\frac{X_1 + \cdots + X_{4800} - 400}{\sqrt{4800 \cdot \tfrac{1}{12} \cdot \tfrac{11}{12}}} \leq u\right\} \approx 0.851$$

since $u = \sqrt{\frac{12}{11}} = 1.044$. It should be recognized that our estimate of $\frac{4}{48}$ can be wrong. Hopefully it is not wrong by much. Furthermore, since the probability which we derived is a continuous function of $p$, if $\frac{4}{48}$ is only slightly different from the correct value of $p$, our answer will be close to the correct answer.

Example 13.2.4 does not constitute a very good way of formulating an inspection problem. Nevertheless, it does show why one would wish to estimate parameters.

### *Example 13.2.5*

Independent sampling of a population whose mathematical model involves a random variable $X$ gives values of 7, 4.3, 8.1, 5.8, 6.2. Estimate $m = E(X)$ by $\bar{X}$ and by $Y = \frac{1}{5}(X_1 + X_2 + X_3) + \frac{3}{10}X_4 + \frac{1}{10}X_5$.

Using $\bar{X}$ we obtain

$$\hat{m} = \tfrac{1}{5}(7 + 4.3 + 8.1 + 5.8 + 6.2) = \tfrac{1}{5}(31.4) + 6.28$$

Using $Y$ and assuming that 7, 4.3, 8.1, 5.8, 6.2 correspond to $X_1$, $X_2$, $X_3$ $X_4$, $X_5$, respectively, we have

$$\hat{m} = \tfrac{1}{5}(19.4) + \tfrac{3}{10}(5.8) + \tfrac{1}{10}(6.8) = 6.30$$

Can we use $Y$ to estimate $m$? Why not? Is it as good an estimator as $X$? We have not discussed the quality of estimators. Perhaps 6.30 obtained from $Y$ is a better estimate than 6.28 obtained from $X$? We can conceive of circumstances when that would be true. But if there is no distinction between good estimators and bad estimators why not use one single estimator for everything, namely zero? In section 13.3 we discuss some theoretical methods of analyzing the quality of estimators. For the present let us trust our intuition to judge whether $\bar{X}$ as an estimator for $E(X)$ is better than zero as an estimator for $E(X)$. We could even leave it to intuition to judge between $\bar{X}$ and $Y$.

### *Example 13.2.6*

On the basis of the data in Example 13.2.3 estimate the variance of $X$ mentioned in that example by the statistics $d^2 = \dfrac{1}{n-1} \sum\limits_{j=1}^{n} (X_j - \bar{X})^2$ and by $m_2 - m_1^2$. Note that $m_1 = \bar{X}$.

To use $d^2$ is to estimate by

$$\tfrac{1}{9}[(3 - 0.9)^2 + 3(-1 - 0.9)^2 + (5 - 0.9)^2 + 2(0 - 0.9)^2$$
$$+ (-2 - 0.9)^2 + (2 - 0.9)^2]$$

after certain identical terms are grouped. This is $5.8777\ldots$. To estimate by $m_2 - m_1^2$ is to compute

$$\tfrac{1}{10}[9 + 1 + 25 + 0 + 0 + 1 + 16 + 1 + 4 + 4] - (0.9)^2$$
$$= 6.1 - 0.81 = 5.29$$

Which of the above is preferable? The estimate by $d^2$ seems less direct than that by $m_2 - m_1^2$. However, we shall see that $d^2$ is preferable in certain respects.

There is nothing that requires a parameter to be related to a moment. Furthermore, if a parameter $\theta = g(E(X^r))$ it does not follow that $g\,(\hat{m}_r)$ is a good estimate of $\theta$. We now study another method of obtaining estimators.

### Example 13.2.7

A numerical population is modeled by a Poisson random variable $X$. Independent sampling of that population yields 3, 0, 1, 1, 2 in a sample of size 5. Find the probability of this happening in $S$ when the Poisson parameter is $\lambda$. For what value of $\lambda$ is this probability largest?

The Poisson probability function has $f_X(k) = e^{-\lambda}\lambda^k/k!$ for $k = 0, 1, 2, \ldots$. Since sampling is independent,

$$P_\lambda(X_1 = 3, \quad X_2 = 0, \quad X_3 = 1, \quad X_4 = 1, \quad X_5 = 2)$$

$$= \left(e^{-\lambda}\frac{\lambda^3}{3!}\right)\left(e^{-\lambda}\frac{\lambda^0}{0!}\right)\left(e^{-\lambda}\frac{\lambda}{1!}\right)\left(e^{-\lambda}\frac{\lambda}{1!}\right)\left(e^{-\lambda}\frac{\lambda^2}{2!}\right)$$

$$= \frac{1}{3!0!1!1!2!}\,e^{-5\lambda}\lambda^7$$

If we graph this probability as a function of $\lambda$ we get a graph such as that in Figure 1. Since this is a differentiable function we can find the maximum by solving

$$\frac{d}{d\lambda}\frac{1}{3!0!1!1!2!}\,e^{-5\lambda}\lambda^7 = 0$$

This is equivalent to solving

$$\frac{d}{d\lambda}e^{-5\lambda}\lambda^7 = 0$$

or

$$-5e^{-5\lambda}\lambda^7 + 7\lambda^6 e^{-5\lambda} = 0$$

Since $e^{-5\lambda} > 0$ we have $-5\lambda^7 + 7\lambda^6 = (-5\lambda + 7)\lambda^6 = 0$.

The value $\lambda = 0$ does not mark the maximum. It is obvious that $-5\lambda + 7 = 0$ or $\lambda = \frac{7}{5}$ marks the maximum. Notice that this value of $\lambda$ is $\frac{1}{5}(3 + 0 + 1 + 1 + 2) = \hat{\bar{X}}$ and that $E(X) = \lambda$.

Example 13.2.7 illustrates the method of maximum-likelihood estimation. We will prove that the above result is a special case of a more general result. But first, another example.

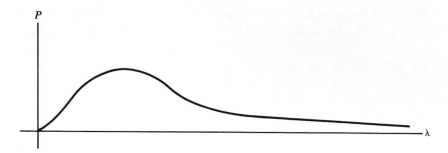

**Figure 1**

### Example 13.2.8

A population has a numerical characteristic which simulates a random variable
$X$ whose distribution is given by

$$P(X = n) = (1 - q)^3 \binom{-3}{n} q^n$$

for $n = 0, 1, 2, \ldots$ and $1 > q > 0$. Independent sampling gives data $n_1, n_2, \ldots, n_r$. Find the probability in $S$ that this data will occur and find the value of $q$ which maximizes this probability.

By independence

$$P_q(X_1 = n_1, \quad X_2 = n_2, \ldots, X_r = n_r)$$

$$= (1 - q)^3 \binom{-3}{n_1} q^{n_1} (1 - q)^3 \binom{-3}{n_2} q^{n_2} \cdots (1 - q)^3 \binom{-3}{n_r} q^{n_r}$$

$$= K(1 - q)^{3r} q^{n_1} q^{n_2} \ldots q^{n_r}$$

where $K = \binom{-3}{n_1} \cdots \binom{-3}{n_r}$ does not depend on $q$.

The maximum of $(1 - q)^{3r} q^{n_1 + n_2 + \cdots + n_r}$ has the same value of $q$ for which $K(1 - q)^{3r} q^{n_1 + \cdots + n_r}$ takes its maximum. Unless $n_1 + n_2 + \cdots + n_r = 0$ the maximum can be found by differentiation and setting the derivative equal to zero.

However, we can analyze the maximum of

$$\log [(1 - q)^{3r} q^{n_1 + n_2 + \cdots + n_r}] = g(q)$$
$$= 3r \log (1 - q) + (n_1 + n_2 + \cdots + n_r) \log q$$

Upon differentiating we obtain

$$g'(q) = \frac{3r}{1 - q} \frac{d}{dq} (1 - q) + \frac{n_1 + n_2 + \cdots + n_r}{q}$$

$$= -\frac{3r}{1 - q} + \frac{n_1 + \cdots + n_r}{q}$$

$$\frac{-3rq - q(n_1 + n_2 + \cdots + n_r) + n_1 + n_2 + \cdots + n_r}{q(1 - q)}$$

Therefore, $g'(q) = 0$ precisely when

$$q = \frac{n_1 + n_2 + \cdots + n_r}{3r + n_1 + n_2 + \cdots + n_r}$$

### Example 13.2.9

Assume that lifetimes of certain electronic components simulate an exponential random variable $X$ with parameter $\lambda$. A sample of lifetimes yields 3.7, 8.3, 5.4, 6.2, 6.8. Find the density on $S$ that a fivefold independent sample would yield those values. Find $\lambda$ which makes that density a maximum.

The density in question is

$$\lambda e^{-\lambda(3.7)}\lambda e^{-\lambda(8.3)}\lambda e^{-\lambda(5.4)}\lambda e^{-\lambda(6.2)}\lambda e^{-\lambda(6.8)}$$

$$= \lambda^5 \exp\left[-\lambda(3.7 + 8.3 + 5.4 + 6.2 + 5.8)\right]$$

$$= \lambda^5 e^{-29.4\lambda}$$

As before $\lambda^5 e^{-29.4\lambda}$ takes its maximum for the same $\lambda$ value as does log $\lambda^5 e^{-29.4\lambda} = 5 \log \lambda - 29.4\lambda$. Furthermore the maximum is marked by a horizontal tangent. Therefore we set

$$0 = \frac{d}{d\lambda}\left(5 \log \lambda - 29.4\lambda\right) = \frac{5}{\lambda} - 29.4$$

It is easy to find $\hat{\lambda} = 5/29.4$ and $1/\hat{\lambda} = 29.4/5$

**Definition**    Let $X$ be one of a family of random variables whose probability function (or density function as the case may be) depends on a parameter $\theta$. Let $f(x; \theta)$ be that probability (or density) function. The *likelihood function L* for $n$-fold independent sampling of $X$ is given by

**(13.2.2)**    $L(x_1, x_2, \ldots, x_n; \theta) = f(x_1; \theta)f(x_2; \theta) \cdots f(x_n; \theta)$

It should be noted that $L$ is nothing more than the joint probability or density function for the random vector $(X_1, X_2, \ldots, X_n)$ used in independent sampling of $X$. It is a probability or density function according to whether $f$ is a probability function or density function.

For the present we assume that $\theta$ is a real number.

**Definition**    Suppose that for each fixed $n$-tuple $(x_1, x_2, \ldots, x_n)$ the likelihood function $g(\theta) = L(x_1, \ldots, x_n; \theta)$ takes on its absolute maximum at the unique point $\bar{\theta}$ which depends on $(x_1, \ldots, x_n)$. The statistic given by

**(13.2.3)**    $k(x_1, x_2, \ldots, x_n) = \bar{\theta}$

is called the *maximum-likelihood estimator* of $\theta$.

Examples 13.2.7–13.2.9 illustrate maximum-likelihood estimation.

### Example 13.2.10

Let $X$ have a uniform density on $0 \leq x \leq a$ for $a > 0$. Find the maximum-likelihood estimator for $a$.

The density function is

$$f(x) = \begin{cases} \dfrac{1}{a}, & 0 \leq x \leq a \\ 0, & \text{otherwise} \end{cases}$$

For a sample of size $n$ with data $(x_1, x_2, \ldots, x_n)$ we have

$$L(x_1, x_2, \ldots, x_n; a) = \begin{cases} \dfrac{1}{a^n}, & 0 \leq x_j \leq a \text{ for each } x_j \\ 0, & a < x_j \quad \text{for some } j \end{cases}$$

If we plot $L$ as a function of $a$ we get Figure 2. The jump occurs at $a = \max(x_1, x_2, \ldots, x_n)$ since for smaller $a$ some $x_j > a$ and $L = 0$. The function $1/a^n$ is monotonically decreasing as $a \to \infty$. Therefore $\bar{a} = \max(x_1, x_2, \ldots, x_n)$. That is the maximum-likelihood estimator of $a$.

Note that in this the maximum was not found by differentiating $L$ and setting the derivative equal to zero. In this problem $L$ is not continuous and it takes its maximum at the discontinuity. In other situations the parameter takes on discrete values. There, too, differentiation may have no value.

Next we consider an example involving the normal distribution.

### Example 13.2.11

Find the values of $m$ and $\sigma$ which maximize the likelihood function for $n$-fold independent sampling of a general normal random variable

$$L(x_1, x_2, \ldots, x_n; m, \sigma) = \prod_{j=1}^{n} \frac{1}{\sqrt{2\pi}\,\sigma} \exp\left[-\frac{1}{2\sigma^2}(x_j - m)^2\right]$$

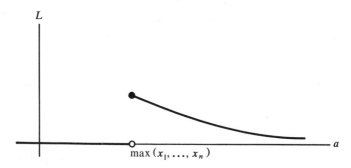

**Figure 2**

where $\prod$ represents product in the way that $\sum$ represents sum. We can see that

$$\log L = \sum_{j=1}^{n} \log \left\{ \frac{1}{\sqrt{2\pi}\,\sigma} \exp \left[ -\frac{1}{2\sigma^2}(x_j - m)^2 \right] \right\}$$

$$= \sum_{j=1}^{n} \left[ -\tfrac{1}{2}\log 2\pi - \log \sigma - \frac{1}{2\sigma^2}(x_j - m)^2 \right]$$

It is easy to see that as $m \to \pm\infty$, $L \to 0$. Furthermore as $\sigma \to +\infty$ or $\sigma \to 0$, $L \to 0$. Therefore the graph of $g(m, \sigma) = L(x_1, x_2, \ldots, x_n; m, \sigma)$ will have a horizontal tangent plane at the maximum which will occur for "finite" values of $m$ and $\sigma$. Furthermore, $\log g$ will take its maximum at the same values of $m$ and $\sigma$ that $g$ takes its maximum. Since $\log g$ is differentiable we must have

$$\frac{\partial}{\partial m} \log g = 0 \quad \text{and} \quad \frac{\partial}{\partial \sigma} \log g = 0 \text{ at the point } (\hat{m}, \hat{\sigma}) \text{ at which } \log g \text{ takes its}$$

maximum. These satisfy

$$\sum_{j=1}^{n} -\frac{1}{2\sigma^2} 2(x_j - m)(-1) = 0$$

and

$$\sum_{j=1}^{n} \left[ -\frac{1}{\sigma} + \frac{1}{\sigma^3}(x_j - m)^2 \right] = 0$$

or

(13.2.4)     $\sum_{j=1}^{n} (x_j - m) = 0$

and

(13.2.5)     $-\dfrac{n}{\sigma} + \dfrac{1}{\sigma^3} \sum_{j=1}^{n} (x_j - m)^2 = 0$

Since $m$ is independent of $j$, (13.2.4) states

$$\sum_{j=1}^{n} x_j - nm = 0 \quad \text{or} \quad \hat{m} = \frac{1}{n} \sum_{j=1}^{n} x_j$$

We may solve (13.2.5) as $n\sigma^2 = \sum_{j=1}^{n} (x_j - m)^2$ after multiplying by $\sigma^3$. But at the maximum, both (13.2.4) and (13.2.5) must be valid. Therefore the value of $m$ in $n\sigma^2 = \sum_{j=1}^{n} (x_j - m)^2$ is just $\hat{m}$. We summarize as

(13.2.6)     $\hat{m} = \dfrac{1}{n} \sum_{j=1}^{n} x_j \quad \text{and} \quad \hat{\sigma}^2 = \dfrac{1}{n} \sum_{j=1}^{n} (x_j - \hat{m})^2$

Although the original definition of maximum-likelihood estimator was made in the case $\theta$ is a real number, Example 13.2.11 shows how to find a pair of statistics which maximize $L$ in the case $\theta = (m, \sigma^2)$. It seems clear that this can be done when $\theta$ is a $k$-tuple of real numbers. We will consider statistics obtained in that fashion to be maximum-likelihood estimators.

In this section we have proposed methods of obtaining intuitively reasonable estimators for certain parameters. No attempt was made to justify their usefulness. It should be noted that there is some duplication (actually a good deal of duplication) between the estimators which were arrived at by matching moments and those arrived at by the maximum-likelihood approach. Furthermore, the maximum-likelihood method has, inherently, a certain probabilistic justification in that it chooses a parameter to maximize the probability or density of the data of the sample. After all, why estimate a parameter by a value under which the data which presumably occurred are extremely unlikely? Furthermore, if one is to choose an estimate under which the data are quite likely, why not choose an estimate under which they are most likely?

## EXERCISES 13.2

1  Let $X$ have the probability function $f(k) = (1 - p)p^k$ for $k = 0, 1, 2, 3, \ldots$ and $0 < p < 1$. Find the maximum-likelihood estimator for $p$.

2  Let $X = (U, V, W)$ have the multinomial probability function

$$f_X(a, b, c) = \frac{k!}{a!b!c!}\, p^a q^b r^c$$

for integers $a$, $b$, $c \geq 0$ and $a + b + c = k$. Naturally, $p + q + r = 1$. Find the maximum-likelihood estimators for $p$, $q$, $r$. Discuss this intuitively, too.

3  Let $X$ have density function $f(x) = Kx^{r-1}e^{-bx}$ for $x > 0$ and $f(x) = 0$ for $x \leq 0$. Here, $r > 0$ is some fixed number and $K$ is chosen so that $\int_{-\infty}^{\infty} f(x)\, dx = 1$. Find a maximum-likelihood estimator for $b > 0$.

4  What is wrong with the following argument for finding the maximum-likelihood estimator for $\theta$ when $f(x; \theta) = 1/\pi[1 + (x - \theta)^2]$?

$$\frac{\partial f}{\partial \theta}(x; \theta) = \frac{-2(x - \theta)}{\pi[1 + (x - \theta)^2]^2}$$

so $\theta = x$.

In exercises 5–10 suppose that $X$ has a density function and that there is exactly one $\theta$ such that $P(X \leq \theta) = \frac{1}{2}$. Consider the estimator $Y$ of $\theta$ as follows: For $n = 2m + 1$, and an independent sample of size $n$ $(X_1, X_2, \ldots, X_n)$ let $Y(s) = X_k(s)$ if there are $m$ values $X_j(s) < X_k(s)$ and $m$ (other) values $X_j(s) > X_k(s)$. See exercise 8.3.5.

5  What is the advantage of $n = 2m + 1$? Show that $Y(s)$ is the middle number when the numbers $X_1(s), X_2(s), \ldots, X_n(s)$ have been put in increasing order.

6  Find $\hat{Y}$ if $n = 5$ and the observations are 2.1, 3.7, 0, 1, $-4.2$.

7  Find $\hat{Y}$ if $n = 9$ and the observations are 2.8, $-2.1$, $-3.6$, $-1.9$, 2.1, $-1.2$, 0.032, $-0.07$, $-2.3$.

**8** Is it true that $\theta = E(X)$? Explain.

**9** Where does the assumption that $X$ has a density function enter? Where does the assumption that there is only one such $\theta$ enter?

**10** Discuss, intuitively, the advantages of $Y$ as an estimator of $\theta$.

**11** We now consider a variant of Example 13.2.4. Suppose that it is in the nature of the manufacturing process that the probability $p$ of an item being defective is a random variable which is constant for any shipment of 4800 of the items. It varies from shipment to shipment with the following probability function

| $p$ IN THE SHIPMENT | 0.005 | 0.01 | 0.015 | 0.02 | 0.04 | 0.06 | 0.08 |
|---|---|---|---|---|---|---|---|
| PROBABILITY OF THAT $p$ | 0.70 | 0.20 | 0.05 | 0.02 | 0.01 | 0.01 | 0.01 |

Thus we know that "most" shipments have $p \leq 0.01$. We might use a different type of "estimator" for $p$ when we have this additional information. We continue to use a sample of size 48. We let $N$ denote the number of defectives found in the sample of size 48. We say that

$P(p = 0.005 \mid N = 0)$
$= (0.995)^{48}(0.70) \div [(0.995)^{48}(0.70) + (0.99)^{48}(0.20) + (0.985)^{48}(0.05) + \cdots]$

(a) Guess at the missing terms in the denominator of $P(p = 0.005 \mid N = 0)$.

(b) What does this remind you of?

(c) Is this a legitimate assertion on the basis of our assumptions?

(d) Compute $P(p = 0.08 \mid N = 0)$ and $P(p = 0.06 \mid N = 4)$.

# 13.3 Quality of estimators

One way to determine which of two estimators is preferable is to subject both to extensive empirical analysis; that is, to use both of them in real-world problems and to choose the one which gives better results. However, extensive work of this sort can be prohibitive both in time and monetary cost. It is desirable to have some theoretical material to assist one in choosing between two estimators even though ultimately one must resort to empirical confirmation.

### Example 13.3.1
Twenty six cards from a standard deck are prepared by an assistant. The object is to estimate $\theta$, the number of red (hearts or diamonds) cards in the prepared twenty six. Use independent sampling of sample size 6 but compare two statistics

as estimators. The first statistic is $26\overline{X} = 26 \cdot \frac{1}{6}$(number of red cards in the sample). The second statistic is defined as follows:

$$Y = \begin{cases} \text{the stage of the first red card in the sample if any red card appears} \\ 13 \text{ if no red card appears} \end{cases}$$

Use the statistic $26/Y$.

Give some special cases and discuss what would happen if each method were used many times and the results averaged.

To get the independent sampling imagine choosing a card, noting its color (R or B), returning it to the 26 cards and reshuffling those cards. Then repeat for a total of six cards. If the list of colors in their order of appearance is RBRRRB we have

$$26\overline{X}(\text{RBRRRB}) = 26 \cdot \frac{4}{6} = 17\tfrac{1}{3} \text{ and } \frac{26}{Y}(\text{RBRRRB}) = \frac{26}{1} = 26$$

Similarly

$$26\overline{X}(\text{BRBRRB}) = 26 \cdot \frac{3}{6} = 13 \text{ and } \frac{26}{Y}(\text{BRBRRB}) = \frac{26}{2} = 13$$

Also,

$$26\overline{X}(\text{BBBRRR}) = 26 \cdot \frac{3}{6} = 13 \text{ and } \frac{26}{Y}(\text{BBBRRR}) = \frac{26}{4} = 6\tfrac{1}{3}$$

Finally,

$$26\overline{X}(\text{RBBBBB}) = 26 \cdot \frac{1}{6} = 4\tfrac{1}{3} \text{ and } \frac{26}{Y}(\text{RBBBBB}) = \frac{26}{1} = 26$$

If we take $N$ samples, where $N$ is a large number we would expect to get about $N \cdot P(s)$ cases of sample $s$. To be more specific, suppose there is one red card only. Then

$$P(\text{BBBBBB}) = \left(\frac{25}{26}\right)^6 \quad P(\text{RBBBBB}) = \frac{1}{26}\left(\frac{25}{26}\right)^5$$

$$P(\text{RRRBBB}) = \left(\frac{1}{26}\right)^3\left(\frac{25}{26}\right)^3$$

(We replace the chosen card so RRRBBB is conceivable!)

Now we expect about $\left(\frac{25}{26}\right)^6 N$ cases of BBBBBB. In these cases $26\overline{X}$ gives an estimate of zero while $26Y$ gives an estimate of two. For the moment, forget that $26/Y$ gives a bad estimate. Imagine taking the arithmetic mean of the estimates given by $26\overline{X}$ in the $N$ samples. That is $(1/N)$[sum of the various estimates] which in good circumstances should be approximately

$(1/N)$[(estimate due to BBBBBB)$NP$(BBBBBB)

$$+ \text{(estimate due to RBBBBR)}NP(\text{RBBBBR})] = \sum_{s \in S} 26\overline{X}(s)P(s)$$

We will not make the same computation for $26/Y$ at this stage.

The object of Example 13.3.1 is to suggest that if we were to repeatedly use an estimator $Z$, under good circumstances the average of the estimates is close to $\sum\limits_{s \in S} Z(s)P(s)$. But (at least in the discrete case) this is $E(Z)$. However, due to cost we do not want to simulate repeated use of an estimator on real-world data to determine the desirability of that estimator. Therefore we just look at $E(Z)$ instead of averaging the values gotten from $Z$ in $N$ repetitions of the sampling procedure. We make a definition.

**Definition**     A statistic $Z$ is said to be an *unbiased estimator* of a parameter $\theta$ if $E_\theta(Z) = \theta$. Naturally we assume that $E_\theta(Z)$ exists for the parameters of interest.

This is our first official criterion for distinguishing between better estimators and poorer estimators. In essence, taking an expectation is like averaging over many ideal trials. We would hope that discrepancies balance out in the averaging. Since we wish to estimate $\theta$, it would seem silly to use a statistic $W$ for which $E_\theta(W) = \theta - 10^{10}$. For that reason we define unbiased estimators this way.

**Theorem 13.3.1**     The moment estimator $m_r$ of (13.2.1) is an unbiased estimator of $E(X^r)$ when it exists. The sample variance

$$d^2 = \frac{1}{n-1} \sum_{j=1}^{n} (X_j - \bar{X})^2$$

is an unbiased estimator of $\sigma^2(X)$.

PROOF

$$E_\theta\left(\frac{X_1^r + X_2^r + \cdots + X_n^r}{n}\right) = \frac{1}{n} E_\theta(\sum_{j=1}^{n} X_j^r)$$

$$= \frac{1}{n} \sum_{j=1}^{n} E_\theta(X_j^r)$$

However, the $X_j$ are identically distributed with $X$ so $E_\theta(X_j^r) = E(X^r)$ for $j = 1, 2, \ldots, n$. Therefore

$$\frac{1}{n} \sum_{j=1}^{n} E_\theta(X_j^r) = \frac{1}{n}[nE(X^r)] = E(X^r)$$

This is the assertion that $m_r$ is an unbiased estimator of $E(X^r)$.

In Theorem 12.2.1 we showed that

$$E_\theta\left(\frac{1}{n-1} \sum_{j=1}^{n} (X_j - \bar{X})^2\right) = \sigma^2(X)$$

That is, $d^2$ is an unbiased estimator of $\sigma^2(X)$.

Note that a maximum-likelihood estimator may or may not be unbiased.

According to (13.2.6) $m_1 = \bar{X}$ is an estimator of $m$ and $\frac{1}{n}\sum_{j=1}^{n}(X_j - \bar{X})^2$ is an estimator for $\sigma^2$. By our definitions both are maximum-likelihood estimators for the normal distribution but the first is unbiased and the second is not.

### Example 13.3.2

Ascertain whether the maximum-likelihood estimator $U = \max(X_1, X_2, \ldots, X_n)$ is an unbiased estimator of $a$ which appears as

$$f_X(x) = \begin{cases} \dfrac{1}{a}, & 0 \le x \le a \\ 0, & \text{otherwise} \end{cases}$$

We compute $E_a(\max(X_1, X_2, \ldots, X_n))$. In exercise 8.3.1 we saw that $F_U(u) = [F_X(u)]^n = \left(\dfrac{u}{a}\right)^n$ if $0 \le u \le a$. The density function $f_U(u) = n\left(\dfrac{u}{a}\right)^{n-1}\dfrac{1}{a}$ if $0 \le u \le a$. Therefore

$$E_a(u) = \int_0^a un\left(\frac{u}{a}\right)^{n-1}\frac{1}{a}\,du$$

$$= \frac{n}{a^n}\int_0^a u^n\,du = \frac{n}{a^n}\frac{u^{n+1}}{n+1}\bigg|_0^a = \frac{na^{n+1}}{a^n(n+1)} = \frac{n}{n+1}a$$

Therefore $U$ is not an unbiased estimator of $a$. Note that $U < a$ with probability 1 so $E_a(U) < a$ is obvious. Note that for large $n$, $E_a(U)$ is very close to $a$.

It could be argued that $E_\theta(Z) = \theta$ is not a strong requirement. After all, you want $Z$ to give values close to $\theta$. Perhaps $E_\theta(Z) = \theta$ because the estimates given by $Z$ which are much larger than $\theta$ are balanced by those which are much smaller than $\theta$ and both of these types occur with high probability. It seems appropriate to require some other criteria.

### Example 13.3.3

A certain class of random variables has variance equal to 10. How large should the sample size be in order that $P_\theta(|\bar{X} - E(X)| \ge 2) \le 0.05$? This gives probability 0.95 or less that your estimate is within 2 of the value of $E(X)$.

This looks like a Chebyshev inequality problem. After all,

$$P_\theta(|\bar{X} - E(X)| \ge 2) = P_\theta(|\bar{X} - E_\theta(\bar{X})| \ge 2) \le \frac{\sigma_\theta^2(\bar{X})}{4}$$

since $E(X) = E_\theta(\bar{X})$.

Furthermore $\sigma_\theta^2(\bar{X}) = \sigma^2(X)/n = 10/n$. We need only find $n$ so that $\frac{1}{4}(10/n) = \frac{1}{20} = 0.05$. This gives $200/4n = 1$ or $50 = n$.

Note that we could have assumed $\sigma^2(X) \le 10$ and $n = 50$ and also obtained our result. If we wish to have $P_\theta(|\bar{X} - E(X)| \ge \frac{1}{2}) \le 0.05$ we could use $n = 800$. If we wish $P_\theta(|\bar{X} - E(X)| \ge 2) \le 0.01$ use $n = 250$.

It should come as no surprise that our good estimators give increasing accuracy for larger and larger $n$. We therefore make another definition.

**Definition**    An estimator $Z$ (which is defined for all $n$) of the parameter $\theta$ is said to be *consistent* if for each $\epsilon > 0$

$$\lim_{n \to \infty} P_\theta(|Z - \theta| \geq \epsilon) = 0$$

This definition is certainly reminiscent of the weak law of large numbers. Furthermore one can prove the consistency of many estimators by using Chebyshev's inequality.

**Definition**    An estimator $Z$ (which is defined for all $n$) of the parameter $\theta$ is said to be *asymptotically unbiased* if $\lim_{n \to \infty} E_\theta(Z) = \theta$.

**Theorem 13.3.2**    Let $Z$ be an asymptotically unbiased estimator of $\theta$ such that $\lim_{n \to \infty} \sigma_\theta^2(Z) = 0$. Then $Z$ is a consistent estimator of $\theta$.

PROOF    By the assumption of asymptotic unbiasedness, for a fixed $\epsilon > 0$ and fixed $\theta$ there is an $n_0$ such that if $n \geq n_0$, $|E_\theta(Z) - \theta| < \epsilon/2$. We show next that if $n \geq n_0$

$$\{s \in S; |Z(s) - \theta| \geq \epsilon\} \subset \{s \in S; |Z(s) - E(Z)| \geq \epsilon/2\}$$

We show that if $s \in \{|Z(s) - \theta| \geq \epsilon\}$, then $s \in \{|Z(s) - E(Z)| \geq \epsilon/2\}$. Let $s \in \{|Z(s) - \theta| \geq \epsilon\}$, $Z(s) - \theta = Z(s) - E(Z) + E_\theta(Z) - \theta$. By the fact that $|a + b| \leq |a| + |b|$,

$$\epsilon \leq |Z(s) - \theta| \leq |Z(s) - E_\theta(Z)| + |E_\theta(Z) - \theta|$$

so $\epsilon - |E_\theta(Z) - \theta| \leq |Z(s) - E_\theta(Z)|$. But for $n \geq n_0$, $\epsilon/2 \geq |E_\theta(Z) - \theta|$ so $-\epsilon/2 \leq -|E_\theta(Z) - \theta|$ and then $\epsilon - \epsilon/2 \leq \epsilon - |E_\theta(Z) - \theta|$ so $\epsilon/2 = \epsilon - \epsilon/2 \leq \epsilon - |E_\theta(Z) - \theta| \leq |Z(s) - E_\theta(Z)|$ and $s \in \{|Z(s) - E_\theta(Z)| \geq \epsilon/2\}$. Therefore for $n \geq n_0$

$$P_\theta(|Z - \theta| \geq \epsilon) \leq P_\theta\left(|Z - E_\theta(Z)| \geq \frac{\epsilon}{2}\right) \leq \frac{\sigma^2(Z)}{(\epsilon/2)^2}$$

But since $\lim_{n \to \infty} \sigma_\theta^2(Z) = 0$ we have $\lim_{n \to \infty} P_\theta(|Z - \theta| \geq \epsilon) = 0$ which says that $Z$ is a consistent estimator of $\theta$.

This theorem reduces the problem of ascertaining the consistency of certain estimators to that of computing expectations and variances. It should be clear from the proof that if $E_\theta(Z) \to \beta \neq \theta$ as $n \to \infty$, and if the variance of $Z$ is reasonable, you will not have a consistent estimator. Furthermore, if the variance of $Z$ is unreasonable you would probably not be interested in $Z$.

Although one may need extremely large values of $n$ to insure that

$$P_\theta(|Z - \theta| \geq \epsilon) \leq \alpha$$

in case $\epsilon$ is very small and $\alpha$ is also very small, and although such large samples are prohibitive in cost, it seems reassuring to use a consistent estimator. Our next criterion for quality of an estimator compares estimators related to samples of the same size.

Suppose that $Z_1$ and $Z_2$ are unbiased estimators of the parameters $\theta$ from some set of parameters. One way to measure the spread of values of the $Z_j$ is to look at their variances. This suggests the following.

**Definition**    For fixed sample size and two statistics $Z_1$ and $Z_2$ with $E_\theta(Z_1) = E_\theta(Z_2) = \theta$, $Z_1$ is said to be a *better* estimator than $Z_2$ if $\sigma_\theta^2(Z_1) \le \sigma_\theta^2(Z_2)$ and strict inequality holds for some $\theta$.

Note that we are comparing only unbiased estimators. We are not choosing to make any comparison between biased estimators. Naturally, we assume that all expectations and variances exist.

This is an official definition of "better." In cases when we wish to make only intuitive judgments of quality of estimators we use the term "preferable."

**Definition**    For fixed sample size, a statistic is said to be the *best* estimator of $\theta$ if $E_\theta(Z) = \theta$ and $Z$ is a better estimator than any other estimator.

Best estimators are called *minimum variance, unbiased estimators*. It can be shown that there cannot be two essentially different unbiased estimators with equal minimum variance in any interesting situation. We use this fact often in our work.

### Example 13.3.4

For independent sampling with $n = 5$ show that $\bar{X}$ is a better estimator of $E(X)$ than is $Z_2 = \frac{1}{5}(X_1 + X_2 + X_3) + \frac{3}{10}X_4 + \frac{1}{10}X_5$.

Clearly $E_\theta(\bar{X}) = E(X)$. Now

$$E_\theta(Z_2) = E_\theta(\tfrac{1}{5}X_1 + X_2 + X_3) + \tfrac{3}{10}X_4 + \tfrac{1}{10}X_5)$$
$$= \tfrac{1}{5}[E_\theta(X_1) + E_\theta(X_2) + E_\theta(X_3)] + \tfrac{3}{10}E_\theta(X_4) + \tfrac{1}{10}E_\theta(X_5)$$

But $E_\theta(X_j) = E(X)$ so

$$E_\theta(Z_2) = \tfrac{1}{5}[3E(X)] + \tfrac{3}{10}E(X) + \tfrac{1}{10}E(X) = E(X)$$

By (12.2.14) $\sigma_\theta^2(\bar{X}) = \frac{1}{5}\sigma^2(X)$. Since $\sigma_\theta^2(X_j) = \sigma^2(X)$, independence guarantees that

$$\sigma_\theta^2(Z_2) = \frac{\sigma^2(X)}{25} + \frac{\sigma^2(X)}{25} + \frac{\sigma^2(X)}{25} + \frac{9}{100}\sigma^2(X) + \frac{1}{100}\sigma^2(X)$$

$$= \frac{12 + 9 + 1}{100}\sigma^2(X) = \frac{22}{100}\sigma^2(X) > \frac{\sigma^2(X)}{5}$$

Since the expectations are the same, the better estimator is the one with the smaller variance. Note that both are unbiased estimators.

**Example 13.3.5**
Show that $2\bar{X}$ is an unbiased estimator of $a$, the uniform distribution parameter of Example 13.3.2. Show that $[(n + 1)/n]U$ (which is unbiased by exercise 13.3.2) is better than $2\bar{X}$. Sampling is independent.

For a random variable $X$ whose density is

$$f(x) = \begin{cases} 1/a, & 0 \leq x \leq a \\ 0, & \text{otherwise} \end{cases}$$

we have

$$E(X) = \int_0^a x \frac{1}{a} \, dx = \frac{1}{a} \int_0^a x \, dx = \frac{1}{a} \frac{x^2}{2} \Big|_0^a = \frac{a^2}{2a} = \frac{a}{2}$$

Therefore

$$E(2\bar{X}) = 2E(\bar{X}) = 2E(X) = 2\left(\frac{a}{2}\right) = a$$

$$E(X^2) = \int_0^a x^2 \frac{1}{a} \, dx = \frac{1}{a} \frac{x^3}{3} \Big|_0^a = \frac{a^2}{3}$$

$$\sigma^2(X) = \frac{a^2}{3} - \left(\frac{a}{2}\right)^2 = a^2(\tfrac{1}{3} - \tfrac{1}{4}) = \frac{a^2}{12}$$

$$\sigma_a^2(2\bar{X}) = 4\sigma_a^2(\bar{X}) = 4\frac{\sigma^2(X)}{n} = \frac{4}{n}\frac{a^2}{12} = \frac{a^2}{3n}$$

$$E_a(U^2) = \int_0^a u^2 f_U(u) \, du = \int_0^a u^2 n \frac{u^{n-1}}{a^n} \, du$$

$$= \frac{n}{a^n} \int_0^a u^{n+1} \, du = \frac{n}{a^n} \frac{a^{n+2}}{n+2} = \frac{n}{n+2} a^2$$

$$\sigma_a^2\left(\frac{n+1}{n} U\right) = E_a\left(\left(\frac{n+1}{n} U\right)^2\right) - \left[E_a\left(\frac{n+1}{n} U\right)\right]^2 = \left(\frac{n+1}{n}\right)^2 E_a(U^2) - a^2$$

$$= \frac{(n+1)^2}{n^2} \frac{n}{n+2} a^2 - a^2 = \left[\frac{(n+1)^2}{n(n+2)} - 1\right] a^2$$

$$= \frac{n^2 + 2n + 1 - n(n+2)}{n(n+2)} a^2 = \frac{1}{n(n+2)} a^2 < \frac{a^2}{3n} = \sigma_a^2(2\bar{X})$$

if $n > 1$

If $n = 1$ the estimators are the same.

**Theorem 13.3.3** (Cramér–Rao Inequality.)    Let $S_0$ be a countable set of real numbers. Let $X(x) = x$ for each $x \in S_0$. Let $f(x; \theta)$ for $x \in S_0$ and $\theta$ in some interval of real numbers be such that

(1)  $f(x; \theta) > 0$,   for each $x \in S_0$ and each $\theta$

(2)  $\sum_{x \in S_0} f(x; \theta) = 1$   for each $\theta$

Then $f(x; \theta)$ determines a probability function for $X$. Suppose that for each $x \in S_0$, $\dfrac{\partial f}{\partial \theta}(x; \theta)$ exists and the series for it is absolutely convergent. Let $Z$ be a statistic on $S$ based on independent sampling of sample size $n$. Suppose that $Z$ is an unbiased estimator of $\theta$ for each $\theta$ in its interval. Then for each $\theta$

**(13.3.1)**    $\sigma_\theta^2(Z) \geq \left\{ n \sum_{x \in S_0} \left[ \dfrac{\partial}{\partial \theta} \log f(x; \theta) \right]^2 f(x; \theta) \right\}^{-1}$

Before proving this theorem we give an example of its use.

### Example 13.3.6

Let $S_0 = \{0, 1\}$ and let $X(x) = x$. Let $f(0; p) = 1 - p$ and $f(1; p) = p$ for $0 < p < 1$. Show that $\overline{X}$ is the best estimator of $p$.

By definition $X(0) = 0$ and $X(1) = 1$ so $E(X) = 0(1 - p) + 1 \cdot p = p$. It is known that $E_p(\overline{X}) = E(X) = p$. Furthermore

$$\sigma_p^2(\overline{X}) = \frac{1}{n^2} \sigma^2(X) = \frac{p(1 - p)}{n}$$

We compute

$$n \sum_{x=0}^{1} \left[ \frac{\partial}{\partial p} \log f(x; p) \right]^2 f(x; p)$$

$$\frac{\partial}{\partial p} \log f(0; p) = \frac{\partial}{\partial p} \log (1 - p) = \frac{-1}{1 - p}$$

$$\frac{\partial}{\partial p} \log f(1, p) = \frac{\partial}{\partial p} \log p = \frac{1}{p}$$

$$n \sum_{x=0}^{1} \left[ \frac{\partial}{\partial p} \log f(x; p) \right]^2 f(x; p) = n \left[ \frac{1}{(1 - p)^2} (1 - p) + \frac{1}{p^2} p \right]$$

$$= n \left( \frac{1}{1 - p} + \frac{1}{p} \right) = n \frac{p + (1 - p)}{(1 - p)p}$$

$$= \frac{n}{p(1 - p)}$$

Now any unbiased estimator of $p$ has a variance greater than or equal to

$$\left\{ n \sum_{x=0}^{1} \left[ \frac{\partial}{\partial p} \log f(x; p) \right]^2 f(x; p) \right\}^{-1} = \left( \frac{n}{p(1-p)} \right)^{-1} = \frac{p(1-p)}{n}$$

But this is precisely the variance of $\bar{X}$. Therefore $\bar{X}$ has the smallest variance of any unbiased estimator of $p$. By the remark after the definition, there cannot be two materially different minimum variance unbiased estimators. Therefore, $\bar{X}$ is the best estimator of $p$.

Example 13.3.6 is typical of a number of cases in which the Cramér–Rao inequality can be exploited to ascertain that an estimator is the best estimator. Admittedly, one must have a statistic to test.

PROOF OF THEOREM 13.3.3     We form

$$L(x_1, x_2, \ldots, x_n; \theta) = f(x_1; \theta)f(x_2; \theta) \cdots f(x_n; \theta)$$

For convenience, we write $x = (x_1, x_2, \ldots, x_n)$. Then on the one hand $L(x, \theta) = P_\theta(X_1 = x_1, \ X_2 = x_2, \ldots, \ X_n = x_n)$ for each $\theta$ and also $L(x, \theta) = P_\theta(x_1, x_2, \ldots, x_n)$ for each $(x_1, x_2, \ldots, x_n) \in S$. Therefore, for each $\theta$

(13.3.2)     $\displaystyle\sum_{x \in S} L(x, \theta) = 1$

Since $Z$ is an unbiased estimator of each $\theta$

(13.3.3)     $\displaystyle\sum_{x \in S} Z(x_1, x_2, \ldots, x_n)L(x_1, x_2, \ldots, x_n; \theta) = \theta$

Now for fixed $x$

$$\frac{\partial}{\partial \theta} \log L(x; \theta) = \frac{\partial}{\partial \theta} \log \prod_{j=1}^{n} f(x_j; \theta)$$

$$= \frac{\partial}{\partial \theta} \sum_{j=1}^{n} \log f(x_j; \theta) = \sum_{j=1}^{n} \frac{\frac{\partial f}{\partial \theta}(x_j; \theta)}{f(x_j; 0)}$$

Let $Y: S_0 \to R$ be defined by

$$Y(x; \theta) = \begin{cases} \dfrac{\frac{\partial f}{\partial \theta}(x; \theta)}{f(x; \theta)}, & f(x; \theta) \neq 0 \\[2mm] 0, & f(x; \theta) = 0 \end{cases}$$

But

$$\frac{\partial}{\partial \theta} \log L(x; \theta) = \frac{\frac{\partial L}{\partial \theta}(x; \theta)}{L(x; \theta)}$$

Therefore,

$$\frac{\partial L}{\partial \theta}(x; \theta) = L(x; \theta) \sum_{j=1}^{n} Y(x_j; \theta)$$

Under reasonable hypotheses we have

$$\frac{\partial}{\partial \theta} \sum_{x \in S} L(x; \theta) = \sum_{x \in S} \frac{\partial}{\partial \theta} L(x; \theta)$$

Therefore, if we differentiate both sides of (13.3.2) with respect to $\theta$ we have

$$\sum_{x \in S} \frac{\partial}{\partial \theta} L(x; \theta) = \frac{\partial}{\partial \theta} 1 = 0$$

or

**(13.3.4)**     $\sum_{x \in S} L(x; \theta) \sum_{j=1}^{n} Y(x_j; \theta) = 0$

Similarly if we differentiate both sides of (13.3.3) and keep in mind that $Z$ does not involve $\theta$ we obtain

**(13.3.5)**     $\displaystyle\sum_{x \in S} Z(x) \frac{\partial L}{\partial \theta}(x; \theta) = \sum_{x \in S} Z(x) L \sum_{j=1}^{n} Y(x_j; \theta) = \frac{\partial}{\partial \theta} \theta = 1$

But since $L$ defines the probability measure $P_\theta$ on $S$ we may interpret (13.3.4) as $E_\theta(\sum_{j=1}^{n} Y_j) = 0$ if we make the standard definition of $Y_j(x) = Y(x_j)$. Similarly (13.3.5) can be interpreted as $E_\theta(Z(\sum_{j=1}^{n} Y_j)) = 1$. Now the $Y_1, Y_2, \ldots, Y_n$ are independent and identically distributed for each $\theta$. Therefore

$$E_\theta(\sum_{j=1}^{n} Y_j) = nE(Y)$$

So if $E_\theta(\sum_{j=1}^{n} Y_j) = 0$, $E(Y) = 0$. Similarly $\sigma_\theta^2(\sum_{j=1}^{n} Y_j) = n\sigma^2(Y)$. Now

$$\text{cov}_\theta(Z, \sum_{j=1}^{n} Y_j) = E_\theta(Z(\sum_{j=1}^{n} Y_j)) - E_\theta(Z)(E_\theta(\sum_{j=1}^{n} Y_j)$$

$$= E_\theta(Z(\sum_{j=1}^{n} Y_j)) - E_\theta(Z) \cdot 0 = 1$$

for each $\theta$. Furthermore by Example 8.7.2

$$1 = \text{cov}_\theta(Z, \sum_{j=1}^{n} Y_j) \le \sigma_\theta(Z)\sigma_\theta(\sum_{j=1}^{n} Y_j)$$

for each $\theta$. Therefore,

$$[\sigma_\theta(\sum_{j=1}^{n} Y_j)]^{-1} \le \sigma_\theta(Z) \quad \text{and} \quad [\sigma_\theta^2(\sum_{j=1}^{n} Y_j)]^{-1} \le \sigma_\theta^2(Z)$$

But since $E_\theta(Y_j) = 0$ and since $\sigma_\theta^2(\sum_{j=1}^{n} Y_j) = n\sigma^2(Y)$ we have, for each $\theta$,

$$\sigma_\theta^2(Z) \ge [n\sigma^2(Y)]^{-1} = [nE(Y^2)]^{-1}$$

$$= \left[ n \sum_{x \in S_0} \left( \frac{\partial}{\partial \theta} \log f(x; \theta) \right)^2 f(x; \theta) \right]^{-1}$$

But that is (13.3.1).

The Cramér–Rao inequality is not restricted to a discrete subset $S_0$. The computations in the proof are valid in case $S_0$ is the real line with $f(x; \theta)$ a suitably smooth mass density on the real line. In that case we have

(13.3.6)   $$\sigma_\theta^2(Z) \ge \left\{ n \int_{-\infty}^{\infty} \left[ \frac{\partial}{\partial \theta} \log f(x; \theta) \right]^2 f(x; \theta)\, d\theta \right\}^{-1}$$

### Example 13.3.7

Suppose $S_0$ is the real line with $X(x) = x$ and with $f(x; m) = (1/\sqrt{2\pi})e^{-(x-m)^2/2}$. Show that $\bar{X}$ is the best estimator.

Since $\sigma^2(X) = 1$, $\sigma_m^2(\bar{X}) = 1/n$. Now

$$\frac{\partial}{\partial m} \left( \log \frac{1}{\sqrt{2\pi}} e^{-(x-m)^2/2} \right) = \frac{\partial}{\partial m} [-\tfrac{1}{2} \log 2\pi - \tfrac{1}{2}(x - m)^2]$$

$$= -\tfrac{1}{2}[2(x - m)(-1)] = (x - m)$$

and

$$\int_{-\infty}^{\infty} (x - m)^2 \frac{1}{\sqrt{2\pi}} e^{-(x-m)^2/2} = \sigma^2(X)$$

since our work on normal random variables has $E(X) = m$ and

$$\int_{-\infty}^{\infty} (x - m)^2 f(x; m)\, dx = \sigma^2(X)$$

Therefore,

$$\left\{ n \int_{-\infty}^{\infty} \left[ \frac{\partial}{\partial \theta} \log f(x; \theta) \right]^2 f(x; \theta)\, dx \right\}^{-1} = (n \cdot 1)^{-1} = \frac{1}{n}$$

We reason again that $\sigma_m^2(\bar{X}) = 1/n$ which is the smallest variance possible among unbiased estimators of $m$. Obviously $\bar{X}$ is an unbiased estimator of $m$.

We cite the result that there cannot be two materially different unbiased estimators both having the minimum variance.

The discrete and continuous cases can be unified if we define a random variable $Y$ (which changes for different $\theta$'s) by

$$Y(x) = \frac{\partial}{\partial \theta} \log f(x; \theta)$$

In this case the sum or integral on the right-hand side of (13.3.1) is just $E(Y^2)$. We then have for each unbiased estimator $Z$ and for each parameter $\theta$

$$\sigma_\theta^2(Z) \geq [nE(Y^2)]^{-1}$$

A bit of algebra gives

$$1 \geq \frac{1}{nE(Y^2)\sigma^2(Z)} > 0$$

Moreover, we have equality to 1 only if $Z$ is the best unbiased estimator.

**Definition**    Let $\theta$ vary over an interval and satisfy the conditions of Theorem 13.3.3. Let $Z$ be an unbiased estimator of $\theta$ based on independent sampling of size $n$. We define the *efficiency* of $Z$, $e(Z)$ by

$$e(Z) = \frac{1}{nE(Y^2)\sigma^2(Z)}$$

## Example 13.3.8

Compute the efficiency of the maximum-likelihood estimator for $\theta = \sigma^2$ when $X$ is $N(0, \theta)$.

We form

$$L(x_1, x_2, \ldots, x_n; \sigma) = \prod_{j=1}^{n} \frac{1}{\sqrt{2\pi\theta}} e^{-x_j^2/2\theta}$$

and compute

$$\frac{\partial \log L}{\partial \theta} = \frac{\partial}{\partial \theta}\left(-\frac{n}{2} \log 2\pi - \frac{n}{2} \log \theta - \frac{1}{2\theta} \sum_{j=1}^{n} x_j^2\right)$$

(13.3.7)    $$= -\frac{n}{2\theta} + \frac{1}{2\theta^2} \sum_{j=1}^{n} x_j^2$$

This is zero precisely if $\theta = \frac{1}{n} \sum_{j=1}^{n} x_j^2$ so $Z(x) = \frac{1}{n} \sum_{j=1}^{n} x_j^2$. Note that if $n = 1$,

(13.3.7) is just $\frac{\partial}{\partial \theta} \log f(x_1; \theta)$. We compute

$$\int_{-\infty}^{\infty} \left(-\frac{1}{2\theta} + \frac{1}{2\theta^2} x_1^2\right)^2 \frac{1}{\sqrt{2\pi\theta}} e^{-x_1^2/2\theta} \, dx_1 = E(Y^2)$$

If we let $x_1 = \theta^{1/2} x$ we obtain

$$\int_{-\infty}^{\infty} \left(-\frac{1}{2\theta} + \frac{1}{2\theta} x^2\right)^2 \frac{1}{\sqrt{2\pi}} e^{-x^2/2}\, dx = \frac{1}{4\theta^2} \int_{-\infty}^{\infty} (-1 + x^2)^2 \frac{1}{\sqrt{2\pi}} e^{-x^2/2}\, dx$$

From our knowledge of the moments of an $N(0, 1)$ random variable

$$E(Y^2) = \frac{1}{4\theta^2} (1 - 2 + 3) = \frac{1}{2\theta^2}$$

We now compute

$$\sigma_\theta^2(Z) = \frac{1}{n^2} \sigma_\theta^2(X_1^2 + X_2^2 + \cdots + X_n^2)$$

$$= \frac{1}{n^2} n\sigma^2(X_1^2)$$

since the $X_j$ are independent and identically distributed. Also,

$$\int_{-\infty}^{\infty} x^4 \frac{1}{\sqrt{2\pi\theta}} e^{-x^2/2\theta}\, dx = \theta^2 \int_{-\infty}^{\infty} y^4 \frac{1}{\sqrt{2\pi}} e^{-y^2/2}\, dy = 3\theta^2$$

which can be seen by substituting $x = \sqrt{\theta}\, y$ and noting that the fourth moment of an $N(0, 1)$ distribution is 3. Now $\int_{-\infty}^{\infty} x^2 \frac{1}{\sqrt{2\pi\theta}} e^{-x^2/2\theta}\, dx = \theta$ since $X$ is $N(0, \theta)$. Therefore

$$\sigma_\theta^2(Z) = \frac{1}{n^2} (3\theta^2 - \theta^2) = \frac{1}{n^2} 2n\theta^2 = \frac{2\theta^2}{n}; \qquad nE(Y^2) = \frac{n}{2\theta^2}$$

It follows that

$$e(Z) = \frac{1}{\dfrac{n}{2\theta^2} \dfrac{2\theta^2}{n}} = 1$$

## Example 13.3.9

Discuss the efficiency of

$$d^2 = \frac{1}{n-1} \sum_{j=1}^{n} (X_j - \bar{X})^2$$

as an estimator of $\theta$ under the same conditions as those in Example 13.3.8.

We know $E_\theta(d^2) = \theta$ from Theorem 13.3.1 so $d^2$ is an unbiased estimator. Now

$$\sigma_\theta^2(s^2) = \frac{m_4}{n} - \frac{1}{n} m_2^2 + \frac{2}{n(n-1)} m_2^2 = \frac{2\theta^2}{n-1}$$

by (12.2.15) and by virtue of $m_4 = 3\theta^2$ and $m_2 = \theta$. Therefore

$$e(d^2) = \frac{1}{\dfrac{n}{2\theta^2}\dfrac{2\theta^2}{n-1}} = \frac{n-1}{n} < 1$$

**Definition**    A statistic $Z$ is said to be an *asymptotically most efficient* estimator for a parameter $\theta$ if it arises from a density function satisfying the hypothesis of Theorem 13.3.3 and if

$$\lim_{n\to\infty} e(Z) = 1$$

Example 13.3.9 displayed an asymptotically most efficient estimator for $\sigma^2$.
 We close with an example of the failure of (13.3.6) because the hypothesis of $f(x;\theta) > 0$ for each $x \in S_0$ is not satisfied.

### Example 13.3.10

Show that $[(n+1)/n]V = [(n+1)/n]\max(X_1, X_2, \ldots, X_n)$ where

$$f(x;a) = \begin{cases} 1/a, & 0 \le x \le a \\ 0, & \text{otherwise} \end{cases}$$

is an unbiased estimator but that (13.3.6) does not apply.
 In Example 13.3.5, we saw that $\sigma_a^2(V) = [1/n(n+2)]a^2$. The Cramér–Rao inequality implies that $\sigma_a^2(Z) \ge 1/nE(Y^2)$ where $E(Y^2)$ is fixed for each fixed parameter value. It is impossible that

$$\frac{1}{n(n+2)}a^2 \ge \frac{1}{nE(Y^2)}$$

for all $n$ because then $a^2 \ge (n+2)/E(Y^2)$.
 Note that we could take $S_0 = \{\text{positive reals}\}$ but there will always be points $x$ in $S_0$ such that $f(x;a) = 0$ for some $a$.

### EXERCISES 13.3

1    Show that the product of moment estimators $m_j m_k$ may not be an unbiased estimator of $E(X^j)E(X^k)$.

◆ 2    Suppose that $\sigma^2(X)$ exists. Show that $a_1 X_1 + a_2 X_2 + \cdots + a_n X_n$ is an unbiased estimator of $E(X)$ if and only if $\sum a_k = 1$. Show that the estimator of this type with smallest variance has $a_k = 1/n$ for $k = 1, 2, \ldots, n$.

3    Show that the estimator $U$ of Example 13.3.2 is a consistent estimator of $a$.

4    If $\theta$ is the fraction of red cards in 26 cards find $E_\theta(26/Y)$ of Example 13.3.1.

Exercises 5–9 deal with a random variable $X$ whose density function is $f(x) = \lambda e^{-\lambda x}$ for $x > 0$.

5  Show that the density function of $X_1 + X_2 + \cdots + X_n$ is $\lambda^n u^{n-1} e^{-\lambda u}/(n-1)!$ for $u > 0$ by using the extended versions of (9.2.2) and Example 9.2.6 on the one hand and the technique of (9.3.4) and (9.3.5) on the other. Recall by exercise 13.2.3 that $1/\bar{X}$ is an estimator for $\lambda$.

6  Use the result of exercise 5 to show that $(n-1)/n\bar{X} = Y$ is an unbiased estimator for $\lambda$.

7  Compute $E(Y^2)$ and $\sigma^2(Y)$ when $n > 2$.

8  Let $Z$ be an unbiased estimator of $\lambda$ based on a sample of size $n$. Show that $\sigma_\lambda^2(Z) \geq \lambda^2/n$.

9  Discuss consistency, efficiency, and asymptotic efficiency of $Y$ of exercise 6 as an estimator of $\lambda$. Note exercises 5–8.

10  Let $X$ have the density function

$$f(x) = \frac{1}{\pi[1 + (x - \theta)^2]}$$

Show that $f(\theta - u) = f(\theta + u)$ and that $P(X \leq u) = \frac{1}{2}$ if and only if $u = \theta$. Show that the statistic $Y$ defined in the material preceding exercise 13.2.5 is an unbiased estimator of $\theta$ when $n = 2m + 1 > 1$. Notice that we are getting an unbiased estimator even though $E(X)$ does not exist.

11  Why does the proof of the Cramér–Rao inequality not extend to Euclidean densities such as the uniform density by merely using integration instead of sums? Example 13.3.10 shows that it cannot extend without some modification, at least.

12  As an employee of a government bureau you are required to estimate the fraction of pennies in a large shipment of mixed coins as follows: Choose a coin, look at it, record penny or not, and then return it to the shipment. You do this until you have recorded $m$ pennies. Let $N$ be the total number of coins which you have recorded. Show that

$$P(N = k) = \binom{k-1}{m-1} p^m (1-p)^{k-m}$$

for $k = m, m+1, m+2, \ldots$, a negative binomial distribution. Note that we are not dealing with a sample of fixed size. Show that $(m-1)/(N-1)$ is an unbiased estimator for $p$.

# 13.4  Confidence intervals

It is more proper to describe the estimators of sections 13.2 and 13.3 as *point estimators*. Using such estimators, one seeks specific numbers as estimates of parameters. In this section we will make a different kind of estimate.

### Example 13.4.1

A population of numbers is modeled by a random variable which is $N(m, 4)$. One obtains the data 3.1842, 3.9883, 3.6521 by a method which can reasonably be thought of as independent sampling. Estimate $m$ and discuss.

It seems inappropriate to use any estimator other than $\bar{X}$. It is unbiased, consistent, and best in the sense of minimum variance. We then get

$$m = \tfrac{1}{3}(3.1842 + 3.9883 + 3.6521) = 3.6082$$

Could 3.6846 be closer to the true value of $m$? Of course. An estimate of this sort can be quite a bit different from the true value. Could 3.2187 be the true value of $m$? Yes, it could. We must recognize that chance fluctuations can lead to bad estimates and 3.6082 may result from an unlikely event when 3.2187 is the true value of $m$.

However, does not the offering of a single number, 3.6082 seem rather arrogant? Surely we do not expect our estimate to be exactly equal to $m$. Perhaps we should be more modest in our claims for our estimates. To this end we recall from 12.2.18 that $\bar{X}$ is $N(m, \tfrac{4}{3})$. Therefore $(\bar{X} - m)/\sqrt{\tfrac{4}{3}}$ is $N(0, 1)$. Up to the accuracy of our tables

$$P\left(-1.65 \leq \frac{\bar{X} - m}{\sqrt{\tfrac{4}{3}}} \leq 1.65\right) = 0.90$$

Now $\sqrt{\tfrac{4}{3}}\,(1.65) = 1.90$ so

$$0.90 = P(-1.90 \leq \bar{X} - m \leq 1.90) = P(1.90 \geq m - \bar{X} \geq -1.90)$$

(13.4.1) $$= P(\bar{X} + 1.90 \geq m \geq \bar{X} - 1.90)$$

To restate this assertion, the probability that the parameter $m$ lies in the interval $[\bar{X} - 1.90, \bar{X} + 1.90]$ is 0.90 up to tabular accuracy. With probability 0.10 the parameter $m$ will not lie in that interval of length 3.80. If we substitute 3.6082 for $\bar{X}$ we can consider the interval from $3.6082 - 1.90 = 1.7082$ to $3.6082 + 1.90 = 5.5082$ as an interval estimate of $m$ and feel fairly certain that $m$ is in that interval.

It should not be construed that

$$P(1.7082 \leq m \leq 5.5082) = 0.90$$

The parameter $m$ does not lead to an event in some sample space upon which there is a probability measure, i.e., $\{s; 1.7082 \leq m \leq 5.5082\}$ is not the definition of any set in any sample space which we have agreed upon so far. It has been our viewpoint that in any specific problem, the relevant parameter or parameters had definite values. Therefore in any specific problem either $1,7082 \leq m \leq 5.5082$ or $1.7082 > m$ or $5.5082 < m$. What, then, is the significance of the 0.90? If we look at things from a frequency viewpoint and imagine using this *procedure* many times and asserting $\hat{m} + 1.90 \geq m \geq \hat{m} - 1.90$ where $\hat{m}$ is the estimate of $m$ from the data via substitution in $\bar{X}$ then about 90% of those assertions will be correct. Moreover, the actual parameter value makes no difference. You could make these assertions in all kinds of $N(m, 4)$ problems with sample size 3 and expect 90% of them to be correct.

### Example 13.4.2

A numerical population is modeled in such a way that the density function is $(1/\sqrt{2\pi}\,\sigma)e^{-x^2/2\sigma^2}$. Data obtained from independent sampling are 1.3002, $-2.7811$, 0.0564, 1.9783. Discuss an interval estimate of $\sigma^2$.

An estimator for $\sigma^2$ in this case is $Z = \frac{1}{4}(X_1^2 + X_2^2 + X_3^2 + X_4^2)$ if we use $X(x) = x$ to convert this to a legitimate random variable problem. We know $X_j$ is $N(0, \sigma^2)$. However we could write $X_j = \sigma Y_j$ where $Y_j$ is $N(0, 1)$. Therefore $Z = \dfrac{\sigma^2}{4}\chi_4^2$ where $\chi_4^2$ is a chi-squared random variable with four degrees of freedom. Furthermore $\dfrac{4}{\sigma^2}Z = \chi_4^2$. For definiteness, we use 0.98 as we just used 0.90. From Table 2 (appendix) we note that $P(\chi_4^2 \le 0.297) = 0.001$ and $P(\chi_4^2 \ge 13.277) = 0.001$. Therefore

$$P\left(0.297 \le \frac{4}{\sigma^2}Z \le 13.277\right) = 0.98$$

This states that

$$P\left(\frac{1}{0.297} \ge \frac{\sigma^2}{4Z} \ge \frac{1}{13.277}\right) = 0.98$$

Now $4\hat{Z} = 13.5052$. Our interval estimate has end points $13.5052/0.297$ and $13.5052/13.277$ or 27.903 and 1.2120 and the interval estimate would be [1.017, 45.472]. If we wanted to use 90% we would have

$$P\left(\frac{1}{0.711} \ge \frac{\sigma^2}{4Z} \ge \frac{1}{9.488}\right) = 0.90$$

Note that we get a large confidence interval because the sample size is small and we ask for a high confidence.

### Example 13.4.3

A numerical population whose density is $f(x; m, \sigma^2) = (1/\sqrt{2\pi\sigma^2})e^{-(x-m)^2/2\sigma^2}$ where $m$ and $\sigma > 0$ are unknown yields data in independent sampling as follows: 3.814, 6.133, 5.311, 4.872, 3.015, 4.362, 5.547. Find an interval estimate for $m$.

We note that

$$\hat{m} = \tfrac{1}{7}(3.814 + 6.133 + 5.311 + 4.872 + 3.015 + 4.362 + 5.547) \doteq 4.722$$

via a standard estimate. Using the estimator $\frac{1}{6} \sum_{K=1}^{7} (X_j - \bar{X})^2$ we obtain $\hat{\sigma}^2 =$ 1.151. In this example we know about $\sigma^2$ only through the data in contrast to our situation in Example 13.4.1 in which $\sigma^2$ was known. There we took advantage of our knowledge of the distribution of $\frac{1}{n} \sum_{j=1}^{n} X_j^2$. We could arbitrarily say that $(\bar{X} - m)/\sqrt{1.151}$ should be approximately $N(0, 1)$. That would constitute allowing $\bar{X}$ to give a bad estimate of $m$ yet assuming that $d^2$ gives a good estimate of $\sigma^2$. This would be unreasonable. Recall from exercise 12.3.2 that if $X_1, X_2, \ldots, X_n$ are independent, identically distributed and $N(m, \sigma^2)$ then

$$\frac{(\bar{X} - m)}{\sqrt{d^2}} \sqrt{n}$$

has Student's $t$-distribution with $n - 1$ degrees of freedom. In this case $n - 1 = 6$.

Let us work at the 90% level for definiteness. Using Table 3, we find $P(\text{Student's } t\text{-random variable} \leq 1.9432) = 0.95$. Therefore $P(-1.9432 \leq \text{Student's } t\text{-random variable} \leq 1.9432) = 0.90$. Therefore

$$P\left(-1.9432 \leq \frac{\bar{X} - m}{\sqrt{d^2}} \sqrt{7} \leq 1.9432\right) = 0.90$$

We may write the inequalities

$$-1.9432 \leq \frac{\bar{X} - m}{\sqrt{d^2}} \sqrt{7} \leq 1.9432$$

as

$$-1.9432 \sqrt{\frac{d^2}{7}} \leq \bar{X} - m \leq 1.9432 \sqrt{\frac{d^2}{7}}$$

or

$$+1.9432 \sqrt{\frac{d^2}{7}} \geq m \geq \bar{X} - 1.9432 \sqrt{\frac{d^2}{7}}$$

which is

$$\bar{X} + 1.9432 \sqrt{\frac{d^2}{7}} \geq m \geq \bar{X} - 1.9432 \sqrt{\frac{d^2}{7}}$$

we take as the interval estimate, the interval from $4.722 - 1.9432 \sqrt{\frac{1.151}{7}}$ to $4.722 + 1.9432 \sqrt{\frac{1.151}{7}}$ or from $4.722 - 0.851$ to $4.722 + 0.851$. That is, our interval estimate is $[3.871, 5.573]$. At the 95% level we would have $[4.722 - 2.4469 \sqrt{\frac{1.151}{7}}, 4.722 + 2.4469 \sqrt{\frac{1.151}{7}}]$.

Note that had we merely taken 1.151 to be $\sigma^2$ and used the method of Example 13.4.1, we would have obtained the interval

$$\left[4.722 - 1.65\sqrt{\frac{1.151}{6}}, \; 4.722 + 1.65\sqrt{\frac{1.151}{6}}\right]$$

which is a shorter interval than

$$\left[4.722 - 1.9432\sqrt{\frac{1.151}{6}}, \; 4.722 + 1.9432\sqrt{\frac{1.151}{6}}\right]$$

We are paying a price for our uncertainty of the variance. This is just, however.

To unify the ideas of these examples we make some definitions.

**Definitions**    Let $Z_1, Z_2$ be statistics on $S$ with $Z_1(s) \le Z_2(s)$ for all $s \in S$. We say that the interval $[Z_1(s), Z(s)]$ is a *confidence interval for a parameter* $\theta$ *of confidence level q or* $(100q\%)$ if

$$\min_\theta P_\theta(Z_1 \le \theta \le Z_2) = q$$

We call an interval $[\hat{Z}_1, \hat{Z}_2]$ a *confidence interval estimate of* $\theta$ *with confidence level q* if $[\hat{Z}_1, \hat{Z}_2]$ are the numbers obtained by substituting data into $Z_1$ and $Z_2$.

As noted before $[Z_1(s), Z_2(s)]$ is a "random interval," whereas $[\hat{Z}_1, \hat{Z}_2]$ is a definite interval. We may say that "$\theta$ is in $[\hat{Z}_1, \hat{Z}_2]$ with confidence $q$" but we will *not* say "with probability $q$" since "probability" has a technical meaning.

In Examples 13.4.1–13.4.3 we allotted equal probability to $P_\theta(Z_1 > \theta)$ and $P_\theta(Z_2 < \theta)$. This is not necessary. Other choices can and should be made when the question requires such choices. Were we to allot probability so that $P_\theta(Z_1 < \theta) = 0$ or $P_\theta(Z_2 > \theta) = 0$ we would have so-called one-sided confidence intervals.

### Example 13.4.4

With the data of Example 13.4.3, find a confidence interval estimate for $\sigma^2$ with confidence 95% where $P_\theta(Z_1 > \theta) = 0$.

We need a right-hand limit only. From Table 2

$$P_\sigma\left(\frac{4Z}{\sigma^2} \ge 0.711\right) = 0.95$$

since $4Z/\sigma^2$ has a $\chi^2$ distribution with four degrees of freedom. Therefore,

$$0.95 = P_\sigma\left(\frac{\sigma^2}{4Z} \le 0.711\right) = P_\sigma(\sigma^2 \le 0.711(4Z))$$

Since $4Z = 13.5052$ we would take the confidence interval estimate to be $[0, 0.711(13.5052)] = [0, 9.602]$.

Another problem which arises is that of finding a confidence interval for the difference $m_1 - m_2$ when we have two populations modeled by an $N(m_1, \sigma^2)$ and $N(m_2, \sigma^2)$. Note that the variances are the same in each case.

### Example 13.4.5

Two classes of students take a test. Class A had been prepared in a certain way for this test while class B had not. The students were chosen from a common pool of 24 students but class A has only eight students whereas class B has 16. The average score of the class A students was 82.3 and the average score of the class B students was 77.6. The computed variance of the scores were 12.37 and 11.34 for classes A and B, respectively. These computed variances are

$$\frac{1}{8} \sum_{j=1}^{8} (x_j - \bar{x})^2, \frac{1}{16} \sum_{j=1}^{16} (y_j - \bar{y})^2$$ for class A and class B, respectively. If we

assume that the class A scores can be thought of as $N(m_1, \sigma^2)$ and those from class B as $N(m_2, \sigma^2)$ and if different students are independent, find 95% confidence interval estimate for $m_1 - m_2$.

If we use $X_1, X_2 \ldots, X_8$ to model class A and $Y_1, Y_2, \ldots, Y_{16}$ to model class B we have that

$$\frac{(\bar{X} - m_1) - (\bar{Y} - m_2)}{\sigma(\bar{X} - \bar{Y})}$$

if $N(0, 1)$. Furthermore $\frac{1}{\sigma^2} \sum_{j=1}^{8} (X_j - \bar{X})^2$ has a $\chi^2$ distribution with seven

degrees of freedom and $\frac{1}{\sigma^2} \sum_{j=1}^{16} (Y_j - \bar{Y})^2$ has a $\chi^2$ distribution with 15 degrees

of freedom by corollary 12.3.2 to Theorem 12.3.1 with a slight modification.

Therefore $\frac{1}{\sigma^2} \sum_{j=1}^{8} (X_j - \bar{X})^2 + \frac{1}{\sigma^2} \sum_{j=1}^{16} (Y_j - \bar{Y})^2$ is $\chi_{22}^2$ since the $X$'s and $Y$'s are

independent. Now $\sigma^2(\bar{X} - \bar{Y}) = \frac{\sigma^2}{8} + \frac{\sigma^2}{16}$. Therefore by section 12.3

$$Z = \frac{(\bar{X} - m_1) - (\bar{Y} - m_2)}{\left(\frac{\sigma^2}{8} + \frac{\sigma^2}{16}\right)^{1/2}} \sqrt{22} \frac{1}{\left[\frac{1}{\sigma^2} \sum_{j=1}^{8} (X_j - \bar{X})^2 + \frac{1}{\sigma^2} \sum_{j=1}^{16} (Y_j - \bar{Y})^2\right]^{1/2}}$$

has a Student's $t$-distribution with 22 degrees of freedom.

If $W$ is such a Student's $t$ random variable, then Table 3 gives

$$P(-2.074 \le W \le 2.074) = 0.95$$

We let

$$V = \sqrt{22} \left(\frac{\sigma^2}{8} + \frac{\sigma^2}{16}\right)^{-1/2} \left[\frac{1}{\sigma^2} \sum_{j=1}^{8} (X_j - \bar{X})^2 + \frac{1}{\sigma^2} \sum_{j=1}^{16} (Y_j - \bar{Y})^2\right]^{-1/2}$$

This can be simplified to

$$V = \sqrt{22} \left(\frac{3}{16}\right)^{-1/2} \left[\sum_{j=1}^{8} (X_j - \bar{X})^2 + \sum_{j=1}^{16} (Y_j - \bar{Y})^2\right]^{-1/2}$$

Therefore since $Z = [\bar{X} - \bar{Y} - (m_1 - m_2)]V$,

$$P_\theta(-2.074 \leq Z \leq 2.074) = P_\theta\left(-\frac{2.074}{V} \leq \bar{X} - \bar{Y} - (m_1 - m_2) \leq \frac{2.074}{V}\right)$$

$$= P_\theta\left(+\frac{2.074}{V} \geq (m_1 - m_2) - (\bar{X} - \bar{Y}) \geq -\frac{2.074}{V}\right)$$

where we have changed our way of describing the event by multiplying the inequalities by $-1$ and changing the senses of the inequality signs. We may take $Z_1 = \bar{X} - \bar{Y} - 2.074/V$ and $Z_2 = \bar{X} - \bar{Y} + 2.074/V$. Then

$$\hat{Z}_1 = (82.3) - (77.6) - \frac{2.074}{\sqrt{22}} \sqrt{\frac{3}{16}} [8(12.37) + 16(11.34)]^{1/2} = 4.7 - 3.2$$

and

$$\hat{Z}_2 = (82.3 - 77.6) - \frac{2.074}{\sqrt{22}} \sqrt{\frac{3}{16}} [8(12.37) + 16(11.34)]^{1/2} \approx 4.7 + 3.2$$

Thus [1.5, 7.9] is the 0.95 confidence interval estimate. One might confidently state that between 1.5 and 7.9 points on the test were added onto the typical student by preparation.

It should be clear by now that one chooses the confidence level to suit himself or to suit others to whom the problem is important. Generally it is chosen by convention and may vary among the different fields to which statistics applies. Certainly, the composition of the statistical tables is relevant to the confidence level chosen. (It would be foolish to use a confidence level of 92.8146.) Generally, confidence levels of 0.90, 0.95, 0.98, and 0.99 are used.

## EXERCISES 13.4

1   Simulate the finding of confidence intervals for $p = P(\text{red})$ as follows: Prepare a deck of cards by setting aside 12 cards without looking at them or have a partner do this. Draw a card from the reduced deck of 40 cards and note its color. Replace it. Reshuffle the 40 cards. Repeat nine more times. Let $k$ be the number of red cards which you found. Look at the 40 cards. Let $R$ be the number of red cards among them. Note whether $R/40$ is in the interval $[k/10 - 1/8, k/10 + 1/8]$. Repeat the procedure several times. Compare your results with those of others doing the same experiment.

**2** Let $X$ be $N(m, D^2)$ where we consider $D$ to be definitely known. Except for the accuracy of the table used show that $[Z_1, Z_2]$ is a 90% confidence interval for $m$ in the following cases.

(a) $Z_1 = \bar{X} - 1.655 D/\sqrt{n}, \qquad Z_2 = \bar{X} + 1.655 D/\sqrt{n}$

(b) $Z_1 = \bar{X} - 1.41 D/\sqrt{n}, \qquad Z_2 = \bar{X} + 2.04 D/\sqrt{n}$

(c) $Z_1 = \bar{X} - 1.88 D/\sqrt{n}, \qquad Z_2 = \bar{X} + 1.477 D/\sqrt{n}$

**3** Under the conditions of exercise 2 show that 90% confidence intervals for $m$ of the form $Z_1 = \bar{X} - \xi, \quad Z_2 = \bar{X} + \eta$ are of smallest length when $\xi = \eta$.

**4** Obtain a 98% confidence interval for $\sigma^2$ of Example 13.4.2 which is shorter than the one given in the example.

**5** Let $X$ have the uniform distribution on $\{x; 0 \le x \le a\}$. Recall that $E(X) = a/2 = $ median $X$. For a sample of size 5 let $Y = $ median$(X_1, X_2, X_3, X_4, X_5)$ and let $\bar{X}$ have its usual meaning. Discuss how to find confidence intervals for $a/2$ if

(a) $Z_1 = Y/(1 + \delta), \qquad Z_2 = Y/1 - \delta$

(b) $Z_1 = \bar{X}/2(1 + \epsilon), \qquad Z_2 = \bar{X}/(1 - E)$

**6** Work Example 13.4.5 if the average scores for classes A and B are the same but the computed variances are 28.09 and 31.36 for classes A and B, respectively.

**7** What would the confidence interval be in Example 13.4.5 if there were 16 students in class A, the average score remained 82.3, and the computed variance corresponding to $\frac{1}{16} \sum_{k=1}^{16} (X_j - \bar{X})^2$ were 12.32? Everything about class $B$ is the same as in the example.

# CHAPTER 14

# Hypothesis Testing

## 14.1 Introduction

A number of examples in section 12.1 suggest ideas which do not fall under the heading of estimation but which seem like prime material for some type of statistical treatment. In a sense, some of the ideas are more basic than estimation. We may summarize topics of this type by questions as follows:

(a)   Is a certain probability measure appropriate for a population?

(b)   Which of a number of probability measures is appropriate for a population?

(c)   Are two characteristics of a population independent?

(d)   Are certain characteristics of two different populations statistically the same?

Frequently these questions are mathematical versions of extremely important problems. Moreover, they are related to problems of estimation. However, they are phrased in such a fashion that one can more easily study how well he is solving these problems. This is best seen in some examples.

### Example 14.4.1
A manufacturer of canned goods chooses to price his goods lower if there is a high percentage of damaged cans in a shipment. For convenience he assumes 5% versus 15% to represent an acceptable damage rate versus an unacceptable rate (to be priced lower). Furthermore, he does not wish to inspect all cans in a

shipment in order to assess the damage. Discuss an inspection procedure consisting of inspecting 100 randomly chosen cans from a shipment so large that we can assume independent sampling as a good approximation.

Let $p$ be the fraction of damaged cans in the shipment. With the assumption of independence

$$P(k \text{ damaged cans in } 100) = \binom{100}{k} p^k (1 - p)^{100-k}$$

Table 1 is an abridged table giving approximate values obtained from using a Poisson approximation. We use the missing entries as well as the entries in the table in further computation.

**Table 1**

| $k$ | 4 | 6 | 8 | 9 | 10 | 12 | 14 | 16 |
|---|---|---|---|---|---|---|---|---|
| $\binom{100}{k}(0.05)^k(0.95)^{100-k}$ | 0.175 | 0.146 | 0.065 | 0.036 | 0.018 | 0.003 | 0.001 | 0.000 |
| $\binom{100}{k}(0.15)^k(0.85)^{100-k}$ | 0.001 | 0.010 | 0.019 | 0.033 | 0.048 | 0.083 | 0.103 | 0.098 |

The percentage error resulting from the Poisson approximation is on the order of 10% or less.

On the basis of the entries in Table 1, the manufacturer may decide upon a cutoff level $k_0$ such that if $k_0$ or more damaged cans appear in the 100 cans which are sampled, the lower price will be charged. If fewer than $k_0$ damaged cans appear, the regular price will be charged for the shipment.

Let us discuss some consequences of certain values of $k_0$. If $k_0 = 12$, the manufacturer is protecting himself well. He will rarely (probability about 0.005) find more than 11 damaged cans in a shipment where $p = 0.05$. Thus he will rarely lower the price on a good lot. However, it seems quite clear that he will frequently be charging the regular price for a bad lot ($p = 0.15$). For example, if we take into account the missing entries $k = 11$ and 7, $P(11$ or fewer damaged cans show when $p = 0.15) \approx 0.18$. This may result in a fair amount of customer dissatisfaction. The manufacturer may give rebates but dissatisfaction of one sort or another may remain. If $k_0 = 8$, he will more frequently (probability about 0.07) be lowering the price of a good shipment. However, he will rarely be overpricing a bad shipment. He can choose between these or use other cutoff levels $k_0$. He must decide and take his chances.

He may also use some other method of inspecting and pricing. If he plans to inspect more cans, he must pay the additional inspection costs.

## Example 14.1.2

A fertilizer used for a particular crop is known to result in an average yield per acre of 500 bushels. To test other fertilizers, 20 randomly chosen plots are fertilized and that crop is grown. Discuss a procedure for determining whether the newly tested fertilizers are better.

Yield per acre is not the only criterion for effectiveness of a fertilizer, but we will focus on that aspect.

There is good evidence to use a normal random variable $X$ as a model for the yield per acre of a randomly chosen plot that has been fertilized with a particular fertilizer. For a fertilizer to be better we would want $m = E(X) > 500$. We note that $\bar{X}$ models the average yield per acre if $X_1, X_2, \ldots, X_{20}$ model the yield per acre of the experimental plots. By section 12.3, if $d^2$ is as usual, then $Z = [(\bar{X} - m)/\sqrt{d^2}]\sqrt{20}$ has a Student's $t$-distribution with 19 degrees of freedom. From Table 3 (appendix) $P(Z < 1.73) = 0.95$. That is,

$$0.95 = P\left(\frac{\bar{X} - m}{\sqrt{d^2}}\sqrt{20} \le 1.73\right) = P\left(\bar{X} - \frac{1.73}{\sqrt{20}}d \le m\right)$$

In order to establish a procedure, if $x_1, x_2, \ldots, x_{20}$ are the observed yields of the 20 experimental plots and $\bar{x} = (x_1 + \cdots + x_{20})/20$, then we accept the view that the fertilizer being tested is better than the standard fertilizer if

$$\bar{x} - \frac{1.73}{\sqrt{20 \cdot 19}}\left[\sum_{j=1}^{20} (x_j - \bar{x})^2\right]^{1/2} > 500$$

If we were to carry out this procedure often with a fertilizer whose $m = 500$ we would expect to be wrong in about 5% of our tests, that is about that often $\bar{X} - (1.73/\sqrt{20})d > 500$.

We can calculate $P(\bar{X} - 1.73(20)^{-1/2} \le 500)$ when $X$ is $N(m, \sigma^2)$, but only when we know $m = 500$ can we use the $t$-distribution. In other cases the computation is complicated. However, arbitrarily good approximations can be obtained. If $W = (1/\sqrt{20})d^2$ and $m > 500$, $P(\bar{X} - 1.73W < 500)$ represents the probability of a better fertilizer showing up no better than the original fertilizer due to chance. Similarly $P(\bar{X} - 1.73W > 500)$ in the case $m < 500$ would represent the probability of a poorer fertilizer showing up as a better fertilizer due to chance.

Finally, if we do not want to get involved with fertilizers which are only slightly better we may work according to

$$\bar{x} - \frac{1.73}{\sqrt{20 \cdot 19}}\left[\sum_{j=1}^{20} (x_j - \bar{x})^2\right]^{1/2} > 500 + \epsilon$$

where $\epsilon$ is chosen according to our needs.

These examples do not necessarily represent the best way to solve problems of this sort. They do suggest a mode of approach and they show that the numerical aspects of the risks and gains can be ascertained within the framework of statistics. However, no amount of mathematics will determine whether a particular customer will be incensed at being overcharged for a 15% damage rate as in Example 14.1.1.

## Example 14.1.3

A die is to be rolled 600 times to determine whether it is biased. Set up a scheme to decide this matter.

An unbiased die would have $P(j$th face shows$) = \frac{1}{6}$. Let $n_j =$ number of times in the 600 rolls that face $j$ shows. Let $X_1, X_2, \ldots, X_{600}$ be independent random variables with $P(X_k = j) = \frac{1}{6}$ for $j = 1, 2, 3, 4, 5, 6; k = 1, 2, \ldots, 600$. These model the rolls of the die. The statistic $N_j(s) = [$number of $X_1(s),$ $X_2(s), \ldots, X_{600}(s)$ which equal $j]$ models $n_j$.

For an unbiased die Theorem 12.3.2 implies that the distribution of $\sum_{j=1}^{6} \frac{(N_j - 100)^2}{100}$ is approximately that of a $\chi^2$ random variable with five degrees of freedom. We could then compute $\sum_{j=1}^{6} \frac{(n_j - 100)^2}{100}$ and if it is greater than a fixed quantity $u_0$, we would assume that the die is biased. The number $u_0$ can be chosen as follows: Suppose you are willing to allow about 10% of your cases in which a fair die is judged to be biased. Then let $u_0$ be such that

$$P(\chi^2 \text{ random variable with five degrees of freedom} \leq u_0) = 0.90$$

We can therefore ascertain the probability of making a certain type of error. Another type of error is that in which a biased die is judged to be unbiased. If we specify, for example $P(X_j = 1) = \frac{1}{5}$, $P(X_j = 2) = \frac{1}{5}$, $P(X_j = 3) = \frac{1}{5}$, $P(X_j = 4) = \frac{2}{15}$, $P(X_j = 5) = \frac{2}{15}$, $P(X_j = 6) = \frac{2}{15}$, or any other specific state of biasedness we can calculate

$$P\left(\sum_{j=1}^{6} \frac{(N_j - 100)^2}{100} < u_0\right)$$

although this might be a complicated computation. It can be approximated by means of the noncentral $\chi^2$ distribution.

We now describe a pattern which includes these examples as special cases and which describes a good many testing and judgment problems. Generally one will have a judgment or hypothesis toward which he is well disposed, perhaps that the shipment of canned goods is satisfactory, the die is unbiased, or the fertilizer is better. We call this hypothesis $H_0$, the *null* hypothesis. In contrast to this is $H_1$, the so-called *alternative* hypothesis. In a sense $H_1$ is what you would rather not have happen (the shipment of canned goods is badly damaged, the die is biased, or the fertilizer is not better). To say that you would rather not have $H_1$ happen is to make a nonmathematical value judgment. Admittedly, what you wish to happen depends on your situation. The storeowner would like to get canned goods at a lower price—especially if they are good but are judged to be bad by sampling fluctuations. However, although there is no logical difference between $H_0$ and $H_1$, they are treated differently for reasons of mathematical convenience. This should be looked for in the development to come.

Associated with $H_0$ are one or more probability measures for $S_0$, the population model. These are the probability measures under which we consider $H_0$ to be correct. As a matter of fact, we set $H_0 = \{$those probability measures pertaining to the phrasing of $H_0\}$. Similarly, $H_1 = \{$those probability measures pertaining to the phrasing of $H_1\}$. Naturally, $H_0 \cap H_1 = \varnothing$ but we do not assume that $H_0 \cup H_1 = \{$all probability measures on $S_0\}$. Frequently it is just parameters which are at issue so we may think of $H_0$ and $H_1$ as being the sets of appropriate parameters.

A hypothesis is called *simple* if it contains precisely one element. Otherwise it is *composite*.

### Example 14.1.4

Discuss reasonable choices of $H_0$ and $H_1$ for Examples 14.1.1, 14.1.2, and 14.1.3.

Considering how Example 14.1.1 was formulated, we could let $H_0 = \{0.05\}$ and $H_1 = \{0.15\}$. Both hypotheses are simple.

In Example 14.1.2 we could take $H_0 = \{(m, \sigma^2); m > 500\}$, $H_1 = \{(500, D^2)\}$ where $D^2$ is the variance of yields resulting from the fertilizer commonly used. $H_0$ is composite; $H_1$ is simple.

In Example 14.1.3, $H_0 = \{$the die is unbiased$\}$, $H_1 = \{$the die is biased$\}$. $H_0$ is simple; $H_1$ is composite.

Our method of testing hypotheses can be described as follows: For a particular method of sampling and for a particular sample size $n$ find a suitable statistic $Z$ and reject $H_0$ (accept $H_1$) if $Z(s) \in A$, a suitable set of real numbers. If $Z(s) \notin A$ accept $H_0$ (reject $H_1$).

**Definition**   Let $H_0, H_1$ be the null hypothesis and alternative hypothesis. Let $S$ be the sample space of the sample and let $Z$ be a statistic. Let $A$ be the set of real numbers whereby $Z(s) \in A$ results in the rejection of $H_0$. The set $C = \{s \in S; Z(s) \in A\}$ is called the *critical region* of the test under discussion. The rejection of $H_0$ when it is true is called an *error of type* 1. The acceptance of $H_0$ when it is false is called an *error of type* 2.

It should be clear that in many interesting situations we must run the risk of making a type-1 or type-2 error. However, we would wish that the probability of such errors is small.

### Example 14.1.5

Describe some of the sample points in the critical region of a test relating to Example 14.1.1 if $H_0 = \{0.05\}$, $H_1 = \{0.15\}$, $Z(s) =$ number of damaged cans in $s$, and $A = \{10, 11, 12, \ldots, 100\}$.

Since $C$ is the set in $S$ in which $Z(s) \in A$ we see immediately that any sample point $s \in S$ which indexes an instance of ten or more damaged cans is in $C$. Thus if we let $D$ denote damage and $N$ denote no damage, $(DNDDDDDDDDDDNNN \cdots N) \in C$. Note that $C \subset S$ while $A$ is a set of real numbers.

### Example 14.1.6

In reference to Example 14.1.1, suppose a critical region $C$ for testing whether the shipment is good or not is as follows: $s \in C$ if and only if $s$ represents no damaged cans in the first ten inspected, and at most one damaged can in the next 20 inspected. Find the probability of an error of type 1.

We want to find the probability of $C$ under the null hypothesis.

$$P(C) = 1 - P(C')$$

$$C' = \{s; s = (\underbrace{NN \cdots N}_{10} \underbrace{\cdots}_{1 \, D \, in \, 20} )\}$$

By independence

$$P(C') = (0.95)^{10}\binom{20}{1}(0.05)(0.95)^{19}$$

$$= 20(0.95)^{29}(0.05) = (0.95)^{29}$$

Therefore,

$$P(C) = 1 - (0.95)^{29}$$

Note that there is no statistic being used in Example 14.1.6. All we really need, in order to perform a test of hypotheses, is a critical region. If the data are indexed by a sample point in the critical region, $H_0$ is rejected. Otherwise it is accepted.

We are actually dealing with two or more probability measures $P_0$ on $S_0$ and with the corresponding probability measures $P$ which they induce on $S$ when the mode of sampling is taken into account. Normally we use independent sampling but other modes could be used.

Since a test is essentially a critical region $C \subset S$ we could begin to describe the probabilistic aspects of the test by giving the probability of $C$ under various probability measures on $S$. Of greatest interest is $q$, a function from the probability measures $P_0$ and $S_0$ to the real numbers given by

**(14.1.1)** $\quad q(P_0) = P(C)$

where $P$ is the probability measure on $S$ induced by $P_0$ on $S_0$.

Frequently, we index our probability measures by a parameter $\theta$ and ignore all other probability measures. In this case we write (14.1.1) as

**(14.1.2)** $\qquad q(\theta) = P_\theta(C) =$ probability of $C$ under the probability measure whose parameter is $\theta$

**Definition** The function $q$ defined in (14.1.1) or (14.1.2) is called the *power* (function) of the test $C$. The function $q^* = 1 - q$ is called the *operating characteristic* (O.C.) function.

We plot a typical O.C. function $q^*$ in the case

$$S = \{(\xi_1, \xi_2, \ldots, \xi_8); \xi_i = \text{H or T}\} \text{ and } C = \{s \in S; s \text{ has five or more H's}\}$$

If $p$ is the probability of heads

$$q^*(p) = 1 - q(p) = 1 - \sum_{k=5}^{8} \binom{8}{k} p^k (1-p)^{8-k}$$

$$= \sum_{k=0}^{4} \binom{8}{k} p^k (1-p)^{8-k}$$

This is plotted in Figure 1.

We note that $q^*(p)$ is the probability of accepting $H_0$ when $p \in H_0$. If $p \in H_1$, $q^*(p)$ is the probability of rejecting $H_1$.

A good test should have $q^*(p)$ close to 1 if $p \in H_0$ and $q^*(p)$ close to zero if $p \in H_1$.

Also plotted in Figure 1 is

$$q_1^*(p) = 1 - \sum_{k=7}^{8} \binom{8}{k} p^k (1-p)^{8-k} = \sum_{k=0}^{6} \binom{8}{k} p^k (1-p)^{8-k}$$

which applies when the critical region $C_1 = \{s \in S; s \text{ has seven or more H's}\}$. Notice that $q_1^*(p)$ will give values closer to 1 than will $q^*(p)$.

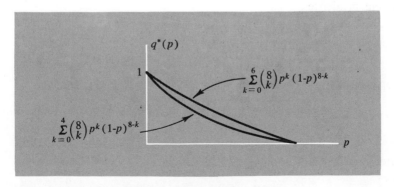

**Figure 1**

We have pointed out that a test is really completely determined by its critical region. Namely, the critical region $C$ is precisely that portion of the sample space of the sample such that, if a list of data $s \in C$, $H_0$ is rejected. If $s \notin C$, $H_0$ is accepted. Therefore, the statistic $Z$ is not the crux of a statistical test; the critical region is.

We now consider the basic case in which $H_0$ and $H_1$ are simple hypotheses. Suppose that $f$ and $g$ are density functions on $S$ which pertain to $H_0$ and $H_1$, respectively. That is, when $H_0$ is true, the probability measure on $S$, the sample space of the sample, is given by a density function $f$. Similarly, when $H_1$ is true, the probability measure on $S$ is given by a density function $g$.

If $c$ is any positive constant we let

**(14.1.3)**    $C_c = \{s \in S; g(s) \geq cf(s)\}$

Note that $f$ and $g$ are likelihood functions. The tests corresponding to (14.1.3) are called *likelihood ratio tests* since we can think of $g(s) \geq cf(s)$ as $g(s)/f(s) \geq c$.

### Example 14.1.7

What is the nature of the regions $C_c$, which one obtains in testing the lifetime of semiconductors when a sample of size 2 is used and when the null hypothesis has the density of lifetimes as $\frac{1}{6}e^{-x/6}$ while the alternative hypothesis has the density of lifetimes as $\frac{1}{2}e^{-x/2}$? Find $C_c$ so that the probability of an error of type 1 is 0.10.

We assume independence. Accordingly, with $S_0 = \{x; x > 0\}$ we can think of $S$ as the first quadrant in 2-dimensional space. Before examining the problem in detail let us think intuitively. As we know, an exponential density $\lambda e^{-\lambda x}$ has expected value $1/\lambda$. We are choosing between $H_0$ with expected life $\frac{1}{\frac{1}{6}} = 6$ and $H_1$ whose expected life is $\frac{1}{\frac{1}{2}} = 2$. Therefore, we would expect a good critical region to have two points $(x_1, x_2)$ where both are fairly small rather than fairly large since we wish to reject $H_0$ if the lifetimes are small.

By independence

$$f(x_1, x_2) = \tfrac{1}{6}e^{-x_1/6}\tfrac{1}{6}e^{-x_2/6} = \tfrac{1}{36}e^{-(x_1+x_2)/6}$$

and

$$g(x_1, x_2) = \tfrac{1}{2}e^{-x_1/2}\tfrac{1}{2}e^{-x_2/2} = \tfrac{1}{4}e^{-(x_1+x_2)/2}$$

For $g(x_1, x_2) \geq cf(x_1, x_2)$ it is necessary and sufficient that

$$\log g(x_1, x_2) \geq \log c + \log f(x_1, x_2)$$

or

$$-\log 4 - \tfrac{1}{2}(x_1 + x_2) \geq \log c - \log 36 - \tfrac{1}{6}(x_1 + x_2)$$

or, after some simplification,

$$x_1 + x_2 \leq 3(\log 36 - \log 4 - \log c) = 3 \log \frac{9}{c}$$

Therefore, $C_c = \left\{(x_1, x_2); x_1 + x_2 < 3 \log \dfrac{9}{c}\right\}$. Such a region is shaded in Figure 2. The region is an isosceles triangle. If $c > 9$ it is the empty set because no sample points in $S$ satisfy $x_1 + x_2 < 0$.

An error of type 1 is a rejection of $H_0$ when it is true. We would then want the probability of $C_c$ under the density $f$. That is,

$$\int_0^{3 \log 9/c} dx_1 \int_0^{3 \log 9/c - x_1} \frac{1}{36} e^{-x_1/6} e^{-x_2/6} \, dx_2$$

It can be seen that

**(14.1.4)**   $\displaystyle\int_0^k dx_1 \int_0^{k-x_1} a^2 e^{-a(x_1 + x_2)} \, dx_2 = 1 - e^{-ak} - ake^{-ak}$

With $a = \frac{1}{6}$, $k = 3 \log (9/c)$, and $b = ak$, we want $0.1 = 1 - e^{-ak} - ake^{-ak}$ or $e^{-b} + be^{-b} = 0.9$. Tables of the exponential function give $b \approx 0.53$. The critical region $C_* = \{(x_1, x_2); x_1 + x_2 \le 3.18\}$ is what we seek since

$$3 \log (9/c) = 6b = 6(0.53)$$

within our approximation. We do not reject $H_0$ unless $x_1$ and $x_2$ are quite small.

It appears as though we frequently accept $H_0$ when the lifetimes are small. For example if $x_1 = 1.2$ and $x_2 = 2.3$ we accept $H_0$. To appreciate this more fully we calculate the probability of an error of type 2 for $C_*$. This is

**(14.1.5)**   $\displaystyle 1 - \int\int_{C_*} \frac{1}{4} e^{-(x_1 + x_2)/2} \, dx_1 \, dx_2 = P(C_*')$

under $g$. By (14.1.4) with $a = \frac{1}{2}$ and $k = 3.18$, (14.1.5) is

$$1 - (1 - e^{-1.56} - 1.56e^{-1.56}) \approx 0.52$$

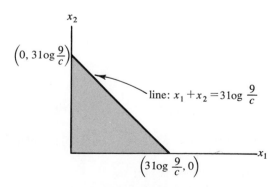

**Figure 2**

This is a large probability of an error of type 2. On the other hand, the sample size is only 2. It would be a miracle if one could get good discrimination between $H_0$ and $H_1$ with such a small sample.

A critical region of the form (14.1.3) makes sense when $f$ and $g$ are probability functions as well as when they are density functions. Certain difficulties arise, however.

### Example 14.1.8

Regarding a biased coin suppose $H_0$ states that $P(\text{H}) = 0.4$ and $H_1$ states that $P(\text{H}) = 0.2$. Set up a test based on a sample of size 20 whereby the probability of an error of type 1 is 0.05.

For convenience we may let $S_0 = \{0, 1\}$ wherein 0 corresponds to T and 1 to H. Then, assuming independent sampling

$$f(x_1, x_2, \ldots, x_{20}) = \left(\frac{2}{5}\right)^{x_1 + x_2 + \cdots + x_{20}} \left(\frac{3}{5}\right)^{20 - (x_1 + x_2 + \cdots + x_{20})}$$

and

$$g(x_1, x_2, \ldots, x_{20}) = \left(\frac{1}{5}\right)^{x_1 + x_2 + \cdots + x_{20}} \left(\frac{4}{5}\right)^{20 - (x_1 + x_2 + \cdots + x_{20})}$$

since $x_1 + x_2 + \cdots + x_{20} =$ [number of 1's (heads)] and $20 -$ (number of 1's) $=$ [number of 0's (tails)]. Then, if we set $\nu = x_1 + x_2 + \cdots + x_{20}$ $g(x_1, x_2, \ldots, x_{20}) \geq cf(x_1, x_2, \ldots, x_{20})$ means $(\frac{1}{5})^\nu (\frac{4}{5})^{20-\nu} \geq c(\frac{2}{5})^\nu (\frac{3}{5})^{20-\nu}$ or

$$\left(\tfrac{1}{5}\right)^\nu \left(\tfrac{4}{5}\right)^{-\nu} \left(\tfrac{3}{5}\right)^\nu \left(\tfrac{2}{5}\right)^{-\nu} > c\left(\tfrac{3}{5}\right)^{20} \left(\tfrac{4}{5}\right)^{-20}$$

or $(\frac{1}{5} \cdot \frac{5}{4} \cdot \frac{3}{5} \cdot \frac{5}{2})^\nu \geq c(\frac{3}{4})^{20}$ which is $(\frac{3}{8})^\nu \geq c(\frac{3}{4})^{20}$ or $\nu \leq b$ since $\frac{3}{8} < 1$.

How do we know that there is a $b$ such that

$$\sum_{0 \leq k \leq b} \binom{20}{k} \left(\frac{2}{5}\right)^k \left(\frac{3}{5}\right)^{20-k} = 0.05$$

As a matter of fact,

$$\sum_{k=0}^{3} \binom{20}{k} \left(\frac{2}{5}\right)^k \left(\frac{3}{5}\right)^{20-k} = 0.01597$$

and

$$\sum_{k=0}^{4} \binom{20}{k} \left(\frac{2}{5}\right)^k \left(\frac{3}{5}\right)^{20-k} = 0.05096$$

Thus there is no value of $c$ for which $P(C_c) = 0.05$ under the probability measure of $H_0$.

We may elect to use the critical region

$$C_* = \{(x_1, x_2, \ldots, x_{20}); x_1 + x_2 + \cdots + x_{20} \leq 4\}$$

while understanding that $P(C_*) = 0.05096 > 0.05$ under the probability measure of $H_0$. After all, 0.05096 is very close to 0.05. On the other hand, what if we wanted the probability of an error of type 1 to be 0.03? This could pose some difficulties. In section 14.2, we present a method to deal with this situation.

If we take $\{(x_1, x_2, \ldots, x_{20}); x_1 + x_2 + \cdots + x_{20} \leq 4\}$ as the critical region, the probability of an error of type 2 is

$$1 - \sum_{k=0}^{4} \binom{20}{k} \left(\frac{1}{5}\right)^k \left(\frac{4}{5}\right)^{20-k} = 1 - 0.63 \approx 0.37$$

This is a high probability of error but it must be realized that the expected number of heads in 20 tosses when $P(\text{H}) = 0.2$ is five. It should be no surprise that the probability of five or more heads in 20 tosses is high when $P(\text{H}) = 0.2$. If we are not happy about such a high probability of an error of type 2 but we want the probability of an error of type 1 to be about 0.05, we will have to use a larger sample.

### Example 14.1.9

If $H_0$ has an $N(6, 9)$ density and $H_1$ has an $N(8, 9)$ density where $S_0$ is the set of real numbers, find the sample size $n$ for independent sampling and find a critical region such that the probability of an error of type 1 is close to 0.05 and the error of type 2 is close to 0.05.

The density functions $f$ and $g$ are

$$g(x_1, x_2, \ldots, x_n) = \prod_{j=1}^{n} \frac{1}{\sqrt{2\pi \cdot 9}} e^{-(1/2 \cdot 9)(x_j - 8)^2}$$

and

$$f(x_1, x_2, \ldots, x_n) = \prod_{j=1}^{n} \frac{1}{\sqrt{2\pi \cdot 9}} e^{-(1/2 \cdot 9)(x_j - 6)^2}$$

The inequality

$$g(x_1, x_2, \ldots, x_n) \geq cf(x_1, x_2, \ldots, x_n)$$

is equivalent to

$$\exp\left[-\frac{1}{18} \sum_{j=1}^{n} (x_j - 8)^2\right] \geq c \exp\left[-\frac{1}{18} \sum_{j=1}^{n} (x_j - 6)^2\right]$$

Taking logarithms gives

$$-\frac{1}{18}\left[\sum_{j=1}^{n} (x_j^2 - 16x_j + 64)\right] \geq \log c - \frac{1}{18}\left[\sum_{j=1}^{n} (x_j^2 - 12x_j + 36)\right]$$

or

$$\frac{1}{18} \sum_{j=1}^{n} [(16 - 12)x_j - 64 + 36] \geq \log c$$

which is

**(14.1.6)** $\quad 4 \sum_{j=1}^{n} x_j > 18 \log c + n \cdot 28$

Since $c$ can be any positive number, the form of (14.1.6) is

$$\sum_{j=1}^{n} x_j \geq b \quad \text{or} \quad \frac{1}{n} \sum_{j=1}^{n} x_j \geq a$$

Therefore our critical regions are of the form $\{\bar{X} \geq a\}$ if we translate into standard random variable form. We note that $E(\bar{X}) = 6$ for $H_0$ and $E(\bar{X}) = 8$ for $H_1$. In both cases $\sigma^2(\bar{X}) = 9/n$ and by our assumptions $\bar{X}$ is normal. We want

$$0.05 \approx \int_{a}^{\infty} \frac{1}{\sqrt{2\pi(9/n)}} e^{-(n/18)(x-6)^2} \, dx$$

and

$$0.05 \approx \int_{-\infty}^{a} \frac{1}{\sqrt{2\pi(9/n)}} e^{-(n/18)(x-8)^2} \, dx$$

These integrals can be changed to

$$\int_{\sqrt{n/9}(a-6)}^{\infty} \frac{1}{\sqrt{2\pi}} e^{-z^2/2} \, dz$$

and

$$\int_{-\infty}^{\sqrt{n/9}(a-8)} \frac{1}{\sqrt{2\pi}} e^{-z^2/2} \, dz$$

respectively, with the substitutions $\sqrt{n/9}\,(x-6) = z$ and $\sqrt{n/9}\,(x-8) = z$, respectively. By Table 1 (appendix) $\int_{1.645}^{\infty} \frac{1}{\sqrt{2\pi}} e^{-z^2/2} \, dz = 0.05$. Using the symmetry of $e^{-z^2/2}$ we have $\sqrt{n/9}\,(a-6) = 1.645$ and $\sqrt{n/9}\,(a-8) = -1.645$. The simultaneous solution of these equations has $n = (4.935)^2$. The nearest integer to $(4.935)^2$ is 25. We could then use $a = 6 + \sqrt{9/25} \cdot 1.645 = 6.987$.

Note that it is easy to handle this problem because $\bar{X}$ is normal when $X$ is normal and because we have a density function rather than a probability function. Furthermore, elaborate tables are available.

In this section we have examined some tests (14.13) which can be used to judge two simple hypotheses. The tests can be arranged to fit certain specifications if not exactly, then approximately. In the next section we see that these tests are the best tests within a certain concept of best.

**EXERCISES 14.1**

1  With reference to Example 14.4.1, suppose that $k_0 = 10$ (that is, if ten or more damaged cans appear in a sample of 100 cans the lower price is charged). Find the probability of the following:
   (a) the probability of a can being damaged is actually 0.20 but the sample leads to a lower price being charged
   (b) the probability of a can being damaged is actually 0.10 but the sample leads to a lower price being charged
   (c) the probability of a can being damaged is actually 0.12 but the sample leads to a higher price being charged.

2  What products other than canned goods can have their quality inspected as in Example 14.1.1?

3  Suppose the manufacturer of Example 14.1.1 loses $300 per shipment if a shipment of good quality (5% damaged) is judged to be of poor quality and loses $400 per shipment if a shipment of poor quality (15% damaged) is judged to be of good quality. With $p_1$ and $p_2$ the probability of judging a good shipment to be bad and of judging a bad shipment to be good, respectively, assume we have the following information about sampling plans of 100 cans and of 200 cans.

| | COST/SHIPMENT | $p_1$ | $p_2$ |
|---|---|---|---|
| INSPECTION OF 100 | $12 | 0.03 | 0.08 |
| INSPECTION OF 200 | $23 | 0.0035 | 0.0116 |

Suppose that an ideal case has only 5% damage or 15% damage rates. Suppose that in this ideal case a 15% damage rate occurs $\frac{1}{3}$ of the time. Is the 200-can sampling plan preferable to the 100-can sampling program? Would a 300-can sampling plan costing $34 be preferable to the other plans?

4  Referring to Example 14.1.2 ascertain whether it would be advisable to put the following fertilizers into use if study of them on 20 plots yielded the tabulated data. Assume that the cost of the new fertilizers can be brought down to the cost of the current standard fertilizer.

| FERTILIZER | $\bar{x}$ | $(\sum (x_j - \bar{x}))^{1/2}$ |
|---|---|---|
| I | 510 | 170 |
| II | 520 | 200 |
| III | 530 | 400 |

5  Referring to Example 14.1.2, prove that the given $Z$ has the Student's $t$-distribution as claimed. Suppose that a new fertilizer actually has an expected

yield of 510 bushels per acre. Suppose that its variance is 1600; i.e., the normal random variable modeling it is $N(510, 1600)$. Set up an integral for the probability of the event that it will be accepted as better than the current standard. Note that if $X_1, X_2, \ldots, X_{20}$ are independent and $N(510, 1600)$, $\bar{X}$ is $\sqrt{80}\, Y + 510$ where $Y$ is $N(0, 1)$. Use the joint density function of $Y$ and $d$ which can be derived from formulas of section 12.3.

6   Make some reasonable objections to the assumptions of Example 14.1.2

7   Referring to Example 14.1.3 determine $u_0$ if we wish to allow about 10% of the instances in which we have a fair die to be judged as a biased die. With that value of $u_0$ how would we judge a die which showed 89 ones, 96 twos, 124 threes, 93 fours, 78 fives, and 120 sixes?

In exercises 8–14 we take up testing hypotheses about the proper probability measure for describing the "breaking of a stick" in which it is more probable that it breaks close to $x = 1$. Specifically, we take $S_0 = \{0 \le x \le 1\}$ and $f(x; \theta) = (\theta + 1)x^\theta$ for $x \in S_0$ and $\theta > 0$. We will use $H_0: \theta = 1$ and $H_1: \theta = 2$. To prevent the calculations from getting unwieldy we take sample size $n = 2$. We assume independent sampling and refer to data as follows: $(0.5, 0.8), (0.8, 0.95), (0.83, 0.76), (0.89, 0.65), (0.77, 0.72)$.

8   What is the density function on the sample space of the sample resulting from $f(x; \theta)$?

9   Consider the critical region $\{(x_1, x); \text{ minimum } (x_1, x_2) \ge 0.715\}$. Under which of the above data should $H_0$ be rejected?

10   Consider the critical region $\{(x_1, x_2); x_1 + x_2 \ge 1.6\}$. Under which of the above data should $H_0$ be rejected?

11   Consider the critical region $\{(x_1, x_2); x_1 x_2 \ge b = 0.6276\}$. Under which of the above data should $H_0$ be rejected?

12   Compute the probability of an error of type 1 for each of the critical regions in exercises 9, 10, and 11. Note that $b = 0.6278$ satisfies $1 - b^2 + 2b^2 \log b = 0.2389$ to four decimal places.

13   Sketch the critical regions of exercises 10, 11, and 12 on a single graph. Compute the probability of an error of type 2 for each of these tests.

14   Do any of the critical regions of exercises 10, 11, and 12 arise from the likelihood ratio test of $H_0$ against $H_1$?

# 14.2   Quality of tests of hypotheses

In section 14.1 we saw examples of tests of two simple hypotheses where we tested hypotheses based on famous distributions. The critical regions of the tests were of the form

$$\{s \in S; \quad \bar{X} \ge c\} \quad \text{or} \quad \{s \in S; \quad \bar{X} \le c\}$$

Not all critical regions have that form but often those which do are very good. It is a fact that our eye and mind seem to pick on averages or expected values as very interesting things. Furthermore, they seem to be easier to work with than many other entities. We shall see cases in which the critical region is not determined by a condition on $\bar{X}$. Another question which arises in connection with the critical regions observed in section 14.1 is whether these are the best critical regions for testing the given hypotheses. To answer this one must have a definition of "best."

It is convenient to have some additional notation and terminology.

**Definition**      Let $S_0$ be a sample space and let $H_0$ and $H_1$ be simple hypotheses about probability measures on $S_0$. Let a fixed sample size and method of sampling be established (usually it is independent sampling). If $C$ is a critical region (of a test), we call $\alpha = \alpha(C) = P(C)$ under $H_0 = $ probability of an error of type 1 the *first deficiency* of $C$. We call $\beta = \beta(C) = P(C)$ under $H_1 = $ probability of an error of type 2 the *second deficiency* of $C$. We say that $C_*$ is a *most powerful test of $H_0$ against $H_1$* if $\beta(C_*) \le \beta(C)$ whenever $\alpha(C_*) = \alpha(C)$. That is, $C_*$ has the lowest second deficiency among all tests $C$ (critical regions) which have the same first deficiency as $C_*$.

A more delicate theory can even allow $\beta(C_*) \le \beta(C)$ whenever $\alpha(C_*) \le \alpha(C)$. We will not use this.

In other books, $\alpha$ is called the size of the test. We have avoided this term because of possible confusion with the sample size. It is proper to ask "why not hold $\beta$ fixed and minimize $\alpha$?" or "why not minimize the sum of $\alpha$ and $\beta$?" or "why not minimize $\frac{1}{2}\alpha + \frac{3}{4}\beta$?" etc. There is no good answer to these questions. One can do these other things. We chose to hold $\alpha$ fixed and minimize $\beta$ because that is relatively easy (fixing $\beta$ and minimizing $\alpha$ is equally easy but the proof is essentially gotten by interchanging $H_1$ and $H_0$) and we do not want to get bogged down in too much theory.

*Notice that we are discussing the most powerful test for a fixed mode of sampling.*

**Theorem 14.2.1**      Let $S_0$ be a sample space and let $H_0$ and $H_1$ be simple hypotheses about probability measures on $S_0$. Let a fixed mode of sampling be established and let $f$ and $g$ determine the probability measures on $S$ associated with $H_0$ and $H_1$, respectively. Then, a critical region of the form $C = \{s \in S;\ g(s) \ge cf(s)\}$ is a most powerful test among all tests of equal first deficiency.

**Proof**      Let $D$ be any critical region with $\alpha(D) = \alpha(C) = \alpha$. We will deal with $f$ and $g$ as though they are density functions even though they may be probability functions.

For any event $A \subset S$ let

$$I_A(s) = \begin{cases} 1, & s \in A \\ 0, & s \notin A \end{cases}$$

If $h$ is any function on $S$,

$$\int_A h(s)\,ds = \int_S I_A(s)h(s)\,ds$$

By assumption

**(14.2.1)** $\qquad \int_S I_C(s)f(s)\,ds = \alpha(C) = \alpha(D) = \int_S I_D(s)f(s)\,ds = \alpha$

Note that

$$\int_S I_C(s)g(s)\,ds = 1 - \int_{C'} g(s)\,ds = 1 - \beta(C)$$

Consider

**(14.2.2)** $\qquad \int_S [I_C(s) - I_D(s)][g(s) - cf(s)]\,ds$

If $s \in C$, $I_C(s) - I_D(s) = 1 - I_D(s) \geq 0$ and $g(s) \geq cf(s)$; i.e., $g(s) - cf(s) \geq 0$.
If $s \notin C$, $I_C(s) - I_D(s) = I_C(s) - 1 \leq 0$ and $g(s) < cf(s)$; i.e., $g(s) - cf(s) < 0$.
Thus for any $s \in S$ the integrand and hence the integral in (14.2.2) is non-negative. This is equivalent to

$$\int_S [I_C(s) - I_D(s)]g(s)\,ds \geq \int_S [I_C(s) - I_D(s)]cf(s)\,ds$$

By (14.2.1)

$$\int_S [I_C(s) - I_D(s)]cf(s)\,ds = c\int_S I_C(s)f(s)\,ds - c\int_S I_D(s)f(s)\,ds$$

$$= c\alpha - c\alpha = 0$$

Then

$$0 \leq \int_S [I_C(s) - I_D(s)]g(s)\,ds = \int_S I_C(s)g(s)\,ds - \int_S I_D(s)g(s)\,ds$$

$$= 1 - \beta(C) - [1 - \beta(D)]$$

$$= \beta(D) - \beta(C)$$

This implies that $\beta(D) \geq \beta(C)$.

Therefore, $C$ is a most powerful test among those tests having first deficiency $\alpha$. Note that we could have used sums instead of integrals and assumed that $f$ and $g$ are probability functions.

Theorem 14.2.1 is a special case of a lemma due to Neyman and Pearson.

Suppose that $0 < \alpha < 1$. Is there always a test whose first deficiency is exactly $\alpha$? Let us look at

$$C_c = \{s \in S; \ g(s) \geq cf(s)\}, \qquad c \geq 0$$

If $c_1 < c_2$, every $s' \in S$ such that $g(s') \geq c_2 f(s')$ also satisfies $g(s') \geq c_1 f(s')$ since $f(s) \geq 0$. Thus for those $c_1, c_2$, $C_{c_1} \supset C_{c_2}$. Therefore under $H_0$, $\alpha(C_c) =$

$P(C_c)$ as a function of $c$ must be monotonically decreasing as $c \to \infty$. Note that $C_0 = S$ so $\alpha(C_0) = 1$. It can be shown that $\alpha(C_c) \to 0$ as $c \to \infty$. Therefore, the graph of $\alpha(C_c)$ looks like Figure 3.

The only difficulties occur if we want a first deficiency $\alpha$ for which $\alpha(C_c) \neq \alpha$ for any $c \geq 0$. Since $0 < \alpha < 1$, this can occur only if $\alpha$ corresponds to a jump in the graph as in Figure 3. If this is the case there are two numbers $0 \leq c_1 < c_2$ such that $\alpha(C_{c_1}) > \alpha > \alpha(C_{c_2})$. If $f$ is given by a density function we have

$$\int_{C_{c_1}} f(s)\, ds > \alpha > \int_{C_{c_2}} f(s)\, ds$$

It is intuitively clear, and we will not prove it, that for some set $K$ such that $C_{c_1} \supset K \supset C_{c_2}$

**(14.2.3)**    $$\int_K f(s)\, ds = \alpha$$

since there is no *point* in Euclidean space with non-zero probability under a density such as $f$.

As a matter of fact, it can be shown that $\alpha(C_c)$ is continuous from the left. Thus we can choose $c_1$ to be the largest $c$ with $\alpha(C_c) > \alpha$. It can be shown that

**(14.2.4)**    $g(s) \geq c_1 f(s)$   for all $s \in K$

We may choose $K$ as a critical region for a test of $H_0$ versus $H_1$. It has first deficiency equal to $\alpha$.

If we do not assume that $f$ is a density function but assume that it is a probability function, all of the above can be carried out by replacing integrals with sums except for the existence of a set $K$ satisfying (14.2.3). However, we can perform randomized tests.

### Example 14.2.1
With the setup of Example 14.1.8 let

$$A = \{(x_1, x_2, \ldots, x_{20}); x_1 + x_2 + \cdots + x_{20} \leq 3\}$$

and

$$B = \{(x_1, x_2, \ldots, x_{20}); x_1 + x_2 + \cdots + x_{20} \leq 4\}$$

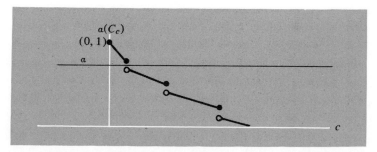

**Figure 3**

Discuss the testing procedure of $H_0$ versus $H_1$ whereby $H_0$ is rejected if $s \in A$ and if $s \in B - A$, then play a game of chance independent of the original experiment and reject or accept $H_0$ according to the outcome of this game of chance.

In Example 14.1.8 we saw that $\alpha(A) = 0.01597$ and $\alpha(B) = 0.05096$ to five decimal places. It is easy to show that for any number $\alpha$ between $\alpha(A)$ and $\alpha(B)$ there is a $\gamma$ with $0 < \gamma < 1$ such that $\alpha = (1 - \gamma)\alpha(A) + \gamma\alpha(B)$. In particular there is a $\gamma_1$ such that $0.05 = (1 - \gamma_1)\alpha(A) + \gamma_1\alpha(B)$. Suppose that we flip a coin with bias $P(H) = \gamma$ whenever $s \in B - A$ and reject $H_0$ if a head appears. Otherwise we accept $H_0$. The probability of rejecting a true $H_0$ under this procedure is as follows:

$$P(\text{rejecting } H_0) = \alpha(A) + P(s \in B - A \text{ and H appears})$$
$$= \alpha(A) + P(B - A)\gamma = \alpha(A) + [P(B) - P(A)]\gamma$$
$$= \alpha(A) + [\alpha(B) - \alpha(A)]\gamma$$

But $\alpha(A) + [\alpha(B) - \alpha(A)]\gamma = (1 - \gamma)\alpha(A) + \gamma\alpha(B) = \alpha$.

Thus, although we do not have a test of $H_0$ against $H_1$ in the original sense, whereby the judgment was based only on the data, we have a slightly supplemented situation. Although one cannot easily obtain biased coins (where the bias is known), one can simulate a biased coin by the use of random numbers. This can give a very good approximation.

Suppose that $H_0$ is a simple hypothesis and $H_1$ is a composite hypothesis. Let $H_2$ be a simple hypothesis consisting of a specific $g \in H_1$. It may be true that $C$, a critical region, furnishes the best test of $H_0$ versus $H_2$ regardless of which $g \in H_1$ comprises $H_2$. In that case the test based on $C$ is called *uniformly most powerful*. It is assumed that $\alpha(C)$ is specified.

### Example 14.2.2

Suppose we are dealing with a biased coin. Then $H_0$ states that $p = P(H) = 0.6$. $H_1$ states that $p < 0.5$. Independent sampling of sample size 10 is to be used. Arrange for $\alpha \leq 0.10$. Discuss.

Suppose we consider $p = 0.5$ and consider

$$\{(x_1, x_2, \ldots, x_{10}); p^\nu(1 - p)^{10 - \nu} \geq c(0.6)^\nu(0.4)^{10 - \nu}\}$$

where $\nu = x_1 + x_2 + \cdots + x_{10}$ as in Example 14.1.8. Note that $\nu$ is the number of heads with the conventions of Example 14.1.8. Upon taking logarithms we obtain

$$\nu \log \frac{p}{1 - p} + 10 \log (1 - p) \geq \log c + \nu \log \frac{0.6}{0.4} + 10 \log 0.4$$

or

$$\nu \left( \log \frac{p}{1 - p} - \log \frac{0.6}{0.4} \right) \geq \log c - 10 \log (1 - p) + 10 \log 0.4$$

or

$$\nu\left[\log \frac{4p}{6(1-p)}\right] \geq \log c - 10 \log (1-p) + 10 \log 0.4$$

since

$$\log \frac{p}{1-p} - \log \frac{0.6}{0.4} = \log \frac{p}{1-p} - \log \tfrac{6}{4}$$

$$= \log p - \log (1-p) - \log 6 + \log 4 = \log \frac{4p}{6(1-p)}$$

But since $p < 0.5$, $10p < 6$ so $4p < 6 - 6p = 6(1-p)$ and $\log \dfrac{4p}{6(1-p)}$ is always negative. Therefore,

**(14.2.5)**     $\nu \leq \dfrac{1}{\log \dfrac{4p}{6(1-p)}}$ $[\log c - 10 \log (1-p) + 10 \log 0.4]$

is the criterion for a most powerful test of $H_0$ against $H_2 = \{p\}$ where $p < 0.05$. Note that any region satisfying (14.2.5) is of the form

$$C_b = \{(x_1, x_2, \ldots, x_{10}); x_1 + \cdots + x_{10} \leq b\}$$

We are asked to find such a region for which $\alpha(C_b) \leq 0.10$. Now

$$\alpha(C_b) = \sum_{k=0}^{b} \binom{10}{k}(0.6)^k(0.4)^{10-k}$$

Binomial probability tables give

$$0.05476 = \sum_{k=7}^{10} \binom{10}{k}(0.4)^k(0.6)^{10-k} = \sum_{l=0}^{3} \binom{10}{l}(0.4)^{10-l}(0.6)^l$$

where $l = 10 - k$ so $\binom{10}{k} = \binom{10}{10-k} = \binom{10}{l}$. Therefore, $C_3 = \{x_1 + x_2 + \cdots + x_{10} \leq 3\}$ is a best test of $H_0 = \{0.6\}$ against $H_2 = \{p\}$ for any $p < 0.5$. Therefore, $C_3$ is a uniformly most powerful test of $H_0 = \{0.6\}$ against $H_1 = \{p; p < 0.5\}$. Note that there is no single value of $\beta$; the value depends on $p \in H_1$.

The O.C. function of $C_3$ is graphed in Figure 4. That is

$$q^*(p) = \sum_{k=4}^{10} \binom{10}{k}p^k(1-p)^{10-k}$$

Note that $q^*$ is monotonic and that all parameters in $H_1$ lie to the left of $\frac{6}{10} \in H_0$.

It should be fairly clear that if $H_1$ has certain geometrical properties in relation to $H_0$ and with respect to $C$, a most powerful test of $H_0$ against $g_1$, a specific density in $H_1$, is a uniformly most powerful test of $H_0$ versus $H_1$.

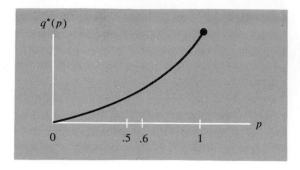

**Figure 4**

### Example 14.2.3

Suppose $S_0$ is the real line and $H_0 = \{N(0, 1)\}$ while $H_1 = \{N(1, 1), N(-1, 1)\}$. Devise a test that treats $N(1, 1)$ and $N(-1, 1)$ equally and such that $\alpha = 0.05$. Independent sampling of size $n$ is used.

It should be fairly clear that if $x_1 + x_2 + \cdots + x_n$ is large in absolute value, then $H_0$ should be rejected. We could choose

$$C = \{(x_1, x_2, \ldots, x_n); |x_1 + x_2 + \cdots + x_n| \geq c\}$$

Since we are using independent sampling and since $S_0$ has normal densities on it from $H_0$ and $H_1$ we can talk in terms of $X(x) = x$ as our random variable. In this case $n\bar{X}(s) = x_1 + x_2 + \cdots + x_n$ for $s = (x_1, \ldots, x_n) \in S$. By known results $n\bar{X}$ is $N(nm, n)$ if $X$ is $N(m, 1)$.

To calculate $\alpha$, $m = 0$ so we want

$$0.05 = P(|n\bar{X}| \geq c) = P(|\sqrt{n}\, Y| \geq c)$$

where $Y$ is $N(0, 1)$. That is, $0.05 = P(|Y| \geq c/\sqrt{n})$. By Table 1 (appendix), $c/\sqrt{n} = 1.96$.

Let us now pass to the O.C. function $q^*(m)$ for the region $C$. We note that

$$C = \{|n\bar{X}| \geq 1.96\sqrt{n}\} = \{n\bar{X} \leq -1.96\sqrt{n}\} \cup \{n\bar{X} \geq 1.96\sqrt{n}\}$$

By definition, the O.C. function gives the probability of the complement of the critical region. Therefore,

$$q^*(m) = P(-1.96\sqrt{n} < n\bar{X} < 1.96\sqrt{n})$$

when $X$ is $N(m, 1)$. But since $n\bar{X}$ is $N(nm, n)$ we may use exercise 6.4.16 to obtain

$$q^*(m) = \int_{-1.96 - \sqrt{n}m}^{1.96 + \sqrt{n}m} \frac{1}{\sqrt{2\pi}} e^{-x^2/2}\, dx$$

This integral may be interpreted as the area under the curve $Y = (1/\sqrt{2\pi})e^{-x^2/2}$ and above the interval $[-1.96 - \sqrt{n}\, m, 1.96 - \sqrt{n}\, m]$, an interval of length 3.92 with midpoint at $x = -\sqrt{n}\, m$.

Since the curve $y = e^{-x^2/2}$ is symmetric with respect to the $y$ axis and takes its maximum at $x = 0$, the area under it and above $[-1.96 - \sqrt{n}\, m, 1.96 -$

$\sqrt{n}\,m]$ is largest when $m = 0$ and by symmetry the areas corresponding to $m = 1$ and $m = -1$ are equal.

If we graph the O.C. functions of $C = \{|n\bar{X}| \geq 1.96\sqrt{n}\}$, $C_1 = \{n\bar{X} \geq 1.65\sqrt{n}\}$, $C_{-1} = \{nX \leq -0.165\sqrt{n}\}$ we will obtain some insight into our problems. Note that $\alpha(C) = \alpha(C_1) = \alpha(C_{-1}) = 0.05$ up to the accuracy of our tables. Furthermore, $C$ and $C_{-1}$ are the most powerful tests of $H_0:m = 0$ against $H_1:m = 1$, and $H_0:m = 0$ against $H_1:m = -1$, respectively.

In Figure 5 we graph $q^*$, $q_1^*$, and $q_{-1}^*$, the O.C. functions of $C$, $C_1$, and $C_{-1}$, respectively. Naturally $q^*(1) > q_1^*(1)$ and $q^*(-1) > q_{-1}^*(-1)$ since $C_1$ and $C_{-1}$ are most powerful tests as previously mentioned. However, $C_1$ is a bad test for $H_0:m = 0$ against $H_1:m = -1$. Therefore, if one wishes to test $H_0:m = 0$ against $H_1:m = \pm 1$ then he must plan to give up some power against $m = 1$ in order to be able to work against $m = -1$.

The ideas relating to Example 14.2.3 suggest a certain class of tests which are very useful in certain cases.

**Definition**   Let $I$ be a set of parameters and let $H_0 \subset I$, $H_1 \subset I$. For convenience let $H_0$ be a simple hypothesis. A test $C$ of $H_0$ against $H_1$ is said to be *unbiased* on $I$ if the O.C. function defined on $I$ takes its maximum at $\theta_0 \in H_0$; i.e., if $q^*(\theta_0) \geq q^*(\theta)$ for all $\theta \in I$. We say that $C$ is a *most powerful unbiased* test (on $I$) of $H_0$ against $H_1$ if whenever $C_1$ is an unbiased test on $I$ with O.C. function $q_1^*$ and $q_1^*(\theta_0) = q^*(\theta_0)$, then $q^*(\theta) \leq q_1^*(\theta)$ for $\theta \in H_1$. That is, $C$ is an unbiased test and has smallest second deficiency at all $\theta \in H_1$ among all unbiased tests with the same first deficiency.

It is not our intention to study the theory of unbiased tests. We state a theorem (without proof) which is useful for finding most powerful unbiased tests.

Let $S_0$, $I$, $H_0$, $H_1$ satisfy

**(14.2.6a)**   $S_0$ is a set of real numbers.

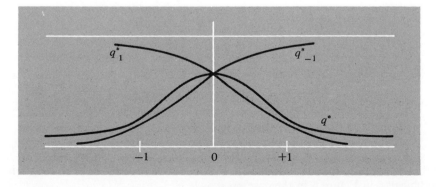

**Figure 5**

**(14.2.6b)**   $I$, the parameter set, is a bounded or unbounded interval of real numbers.

**(14.2.6c)**   $H_0$ and $H_1$ are simple hypotheses with $\theta_0 \in H_0$ in the interior of $I$.

**(14.2.6d)**   $f(s; \theta)$, the likelihood function on $S$ corresponding to parameter $\theta$, is such that $\dfrac{\partial f}{\partial \theta}(s; \theta)$ exists for each $s \in S$ and each $\theta \in I$ and is of sufficiently slow growth that $\dfrac{\partial}{\partial \theta} \displaystyle\int_S f(s; \theta)\, ds = \int_S \dfrac{\partial f}{\partial \theta}(s; \theta)\, ds.$

**Theorem 14.2.2**   Under conditions (14.2.6) any critical region

**(14.2.7)**   $C = \left\{ s \in S; f(s; \theta_1) \geq cf(s; \theta_0) + c_1 \dfrac{\partial f}{\partial \theta}(s; \theta_0) \right\}$

where $c > 0$ which is an unbiased test of $H_0 : \theta = \theta_0$ against $H_1 : \theta = \theta_1$ is a most powerful unbiased test on $I$ of $H_0$ against $H_1$.

### Example 14.2.4

Show that the critical region $C$ of Example 14.2.3 is a most powerful unbiased test of $H_0 : m = 0$ against $H_1 : m = 1$ or $H_1 : m = -1$ where the parameter set $I = \{m; m \in R\}$. We are taking the fixed $\sigma^2 = 1$.

In the discussion, all sums and products range from 1 to $n$. We know

$$f(s; m) = \prod \frac{1}{\sqrt{2\pi}} e^{-(x_j - m)^2/2} = (2\pi)^{-n/2} \exp\left[-\tfrac{1}{2} \sum (x_j - m)^2\right]$$

$$\frac{\partial}{\partial m} f(s; m) = f(s; m)\left[-\tfrac{1}{2} \sum 2(x_j - m)(-1)\right] = f(s; m)\left[\sum (x_j - m)\right]$$

Now $f(s; m) \geq cf(s; 0) + c_1 \dfrac{\partial f}{\partial m}(s; 0)$ is equivalent to

$$(2\pi)^{-n/2} \exp\left(-\tfrac{1}{2} \sum (x_j - m)^2\right) \geq (2\pi)^{-n/2} \exp\left(-\tfrac{1}{2} \sum x_j^2\right)(c + c_1 \sum x_j)$$

After division by the factors $(2\pi)^{-n/2}$ and $\exp\left(-\tfrac{1}{2} \sum x_j^2\right)$, which does not change the sense of the inequality, we have

$$\exp\left(-\tfrac{1}{2}nm^2 + \sum x_j m\right) \geq (c + c_1 \sum x_j)$$

Since the constants $c$ and $c_1$ are restricted only by $c > 0$ and since $\exp\left(-\tfrac{1}{2}nm^2\right)$ is positive and constant with respect to the $s \in S$, we can divide both sides by it and retain the notation $c$ and $c_1$ for the resulting constants. Our key condition, equivalent to (14.2.7), is now

$$\exp\left(m \sum x_j\right) \geq c + c_1 \sum x_j$$

With $x^* = \sum x_j$, this is just $e^{mx^*} \geq c + c_1 x^*$. We analyze this inequality, graphically, in Figure 6 in the case $m = -1$.

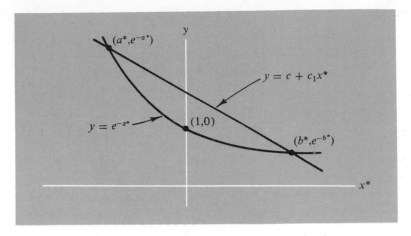

**Figure 6**

The figure indicates how a line $y = c + c_1 x^*$ with $c > 1$ crosses the curve $y = e^{-x^*}$. Furthermore, by choosing $c$ and $c_1$ properly, we can arrange that the intersections of these graphs have horizontal coordinates $a^* = -1.96\sqrt{n}$ and $b^* = 1.96\sqrt{n}$, respectively. Obviously, $e^{-x^*} \geq c + c_1 x^*$ if $x_1 + x_2 + \cdots + x_n = x^* \leq -1.96\sqrt{n}$ or if $x_1 + \cdots + x_n \geq 1.96\sqrt{n}$. This is the critical region $C$ of Example 14.2.3. It was shown to give an unbiased test. By Theorem 14.2.2 the test is best unbiased of $m = 0$ against $m = -1$. Similarly, but with different choices of $c$, $c_1$ we can show that it is the best unbiased test of $m = 0$ against $m = 1$. However, it is the set $C \subset S$ that is the critical region. That we need two equivalent descriptions of it to prove it is a best unbiased test of a simple hypothesis against a composite hypothesis is perfectly acceptable, mathematically.

It should be quite clear from the preceding examples that it is difficult to get something for nothing. If we know very little about the alternatives it is hard to design a good test. Admittedly, the likelihood-ratio tests are uniformly most powerful if $H_0$ and $H_1$ are simple or if $H_1$ is located conveniently with respect to $H_0$. Furthermore, there are many practical situations in industrial problems such as quality control, in which the Neyman–Pearson regions are appropriate because $H_1$ is on "one side" of $H_0$. In many other problems uniformly most powerful tests are not available and we must compromise. The most powerful unbiased tests are a useful compromise. However, when specific information is available one may wish to choose a biased test rather than an unbiased test.

### Example 14.2.5
An independent sample of size 4 is used to test the hypothesis $H_0 : m = 0$ against $H : m = 1$ or $m = -2$ where we assume that an $N(m, 1)$ distribution is the

correct one. Determine the values of the O.C. function belonging to the regions

$$C_0 = \{-1.96 \le \tfrac{1}{2}(x_1 + x_2 + x_3 + x_4) \le 1.96\}$$
$$C_1 = \{-2.33 \le \tfrac{1}{2}(x_1 + x_2 + x_3 + x_4) \le 1.75\}$$
$$C_2 = \{-3.03 \le \tfrac{1}{2}(x_1 + x_2 + x_3 + x_4) \le 1.65\}$$

at $m = 0$, $m = 1$, and $m = -2$. Discuss the merits of these tests.

We let $X(x) = x$ and study its distribution according to $m = 0, 1,$ or $-2$. In all cases we take $\sigma^2(X) = 1$. Now let $Y = \tfrac{1}{2}(X_1 + X_2 + X_3 + X_4)(s) = \tfrac{1}{2}(x_1 + x_2 + x_3 + x_4)$ where $s = (x_1, x_2, x_3, x_4)$. Then $\sigma_m^2(Y) = 1$ while $E_m(Y) = 2E(X)$.

When $m = 0$,

$$P(C_0) = P(-1.96 \le N(0, 1) \le 1.96) = 1 - 2P(N(0, 1) \ge 1.96)$$
$$= 1 - 2(0.025) = 0.95$$

$$P(C_1) = P(-2.33 \le N(0, 1) \le 1.75) = 1 - 0.0099 - 0.0401$$
$$= 0.95$$

$$P(C_2) = P(-3.03 \le N(0, 1) \le 1.65) + 1 - 0.0005 - 0.0495 = 0.95$$

Therefore the O.C. functions of $C_0$, $C_1$, and $C_2$ take the value 0.95 at $m = 0$. When $m = 1$; $E_m(Y) = 2$. Then

$$P(-1.96 \le N(2, 1) \le 1.96) = P(-3.96 \le N(0, 1) \le -0.04) = 0.4840$$

Next

$$P(-2.33 \le N(2, 1) \le 1.75) = P(-4.33 \le N(0, 1) \le -0.25) = 0.4013$$

and

$$P(-3.03 \le N(2, 1) \le 1.65) = P(5.03 \le N(0, 1) \le -0.35) = 0.3632$$

When $m = -2$, $E_m(Y) = -4$. Then

$$P(-1.96 \le N(-4, 1) \le 1.96) = P(2.04 \le N(0, 1) \le 5.96) = 0.0207$$
$$P(-2.33 \le N(-4, 1) \le 1.75) = P(1.67 \le N(0, 1) \le 5.75) = 0.0475$$

and

$$P(-3.03 \le N(-4, 1) \le 1.65) = P(0.97 \le N(0, 1) \le 5.65) = 0.1660$$

We summarize these results in Table 2.

**Table 2**

|  | $m$ | 0 | 1 | $-2$ |
|---|---|---|---|---|
|  | $C_0$ | 0.95 | 0.4840 | 0.0207 |
| O.C. | $C_1$ | 0.95 | 0.4013 | 0.0475 |
|  | $C_2$ | 0.95 | 0.3632 | 0.1660 |

Each of these tests has the same first deficiency, namely 0.05. Test $C_0$ is an unbiased test but it has a high probability, 0.4840, of rejecting $m = 1$ when it is true. Test $C_2$ has about as small a value as the tables permit for the probability of rejecting $m = 1$ when it is true. Test $C_1$ has the smallest total error of the three tests with respect to the three parameters.

Which test we choose depends on our objectives. If we fear rejecting $m = 1$ when it is true but are less concerned about rejecting $m = -2$ when it is true, we may wish to choose $C_2$. It should be clear that our choice of a test depends on external objectives which are not given in the statement of the example. To the extent that we can precisely formulate these objectives we can choose a good test to attain them. If we cannot formulate them too precisely, we may be wise to choose a test by convention and not worry whether it is the best test.

We have been limiting ourselves to simple hypotheses because we could then arrange for tests to have given deficiencies—mainly to have given first deficiency. If $H_0$ contains more than one distribution, the O.C. function may not be constant on $H_0$. On the other hand it may be constant in certain special cases. Consider the next example.

### Example 14.2.6

Let $S_0$ be the set of real numbers. We use independent sampling with sample size $n$. Discuss testing $H_0 = \{N(0, \sigma^2) \text{ distributions}; \sigma > 0\}$ against $H_1 = \{(N(1, \tau^2) \text{ distributions}; \tau > 0\}$ using the critical region

$$\{x_1 + x_2 + \cdots + x_n \geq c(\sum_{j=1}^{n} (x_j - \bar{x})^2)^{1/2}\}$$

where $\bar{x} = (1/n)(x_1 + x_2 + \cdots + x_n)$. Note that $\sigma^2$ and $\tau^2$ vary over all positive real numbers.

The critical region in question may be described as the set of points satisfying

$$(14.2.8) \qquad \frac{x_1 + x_2 + \cdots + x_n}{[\sum\limits_{j=1}^{n} (x_j - \bar{x})^2]^{1/2}} \sqrt{\frac{n-1}{n}} \geq c\sqrt{\frac{n-1}{n}}$$

If we introduce a standard random variable $X$ on $S_0$ with $X(x) = x$ we may interpret the left-hand side of inequality (14.2.8) as

$$\frac{X_1 + X_2 + \cdots + X_n}{[\sum\limits_{j=1}^{n} (X_j - \bar{X})^2]^{1/2}} \sqrt{\frac{n-1}{n}} = \frac{n(\bar{X} - 0)}{\sqrt{n-1}\,d} \sqrt{\frac{n-1}{n}} = \frac{\bar{X} - 0}{d} \sqrt{n}$$

Now, if $X$ is $N(0, \sigma^2)$ for any $\sigma^2 > 0$ the aftermath of Theorem 12.3.1 is that $[(\bar{X} - 0)/d]\sqrt{n}$ has a Student's $t$-distribution with $n - 1$ degrees of freedom. Thus for various values of $c\sqrt{\dfrac{n-1}{n}} = k$ we may find

$$P(\{X_1 + \cdots + X_n \geq c[\sum_{j=1}^{n} (X_j - \bar{X})^2]^{1/2}\})$$

by consulting tables for the above-mentioned distribution. But that probability is the first deficiency when $P$ is taken relative to $N(0, \sigma^2)$ on $S_0$.

Note that $\sigma^2$ is divided out in the formation of $(\bar{X} - 0)/d$ and so the first deficiency is independent of $\sigma^2$. However, the distribution of $(\bar{X}/d)\sqrt{n}$ does depend on $\sigma^2$ if $X$ is $N(1, \sigma^2)$ so the second deficiency is not constant on $H_1$. We will not go into any more detail on these matters.

### EXERCISES 14.2

In exercises 1–5 find the form of the critical regions which give the most powerful tests. Describe this form in terms of inequalities which the points of the region must satisfy. Assume independent sampling.

**1**  Let $X$ be $N(m, 4)$, sample size $n = 36$, $H_0:m = 3$ and $H_1:m = -5$.

**2**  Let $X$ have density function $b^{3/2}x^{1/2}e^{-bx}/\Gamma(\frac{3}{2})$ for $x > 0$, sample size $n = 25$, $H_0:b = 1$ and $H_1:b = 3$.

**3**  Let $X$ have probability function $(-1)^k \binom{-r}{k}(\frac{1}{3})^k(\frac{2}{3})^r$ for $k = 0, 1, 2, 3, \ldots,$ sample size $n = 20$, $H_0:r = 4$ and $H_1:r = 1$. See exercise 1.5.4.

**4**  Let $X$ be $N(m, \sigma^2)$, sample size $n = 100$, $H_0:m = 3$, $\sigma^2 = 4$ and $H_1:m = -5$, $\sigma^2 = 8$.

**5**  Let $X$ have density function $b^{\gamma}x^{\gamma-1}e^{-bx}/\Gamma(\gamma)$ for $x > 0$, $n = 25$, $H_0:\gamma = 1$, $b = 2$, and $H_1:\gamma = 2$, $b = 1$.

**♦ 6**  In Example 14.2.1 we had $\alpha(A) = 0.01597$ and $\alpha(B) = 0.05096$. We used a randomized test to get $\alpha = 0.05$. Each sample point with exactly four 1's and sixteen 0's has probability $(0.4)^4(0.6)^{16} \approx 7.222 \times 10^{-6}$. We could choose a fixed set $L$ of exactly 133 such sample points so $P(L) = 7.22 \times 10^{-6} \times 133 \approx 0.00096$. Since $L \subset B$, $P(B - L) = 0.05000$ to five decimal places. Discuss $B - L$ as a test of $H_0$ against $H_1$ with $\alpha(B - L) = 0.05$.

**7**  How does the method in exercise 6 fail in some problems which can be handled by randomized tests?

**8**  Show that the tests of exercises 1 and 2 are uniformly most powerful tests of $H_0:m = 3$ against $H_1:m < 3$ and $H_0:b = 1$ against $H_1:b > 1$, respectively.

**9**  Show that the test of exercise 4 is a uniformly most powerful test of $H_0:m = 3$, $\sigma^2 = 4$ against $H_1:m = 11 - 2D^2$, $\sigma^2 = D^2$ for $D^2 > 4$.

**10**  Which test of exercise 1 has $\alpha = 0.10$?

**11**  Which test of exercise 2 has $\alpha = 0.20$? Recall that the sum of independent gamma random variables can be a gamma random variable.

**12**  Find $\beta$ of exercises 10 and 11.

**♦ 13**  Let $C$ be an unbiased test for $H_0 = \theta_0$, where $\theta_0$ is in the interior of an interval. Show, intuitively, that $\dfrac{\partial}{\partial \theta} \displaystyle\int_C f(s; \theta_0)\, ds = 0$. Under reasonable hypotheses this means that $\displaystyle\int_C \dfrac{\partial f}{\partial \theta}(s; \theta_0)\, ds = 0$.

# 14.3  The $\chi^2$ test

In section 12.3 we studied some examples and provided a theorem related to the following problem. Let $S_0$ be the sample space for a population. We believe that $P_0$ is the appropriate probability measure for this population. What kind of data confirms this belief? Assume independent sampling.

We approached this problem by partitioning $S_0$ into disjoint events $A_1$, $A_2$, ..., $A_r$ with $S_0 = A_1 \cup A_2 \cup \cdots \cup A_r$ and with $P_0(A_j) = \pi_j > 0$. We then defined the random variables $N_j$ by $N_j(s) = $ number of entries of $s = (s_1, s_2, \ldots, s_r)$ which belong to $A_j$. Note that $s \in S$, the sample space of the sample, so $N_j$ is a statistic. However, we then considered the other statistic

$$Y_n = \sum_{j=1}^{r} \frac{(N_j - n\pi_j)^2}{n\pi_j}$$

This has a certain limiting distribution. We next obtained experimental values $\hat{N}_j$ from the actual data and computed

$$\hat{Y}_n = \sum_{j=1}^{r} \frac{(\hat{N}_j - n\pi_j)^2}{n\pi_j}$$

If this is small based on the limiting distribution we accept $P_0$ as appropriate.

If $\hat{Y}_n$ is large we reject the assumption that $P_0$ is the appropriate probability measure. What should be considered to be large depends on the risk of error which we are willing to accept.

To be more specific, suppose that we are willing to accept a probability of $\alpha$ of deciding against $P_0$ when it is appropriate in order to get a reasonable chance of deciding against $P_0$ when it is *inappropriate*. We look in Table 2 (appendix) for $P(Z_{r-1} > u_0) = \alpha$ where $Z_{r-1}$ is a $\chi^2$ random variable with $r - 1$ degrees of freedom. We then consider $\hat{Y}_n > u_0$ to be large enough to reject $P_0$. Thus we accept $P_0$ if $\hat{Y}_n \leq u_0$.

This procedure is used because Theorem 12.3.2 states that

$$\lim_{n \to \infty} P\left( Y_n = \sum_{j=1}^{n} \frac{(N_j - n\pi_j)^2}{n\pi_j} \leq u \right) = P(Z_{r-1} \leq u)$$

Note that we are again using a limit theorem as an approximation theorem for large $n$. Ideally we would wish to use the exact distribution of $Y_n$ but that is usually complicated.

Throughout the remainder of this section we let $Z_q$ denote a $\chi^2$ random variable with $q$ degrees of freedom.

### Example 14.3.1
We wish to test whether equal likelihood is a reasonable assumption for $S_0 = \{1, 2, 3, 4\}$ when a sample of size 40 is to be taken. How should we test?

It is natural to let $A_1 = \{1\}$, $A_2 = \{2\}$, $A_3 = \{3\}$, $A_4 = \{4\}$. Let $\hat{N}_j$ be the number of $j$'s which appear in the data. For definiteness let $\alpha = 0.10$. Then, we reject equal likelihood if

$$\hat{Y}_{40} = \sum_{j=1}^{4} \frac{(\hat{N}_j - 10)^2}{10} > 6.25$$

The number 6.25 is used because $P(Z_3 > 6.25) = 0.10$.

Suppose we observe $\hat{N}_1 = 15$, $\hat{N}_2 = 5$, $\hat{N}_3 = 13$, $\hat{N}_4 = 7$. Then

$$\hat{Y}_{40} = \frac{(15 - 10)^2}{10} + \frac{(5 - 10)^2}{10} + \frac{(13 - 10)^2}{10} + \frac{(7 - 10)^2}{10}$$

$$= \frac{25 + 25 + 9 + 9}{10} = 6.8$$

This is greater than 6.25 so we reject the hypothesis of equal likelihood.

### Example 14.3.2

Suppose we believe that the correct probability measure for $S_0$ in Example 14.3.1 is $P_0(A_1) = 0.30$, $P_0(A_2) = 0.20$, $P_0(A_3) = 0.25$, and $P_0(A_4) = 0.25$. Test this using our $\chi^2$ method with $\alpha = 0.10$ and with the $\hat{N}_j$ given in the discussion of Example 14.3.1.

We need only consider

$$\frac{[\hat{N}_1 - 40(0.30)]^2}{40(0.30)} + \frac{[\hat{N}_2 - 40(0.20)]^2}{40(0.20)} + \frac{[\hat{N}_3 - 40(0.25)]^2}{40(0.25)} + \frac{[\hat{N}_4 - 40(0.25)]^2}{40(0.25)}$$

$$= \frac{(15 - 12)^2}{12} + \frac{(5 - 8)^2}{8} + \frac{(13 - 10)^2}{10} + \frac{(7 - 10)^2}{10}$$

$$= 0.75 + 1.125 + 0.9 + 0.9 = 3.675 < 6.25$$

As a matter of fact, the data $\hat{N}_1 = 15$, $\hat{N}_2 = 5$, $\hat{N}_3 = 13$, $\hat{N}_4 = 7$ was obtained by choosing with replacement from a set of 20 playing cards consisting of six clubs, four diamonds, five hearts, and five spaces. If we had chosen $\alpha = 0.30$ we would have rejected a correct hypothesis using this test because then $u_0 = 3.66$.

This method of testing is a simple case of the "$\chi^2$ test for goodness of fit." It was originated by Karl Pearson (1857–1936). Frequently one will see

$$\chi^2 = \sum_{j=1}^{r} \frac{(N_j - n\pi_j)^2}{n\pi_j}$$

This is fairly suggestive. It is called *Pearson's* $\chi^2$ to distinguish it from our "official" $\chi^2$. This freewheeling use of notation has its advantages as well as its disadvantages.

It is conventional to choose $A_j$ and $n$ so that $n\pi_j \geq 5$. However, if $r \geq 10$ one or two cases of $n\pi_j$ even as small as 1 are acceptable. Generally, when possible, we try to make the $\pi_j$ equal or approximately so. Also by convention the values of $\alpha$ are usually 0.30, 0.10, 0.5, 0.01 although other values are tabulated.

The events $A_j$, which we will call *cells*, are chosen at the discretion of the statistician. The limit theorem is valid regardless of the choice of these cells. However, in the case of a discrete sample space $S_0$, some of the cells are natural in that they consist of one sample point with high probability. In other cases the cells are chosen so that $n\pi_j \geq 5$ and so that the characteristics of the distribution in question will be distinguished by the cells. In this regard, we would not wish to choose two cells $A_1 = \{x; x < 0\}$ and $A_2 = \{x; x \geq 0\}$ when we have $n = 300$ and are testing whether the underlying distribution is $N(0, 1)$. It is quite clear that there should be numerous cells unless $S_0$ is a discrete sample space with only a few sample points.

### Example 14.3.3

It is desired to test data about the lengths of 100 telephone calls to ascertain whether an exponental distribution with density $\frac{2}{5}e^{-2x/5}$ (i.e., $\lambda = \frac{2}{5}$) is the correct probability measure. Find some useful $A_j$'s and the corresponding $\pi_j$'s.

The natural sample space is $S_0 = \{x; x \geq 0\}$. It seems appropriate to choose the $A_j$ as intervals although the theorem does not require this. If $A_j = \{x; a \leq x < b\}$

$$P_0(A_j) = \int_a^b \tfrac{2}{5}e^{-2x/5}\, dx = -e^{-2x/5}\big|_a^b = e^{-2a/5} - e^{-2b/5}$$

Note that the right-hand end points of $A_j$ is not in $A_j$. This has no effect on the probabilities $\pi_j$ but it insures the requirement of the disjointness of the $A_j$ in the discussion. More precisely, let

$$A_1 = \{x; 0 < x < 1\}, \quad A_2 = \{x; 1 \leq x < 2\}, \quad A_3 - \{x; 2 \leq x < 3\}, \ldots,$$
$$A_{10} = \{x; 9 \leq x < 10\}, \quad \text{and } A_{11} = \{x; 10 \leq x\}$$

From tables of exponentials we find

$$\pi_1 = 0.330, \qquad \pi_5 = 0.067, \qquad \pi_9 = 0.014,$$
$$\pi_2 = 0.221, \qquad \pi_6 = 0.044, \qquad \pi_{10} = 0.009,$$
$$\pi_3 = 0.148, \qquad \pi_7 = 0.030, \qquad \pi_{11} = 0.018$$
$$\pi_4 = 0.099, \qquad \pi_8 = 0.020,$$

It is easy to compute $n\pi_j = 100\pi_j$ by inspection. Note that $100\pi_j < 5$ if $j \geq 6$. It may be appropriate to pool $A_9$, $A_{10}$, and $A_{11}$ if not $A_7$ and $A_8$. What is more important, perhaps, is to split $A_1$. It has "too much" probability; it could be split into two cells. Perhaps the same is true for $A_2$.

If we are planning to apply the $\chi^2$ test to a population we can split and pool before the data have been gathered. Moreover, we may then simplify the actual gathering of data. However, if data have already been gathered and processed, we may not be able to split, but we can always pool. Splitting may be impossible because the data may state only that there were seven telephone calls whose length was between three minutes and four minutes, without stating the number whose length was between three and three and a half minutes.

In some problems one may wish to test whether a *family* of distributions is appropriate for a population when he is not certain which parameter or parameters are appropriate. For example, Rutherford and Geiger accumulated data on the number of instances of 0, 1, 2, ... firings of a Geiger counter in various time intervals of fixed length due to certain radiation. The results appear in Table 3.

**Table 3**

| $j$ | 0 | 1 | 2 | 3 | 4 | 5 | 6 | 7 | 8 | 9 | 10 |
|---|---|---|---|---|---|---|---|---|---|---|---|
| $\hat{N}_j$ | 57 | 203 | 383 | 525 | 532 | 408 | 273 | 139 | 45 | 27 | 16 |
| $n\hat{\pi}_j$ | 54.4 | 210.5 | 407.4 | 525.5 | 508.4 | 393.5 | 253.8 | 140.3 | 67.9 | 29.2 | 17.1 |

The index $j$ is the number of firings in an interval. Thus $\hat{N}_2 = 383$ means that there were 383 cases (intervals) in which two firings were observed. Furthermore, $\hat{N}_0 = 57$ means that in 57 intervals the counter did not fire. The theory applied to these problems states that a Poisson distribution describes such a population. We could estimate the parameter on the basis of these data. It was found that the total number of particles was 10 094 and the total number of time intervals was 2608. The usual estimate for $\lambda$ would be 10 094 ÷ 2608 = 3.87. This value for $\lambda$ gives values for $n\pi_j$ listed as $n\hat{\pi}_j$ in Table 3.

We next compute

$$\hat{Y} = \sum_{j=1}^{10} \frac{(\hat{N}_j - n\hat{\pi}_j)^2}{n\hat{\pi}_j}$$

We would wish to choose $u_0$ to compare $\hat{Y}$ to $u_0$. However, it should be kept in mind that the $\hat{\pi}_j$ are not deduced from a definite distribution but are deduced from an estimated parameter. It is customary to deduct one degree of freedom for each parameter estimated. Therefore, since there are 11 cells we would obtain $u_0$ from the $\chi^2$ table with nine degrees of freedom. A computation reveals $\hat{Y} = 12.9$. This compares with $P(\chi^2$ with nine degrees of freedom $\geq 14.7) = 0.10$. That is, intuitively, 10% of the time *by chance* a Poisson random variable will lead to an estimate of $\lambda = 3.87$ and a $Y$ value exceeding 14.7.

One degree of freedom is deducted for every parameter estimated from the data because there are theorems which prove that this gives the proper limiting relationship. Unfortunately, the methods of estimation under which these theorems are valid are more complicated than the methods we have studied. Moreover, if one uses some of the standard estimators the limiting assertions can be false. Nevertheless, we propose that the student get some practice in using the $\chi^2$ test with these standard estimators even though that may be mathematically unethical.

### Example 14.3.4

Test at the level of $\alpha = 0.10$, the hypothesis that a normal distribution is applicable to the following data:

$$-7.32, \ -6.97, \ -6.15, \ -4.88, \ -4.75, \ -4.32, \ -4.16, \ -4.01,$$
$$-3.88, \ -3.79, \ -3.64, \ -3.44, \ -3.21, \ -3.17, \ -3.06, \ -2.94,$$
$$-2.81, \ -2.69, \ -2.44, \ -2.38, \ -2.33, \ -2.28, \ -2.26, \ -2.19,$$
$$-2.15, \ -2.12, \ -2.07, \ -2.05, \ -2.04, \ -2.01, \ -1.91, \ -1.86,$$
$$-1.69, \ -1.64, \ -1.51, \ -1.46, \ -1.30, \ -1.19, \ -1.08, \ -1.02,$$
$$-0.94, \ -0.76, \ -0.65, \ -0.38, \ -0.20, \ -0.17, \ -0.02, \ \ 0.17,$$
$$0.38, \quad 0.42, \quad 0.67, \quad 0.94, \quad 1.36, \quad 1.52, \quad 1.95, \quad 2.13,$$
$$2.61, \quad 2.95, \quad 3.36, \quad 3.84$$

If we average the data we obtain $\hat{m} = -1.61$ to two places. We estimate $\sigma^2$ by

$$\tfrac{1}{60} \sum_{j=1}^{60} (x_j - 1.61)^2 = 4.72$$

and hence $\hat{\sigma} = 2.18$. We choose cells as follows:

$$[\hat{m}, \hat{m} + 0.2\hat{\sigma}], \quad [\hat{m} + 0.2\hat{\sigma}, \hat{m} + 0.4\hat{\sigma}], \quad [\hat{m} + 0.4\hat{\sigma}, \hat{m} + 0.7\hat{\sigma}],$$

$$[\hat{m} + 0.7\hat{\sigma}, \hat{m} + \hat{\sigma}], \quad [\hat{m} + \hat{\sigma}, \hat{m} + 1.5\hat{\sigma}], \quad \{x; \hat{m} + 1.5\hat{\sigma} < x\}$$

and their reflections in the line $x = \hat{m}$. Thus we would want the number of items of data in the following intervals:

| | |
|---|---|
| $x \leq -4.88$ | $-1.61 < x \leq -1.174$ |
| $-4.88 < x \leq -3.79$ | $-1.174 < x \leq -0.738$ |
| $-3.79 < x \leq -3.136$ | $-0.738 < x \leq -0.084$ |
| $-3.136 < x \leq -2.482$ | $-0.084 < x \leq 0.57$ |
| $-2.482 < x \leq -2.046$ | $0.57 < x \leq 1.66$ |
| $-2.046 < x \leq -1.61$ | $1.66 < x$ |

We tabulate our results by numbering the cells $1, 2, \ldots, 12$ in the order presented and by ascertaining the probability of the cells from Table 4.

### Table 4

| CELL $j$ | 1 | 2 | 3 | 4 | 5 | 6 |
|---|---|---|---|---|---|---|
| $n\hat{\pi}_j$ | 4.008 | 5.514 | 4.998 | 6.156 | 4.566 | 4.758 |
| OBSERVED IN CELL $j$ | 4 | 6 | 4 | 4 | 10 | 6 |

| CELL $j$ | 7 | 8 | 9 | 10 | 11 | 12 |
|---|---|---|---|---|---|---|
| $n\hat{\pi}_j$ | 4.758 | 4.566 | 6.156 | 4.998 | 5.514 | 4.003 |
| OBSERVED IN CELL $j$ | 4 | 4 | 4 | 4 | 4 | 6 |

A calculation gives

$$\sum_{j=1}^{12} \frac{(\hat{N}_j - n\hat{\pi}_j)^2}{n\hat{\pi}_j} = 10.41$$

where $\hat{N}_j$ is the number observed in the $j$th cell. Since we have estimated two parameters and since there are 12 cells, we want $12 - 1 - 2 = 9$ degrees of freedom. Since $P(Z_9 \leq 12.2) = 0.90$, $P(Z_9 > 12.2) = 0.10$. Since $10.41 < 12.2$ we accept the hypothesis that the normal distribution is applicable.

Although a number of $n\hat{\pi}_j$ are somewhat less than five, there are quite a few cells. Moreover, they are not much less than five.

Note that the parameters were estimated by the maximum-likelihood estimates. This is not strictly legal but we will use this procedure nonetheless.

Another problem about which the $\chi^2$ provides information is illustrated by the following example.

### Example 14.3.5

Two populations can be classed into four cells each. Independent sampling is done on each with the data displayed in Table 5. Are the statistical characteristics of these two populations the same?

First of all it seems that we should not come to the conclusion that the statistical characteristics of these two populations are the same. Cells 2 and 4 are quite disparate. On the other hand, chance variations may account for these disparities.

On the basis of the sampling it would seem that the probability of cell 1 is $\frac{11}{60}$, that of cell 2 is $\frac{13}{60}$, that of cell 3 is $\frac{14}{60}$, and that of cell 4 is $\frac{22}{60}$. On this basis, keeping in mind that population 1 is sampled 32 times while population 2 is

**Table 5**

| CELL | 1 | 2 | 3 | 4 | total |
|------|---|---|---|---|-------|
| POPULATION 1 | 5 | 3 | 9 | 15 | 32 |
| POPULATION 2 | 6 | 10 | 5 | 7 | 28 |
| TOTAL | 11 | 13 | 14 | 22 | 60 |

**Table 6**

| CELL | 1 | 2 | 3 | 4 |
|------|---|---|---|---|
| POPULATION 1 | $\frac{88}{15}$ | $\frac{104}{15}$ | $\frac{112}{15}$ | $\frac{176}{15}$ |
| POPULATION 2 | $\frac{77}{15}$ | $\frac{91}{15}$ | $\frac{98}{15}$ | $\frac{154}{15}$ |

sampled only 28 times, the expected number of occurrences are given in Table 6 in corresponding positions. We form

$$\frac{(-\frac{13}{15})^2}{\frac{88}{15}} + \frac{(\frac{13}{15})^2}{\frac{77}{15}} + \frac{(-\frac{59}{15})^2}{\frac{104}{15}} + \frac{(\frac{59}{15})^2}{\frac{91}{15}} + \frac{(\frac{23}{15})^2}{\frac{112}{15}} + \frac{(-\frac{23}{15})^2}{\frac{98}{15}} + \frac{(\frac{49}{15})^2}{\frac{176}{15}} + \frac{(-\frac{49}{15})^2}{\frac{154}{15}} \approx 7.68$$

Now to what should we compare this number? Note that there are eight positions. There are two constraints from the fact that the sum of the entries for population 1 is 32 and the sum of the entries for population 2 is 28. Then, three parameters were estimated since the sum of the probabilities is 1 and the probability of cell 4 could be derived from those of cells 1, 2, and 3. Therefore we should subtract three more degrees of freedom.

As a matter of fact, the theory justifies exactly this. Thus, $8 - 2 - 3 = 3 = (4 - 1)(2 - 1)$ is the number of degrees of freedom. Now $P(Z_3 > 6.25) = 0.10$. Therefore, we would reject the hypothesis that the distributions are the same at the 0.10 level. We would not reject it at the 0.05 level and we certainly would not reject it at the 0.02 level.

Tables such as Table 5 are called *contingency tables*. Testing whether two (or more) populations have the same distribution (statistical characteristics) is called testing for *homogeneity*. We will see another use for contingency tables shortly.

We may describe the general contingency table for homogeneity by Table 7. Let $n_{j1} + n_{j2} + \cdots + n_{jk} = n_{j.}$ for $j = 1, 2, \ldots, v$. Let $n_{11} + n_{21} + \cdots + n_{vl} = n_{.l}$ for $l = 1, 2, \ldots, k$. Let the sum of all the entries be $n$.

To test for homogeneity we form

**(14.3.1)** $$\sum_{j=1}^{v} \sum_{l=1}^{k} \frac{\left(n_{jl} - \dfrac{n_{.l}n_{j.}}{n}\right)^2}{\dfrac{n_{.l}n_{j.}}{n}}$$

### Table 7

| CELL | 1 | 2 | 3 | $\cdots$ | $k$ | total |
|------|---|---|---|------|-----|-------|
| POPULATION 1 | $n_{11}$ | $n_{12}$ | $n_{13}$ | | $n_{1k}$ | $n_{1.}$ |
| POPULATION 2 | $n_{21}$ | $n_{22}$ | $n_{23}$ | | $n_{2k}$ | $n_{2.}$ |
| $\vdots$ | $\vdots$ | | | | | |
| POPULATION $v$ | $n_{v1}$ | $n_{v2}$ | $n_{v3}$ | | $n_{vk}$ | $n_{v.}$ |
| TOTAL | $n_{.1}$ | $n_{.2}$ | $n_{.3}$ | | $n_{.k}$ | $n$ |

and deal with it as though its distribution is approximately that of a $\chi^2$ random variable with

$$vk - v - (k - 1) = (v - 1)(k - 1)$$

degrees of freedom.

Note that $n_{.l}/n$ is the *estimate* for the probability of the $l$th cell.

One could envision that $2 \times 2$ contingency tables would be very common. They would apply to two populations and to a *yes* or *no* (defective or acceptable, heads or tails) breakdown of the individuals of the population. By (14.3.2) $(2 - 1)(2 - 1) = 1$ degree of freedom is appropriate. However, the expression (14.3.1) is not too good an approximation of $Z_1$. Therefore, a correction, called *Yates' correction*, is frequently used. This correction is explained in the next example.

### Example 14.3.6

Two shipments of fruit are inspected. A piece of fruit is classified as good quality or inferior quality. The results are displayed in Table 8. Is it reasonable to assume that both shipments have the same underlying quality characteristics? Test on the 0.10 level.

Following the notation of (14.3.1) we have $n_{1.} = 237$, $n_{2.} = 133$, $n_{.1} = 107$, $n_{.2} = 263$, and $n = 370$. We will need

$$73 - \frac{237 \cdot 107}{370}, \qquad 164 - \frac{237 \cdot 263}{370}, \qquad 34 - \frac{133 \cdot 107}{370}, \qquad 99 - \frac{133 \cdot 263}{370}$$

**Table 8**

|  | INFERIOR | GOOD |
|---|---|---|
| SHIPMENT 1 | 73 | 164 |
| SHIPMENT 2 | 34 | 99 |

Now

$$\frac{237 \cdot 107}{370} \approx 68.54, \qquad \frac{237 \cdot 263}{370} \approx 168.46,$$

$$\frac{133 \cdot 107}{370} \approx 38.46, \qquad \frac{113 \cdot 263}{370} \approx 94.54$$

Normally we would compute $73 - 68.54$, $164 - 168.46$, $34 - 38.46$, and $99 - 94.54$, but the Yates correction requires a subtraction of $\frac{1}{2}$ from the absolute value of these numbers. We obtain

$$|73 - 68.54| - \tfrac{1}{2} = 4.46 - 0.50 = 3.96$$
$$|164 - 168.46| - \tfrac{1}{2} = 4.46 - 0.50 = 3.96$$
$$|34 - 38.46| - \tfrac{1}{2} = 4.46 - 0.50 = 3.96$$
$$|99 - 94.56| - \tfrac{1}{2} = 4.46 - 0.50 = 3.96$$

We then compute

$$\frac{(3.96)^2}{68.54} + \frac{(3.96)^2}{168.46} + \frac{(3.96)^2}{38.46} + \frac{(3.96)^2}{94.56} \approx 0.8$$

This is much less than the rejection level 2.71 arising from $P(Z_1 > 2.71) = 0.10$.

Another natural problem for the $\chi^2$ test is that of "determining" independence of two (or more) characteristics of the same population. Specifically, suppose $X$ and $Y$ are two random variables on $S_0$, the model of a population. Presumably for each $w \in S_0$, $X(w)$ and $Y(w)$ refer to some real-world characteristics. Is it reasonable to assume that $X$ and $Y$ are independent in some particular case?

### Example 14.3.7

The following data* were obtained concerning price changes of certain stocks and the second vowel in the corporate name. Are these two characteristics independent?

One should be suspicious of independence here. Line 0 has many changes of $-\frac{5}{8}$ or less. Line I has few changes of large absolute value. To obey the official definition of random variable, let us code A as 1, E as 2, I as 3, O as 4, and U

*N.Y. Stock Exchange Transactions, April 18, 1968.

## Table 9

PRICE CHANGE FROM PREVIOUS DAY

|  |  | $-\frac{5}{8}$ or less | $-\frac{1}{2}$ to $-\frac{1}{8}$ | $\frac{1}{8}$ to $\frac{1}{2}$ | $\frac{5}{8}$ to $1\frac{1}{2}$ | more than $1\frac{1}{2}$ |  |
|---|---|---|---|---|---|---|---|
|  | A | 3 | 16 | 16 | 6 | 5 | 46 |
| SECOND | | | | | | | |
| VOWEL | E | 3 | 13 | 17 | 19 | 6 | 58 |
| IN | | | | | | | |
| CORPORATE | I | 1 | 9 | 15 | 11 | 1 | 37 |
| NAME | | | | | | | |
|  | O | 8 | 11 | 10 | 8 | 6 | 43 |
|  | U | 3 | 5 | 5 | 2 | 1 | 16 |
|  |  | 18 | 54 | 63 | 46 | 19 | 200 |

as 5. Thus $X(w) = \{1, 2, 3, 4, 5\}$ where $w$ is a listed stock and $X(w)$ is the code for the second vowel of the corporate title. We can take $Y(w)$ to be the price change of this stock from the previous day. If $X$ and $Y$ are independent, we should have $P(X = j$ and $Y = k) = P(X = j)P(Y = k)$. Thus, for example, $P(X = 2$ and $\frac{1}{8} \leq Y \leq \frac{1}{2}) = P(X = 2)P(\frac{1}{8} \leq Y \leq \frac{1}{2})$. Since we have no commitments to the underlying distribution of price changes and second vowel of the corporate title, we must estimate the probabilities $P(X = j)$ for $j = 1, 2, \ldots, 5$ and $P(Y \leq -\frac{5}{8}), P(-\frac{1}{2} \leq Y \leq -\frac{1}{8}), P(\frac{1}{8} \leq Y \leq \frac{1}{2}), P(\frac{5}{8} \leq Y \leq 1\frac{1}{2}), P(1\frac{5}{8} \leq Y)$ which are the only germane categories in view of the given data. Such estimates are completely natural. We can take $P(Y \leq -\frac{5}{8}) = \frac{18}{200}$, $P(\frac{1}{8} \leq Y \leq \frac{1}{2}) = \frac{63}{200}$, $P(1\frac{5}{8} \leq Y) = \frac{19}{200}$, $P(x = 2) = \frac{58}{200}$ and so on. We are merely using the observed relative frequencies. These can be shown to be maximum-likelihood estimates.

On the basis of the assumed independence and the estimates made, it seems natural that, e.g., $P(x = 2$ and $\frac{1}{8} \leq Y \leq \frac{1}{2}) = \frac{58}{200} \cdot \frac{63}{200}$, $P(x = 4$ and $1\frac{5}{8} \leq Y) = \frac{43}{200} \cdot \frac{19}{200}$ etc. Hence, we multiply the estimates for the events whose intersection is under consideration. This, too, gives a maximum-likelihood estimate.

Now what are the $n\hat{\pi}_j$? Three examples should suffice. For $X = 2$, $\frac{5}{8} \leq Y \leq 1\frac{1}{2}$ we should have $200 \cdot \frac{58}{200} \cdot \frac{46}{200} = n\hat{\pi}_j$. For $X = 5$, $Y \leq -\frac{5}{8}$ we should have $200 \cdot \frac{16}{200} \cdot \frac{18}{200}$. For $x = 3$, $-\frac{1}{2} \leq Y \leq -\frac{1}{8}$ we should have $200 \cdot \frac{37}{200} \cdot \frac{54}{200}$. In each of these cases $n\hat{\pi}_j$ is exactly the number which one obtains in the test for homogeneity. Thus, the Pearson's $\chi^2$ quantity should be identical to (14.3.1). This is not all, however. If we consider a $v$ by $k$ contingency table which we are testing for independence, we may compute the number of degrees of freedom as

$$vk - (v - 1) - (k - 1) - 1 = vk - v - k + 1 = (v - 1)(k - 1)$$

because we estimate $v - 1 + k - 1$ parameters and there is a sample size of $n$.

All computation is identical to that in the test for homogeneity although the results we seek are quite different.

In the present problem we found Pearson's $\hat{\chi}^2$ to be 25.18. There are $(5 - 1)(5 - 1) = 16$ degrees of freedom. The assumption of independence would be rejected at the level $\alpha = 0.10$ but not at the level $\alpha = 0.05$. Nonetheless, the reader should not rush out and buy stock whose corporate title has second vowel I in preference to that whose second vowel is U.

It should be clear that the $\chi^2$ test should be used with some caution.

In essence the $\chi^2$ test has a null hypothesis $H_0$ which may be simple or composite. The alternative $H_1$ has the form $H_0'$. However, traditionally, little attention is paid to the second deficiency because it varies considerably according to which $P_1 \in H_1$ is assumed to be in effect. For example, a biased coin whose head bias $P(\text{H}) = 0.4999$ will invariably escape detection in a $\chi^2$ test for a fair coin. But this is quite reasonable. For most practical purposes such a coin is indistinguishable from a fair coin. A coin whose head bias is $P(\text{H}) = \frac{4}{10}$ will be detected with probability about 0.75, when $n \approx 400$ in a $\chi^2$ test for a fair coin with $\alpha = 0.10$. If $\alpha = 0.20$, $n \approx 270$ will do as good a job. If we consider a bias with $P(\text{H}) = \frac{3}{10}$ and $\alpha = 0.10$, $n \approx 100$ will suffice to detect a biased coin with probability of about 0.75.

These remarks are not made to disparage the $\chi^2$ tests, especially so because there is no difference between the $\chi^2$ test for $H_0:P(\text{H}) = 0.5$ and $H_1:P(\text{H}) \neq 0.5$ and the best unbiased test, but to point to the difficulties inherent in testing when $H_1$ is the complement of $H_0$.

**EXERCISES 14.3** _____

Exercises 1–3 refer to the following table.

**Table 10**

| $n$ | 0 | 1 | 2 | 3 | 4 | 5 | 6 | 7 | 8 | 9 | 10 | 11 | 12 |
|---|---|---|---|---|---|---|---|---|---|---|---|---|---|
| FREQUENCY | 3 | 12 | 18 | 14 | 8 | 3 | 1 | 0 | 1 | 0 | 0 | 0 | 0 |

1   Suppose that the $\chi^2$ test is to be used to decide whether a negative binomial probability function

$$P(X = n) = (-1)^n \binom{-r}{n} p^n (1 - p)^r$$

for $n = 0, 1, 2, 3, \ldots$ would apply to the data of Table 10. Both $0 < p < 1$ and $r > 0$ are to be estimated. Discuss pooling and the number of degrees of freedom. Do not estimate $p$ or $r$. Assume that the observations for $n \geq 12$ are all zero. See exercise 1.5.4.

2 Discuss the problem in exercise 1 if a regular binomial distribution of the form $\binom{12}{n} p^n (1 - p)^{12-n}$ for $0 \leq n \leq 12$ is assumed. Carry out the test.

3 Suppose that the experiments leading to the data in Table 10 were repeated and essentially the same results were found. How would this affect the $\chi^2$ test when all of the data are pooled? Assume that exactly the same results were obtained while recognizing that small variations would not affect your conclusion much.

4 Use the $\chi^2$ test with $\alpha = 0.10$ to decide whether the data of Example 14.3.4 come from a random variable whose density is $\pi^{-1}[1 + (x - \theta)^2]^{-1}$ for all $x$. Estimate $\theta$ by the median. The density is a generalized Cauchy density.

5 Explain why $|n_{jl} - n_j.n._l/n|$ is independent of $j$ and $l$ in a $2 \times 2$ contingency table.

## Table 11

| State | ELEMENTARY | | | HIGH SCHOOL | | COLLEGE | |
|---|---|---|---|---|---|---|---|
| | less than 5 | 5–7 | 8 | 1–3 | 4 | 1–3 | 4 or more |
| N.H. | 15 | 41 | 76 | 65 | 94 | 30 | 25 |
| N.J. | 251 | 489 | 661 | 734 | 885 | 277 | 303 |
| N.Y. | 785 | 1187 | 1907 | 2111 | 2431 | 804 | 902 |
| Ohio | 292 | 661 | 1025 | 1144 | 1469 | 412 | 375 |
| Ore. | 33 | 88 | 191 | 202 | 286 | 112 | 84 |
| Minn. | 74 | 195 | 486 | 279 | 480 | 192 | 138 |

Table 11 gives the number in 1000's of people aged 25 or over in various education attainment levels. Exercises 6–10 ask that the table be treated as a contingency table as is.

6 Test rows Ore. and Minn. for homogeneity with $\alpha = 0.10$.

7 Test rows N.J. and N.Y. for homogeneity with $\alpha = 0.30$.

8 Test rows N.J. and Ohio for homogeneity with $\alpha = 0.10$.

**9**   Test rows N.H., Ore., and Minn. for homogeneity with $\alpha = 0.10$.

**10**   On what real-world bases would it be rather ridiculous to think of data of this kind as forming a contingency table to which the $\chi^2$ test may be applied?

**11**   Suppose that a population has three characteristics which we wish to test for independence. Extend the concept of the $\chi^2$ test for independence of two characteristics. Assume that characteristics 1, 2, 3 have $r_1$, $r_2$, and $r_3$ cells respectively. How many degrees of freedom should be used?

# 14.4   Other tests of hypotheses

In this section we discuss certain tests of hypotheses whose theory is outside the scope of this text. However, the procedures for using them are fairly direct.

The critical regions of the likelihood ratio tests of section 2 can be described as follows: Let $L_0$ be the likelihood function on $S$ associated with the (simple) null hypothesis $H_0$ and let $L_1$ be the likelihood function for $H_1$ (also assumed simple). The critical regions are of the form

$$C_c = \left\{ (x_1, x_2, \ldots, x_n) \in S; \; \frac{L_1(x_1, \ldots, x_n)}{L_0(x_1, \ldots, x_n)} \geq c \right\}$$

The quantity $L_1(x_1, \ldots, x_n)/L_0(x_1, \ldots, x_n)$ measures how much better the "data point" $x = (x_1, x_2, \ldots, x_n)$ satisfies $H_1$ than $H_0$. Thus, if $L_1/L_0$ is large, we feel that $x$ is much more plausible under $H_1$ than under $H_0$. Similarly, an expression of the form $L_0/L_1$ measures the plausibility of $H_0$ compared to $H_1$ for $x$.

A suitable modification of these ratios provides a test of composite hypotheses. For convenience we assume that the likelihood functions are probability or density functions which depend on a parameter $\theta$. We write $L(x_1, \ldots, x_n; \theta) = L(x; \theta)$ to denote them.

Let $Q$ be the set of all parameters which are being considered in a problem. This is a vague description but it has advantages. Let

**(14.4.1)**   $\displaystyle \lambda(x) = \frac{\underset{\theta \in H_0}{\text{l.u.b.}}\, L(x; \theta)}{\underset{\theta \in Q}{\text{l.u.b.}}\, L(x; \theta)}$

As usual, l.u.b. denotes the least upper bound. The numerator measures how the data point $x$ fits "best" under $H_0$. The denominator measures how $x$ fits "best" under $Q \supset H_0$. Since $Q \supset H_0$, $0 \leq \lambda \leq 1$. Furthermore, $\lambda$ is small when $x$ is implausible under all $\theta \in H_0$. We use the test with critical region. $\{ x \in S; \; \lambda(x) \leq c \}$ where $c > 0$ is a suitable constant chosen so that $P(\text{reject})$ under $H_0$ is $\leq$ some prescribed $\alpha$.

### Example 14.4.1

Assuming only normal densities, suppose $H_0$ is the hypothesis $m = 0$, $\sigma^2 = 1$ and $H_1$ is $m = \pm 1$, $\sigma^2 = 1$. What is the nature of $\lambda$ in (14.4.1) for this problem? Assume no other parameters than those from $H_0$ and $H_1$.

In this problem

$$L(x; m) = \prod_{j=1}^{n} (2\pi)^{-1/2} [\exp(-\tfrac{1}{2}(x_j - m)^2]$$

$$= (2\pi)^{-n/2} \exp \left[ -\frac{1}{2} \sum_{j=1}^{n} (x_j - m)^2 \right]$$

We let

$$\bar{x} = \frac{1}{n}(x_1 + \cdots + x_n) \quad \text{and} \quad v^2 = \frac{1}{n} \sum_{j=1}^{n} (x_j - \bar{x})^2$$

It is easy to verify that

$$v^2 + \bar{x}^2 = \frac{1}{n} \sum_{j=1}^{n} x_j^2$$

Now

$$\sum_{j=1}^{n} (x_j - m)^2 = \sum_{j=1}^{n} (x_j^2 - 2mx_j + m^2) = n(v^2 + \bar{x}^2) - 2nm\bar{x} + nm^2$$

$$= n[v^2 + (\bar{x}^2 - 2m\bar{x} + m^2)] = n[v^2 + (m - \bar{x})^2]$$

Therefore,

$$L(x; m) = (2\pi)^{-n/2} \exp \left[ -\frac{n}{2} v^2 - \frac{n}{2}(m - \bar{x}^2) \right]$$

Note that $(2\pi)^{-n/2} \exp \left( -\dfrac{n}{2} v^2 \right)$ does not depend on $m$. Therefore in forming $\lambda(x)$ this will appear both in the numerator and denominator. It can be ignored if it is ignored in both places. Now

$$\underset{m \in H_0}{\text{l.u.b.}} \exp \left[ -\frac{n}{2}(m - \bar{x})^2 \right] = \exp \left( -\frac{n}{2} \bar{x}^2 \right)$$

To obtain $\underset{m \in Q}{\text{l.u.b.}} \exp \left[ -\dfrac{n}{2}(m - \bar{x})^2 \right]$ note that we must minimize $(m - \bar{x})^2$. Thus, if $\bar{x} > \tfrac{1}{2}$ we get $(1 - \bar{x})^2$; if $\bar{x} < -\tfrac{1}{2}$ we get $(-1 - \bar{x})^2$; if $-\tfrac{1}{2} \le \bar{x} \le \tfrac{1}{2}$ we get $(\bar{x})^2$. Thus

$$\underset{m \in Q}{\text{l.u.b.}} \exp \left[ -\frac{n}{2}(m - \bar{x})^2 \right] = \begin{cases} \exp \left( -\dfrac{n}{2} \bar{x}^2 \right), & -\tfrac{1}{2} \le \bar{x} \le \tfrac{1}{2} \\[2ex] \exp \left[ -\dfrac{n}{2}(1 - \bar{x})^2 \right], & \tfrac{1}{2} < \bar{x} \\[2ex] \exp \left[ -\dfrac{n}{2}(-1 - \bar{x})^2 \right], & -\tfrac{1}{2} > \bar{x} \end{cases}$$

It can be verified that

$$\lambda(x) = \begin{cases} 1, & -\tfrac{1}{2} \le \bar{x} \le \tfrac{1}{2} \\ \exp\left[\dfrac{n}{2}(1 - 2\bar{x})\right], & \tfrac{1}{2} < \bar{x} \\ \exp\left[\dfrac{n}{2}(1 + 2\bar{x})\right], & -\tfrac{1}{2} > \bar{x} \end{cases}$$

It is not always so easy to compute $\lambda$ as in this example but it can often be done with a bit more work. To some extent we are maximizing likelihood functions as in maximum-likelihood estimation, but here we may have more constraints on the parameters than previously. This does not change matters theoretically, only practically.

**Definition**    The statistic

$$\lambda(x) = \frac{\underset{\theta \in H_0}{\text{l.u.b.}}\ L(x;\theta)}{\underset{\theta \in Q}{\text{l.u.b.}}\ L(x;\theta)}$$

is called a *likelihood ratio* statistic. A test of the form

$$A_c\{x \in S;\ \lambda(x) \le c\} \quad \text{where } 0 \le c \le 1$$

is called a *likelihood ratio test*.

There are three advantages to likelihood ratio tests. First, they always exist. Second, in the proper circumstances they are identical with most powerful or best, unbiased tests. Third, the limiting distribution of $\lambda$ for large $n$ under $H_0$ is known in good circumstances.

We can determine the constant $c$ by arranging for

$$\alpha - \underset{\theta \in H_0}{\text{l.u.b.}}\ P_\theta(A_c) = \underset{\theta \in H_0}{\text{l.u.b.}} \int_{A_c} L(x;\theta)\, dx$$

The quantity $\alpha$ is the largest first deficiency which occurs for $\theta \in H_0$ and $A_c$.

## Example 14.4.2

We consider densities on the positive real axis of the form

$$\frac{b^k x^{k-1} e^{-bk}}{\Gamma(k)}$$

for $k = 1, 2$ and $b > c$. For $n = 2$ find a likelihood ratio test with $\alpha = 0.20$ when $H_0$ is $k = 1$ and $H_1$ is $k = 2$.

Note that for $k = 1$ we have $be^{-bx}$, the usual exponential density with $b$

instead of $\lambda$. Its likelihood function is $b^n e^{-b(x_1 + x_2 + \cdots + x_n)} = b^n e^{-bn\bar{x}}$. For $b > 0$ this is largest when $b = (\bar{x})^{-1}$. Therefore,

$$\underset{(k,b) \in H_0}{\text{l.u.b.}} \ L(x; k, b) = (\bar{x})^{-n} e^{-n}$$

We study

$$L(x; 2, b) = b^{2n}(x_1 x_2 \cdots x_n) e^{-bn\bar{x}}$$

since $\Gamma(2) = 1$. We let $x_1 x_2 \cdots x_n = x^*$. We maximize $b^{2n} e^{-bn\bar{x}}$ by maximizing $\log b^{2n} e^{-2n\bar{x}}$. This gives $2n \log b - bn\bar{x}$. Its derivative is $2n/b - n\bar{x}$. Therefore it is largest when $b = 2(\bar{x})^{-1}$. Therefore $\underset{(k,b) \in H_1}{\max} \ L(x; k, b) = 2^{2n}(\bar{x})^{-2n} x^* e^{-2n}$. We have

$$\lambda(x) = \begin{cases} 1, & (\bar{x})^{-n} e^{-n} \geq 2^{2n}(\bar{x})^{-2n} x^* e^{-2n} \\ \dfrac{(\bar{x})^{-n} e^{-n}}{2^{2n}(\bar{x})^{-2n} x^* e^{-2n}} & (\bar{x})^{-n} e^{-n} < 2^{2n}(\bar{x})^{-2n} x^* e^{-2n} \end{cases}$$

since $Q = H_0 \cup H_1$.

We now investigate

$$\frac{(\bar{x})^{-n} e^{-n}}{2^{2n}(\bar{x})^{-2n} x^* e^{-2n}} = 2^{-2n} e^n (\bar{x})^n (x^*)^{-1}$$

We want $2^{-2n} e^n (\bar{x})^n (x^*)^{-1} \leq c < 1$. We now take $n = 2$ and let $x_2 = mx_1$. This gives

**(14.4.2)** $\quad 2^{-4} e^2 \left( \dfrac{x_1 + mx_1}{2} \right)^2 \dfrac{1}{x_1 mx_1} \leq c < 1 \quad$ or $\quad e^2 (1 + m)^2 \leq 2^5 mc$

This is

**(14.4.3)** $\quad \dfrac{1}{m} + 2 + m = \dfrac{(1 + m)^2}{m} \left( \dfrac{2^b}{e^2} \right) c$

Since $1/m + 2 + m$ takes its minimum for $m = 1$ we must have $4 \leq (2^6/e^2)c$ or $e^2/16 \leq c$ if there are to be any $m$ to satisfy (14.4.2). If $e^2/16 \leq c$ we note that $1/m + 2 + m$ is symmetric in $m$ and $1/m$ so the numbers $m$ satisfying (14.4.3) must lie in an interval $[m_1, 1/m_1]$ for $0 < m_1 \leq 1$ and $m_1$ depends on $c$. But $mx_1 = x_2$ so we have $m_1 \leq x_2/x_1 \leq 1/m_1$ as the criteria for our critical regions.

We now compute the first deficiency of the critical region $\{(x_1, x_2);$ $m_1 < x_2/x_1 < 1/m_1\}$ as a function of $\theta = (k, b) \in H_0$. That is, we compute

$$\int_0^\infty dx_1 \int_{m_1 x}^{x_1/m_1} b^2 e^{-bx_1} e^{-bx_2} \, dx_2$$

the integral of the likelihood function $L(x; 1, b)$ over the critical region. Since the inner integration is on $x_2$ we may write this as

$$\int_0^\infty be^{-bx_1}\left(\int_{m_1x_1}^{x_1/m_1} be^{-bx_2}\,dx_2\right)dx_1 = \int_0^\infty be^{-bx_1}(-e^{-bx_2})\big|_{m_1x_1}^{x_1/m_1}$$

$$= \int_0^\infty be^{-bx_1}(e^{-bm_1x_1} - e^{-b(1/m_1)x_1})\,dx_1$$

$$= \left(-\frac{b}{b+bm_1}e^{-b(1+m_1)x_1} + \frac{be^{-b(1+1/m_1)x_1}}{b+b\frac{1}{m_1}}\right)\bigg|_0^\infty$$

$$= \frac{1}{1+m_1} - \frac{1}{1+\frac{1}{m_1}}$$

$$= \frac{1}{1+m_1} - \frac{m_1}{m_1+1} = \frac{1-m_1}{1+m_1}$$

This is independent of $b$ so its l.u.b. is exactly $(1 - m_1)/(1 + m_1)$. Since this is to be $\alpha = 0.20$ we have $(1 - m_1)/(1 + m_1) = 0.20$. Therefore $m_1 = \frac{2}{3}$. The critical region is the set of points in the $(x_1, x_2)$ first quadrant above the line $x_2 = \frac{2}{3}x_1$ and below the line $x_2 = \frac{3}{2}x_1$.

Examples 14.4.1 and 14.4.2 give some idea of the difficulties of constructing likelihood ratio tests. Example 14.4.2 could have been more difficult if the first deficiency depended on $b$, but then we would have taken the l.u.b. and set it equal to $\alpha$. That would determine $m_1$.

We finally point out without proof that for suitable $H_0$ and $Q$

**(14.4.4)**     $\lim_{n \to \infty} P(-2 \log \lambda \le u) = P(Z_r \le u)$

where $Z_r$ is $\chi^2$ with $r$ degrees of freedom. A thorough discussion of (14.4.4) is difficult at this level. However if $\theta = (\theta_1, \theta_2, \ldots, \theta_k)$ is a point in $k$-dimensional space and if $L(x, \theta)$ is a suitably smooth function of $x$ and $\theta$ we expect (14.4.4) to hold. The number of degrees of freedom $r$ is determined as follows: Assuming that $Q = \{\theta\}$ has the structure of $k$-dimensional space and $H_0 \subset Q$ has the structure of $l$-dimensional space, the degrees of freedom is then the difference in dimension, namely $r = k - l$.

As an illustration of the varied uses of the likelihood ratio test consider the following example:

### Example 14.4.3

Groups of people in six geographic areas are sampled as to I.Q. on a fixed intelligence test. Assume that I.Q. score is $N(m_j, \sigma_j^2)$ for the $j$th group. Find likelihood ratio tests for $H_0 : \sigma_1^2 = \sigma_2^2 = \sigma_3^2 = \sigma_4^2 = \sigma_5^2 = \sigma_6^2$ against $H_1$ which says that $H_0$ is false.

Let $n_j$ be the number of people tested in the $j$th region. There is no reason to assume the $n_j$ to be equal. Our likelihood function is

**(14.4.5)**
$$\prod_{j=1}^{6} (2\pi\sigma_j^2)^{-n_j/2} \prod_{k=1}^{n_j} \exp\left[-\frac{1}{2\sigma_j^2}(x_{jk} - m_j)^2\right]$$

If we maximize subject to $\sigma_1^2 = \sigma_2^2 = \cdots = \sigma_6^2 = \sigma^2$ we get

$$-(n_1 + \cdots + n_6)\frac{1}{\sigma} - \frac{1}{\sigma^3}\sum_{j=1}^{6}\sum_{k=1}^{n_j}(x_{jk} - m_j)^2 = 0$$

after taking logarithms and differentiating with respect to $\sigma$. We also get

$$\sum_{k=1}^{n_j}\frac{2}{2\sigma_j^2}(x_{jk} - m_j) = 0, \qquad j = 1, 2, \ldots, 6$$

after differentiating the logarithm of (14.4.5) with respect to $m_j$. These yield

$$\hat{m}_j = \frac{\sum_{k=1}^{n_j} x_{jk}}{n_j}$$

and

$$\hat{\sigma}^2 = \frac{1}{n_1 + n_2 + \cdots + n_6}\sum_{j=1}^{6}\sum_{k=1}^{n_j}(x_{jk} - \hat{m}_j)^2$$

Let $n_1 + n_2 + \cdots + n_6 = n$. Thus

$$\underset{\theta \in H_0}{\text{l.u.b.}}\ L(x; \theta) = (2\pi)^{-n/2}(\hat{\sigma}^2)^{-n/2}e^{-n/2}$$

It is easy to see that if we make no restriction on the $\sigma_j^2$ we get

$$\underset{\theta \in \Omega}{\text{l.u.b.}}\ L(x; \theta) = (2\pi)^{-n/2}\prod_{j=1}^{6}(\hat{\sigma}_j^2)^{-n_j/2}e^{-n_j/2}$$

where

$$\hat{\sigma}_j^2 = \frac{1}{n_j}\sum_{k=1}^{n_j}(x_{jk} - \hat{m}_j)^2$$

Our tests are therefore of the form

$$\lambda(x) = \frac{(\hat{\sigma}^2)^{-n/2}}{\prod_{j=1}^{6}(\hat{\sigma}_j^2)^{-n_j/2}} \le c < 1$$

or

$$\lambda(x) = \frac{\hat{\sigma}_1^{n_1}\hat{\sigma}_2^{n_2}\cdots\sigma_6^{n_2}}{\left(\dfrac{n_1\hat{\sigma}_1^2 + n_2\hat{\sigma}_2^2 + \cdots + n_6\hat{\sigma}_6^2}{n}\right)^{n/2}} \le c < |$$

since

$$\hat{\sigma}^2 = \frac{1}{n} \sum_{j=1}^{6} n_j \hat{\sigma}_j^2$$

For small values of $n_j$ this is not the best approach but we will not go into that issue. For $n_1, n_2, \ldots, n_6$ large we would argue that $-2 \log \lambda$ has, approximately, a $\chi^2$ distribution with $2 \cdot 6 - 7 = 5$ degrees of freedom. We obtain $2 \cdot 6$ because $Q$ involves parameters of the form $(\sigma_1^2, m_1, \sigma_2^2, m_2, \sigma_3^2, m_3, \ldots, \sigma_6^2, m_6)$, a vector in 12-dimensional space. The set of points of this form satisfying $\sigma_1^2 = \sigma_2^2 = \cdots = \sigma_6^2$ can be shown to be 7-dimensional.

We now turn to a test which replaces the $\chi^2$ test in certain cases. We restrict ourselves to random variables on $S_0$ with continuous distribution. As a first step, we approximate the distribution function.

### Example 14.4.4

Independent sampling of size 8 yields the data $1.87, -2.62, -1.03, 0.21, -2.81, -0.86, -0.41, 0.45$. Find a plausible approximation to the distribution function $F$ of the random variable $X$ modeling the characteristic which yields this data.

In reasonable circumstances the data give a good idea of the structure of a random variable. Since we have nothing to go on except these data let us make commitments and express reservations later. For convenience we reorder the data in an increasing sequence; $-2.81, -2.62, -1.03, -0.86, -0.41, 0.21, 0.45, 1.87$. A random variable which takes these values with equal likelihood would have the distribution function

$$\hat{G}(u) = \begin{cases} 0, & u < -2.81 \\ 1/8, & -2.81 \le u < -2.62 \\ 1/4, & -2.62 \le u < -1.03 \\ 3/8, & -1.03 \le u < -0.86 \\ 1/2, & -0.86 \le u < -0.41 \\ 5/8, & -0.41 \le u < 0.21 \\ 3/4, & 0.21 \le u < 0.45 \\ 7/8, & 0.45 \le u < 1.87 \\ 1, & 1.87 \le u \end{cases}$$

This function is graphed in Figure 7.

Needless to say, we would not expect $\hat{G}$ to fit as close to $F$ as in the figure but it would be nice. Furthermore, if we took a much larger sample we would expect a good fit much of the time. We can easily prove that for large $n$, $(1/n)$ (number of random variables from $\{X_1, X_2, \ldots, X_n\} \le u$) is close to $P(X \le u)$ with high probability. But $F(u) = P(X \le u)$ and $(1/n)$ (number of random variables from $\{X_1, X_2, \ldots, X_n\} \le u$) is the mathematical abstraction of $\hat{G}(u)$. We therefore take $\hat{G}$ to be a plausible approximation of $F$.

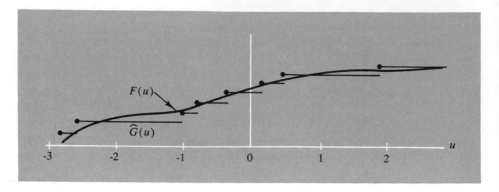

**Figure 7**

**Definition**    Let $S_0$ be the real numbers with a probability measure $P_0$. Let $S$ be the usual sample space of the sample with independent sampling and probability measure $P$. For $x = (x_1, x_2, \ldots, x_n) \in S$, the (random) function $G_x$ given by

**(14.4.6)**    $G_x(u) = \dfrac{1}{n}$ (number of $x_j$ in $(x_1, x_2, \ldots, x_n)$ such that $x_j \leq u$)

is called the *empirical distribution function* of $X(x) = x$.

The term "empirical" is a misnomer in this definition. The function $G_x$ is not empirical in the sense of its being determined by data. However, it models something which can be determined by data.

We now let

**(14.4.7)**    $D_n(x) = \underset{-\infty < u < \infty}{\text{l.u.b.}} \ |G_x(u) - F(u)|$

where $F$ is the distribution function of $X$ in the above definition and $G_x$ is defined by (14.4.6).

Since $F$ is monotonically increasing and $G_x$ is constant on intervals it follows that $F$ is farthest from $G_x$ at the endpoint of intervals of constancy of $G_x$ (see Figure 8) which illustrates the three relationships which a horizontal line segment can have with a (continuous) monotonic curve. Thus if $F$ is continuous and if the $x_1, x_2, \ldots, x_n$ in $x$ are all different and we let $G_x = G$,

$$D_n(x) = \max \{ \lim_{\substack{u \to x_j \\ u < x_j}} (|G(u) - F(u)|, \ \lim_{\substack{u \to x_j \\ u > x_j}} |G(u) - F(u)|\}$$

But

$$\lim_{u \to x_j} F(u) = F(x_j) \quad \text{and} \quad \lim_{\substack{u \to x_j \\ u > x_j}} G(u) = G(x_j)$$

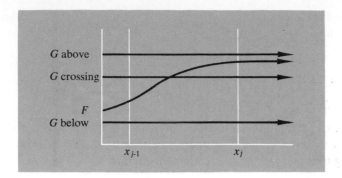

**Figure 8**

while

$$\lim_{\substack{u \to x_j \\ u < x_j}} G(u) = \begin{cases} G(x_{j-1}), & j > 1 \\ 0, & j = 1 \end{cases}$$

Therefore

$$D_n(x) = \max\{|0 - F(x_1)|, |G(x_1) - F(x_1)|, |G(x_1) - F(x_2)|, |G(x_2) - F(x_2)|, \ldots\}$$

A result of A. N. Kolmogorov states that if $F$ is continuous

**(14.4.8)**   $$\lim_{n \to \infty} P(\sqrt{n}\, D_n(x) \le u) = \begin{cases} \sum_{k=-\infty}^{\infty} (-1)^k e^{-2k^2 u^2}, & u > 0 \\ 0, & u \le 0 \end{cases}$$

This is another limit theorem and the function on the right-hand side is tabulated for various values of $u$.

We turn to an example of the use of Kolmogorov's result. For purposes of illustration we use a smaller value of $n$ than may be justified.

### Example 14.4.5

A telephone exchange records some lengths of telephone calls as 0.12, 0.42, 0.81, 1.59, 1.83, 2.22, 3.51, 4.14, 7.62 minutes. How does this compare with an assumption that call lengths are modeled by a random variable which is exponential with $\lambda = \frac{1}{3}$?

The distribution function of the random variable is $F(u) = 1 - e^{-u/3}$. Tables of $e^{-x}$ give

$$1 - e^{-0.12/3} = 0.039, \quad 1 - e^{-0.42/3} = 0.131, \quad 1 - e^{-0.81/3} = 0.237$$
$$1 - e^{-1.59/3} = 0.413, \quad 1 - e^{-1.83/3} = 0.457, \quad 1 - e^{-2.22/3} = 0.523$$
$$1 - e^{-3.51/3} = 0.680, \quad 1 - e^{-4.14/3} = 0.748, \quad 1 - e^{-7.62/3} = 0.921$$

and

$$\hat{G}(0.12) = 0.111, \quad \hat{G}(0.42) = 0.222, \quad \hat{G}(0.81) = 0.333,$$
$$\hat{G}(1.59) = 0.444, \quad \hat{G}(1.83) = 0.556, \quad \hat{G}(2.22) = 0.667,$$
$$\hat{G}(3.51) = 0.778, \quad \hat{G}(4.14) = 0.889, \quad \hat{G}(7.62) = 1.00$$

where $\hat{G}$ is the experimental empirical distribution function. It can be seen that $\hat{D}_9$, the result of substituting our data into the formulas for $D_9$, is

$$\hat{D}_9 = 0.144 = |\hat{G}(2.22) - F(2.22)|$$

Now $\sqrt{9}\,\hat{D}_9 = 0.432$. Tables show that $\lim_{n \to \infty} P(\sqrt{n}\,D_n(x) \le 0.432) = 0.0059$ or $\lim_{n \to \infty} P(\sqrt{n}\,D_n(x) > 0.432) = 0.9941$.

Thus, except for the reservation about $n = 9$ being large enough (and it certainly is not large enough) we should be extremely satisfied with our data as confirming that particular exponential distribution.

## EXERCISES 14.4

1  Derive the likelihood ratio tests for normal random variables where $H_0 : m = 0$ and $H_1 : m \ne 0$ are the hypotheses. Take $\sigma^2$ to be unspecified and $Q = H_0 \cup H_1$.

2  Let $\lambda$ be as in Example 14.4.1. Show that tests of the form $\{x;\ \lambda(x) \le c < 1\}$ are of the form $\{x;\ |\bar{x}| \ge a > \frac{1}{2}\}$. Find the test of this form whose first deficiency is 0.05 when $n = 25$.

3  Find the likelihood ratio tests for normal random variables where $H_0 : m = 0$, $\sigma^2 = 1$, $H_1 : m = 1$, $\sigma^2 = 1$, or $m = -2$, $\sigma^2 = 1$, and $Q = H_0 \cup H_1$.

4  Let $H_0$ and $H_1$ be simple hypotheses. Let $Q = H_0 \cup H_1$. Show that the most powerful tests are essentially the likelihood ratio tests of these hypotheses.

5  If $Q$ is the collection of all parameters for a family of density functions, show that

$$\text{l.u.b.}_{\theta \in Q} L(x_1, x_2, \ldots, x_n;\ \theta) = L(x_1, x_2, \ldots, x_n;\ \hat{\theta})$$

where $\hat{\theta}$ is the maximum likelihood estimator of $\theta$ at the point $(x_1, x_2, \ldots, x_n)$.

The next two exercises compare the likelihood ratio and $\chi^2$ test in the simplest case.

6  Let $P(X = j) = p_j \ge 0$ for $j = 1, 2, 3$ and $p_1 + p_2 + p_3 = 1$. Let $X_1, X_2, \ldots, X_n$ be $n$ independent random variables identically distributed with $X$. Let $H_0 : p_j = \pi_j$ for $j = 1, 2, 3$ where $\pi_j > 0$ are fixed for the problem be the null hypothesis. Let $H_1$ consist of all other choices of $p_j$. Let $N_j = $ number of $(X_1, \ldots, X_n)$ which are $j$. Show that

$$\lambda = \frac{\pi_1^{N_1} \pi_2^{N_2} \pi_3^{N_3}}{\left(\dfrac{N_1}{n}\right)^{N_1} \left(\dfrac{N_2}{n}\right)^{N_2} \left(\dfrac{N_3}{n}\right)^{N_3}}$$

Compute $\lambda$ in the case the $\pi_j$ are equal and $X(s) = (1, 2, 1, 1, 3, 2, 1, 2, 2, 3, 1, 1, 3, 3, 1)$.

7  Show that the $\lambda$ of exercise 6 satisfies

$$-2 \log \lambda = \sum_{j=1}^{3} 2N_j \log \frac{N_j}{n\pi_j}$$

Let $v_j = N_j - n\pi_j$ and $n\pi_j = e_j$. Show that $v_1 + v_2 + v_3 = 0$. Use the MacLaurin series $\log(1 + x) = x - x^2/2 + x^3/3 - x^4/4 \cdots$ to show that

$$-2 \log \lambda = \frac{v_1^2}{e_1} + \frac{v_2^2}{e_2} + \frac{v_3^2}{e_3} + v_1^3 B_1 + v_2^3 B_2 + v_3^3 B_3$$

$$= \text{(quantity used in a } \chi^2 \text{ test)}$$
$$+ \text{(term considerably smaller than the first term if } H_0 \text{ is true)}$$

Note the relationship between our result here and equation (14.4.4).

8  Find the likelihood ratio tests for normal random variables where $H_0 : m = 0$, $\sigma^2 = 1$, $H_1 : m = 1$, $\sigma^2 = 1$, or $m = -1$, $\sigma^2 = 2$, and $Q = H_0 \cup H_1$. How does the test "protect" against the larger variance $\sigma^2 = 2$ when $m = -1$.

9  Use Kolmogorov's test to decide whether the data 0.019, 0.035, 0.062, 0.068, 0.118, 0.193, 0.198, 0.222, 0.238, 0.279, 0.296, 0.317, 0.384, 0.431, 0.453, 0.470, 0.521, 0.543, 0.602, 0.668, 0.723, 0.741, 0.795, 0.803, 0.814, 0.861, 0.867, 0.893, 0.913, 0.956 can be thought of as arising from a uniform distribution on $0 \le x \le 1$. Work at the 0.10 level.

10  Use Kolmogorov's test to decide whether the data of exercise 9 can be thought of as arising from a random variable whose distribution function $F(x) = \frac{3}{2}x$ for $0 \le x \le \frac{1}{2}$ and $F(x) = \frac{1}{2} + \frac{1}{2}x$ for $\frac{1}{2} < x \le 1$; $F(x) = 0$ if $x < 0$ and $F(x) = 1$ if $x > 1$. Work at the 0.10 level.

11  Contrast the results of exercises 14.4.9 and 14.4.10 with those of Example 14.3.4 and exercise 14.3.4.

12  Let $u_1 < u_2 < \cdots < u_k$ be real numbers. Let $X$ be a random variable on $S_0$ with distribution function $F$. For $G_x(u)$ of (14.4.6) and $\epsilon > 0$ show that as the sample size $n \to \infty$ $P(|F(u_j) - G_x(u_j)| < \epsilon$ for $j = 1, 2, \ldots, k) \to 1$.

13  Suppose that we have a random variable on $S_0$ whose distribution function $F$ is continuous. Show that for any $\epsilon > 0$, $P(\max_{-\infty < u < \infty} |F(u) - G_x(u)| < \epsilon) \to 1$ as $n \to \infty$. [*Hint:* Note $F$ and $G$ are monotonic, $0 \le F(u)$, $G(u) \le 1$ and see exercise 12.]

# CHAPTER 15

# Some Topics in Sampling

## 15.1 Analysis of variance

A fundamental problem in statistics is illustrated by the following example.

**Example 15.1.1**
It is desired to ascertain whether students who are given lunch at school perform better than those who are not given lunch at school. Furthermore, is the amount of lunch they receive relevant?

It is conceivable that the nature of the school might be relevant in this problem. For example, eating lunch in a dingy lunchroom might have a different effect than eating in a pleasant lunchroom. In any event, suppose that a number of children are chosen at random in each of ten schools with varying facilities to receive sandwiches or a hot lunch. We number the schools 1, 2, ..., 10 and we use 1 to denote no lunch at school, 2 to denote a sandwich lunch, and 3 to denote hot lunch. We let $x_{jk}$ be the model for the average score on a standard test attained by the children in the experiment. From here on we can make a number of assumptions of differing degrees of controversy. We assume that the $x_{jk}$ are independent normal random variables.

The normality would not be too controversial since we are averaging what we could consider independent scores chosen at random from a fixed population of scores. Independence of the $x_{jk}$ seems reasonable because the children in the $j, k$ classification are different from those in the $j', k'$ classification if $(j, k) \neq$

$(j', k')$. Now come more controversial assumptions. Let $x_{jk}$ be $N(m_{jk}, \sigma_{jk}^2)$. We assume that $\sigma_{jk}^2 = \sigma^2$ does not depend on $j, k$. This is quite an assumption. We would certainly require that the number of children in the $j, k$ classification is the same for all $j$ and $k$. It could be hard to justify the assumption that each school would have the same variance in test taking potential. In addition to this we assume that

$$m_{jk} = a_j + b_k + c$$

where $c$ is an expected test average for all $j, k$, $a_j$ represents an addition (or subtraction), due to receiving a certain type of food, and $b_k$ represents an addition due to the variation of circumstances from one school to another. We assume $\sum\limits_{j=1}^{3} a_j = 0$ and $\sum\limits_{k=1}^{10} b_k = 0$. Suppose we wish to test the hypothesis that $a_1 = a_2 = a_3 = 0$; i.e., that receipt of lunch at school has no effect on performance on the test.

Let

$$\bar{x}_{.k} = \tfrac{1}{3}(x_{1k} + x_{2k} + x_{3k}) = \frac{1}{3} \sum_{j=1}^{3} x_{jk}$$

(15.1.1)    $$\bar{x}_{j.} = \frac{1}{10} \sum_{k=1}^{10} x_{jk} = \tfrac{1}{10}(x_{j1} + x_{j2} + \cdots + x_{j10})$$

and

$$\bar{x} = \frac{1}{10} \sum_{k=1}^{10} \bar{x}_{.k} = \frac{1}{3} \sum_{j=1}^{3} \bar{x}_{j.} = \frac{1}{30} \sum_{j=1}^{3} \sum_{k=1}^{10} x_{jk}$$

These are similar to quantities which we have used before. The $\bar{x}_{.k}$ are averages for a fixed school with different lunch situations. The $\bar{x}_{j.}$ are averages for a fixed lunch situation but for different schools.

We note that $\sum\limits_{k=1}^{10} (x_{jk} - \bar{x}_{j.}) = 0$ and $\sum\limits_{k=1}^{10} (\bar{x}_{.k} - \bar{x}) = 0$. Therefore

$$\sum_{k=1}^{10} \sum_{j=1}^{3} (x_{jk} - \bar{x}_{j.} - \bar{x}_{.k} + \bar{x})(\bar{x}_{.j} - \bar{x})$$

$$= \sum_{j=1}^{3} \sum_{k=1}^{10} (x_{jk} - \bar{x}_{j.} - \bar{x}_{.k} + \bar{x})(\bar{x}_{.j} - \bar{x})$$

(15.1.2)    $$= \sum_{j=1}^{3} (\bar{x}_{.j} - \bar{x}) \sum_{k=1}^{10} [x_{jk} - \bar{x}_{j.} - (\bar{x}_{.k} - \bar{x})]$$

$$= \sum_{j=1}^{3} (\bar{x}_{.j} - \bar{x})[\sum_{k=1}^{10} (x_{jk} - \bar{x}_{j.}) - \sum_{k=1}^{10} (\bar{x}_{.k} - \bar{x})]$$

$$= \sum_{j=1}^{3} (\bar{x}_{.j} - \bar{x})(0 - 0) = 0$$

Similarly

$$\sum_{k=1}^{10} \sum_{j=1}^{3} (x_{jk} - \bar{x}_{j.} - \bar{x}_{.k} + \bar{x})(\bar{x}_{.k} - \bar{x}) = 0$$

and

$$\sum_{k=1}^{10} \sum_{j=1}^{3} (x_{j.} - x)(x_{.k} - x) = 0$$

Therefore, since $x_{jk} - \bar{x} = \bar{x}_{j.} - \bar{x} + \bar{x}_{.k} - \bar{x} + x_{jk} - \bar{x}_{j.} - \bar{x}_{.k} + \bar{x}$ we have

$$\sum_{k=1}^{10} \sum_{j=1}^{3} (x_{jk} - \bar{x})^2 = \sum_{k=1}^{10} \sum_{j=1}^{3} (\bar{x}_{j.} - \bar{x} + \bar{x}_{.k} - \bar{x} + \bar{x}_{jk} - \bar{x}_{j.} - \bar{x}_{.k} + \bar{x})^2$$

$$= \sum\sum (\bar{x}_{j.} - \bar{x})^2 + \sum\sum (\bar{x}_{.k} - \bar{x})^2 + \sum (x_{jk} - \bar{x}_{j.} - \bar{x}_{.k} + \bar{x})^2$$
$$+ 2\sum\sum (\bar{x}_{j.} - \bar{x})(\bar{x}_{.k} - \bar{x})$$
$$+ 2\sum\sum (\bar{x}_{j.} - \bar{x})(\bar{x}_{jk} - \bar{x}_{j.} - \bar{x}_{.k} + \bar{x})$$
$$+ 2\sum\sum (\bar{x}_{.k} - \bar{x})(x_{jk} - \bar{x}_{j.} - \bar{x}_{.k} + \bar{x})$$

The last three sums were observed to be zero. Therefore

**(15.1.3)**   $$\sum_{k=1}^{10} \sum_{j=1}^{3} (x_{jk} - \bar{x})^2 = \sum\sum (\bar{x}_{j.} - \bar{x})^2 + \sum\sum (\bar{x}_{.k} - \bar{x})^2$$
$$+ \sum\sum (x_{jk} - \bar{x}_{j.} - \bar{x}_{.k} + \bar{x})^2$$

The term $\sum_{k=1}^{10} \sum_{j=1}^{3} (x_{jk} - \bar{x})^2$ suggests the variation of average test performances of the samples in different schools and different lunch situations. It is decomposed into three terms. Intuitively, $\sum_{k=1}^{10} \sum_{j=1}^{3} (\bar{x}_{j.} - \bar{x})^2$ is a measure of the variation of test performances due to different lunch arrangements (the different schools have been averaged out) while $\sum_{k=1}^{10} \sum_{j=1}^{3} (\bar{x}_{.k} - \bar{x})^2$ is the variation due to different schools (lunch arrangements have been averaged out). The other term on the right-hand side of (15.1.3) accounts for more individual variations after school and lunch effects have been eliminated.

By our assumptions

$$E(\bar{x}_{j.}) = \frac{1}{10} \sum_{k=1}^{10} E(x_{jk}) = a_j + \tfrac{1}{10}(b_1 + \cdots + b_{10}) + c = a_j + c$$

since $\sum b_k = 0$. Also

$$E(\bar{x}) = \frac{1}{3} \sum_{j=1}^{3} E(\bar{x}_{j.}) = \tfrac{1}{3}(a_1 + c + a_2 + c + a_3 + c) = c$$

since $a_1 + a_2 + a_3 = 0$. Under $H_0$, $a_j = 0$ so $E(\bar{x}_{j.}) = E(\bar{x})$. Therefore if lunch arrangement has no effect $\sum\limits_{j=1}^{3} (\bar{x}_{j.} - \bar{x})^2$ would be smaller than if lunch arrangement is a factor in test performance—i.e., $E(\bar{x}_{j.}) \neq E(\bar{x})$.

The $x_{jk}$ are independent $N(m_{jk}, \sigma^2)$ so the $\bar{x}_{j.}$ are independent and $N(a_j + c, \frac{1}{10}\sigma^2)$ random variables. If $a_j = 0$, $j = 1, 2, 3$ a slight modification of corollary 1 of Theorem 12.3.1 shows that

$$A = \sum_{j=1}^{3} \frac{(\bar{x}_{j.} - \bar{x})^2}{\frac{1}{10}\sigma^2} = \sum_{k=1}^{10} \sum_{j=1}^{3} \frac{(\bar{x}_{j.} - \bar{x})^2}{\sigma^2}$$

has the $\chi^2$ distribution with $3 - 1 = 2$ degrees of freedom since $\frac{1}{3}(\bar{x}_1 + \bar{x}_2 + \bar{x}_3)$ $= \bar{x}$ and the summation $\sum\limits_{k=1}^{10}$ merely adds ten copies of $\sum\limits_{j=1}^{3} \frac{(\bar{x}_{j.} - \bar{x})^2}{\sigma^2}$ together.

Furthermore we will see that $C = \sum\limits_{k=1}^{10} \sum\limits_{j=1}^{3} (x_{jk} - \bar{x}_{j.} - \bar{x}_{.k} + \bar{x})^2$ is independent of $A$ and that under $H_0$, $\frac{1}{\sigma^2} C$ has a $\chi^2$ distribution with $(3 - 1)(10 - 1) = 18$ degrees of freedom. Thus under $H_0$

$$V = \frac{9 \sum\limits_{j=1}^{3} \sum\limits_{k=1}^{10} (\bar{x}_{j.} - \bar{x})^2}{\sum\limits_{j=1}^{3} \sum\limits_{k=1}^{10} (x_{jk} - \bar{x}_{j.} - \bar{x}_{.k} + \bar{x})^2} = \frac{\dfrac{A}{2\sigma^2}}{\dfrac{C}{2 \cdot 9\sigma^2}}$$

has an $F$-distribution with 2, 18 as its indices. See exercise 15.1.9.

We would reject $H_0$ if $\hat{V} \geq k_0$ where $\hat{V}$ is the experimental value obtained by substituting the experimental test averages for $x_{jk}$ and $k_0$ is determined so that for $V$, an $F$ random variable with indices 2, 18, $P(V \geq k_0) = \alpha$, an acceptable first deficiency.

The method used in Example 15.1.1 is called *analysis of variance*. It breaks up the "sample variance," $\sum\limits_{k=1}^{10} \sum\limits_{j=1}^{3} (x_{jk} - \bar{x})^2$, into a crucial component such as $A$ and an interfering component such as $B = \sum\limits_{k=1}^{10} \sum\limits_{j=1}^{3} (\bar{x}_{.k} - \bar{x})^2$ as well as $C$ in order to obtain a more sensitive analysis of the effect of the crucial variable (school lunch).

We could contrast this experiment with certain other experiments. If we tested in one school only, the results could be criticized as applicable to the one school only. If we tested with the same groups of children in the ten schools we would have an improper experiment because we would be using the same number of children in each school and therefore not choosing children at random. Thus a pattern in which the effects of lunch given at school depends upon the character of the school could be obscured.

To see how the current method is more sensitive consider the information in Table 1, tabulated for six schools instead of ten. The numbers in the body of the table represent the average test score attained by the fixed number of children in each category.

**Table 1**

| | SCHOOLS | | | | | | AVERAGE |
| | 1 | 2 | 3 | 4 | 5 | 6 | $(\bar{x}_{j.})$ |
|---|---|---|---|---|---|---|---|
| NO LUNCH AT SCHOOL | 67 | 84 | 62 | 69 | 70 | 68 | 70 |
| SANDWICHES AT SCHOOL | 70 | 81 | 68 | 74 | 68 | 71 | 72 |
| HOT LUNCH AT SCHOOL | 75 | 83 | 69 | 73 | 72 | 72 | 74 |
| AVERAGE $(\bar{x}_{.k})$ | 70.7 | 82.7 | 66.3 | 72 | 70 | 70.3 | 72 $(\bar{x})$ |

The $V$ statistic is

$$\frac{5 \sum\limits_{j=1}^{3} \sum\limits_{k=1}^{6} (\bar{x}_{j.} - \bar{x})^2}{\sum\limits_{j=1}^{3} \sum\limits_{k=1}^{6} (x_{jk} - \bar{x}_{j.} - \bar{x}_{.k} + \bar{x})^2}$$

Therefore

$$\hat{V} = \frac{5 \cdot 6[(70 - 72)^2 + (72 - 72)^2 + (74 - 70)^2]}{48.68} = \frac{240}{48.68} \approx 4.9$$

Since $P(V \geq 4.10) = 0.05$ where $V$ has an $F$-distribution with 2, 10 degrees of freedom, we would reject the hypothesis that lunch at school has no effect on test performance at the 5% first deficiency level.

If we had not classified both as to school and lunch arrangement, and viewed the data of the table as resulting from diverse experiments, it seems quite clear that the deviations of 70 and 74 from 72 would be considered very small in comparison to the deviations of the entries from their averages. The latter deviations would suggest a value of the variance so large as to easily absorb the deviations $70 - 72$ and $74 - 72$.

To be more specific we note that

**(15.1.4)** $\sum\limits_{j=1}^{3} \sum\limits_{k=1}^{6} (x_{jk} - \bar{x})^2 = \sum \sum (x_{jk} - x_{j.} + x_{j.} - \bar{x})^2$

$$= \sum \sum (x_{jk} - x_{j.})^2 + 2 \sum \sum (x_{jk} - x_{j.})(x_{j.} - \bar{x})$$
$$+ \sum \sum (x_{j.} - \bar{x})^2$$

Just as we argued in the "two-way" classification above we can show that the middle term on the right is zero and that the two end terms on the right are independent under $H_0$. Furthermore

$$\frac{\sum\limits_{j=1}^{3}\sum\limits_{k=1}^{6}(\bar{x}_{j.} - \bar{x})^2/2}{\sum\limits_{j=1}^{3}\sum\limits_{k=1}^{6}(x_{jk} - \bar{x}_{j.})^2/15}$$

has the $F$-distribution with 2, 15 degrees of freedom. Now the experimental values analogous to $\sum\sum(x_{jk} - \bar{x})^2$ and $\sum\sum(x_{j.} - \bar{x})^2$ are 552 and 48, respectively. By subtraction the experimental analog to $\sum\limits_{j=1}^{3}\sum\limits_{k=1}^{6}(x_{jk} - \bar{x}_{j.})^2$ is $552 - 48 = 504$. Now $\frac{48}{2} \div \frac{504}{15} = \frac{360}{504} < 1$. This would lead to an acceptance of the hypothesis that the various lunch categories have no influence on test scores.

We now turn to the theory behind the assertion that the statistic $V$ has the $F$-distribution when the $a_j = 0$. We assume $j = 1, 2, \ldots, m$ and $k = 1, 2, \ldots, n$.

We first assume that $\sigma^2 = 1$. This leads to no loss of generality because any value of $\sigma^2 \neq 1$ appears both in the numerator and denominator of $V$ as a common factor.

Since the $x_{jk}$ are assumed to be independent, $\bar{x}_{1.}, \bar{x}_{2.}, \ldots, \bar{x}_{m.}$ are independent and they are $N\left(c, \dfrac{1}{n}\right)$ when $a_j = 0$ for $j = 1, 2, \ldots, m$. Since $\bar{x} = \dfrac{1}{m}\sum\limits_{j=1}^{m}\bar{x}_{j.}$ by corollary 1 of Theorem 12.3.1

$$m\sum\limits_{j=1}^{m}(\bar{x}_{j.} - \bar{x})^2 = \sum\limits_{k=1}^{n}\sum\limits_{j=1}^{m}(\bar{x}_{j.} - \bar{x})^2$$

is $\chi^2$ with $m - 1$ degrees of freedom. Similarly

$$\sum\limits_{j-1}^{m}\sum\limits_{k=1}^{n}(\bar{x}_{.k} - b_k - \bar{x})^2$$

is $\chi^2$ with $n - 1$ degrees of freedom. We now examine three random variables; namely,

**(15.1.5)**   $\bar{x}_{j.} - \bar{x}, \qquad \bar{x}_{.k} - b_k - \bar{x}, \quad$ and $\quad x_{lv} - \bar{x}_{.v} - \bar{x}_{l.} + \bar{x}$

These have a multidimensional normal distribution because they are linear combinations of the independent normal random variables $x_{jk}, j = 1, 2, \ldots, m$ and $k = 1, 2, \ldots, n$.

We show that the random variables of (15.1.5) are independent by showing that the covariance of any pair is zero. Let $y_{jk} = x_{jk} - b_k - c$. It is clear that $E(y_{jk}) = 0$ so $y_{jk}$ is $N(0, 1)$ and the set $\{y_{jk}\}$ is a set of independent random variables. Also

$$\bar{x}_{j.} - \bar{x} = \bar{y}_{j.} - \bar{y}, \qquad \bar{x}_{.k} - \bar{x} - b_k = \bar{y}_{.k} - \bar{y}$$

and

$$x_{lv} - \bar{x}_{.v} - \bar{x}_{l.} + \bar{x} = y_{lv} - \bar{y}_{.v} - \bar{y}_{l.} - \bar{y}$$

Using the techniques of Theorems 12.2.1 and 12.2.2 we have $E(\bar{y}_{j.}\bar{y}_{.k}) = \dfrac{1}{mn} \cdot 1$

since there is only one term of $\bar{y}_{j.}$ common to $\bar{y}_{.k}$, namely, $y_{jk}$. Its second moment is 1. Similarly $E(\bar{y}_{j.}y) = (1/m)(1/n) = 1/mn$ since $\bar{y} = (1/m)[\bar{y}_{1.} + \bar{y}_{2.} + \cdots + \bar{y}_{m.}]$ and $E(\bar{y}_{j.}^2) = n/n^2$. Now

$$E((\bar{y}_{j.} - \bar{y})(\bar{y}_{.k} - \bar{y})) = E(\bar{y}_{j.}\bar{y}_{.k}) - E(\bar{y}\bar{y}_{.k}) - E(\bar{y}_{j.}\bar{y}) + E(\bar{y}^2)$$

**(15.1.6)**
$$= \frac{1}{mn} - \frac{1}{mn} - \frac{1}{mn} + \frac{mn}{(mn)^2} = 0$$

Furthermore

$$E((\bar{y}_{j.} - \bar{y})(y_{lv} - \bar{y}_{.v} - \bar{y}_{l.} + \bar{y}))$$
$$= E((\bar{y}_{j.} - \bar{y})(y_{lv} - \bar{y}_{l.})) - E((\bar{y}_{j.} - \bar{y})(\bar{y}_{.v} - \bar{y}))$$

We proved in (15.1.6) that the second expectation is zero. Now since $E(y_{jk}) = 0$,

$$E((\bar{y}_{j.} - \bar{y})(y_{lv} - \bar{y}_{l.})) = E(\bar{y}_{j.}y_{lv}) - E(\bar{y}y_{lv}) - E(\bar{y}_{j.}\bar{y}_{l.}) + E(\bar{y}\bar{y}_{l.})$$
$$= \frac{1}{n} \delta_{jl} - \frac{1}{mn} - \frac{n}{nn} \delta_{jl} + \frac{1}{mn} = 0$$

since $\delta_{jl}$ appears because $y_{lv}$ enters into $\bar{y}_j$ if and only if $j = l$ and $\bar{y}_{j.}$ and $\bar{y}_{l.}$ are uncorrelated unless $j = l$. Similarly we can prove that

$$E((\bar{y}_{.k} - \bar{y})(y_{lv} - \bar{y}_{l.} - \bar{y}_{.v} + \bar{y})) = 0$$

Under $H_0$ we have

$$\sum\sum [x_{jk} - b_k - c - (\bar{x} - c)]^2 = \sum [(x_{jk} - b_k - \bar{x})^2]$$
$$= \sum\sum [(\bar{x}_{j.} - \bar{x})^2 + (\bar{x}_{.k} - b_k - \bar{x})^2$$

**(15.1.7)**
$$+ (x_{jk} - \bar{x}_{j.} - \bar{x}_{.k} + \bar{x})^2]$$
$$= \sum\sum [(\bar{x}_{j.} - \bar{x})^2 + (\bar{x}_{.k} - b_k - \bar{x})^2$$
$$+ (x_{jk} - \bar{x}_{j.} - \bar{x}_{.k} - \bar{x})^2]$$

because the mixed product terms are zero as in (15.1.2). The random variables $\bar{x}_{j.} - \bar{x}$, $\bar{x}_{.k} - b_k - \bar{x}$, and $x_{lv} - \bar{x}_{l.} - \bar{x}_{.v} + \bar{x}$ are independent for all choices of indices $j$, $k$, $l$, $v$. Therefore

$$A = \sum_{j=1}^{m} \sum_{k=1}^{n} (\bar{x}_{j.} - \bar{x})^2$$

$$B = \sum_{p=1}^{m} \sum_{q=1}^{n} (\bar{x}_{.q} - b_k - \bar{x})^2$$

and

$$C = \sum_{l=1}^{m} \sum_{v=1}^{n} (x_{lv} - \bar{x}_{l.} - \bar{x}_{.v} + \bar{x})^2$$

are independent. $A$ and $B$ are $\chi^2$ with $m - 1$ and $n - 1$ degrees of freedom, respectively. Also $\sum_{j=1}^{m} \sum_{k=1}^{n} (x_{jk} - b_k - \bar{x})^2$ is $\chi^2$ with $mn - 1$ degrees of freedom.

Now, by exercise 9.3.10 a $\chi^2$ random variable with $\nu$ degrees of freedom has characteristic function $(1 - 2it)^{-\nu/2}$. Since $A$, $B$, and $C$ are independent, we can take the characteristic function of the extremes of equation (15.1.7) and obtain

$$(1 - 2it)^{-(mn-1)/2} = (1 - 2it)^{-(m-1)/2}(1 - 2it)^{-(n-1)/2}\phi_C(t)$$

Therefore

$$\phi_C(t) = (1 - 2it)^{-[mn-1-(m-1)-(n-1)]/2}$$
$$= (1 - 2it)^{-[mn-m-n+1]/2} = (1 - 2it)^{-(m-1)(n-1)/2}$$

Thus we see that $C$ is $\chi^2$ with $(m - 1)(n - 1)$ degrees of freedom when $H_0$ is true. Therefore,

$$\frac{\dfrac{1}{m-1}A}{\dfrac{1}{(m-1)(n-1)}C} = \frac{(n-1)A}{C}$$

has an $F$-distribution with $m - 1$, $n - 1$ as its indexes when $H_0$ is true.

The material in this section constitutes the briefest of introductions to analysis of variance.

**EXERCISES 15.1** ———————————————————————————

The data in Table 2 represent the average mileage per gallon attained by cars in four weight classes using brands $X$, $Y$, and $Z$ of gasoline (the data are fictitious). Exercises 1–5 refer to these data.

### Table 2

CLASS OF CAR

| GASOLINE | 1 | 2 | 3 | 4 | AVERAGE |
|----------|-----|------|-----|------|---------|
| $X$ | 20 | 18 | 16 | 15 | 17.25 |
| $Y$ | 19 | 17.1 | 14 | 13.1 | 15.8 |
| $Z$ | 21 | 18 | 15 | 13 | 16.75 |
| AVERAGE | 20 | 17.7 | 15 | 13.7 | 16.6 |

1 How would you set up an experiment of this type? What assumptions would you be making if you intended to use analysis of variance to judge whether the three brands of gasoline are equally good?

2 Decide at the $\alpha = 0.05$ level whether the three brands are equally good using the two-way classification.

3  Under the conditions of exercise 2 decide whether all four weight classes can be assumed to be equal in gasoline economy.

4  Apply the one-way classification indicated in (15.1.5) to decide the issues in exercises 2 and 3.

5  Show that the experimental values 20, 17.7, 15, and 13.7 could reasonably come from independent normal random variables with expectation 15 and equal but unspecified variance. How does this relate to exercise 3?

6  Show that independent normal random variables $x_{11}$, $x_{12}$, $x_{13}$, $x_{21}$, $x_{22}$, $x_{23}$ with $E(x_{1k}) = 2$, $E(x_{2k}) = -2$, $\sigma^2(x_{j1}) = \sigma^2(x_{j2}) = 1$, and $\sigma^2(x_{j3}) = 25$ could, with reasonably large probability, produce data such as

| 2 | 3 | −1 |
|---|---|----|
| −2 | −2 | 0 |

and

| 2 | 2.5 | 0 |
|---|-----|---|
| −2 | −1.5 | −1 |

in the usual ordering. Proceed as though analysis of variance is legitimate and show that the inequality of $E(\bar{x}_{1.})$ and $E(\bar{x}_{2.})$ can frequently be lost.

7  Show, intuitively, that data such as that in the tables of exercise 6 would be rare if the expectations were unchanged but the variances were all equal to 1.

8  Discuss how the $F$-distribution can be applied to test equality of the variances if there are enough data.

♦ 9  Let $X$ and $Y$ be independent random variables having $\chi^2$ distributions with $m$ and $n$ degrees of freedom, respectively. Show that $\dfrac{X/m}{Y/n}$ has an $F$-distribution with $m$, $n$ degrees of freedom by using exercise 8.5.4.

# 15.2  Stratified sampling

Many practical sampling problems involve populations whose characteristics are known to some extent. For example, wealthy people, farmers, and New Englanders are more likely to be Republicans. Thus, if we wish to obtain information about preference for Republican candidates, we might weight our sample to include a disproportionate number of members of such groups. Naturally, if we can arrange to sample registered Republican's only, we might obtain still more significant results. Regardless, strict random sampling of all people in the United States would not be very efficient. Therefore, we study the elementary theory of *stratified sampling*; that is, sampling in which the population is divided into subpopulations in advance. Within these subpopulations we use random sampling.

We assume a finite population of numbers $S = \{x_{jk}; j = 1, \ldots, M$ and $k = 1, 2, \ldots, N_j\}$. The $j$th stratum (subpopulation) is $S_j = \{x_{jk}; k = 1, \ldots, N_j\}$. We discuss sampling for the arithmetic mean $m$. We use the following notation

**(15.2.1)**
$$N = \sum_{j=1}^{M} N_j$$

$$m_j = \frac{1}{N_j} \sum_{k=1}^{N_j} x_{jk}$$

Therefore,

$$m = \frac{1}{N} \sum_{j=1}^{M} \sum_{k=1}^{N_j} x_{jk} = \sum_{j=1}^{M} \frac{N_j}{N} \frac{1}{N_j} \sum_{k=1}^{N_j} x_{jk} = \sum_{j=1}^{M} \frac{N_j}{N} m_j$$

We also let

$$\sigma_j^2 = \frac{1}{N_j} \sum_{k=1}^{N_j} (x_{jk} - m_j)^2$$

**(15.2.2)**
$$\sigma^2 = \frac{1}{N} \sum_{j=1}^{M} \sum_{k=1}^{N_j} (x_{jk} - m)^2$$

$$d^2\dagger = \frac{1}{N} \sum_{j=1}^{M} N_j (m_j - m)^2$$

It should be noted that these quantities are underlying quantities. They are not *defined* by sampling.

As in analysis of variance we observe that

$$\sigma^2 = \frac{1}{N} \sum_{j=1}^{M} \sum_{k=1}^{N_j} (x_{jk} - m)^2 = \frac{1}{N} \sum_{j=1}^{M} \sum_{k=1}^{N_j} (x_{jk} - m_j + m_j - m)^2$$

**(15.2.3)**
$$= \frac{1}{N} \sum_{j=1}^{M} \sum_{k=1}^{N_j} [(x_{jk} - m_j)^2 + (m_j - m)^2]$$

$$= \frac{1}{N} \sum_{j=1}^{M} \sum_{k=1}^{N_j} (x_{jk} - m_j)^2 + \frac{1}{N} \sum_{j=1}^{M} N_j (m_j - m)^2$$

$$= \frac{1}{N} \sum_{j=1}^{M} N_j \sigma_j^2 + d^2$$

† This is a different use of the symbol $d^2$ than in Chapter 12.

We see that partitions of $S$ into strata for which $d^2$ is large result in $N_j \sigma_j^2$ being small since $\sigma^2$ is constant.

Since our population is finite it seems appropriate to study sampling without replacement as well as stratified sampling.

Suppose we have a population of numbers $S_0 = \{a_1, a_2, \ldots, a_n\}$. In sampling without replacement we consider

$$S = \{(y_1, y_2, \ldots, y_r); y_j \in S_0 \text{ and are all different}\}$$

The set $S$ has ${}_nP_r$ elements which we take to be equally likely. In order to keep random variables in the discussion we let $Y_k(s) = y_k$ for $s = (y_1, y_2, \ldots, y_s)$ and $1 \leq k \leq r$. It is easy to see that the $Y_k$ are identically distributed but that they are not independent.

We calculate the expected value and variance of $Y_k$. Let $m = \dfrac{1}{n} \displaystyle\sum_{j=1}^{n} a_j$.

Obviously $E(Y_k) = m$. We note that

$$E(Y_k^2) = \frac{1}{n} \sum_{j=1}^{n} a_j^2$$

If $k \neq l$,

$$P(Y_k = a_j \text{ and } Y_l = a_v) = \frac{{}_{n-2}P_{r-2}}{{}_nP_r}$$

because there are ${}_{n-2}P_{r-2}$ sample points having $a_j$ in the $k$th position and $a_v$ in the $l$th position. Therefore,

$$
\begin{aligned}
E(Y_k Y_l) &= \frac{(n-2)(n-3)\cdots(n-r+1)}{n(n-1)(n-2)\cdots(n-r+1)} \sum_{j \neq v} a_j a_v \\
&= \frac{1}{n(n-1)} [(a_1 + \cdots + a_n)^2 - (a_1^2 + a_2^2 + \cdots + a_n^2)] \\
&= \frac{1}{n(n-1)} [(nm)^2 - nE(Y_k^2)] = \frac{1}{n-1} [nm^2 - (\sigma^2 + m^2)] \\
&= \left[ m^2 - \frac{\sigma^2}{n-1} \right]
\end{aligned}
$$

where $\sigma^2 = E(Y_k^2) - m^2 = \sigma^2(Y_k)$.

By elementary results

(15.2.4)    $E(Y_1 + Y_2 + \cdots + Y_r) = rm$

and $\sigma^2(Y_1 + \cdots + Y_r) = E((Y_1 + \cdots + Y_r)^2) - (rm)^2$.

Therefore,

$$\sigma^2(Y_1 + \cdots + Y_r) = E((Y_1 + \cdots + Y_r)^2) - r^2 m^2$$
$$= rE(Y_1^2) + E(\sum_{j \neq v} Y_j Y_v) - r^2 m^2$$

$$= r(\sigma^2 + m^2)$$

**(15.2.5)**
$$+ (r^2 - r)\frac{1}{n-1}[(n-1)m^2 - \sigma^2] - r^2 m^2$$

$$= r\left(1 - \frac{r-1}{n-1}\right)\sigma^2 + (r + r^2 - r - r^2)m$$

$$= r\left(1 - \frac{r-1}{n-1}\right)\sigma^2$$

We now assume that we sample each stratum $S_j$ using sampling without replacement of size $r_j$. Letting $X_{j1}, X_{j2}, \ldots, X_{jr_j}$ be the random variables of this sampling, we have

$$\bar{X}_j = \frac{1}{r_j}(X_{j1} + X_{j2} + \cdots + X_{jr_j})$$

It seems clear that we can estimate $m$ by $Z = \sum_{j=1}^{M} \frac{N_j}{N}\bar{X}_j$. Moreover by (15.2.4)

$$E\left(\sum_{j=1}^{M} \frac{N_j}{N}\bar{X}_j\right) = \sum_{j=1}^{M} \frac{N_j}{N}E(\bar{X}_j) = \sum_{j=1}^{M} \frac{N_j}{N}\frac{1}{r_j}r_j m_j = \sum_{j=1}^{M} \frac{N_j}{N}m_j = m$$

so this "mixture" of the $\bar{X}_j$ is an unbiased estimator of $m$. Under normal conditions we would want to minimize the variance of $Z$. Now

**(15.2.6)**
$$\sigma^2(Z) = \sum_{j=1}^{M} \left(\frac{N_j}{N}\right)^2 \frac{1}{r_j}\left(1 - \frac{r_j - 1}{N_j - 1}\right)\sigma_j^2$$

$$= \sum_{j=1}^{M} \left(\frac{N_j}{N}\right)^2 \frac{1}{r_j}\left(\frac{N_j - r_j}{N_j - 1}\right)\sigma_j^2$$

since by (15.2.5)

$$\sigma^2(\bar{X}_j) = \frac{1}{r_j^2}\left(1 - \frac{r_j - 1}{N_j - 1}\right)$$

If we consider $S_1, \ldots, S_M$ to be fixed we can influence $\sigma^2(Z)$ by selecting $r_1, \ldots, r_M$. Let us assume that $r = r_1 + r_2 + \cdots + r_M$ is fixed.

Note that $\sigma^2(Z)$ does not involve $d^2$, the variation between strata. Furthermore, if we can arrange for $d^2$ to be close to $\sigma^2$, the number $N_j\sigma_j^2$ will be small and then $\sigma^2(Z)$ will be small.

## Example 15.2.1

Let $S = \{1, 2, 3, \ldots, 20\}$. Discuss stratified sampling using sampling without replacement and sample size 4 with the following strata.

(1)   $S_1 = \{1, 2, 3, \ldots, 10\}$,        $S_2 = \{11, 12, 13, \ldots, 20\}$
(2)   $S_1' = \{1, 3, 5, \ldots, 19\}$,        $S_2' = \{2, 4, 6, \ldots, 20\}$
(3)   $S_1'' = \{1, 2, 3, 4, 5, 16, 17, 18, 19, 20\}$,        $S_2'' = \{6, 7, 8, \ldots, 15\}$

Also discuss unstratified sampling in the above fashion.

In all cases $m = 10.5$. We list some variances which can be verified.

$$\sigma^2 = 33.25$$

(1)   $\sigma_1^2 = 8.25$,        $\sigma_2^2 = 8.25$
(2)   $\sigma_1^2 = 33$,        $\sigma_2^2 = 33$
(3)   $\sigma_1^2 = 58.25$,        $\sigma_2^2 = 8.25$

With unstratified sampling, the variance of $\frac{1}{4}(Y_1 + Y_2 + Y_3 + Y_4)$ is

$$(\tfrac{1}{4})^2 4(1 - \tfrac{3}{19})33.25 = 7$$

In the other cases our statistic is

$$\left(\frac{N_1}{N}\right)^2 \frac{1}{r_1} \left(\frac{N_1 - r_1}{N_1 - 1}\right)\sigma_1^2 + \left(\frac{N_2}{N}\right)^2 \frac{1}{r_2} \left(\frac{N_2 - r_2}{N_2 - 1}\right)\sigma_1^2$$

If we use $r_1 = r_2 = 2$, we obtain

(1)    $\sigma^2(Z) = \frac{1}{4}\cdot\frac{1}{2}\cdot(\frac{8}{9})\cdot 8.25 + \frac{1}{4}\cdot\frac{1}{2}\cdot(\frac{8}{9})\cdot 8.25 = 1.833$
(2)    $\sigma^2(Z') = (\frac{1}{4}\cdot\frac{1}{2}\cdot\frac{8}{9}\cdot 33)\cdot 2 = 4\sigma^2(Z) = 7.333$
(3)    $\sigma^2(Z'') = \frac{1}{4}\cdot\frac{1}{2}\cdot\frac{8}{9}\cdot(58.25 + 8.25) = 7.389$

If we use $r_1 = 3$, $r_2 = 1$ in (3) we get a variance of $\frac{1}{4}\cdot\frac{1}{3}\cdot\frac{7}{9}\cdot 58.25 + \frac{1}{4}\cdot 8.25$ $= 5.83$.

If we are so foolish as to use $r_1 = 1$ and $r_2 = 3$ we obtain $\frac{1}{4}\cdot 58.25 + \frac{1}{4}\cdot\frac{1}{3}\cdot\frac{7}{9}\cdot$ $8.25 = 15.09$.

The example shows the difficulties inherent in stratified sampling. Note that $\sigma^2(Z') > \sigma^2(\frac{1}{4}(Y_1 + \cdots + Y_4))$. Thus, if the strata are badly chosen the estimator for $m$ using stratified sampling is "worse" than that using unstratified sampling. That is, worse in the sense of larger variance. Note that $d^2$, the variation between strata, is $\frac{1}{4}$ and 0 in cases (2) and (3), respectively. Such small values of $d^2$ open up opportunities for bad stratified sampling. Note, further, that $r_1 = 3$ and $r_2 = 1$ gives a smaller variance of the estimator than does $r_1 = 2$ and $r_2 = 2$. Generally, it is desirable to use a larger sample size in a stratum with a larger variance. The reason should be clear, intuitively.

We now find the best choices for the $r_j$ when the strata are specified.

We have $r_1 + r_2 + \cdots + r_M = r$ and

$$\sigma^2(Z) = \sum_{j=1}^{M} \left(\frac{N_j}{N}\right)^2 \frac{1}{r_j} \frac{(N_j - r_j)}{N_j - 1} \sigma_j^2$$

We let $\left(\dfrac{N_j}{N}\right)^2 \dfrac{N_j}{N_j - 1} \sigma_j^2 = \alpha_j^2$ and $Q = \displaystyle\sum_{j=1}^{M} \left(\dfrac{N_j}{N}\right)^2 \dfrac{1}{N_j - 1} \sigma_j^2$. Therefore

$$\sigma^2(Z) = \sum_{j=1}^{M} \left(\frac{N_j}{N}\right)^2 \frac{1}{N_j - 1} \frac{N_j}{r_j} \sigma_j^2 - \sum_{j=1}^{M} \left(\frac{N_j}{N}\right)^2 \frac{1}{N_j - 1} \sigma_j^2$$

$$= \sum_{j=1}^{M} \alpha_j^2 r_j^{-1} - Q$$

Since we are assuming that the strata are fixed, $Q$ is constant for variable $r_j$.

To minimize $\sigma^2(Z)$ we need only minimize $\phi(r) = \displaystyle\sum_{j=1}^{M} \alpha_j^2 r_j^{-1}$. This is to be done subject to $r_1 + r_2 + \cdots + r_M = r$ where the $r_j$ are integers. Suppose, temporarily, that we treat $r_1, \ldots, r_M$ as arbitrary real numbers whose sum is $r$. We note that $r_M = r - r_1 - r_2 - \cdots - r_{M-1}$.

$$\frac{\partial \phi}{\partial r_k} = \alpha_k^2 (-1) r_k^{-2} + \alpha_M^2 (r - r_1 - r_2 - \cdots - r_{M-1})^{-2} (-1)(-1),$$

$$k = 1, 2, \ldots, M - 1$$

where all terms $\alpha_j^2 r_j^{-1}$ for $j \neq k$ have partial derivatives with respect to $r_k$ equal to zero. Furthermore

$$\frac{\partial}{\partial r_k} (r - r_1 - r_2 - \cdots - r_{M-1})^{-1} = (-1)(r - r_1 - \cdots - r_M)^{-2}(-1)$$

At the minimum

$$\frac{\partial \phi}{\partial r_k} = -\left(\frac{\alpha_k}{r_k}\right)^2 + L = 0$$

where $L = \alpha_M^2 (r - r_1 - r_2 - \cdots - r_{M-1})^{-2}$. Thus for $k, l = 1, 2, \ldots, M - 1$

$$\left(\frac{\alpha_k}{r_k}\right)^2 = \left(\frac{\alpha_l}{r_l}\right)^2 = \left(\frac{\alpha_M^2}{r_M}\right)^2$$

Thus at the minimum we have

$$\left(\frac{\alpha_k}{r_k}\right)^2 = \frac{1}{\beta^2}$$

a constant with respect to $k$. Thus $r_k = \alpha_k \beta$. Since $\sum\limits_{k=1}^{M} r_k = r$ we have $\sum\limits_{k=1}^{M} \alpha_k \beta = r$

or $\beta = \dfrac{r}{\sum\limits_{k=1}^{M} \alpha_k}$. Therefore

$$r_k = \frac{\alpha_k r}{\sum\limits_{k=1}^{M} \alpha_k}, \qquad k = 1, 2, \ldots, M$$

minimizes $\sigma^2(Z)$ among real values of the $r_k$. In terms of the $N_j$, etc.,

**(15.2.7)** $\qquad r_k^* = \left( \dfrac{N_k}{N} \sqrt{\dfrac{N_k}{N_k - 1}} \, \sigma_k r \right) \div \left( \sum\limits_{j=1}^{M} \dfrac{N_j}{N} \sqrt{\dfrac{N_j}{N_j - 1}} \, \sigma_j \right)$

or

$$r_k^* \sim \frac{N_k}{N} \sqrt{\frac{N_k}{N_k - 1}} \, \sigma_k$$

with $\sum\limits_{k=1}^{M} r_k^* = r$. For large $N_k$, $\sqrt{\dfrac{N_k}{N_k - 1}} \approx 1$.

To solve our original problem we would examine the nearest integers $r_j$ to $r_j^*$.

### Example 15.2.2

Let $S = \{1, 2, 3, \ldots, 10, 12, 14, 16, \ldots, 30, 34, 38, 42, \ldots, 70\}$. Let $S_1 = \{1, 2, 3, \ldots, 10\}$, $S_2 = \{12, 14, 16, \ldots, 30\}$, $S_3 = \{34, 38, 42, \ldots, 70\}$. Note that $N = 30$ and $N_1 = N_2 = N_3 = 10$. Let $r = 8$. Find the $r_j$ for minimizing $\sigma^2(Z)$ as in the above theory.

We first find $\sigma_k^2$. A simple computation gives $\sigma_1^2 = 8.25$. Obviously $\sigma_2^2 = 33$ and $\sigma_3^2 = 132$. The numbers $r_k^*$ must be proportional to $\dfrac{N_k}{N} \sqrt{\dfrac{N_k}{N_k - 1}} \, \sigma_k$. But since the $N_k$ are all equal, in fact the $r_k^*$ are proportional to $\sigma_k$. Since the $\sigma_k^2$ are proportional to 1, 4, 16, respectively, the $\sigma_k$ are proportional to 1, 2, 4. We therefore let $r_1 = \alpha$, $r_2 = 2\alpha$, and $r_3 = 4\alpha$ with $r_1 + r_2 + r_3 = \alpha + 2\alpha + 4\alpha = 8$ or $\alpha = \frac{8}{7}$. This suggests $r_1 = 1$, $r_2 = 2$, and $r_3 = 5$. With this choice

$$\sum_{j=1}^{3} \left( \frac{N_j}{N} \right)^2 \frac{1}{N_j - 1} (N_j - r_j)\sigma_j^2 = \sum_{j=1}^{3} (\tfrac{10}{30})^2 \tfrac{1}{9}(9 - r_j)\sigma_j^2$$

$$= \tfrac{1}{81}[8(8.25) + 7(33) + 4(132)] = 10.185$$

If we took $r_1 = 1$, $r_2 = 3$, $r_3 = 4$ we would have $\tfrac{1}{81}[8(8.25) + 6(33) + 5(132)] \approx 11.4$.

**EXERCISES 15.2** _____

1  Verify the calculations in Example 15.2.1.

2  At one time it was assumed that statified sampling with $r_j$ proportional to $N_j$ would always give better results than unstratified sampling. What does Example 15.2.1 say about that matter?

3  Discuss the uncertainties about $\sigma_j^2$ which appear in real-world sampling problems but which do not appear in a purely theoretical problem.

4  Suppose that $S_0 = \{1, 2, 4, 8, 16, 32, 64, 128, 256, 512, 1024\}$. Find the best choices for $r_j$ if the strata are $S_1 = \{1, 2, 4, 8, 16\}$, $S_2 = \{32, 64, 128\}$, $S_3 = \{256, 512, 1024\}$, and $r = 6$.

5  Work through the relevant material of section 15.2 under the assumption that sampling with replacement is used in the strata.


# 15.3  Sequential sampling

Our work so far has assumed fixed sample size. That is, any two samples relating to a specific test or estimate had the same number of pieces of data. This assumption makes the mathematical theory easier. However, such is not the most practical mode of sampling in many real-world problems.

**Example 15.3.1**
A biased coin is under study. The null hypothesis, $H_0$, states that $P(\text{H}) = \frac{4}{10}$ and $H_1$ states that $P(\text{H}) = \frac{1}{10}$. The hypotheses are tested by tossing the coin independently until one of the following occurs:

(a)  For some $n \leq 100$,
$$n^{-1}(\text{number of H's thrown so far}) \geq \tfrac{3}{10} + 2/n$$

(b)  For some $n \leq 100$,
$$n^{-1}(\text{number of H's thrown so far}) \leq \tfrac{2}{10} - 2/n$$

(c)  The coin has been tossed 100 times and neither (a) nor (b) has occurred.

If (a) occurs, we accept $H_0$ at the toss at which it occurs and stop tossing. If (b) occurs, we reject $H_0$ at the toss that it occurs and stop tossing. If (c) occurs we stop tossing at the 100th toss and reject $H_0$ if there have been fewer than 22 H's in the 100 tosses. Otherwise we accept $H_0$. Discuss this test.

No decision can be reached before the third toss. If the first three tosses show H, we find $3^{-1} \cdot 3 \geq \tfrac{3}{10} + \tfrac{2}{3}$. Thus (a) occurs and we accept $H_0$. We cannot reject $H_0$ before ten tosses. But if each of the first ten tosses show T it is easy to see that (b) occurs. Note that three H's on the first three tosses would be quite unlikely if $H_1$ were true. Indeed under $H_1$ $P(\text{three H's in three tosses}) = (\tfrac{1}{10})^3$

$= 10^{-3}$. Furthermore, under $H_0$, $P$(ten T's in ten tosses) $= (\frac{6}{10})^{10} \approx 0.006$. It is not likely that we would quickly reject a true hypothesis. A T on one of the first three tosses would force us to make at least five tosses. If those tosses result in one T and four H's, $H_0$ is accepted.

The order of appearance of H's and T's is important. For example, we can imagine, under $H_0$, a situation in which ten consecutive T's would be followed by many H's, except that we stop tossing after the ten T's and reject $H_0$. This is one of the risks inherent in chance phenomena. We feel comforted in the knowledge that this sort of event is quite unlikely in a properly designed test.

A test like that in the example has the advantage of giving a quick decision when one of the hypotheses is well confirmed by early evidence. If the early evidence is inconclusive, the test provides for the gathering of additional data. However, if either $H_0$ or $H_1$ is true it is rare that the full 100 tosses are needed to decide about the hypotheses.

The test in Example 15.3.1 is rather rudimentary. Its object is that of rejecting $H_0$ if the observed bias probability is less than $\frac{2}{10}$ and of accepting $H_0$ if the observed bias probability is greater than $\frac{3}{10}$. Note that the test is keyed to

$$n^{-1}(\text{number of H's thrown up to the } n\text{th toss})$$

This is the type of quantity to which a number of good tests of fixed sample size are keyed. However, the test in the example does not permit a sample of size greater than 100. In some cases it is not very convenient to have such a stopping point. We next suggest a kind of test where additional data may always be gathered if data are inconclusive.

Let $S_0$ be the real numbers as usual. Let $X(x) = x$ so that the distribution of $X$ is essentially the probability measure on $S_0$. Let $f(x; \theta)$ be the density or probability function of $X$ corresponding to the parameter value $\theta$. We assume independent sampling. For each $n$ we have the logarithm of the likelihood ratio for a sample of size $n$ where $H_0 = \{\theta_0\}$ and $H_1 = \{\theta_1\}$

**(15.3.1)**   $\log \lambda_n(x_1, \ldots, x_n) = \log \dfrac{L(x_1, \ldots, x_n; \theta_1)}{L(x_1, \ldots, x_n; \theta_0)} = \sum\limits_{j=1}^{n} \log \dfrac{f(x_j; \theta_1)}{f(x_j; \theta_0)}$

The terms in the sum in equation (15.3.1) are independent, identically distributed random variables on $S$, the sample space of the sample of size $n$. However we will not think of $n$ as being fixed for a specific test.

In the study of hypothesis testing with fixed sample size we use critical regions determined by inequalities of the form

$$\lambda_n(x) \geq k, \quad \text{where } x = (x_1, \ldots, x_n)$$

That is, we reject $H_0$ if $x$ is a good deal more compatible with $H_1$ than with $H_0$. There is no reason not to do the same with samples whose size is not fixed.

We assume that a sample $x = (x_1, \ldots, x_n)$ of size $n$ has been chosen and proceed as follows: For $0 < A < B$

**(15.3.2a)**     If $\lambda_n(x) \geq B$ accept $H_1$

**(15.3.2b)**     If $\lambda_n(x) \leq A$ accept $H_0$

**(15.3.2c)**     If $A < \lambda_n(x) < B$ obtain $x_{n+1}$ and study $\lambda_{n+1}(x_1, \ldots, x_n, x_{n+1})$ from the point of view of (a), (b), and (c).

We call such a test a *sequential likelihood-ratio* test.
    Let

$$Z(x) = \log \frac{f(x; \theta_1)}{f(x; \theta_0)}$$

and let $W_n(x) = Z(x_1) + \cdots + Z(x_n)$. With $a = \log A$ and $b = \log B$ we may recast the inequalities in (15.3.2) as

**(15.3.3a)**     $W_n \geq b$

**(15.3.3b)**     $W_n \leq a$

**(15.3.3c)**     $a < W_n < b$

It is clear that we should take $0 < A < 1 < B$ in order to make it a reasonable test. This is equivalent to $a < 0 < b$.
    A critical question about such a test is whether it will always end. Is it not possible that for every whole number $n$ condition (15.3.3c) will hold and we will be forced to continue sampling without ever making a decision? It seems clear that if $E(Z) > 0$, $W_n$ should drift in the positive direction and we will eventually find $W_n \geq b$ for some $n$. Similarly, if $E(Z) < 0$, we will eventually find $W_n \leq a$ for some $n$. If $E(Z) = 0$, drift does not occur. However, our knowledge of random walks makes it clear that there will be enough oscillation so that only with probability zero will (15.3.3c) hold for all $n$. This is only a theoretical reassurance that testing will stop. However, theoretical reassurance is preferable to theoretical evidence to the contrary.
    Another question is that of the values of $A$ and $B$ which are appropriate for a particular problem. As might be expected $A$ and $B$ are determined by the first and second deficiencies which are desired for the problem.
    Finally, since $n$ is not fixed and the stage of stopping depends on the data we treat this stage as a random variable. Thus if $x_1, x_2, x_3, \ldots$ is an infinite sequence of elements of $S_0$ we let

$$N(x_1, x_2, x_3, \ldots) = \text{smallest } \{n \geq 1; \quad W_n \geq b \quad \text{or} \quad W_n \leq a\}$$

As an idea of how long we must test we may wish to know $E(N)$ or $P(N \geq u)$.
    We next give some approximations for the key quantities $E(N)$, $A$, and $B$. Later we discuss the background of the approximations.

Suppose that we wish to have first and second deficiencies approximately equal to $\alpha$ and $\beta$, respectively. Then the values

**(15.3.4)**   $A = \dfrac{\beta}{1-\alpha}$   and   $B = \dfrac{1-\beta}{\alpha}$

will suffice. Henceforth we assume that $A$ and $B$ are so chosen.

Let $\theta$ be a parameter for the possible distributions on $S_0$ which are germane to the problem. Let

$$q^*(\theta) = P(\text{accepting } H_0 \text{ when } \theta \text{ is the true parameter})$$
$$= P_\theta(W_{N(x_1, x_2, x_3, \ldots)} \le a)$$

The function $q^*$ is the O.C. function of the test as defined in equations (14.1.3). To state that $\theta$ is the parameter is to state that $f(x; \theta)$ is the distribution on $S_0$. We now take up an approximation for $q^*(\theta)$ for those $\theta$ for which $E_\theta(Z) \ne 0$. Let $h(\theta)$ be such that $h(\theta) \ne 0$ and $E_\theta(\lambda_1^{h(\theta)}) = 1$. The theory guarantees that there is only one such number when $E_\theta(Z) \ne 0$. If we are using density functions

$$E_\theta(\lambda_1^{h(\theta)}) = \int_{-\infty}^{\infty} \lambda_1^{h(\theta)} f(x; \theta)\, dx$$

Note that $h(\theta_0) = 1$ and $h(\theta_1) = -1$ by the definition of $\lambda_1$.

With the above definition of $h(\theta)$,

**(15.3.5)**   $q^*(\theta) \approx \dfrac{B^{h(\theta)} - 1}{B^{h(\theta)} - A^{h(\theta)}}$

If $E_\theta(Z) = 0$ a different approximation is needed. We see that the expression on the right-hand side of (15.3.5) gives $1 - \alpha$ and $\beta$ when $h(\theta) = 1$ and $h(\theta) = -1$, respectively.

It can also be shown that if $E_\theta(Z) \ne 0$

**(15.3.6)**   $E_\theta(N) \approx \dfrac{q^*(\theta) \log A + [1 - q^*(\theta)] \log B}{E_\theta(Z)}$

### Example 15.3.2

The phone company wishes to determine whether the expected length of telephone calls originating in a particular area has increased from four to five minutes. A sequential likelihood-ratio test is to be used and an exponential density $f(x; \theta) = \theta e^{-\theta x}$ is to be assumed for the call lengths. For definiteness $H_0 : \theta^{-1} = 5$ and $H_1 : \theta^{-1} = 4$ are the hypotheses. The company wishes to be very careful that $H_0$ is not accepted when $H_1$ is true since it may be obliged to install additional equipment if the expected call length has increased. Therefore it wants $\beta = 0.005$ and $\alpha = 0.05$. Set up the test. Find approximate values for $q^*(\theta)$ and $E_\theta(N)$ when $\theta^{-1} = 4.0, 4.513, 5.0, 5.625$.

We use the standard approximations $A = \beta/(1 - \alpha)$ and $B = (1 - \beta)/\alpha$. This gives $A = 0.005/0.95 = 0.005263$ and $B = 0.995/0.05 = 19.9$. From tables of natural logarithms we obtain $\log A = -5.22575$ and $\log B = 2.99072$. As observed above $q^*(5) \approx 0.95$ and $q^*(4) \approx 0.005$.

Next we turn to the problem of finding $h(\theta)$ for other $\theta$. In the present test

$$\lambda_1(x) = \frac{\frac{1}{4}e^{-x/4}}{\frac{1}{5}e^{-x/5}} = \frac{5}{4}e^{-x/20}$$

and

$$Z(x) = \log \tfrac{5}{4} - x/20 = 0.22315 - x/20$$

From previous results about the exponential distribution, $E_\theta(Z) = 0.22315 - (20\theta)^{-1}$. This is not zero for any $\theta$ in this example.

The defining equation for $h(\theta)$ is $E_\theta(\lambda_1^{h(\theta)}) = 1$. This is

$$\int_0^\infty (\tfrac{5}{4}e^{-x/20})^{h(\theta)}\theta e^{-\theta x}\,dx = \int_0^\infty (\tfrac{5}{4})^{h(\theta)}\theta e^{-[h(\theta)/20 + \theta]x}\,dx$$

$$= (\tfrac{5}{4})^{h(\theta)}\,\frac{\theta}{h(\theta)/20 + \theta} = 1$$

or

$$\theta(\tfrac{5}{4})^{h(\theta)} = h(\theta)/20 + \theta$$

We have already noted that $h(5^{-1}) = 1$ and $h(4^{-1}) = -1$. It can be verified that $h(5.625^{-1}) = 2$ and $h(4.513^{-1}) = 0.10$ to two significant figures. Approximations for $q^*(\theta)$ may be obtained by means of (15.3.5). We can then obtain approximations for $E_\theta(N)$ by means of (15.3.6). We tabulate the results in Table 3.

### Table 3

| $\theta^{-1}$ | $E_\theta(Z)$ | $h(\theta)$ | $q^*(\theta)$ | $E_\theta(N)$ |
|------|------------|-------|---------|-------|
| 4.0   | 0.02315   | $-1$  | 0.005   | 128   |
| 4.513 | $-0.00350$ | 0.10 | 0.45003 | 202   |
| 6.0   | $-0.02685$ | 1    | 0.95    | 179   |
| 5.625 | $-0.05810$ | 2    | 0.9975  | 89.5  |

It is only fair to say that $h(4.513^{-1})$ and $h(5.625^{-1})$ were not found by solving the transcendental equation

$$\theta(\tfrac{5}{4})^h = h/20 + \theta$$

for $h$ while $\theta$ is fixed. Rather, $h$ was chosen and then $\theta$ was determined. This is elementary. However, in many cases this represents a realistic approach to solving this equation. We choose many values for $h$; determine the corresponding values of $\theta$ and interpolate for others.

Much more interesting from a probabilistic viewpoint is the behavior when $\theta^{-1} = 4.513$. Although there is a drift in the negative direction because of $E_\theta(Z) < 0$, $H_0$ is accepted only about 45% of the time. This is because the negative barrier $\log A = -5.22575$ is much farther away from zero than the positive barrier $\log B = 2.99072$. Chance fluctuations carry $\lambda_n$ toward $B$ quite often. These are enough to get $\lambda_n \geq B$ often.

Note how much more quickly we decide on $H_0$ when $\theta^{-1} = 5.625$ than when $\theta^{-1} = 5.0$. This is an indication of the strength of a sequential test. When something is obviously true, you find out about it quickly and the test ends.

We close with a few brief remarks about the theoretical aspects of the sequential likelihood-ratio test. The definition of $W_n$ shows that such a test is keyed to a random walk in which steps are determined by the distribution of $Z$. The test is reminiscent of the gambler's ruin (Theorem 11.2.3) except the steps can be much more complicated.

**EXERCISES 15.3** _____

Exercises 1–4 refer to Example 15.3.2.

**1** Suppose that a sequence of ten calls of lengths 0.350, 10.628, 2.734, 3.841, 0.534, 0.721, 15.321, 4.152, 6.418, 1.137 arises. Does the sequential likelihood-ratio test reach a decision thus far?

**2** Show that a sequence whose first 14 calls had essentially zero length would just suffice for acceptance of $H_1$ at the fourteenth call.

**3** Show that if the first three calls lasted 63, 48, 23 minutes, respectively, then $H_0$ would be accepted after the third call but not after the second call.

**4** Discuss the action on a sequence of calls whose lengths are 63, 48, 2.5, 1.3, 4.2, 13.06 at the sixth stage. Find the maximum-likelihood estimate of $\theta$ in $\theta e^{-\theta x}$ for such data.

**5** Let $\alpha = 0.05$, $\beta = 0.05$, $A = \beta/(1 - \alpha)$, and $B = (1 - \beta)/\alpha$. Describe sequential likelihood-ratio tests for normal random variables with

   (a)  $H_0: m = 1$,   $\sigma^2 = 1$   and   $H_1: m = -1$,   $\sigma^2 = 1$
   (b)  $H_0: m = 1$,   $\sigma^2 = 4$   and   $H_1: m = -1$,   $\sigma^2 = 4$

Since $A$ and $B$ are the same for both tests how does the difference in the variances manifest itself?

**6** Set up the sequential likelihood-ratio test for the hypotheses of Example 15.3.1 with $\alpha = 0.05$ and $\beta = 0.05$. Plot the O.C. function of the test. Find $E(N)$ when $P(\text{H}) = 0.1$, 0.4, and 0.25.

**7** Consider the hypotheses $H_0: P(\text{H}) = \frac{1}{3}$ and $H_1: P(\text{H}) = \frac{2}{3}$ for a biased coin. Let $\alpha = \beta = 0.05$. Show how the sequential likelihood-ratio test is essentially the problem of passage of a random walk of step length $\log 2$ to one of two barriers as in the last part of section 11.2. Compare the O.C. function to $\sum_{n=1}^{\infty} r_n(a, b)$ of (11.2.3). Note the relationship between $a, b$ of this chapter and $a, b$ of Chapter 11.

8   Consider normal distributions. Let $H_0 : m = m_0, \quad \sigma^2 = K$ be tested against $H_1 : m = m_1, \quad \sigma^2 = K$ by our sequential likelihood-ratio test. Let $\theta$ refer to an $N(m, K)$ distribution. Show that

$$E(\lambda_1^u) = \exp\left\{-\frac{1}{2K}\left[m_1^2 u - m_0^2 u + m^2 - (m - m_0 u + m_1 h)^2\right]\right\}$$

where, of course, the expectation is taken with respect to the $N(m, K)$ distribution.

[*Hint:* $\int_{-\infty}^{\infty} (2\pi K)^{-1/2} e^{-(x-c)^2/2K} \, dx = 1.$]

9   Use the result of exercise 8 to find our standard approximation to $q^*(\theta)$ for the test in exercise 5(a) when $\theta$ signifies $N(m, 1)$ and $m = -2, -1.5, -1$ $-\frac{3}{4}, -0.1, 0.1, \frac{1}{4}, \frac{1}{2}, 0.8, 1, 1.5.$

# References

Buck, R. C. *Advanced Calculus*, 2nd ed. (New York: McGraw-Hill, 1965).

Reference for sections 1.4 and 8.5 and for general topics in calculus in support of this text.

Conant, James. *On Understanding Science: An Historical Approach*—Terry Lecture (New Haven: Yale University Press, 1947).

Reference for Chapter 4.

Feller, W. *Introduction to Probability Theory and its Applications*, Vol. 1, 2nd ed. (New York: Wiley, 1957).

A classic; reference for Chapters 1,3,5 and 11; contains many excellent problems and examples.

Fisz, Marek. *Probability Theory and Mathematical Statistics*, 3rd ed. (New York: Wiley, 1963).

Its topics parallel this text but go further; material entitled "Problems and Complements" comprise a useful introduction to the literature of probability and statistics.

Rahman, N. A. *Exercises in Probability and Statistics* (New York: Hafner, 1967).

Contains problems for those who find the exercises of this text too easy.

Sveshnikov, A. A., ed., translated by Scripta Technica. *Problems in Probability Theory, Mathematical Statistics and Theory of Random Functions* (Philadelphia: W. B. Saunders, 1968).

See remark about reference 5.

Uspensky, J. V. *Introduction to Mathematical Probability* (New York: McGraw-Hill, 1937).

Reference for section 1.4; contains a number of analytic sidelights, e.g. Weierstrass approximation theorem via Chebyshev's inequality; another classic.

# Appendix

# Table 1.   Normal Areas and Ordinates*

| $t$ | $\phi(t)$ | $\int_0^t \phi(t)\,dt$ | $t$ | $\phi(t)$ | $\int_0^t \phi(t)\,dt$ | $t$ | $\phi(t)$ | $\int_0^t \phi(t)\,dt$ |
|---|---|---|---|---|---|---|---|---|
| 0.00 | 0.39894 | 0.00000 | 0.45 | 0.36053 | 0.17364 | 0.90 | 0.26609 | 0.31594 |
| 0.01 | 0.39892 | 0.00399 | 0.46 | 0.35889 | 0.17724 | 0.91 | 0.26369 | 0.31859 |
| 0.02 | 0.39886 | 0.00798 | 0.47 | 0.35723 | 0.18082 | 0.92 | 0.26129 | 0.32121 |
| 0.03 | 0.39876 | 0.01197 | 0.48 | 0.35553 | 0.18439 | 0.93 | 0.25888 | 0.32381 |
| 0.04 | 0.39862 | 0.01595 | 0.49 | 0.35381 | 0.18793 | 0.94 | 0.25647 | 0.32639 |
| 0.05 | 0.39844 | 0.01994 | 0.50 | 0.35207 | 0.19146 | 0.95 | 0.25406 | 0.32894 |
| 0.06 | 0.39822 | 0.02392 | 0.51 | 0.35029 | 0.19497 | 0.96 | 0.25164 | 0.33147 |
| 0.07 | 0.39797 | 0.02790 | 0.52 | 0.34849 | 0.19847 | 0.97 | 0.24923 | 0.33398 |
| 0.08 | 0.39767 | 0.03188 | 0.53 | 0.34667 | 0.20194 | 0.98 | 0.24681 | 0.33646 |
| 0.09 | 0.39733 | 0.03586 | 0.54 | 0.34482 | 0.20540 | 0.99 | 0.24439 | 0.33891 |
| 0.10 | 0.39695 | 0.03983 | 0.55 | 0.34294 | 0.20884 | 1.00 | 0.24197 | 0.34134 |
| 0.11 | 0.39654 | 0.04380 | 0.56 | 0.34105 | 0.21226 | 1.01 | 0.23955 | 0.34375 |
| 0.12 | 0.39608 | 0.04776 | 0.57 | 0.33912 | 0.21566 | 1.02 | 0.23713 | 0.34614 |
| 0.13 | 0.39559 | 0.05172 | 0.58 | 0.33718 | 0.21904 | 1.03 | 0.23471 | 0.34850 |
| 0.14 | 0.39505 | 0.05567 | 0.59 | 0.33521 | 0.22240 | 1.04 | 0.23230 | 0.35083 |
| 0.15 | 0.39448 | 0.05962 | 0.60 | 0.33322 | 0.22575 | 1.05 | 0.22988 | 0.35314 |
| 0.16 | 0.39387 | 0.06356 | 0.61 | 0.33121 | 0.22907 | 1.06 | 0.22747 | 0.35543 |
| 0.17 | 0.39322 | 0.06749 | 0.62 | 0.32918 | 0.23237 | 1.07 | 0.22506 | 0.35769 |
| 0.18 | 0.39253 | 0.07142 | 0.63 | 0.32713 | 0.23565 | 1.08 | 0.22265 | 0.35993 |
| 0.19 | 0.39181 | 0.07535 | 0.64 | 0.32506 | 0.23891 | 1.09 | 0.22025 | 0.36214 |
| 0.20 | 0.39104 | 0.07926 | 0.65 | 0.32297 | 0.24215 | 1.10 | 0.21785 | 0.36433 |
| 0.21 | 0.39024 | 0.08317 | 0.66 | 0.32086 | 0.24537 | 1.11 | 0.21546 | 0.36650 |
| 0.22 | 0.38940 | 0.08706 | 0.67 | 0.31874 | 0.24857 | 1.12 | 0.21307 | 0.36864 |
| 0.23 | 0.38853 | 0.09095 | 0.68 | 0.31659 | 0.25175 | 1.13 | 0.21069 | 0.37076 |
| 0.24 | 0.38762 | 0.09483 | 0.69 | 0.31443 | 0.25490 | 1.14 | 0.20831 | 0.37286 |
| 0.25 | 0.38667 | 0.09871 | 0.70 | 0.31225 | 0.25804 | 1.15 | 0.20594 | 0.37493 |
| 0.26 | 0.38568 | 0.10257 | 0.71 | 0.31006 | 0.26115 | 1.16 | 0.20357 | 0.37698 |
| 0.27 | 0.38466 | 0.10642 | 0.72 | 0.30785 | 0.26424 | 1.17 | 0.20121 | 0.37900 |
| 0.28 | 0.38361 | 0.11026 | 0.73 | 0.30563 | 0.26730 | 1.18 | 0.19886 | 0.38100 |
| 0.29 | 0.38251 | 0.11409 | 0.74 | 0.30339 | 0.27035 | 1.19 | 0.19652 | 0.38298 |
| 0.30 | 0.38139 | 0.11791 | 0.75 | 0.30114 | 0.27337 | 1.20 | 0.19419 | 0.38493 |
| 0.31 | 0.38023 | 0.12172 | 0.76 | 0.29887 | 0.27637 | 1.21 | 0.19186 | 0.38686 |
| 0.32 | 0.37903 | 0.12552 | 0.77 | 0.29659 | 0.27935 | 1.22 | 0.18954 | 0.38877 |
| 0.33 | 0.37780 | 0.12930 | 0.78 | 0.29431 | 0.28230 | 1.23 | 0.18724 | 0.39065 |
| 0.34 | 0.37654 | 0.13307 | 0.79 | 0.29200 | 0.28524 | 1.24 | 0.18494 | 0.39251 |

*(continued)*

* From *Mathematics of Statistics*, Part One, 3rd ed., by J. F. Kenney and E. S. Keeping, Copyright © 1939, 1946, 1954, by Litton Educational Publishing, Inc., by permission of Van Nostrand Reinhold Company.

## Table 1. (*Continued*)

| $t$ | $\phi(t)$ | $\int_0^t \phi(t)\,dt$ | $t$ | $\phi(t)$ | $\int_0^t \phi(t)\,dt$ | $t$ | $\phi(t)$ | $\int_0^t \phi(t)\,dt$ |
|---|---|---|---|---|---|---|---|---|
| 0.35 | 0.37524 | 0.13683 | 0.80 | 0.28969 | 0.28814 | 1.25 | 0.18265 | 0.39435 |
| 0.36 | 0.37391 | 0.14058 | 0.81 | 0.28737 | 0.29103 | 1.26 | 0.18037 | 0.39617 |
| 0.37 | 0.37255 | 0.14431 | 0.82 | 0.28504 | 0.29389 | 1.27 | 0.17810 | 0.39796 |
| 0.38 | 0.37115 | 0.14803 | 0.83 | 0.28269 | 0.29673 | 1.28 | 0.17585 | 0.39973 |
| 0.39 | 0.36973 | 0.15173 | 0.84 | 0.28034 | 0.29955 | 1.29 | 0.17360 | 0.40147 |
| 0.40 | 0.36827 | 0.15542 | 0.85 | 0.27798 | 0.30234 | 1.30 | 0.17137 | 0.40320 |
| 0.41 | 0.36678 | 0.15910 | 0.86 | 0.27562 | 0.30511 | 1.31 | 0.16915 | 0.40490 |
| 0.42 | 0.36526 | 0.16276 | 0.87 | 0.27324 | 0.30785 | 1.32 | 0.16694 | 0.40658 |
| 0.43 | 0.36371 | 0.16640 | 0.88 | 0.27086 | 0.31057 | 1.33 | 0.16474 | 0.40824 |
| 0.44 | 0.36213 | 0.17003 | 0.89 | 0.26848 | 0.31327 | 1.34 | 0.16256 | 0.40988 |
| 1.35 | 0.16038 | 0.41149 | 1.80 | 0.07895 | 0.46407 | 2.25 | 0.03174 | 0.48778 |
| 1.36 | 0.15882 | 0.41309 | 1.81 | 0.07754 | 0.46485 | 2.26 | 0.03103 | 0.48809 |
| 1.37 | 0.15608 | 0.41466 | 1.82 | 0.07614 | 0.46562 | 2.27 | 0.03034 | 0.48840 |
| 1.38 | 0.15395 | 0.41621 | 1.83 | 0.07477 | 0.46638 | 2.28 | 0.02965 | 0.48870 |
| 1.39 | 0.15183 | 0.41774 | 1.84 | 0.07341 | 0.46712 | 2.29 | 0.02898 | 0.48899 |
| 1.40 | 0.14973 | 0.41924 | 1.85 | 0.07206 | 0.46784 | 2.30 | 0.02833 | 0.48928 |
| 1.41 | 0.14764 | 0.42073 | 1.86 | 0.07074 | 0.46856 | 2.31 | 0.02768 | 0.48956 |
| 1.42 | 0.14556 | 0.42220 | 1.87 | 0.06943 | 0.46926 | 2.32 | 0.02705 | 0.48983 |
| 1.43 | 0.14350 | 0.42364 | 1.88 | 0.06814 | 0.46995 | 2.33 | 0.02643 | 0.49010 |
| 1.44 | 0.14146 | 0.42507 | 1.89 | 0.06687 | 0.47062 | 2.34 | 0.02582 | 0.49036 |
| 1.45 | 0.13943 | 0.42647 | 1.90 | 0.06562 | 0.47128 | 2.35 | 0.02522 | 0.49061 |
| 1.46 | 0.13742 | 0.42786 | 1.91 | 0.06439 | 0.47193 | 2.36 | 0.02463 | 0.49086 |
| 1.47 | 0.13542 | 0.42922 | 1.92 | 0.06316 | 0.47257 | 2.37 | 0.02406 | 0.49111 |
| 1.48 | 0.13344 | 0.43056 | 1.93 | 0.06195 | 0.47320 | 2.38 | 0.02349 | 0.49134 |
| 1.49 | 0.13147 | 0.43189 | 1.94 | 0.06077 | 0.47381 | 2.39 | 0.02294 | 0.49158 |
| 1.50 | 0.12952 | 0.43319 | 1.95 | 0.05959 | 0.47441 | 2.40 | 0.02239 | 0.49180 |
| 1.51 | 0.12758 | 0.43448 | 1.96 | 0.05844 | 0.47500 | 2.41 | 0.02186 | 0.49202 |
| 1.52 | 0.12566 | 0.43574 | 1.97 | 0.05730 | 0.47558 | 2.42 | 0.02134 | 0.49224 |
| 1.53 | 0.12376 | 0.43699 | 1.98 | 0.05618 | 0.47615 | 2.43 | 0.02083 | 0.49245 |
| 1.54 | 0.12188 | 0.43822 | 1.99 | 0.05508 | 0.47670 | 2.44 | 0.02033 | 0.49266 |
| 1.55 | 0.12001 | 0.43943 | 2.00 | 0.05399 | 0.47725 | 2.45 | 0.01984 | 0.49286 |
| 1.56 | 0.11816 | 0.44062 | 2.01 | 0.05292 | 0.47778 | 2.46 | 0.01936 | 0.49305 |
| 1.57 | 0.11632 | 0.44179 | 2.02 | 0.05186 | 0.47831 | 2.47 | 0.01889 | 0.49324 |
| 1.58 | 0.11450 | 0.44295 | 2.03 | 0.05082 | 0.47882 | 2.48 | 0.01842 | 0.49343 |
| 1.59 | 0.11270 | 0.44408 | 2.04 | 0.04980 | 0.47932 | 2.49 | 0.01797 | 0.49361 |
| 1.60 | 0.11092 | 0.44520 | 2.05 | 0.04879 | 0.47982 | 2.50 | 0.01753 | 0.49379 |
| 1.61 | 0.10915 | 0.44630 | 2.06 | 0.04780 | 0.48030 | 2.51 | 0.01709 | 0.49396 |
| 1.62 | 0.10741 | 0.44738 | 2.07 | 0.04682 | 0.48077 | 2.52 | 0.01667 | 0.49413 |
| 1.63 | 0.10567 | 0.44845 | 2.08 | 0.04586 | 0.48124 | 2.53 | 0.01625 | 0.49430 |
| 1.64 | 0.10396 | 0.44950 | 2.09 | 0.04491 | 0.48169 | 2.54 | 0.01585 | 0.49446 |

# Table 1. *(Continued)*

| $t$ | $\phi(t)$ | $\int_0^t \phi(t)\,dt$ | $t$ | $\phi(t)$ | $\int_0^t \phi(t)\,dt$ | $t$ | $\phi(t)$ | $\int_0^t \phi(t)\,dt$ |
|---|---|---|---|---|---|---|---|---|
| 1.65 | 0.10226 | 0.45053 | 2.10 | 0.04398 | 0.48214 | 2.55 | 0.01545 | 0.49461 |
| 1.66 | 0.10059 | 0.45154 | 2.11 | 0.04307 | 0.48257 | 2.56 | 0.01506 | 0.49477 |
| 1.67 | 0.09893 | 0.45254 | 2.12 | 0.04217 | 0.48300 | 2.57 | 0.01468 | 0.49492 |
| 1.68 | 0.09728 | 0.45352 | 2.13 | 0.04128 | 0.48341 | 2.58 | 0.01431 | 0.49506 |
| 1.69 | 0.09566 | 0.45449 | 2.14 | 0.04041 | 0.48382 | 2.59 | 0.01394 | 0.49520 |
| 1.70 | 0.09405 | 0.45543 | 2.15 | 0.03955 | 0.48422 | 2.60 | 0.01358 | 0.49534 |
| 1.71 | 0.09246 | 0.45637 | 2.16 | 0.03871 | 0.48461 | 2.61 | 0.01323 | 0.49547 |
| 1.72 | 0.09089 | 0.45728 | 2.17 | 0.03788 | 0.48500 | 2.62 | 0.01289 | 0.49560 |
| 1.73 | 0.08933 | 0.45818 | 2.18 | 0.03706 | 0.48537 | 2.63 | 0.01256 | 0.49573 |
| 1.74 | 0.08780 | 0.45907 | 2.19 | 0.03626 | 0.48574 | 2.64 | 0.01223 | 0.49585 |
| 1.75 | 0.08628 | 0.45994 | 2.20 | 0.03547 | 0.48610 | 2.65 | 0.01191 | 0.49598 |
| 1.76 | 0.08478 | 0.46080 | 2.21 | 0.03470 | 0.48645 | 2.66 | 0.01160 | 0.49609 |
| 1.77 | 0.08329 | 0.46164 | 2.22 | 0.03394 | 0.48679 | 2.67 | 0.01130 | 0.49621 |
| 1.78 | 0.08183 | 0.46246 | 2.23 | 0.03319 | 0.48713 | 2.68 | 0.01100 | 0.49632 |
| 1.79 | 0.08038 | 0.46327 | 2.24 | 0.03246 | 0.48745 | 2.69 | 0.01071 | 0.49643 |
| 2.70 | 0.01042 | 0.49653 | 3.15 | 0.00279 | 0.49918 | 3.60 | 0.00061 | 0.49984 |
| 2.71 | 0.01014 | 0.49664 | 3.16 | 0.00271 | 0.49921 | 3.61 | 0.00059 | 0.49985 |
| 2.72 | 0.00987 | 0.49674 | 3.17 | 0.00262 | 0.49924 | 3.62 | 0.00057 | 0.49985 |
| 2.73 | 0.00961 | 0.49683 | 3.18 | 0.00254 | 0.49926 | 3.63 | 0.00055 | 0.49986 |
| 2.74 | 0.00935 | 0.49693 | 3.19 | 0.00246 | 0.49929 | 3.64 | 0.00053 | 0.49986 |
| 2.75 | 0.00909 | 0.49702 | 3.20 | 0.00238 | 0.49931 | 3.65 | 0.00051 | 0.49987 |
| 2.76 | 0.00885 | 0.49711 | 3.21 | 0.00231 | 0.49934 | 3.66 | 0.00049 | 0.49987 |
| 2.77 | 0.00861 | 0.49720 | 3.22 | 0.00224 | 0.49936 | 3.67 | 0.00047 | 0.49988 |
| 2.78 | 0.00837 | 0.49728 | 3.23 | 0.00216 | 0.49938 | 3.68 | 0.00046 | 0.49988 |
| 2.79 | 0.00814 | 0.49736 | 3.24 | 0.00210 | 0.49940 | 3.69 | 0.00044 | 0.49989 |
| 2.80 | 0.00792 | 0.49744 | 3.25 | 0.00203 | 0.49942 | 3.70 | 0.00042 | 0.49989 |
| 2.81 | 0.00770 | 0.49752 | 3.26 | 0.00196 | 0.49944 | 3.71 | 0.00041 | 0.49990 |
| 2.82 | 0.00748 | 0.49760 | 3.27 | 0.00190 | 0.49946 | 3.72 | 0.00039 | 0.49990 |
| 2.83 | 0.00727 | 0.49767 | 3.28 | 0.00184 | 0.49948 | 3.73 | 0.00038 | 0.49990 |
| 2.84 | 0.00707 | 0.49774 | 3.29 | 0.00178 | 0.49950 | 3.74 | 0.00037 | 0.49991 |
| 2.85 | 0.00687 | 0.49781 | 3.30 | 0.00172 | 0.49952 | 3.75 | 0.00035 | 0.49991 |
| 2.86 | 0.00668 | 0.49788 | 3.31 | 0.00167 | 0.49953 | 3.76 | 0.00034 | 0.49992 |
| 2.87 | 0.00649 | 0.49795 | 3.32 | 0.00161 | 0.49955 | 3.77 | 0.00033 | 0.49992 |
| 2.88 | 0.00631 | 0.49801 | 3.33 | 0.00156 | 0.49957 | 3.78 | 0.00031 | 0.49992 |
| 2.89 | 0.00613 | 0.49807 | 3.34 | 0.00151 | 0.49958 | 3.79 | 0.00030 | 0.49992 |
| 2.90 | 0.00595 | 0.49813 | 3.35 | 0.00146 | 0.49960 | 3.80 | 0.00029 | 0.49993 |
| 2.91 | 0.00578 | 0.49819 | 3.36 | 0.00141 | 0.49961 | 3.81 | 0.00028 | 0.49993 |
| 2.92 | 0.00562 | 0.49825 | 3.37 | 0.00136 | 0.49962 | 3.82 | 0.00027 | 0.49993 |
| 2.93 | 0.00545 | 0.49831 | 3.38 | 0.00132 | 0.49964 | 3.83 | 0.00026 | 0.49994 |
| 2.94 | 0.00530 | 0.49836 | 3.39 | 0.00127 | 0.49965 | 3.84 | 0.00025 | 0.49994 |

## Table 1. (*Continued*)

| $t$ | $\phi(t)$ | $\int_0^t \phi(t)\, dt$ | $t$ | $\phi(t)$ | $\int_0^t \phi(t)\, dt$ | $t$ | $\phi(t)$ | $\int_0^t \phi(t)\, dt$ |
|---|---|---|---|---|---|---|---|---|
| 2.95 | 0.00514 | 0.49841 | 3.40 | 0.00123 | 0.49966 | 3.85 | 0.00024 | 0.49994 |
| 2.96 | 0.00499 | 0.49846 | 3.41 | 0.00119 | 0.49968 | 3.86 | 0.00023 | 0.49994 |
| 2.97 | 0.00485 | 0.49851 | 3.42 | 0.00115 | 0.49969 | 3.87 | 0.00022 | 0.49995 |
| 2.98 | 0.00471 | 0.49856 | 3.43 | 0.00111 | 0.49970 | 3.88 | 0.00021 | 0.49995 |
| 2.99 | 0.00457 | 0.49861 | 3.44 | 0.00107 | 0.49971 | 3.89 | 0.00021 | 0.49995 |
| | | | | | | | | |
| 3.00 | 0.00443 | 0.49865 | 3.45 | 0.00104 | 0.49972 | 3.90 | 0.00020 | 0.49995 |
| 3.01 | 0.00430 | 0.49869 | 3.46 | 0.00100 | 0.49973 | 3.91 | 0.00019 | 0.49995 |
| 3.02 | 0.00417 | 0.49874 | 3.47 | 0.00097 | 0.49974 | 3.92 | 0.00018 | 0.49996 |
| 3.03 | 0.00405 | 0.49878 | 3.48 | 0.00094 | 0.49975 | 3.93 | 0.00018 | 0.49996 |
| 3.04 | 0.00393 | 0.49882 | 3.49 | 0.00090 | 0.49976 | 3.94 | 0.00017 | 0.49996 |
| | | | | | | | | |
| 3.05 | 0.00381 | 0.49886 | 3.50 | 0.00087 | 0.49977 | 3.95 | 0.00016 | 0.49996 |
| 3.06 | 0.00370 | 0.49889 | 3.51 | 0.00084 | 0.49978 | 3.96 | 0.00016 | 0.49996 |
| 3.07 | 0.00358 | 0.49893 | 3.52 | 0.00081 | 0.49978 | 3.97 | 0.00015 | 0.49996 |
| 3.08 | 0.00348 | 0.49897 | 3.53 | 0.00079 | 0.49979 | 3.98 | 0.00014 | 0.49997 |
| 3.09 | 0.00337 | 0.49900 | 3.54 | 0.00076 | 0.49980 | 3.99 | 0.00014 | 0.49997 |
| | | | | | | | | |
| 3.10 | 0.00327 | 0.49903 | 3.55 | 0.00073 | 0.49981 | | | |
| 3.11 | 0.00317 | 0.49906 | 3.56 | 0.00071 | 0.49981 | | | |
| 3.12 | 0.00307 | 0.49910 | 3.57 | 0.00068 | 0.49982 | | | |
| 3.13 | 0.00298 | 0.49913 | 3.58 | 0.00066 | 0.49983 | | | |
| 3.14 | 0.00288 | 0.49916 | 3.59 | 0.00063 | 0.49983 | | | |

## Table 2.  $\chi^2$ Distribution

| Degrees of freedom | P = 0.99 | 0.98 | 0.95 | 0.90 | 0.80 | 0.70 | 0.50 | 0.30 | 0.20 | 0.10 | 0.05 | 0.02 | 0.01 |
|---|---|---|---|---|---|---|---|---|---|---|---|---|---|
| 1 | 0.000157 | 0.000628 | 0.00393 | 0.0158 | 0.0642 | 0.148 | 0.455 | 1.074 | 1.642 | 2.706 | 3.841 | 5.412 | 6.635 |
| 2 | 0.0201 | 0.0404 | 0.103 | 0.211 | 0.446 | 0.713 | 1.386 | 2.408 | 3.219 | 4.605 | 5.991 | 7.824 | 9.210 |
| 3 | 0.115 | 0.185 | 0.352 | 0.584 | 1.005 | 1.424 | 2.366 | 3.665 | 4.642 | 6.251 | 7.815 | 9.837 | 11.341 |
| 4 | 0.297 | 0.429 | 0.711 | 1.064 | 1.649 | 2.195 | 3.357 | 4.878 | 5.989 | 7.779 | 9.488 | 11.668 | 13.277 |
| 5 | 0.554 | 0.752 | 1.145 | 1.610 | 2.343 | 3.000 | 4.351 | 6.064 | 7.289 | 9.236 | 11.070 | 13.388 | 15.086 |
| 6 | 0.872 | 1.134 | 1.635 | 2.204 | 3.070 | 3.828 | 5.348 | 7.231 | 8.558 | 10.645 | 12.592 | 15.033 | 16.812 |
| 7 | 1.239 | 1.564 | 2.167 | 2.833 | 3.822 | 4.671 | 6.346 | 8.383 | 9.803 | 12.017 | 14.067 | 16.622 | 18.475 |
| 8 | 1.646 | 2.032 | 2.733 | 3.490 | 4.594 | 5.527 | 7.344 | 9.524 | 11.030 | 13.362 | 15.507 | 18.168 | 20.090 |
| 9 | 2.088 | 2.532 | 3.325 | 4.168 | 5.380 | 6.393 | 8.343 | 10.656 | 12.242 | 14.684 | 16.919 | 19.679 | 21.666 |
| 10 | 2.558 | 3.059 | 3.940 | 4.865 | 6.179 | 7.267 | 9.342 | 11.781 | 13.442 | 15.987 | 18.307 | 21.161 | 23.209 |
| 11 | 3.053 | 3.609 | 4.575 | 5.578 | 6.989 | 8.148 | 10.341 | 12.899 | 14.631 | 17.275 | 19.675 | 22.618 | 24.725 |
| 12 | 3.571 | 4.178 | 5.226 | 6.304 | 7.807 | 9.034 | 11.340 | 14.011 | 15.812 | 18.549 | 21.026 | 24.054 | 26.217 |
| 13 | 4.107 | 4.765 | 5.892 | 7.042 | 8.634 | 9.926 | 12.340 | 15.119 | 16.985 | 19.812 | 22.362 | 25.472 | 27.688 |
| 14 | 4.660 | 5.368 | 6.571 | 7.790 | 9.467 | 10.821 | 13.339 | 16.222 | 18.151 | 21.064 | 23.685 | 26.873 | 29.141 |
| 15 | 5.229 | 5.985 | 7.261 | 8.547 | 10.307 | 11.721 | 14.339 | 17.322 | 19.311 | 22.307 | 24.996 | 28.259 | 30.578 |
| 16 | 5.812 | 6.614 | 7.962 | 9.312 | 11.152 | 12.624 | 15.338 | 18.418 | 20.465 | 23.542 | 26.296 | 29.633 | 32.000 |
| 17 | 6.408 | 7.255 | 8.672 | 10.085 | 12.002 | 13.531 | 16.338 | 19.511 | 21.615 | 24.769 | 27.587 | 30.995 | 33.409 |
| 18 | 7.015 | 7.906 | 9.390 | 10.865 | 12.857 | 14.440 | 17.338 | 20.601 | 22.760 | 25.989 | 28.869 | 32.246 | 34.805 |
| 19 | 7.633 | 8.567 | 10.117 | 11.651 | 13.716 | 15.352 | 18.338 | 21.689 | 23.900 | 27.204 | 30.144 | 33.687 | 36.191 |
| 20 | 8.260 | 9.237 | 10.851 | 12.443 | 14.578 | 16.266 | 19.337 | 22.775 | 25.038 | 28.412 | 31.410 | 35.020 | 37.566 |
| 21 | 8.897 | 9.915 | 11.591 | 13.240 | 15.445 | 17.182 | 20.337 | 23.858 | 26.171 | 29.615 | 32.671 | 36.343 | 38.932 |
| 22 | 9.542 | 10.600 | 12.338 | 14.041 | 16.314 | 18.101 | 21.337 | 24.939 | 27.301 | 30.813 | 33.924 | 37.659 | 40.289 |
| 23 | 10.196 | 11.293 | 13.091 | 14.848 | 17.187 | 19.021 | 22.337 | 26.018 | 28.429 | 32.007 | 35.172 | 38.968 | 41.638 |
| 24 | 10.856 | 11.992 | 13.848 | 15.659 | 18.062 | 19.943 | 23.337 | 27.096 | 29.553 | 33.196 | 36.415 | 40.270 | 42.980 |
| 25 | 11.524 | 12.697 | 14.611 | 16.473 | 18.940 | 20.867 | 24.337 | 28.172 | 30.675 | 34.382 | 37.652 | 41.566 | 44.314 |
| 26 | 12.198 | 13.409 | 15.379 | 17.292 | 19.820 | 21.792 | 25.336 | 29.246 | 31.795 | 35.563 | 38.885 | 42.856 | 45.642 |
| 27 | 12.879 | 14.125 | 16.151 | 18.114 | 20.703 | 22.719 | 26.336 | 30.319 | 32.912 | 36.741 | 40.113 | 44.140 | 46.963 |
| 28 | 13.565 | 14.847 | 16.928 | 18.939 | 21.588 | 23.647 | 27.336 | 31.391 | 34.027 | 37.916 | 41.337 | 45.419 | 48.278 |
| 29 | 14.256 | 15.574 | 17.708 | 19.768 | 22.475 | 24.577 | 28.336 | 32.461 | 35.139 | 39.087 | 42.557 | 46.693 | 49.588 |
| 30 | 14.953 | 16.306 | 18.493 | 20.599 | 23.364 | 25.508 | 29.336 | 33.530 | 36.250 | 40.256 | 43.773 | 47.962 | 50.892 |

For degrees of freedom greater than 30, the expression $\sqrt{2\chi^2} - \sqrt{2n' - 1}$ may be used as a normal deviate with unit variance, where $n'$ is the number of degrees of freedom. Table 2 is taken from Table III of Fisher: *Statistical Methods for Research Workers*, published by Oliver and Boyd Limited, Edinburgh and by permission of the authors and publishers.

## Table 3.  Student's $t$-Distribution*

| Degrees of freedom $n$ | Probability of a deviation greater than $t$ | | | | | |
|:---:|:---:|:---:|:---:|:---:|:---:|:---:|
| | 0.005 | 0.01 | 0.025 | 0.05 | 0.1 | 0.15 |
| 1 | 63.657 | 31.821 | 12.706 | 6.314 | 3.078 | 1.963 |
| 2 | 9.925 | 6.965 | 4.303 | 2.920 | 1.886 | 1.386 |
| 3 | 5.841 | 4.541 | 3.182 | 2.353 | 1.638 | 1.250 |
| 4 | 4.604 | 3.747 | 2.776 | 2.132 | 1.533 | 1.190 |
| 5 | 4.032 | 3.365 | 2.571 | 2.015 | 1.476 | 1.156 |
| 6 | 3.707 | 3.143 | 2.447 | 1.943 | 1.440 | 1.134 |
| 7 | 3.499 | 2.998 | 2.365 | 1.895 | 1.415 | 1.119 |
| 8 | 3.355 | 2.896 | 2.306 | 1.860 | 1.397 | 1.108 |
| 9 | 3.250 | 2.821 | 2.262 | 1.833 | 1.383 | 1.100 |
| 10 | 3.169 | 2.764 | 2.228 | 1.812 | 1.372 | 1.093 |
| 11 | 3.106 | 2.718 | 2.201 | 1.796 | 1.363 | 1.088 |
| 12 | 3.055 | 2.681 | 2.179 | 1.782 | 1.356 | 1.083 |
| 13 | 3.012 | 2.650 | 2.160 | 1.771 | 1.350 | 1.079 |
| 14 | 2.977 | 2.624 | 2.145 | 1.761 | 1.345 | 1.076 |
| 15 | 2.947 | 2.602 | 2.131 | 1.753 | 1.341 | 1.074 |
| 16 | 2.921 | 2.583 | 2.120 | 1.746 | 1.337 | 1.071 |
| 17 | 2.898 | 2.567 | 2.110 | 1.740 | 1.333 | 1.069 |
| 18 | 2.878 | 2.552 | 2.101 | 1.734 | 1.330 | 1.067 |
| 19 | 2.861 | 2.539 | 2.093 | 1.729 | 1.328 | 1.066 |
| 20 | 2.845 | 2.528 | 2.086 | 1.725 | 1.325 | 1.064 |
| 21 | 2.831 | 2.518 | 2.080 | 1.721 | 1.323 | 1.063 |
| 22 | 2.819 | 2.508 | 2.074 | 1.717 | 1.321 | 1.061 |
| 23 | 2.807 | 2.500 | 2.069 | 1.714 | 1.319 | 1.060 |
| 24 | 2.797 | 2.492 | 2.064 | 1.711 | 1.318 | 1.059 |
| 25 | 2.787 | 2.485 | 2.060 | 1.708 | 1.316 | 1.058 |
| 26 | 2.779 | 2.479 | 2.056 | 1.706 | 1.315 | 1.058 |
| 27 | 2.771 | 2.473 | 2.052 | 1.703 | 1.314 | 1.057 |
| 28 | 2.763 | 2.467 | 2.048 | 1.701 | 1.313 | 1.056 |
| 29 | 2.756 | 2.462 | 2.045 | 1.699 | 1.311 | 1.055 |
| 30 | 2.750 | 2.457 | 2.042 | 1.697 | 1.310 | 1.055 |
| $\infty$ | 2.576 | 2.326 | 1.960 | 1.645 | 1.282 | 1.036 |

*(continued)*

The probability of a deviation *numerically* greater than $t$ is twice the probability given at the head of the table.

* Table 3 is taken from Table IV of Fisher: *Statistical Methods for Research Workers,* published by Oliver & Boyd Limited, Edinburgh and by permission of the author and publishers.

## Table 3.    (*Continued*)

| Degrees of freedom $n$ | Probability of a deviation greater than $t$ | | | | | |
|---|---|---|---|---|---|---|
| | 0.2 | 0.25 | 0.3 | 0.35 | 0.4 | 0.45 |
| 1 | 1.376 | 1.000 | 0.727 | 0.510 | 0.325 | 0.158 |
| 2 | 1.061 | 0.816 | 0.617 | 0.445 | 0.289 | 0.142 |
| 3 | 0.978 | 0.765 | 0.584 | 0.424 | 0.277 | 0.137 |
| 4 | 0.941 | 0.741 | 0.569 | 0.414 | 0.271 | 0.134 |
| 5 | 0.920 | 0.727 | 0.559 | 0.408 | 0.267 | 0.132 |
| 6 | 0.906 | 0.718 | 0.553 | 0.404 | 0.265 | 0.131 |
| 7 | 0.896 | 0.711 | 0.549 | 0.402 | 0.263 | 0.130 |
| 8 | 0.889 | 0.706 | 0.546 | 0.399 | 0.262 | 0.130 |
| 9 | 0.883 | 0.703 | 0.543 | 0.398 | 0.261 | 0.129 |
| 10 | 0.879 | 0.700 | 0.542 | 0.397 | 0.260 | 0.129 |
| 11 | 0.876 | 0.697 | 0.540 | 0.396 | 0.260 | 0.129 |
| 12 | 0.873 | 0.695 | 0.539 | 0.395 | 0.259 | 0.128 |
| 13 | 0.870 | 0.694 | 0.538 | 0.394 | 0.259 | 0.128 |
| 14 | 0.868 | 0.692 | 0.537 | 0.393 | 0.258 | 0.128 |
| 15 | 0.866 | 0.691 | 0.536 | 0.393 | 0.258 | 0.128 |
| 16 | 0.865 | 0.690 | 0.535 | 0.392 | 0.258 | 0.128 |
| 17 | 0.863 | 0.689 | 0.534 | 0.392 | 0.257 | 0.128 |
| 18 | 0.862 | 0.688 | 0.534 | 0.392 | 0.257 | 0.127 |
| 19 | 0.861 | 0.688 | 0.533 | 0.391 | 0.257 | 0.127 |
| 20 | 0.860 | 0.687 | 0.533 | 0.391 | 0.257 | 0.127 |
| 21 | 0.859 | 0.686 | 0.532 | 0.391 | 0.257 | 0.127 |
| 22 | 0.858 | 0.686 | 0.532 | 0.390 | 0.256 | 0.127 |
| 23 | 0.858 | 0.685 | 0.532 | 0.390 | 0.256 | 0.127 |
| 24 | 0.857 | 0.685 | 0.531 | 0.390 | 0.256 | 0.127 |
| 25 | 0.856 | 0.684 | 0.531 | 0.390 | 0.256 | 0.127 |
| 26 | 0.856 | 0.684 | 0.531 | 0.390 | 0.256 | 0.127 |
| 27 | 0.855 | 0.684 | 0.531 | 0.389 | 0.256 | 0.127 |
| 28 | 0.855 | 0.683 | 0.530 | 0.389 | 0.256 | 0.127 |
| 29 | 0.854 | 0.683 | 0.530 | 0.389 | 0.256 | 0.127 |
| 30 | 0.854 | 0.683 | 0.530 | 0.389 | 0.256 | 0.127 |
| ∞ | 0.842 | 0.674 | 0.524 | 0.385 | 0.253 | 0.126 |

The probability of a deviation *numerically* greater than $t$ is twice the probability given at the head of the table.

# Table 4.   F-Distribution*

5% (Roman Type) and 1% (Bold-Face Type) Points for the Distribution of $F$

Degrees of freedom for numerator ($\nu_1$)

| Degrees of freedom for denominator ($\nu_2$) | 1 | 2 | 3 | 4 | 5 | 6 | 7 | 8 | 9 | 10 | 11 | 12 | 14 | 16 | 20 | 24 | 30 | 40 | 50 | 75 | 100 | 200 | 500 | ∞ |
|---|---|---|---|---|---|---|---|---|---|---|---|---|---|---|---|---|---|---|---|---|---|---|---|---|
| 1 | 161 / **4052** | 200 / **4999** | 216 / **5403** | 225 / **5625** | 230 / **5764** | 234 / **5859** | 237 / **5928** | 239 / **5981** | 241 / **6022** | 242 / **6056** | 243 / **6082** | 244 / **6106** | 245 / **6142** | 246 / **6169** | 248 / **6208** | 249 / **6234** | 250 / **6258** | 251 / **6286** | 252 / **6302** | 253 / **6323** | 253 / **6334** | 254 / **6352** | 254 / **6361** | 254 / **6366** |
| 2 | 18.51 / **98.49** | 19.00 / **99.01** | 19.16 / **99.17** | 19.25 / **99.25** | 19.30 / **99.30** | 19.33 / **99.33** | 19.36 / **99.34** | 19.37 / **99.36** | 19.38 / **99.38** | 19.39 / **99.40** | 19.40 / **99.41** | 19.41 / **99.42** | 19.42 / **99.43** | 19.43 / **99.44** | 19.44 / **99.45** | 19.45 / **99.46** | 19.46 / **99.47** | 19.47 / **99.48** | 19.47 / **99.48** | 19.48 / **99.49** | 19.49 / **99.49** | 19.49 / **99.49** | 19.50 / **99.50** | 19.50 / **99.50** |
| 3 | 10.13 / **34.12** | 9.55 / **30.81** | 9.28 / **29.46** | 9.12 / **28.71** | 9.01 / **28.24** | 8.94 / **27.91** | 8.88 / **27.67** | 8.84 / **27.49** | 8.81 / **27.34** | 8.78 / **27.23** | 8.76 / **27.13** | 8.74 / **27.05** | 8.71 / **26.92** | 8.69 / **26.83** | 8.66 / **26.69** | 8.64 / **26.60** | 8.62 / **26.50** | 8.60 / **26.41** | 8.58 / **26.30** | 8.57 / **26.27** | 8.56 / **26.23** | 8.54 / **26.18** | 8.54 / **26.14** | 8.53 / **26.12** |
| 4 | 7.71 / **21.20** | 6.94 / **18.00** | 6.59 / **16.69** | 6.39 / **15.98** | 6.26 / **15.52** | 6.16 / **15.21** | 6.09 / **14.98** | 6.04 / **14.80** | 6.00 / **14.66** | 5.96 / **14.54** | 5.93 / **14.45** | 5.91 / **14.37** | 5.87 / **14.24** | 5.84 / **14.15** | 5.80 / **14.02** | 5.77 / **13.93** | 5.74 / **13.83** | 5.71 / **13.74** | 5.70 / **13.69** | 5.68 / **13.61** | 5.66 / **13.57** | 5.65 / **13.52** | 5.64 / **13.48** | 5.63 / **13.46** |
| 5 | 6.61 / **16.26** | 5.79 / **13.27** | 5.41 / **12.06** | 5.19 / **11.39** | 5.05 / **10.97** | 4.95 / **10.67** | 4.88 / **10.45** | 4.82 / **10.27** | 4.78 / **10.15** | 4.74 / **10.05** | 4.70 / **9.96** | 4.68 / **9.89** | 4.64 / **9.77** | 4.60 / **9.68** | 4.56 / **9.55** | 4.53 / **9.47** | 4.50 / **9.38** | 4.46 / **9.29** | 4.44 / **9.24** | 4.42 / **9.17** | 4.40 / **9.13** | 4.38 / **9.07** | 4.37 / **9.04** | 4.36 / **9.02** |
| 6 | 5.99 / **13.74** | 5.14 / **10.92** | 4.76 / **9.78** | 4.53 / **9.15** | 4.39 / **8.75** | 4.28 / **8.47** | 4.21 / **8.26** | 4.15 / **8.10** | 4.10 / **7.98** | 4.06 / **7.87** | 4.03 / **7.79** | 4.00 / **7.72** | 3.96 / **7.60** | 3.92 / **7.52** | 3.87 / **7.39** | 3.84 / **7.31** | 3.81 / **7.23** | 3.77 / **7.14** | 3.75 / **7.09** | 3.72 / **7.02** | 3.71 / **6.99** | 3.69 / **6.94** | 3.68 / **6.90** | 3.67 / **6.88** |
| 7 | 5.59 / **12.25** | 4.74 / **9.55** | 4.35 / **8.45** | 4.12 / **7.85** | 3.97 / **7.46** | 3.87 / **7.19** | 3.79 / **7.00** | 3.73 / **6.84** | 3.68 / **6.71** | 3.63 / **6.62** | 3.60 / **6.54** | 3.57 / **6.47** | 3.52 / **6.35** | 3.49 / **6.27** | 3.44 / **6.15** | 3.41 / **6.07** | 3.38 / **5.98** | 3.34 / **5.90** | 3.32 / **5.85** | 3.29 / **5.78** | 3.28 / **5.75** | 3.25 / **5.70** | 3.24 / **5.67** | 3.23 / **5.65** |
| 8 | 5.32 / **11.26** | 4.46 / **8.65** | 4.07 / **7.59** | 3.84 / **7.01** | 3.69 / **6.63** | 3.58 / **6.37** | 3.50 / **6.19** | 3.44 / **6.03** | 3.39 / **5.91** | 3.34 / **5.82** | 3.31 / **5.74** | 3.28 / **5.67** | 3.23 / **5.56** | 3.20 / **5.48** | 3.15 / **5.36** | 3.12 / **5.28** | 3.08 / **5.20** | 3.05 / **5.11** | 3.03 / **5.06** | 3.00 / **5.00** | 2.98 / **4.96** | 2.96 / **4.91** | 2.94 / **4.88** | 2.93 / **4.86** |
| 9 | 5.12 / **10.56** | 4.26 / **8.02** | 3.86 / **6.99** | 3.63 / **6.42** | 3.48 / **6.06** | 3.37 / **5.80** | 3.29 / **5.62** | 3.23 / **5.47** | 3.18 / **5.35** | 3.13 / **5.26** | 3.10 / **5.18** | 3.07 / **5.11** | 3.02 / **5.00** | 2.98 / **4.92** | 2.93 / **4.80** | 2.90 / **4.73** | 2.86 / **4.64** | 2.82 / **4.56** | 2.80 / **4.51** | 2.77 / **4.45** | 2.76 / **4.41** | 2.73 / **4.36** | 2.72 / **4.33** | 2.71 / **4.31** |
| 10 | 4.96 / **10.04** | 4.10 / **7.56** | 3.71 / **6.55** | 3.48 / **5.99** | 3.33 / **5.64** | 3.22 / **5.39** | 3.14 / **5.21** | 3.07 / **5.06** | 3.02 / **4.95** | 2.97 / **4.85** | 2.94 / **4.78** | 2.91 / **4.71** | 2.86 / **4.60** | 2.82 / **4.52** | 2.77 / **4.41** | 2.74 / **4.33** | 2.70 / **4.25** | 2.67 / **4.17** | 2.64 / **4.12** | 2.61 / **4.05** | 2.59 / **4.01** | 2.56 / **3.96** | 2.55 / **3.93** | 2.54 / **3.91** |
| 11 | 4.84 / **9.65** | 3.98 / **7.20** | 3.59 / **6.22** | 3.36 / **5.67** | 3.20 / **5.32** | 3.09 / **5.07** | 3.01 / **4.88** | 2.95 / **4.74** | 2.90 / **4.63** | 2.86 / **4.54** | 2.82 / **4.46** | 2.79 / **4.40** | 2.74 / **4.29** | 2.70 / **4.21** | 2.65 / **4.10** | 2.61 / **4.02** | 2.57 / **3.94** | 2.53 / **3.86** | 2.50 / **3.80** | 2.47 / **3.74** | 2.45 / **3.70** | 2.42 / **3.66** | 2.41 / **3.62** | 2.40 / **3.60** |
| 12 | 4.75 / **9.33** | 3.88 / **6.93** | 3.49 / **5.95** | 3.26 / **5.41** | 3.11 / **5.06** | 3.00 / **4.82** | 2.92 / **4.65** | 2.85 / **4.50** | 2.80 / **4.39** | 2.76 / **4.30** | 2.72 / **4.22** | 2.69 / **4.16** | 2.64 / **4.05** | 2.60 / **3.98** | 2.54 / **3.86** | 2.50 / **3.78** | 2.46 / **3.70** | 2.42 / **3.61** | 2.40 / **3.56** | 2.36 / **3.49** | 2.35 / **3.46** | 2.32 / **3.41** | 2.31 / **3.38** | 2.30 / **3.36** |
| 13 | 4.67 / **9.07** | 3.80 / **6.70** | 3.41 / **5.74** | 3.18 / **5.20** | 3.02 / **4.86** | 2.92 / **4.62** | 2.84 / **4.44** | 2.77 / **4.30** | 2.72 / **4.19** | 2.67 / **4.10** | 2.63 / **4.02** | 2.60 / **3.96** | 2.55 / **3.85** | 2.51 / **3.78** | 2.46 / **3.67** | 2.42 / **3.59** | 2.38 / **3.51** | 2.34 / **3.42** | 2.32 / **3.37** | 2.28 / **3.30** | 2.26 / **3.27** | 2.24 / **3.21** | 2.22 / **3.18** | 2.21 / **3.16** |
| 14 | 4.60 / **8.86** | 3.74 / **6.51** | 3.34 / **5.56** | 3.11 / **5.03** | 2.96 / **4.69** | 2.85 / **4.46** | 2.77 / **4.28** | 2.70 / **4.14** | 2.65 / **4.03** | 2.60 / **3.94** | 2.56 / **3.86** | 2.53 / **3.80** | 2.48 / **3.70** | 2.44 / **3.62** | 2.39 / **3.51** | 2.35 / **3.43** | 2.31 / **3.34** | 2.27 / **3.26** | 2.24 / **3.21** | 2.21 / **3.14** | 2.19 / **3.11** | 2.16 / **3.06** | 2.14 / **3.02** | 2.13 / **3.00** |
| 15 | 4.54 / **8.68** | 3.68 / **6.36** | 3.29 / **5.42** | 3.06 / **4.89** | 2.90 / **4.56** | 2.79 / **4.32** | 2.70 / **4.14** | 2.64 / **4.00** | 2.59 / **3.89** | 2.55 / **3.80** | 2.51 / **3.73** | 2.48 / **3.67** | 2.43 / **3.56** | 2.39 / **3.48** | 2.33 / **3.36** | 2.29 / **3.29** | 2.25 / **3.20** | 2.21 / **3.12** | 2.18 / **3.07** | 2.15 / **3.00** | 2.12 / **2.97** | 2.10 / **2.92** | 2.08 / **2.89** | 2.07 / **2.87** |
| 16 | 4.49 / **8.53** | 3.63 / **6.23** | 3.24 / **5.29** | 3.01 / **4.77** | 2.85 / **4.44** | 2.74 / **4.20** | 2.66 / **4.03** | 2.59 / **3.89** | 2.54 / **3.78** | 2.49 / **3.69** | 2.45 / **3.61** | 2.42 / **3.55** | 2.37 / **3.45** | 2.33 / **3.37** | 2.28 / **3.25** | 2.24 / **3.18** | 2.20 / **3.10** | 2.16 / **3.01** | 2.13 / **2.96** | 2.09 / **2.89** | 2.07 / **2.86** | 2.04 / **2.80** | 2.02 / **2.77** | 2.01 / **2.75** |

* Reproduced by permission from *Statistical Methods*, 6th ed., by George W. Snedecor and William C. Cochran, © 1967 by The Iowa State University Press, Ames, Iowa., U.S.A.

| df | | | | | | | | | | | | | | | | | | | | | | | | |
|----|----|----|----|----|----|----|----|----|----|----|----|----|----|----|----|----|----|----|----|----|----|----|----|----|
| 17 | 1.96 / 2.65 | 1.97 / 2.67 | 1.99 / 2.70 | 2.02 / 2.76 | 2.04 / 2.79 | 2.08 / 2.86 | 2.11 / 2.92 | 2.15 / 3.00 | 2.19 / 3.08 | 2.23 / 3.16 | 2.29 / 3.27 | 2.33 / 3.35 | 2.38 / 3.45 | 2.41 / 3.52 | 2.45 / 3.59 | 2.50 / 3.68 | 2.55 / 3.79 | 2.62 / 3.93 | 2.70 / 4.10 | 2.81 / 4.34 | 2.96 / 4.67 | 3.20 / 5.18 | 3.59 / 6.11 | 4.45 / 8.40 |
| 18 | 1.92 / 2.57 | 1.93 / 2.59 | 1.95 / 2.62 | 1.98 / 2.68 | 2.00 / 2.71 | 2.04 / 2.78 | 2.07 / 2.83 | 2.11 / 2.91 | 2.15 / 3.00 | 2.19 / 3.07 | 2.25 / 3.19 | 2.29 / 3.27 | 2.34 / 3.37 | 2.37 / 3.44 | 2.41 / 3.51 | 2.46 / 3.60 | 2.51 / 3.71 | 2.58 / 3.85 | 2.66 / 4.01 | 2.77 / 4.25 | 2.93 / 4.58 | 3.16 / 5.09 | 3.55 / 6.01 | 4.41 / 8.28 |
| 19 | 1.88 / 2.49 | 1.90 / 2.51 | 1.91 / 2.54 | 1.94 / 2.60 | 1.96 / 2.63 | 2.00 / 2.70 | 2.02 / 2.76 | 2.07 / 2.84 | 2.11 / 2.92 | 2.15 / 3.00 | 2.21 / 3.12 | 2.26 / 3.19 | 2.31 / 3.30 | 2.34 / 3.36 | 2.38 / 3.43 | 2.43 / 3.52 | 2.48 / 3.63 | 2.55 / 3.77 | 2.63 / 3.94 | 2.74 / 4.17 | 2.90 / 4.50 | 3.13 / 5.01 | 3.52 / 5.93 | 4.38 / 8.18 |
| 20 | 1.84 / 2.42 | 1.85 / 2.44 | 1.87 / 2.47 | 1.90 / 2.53 | 1.92 / 2.56 | 1.96 / 2.63 | 1.99 / 2.69 | 2.04 / 2.77 | 2.08 / 2.86 | 2.12 / 2.94 | 2.18 / 3.05 | 2.23 / 3.13 | 2.28 / 3.23 | 2.31 / 3.30 | 2.35 / 3.37 | 2.40 / 3.45 | 2.45 / 3.56 | 2.52 / 3.71 | 2.60 / 3.87 | 2.71 / 4.10 | 2.87 / 4.43 | 3.10 / 4.94 | 3.49 / 5.85 | 4.35 / 8.10 |
| 21 | 1.81 / 2.36 | 1.82 / 2.38 | 1.84 / 2.42 | 1.87 / 2.47 | 1.89 / 2.51 | 1.93 / 2.58 | 1.96 / 2.63 | 2.00 / 2.72 | 2.05 / 2.80 | 2.09 / 2.88 | 2.15 / 2.99 | 2.20 / 3.07 | 2.25 / 3.17 | 2.28 / 3.24 | 2.32 / 3.31 | 2.37 / 3.40 | 2.42 / 3.51 | 2.49 / 3.65 | 2.57 / 3.81 | 2.68 / 4.04 | 2.84 / 4.37 | 3.07 / 4.87 | 3.47 / 5.78 | 4.32 / 8.02 |
| 22 | 1.78 / 2.31 | 1.80 / 2.33 | 1.81 / 2.37 | 1.84 / 2.42 | 1.87 / 2.46 | 1.91 / 2.53 | 1.93 / 2.58 | 1.98 / 2.67 | 2.03 / 2.75 | 2.07 / 2.83 | 2.13 / 2.94 | 2.18 / 3.02 | 2.23 / 3.12 | 2.26 / 3.18 | 2.30 / 3.26 | 2.35 / 3.35 | 2.40 / 3.45 | 2.47 / 3.59 | 2.55 / 3.76 | 2.66 / 3.99 | 2.82 / 4.31 | 3.05 / 4.82 | 3.44 / 5.72 | 4.30 / 7.94 |
| 23 | 1.76 / 2.26 | 1.77 / 2.28 | 1.79 / 2.32 | 1.82 / 2.37 | 1.84 / 2.41 | 1.88 / 2.48 | 1.91 / 2.53 | 1.96 / 2.62 | 2.00 / 2.70 | 2.04 / 2.78 | 2.10 / 2.89 | 2.14 / 2.97 | 2.20 / 3.07 | 2.24 / 3.14 | 2.28 / 3.21 | 2.32 / 3.30 | 2.38 / 3.41 | 2.45 / 3.54 | 2.53 / 3.71 | 2.64 / 3.94 | 2.80 / 4.26 | 3.03 / 4.76 | 3.42 / 5.66 | 4.28 / 7.88 |
| 24 | 1.73 / 2.21 | 1.74 / 2.23 | 1.76 / 2.27 | 1.80 / 2.33 | 1.82 / 2.36 | 1.86 / 2.44 | 1.89 / 2.49 | 1.94 / 2.58 | 1.98 / 2.66 | 2.02 / 2.74 | 2.09 / 2.85 | 2.13 / 2.93 | 2.18 / 3.03 | 2.22 / 3.09 | 2.26 / 3.17 | 2.30 / 3.25 | 2.36 / 3.36 | 2.43 / 3.50 | 2.51 / 3.67 | 2.62 / 3.90 | 2.78 / 4.22 | 3.01 / 4.72 | 3.40 / 5.61 | 4.26 / 7.82 |
| 25 | 1.71 / 2.17 | 1.72 / 2.19 | 1.74 / 2.23 | 1.77 / 2.29 | 1.80 / 2.32 | 1.84 / 2.40 | 1.87 / 2.45 | 1.92 / 2.54 | 1.96 / 2.62 | 2.00 / 2.70 | 2.06 / 2.81 | 2.11 / 2.89 | 2.16 / 2.99 | 2.20 / 3.05 | 2.24 / 3.13 | 2.28 / 3.21 | 2.34 / 3.32 | 2.41 / 3.46 | 2.49 / 3.63 | 2.60 / 3.86 | 2.76 / 4.18 | 2.99 / 4.68 | 3.38 / 5.57 | 4.24 / 7.77 |
| 26 | 1.69 / 2.13 | 1.70 / 2.15 | 1.72 / 2.19 | 1.76 / 2.25 | 1.78 / 2.28 | 1.82 / 2.36 | 1.85 / 2.41 | 1.90 / 2.50 | 1.95 / 2.58 | 1.99 / 2.66 | 2.05 / 2.77 | 2.10 / 2.86 | 2.15 / 2.96 | 2.18 / 3.02 | 2.22 / 3.09 | 2.27 / 3.17 | 2.33 / 3.29 | 2.39 / 3.42 | 2.47 / 3.59 | 2.59 / 3.82 | 2.74 / 4.14 | 2.96 / 4.64 | 3.37 / 5.53 | 4.22 / 7.72 |
| 27 | 1.67 / 2.10 | 1.68 / 2.12 | 1.71 / 2.16 | 1.74 / 2.21 | 1.76 / 2.25 | 1.80 / 2.33 | 1.84 / 2.38 | 1.88 / 2.47 | 1.93 / 2.55 | 1.97 / 2.63 | 2.03 / 2.74 | 2.08 / 2.83 | 2.13 / 2.93 | 2.16 / 2.98 | 2.20 / 3.06 | 2.25 / 3.14 | 2.30 / 3.26 | 2.37 / 3.39 | 2.46 / 3.56 | 2.57 / 3.79 | 2.73 / 4.11 | 2.95 / 4.60 | 3.35 / 5.49 | 4.21 / 7.68 |
| 28 | 1.65 / 2.06 | 1.67 / 2.09 | 1.69 / 2.13 | 1.72 / 2.18 | 1.75 / 2.22 | 1.78 / 2.30 | 1.81 / 2.35 | 1.87 / 2.44 | 1.91 / 2.52 | 1.96 / 2.60 | 2.02 / 2.71 | 2.06 / 2.80 | 2.12 / 2.90 | 2.15 / 2.95 | 2.19 / 3.03 | 2.24 / 3.11 | 2.29 / 3.23 | 2.36 / 3.36 | 2.44 / 3.53 | 2.56 / 3.76 | 2.71 / 4.07 | 2.93 / 4.57 | 3.34 / 5.45 | 4.20 / 7.64 |
| 29 | 1.64 / 2.03 | 1.65 / 2.06 | 1.68 / 2.10 | 1.71 / 2.15 | 1.73 / 2.19 | 1.77 / 2.27 | 1.80 / 2.32 | 1.85 / 2.41 | 1.90 / 2.49 | 1.94 / 2.57 | 2.00 / 2.68 | 2.05 / 2.77 | 2.10 / 2.87 | 2.14 / 2.92 | 2.18 / 3.00 | 2.22 / 3.08 | 2.28 / 3.20 | 2.35 / 3.33 | 2.43 / 3.50 | 2.54 / 3.73 | 2.70 / 4.04 | 2.92 / 4.54 | 3.33 / 5.42 | 4.18 / 7.60 |
| 30 | 1.62 / 2.01 | 1.64 / 2.03 | 1.66 / 2.07 | 1.69 / 2.13 | 1.72 / 2.16 | 1.76 / 2.24 | 1.79 / 2.29 | 1.84 / 2.38 | 1.89 / 2.47 | 1.93 / 2.55 | 1.99 / 2.66 | 2.04 / 2.74 | 2.09 / 2.84 | 2.12 / 2.90 | 2.16 / 2.98 | 2.21 / 3.06 | 2.27 / 3.17 | 2.34 / 3.30 | 2.42 / 3.47 | 2.53 / 3.70 | 2.69 / 4.02 | 2.92 / 4.51 | 3.32 / 5.39 | 4.17 / 7.56 |
| 32 | 1.59 / 1.96 | 1.61 / 1.98 | 1.64 / 2.02 | 1.67 / 2.08 | 1.69 / 2.12 | 1.74 / 2.20 | 1.76 / 2.25 | 1.82 / 2.34 | 1.86 / 2.42 | 1.91 / 2.51 | 1.97 / 2.62 | 2.02 / 2.70 | 2.07 / 2.80 | 2.10 / 2.86 | 2.14 / 2.94 | 2.19 / 3.01 | 2.25 / 3.12 | 2.32 / 3.25 | 2.40 / 3.42 | 2.51 / 3.66 | 2.67 / 3.97 | 2.90 / 4.46 | 3.30 / 5.34 | 4.15 / 7.50 |
| 34 | 1.57 / 1.91 | 1.59 / 1.94 | 1.61 / 1.98 | 1.64 / 2.04 | 1.67 / 2.08 | 1.71 / 2.15 | 1.74 / 2.21 | 1.80 / 2.30 | 1.84 / 2.38 | 1.89 / 2.47 | 1.95 / 2.58 | 2.00 / 2.66 | 2.05 / 2.76 | 2.08 / 2.82 | 2.12 / 2.89 | 2.17 / 2.97 | 2.23 / 3.08 | 2.30 / 3.21 | 2.38 / 3.38 | 2.49 / 3.61 | 2.65 / 3.93 | 2.88 / 4.42 | 3.28 / 5.29 | 4.13 / 7.44 |
| 36 | 1.55 / 1.87 | 1.56 / 1.90 | 1.59 / 1.94 | 1.62 / 2.00 | 1.65 / 2.04 | 1.69 / 2.12 | 1.72 / 2.17 | 1.78 / 2.26 | 1.82 / 2.35 | 1.87 / 2.43 | 1.93 / 2.54 | 1.98 / 2.62 | 2.03 / 2.72 | 2.06 / 2.78 | 2.10 / 2.86 | 2.15 / 2.94 | 2.21 / 3.04 | 2.28 / 3.18 | 2.36 / 3.35 | 2.48 / 3.58 | 2.63 / 3.89 | 2.86 / 4.38 | 3.26 / 5.25 | 4.11 / 7.39 |
| 38 | 1.53 / 1.84 | 1.54 / 1.86 | 1.57 / 1.90 | 1.60 / 1.97 | 1.63 / 2.00 | 1.67 / 2.08 | 1.71 / 2.14 | 1.76 / 2.22 | 1.80 / 2.32 | 1.85 / 2.40 | 1.92 / 2.51 | 1.96 / 2.59 | 2.02 / 2.69 | 2.05 / 2.75 | 2.09 / 2.82 | 2.14 / 2.91 | 2.19 / 3.02 | 2.26 / 3.15 | 2.35 / 3.32 | 2.46 / 3.54 | 2.62 / 3.86 | 2.85 / 4.34 | 3.25 / 5.21 | 4.10 / 7.35 |
| 40 | 1.51 / 1.81 | 1.53 / 1.84 | 1.55 / 1.88 | 1.59 / 1.94 | 1.61 / 1.97 | 1.66 / 2.05 | 1.69 / 2.11 | 1.74 / 2.20 | 1.79 / 2.29 | 1.84 / 2.37 | 1.90 / 2.49 | 1.95 / 2.56 | 2.00 / 2.66 | 2.04 / 2.73 | 2.07 / 2.80 | 2.12 / 2.88 | 2.18 / 2.99 | 2.25 / 3.12 | 2.34 / 3.29 | 2.45 / 3.51 | 2.61 / 3.83 | 2.84 / 4.31 | 3.23 / 5.18 | 4.08 / 7.31 |

## Table 4. (Continued)

### 5% (Roman Type) and 1% (Bold-Face Type) Points for the Distribution of $F$

Each cell shows the 5% point (Roman) and 1% point (Bold-Face). Values given as "5% / 1%".

| $v_2$ \\ $v_1$ | 1 | 2 | 3 | 4 | 5 | 6 | 7 | 8 | 9 | 10 | 11 | 12 | 14 | 16 | 20 | 24 | 30 | 40 | 50 | 75 | 100 | 200 | 500 | ∞ |
|---|---|---|---|---|---|---|---|---|---|---|---|---|---|---|---|---|---|---|---|---|---|---|---|---|
| 42 | 4.07 / 7.27 | 3.22 / 5.15 | 2.83 / 4.29 | 2.59 / 3.80 | 2.44 / 3.49 | 2.32 / 3.26 | 2.24 / 3.10 | 2.17 / 2.96 | 2.11 / 2.86 | 2.06 / 2.77 | 2.02 / 2.70 | 1.99 / 2.64 | 1.94 / 2.54 | 1.89 / 2.46 | 1.82 / 2.35 | 1.78 / 2.26 | 1.73 / 2.17 | 1.68 / 2.08 | 1.64 / 2.02 | 1.60 / 1.94 | 1.57 / 1.91 | 1.54 / 1.85 | 1.51 / 1.80 | 1.49 / 1.78 |
| 44 | 4.06 / 7.24 | 3.21 / 5.12 | 2.82 / 4.26 | 2.58 / 3.78 | 2.43 / 3.46 | 2.31 / 3.24 | 2.23 / 3.07 | 2.16 / 2.94 | 2.10 / 2.84 | 2.05 / 2.75 | 2.01 / 2.68 | 1.98 / 2.62 | 1.92 / 2.52 | 1.88 / 2.44 | 1.81 / 2.32 | 1.76 / 2.24 | 1.72 / 2.15 | 1.66 / 2.06 | 1.63 / 2.00 | 1.58 / 1.92 | 1.56 / 1.88 | 1.52 / 1.82 | 1.50 / 1.78 | 1.48 / 1.75 |
| 46 | 4.05 / 7.21 | 3.20 / 5.10 | 2.81 / 4.24 | 2.57 / 3.76 | 2.42 / 3.44 | 2.30 / 3.22 | 2.22 / 3.05 | 2.14 / 2.92 | 2.09 / 2.82 | 2.04 / 2.73 | 2.00 / 2.66 | 1.97 / 2.60 | 1.91 / 2.50 | 1.87 / 2.42 | 1.80 / 2.30 | 1.75 / 2.22 | 1.71 / 2.13 | 1.65 / 2.04 | 1.62 / 1.98 | 1.57 / 1.90 | 1.54 / 1.86 | 1.51 / 1.80 | 1.48 / 1.76 | 1.46 / 1.72 |
| 48 | 4.04 / 7.19 | 3.19 / 5.08 | 2.80 / 4.22 | 2.56 / 3.74 | 2.41 / 3.42 | 2.30 / 3.20 | 2.21 / 3.04 | 2.14 / 2.90 | 2.08 / 2.80 | 2.03 / 2.71 | 1.99 / 2.64 | 1.96 / 2.58 | 1.90 / 2.48 | 1.86 / 2.40 | 1.79 / 2.28 | 1.74 / 2.20 | 1.70 / 2.11 | 1.64 / 2.02 | 1.61 / 1.96 | 1.56 / 1.88 | 1.53 / 1.84 | 1.50 / 1.78 | 1.47 / 1.73 | 1.45 / 1.70 |
| 50 | 4.03 / 7.17 | 3.18 / 5.06 | 2.79 / 4.20 | 2.56 / 3.72 | 2.40 / 3.41 | 2.29 / 3.18 | 2.20 / 3.02 | 2.13 / 2.88 | 2.07 / 2.78 | 2.02 / 2.70 | 1.98 / 2.62 | 1.95 / 2.56 | 1.90 / 2.46 | 1.85 / 2.39 | 1.78 / 2.26 | 1.74 / 2.18 | 1.69 / 2.10 | 1.63 / 2.00 | 1.60 / 1.94 | 1.55 / 1.86 | 1.52 / 1.82 | 1.48 / 1.76 | 1.46 / 1.71 | 1.44 / 1.68 |
| 55 | 4.02 / 7.12 | 3.17 / 5.01 | 2.78 / 4.16 | 2.54 / 3.68 | 2.38 / 3.37 | 2.27 / 3.15 | 2.18 / 2.98 | 2.11 / 2.85 | 2.05 / 2.75 | 2.00 / 2.66 | 1.97 / 2.59 | 1.93 / 2.53 | 1.88 / 2.43 | 1.83 / 2.35 | 1.76 / 2.23 | 1.72 / 2.15 | 1.67 / 2.06 | 1.61 / 1.96 | 1.58 / 1.90 | 1.52 / 1.82 | 1.50 / 1.78 | 1.46 / 1.71 | 1.43 / 1.66 | 1.41 / 1.64 |
| 60 | 4.00 / 7.08 | 3.15 / 4.98 | 2.76 / 4.13 | 2.52 / 3.65 | 2.37 / 3.34 | 2.25 / 3.12 | 2.17 / 2.95 | 2.10 / 2.82 | 2.04 / 2.72 | 1.99 / 2.63 | 1.95 / 2.56 | 1.92 / 2.50 | 1.86 / 2.40 | 1.81 / 2.32 | 1.75 / 2.20 | 1.70 / 2.12 | 1.65 / 2.03 | 1.59 / 1.93 | 1.56 / 1.87 | 1.50 / 1.79 | 1.48 / 1.74 | 1.44 / 1.68 | 1.41 / 1.63 | 1.39 / 1.60 |
| 65 | 3.99 / 7.04 | 3.14 / 4.95 | 2.75 / 4.10 | 2.51 / 3.62 | 2.36 / 3.31 | 2.24 / 3.09 | 2.15 / 2.93 | 2.08 / 2.79 | 2.02 / 2.70 | 1.98 / 2.61 | 1.94 / 2.54 | 1.90 / 2.47 | 1.85 / 2.37 | 1.80 / 2.30 | 1.73 / 2.18 | 1.68 / 2.09 | 1.63 / 2.00 | 1.57 / 1.90 | 1.54 / 1.84 | 1.49 / 1.76 | 1.46 / 1.71 | 1.42 / 1.64 | 1.39 / 1.60 | 1.37 / 1.56 |
| 70 | 3.98 / 7.01 | 3.13 / 4.92 | 2.74 / 4.08 | 2.50 / 3.60 | 2.35 / 3.29 | 2.23 / 3.07 | 2.14 / 2.91 | 2.07 / 2.77 | 2.01 / 2.67 | 1.97 / 2.59 | 1.93 / 2.51 | 1.89 / 2.45 | 1.84 / 2.35 | 1.79 / 2.28 | 1.72 / 2.15 | 1.67 / 2.07 | 1.62 / 1.98 | 1.56 / 1.88 | 1.53 / 1.82 | 1.47 / 1.74 | 1.45 / 1.69 | 1.40 / 1.62 | 1.37 / 1.56 | 1.35 / 1.53 |
| 80 | 3.96 / 6.96 | 3.11 / 4.88 | 2.72 / 4.04 | 2.48 / 3.56 | 2.33 / 3.25 | 2.21 / 3.04 | 2.12 / 2.87 | 2.05 / 2.74 | 1.99 / 2.64 | 1.95 / 2.55 | 1.91 / 2.48 | 1.88 / 2.41 | 1.82 / 2.32 | 1.77 / 2.24 | 1.70 / 2.11 | 1.65 / 2.03 | 1.60 / 1.94 | 1.54 / 1.84 | 1.51 / 1.78 | 1.45 / 1.70 | 1.42 / 1.65 | 1.38 / 1.57 | 1.35 / 1.52 | 1.32 / 1.49 |
| 100 | 3.94 / 6.90 | 3.09 / 4.82 | 2.70 / 3.98 | 2.46 / 3.51 | 2.30 / 3.20 | 2.19 / 2.99 | 2.10 / 2.82 | 2.03 / 2.69 | 1.97 / 2.59 | 1.92 / 2.51 | 1.88 / 2.43 | 1.85 / 2.36 | 1.79 / 2.26 | 1.75 / 2.19 | 1.68 / 2.06 | 1.63 / 1.98 | 1.57 / 1.89 | 1.51 / 1.79 | 1.48 / 1.73 | 1.42 / 1.64 | 1.39 / 1.59 | 1.34 / 1.51 | 1.30 / 1.46 | 1.28 / 1.43 |
| 125 | 3.92 / 6.84 | 3.07 / 4.78 | 2.68 / 3.94 | 2.44 / 3.47 | 2.29 / 3.17 | 2.17 / 2.95 | 2.08 / 2.79 | 2.01 / 2.65 | 1.95 / 2.56 | 1.90 / 2.47 | 1.86 / 2.40 | 1.83 / 2.33 | 1.77 / 2.23 | 1.72 / 2.15 | 1.65 / 2.03 | 1.60 / 1.94 | 1.55 / 1.85 | 1.49 / 1.75 | 1.45 / 1.68 | 1.39 / 1.59 | 1.36 / 1.54 | 1.31 / 1.46 | 1.27 / 1.40 | 1.25 / 1.37 |
| 150 | 3.91 / 6.81 | 3.06 / 4.75 | 2.67 / 3.91 | 2.43 / 3.44 | 2.27 / 3.13 | 2.16 / 2.92 | 2.07 / 2.76 | 2.00 / 2.62 | 1.94 / 2.53 | 1.89 / 2.44 | 1.85 / 2.37 | 1.82 / 2.30 | 1.76 / 2.20 | 1.71 / 2.12 | 1.64 / 2.00 | 1.59 / 1.91 | 1.54 / 1.83 | 1.47 / 1.72 | 1.44 / 1.66 | 1.37 / 1.56 | 1.34 / 1.51 | 1.29 / 1.43 | 1.25 / 1.37 | 1.22 / 1.33 |
| 200 | 3.89 / 6.76 | 3.04 / 4.71 | 2.65 / 3.88 | 2.41 / 3.41 | 2.26 / 3.11 | 2.14 / 2.90 | 2.05 / 2.73 | 1.98 / 2.60 | 1.92 / 2.50 | 1.87 / 2.41 | 1.83 / 2.34 | 1.80 / 2.28 | 1.74 / 2.17 | 1.69 / 2.09 | 1.62 / 1.97 | 1.57 / 1.88 | 1.52 / 1.79 | 1.45 / 1.69 | 1.42 / 1.62 | 1.35 / 1.53 | 1.32 / 1.48 | 1.26 / 1.39 | 1.22 / 1.33 | 1.19 / 1.28 |
| 400 | 3.86 / 6.70 | 3.02 / 4.66 | 2.62 / 3.83 | 2.39 / 3.36 | 2.23 / 3.06 | 2.12 / 2.85 | 2.03 / 2.69 | 1.96 / 2.55 | 1.90 / 2.46 | 1.85 / 2.37 | 1.81 / 2.29 | 1.78 / 2.23 | 1.72 / 2.12 | 1.67 / 2.04 | 1.60 / 1.92 | 1.54 / 1.84 | 1.49 / 1.74 | 1.42 / 1.64 | 1.38 / 1.57 | 1.32 / 1.47 | 1.28 / 1.42 | 1.22 / 1.32 | 1.16 / 1.24 | 1.13 / 1.19 |
| 1000 | 3.85 / 6.66 | 3.00 / 4.62 | 2.61 / 3.80 | 2.38 / 3.34 | 2.22 / 3.04 | 2.10 / 2.82 | 2.02 / 2.66 | 1.95 / 2.53 | 1.89 / 2.43 | 1.84 / 2.34 | 1.80 / 2.26 | 1.76 / 2.20 | 1.70 / 2.09 | 1.65 / 2.01 | 1.58 / 1.89 | 1.53 / 1.81 | 1.47 / 1.71 | 1.41 / 1.61 | 1.36 / 1.54 | 1.30 / 1.44 | 1.26 / 1.38 | 1.19 / 1.28 | 1.13 / 1.19 | 1.08 / 1.11 |
| ∞ | 3.84 / 6.64 | 2.99 / 4.60 | 2.60 / 3.78 | 2.37 / 3.32 | 2.21 / 3.02 | 2.09 / 2.80 | 2.01 / 2.64 | 1.94 / 2.51 | 1.88 / 2.41 | 1.83 / 2.32 | 1.79 / 2.24 | 1.75 / 2.18 | 1.69 / 2.07 | 1.64 / 1.99 | 1.57 / 1.87 | 1.52 / 1.79 | 1.46 / 1.69 | 1.40 / 1.59 | 1.35 / 1.52 | 1.28 / 1.41 | 1.24 / 1.36 | 1.17 / 1.25 | 1.11 / 1.15 | 1.00 / 1.00 |

Degrees of freedom for numerator ($v_1$) across top; Degrees of freedom for denominator ($v_2$) down the left.

## Table 5. Critical Values for $D_\alpha$ in the Kolmogorov-Smirnov Test*

| $\alpha$ <br> $n$ | 0.20 | 0.10 | 0.05 | 0.01 |
|---|---|---|---|---|
| 5 | 0.45 | 0.51 | 0.56 | 0.67 |
| 10 | 0.32 | 0.37 | 0.41 | 0.49 |
| 15 | 0.27 | 0.30 | 0.34 | 0.40 |
| 20 | 0.23 | 0.26 | 0.29 | 0.36 |
| 25 | 0.21 | 0.24 | 0.27 | 0.32 |
| 30 | 0.19 | 0.22 | 0.24 | 0.29 |
| 35 | 0.18 | 0.20 | 0.23 | 0.27 |
| 40 | 0.17 | 0.19 | 0.21 | 0.25 |
| 45 | 0.16 | 0.18 | 0.20 | 0.24 |
| 50 | 0.15 | 0.17 | 0.19 | 0.23 |
| Large Values | $\dfrac{1.07}{\sqrt{n}}$ | $\dfrac{1.22}{\sqrt{n}}$ | $\dfrac{1.36}{\sqrt{n}}$ | $\dfrac{1.63}{\sqrt{n}}$ |

The entries $D_\alpha$ in the body of the table are such that $P(D_n(x) \geq D_\alpha) = \alpha$.

* Reproduced from *Introduction to Mathematical Statistics*, 3rd ed., by Paul G. Hoel, © 1962 by John Wiley & Sons, Inc.

# Answers to Selected Exercises

**1.1**

**1** (a) $4^2$, (b) $4^3$, (c) $4^4$, (d) $4 \cdot 3$, (e) $4 \cdot 3 \cdot 2 \cdot 1$

**2** (a) $4^2$, (b) $4^3$, (c) Why not?

**3** $1 + 2 + 4 + 6$

**4** (a) $26^6$, (b) $26 \cdot 25 \cdot 24 \cdot \cdots \cdot 21$

**5** $14 \cdot 13 \cdot 12 \cdot \cdots \cdot 2 \cdot 1$; In no material way.

**6** Nothing.

**7** See Example 1.1.3.

**8** (a) none, (b) 1, (c) 1, (d) 2, (e) 6, (f) 33

**9** $21 \cdot 5$

**11** $m_1 \cdot m_2$

**12** $m_1 \cdot m_2 \cdot m_3$

**13** $m_1 \cdot m_2 \cdot \cdots \cdot m_r$

**14** $10 \cdot 21 \cdot 999$

**15** See exercise 14.

**16** $(f(1) + 1)(f(2) + 1)(f(3) + 1) \cdots (f(17) + 1)$;
$(2f(1) + 1)(2f(2) + 1) \cdots (2f(17) + 1)$

**17** $4 \cdot 3 \cdot 4$

**18** $26^4 - 26 \cdot 25 \cdot 24 \cdot 23$

**1.2**

**2** See exercise 3.

**5** $_6P_4$

**6** $9!$; $5! \cdot 4!$

**7** $_{9980}C_6$; $_{10000}C_6 - {}_{9980}C_6$

**8** $_{100}C_4 - {}_{60}C_4 - {}_{40}C_4$

**9** (a) $\{3, 8, 11\}$, 16, $\{p, b, c\}$, $\{m, c, p\}$, $\{a, b, m\}$, $c$

**10**  $13 \cdot \binom{12}{2} \cdot \binom{4}{3} \cdot \binom{4}{1} \cdot \binom{4}{1}$

**11**  $\binom{13}{5} \cdot \binom{13}{2}$

**13**  $\binom{13}{2} \cdot \binom{11}{2} \cdot \binom{4}{2} \cdot \binom{4}{2} \cdot \binom{4}{3} \cdot \binom{4}{3}$

**14**  See Example 1.2.2.

**17**  $20 \cdot 19 \cdot 18$

**18**  $4 \cdot \binom{13}{5}$

**19**  See exercise 20;  $k^n - \binom{k}{1}(k - 1)^n + \binom{k}{2}(k - 2)^n \ldots$

**1.3**

**1**  $\dfrac{10!}{3!3!2!2!}$

**2**  $\dfrac{7!}{2!2!2!} + \dfrac{7!}{3!2!} + \dfrac{7!}{3!2!} + \dfrac{7!}{3!2!2!}$

**4**  $\dfrac{8!}{4!2!2!}$

**6**  It is the number of words having certain properties.

**8**  $10^{12} \cdot \binom{21}{10}$

**9**  Insert a refuse box.

**10**  $\dfrac{n!}{n_1!n_2! \cdots n_k!}$

**11**  $\binom{14}{7}$

**12**  $\binom{11}{4}$

**13**  See exercise 15.

**15**  $\binom{m + r - 1}{m} \cdot \binom{n + 5 - 1}{n}$

**1.4**

**3**  $\dfrac{2}{\sqrt{6\pi n}} \left(\dfrac{4^4}{3^3}\right)^n$

**4**  $(2\pi)^{-3/2} \cdot 2 \cdot 4^{100}/5^3$

**6**  $f(x) = x + \sqrt{x}, \quad g(x) = x$

**1.5**

**2**  $(1 - x)^{-k-1}$

**6**  $(a_0A + a_1Ax - a_0Bx)/(A - Bx - Cx^2)$   for sufficiently small $|x|$

**8**  $(3x + 2x^2)/(1 - 2x - 2x^2)$   for sufficiently small $|x|$

**9**  $(3x + 3x^2 + 2x^3)/(1 - 2x - 2x^2 - 2x^3)$   for sufficiently small $|x|$

**15**  $(1 - \frac{1}{2}x)/(1 - x)$   for $|x| < 1$;

$$1 - \left[\sum_{j=0}^{\infty} \binom{2j}{j} 2^{-2j} x^j\right]^{-1} = 1 - 1/(1 - 4x)^{-1/2} = 1 - (1 - 4x)^{1/2}$$

**2.1**

1   $M \cap N = \{1, 4, 7\}$, $N - M = \{3, 8\}$
3   $\varnothing$
6   It is $N$.
9   $\{(1, \sqrt{2}), (1, -\sqrt{2}), (-1, \sqrt{2}), (-1, -\sqrt{2})\}$
12   $K \cap L = \{p \in S; p(x) = ax^3 + bx^2 - 2x + 3\}$,
     $K \cap L \cap M \cap N = \{-\frac{26}{3}x^3 + \frac{1}{2}x^2 - 2x + 3\}$

**2.2**

1   $s_{18} = \frac{7}{1}$, $s_{19} = \frac{5}{3}$, $s_{20} = \frac{3}{5}$
3   (c) $0.12499999 \cdots$
7   $\{1, 3, 5, 7, \ldots\}$

**2.3**

3   A rectangle
4   A solid finite cylinder
8   $\{(x, y); x = y\}$

**2.4**

1   Yes;  No
2   No;  No;  Yes
4   No;  Yes
5   $m^n$
7   No
9   (a) $R$,  (c) $R$,  (e) $-1 \le x \le 1$,  (g) $-1 \le x \le 1$,  (j) $x > 1$
11   The function $f$ for which $f(x) = x^2$ is continuous.

**2.5**

3   $\cup$
10   $\displaystyle\bigcap_{n=1}^{\infty}\left\{x; 1 - \frac{1}{n} < x < 1 + \frac{1}{n}\right\}$ is not open.

**3.2**

1   $\{1, 2, 3, \ldots, 12\}$
2   Same as exercise 1.
4   $\{0, 1, 2, \ldots, 6\}$
6   The set of combinations of two flashbulbs from the given flashbulbs;
     $2\binom{10}{2} = 2^{45}$
7   The set of lists of two numbers out of ten numbers;   $10^2$
8   $1 + \dbinom{8}{2}$
11   E1 and E3
14   The set of all combinations of five cards;   $\dbinom{52}{5}$;   $\dbinom{13}{5} \cdot \dbinom{4}{1}^5$;
     $4 \cdot \dbinom{13}{5}$

**3.3**

1   See Example 3.3.2.
2   $\dbinom{30}{14} \div 2^{30}$
3   $\dbinom{30}{14} \cdot 2^{14} \cdot 4^{16} \div 6^{30}$

**5**  $\dbinom{13}{2} \cdot 11 \cdot \dbinom{4}{2}^2 \cdot \dbinom{4}{1} \div \dbinom{52}{5}$

**8**  See exercise 3.2.14.

**10**  $21^3 \div 26^3$;   $3 \cdot 5 \cdot 21^2 \div 26^3$

**11**  $4 \div \dbinom{52}{5}$

**13**  $1 \div \dbinom{13}{5}$ which is greater than the answer to exercise 12.

**15**  See exercise 1.

**16**  $\left[ \dbinom{6}{2} \cdot \dbinom{4}{3} + \dbinom{6}{1} \cdot \dbinom{4}{4} \right] \div \dbinom{10}{5}$   .

**17**  See exercise 3.

**18**  $\dfrac{30!}{(5!)^6} \div 6^{30}$

**19**  $\dbinom{30}{2} \cdot \dbinom{40}{3} \cdot \dbinom{50}{5} \div \dbinom{90}{10}$

**3.4**

**1**  $P(n) = n/21$

**2**  (a)  $n^2/91$

**4**  (a) $P(n) = 5(\tfrac{1}{6})^n/6$,   (b) $P(n) = e^{-3}3^n/n!$

**3.5**

**1**  $\tfrac{1}{2}$

**3**  $\tfrac{1}{4}$;   Probability of an event is proportional to its area.

**4**  $\tfrac{1}{8}$;   Probability of an event is proportional to its volume.

**6**  (a) $(2/\sqrt{3})^3 \div 4\pi/3$

**9**  (a) $p(x) = 6(x - x^2)$,   (b) $p(x) = \dfrac{\pi}{2} \sin \pi x$,

(c)  $p(x) = 1 - 2|x - \tfrac{1}{2}|$

**10**  $p(x) = 12(x - \tfrac{1}{2})^2$

**11**  (f) $p(a, c) = c/4$

**4.1**

**3**  The amplitudes of oscillation do not decrease as time passes.

**4**  $e^{-at} \cos(wt + b)$ behaves like $\cos(wt + b)$ for small $t$.

**5.1**

**1**  $\{1, 2, 3, \ldots, 55\} \times \{1, 2, 3\}$;   Yes

**2**  $\{1, 2, \ldots, 13\} \times \{1, 2, \ldots, 13\} \times \{H, T\}$

**3**  A solid cylinder whose base is of radius one unit and whose height is one unit.

**5.2**

**3**  No

**4**  Yes;   Yes;   Yes

**7**  $6!/6^6$

**8**  $\displaystyle\sum_{k=0}^{3} \dbinom{20}{k}(\tfrac{1}{4})^k(\tfrac{3}{4})^{20-k}$;   $\displaystyle\sum_{k=0}^{9} \dbinom{60}{k}(\tfrac{1}{4})^k(\tfrac{3}{4})^{60-k}$

**12**  $M_1 = \{1, 2, 4, 12\}$,   $M_2 = \{1, 2, 5, 6, 7, 8\}$,   $M_3 = \{1, 5, 6, 9, 10, 12\}$

**5.3**

**1** $\frac{2}{11}$

**3** $\frac{1}{6}\binom{5}{3}2^{-5} \div \frac{1}{6}\left[\binom{3}{3}2^{-3} + \binom{4}{3}2^{-4} + \binom{5}{3}2^{-5} + \binom{6}{3}2^{-6}\right]$

**6** $\binom{50}{3}\Big/\binom{100}{6} \div \left[1 - \binom{50}{6}\cdot\binom{2}{1}^{6}\Big/\binom{100}{6}\right]$

**12** 0.12;  0.18

**5.4**

**3** No;  No;  Yes

**4** $9 - 12\log 2$

**6** $P(M) = \log(13/11)$

**9** $(1 - (\frac{19}{20})^2) \div (1 - (\frac{9}{10})^2)$

**5.5**

**2** $\binom{5}{2}\binom{43}{2}(0.20) \div \left[\binom{2}{2}\binom{46}{2}(0.20) + \binom{5}{2}\binom{43}{2}(0.20)\right.$

$+ \binom{8}{2}\binom{40}{2}(0.05) + \binom{12}{2}\binom{36}{2}(0.05)$

$+ \binom{20}{2}\binom{28}{2}(0.02) + \binom{24}{2}^{2}(0.02) + \cdots$

$\left. + \binom{45}{2}\binom{3}{2}(0.04)\right] \approx 0.11$

**4** $[\sum P(N \mid M \cap H_j)P(M \mid H_j)P(H_j)] \div [\sum P(M \mid H_j)P(H_j)]$

**6** See exercise 7.

**7** Zero

**8** (a) $(0.99)^3/[(0.99)^3 + 3(0.02)(0.99)^2 + 3(0.02)^2(0.99) + (0.02)^3] \approx 0.98$,
(b) $(0.99)^2(0.01)/[(0.99)^2(0.98) + (0.99)^2(0.01) + 2(0.99)(0.02)(0.98)$
$+ 2(0.02)(0.99)(0.01) + (0.02)^2(0.01) + (0.02)^2(0.98)] \approx 0.01$

**6.1**

**1** (In part) $X(1) = 0$,  $X(2) = 1$,  $X(3) = 1$,  $X(4) = 2$,  $X(6) = 0$,
$X(8) = 3$

**2** $P(n) = n^2/385$,  $X(n) = 3 - n$

**5** $X(T) = a$,  $X(H) = b$

**7** This situation is impossible.

**8** (In part) $P(n) = e^{-\lambda}\lambda^n/n!$

**9** Yes, Yes;  $\binom{3500}{60}\Big/\binom{6000}{60}$, $\binom{500}{60}\Big/\binom{6000}{60}$;  300, 60

**15** {samples having only poor corn};  $\binom{500}{60} \div \binom{6000}{60}$;

{samples with one medium ear, 59 poor ears}

**6.2**

**1** $f(4) = \binom{20}{4}K$,  $f(3.5) = \binom{20}{3}\cdot\binom{10}{1}K$,  $f(3) = \binom{20}{2}\cdot\binom{10}{2}K$,

$f(2.5) = \binom{20}{1}\cdot\binom{10}{3}K$,   $f(2) = \binom{10}{4}K$  where $K = \binom{30}{4}^{-1}$

**4** No; $\sum f(n) \neq 1$

**6** (In part) $F(u) = 1 - (u + 1)^{-1}$,  $u = 1, 2, 3, \ldots$.

**10** (a) random variable,  (b) none,  (c) probability function or random variable,  (d) random variable or distribution function, (e) none,  (f) random variable or distribution function

**12** Yes;  No

**6.3**

**1** (In part) $F(u) = u^3$,  $0 \le u \le 1$

**3** $f_Y(u) = -af(au + b)$

**8** $f(x) = e^x(e^x + 1)^{-2}$  for all $x$

**9** Yes;  Yes

**11** Yes

**16** Axiom M5

**6.4**

**1** $(\pi m)^{-1/2}$ by Example 1.4.2

**4** $1 - \binom{5000}{0}(0.9999)^{5000} - \binom{5000}{0}(0.0001)(0.9999)^{4999}$

$- \binom{5000}{2}(0.0001)^2(0.9999)^{4998}$

**9** $\frac{1}{2}$

**11** $\lambda = n$

**12** $\sigma = |x|$

**14** (In part) $F(u) = 1 - e^{-u}$,  $u \ge 0$

**17** It is the Cauchy distribution.

**6.5**

**5** No

**6** Yes

**7.1**

**1** (a) 4930,  (b) 4000,  (c) bimodal

**2**

| INCOME IN $1000 | 1 | 2 | 3 | 4 | 16 |
|---|---|---|---|---|---|
| NUMBER OF FAMILIES | 20 | 20 | 20 | 30 | 10 |

**7.2**

**1** $\frac{37}{8}$

**2** $E(X) = c$

**4** (a) 7,  (b) $\frac{49}{4}$,  (c) $\frac{77}{4}$

**8** $E(X) = \lambda$

**10** $E(X) = np$

**11** $\frac{1}{2}$

**12** $[1 \cdot 3 + 2 \cdot 5 + 3 \cdot 7 + 4 \cdot 9 + \cdots + 10 \cdot 21]/121$

**7.3**

**1** $E(X) = -\frac{1}{2}$,  $E(Y) = 2e - 2$

**3** $p(x)$ is not a probability mass density.

**4** $E(X)$ does not exist.

**5** $E(X) = 1$,  median of $X = \log 2$

**7** Expected area is $\pi$;  $F(u) = u/4\pi + u(4\pi)^{-1}\log 4\pi/u$,  $0 < u < 4\pi$

**9** $E(X) = 1$;  Integrate by parts to get $E(Y_{m+1}) = (2m + 1)E(Y_m)$.

**10** Integrate by parts.

**12** *Hint:* Polar coordinates.

**14** As $x \to \infty$, $x(1 - F(x)) \to 0$ and $\int_0^\infty (1 - F(x))\, dx$ converges

**16** $E(X) = \alpha/b$

**17** *Hint:* Use a square as the sample space.

**7.4**

**4** *This is not a true assertion. It cannot be shown.*

**7** $E(X_j) = \frac{1}{2}$

**7.5**

**6** $\sigma^2 = 1$

**7** $\sigma^2 = \alpha$

**8** $\sigma^2(Y) = \sigma^2(X)$

**9** Moments of order less than $n$

**10** Moments of order less than $n/2$

**7.6**

**2** Yes; No

**3** $X = \pm c$ with probability $\frac{1}{2}$; No

**9** $1 - 900 \cdot \frac{35}{12}/(250)^2 = 0.958$

**8.1**

**2** Sum $P(s)$ over all $s \in S$ which satisfy $X(s) \le u$, $Y(s) \le v$ and $Z(s) \le w$.

**3** (a) $\frac{1}{4}$, (b) $\frac{7}{12} + (4\pi)^{-1}(\sqrt{3} - 1)$

**8.2**

**3** For $u > v > 0$ $\quad F(u, v) = v\left(\log \dfrac{u}{v} - \log \dfrac{1 + u}{1 + v} + \dfrac{1}{1 + u}\right)$

**4** $f(x, y) = e^x e^y$ if both $x < 0$, $y < 0$

**5** (In part) Density $\pi^{-1}(1 - y^2)^{-1/2}$, $\quad 0 < x < 1$, $\quad -1 < y < 1$

**7** (In part) Density $1/(3 \cdot 2\sqrt{x})$, $\quad 0 < x \le 1$, $\quad 0 \le y \le 3$, $-1 \le z \le 0$

**8** (In part) Uniform density on parallelogram with vertices $(0, 0)$, $(1, 0)$, $(1, 1)$ and $(2, 1)$.

**8.4**

**2** $\pi^{-1}(\alpha - 1)$

**3** Ellipses centered at $(0, 0)$

**4** $1 - e^{-u^2/2A}$

**8.5**

**1** $u^{-1/2} e^{-u/2}$

**2** $f_{(X, Y)}(x - m_1, y - m_2)$

**3** $(2\pi |D|)^{-1} \exp\left(-[(c^2 + d^2)x^2 - 2(ac + bd)xy + (a^2 + b^2)y^2]/2D^2\right)$
$D = ad - bc$

**10** $(2\pi)^{-3/2}[t_1(t_2 - t_1)(t_3 - t_2)]^{-1/2} \exp\left(-\frac{1}{2}[x_1^2/t_1 \right.$
$\left. + (x_2 - x_1)^2/(t_2 - t_1) + (x_3 - x_2)^2/(t_3 - t_2)]\right)$

**13** See exercise 14.

**14** See exercise 15.

**8.6**

**1** $f_X(x) = 2\sqrt{1 - x^2}/\pi$ $\quad -1 \le x \le 1$

**2** (In part) $f_X(x) = (2\pi A)^{-1/2} e^{-x^2/2A}$,
$f_{Y|X}(y \mid x) = (2\pi D/A)^{-1/2} \exp\left(-A[y - Bx/A]^2/2D\right)$

**3** $f_Y(y) = 2e^{-2y}, \quad y \geq 0$

**4** $f_Y(n) = e^{-\lambda p}(\lambda p)^n/n!$

**9** $f_{(X, Y)}(x, y) = \lambda e^{-\lambda x}(2\pi x)^{-1/2}e^{-y^2/2x}, \quad x > 0$

**8.7**

**1** $B$

**3** $\frac{1}{4}$

**6** Approximately zero; Fairly close to $+1$

**11** $E(X \mid y) = E(X)$

**13** $E(X \mid y) = \lambda(1 - p) + y, \quad y = 0, 1, 2, 3, \ldots.$

**16** $m_1 + \dfrac{B}{C}(y - m_2)$

**9.1**

**2** *Hint:* A definite integral is a limit of sums of a special type.

**9** $S(x_1, y_1, x_2, y_2) + S(x_2, y_2, x_3, y_3) + S(x_3, y_3, x_1, y_1) = 0$ for those

$n$ since $S(a, b, c, d) = \dfrac{1}{n + 1}((a + ib)^{n+1} - (c + id)^{n+1})$

**11** arctan $r$; See exercise 10.

**12** See exercise 11.

**16** See exercise 15 and exercise 2.

**17** *Hint:* Substitute $\alpha t = u$ and change variables of integration.

**9.2**

**1** $\phi_X(t) = \frac{1}{2}e^{it}/(1 - \frac{1}{2}e^{it}), \quad E(X) = 2,$ and $\sigma^2(X) = 2$

**3** $\phi_Y(t) = e^{ibt}e^{-a^2t^2/2}$ and $f_Y(x) = e^{-(x-b)^2/2a^2}/\sqrt{2\pi} \, |a|$

**5** See exercise 20.

**7** $\phi_X(t) = (1 + t^2)^{-1}, \quad E(X) = 0,$ and $E(X^2) = 2$

**10** $\phi_Y(t) = e^{ibt}(1 - iat)^{-1}$

**14** $(1 - it)^{-3}$

**15** Identical to exercise 14.

**18** $E(X^n) = n!/\lambda^n$

**19** *Hint:* See exercise 17.

**20** *Hint:* $E(\sin tX) = 0.$

**9.3**

**1** (a) $f(-2) = \frac{1}{3}, \quad f(1) = \frac{1}{2}, \quad f(0) = \frac{1}{6}$

(c) This is not a characteristic function; $\phi(0) = \frac{1}{2}$

(e) $f(k - 4) = \dbinom{18}{k}p^k(1 - p)^{18-k}, \quad k = 0, 1, 2, \ldots, 18$

(f) $f(x) = 1, \quad -\frac{1}{2} \leq x \leq \frac{1}{2}$

(h) $X = Y + Z$ where $Y$ is binomial with $n = 6$, $p = \frac{1}{3}$ and $Z$ is Poisson with $\lambda = 3$.

**4** If $\phi(t) = \displaystyle\int_{-\infty}^{\infty} e^{itx}f(x) \, dx$ is non-negative and can be integrated, then

$f(-t)$ gives the characteristic function of a random variable whose

density function is $\dfrac{1}{2\pi} \phi(x)$.

**7** Poisson with parameter $\lambda + \mu$

**9.4**     **1**  $\exp\left(-\lambda + \lambda e^{it_1}(pe^{it_2} + 1 - p)\right)$

**3**  $\phi(t_1, t_2) = \exp\left(-\tfrac{1}{2}[(at_1 + ct_2)^2 + (bt_1 + dt_2)^2]\right)$

**5**  *Hint:* If $a \neq 0$, $d = bc/a$ and $\alpha = a$, $\beta = -c$ suffice.

**6**  $\phi(t_1, \ldots, t_r) = (e^{it_1}p_1 + \cdots + e^{it_r}p_r)^n$

**10.1**    **1**  $P(\text{fewer than five defectives}) \approx e^{-5}\left(1 + \dfrac{5}{1!} + \dfrac{5^2}{2!} + \dfrac{5^3}{3!} + \dfrac{5^4}{4!}\right) \approx 0.44$

**2**  $P(X_{40} \leq 3) \div P(Y \leq 3) \approx 2.5 \times 10^{-3}$

**4**  The estimates are all $\sqrt{2}/2 = 0.707107$ to six places. The exact answer when $n = 1000$ is $707/1000$.

**6**  Converges to $X$ where $P(X = 1) = \tfrac{1}{2}$ and $P(X = 0) = \tfrac{1}{2}$

**7**  Converges to $X$

**10** If $X_n$ is exponentially distributed, it does not converge in distribution since for all real $u$ $P(X_n \leq u) \to 0$.

**10.2**    **1**  Zero

**3**  0.0013;   0.0228

**6**  0.8185

**7**  $\sigma^2(Z) = \sqrt{n}\,\sigma$;   $E(Z) = n\mu$

**10** (a) $-4.5$ and 2,   (c) $-4.5$ and $-1.5$

**10.3**    **4**  The sequence of $U_j$ converges uniformly on $S$.

**6**  $\phi_{X_n}(t)$ converges if and only if $t = 2k$ for some integer $k$.

**7**  If $c > \tfrac{1}{2}$ we have convergence in distribution to the constant zero random variable. If $c = \tfrac{1}{2}$, see exercise 10.2.12.

**8**  The same Cauchy distribution;   Since the $(X_1 + \cdots + X_n)/n$ are identically distributed they must converge in distribution.

**10** $X_n$ converges in probability to a random variable whose density function $f(x) = 2x$,   $0 \leq x \leq 1$.

**10.4**    (See section 10.5 for interpretation of limiting characteristic functions.)

**2**  The limit of the characteristic functions is $\exp\left(-\tfrac{1}{2}[t_1^2 + t_1 t_2 + \tfrac{1}{3}t_2^2]\right)$.

**3**  $\phi_{Y_n}(t_1, t_2) = (\cos t_1/n^{1/2} + \cos t_2/n^{1/2} + \cos(t_1 + t_2)/n^{1/2})^n/3^n$

**5**  The limit of the characteristic functions is $\exp\left(-\tfrac{1}{2}\sum p_j(a_j t_1 + b_j t_2)^2\right)$

**10.5**    **1**  $\phi(t_1, t_2, t_3) = \exp\left(-\tfrac{1}{2}[(3t_1 + t_2 + t_3)^2 + (t_1 + t_2 - 6t_3)^2 + (t_1 - t_2 - 2t_3)^2]\right)$

**3**  If $(X_1, \ldots, X_n)$ is centered,

$$\phi_{(Y_1, \ldots, Y_n)}(t_1, \ldots, t_n) = E(\exp\left(\sum_{k=1}^{n} t_k \sum_{j=1}^{n} c_{kj}X_j\right))$$

$$= E(\exp\left(\sum_{j=1}^{n} \left(\sum_{k=1}^{n} t_k c_{kj}\right)X_j\right))$$

$$= \exp\left(-\tfrac{1}{2}\sum_{j=1}^{n}\sum_{l=1}^{n} a_{jl}\left(\sum_{k=1}^{n} t_k c_{kj}\right)\left(\sum_{m=1}^{n} t_m c_m\right)\right)$$

$$= \exp\left(-\tfrac{1}{2}\sum_{k=1}^{n}\sum_{m=1}^{n}\left(\sum_{j=1}^{n}\sum_{l=1}^{n} c_{kj}a_{jl}c_{ml}\right)t_k t_m\right)$$

**5**  *Hint:* In Theorem 10.5.1, each $b_{jj} = 1$.

**10**    Use $a_{jk} = c_{jk}(\sum_{m=1}^{n} c_{jm}^2)^{-1/2}$   where   the   $c_{jk}$   satisfy   $c_{jk} = \sqrt{\pi_k}$,

$j \leq k \leq n$,   $c_{j,j-1} = (\pi_1 + \pi_2 + \cdots + \pi_{j-1} - 1)/\sqrt{\pi_{j-1}}$,   $c_{jk} = 0$,
$1 \leq k \leq j - 2$.

## 11.1

**2**   (a) $t = 18$,   (c) $t = 12$,   (d) does not exist

**3**   (b) $p^9(1 - p)^{12}$

**5**   (a) 18,   (b) 2

**7**   $P$(walker is at $\pm 3$ at $t = 3$) = $1/18$, $P$(walker is at $\pm 1$ at $t = 3$) = $4/9$

**8**   $P(X_j = 1) = p$, $P(X_j = -1) = 1 - p$

**9**   $N(x_0, n)$

**10**   *Hint:*   Leave it as an integral.

**11**   $e^{it(x_0 + na)}(pe^{itb} + (1 - p)e^{-itb})^n$

**12**   (c) $P(W_n = (j, k)) = \sum \dfrac{n!}{a!b!c!d!} 2^{-n}$ where the sum is taken over

non-negative $a$, $b$, $c$, $d$ with $a - b = j$,   $c - d = k$,   $a + b + c + d = n$

**13**   (b) $1 - e^{-a^2/2n}$

**14**   See exercise 10.4.3.

## 11.2

**3**   (a) $p/q$,   (c) $(p/q)^4$

**5**   *Hint:*   $\alpha_{2n+1} = \sum_{j=1}^{n} v_{2n+1-2j}p_{2j}$

**6**   $v_8^{(2)} = 7/128$ and coincidentally $v_8^{(4)} = 7/128$

**10**   Despite the plausibility of the suggested mode of approach and the obviousness of $\beta(p_1) < \beta(p_2)$ when $p_1 < p_2$ (it is easier to break the bank with a coin more biased in your favor), it is not true that a particular bank breaking sequence will have a higher probability under a larger $p$; e.g. look at $l = 4$, $c = 3$, THTHTHTHTHH with $p = 0.8$ and $0.9$. One must examine $U(1, c)$ and show that it is monotonic in $p$.

**12**   $U(z, 2) = U(z, 1)/zp$,   $U(z, 3) = (1 - z^2pq)U(z, 1)/z^2p^2$,
$U(z, 4) = (1 - 2z^2pq)U(z, 1)/z^3p^3$

**16**   $\dfrac{d}{dz} \sum_{n=0}^{\infty} v_{2n+1}z^{2n+1} = 2p(1 - 4pqz^2)^{-1/2} - [1 - (1 - 4pqz^2)^{1/2}]/2qz^2$

**19**   *Hint:*   Multiply both sides of the equations for $P(T_{2n} = 2k)$ of exercise 18 by $\alpha^{2n}\beta^{2k}$ and sum on $n = 0, 1, 2, \ldots$ and $0 \leq k \leq n$. Use the methods of exercises 1.5.11 and 1.5.14 and $P(x_s(u) < 0$ for $0 < u < 2j$,   $x_s(2j) = 0) = \frac{1}{2}p_{2j}$ when $p = \frac{1}{2}$. After simplification this should give

$Q(\alpha, \beta) = 1 + \frac{1}{2}[R(\alpha) + R(\alpha\beta)]Q(\alpha, \beta) + \frac{1}{2}(1 - \alpha^2)^{-1}[\alpha^2 - R(\alpha)]$
$+ \frac{1}{2}(1 - \alpha^2\beta^2)^{-1}[\alpha^2\beta^2 - R(\alpha\beta)]$

with $R(z) = \sum_{j=1}^{\infty} p_{2j}z^2$ as in (11.2.8a) with $p = q = \frac{1}{2}$.

## 11.3

**2**   $N(x_0, A^2t)$

**4**   $E(Y_n) = 2ABt$,   $\sigma^2(Y_n) = A^2t[1 - 4B^2/n]$, limiting distribution is $N(2ABt, A^2t)$

**6** (a) density $(2\pi t)^{-1} \exp\left(-(x^2 + y^2)/2t\right)$, (b) limiting distribution is a centered normal with $\sigma^2(X) = 2t/3 = \sigma^2(Y)$, $E(XY) = t/3$

**11.4**

**1** (a) $\{\omega; \omega(t) < b\} = \bigcup\limits_{n=1}^{\infty} \{\omega; \omega(t) \le b - 1/n\}$, (c) Try $b = y_0$.

**7** *Hint:* Change the order of integration in the repeated integral.

**11** *Hint:*

$\{\omega; \omega(u) \le b \text{ for } 0 \le u \le t\}$

$$= \bigcap_{n=1}^{\infty} \left\{\omega; \omega(u) < b - \frac{1}{n} \text{ for } 0 \le u \le t\right\}$$

**12** 0.497

**15** $E(W_t) = t^3$

**16** It is identical to the expression in exercise 13.

**19** (b) $N(0, t^5/20)$

**11.5**

**4** $f_{2,3,5}(4, 2, 7) = (1/10)^3$, $F_{2,3,5}(4, 2, 7) = (0.5)(0.3)(0.8)$

**6** $X_t(\frac{1}{3}) = -\frac{3}{2}\sum\limits_{n=1}^{100} n^{-1} \cos nt$; $E(X_t) = 0$,

$\sigma^2(X_t) = \frac{33}{4}\sum\limits_{n=1}^{100} n^{-2} \cos^2 nt$,

$E(X_t X_u) = \frac{33}{4}\sum\limits_{n=1}^{100} n^{-2} \cos nt \cos nu$

**7** $\exp\left(b[-v + te^{i(t_1 + t_2 + t_3)} + (u - t)e^{i(t_2 + t_3)} + (v - t)e^{it_3}]\right)$

**11** The sample functions are monotonically nondecreasing, non-negative, integer-valued functions on $Q$ with $\omega(0) = 0$. The set of sample functions having jumps at elements of $Q$ has probability measure zero.

**11.6**

**2** For all intents and purposes the chain "starts" at the state $k$.

**3** $p_{j,j+1} = p$ and $p_{j,j-1} = 1 - p$ if $1 \le j \le N - 1$, $p_{00} = 1$, $p_{NN} = 1$ and all other $p_{jk} = 0$

**5** $p_{j,j-4}^{(2)} = \left(\frac{j}{2} + N\right)\left(\frac{j}{2} + N - 1\right)/4N^2$, $p_{j,j+4}^{(2)} = \left(N - \frac{j}{2}\right)\left(N - \frac{j}{2} - 1\right)/4N^2$,

$p_{jj}^{(2)} = (4N^2 + 4N - j^2)/8N^2$, other $p_{jk}^{(2)} = 0$

**7** See equation (11.2.10).

**11** Equation (11.2.5) can be generalized to states other than 0 on a line and $P(x_s(2k) = 0, x_s(t) \ne 0$ for $0 < t < 2k$, and $x_s(2n) = 0)$ can be generalized as $P$(chain is in appropriate state at $t = n$ | first return occurred at $t = k)P$(first return occurs at $t = k$). However, even for a Markov chain with stationary transition probabilities, $P$(chain is in state $j$ at $t = n$ | first return occurs at $t = k) = p_{jj}^{(n-k)}$. Equation (11.6.6) is based on properties of generating functions.

**12** Intuitively, $P$(ever returning to $j$ | starts in $j$) = $P$(ever reaching $j$ | starts in $k$)$P$(ever reaching $k$ | starts in $j$).

**16** It is worth working out without help.

**11.7**

**2** Compare to $E(\lambda_\omega(t_3) - \lambda_\omega(t_2))E(\lambda_\omega(t_2) - \lambda_\omega(t_1))$

**4** Then use $Z_j(\omega) = \omega(t_j) - m(t_j)$.

**5** See exercise 11.4.13.
**9** $E(\lambda_\omega(u)) = a(u) + b(u)m(q(u))$, cov $(\lambda_\omega(u)\lambda_\omega(v)) = b(u)b(v)r(q(u), q(v))$

**12.1**

**1** Flip a fair coin 3 times. Declare H if exactly one head shows.
**3** $Y$ and $Z$ are not independent.
**6** $Y$ is approximately $N(2, 0.99)$.
**7** Each head run must be preceded and succeeded by a tail run or nothing. Nothing can occur only for the first or last head run.
**8** $2!3!2^{-20}$
**11** There should be a fair number of runs but not too many.
**12** Yes;   One must be careful.
**13** First procedure: 0.226, second procedure: 0.453
**15** 4.739
**16** 3.429
**17** 5.392, 5.119
**18** 5.523, 6.236; note that second procedure is worse

**12.2**

**4** $\bar{X}(s) \neq 0 = E(X)$ for each $s \in S$
**6** $X_1 = X_3$ identically distributed with $X_2$, $X_1$ and $X_2$ independent
**10** $4!/2!2!$ (that is just Chapter 1)
**12** Take expected values of the expressions in exercise 8, 9, and 11. Use exercise 7 and the comment of exercise 9.
**14** $(n/ita)^n(e^{ita/n} - 1)^n$
**15** $\sqrt{\pi}$, 3, $\pi$
**16** min $(X_1, \ldots, X_n)$, max $(X_1, \ldots, X_n)$
**17** median $(X_1, \ldots, X_n)$
**19** $f_r(c) = (n - 1)\lambda e^{-\lambda c}(1 - e^{-\lambda c})^{n-2}$, $n \geq 2$ in order to define $r$

**12.3**

**1** (a) 3.822,  (d) 0.711,  (f) $-0.389$,  (g) 1.533
**2** *Hint:*  $X = \sigma Y + m$ where $Y$ is $N(0, 1)$.
**3** *Hint:*  If $l > k$, $a_{jl}$ is constant for $j = 1, 2, \ldots, k$.
**4** Not fair at 90% level; fair at 95% level
**7** Assuming independence of the weights of different people and normal distribution of those weights, this is a borderline case whose answer depends heavily on the criteria.
**9** Look at Table 1 (appendix).

**13.1**

**1** (a) $P_\theta(k_1, k_2, \ldots, k_n) = e^{-n\theta}\theta^{k_1 + k_2 + \cdots + k_n}/k_1!k_2! \cdots k_n!$
**2** (b) $E_{\sqrt{2}}(Z) = n\sqrt{2}$

**13.2**

**1** $(k_1 + \cdots + k_n)/(n + k_1 + \cdots + k_n)$
**2** $\bar{p} = (a_1 + \cdots + a_n)/nk$,   $\bar{q} = (b_1 + \cdots + b_n)/nk$,   $\bar{r} = (c_1 + \cdots + c_n)/nk$
**4** It is valid only in case $n = 1$.
**8** Not true. Consider $f_X(x) = 3x^2$ for $0 \leq x \leq 1$ and $\theta = 2^{-1/3}$
**10** $Y$ would seem to be close to $\theta$, the median, with high probability when $m$ is large.
**11** (b) Bayes' formula, hopefully.
(d) $P(p = 0.08 \mid N = 0) \approx 0.0002$

**13.3**  **1**  For example, $m_1 \cdot m_1$

**3**  Use Theorem 13.3.2.

**4**  $26\theta[\sum\limits_{j=1}^{6} (1 - \theta)^{j-1}/j] + 2(1 - \theta)^6$

**7**  $E(Y^2) = (n - 1)\lambda^2/(n - 2)$,   $\sigma^2(Y) = \lambda^2/(n - 2)$

**8**  This is a direct consequence of the Cramér–Rao inequality.

**9**  $e(Y) = (n - 2)/n$

**11**  To a certain extent it is because $\dfrac{\partial}{\partial\theta} \int_S L(x, \theta)\, dx \neq \int_S \dfrac{\partial L}{\partial\theta} (x, \theta)\, dx$

because the true limits of integration depend on $\theta$.

**12**  $\sum\limits_{k=m}^{\infty} \dfrac{m - 1}{k - 1} \dbinom{k - 1}{m - 1} p^m (1 - p)^{k-m} = p^m(1 - (1 - p))^{1-m}$

**13.4**  **4**  $[0, 31.481]$

**5**  (a) By exercise 8.3.5, $f_Y(x) = \dfrac{5!}{2!2!} (x/a)^2(1 - x/a)^2/a$, $0 \le x \le a$.

To find a 90% confidence interval solve
$$0.90 = \int_{a(1-\delta)}^{a(1+\delta)} f_Y(x)\, dx = \int_{1-\delta}^{1+\delta} 30x^2(1 - x)^2\, dx$$
or $0.90 = (15\delta - 10\delta^3 + 3\delta^5)/8$.

(b) The same approach can be used except the density function of $5(\bar{X} - a/2)/a$ is, unfortunately, $115/192 - 5x^2/8 + x^4/4$,  $-\frac{1}{2} \le x \le \frac{1}{2}$;
$55/96 + 5|x|/24 - 5x^2/4 + 5|x|^3/6 - x^4/6$,  $\frac{1}{2} \le |x| \le \frac{3}{2}$;  $(|x| - \frac{5}{2})^4/24$,
$\frac{3}{2} \le |x| \le \frac{5}{2}$;  0, otherwise.

**6**  $[-0.1, 9.5]$

**14.1**  **1**  (a) Approximately 0.002

**3**  Expected cost on 100 is \$28.67, on 200 it is \$25.24, on 300 it is \$34 + .

**5**  $\int_0^{\infty} dv \int_{\sqrt{5}/2 + 1.73v/\sqrt{19}}^{\infty} \dfrac{1}{\sqrt{2\pi}} e^{-y^2/2} 2v[2^{19/2}\Gamma(19/2)]^{-1} v^{37} e^{-v^2/2}\, dy$

**8**  $(\theta + 1)^2(x_1 x_2)^{\theta}$

**10**  $(0.8, 0.95)$

**12**  They are all 0.2389 to four decimal places.

**13**  Respectively, 0.4026, 0.4070, 0.4848

**14.2**  **2**  $\{(x_1, x_2, \ldots, x_{25});\ a \ge \sum x_j\}$

**3**  $\left\{(k_1, k_2, \ldots, k_{20});\ \prod\limits_{j=1}^{20} \dfrac{4 \cdot 5 \cdots (4 + k_j - 1)}{1 \cdot 2 \cdots (1 + k_j - 1)} \ge a\right\}$ subject to $4 \cdot 5 \cdots$
$(4 + k - 1) = 1$ if $k = 0$

**5**  $\sum \log x_j + \sum x_j \ge a$

**6**  It is a valid test but it does not depend only on the number of H's and T's.

**7**  The "extra" sample points may not have probabilities which add up to a useful amount.

**11**  One must solve $\int_0^a x^{73/2} e^{-x}\, dx = \Gamma(\frac{75}{2})/5$ for $a$. It may be best done by machine.

**12**  Regarding exercise 10, $\beta = 0.0000$

**14.3**
2  $\hat{p} = 0.02097$; pool cells 5, 6, 7, ..., 12; $\hat{\chi}^2 \approx 0.31$; the binomial fits very well.

7  $\hat{\chi}^2 \approx 10.192$;   not homogeneous

9  $\hat{\chi}^2$ is much larger than 18.549.

10  First of all, the data came from a complete census. This is not a sample. Secondly, multiplying by $1/1000$ is improper in view of exercise 3—it could distort the test. However, in the other direction, if $n$, the total number of individuals sampled, is large the $\chi^2$ test will deny homogeneity unless we really have homogeneity. Thus, a workable but not perfect theory could be thrown out almost as a result of too much data.

**14.4**
1  $\{x; \sum (x_j - \bar{x})^2 \le c \sum x_j^2\}$ where $x = (x_1, \ldots, x_n)$

3  $\{x; \min [1, e^{n(\bar{x} - 1/2)}, e^{2(\bar{x} + 1)n}] \le c\}$ where $x = (x_1, \ldots, x_n)$

8  $\{x; \min [1, e^{n(1 - 2\bar{x})/2}, 2^{n/2} \exp (-(\sum x_j^2 - 2n\bar{x} - n)/4)] \le c\}$

10  $\hat{D}_{30}(x) = 0.7605 - 16/30 > 0.22$ so we reject $F$.

12  Show that for each $u_j$, $P(|F(u_j) - G_x(u_j)| \ge \epsilon) \to 0$.

13  Since $F$ and $G$ are bounded by zero and one and are monotonic, their closeness on a suitable finite set $\{u_1, \ldots, u_k\}$ guarantees their closeness for all $u$.

**15.1**
2  $\hat{V} = \dfrac{4.34/2}{2.74/6} = 4.8636 < 5.14$;   different brands of gas would be judged to be equally good

4  Even the one-way classification shows that the different car classes do not do equally well in gas mileage;   $\hat{V} = 20.11$

5  *Hint:*  Apply the $t$-distribution.

6  For the second table $\hat{V} = 9 < 18.51$.

**15.2**
4  $r_1 = 1$,   $r_2 = 2$,   $r_3 = 3$

**15.3**
1  Not yet

2  *Hint:*  Act as though the actual call length is zero.

4  $\hat{\theta} = (22.01)^{-1} = 0.045434$

5  (a) $Z(x) = -2x$,   $a = \log \dfrac{0.05}{0.95} = -2.94444$,   $b = -a = 2.94444$

7  $A = 1/19$, $B = 19$ and $Z(x) = (2x - 1) \ln 2$ where $x = 1$ for H and $x = 0$ for T. This problem corresponds to a random walk of step length $\log 2$ starting at zero reaching $5 \log 2$ or $-5 \log 2$ which is the same as a random walk of step length 1 reaching 5 or $-5$. Since $19 = 2^{4.248}$ and $2^{h(\theta)} = \theta^{-1} - 1$, if $\theta = P(\text{H})$ the approximate O.C. function is

$$\frac{((1 - \theta)/\theta)^{4.248} - 1}{((1 - \theta)/\theta)^{4.248} - (\theta/(1 - \theta))^{4.248}} = \frac{(1 - \theta)^{8.496} - (1 - \theta)^{4.248}\theta^{4.248}}{(1 - \theta)^{8.496} - \theta^{8.496}}$$

This should be contrasted with $\dfrac{(1 - \theta)^{10} - \theta^5(1 - \theta)^5}{(1 - \theta)^{10} - \theta^{10}}$ in the random walk. However, you would get this last expression for the approximate O.C. function if $\alpha = \beta = 1/33$.

# Index